Clinical Chemistry

An Overview

Clinical Chemistry

An Overview

Edited by

N. C. den Boer

University Hospital Rotterdam
Rotterdam, The Netherlands

C. van der Heiden

University Hospital for Children and Youth
Utrecht, The Netherlands

B. Leijnse

University Hospital Rotterdam
Rotterdam, The Netherlands

and

J. H. M. Souverijn

University Hospital Leiden
Leiden, The Netherlands

PLENUM PRESS • NEW YORK AND LONDON

Library of Congress Cataloging in Publication Data

International Congress on Clinical Chemistry (13th: 1987: Hague, Netherlands)
Clinical chemistry: an overview / edited by N. C. den Boer . . . [et al.].
p. cm.
"Proceedings of the Thirteenth International Congress of Clinical Chemistry, and the Seventh European Congress of Clinical Chemistry, held June 28–July 3, 1987, in the Hague, the Netherlands" – T.p. verso.
Includes bibliographies and indexes.
ISBN 0-306-43093-2
1. Chemistry, Clinical – Congresses. I. Boer, N. C. den. II. European Congress of Clinical Chemistry (7th: 1987: Hague, Netherlands) III. Title.
[DNLM: 1. Chemistry, Clinical – congresses. QY 90 I583c 1987]
RB40.I6 1987
616.07′5 – dc19
DNLM/DLC 88-31627
for Library of Congress CIP

Proceedings of the Thirteenth International Congress of Clinical Chemistry,
and the Seventh European Congress of Clinical Chemistry,
held June 28–July 3, 1987, in The Hague, The Netherlands

A Division of Plenum Publishing Corporation
233 Spring Street, New York, N.Y. 10013

Printed in the United States of America

PREFACE

The XIIIth International and the VIIth European Congress of Clinical Chemistry took place at the Netherlands Congress Centre in The Hague, from June 28th to July 3rd 1987.

The Organizing Committee and the Scientific Committee for these combined congresses aimed to present the state-of-the-science as well as the state-of-the-art for those fields of clinical chemistry which show a strong progress and which will most probably inflict a great part of all clinical chemists.

"Clinical Chemistry, an Overview" comprises almost all papers which were presented during the congress in 5 plenary lectures and 97 lectures during 24 symposia. The invited speakers, being experts in their fields of clinical chemistry, succeeded very well in presenting an overview over the newest developments in connection to the knowledge already known, thereby demonstrating the progress made in clinical chemistry during the last years.

The Editors take great pleasure in thanking once more the members of the Scientific Committee and of the International Scientific Advisory Board in creating an excellent scientific programme for this congress.
The Editors also take great pleasure in thanking all those whose afforts have made possible the publication of this book. We are most grateful to all speakers who also prepared a manuscript for publication. The Editors also appreciate the most helpful and encouraging attitude of Plenum Press Publishers Corporation.

We would like to express once again our appreciation to our wives who were so kind in encouraging us in our efforts and in accepting our frequent absence to finish this task.

The Editors of "Clinical Chemistry, an Overview" hope that this publication will be widely accepted and that it will stimulate further research in clinical chemistry and be of benefit to the patients.

The Editors

Rotterdam, Utrecht, Leiden: June 1988

CONTENTS

CHAPTER 1

CHAPTER 2
GENES, GENE-TECHNOLOGY AND DNA-PROBES

CHAPTER 3
THERAPEUTIC DRUG MONITORING AND TOXICOLOGY

CHAPTER 4
BIOSENSORS AND BIOCHIPS

CHAPTER 5
AUTOIMMUNE ENDOCRINOLOGICAL AND NEUROLOGICAL DISORDERS

CHAPTER 6
COAGULATION

CHAPTER 7
RHEUMATIC DISEASES

CHAPTER 8
ION SELECTIVE ELECTRODES

CHAPTER 9
BEDSIDE CHEMISTRY AND SELFTESTING

CHAPTER 10
CLINICAL CHEMICAL ASPECTS OF TRANSPLANTATION

CHAPTER 11
ENZYMOLOGY

CHAPTER 12
CHEMI- AND BIOLUMINESCENCE

CHAPTER 13
CLINICAL CHEMICAL ASPECTS OF AGEING

CHAPTER 14
LABORATORY MANAGEMENT AND COST EFFECTIVENESS

CHAPTER 23
THE THYROID AND FUNCTIONING HORMONES

CHAPTER 24
VETERINARY CLINICAL CHEMISTRY

CHAPTER 25
CLINICAL BIOCHEMISTRY IN DEVELOPING COUNTRIES

CHAPTER 1

The interaction between industry and the clinical chemist
B. Leijnse

High resolution nuclear magnetic resonance spectroscopy in clinical chemistry and diseases
J.K. Nicholson

Diagnostic related groups; the influence of refunding of cost on the organization of clinical chemistry laboratories
A.L. Giles

The role of clinical chemistry in preventive medicine
K. Miyai

THE INTERACTION BETWEEN INDUSTRY AND THE CLINICAL CHEMIST

B. Leijnse, Rotterdam, The Netherlands

Department of Clinical Chemistry University Hospital
Department of Chemical Pathology Erasmus University

The clinical chemistry laboratory performs qualitative and quantitative analyses on blood, urine, spinal flood, feces, calculi and other materials. This data-acquisition is indispensable for the determination of the nature of many diseases and the treatment and cure of many patients. Consequently clinical chemistry is an essential part of modern medicine. The clinical chemist is in charge of the workroutine of his department and he is ultimately responsible for the results of the analyses and the information provided by his laboratory. These are well known facts and they give a good outline of our discipline. But a more detailed picture is needed if we wish to analyse the interaction between the clinical chemist and industry. To acquire the necessary facts it is advisable to focus our attention on a famous clinical chemist because every discipline is the result of human effort.

I ask your attention for D.D. van Slyke. I must confess I mention D.D. not only for his brilliant contributions to clinical chemistry but also because his name is very typical for The Netherlands. It means that his ancestors came from het "Slijk" that is the flat mud so abundantly present in the part of our country named Zeeland. Already the well known Roman historian Plinius wondered how it was possible to live in a civilized way below sea level. But the name Van Slyke makes clear the Dutch are even proud of it.

Donald D. van Slyke (1883-1971) was an outstanding analytical chemist. But he was also able to give an excellent interpretation of the analytical results obtained in body fluids because he had a thorough knowledge of biochemistry and medical problems. In his famous book - Quantitative Clinical Chemistry -, published in 1932, he and his co-author, the clinician John P. Peeters, laid down the foundation of modern clinical chemistry. It consists of an amalgation of analytical skill and the ability to translate results into meaningful medical information of Methods and Interpretations. Donald D. van Slyke considered the Norwegian Ivar Bang as the founder of modern clinical chemistry. I think we must give also honour to the Swede Otto Folin. Folin did most of his work in the U.S.A. Bang and Folin combined, like Van Slyke, professional skill as analytical chemists with profound knowledge of medical problems. Johann von Scherer coined in 1843 the term clinical chemistry. Can we find traces of clinical chemistry further back in history? As we confine ourselves to urine-analyses the roots are deep indeed. Uroscopy was practised for centuries. We owe famous Dutch painters gratitude for their paintings giving us a clear insight in the art of uroscopy. Other aspects of medical alchemy are also immortalized by our painters. We must not regard uroscopy as unworthy. Hippocrates found it important enough to mention it in his Aphorism. Of course this is no reason at all to claim Hippocrates as the first clinical chemist. He is rightly called the Father of Medicine. For he (or the members of the school associated with this name) emphasized, for the first time, the importance of the physician's detailed and systematic description of his observations as he attempted to treat a particular individual's illness. Because of this and the fact that common diseases were accepted as entities medicine became a science. It took two thousand years more before chemistry became a science and could contribute so much to medicine that clinical chemistry was a well-founded name. The famous Dutch physician Herman Boerhaave praised the performance of chemistry highly in 1703. But in all fairness we have to admit it was merely a prophecy. Antoine Lavoisier had yet to be born. He detected the keystone of chemistry, the art of weighing. Weighing is a very commonplace operation compared with the flamboyant ideas of the alchemists. But Simplex Sigillum Veri.

Lavoisier was also member of the group scientists who have laid the foundations of the SI. Le Système Métrique was in 1890 presented to the Constituante by Talleyrand and accepted. The new quantities and units were meant "à tous les temps, à tous les peuples - for ever, for all people". Lavoisier's judgement was "Jamais rien de plus grand et de plus simple, de plus cohérent dans toutes ses parties n'est sorti de la main des hommes". Le Système Métrique resulted in Le Système International d'Unites. The SI is used in clinical chemistry for two decades. It is a token of the scientific standing of clinical chemistry. Clinicians were not everywhere enthousiastic about the use of SI-units by the clinical chemist, to put it mildly. But if I am well informed we may expect that the introduction of SI-units in medicine will be completed in the next century in all countries in the world. In the European Community it is already unlawful not to use the SI. Chemistry flourished on the basis of the work of Lavoisier, Dalton and others. A mighty chemical industry was created. The tremendous influence of chemical inventions on the western society is well known. A stream of ideas was flowing steadily from the laboratories to the factories making new production procedures possible. The control of these procedures and the quality of the products was achieved by the analytical methods in the chemical laboratory. What did the laboratories receive from industry? In the beginning of the nineteenth century nothing worth mentioning, all things necessary were fabricated in the laboratories. The chemical laboratory fabricated its own instruments, glassware and chemicals. But gradually industry started to supply the laboratories with these necessities. This development made that laboratories had to rely on industry for essential goods. No laboratory could work without filterpaper to name a simple thing one had to buy. However the degree of freedom of the chemist remains substantial. Up to far in this century he could use the techniques he preferred, he could adapt the techniques according to his wishes. Special glassware and instruments could be fabricated in most laboratories. The chemist was still master in his laboratory because the essential part of his work depended on the combination of his craftmanship and his theoretical knowledge. Industry was dependent on his inventions.

The situation in clinical chemistry was identical with the situation in other fields of chemistry. Donald D. van Slyke developed various ingenious pieces of apparatus and many analytical procedures. Bang and Folin did the same. Of course industry participated in this progress by producing the invented instruments and the chemicals needed. It was the era of the inventive user. Since Jules Duboscq introduced his colorimeter in 1854 several hundreds of this and similar instruments such as the Klett colorimeter were manufactured and sold. But the clinical chemists and the technicians incorporated these instruments in the usual procedures. All work was principally done in the same way as it was done by Lavoisier, that was manually.

In 1950 in advanced clinical chemical laboratories sodium and cholesterol in blood were measured with excellent gravimetric methods. Determinations of calcium by oxalate precipitation and an oxydimetric titration with permanganate developed by Kramer and Tisdall were performed daily in every good equipped clinical chemistry laboratory. The analytical skill of the clinical chemists and the technicians was excellent just like the quality of the routine methods used. The goal was the highest analytical accuracy and analytical precision possible. The standards valid at that time in analytical chemistry were used in judging the performance in clinical chemistry.

However, it was impossible for the clinical chemist to fulfil the ever increasing demands of the clinicians for laboratory investigations with the classical analytical methods. The more the number of analyses was growing the more imprecision and inaccuracy increased. Mechanization of the analytical techniques was the solution. Human manipulative effort had to be replaced by the performance of a given process by mechanical and instrumental devices. Numerous efforts were made to construct mechanized versions of manual laboratory techniques and procedures. A turning point in the history of clinical chemistry was the development by Skeggs of the continuous flow principle. Technicon Instruments Corporation pioneered in this field with the Auto-Analyzer in 1957. From the viewpoint of classical analytical chemistry the continuous flow principle was not a thing of beauty. But for the clinical chemist analysing long series of blood samples it was a joy to get a reasonable reproducibility with the Auto-Analyzer. Besides the Auto-Analyzer was not a black box. On the contrary the apparatus allowed the clinical chemist to improvise. The reagents needed could be made up in the own laboratory from chemicals bought from other companies. The clinical chemist using the continuous flow principle maintained a great influence on the analytical procedures in his laboratory. The same can be said about mechanized versions of basic manual laboratory procedures of other manufacturers such as the centrifugal analyzers and the discrete analyzers.

Instead of mechanization the term automation is often applied in the field of clinical chemistry. This is not entirely correct if the apparatus is not self adjusting by feedback of information. But as automation is the ultimate aim and today a microprocessor is built in many analyzers there is no reason to censure the term. Due to the successful introduction of mechanization-automation in clinical chemistry the workload in the clinical chemistry laboratories in the western world could steadily increase without a comparable increase in number of staff. In this period the activity of industrial research got a firm grip on the development of clinical chemistry. Laboratories were equipped with sophisticated instruments and supplied with corresponding reagents by one and the same manufacturer. And all kind of kits could be bought on the market. Consequently there was a radical change in the status of the inventive user. He lost his dominant position and became a link in a complicated chain of events.

To become fully master of the development in the field of clinical chemistry industry redoubled the efforts. The innovative process was systematically organized. The drawback of this otherwise useful development is the black box character of the modern automated systems. Analytical procedures are hidden in software. The clerical procedures are also computer aided. And the source of this software is often not available. The exact composition of the reagents necessary is often kept secret. The clinical chemist has not only lost every degree of freedom to adapt analytical procedures but sometimes does not even know the details of a determination performed by an automated analyzer in his own laboratory. And the situation is worse in the field of bedside testing and home testing, especially if dry chemistry is used. In the hospital the clinical chemist if often put in charge of the quality assurance of the bedside testing. Not a nice part of his job if the manufacturers prefer to deal about this part of clinical chemistry with members of the hospital staff lacking professional clinical chemical knowledge.

In my experience manufacturers are not eager to listen to objections against this situation. I sometimes fear they like the black box character of modern analyzers and bedside testing. The almalgamation of the manufacturers of instruments with the manufacturers of kits and reagents reinforces this attitude. It is evident that industry shall acquire in this way exclusive control of analytical procedures in clinical chemistry. In my opinion this monopoly will be harmful to the progress of the analytical side of clinical chemistry. And our discipline badly needs fundamental improvement in the analytical sector. Analytical specificity is still lacking in the majority of the routine methods. The well-known matrix effect is the consequence of the lack of analytical accuracy and the variable composition of the body fluids. It is the Achilles heel of clinical chemistry. Of course we wish to remove this weakness. Our ideal must be to use only methods able to determine solely the components they purport to measure.

In my opinion a monopoly of the industry based on black boxes and kits with a partly concealed composition is an unsound basis to improve the analytical side of clinical chemistry. I feel obliged to mention specifically the field of immunoassays. In this branch of clinical chemistry the detection and quantification of chemical substances is performed with antigen-antibody interactions. The world market for immunoassays in biological fluids is large, it amounted in 1985 to 1250 million ECU. The worldwide revenue of diagnostics and diagnostic instruments amounted in 1985 to about 5000 million ECU. The conclusion is that immunoassays already include a quarter of this market. Mechanization-automation is on the moment introduced in this field by manufacturers. And all kind of biological important substances can be estimated with immunoassays. Therefore it is reasonable to expect a great expansion of the market in this branch of clinical chemistry. The manufacturers of instruments and the corresponding reagents will they resist the temptation to build automated analyzers for immunoassays with an extreme black box character? I sincerely hope so because the freedom of action of most clinical chemists is already very limited. Often they have to pay for their automated analyzers by using exclusively the reagents of the manufacturers. Because of the reasons mentioned and the fact that antibody-specificity is a difficult problem it is to be feared that the accuracy of immunoassays in body fluids will not be granted priority. It is a bad omen that the term immunology is frequently used for this field of laboratory methods. Of course there is no reason at all to call it immunology. It is clinical chemistry with antibodies as reagents. There is no need for an indication depending on the reagents used.

To judge immunoassays the criteria of classical analytical chemistry must be used. Considered from this point of view immunoassays can be improved. Both the reagents and the substances to be detected and quantitated are not well-defined. Furthermore the WHO assigned the values in its reference preparation for immunoassays of proteins in International Units. Factors for conversions of IU to mass units have not been defined. Because of these facts each laboratory has to establish its own reference ranges. Accuracy and the SI are ignored. It is a chaotic situation in this very important branch of clinical chemistry. It has to be feared that this primitive state will be semipermanent as a result of the introduction of automated analyzers with a black box character unless the clinical chemist and industry can outline a common strategy for a solution. But industry and the clinical chemist do they have a common interest in this matter?

The clinical chemist must be always aware of the fact that analytical chemistry is a keystone of clinical chemistry. The standards of his profession forbid him to degenerate into a mere consumer of kits and a user of automated black boxes. He is a producer of clinical chemical information. For that purpose he interprets analytical results and must therefore have a profound knowledge of the analytical methods used in his laboratory. His profession is founded by Bang, Folin, Van Slyke and other excellent analytical chemists. Industry must understand, even appreciate this attitude.

The clinical chemist has today to accept the preponderant position of technology and consequently the importance of industry for his profession. He must accept that manufacturers find it profitable to deal with as much people as possible. Therefore it is obvious manufacturers prefer analytical methods requiring no professional knowledge at all. They like black boxes, beside testing, home testing, dry chemistry and patents. A black box is user-friendly for the unskilled and moreover a good hiding place for valuable industrial secrets. The market for bedside testing and home testing is not to be sneezed at. The clinical chemist must not criticize industry in trying to reach as large a market as possible in laboratory medicine. It is a chief feature of industry in the open western society. Thanks to this quality the remarkable expansion of clinical chemistry in the western world was possible. And the benefits of this expansion for numerous patients are significant.

Summarizing: We have to accept the fact that the clinical chemist and industry have a number of conflicting interests. But industry and the clinical chemist have a common goal. It is the development of clinical chemistry into a full proof tool for making diagnosis and aiding therapy. Valid data are essential to get valuable clinical chemical information. Therefore the improvement of analytical precision and analytical accuracy deserves the highest attention. The classical analytical standards must be highly valued. It must be stressed that accuracy is not less important than precision. The true value must be our target. We must defeat the matrix effect. Using uncertain consensus values is unworthy. We must standardize both analytical results and reference ranges among laboratories, in bedside testing and home testing all over the world. It is a difficult task but we can make in this way clinical chemistry very patient-friendly. Consequently the worldmarket for diagnostics and diagnostic instruments will grow and the benefits for the patients will be considerable.

I emphasize the need for a close cooperation between industry and the clinical chemist in order to achieve this common goal.

HIGH RESOLUTION NUCLEAR MAGNETIC RESONANCE SPECTROSCOPY IN CLINICAL CHEMISTRY AND DISEASE DIAGNOSIS

J.K. Nicholson

Department of Chemistry, Birkbeck College, University of London, Gordon House, Gordon Street, London WC1E 6BT, U.K.

INTRODUCTION

For over thirty years nuclear magnetic resonance (NMR) spectroscopy has been an important tool for the chemist to aid identification, characterisation and structural analysis of organic compounds. More recently, major developments in the theory and practice of NMR spectrosopy including the advances in superconducting magnet and computer technology have allowed biological and medical problems of great complexity to be addressed successfully.

There are three main types of biomedical NMR instrumentation which are suitable for generating different kinds of clinical information. These include Magnetic Resonance Imaging (MRI), which can be used to produce a three dimensional image of the body by mapping the distribution of water protons, and Topical NMR spectroscopy, which can allow non-invasive metabolic monitoring in localised tissue regions. Both these techniques have received considerable clinical attention. Indeed, over the next few years MRI will probably replace X-ray tomographic imaging for many applications as it uses no ionising radiation. The subject of this paper is the application of High Resolution proton NMR spectroscopy to problems in clinical chemistry and disease diagnosis. Paradoxically, although this is the "oldest" type of NMR spectroscopy, the instrumentation being identical to that used in classical organic chemical studies, it has only recently been realised that it has a very important role to play in medical research and diagnosis. This is mainly because it has only been since the late 1970's that suitably sensitive high field instrumentation has been available and that certain technical problems associated with the suppression of the strong solvent water signal and extensive overlap of peaks from complex biological matrices have been overcome. Furthermore, NMR spectroscopy is still developing very rapidly with major advances in instrumentation hardware and software being expected within the next few years, further increasing the promise of the technique in a clinical chemical role. It should also be appreciated that high resolution NMR spectroscopy is not simply another analytical technique, as NMR spectra also give data on the dynamic structure and molecular motions of compounds and molecular aggregates in solution. This is a unique type of information which can only be generated by NMR measurements and may carry important clinical messages.

Although several (stable) isotopes can be used in clinical NMR studies, proton NMR spectroscopy has the greatest sensitivity and shows most promise for future developments. It is now possible to use NMR to study the biochemical composition of biological samples including urine, blood plasma, bile, seminal fluid, cerebrospinal fluid as well as tissue and faecal extracts and so obtain useful metabolic information. A wide range of high resolution proton NMR studies of body fluids have now been performed to tackle problems in intermediary metabolism and biochemistry [1-10], screening for inborn errors of metabolism [11-18], testing for malignant tumours [19-31], and drug metabolism, excretion and toxicological investigations of clinical relevance [32-48]. A summary of many of these clinical applications is given in table 1.

Advantages of using proton NMR spectroscopy for use in body fluid analysis include its rapid multicomponent detection facility together with minimal physical or chemical treatment of the sample and rapid metabolic fingerprinting time. The total measurement time in the spectrometer is usually <5 min. Only small volumes (0.3 ml) of biofluid are required for NMR analysis and the technique is non-destructive. NMR is suitable for obtaining many types of diagnostic information, because unlike most conventional analytical procedures, there is no need to make assumptions prior to analysis, as to the nature of the components present in the sample. This contrasts markedly with many analytical methods (particularly chromatography) where sample preparation and detection are deliberately optimised to select a restricted range of closely related compounds. However, it is also possible to use simple and rapid chromatographic clean-up procedures (see below) on crude biofluid samples prior to NMR measurement, which greatly extend the range of compounds that can be detected because problems relating to insensitivity and peak overlap can be minimised. It should be noted that in comparison to most chromatographic and other spectroscopic methods, NMR is relatively insensitive, although NMR instrumentation and data processing software are still improving rapidly. With the type of spectrometers that are currently available, in order for proton NMR to be useful in metabolic studies, analytes must normally be present at concentrations >50 μM. The analyte must also possess suitable resonances, usually from CH, CH_2 or CH_3 protons, that can readily be identified and quantified.

BASIC NMR SPECTROSCOPY

The phenomenon of nuclear magnetic resonance was first observed by Bloch in 1946 [49]. It occurs because the nuclei of some isotopes behave like bar magnets when placed in a magnetic field. These nuclei possess an electric charge and mechanical spin which has an associated angular momentum, the combination of spin and charge gives rise to magnetic properties such that when the nuclei are placed in a strong magnetic field they align with its axis. If a radiofrequency pulse of an appropriate energy is then applied absorption of energy will occur as the nuclei change their orientation in the magnetic field and resonance is said to take place. The resonance frequency of a nucleus is dependant both on the applied magnetic field strength and the gyromagnetic ratio (γ, a measure of the energy difference between the two spin states, corresponding to the nuclei either orientated with or against the applied field) of the nucleus (see below). After the radiofrequency pulse is turned off, a receiver coil in the spectrometer detects the signals from the nuclei as their "excited" spin-state decays back to the thermal equilibrium position. The NMR signals decay with time and the pattern of radiofrequency emissions detected by the spectrometer is called a free induction decay (FID). Chemically distinct nuclei have absorb or resonate at different radiofrequencies, thus appropriate computer analysis of the FID (see below) gives rise to chemical

Table 1: Clinical chemical applications of proton NMR spectroscopy

Application and comments		Biological fluid
INVESTIGATION OF BODY FLUID BIOCHEMISTRY AND METABOLIC DISORDERS		
Normal subjects: Assignment of NMR spectra and measurement of metabolites including: acetate, lactate, valine, alanine, glucose, creatinine, citrate, mobile fatty acids, cholesterol and esters, total calcium and magnesium (as EDTA complexes)		plasma CSF
creatinine, hippurate, sarcosine, indoxyl sulphate		urine
N-acetyl glycoproteins		plasma
LDL, VLDL, chylomicrons		plasma
D-lactic acidosis: Characterisation of disorder in patient following intestinal surgery		plasma
Fasting and diabetes mellitus: Investigation of ketone body production and excretion and related biochemical changes		plasma and urine
Inborn errors of metabolism: Characterisation and metabolic fingerprinting by NMR		
phenylketonuria		urine
glyceroluria		
propionic acidaemia		
methylmalonic aciduria	In these studies a	
maple syrup urine	wide range of abnormal	
isovaleric acidaemia	metabolites detected	
glutaric aciduria types	reflecting specific	
1 and 2	enzymic deficiencies	urine
β-ketothiolase deficiency		
hyperglycinaemia		
biotinidase deficiency		
3-methylcrotonyl-CoA carboxylae deficiency		
3-hydroxy-3 methylglutaconyl-CoA lyase defect		
L-2-hydroxyglutaric aciduria		
"CANCER TESTING"		
screening for malignancy by measurement of composite linewidths of lipoprotein/fucosylated glycolipid signals		plasma
papers reporting "successful" test results		
papers reporting equivocal or negative test results		

information on the spin-active nuclei in the sample. A detailed consideration of the theory and application of general and biological NMR spectroscopy is beyond the scope of this paper, and for this the reader is referred to more detailed texts [50, 51]. However, a basic description of the salient features of modern NMR spectroscopy and the important parameters of the NMR experiment are given below.

All biomedical NMR spectroscopy is of the pulsed Fourier transform type. This involved the irradiation of the sample with an observation pulse of radio energy containing many frequencies needed to excite all the nuclei

(Table 1 cont'd)

Detection of drugs in urine of man and animals		
Ethanol	detection in untreated urine samples	urine
Methanol	including formate as methanol metabolite	
Paracetamol	detection and quantification of major metabolites investigation of overdoses	urine
Ampicillin	detection of parent and major metabolites	urine
Benzyl penicillin	detection of parent compound and penillic acid degradation product	urine
N-methyl formamide	detection/quantification of parent and major metabolites	urine
Metronidazole	detection and quantification of metabolites	urine
Methamphetamine	detection of parent compound and analogues	urine
Oxpentyffline	quantification of major acidic metabolite critical comparison with HPLC	freeze-dried urine
Oxpentyffline Ibuprofen Aspirin Naproxen Paracetamol	use of Solid Phase Extraction to eliminate endogenous interferences and allow detection and characterization of major metabolites by NMR and Mass Spectrometry	extracted and freeze-dried urine

of interest within the sample. After the observation pulse as the nuclei relax to their thermal equilibrium position, radio energy is emitted at all the frequencies previously absorbed by the nuclei. These data, showing the decay of nuclear magnetization with time (i.e. the FID's) are detected in the form of analog voltages in the receiver coils and are digitized and stored in the computer, prior to Fourier transformation which converts the FID time domain spectrum into a frequency domain spectrum. This contains resonance frequency and coupling constant (see below) information and intensity data which can be readily interpreted. The resulting Fourier transform (FT) NMR spectrum is then mathematically equivalent to a data set that would be collected by scanning through all the individual resonance frequencies and measuring their absorption intensities, a so called "continuous wave" (CW) spectrum. There are many advantages of FT over CW NMR spectroscopy. Data on all the protons in a sample are collected simultaneously without the need for scanning which greatly speeds data collection. Also, consecutive pulse NMR scans can easily be stacked and averaged in the computer memory, thus improving signal and noise ratios, as in FT NMR the signal strength increases in direct proportion to the number of scans whereas the noise, being random, increase in proportion to the square root of the number of scans. Other advantages of FT NMR include the ability to use the computer to mathematically manipulate time domain spectra prior to Fourier transformation and so cosmetically improve resolution or signal to noise ratios. In modern FT NMR instruments, most of the spectrometer functions are computer controlled, allowing complex multipulse experiments to be planned and executed.

The exact frequency at which a proton resonates depends not only on γ and the applied field but also on its shielding by orbiting electrons which give rise to small local magnetic fields. The resonance frequency is thus sensitive to the chemical environment of the proton, each type has a unique local magnetic environment so giving slightly different resonance frequencies. So for example the resonance frequencies of the CH_3 protons of ethanol would be several hundred Hz different from the CH_2 protons of ethanol. The first major parameter in NMR spectroscopy is therefore the chemical shift, which is a measure of the electron density about the nucleus and consequently its chemical environment. The practical outcome of this is that all organic molecules give rise to their own fingerprint of resonances, and if suitable sophisticated instrumentation is available complex mixtures of organic compounds can be analysed by interpreting the patterns of NMR signals that are produced. It is for this reason that NMR spectroscopy is very useful in the determination of the composition of biological fluids.

The most widely used reference compound for proton NMR spectroscopy is tetramethylsilane or TMS [$(CH_3)_4Si$] which has 12 chemically and magnetically equivalent methyl protons. The chemical shift of TMS is arbitrarily designated as zero parts per million or $\delta = 0$ PPM. The chemical shift of a sample proton is then defined in terms of d units using the following relationship:

$$\delta = \frac{v_S - v_{TMS}}{\text{operating freqency}} \times 10^6 \text{ PPM}$$

Where v_S and v_{TMS} are the resonance frequencies of the sample and the standard in a given external magnetic field. This gives convenient numbers for the chemical shifts of sample nuclei which are independent of the magnetic field strength at which the spectrum is measured. Sodium tetradeutero trimethylsilyl propionic acid (TSP) is used as a chemical shift reference in aqueous solutions its shift being virtually identical with TMS. For most biological applications of NMR spectroscopy very high field magnetic strengths (e.g. 9.4 T) are employed using superconducting magnets. The main advantages of high field instrumentation being good sensitivity and improved frequency dispersion of chemical shifts from different protons hence minimising peak overlap.

The second important parameter in NMR spectroscopy is spin-spin coupling. This arises because nuclei which are closely connected by electronic orbitals can sense the spin-state of their neighbours i.e. whether they are aligned with or against the field. This leads to characteristic splitting of the NMR spectral lines for nuclei that is dependent on the number of near spin-active-neighbours, the number of connecting bonds and their geometrical relationships. The magnitude of the coupling constants (J) in Hertz is independent of the applied field strength. The majority of small molecules that have adjacent CH, CH_2 and CH_3 groups show extensive proton-proton coupling. Two, three and four bond couplings are usually observable, becoming weaker as the number of bonds between the nuclei increases.

Spin-active nuclei in a magnetic field can exchange energy with their environment by radiationless transitions or relaxation processes. This occurs because of fluctuations of the local magnetic field close to the resonance frequency of the nucleus which help it lose energy to the surroundings. Relaxation is the third important NMR parameter that can be measured for spin-active nuclei. Two main types of relaxation occur and are designated T_1 and T_2 relaxation processes, corresponding to relaxation back in the direction of the applied magnetic field (along the z axis) or at right angles to this (in the xy plane). T_1 and T_2 relaxation times are exponential decay constants and are expressed in seconds. There are various mechanisms whereby T_1 and T_2 relaxation can occur, detailed discussion of which are outside the scope of this paper [see 49]. However, relaxation phenomena are important in biomedical NMR spectroscopy for a variety of reasons. For instance, if quantitative NMR data are to be obtained for metabolites in biological fluids, all spectra must be fully T_1 relaxed. In FT NMR spectroscopy sufficient time must be allowed after the application of the observation pulse otherwise magnetic equilibrium will not be re-established prior to the next pulse. This would cause progressive saturation of the resonances and a reduction in the intensity of the signals, those from nuclei with long T_1 relaxation times experiencing the most severe distortions.

T_1 relaxation times are dependent on factors such as temperature and viscosity, higher temperatures slowing down and higher viscosities speeding up relaxation. The presence of paramagnetic species in solution may greatly reduce relaxation times. Unpaired electrons have large magnetic moments compared with those of the spin-active nuclei and so paramagnetic ions produce the large fluctuations in the local magnetic field about the nucleus and give the nuclei a very efficient relaxation mechanism. T_1 and T_2 relaxation times for a given nucleus are closely related, but T_2 relaxation times are particularly dependent on the molecular mobility of the solute under study and the so called "tumbling time" in solution. In practice this means that most protons on large molecules such as proteins, nucleic acids and some polysaccharides and lipids have efficient T_2 relaxation processes. There is a simple mathematical relationship between the linewidth of an NMR line and the T_2 relaxation time of the nucleus, given by

$$T_2^{obs} = [\pi . \Delta v_{½}]^{-1}$$

where $\Delta v_{½}$ is the half height linewidth of the peak in Hertz and T_2^{obs} is the observed transverse relaxation time in seconds. Nuclei with very short T_2 relaxation times consequently give rise to very broad lines. Therefore, biological matrices containing large amounts of protein and/or nucleic acid such as plasma or cell suspensions will give many broad proton NMR resonances. This may be very troublesome in proton NMR spectroscopy of biofluids, where the macromolecular resonances from thousands of magnetically non-equivalent protons give rise to a broad envelope of signals which mask the signals from low molecular weight compounds of metabolic interest. Metabolites exhibiting extensive protein binding will also have intrinsically broad NMR lines in such matrices, as their molecular motion is constrained by the macromolecules to which they are bound, consequently shortening their T_2 relaxation times. Fortunately, there are spin-echo Fourier transfer NMR methods [52, 53] that allow low molecular weight compounds (usually with long T_2's) to be observed in the presence of macromolecules. Spin-echo pulse sequences include relaxation delays prior to the collection of the FID during which signals from species with fast T_2 relaxation times are significantly attenuated leaving only signals from low molecular weight or high molecular mobility moieties [2, 4]. Occasionally some regions of macromolecular structures are highly mobile and give rise to sharp lines which are not suppressed in spin-echo experiments, e.g. N-acetylated sugar signals from plasma glycoproteins and also from certain mobile aliphatic sidechains on lipoproteins [8, 9].

To summarise, there are several parameters of the NMR spectrum that are important in the detection and identification of metabolites in biological matrices. These include: the chemical shift, giving information on the chemical environment of the nuclei concerned; the multiplicity of the resonance line giving information about the number, distance and geometry of spin-active neighbours; the integrated area of the resonance which is directly proportional to the concentration of the species; the relaxation time which conveys information about molecular mobility and macromolecular binding.

In general, ^{1}H NMR spectroscopy of intact biological materials and biological fluids is hindered by two main problems: Firstly, there is often extensive peak overlap of the many signals from a wide variety of biomolecules appearing in a relatively narrow chemical shift range. This is minimised by use of the high field strength spectrometers e.g. of 9.4 T (400 Hz, ^{1}H frequency) and above, giving wide spectral dispersion of signals [6]. However, the second and more important problem is associated with the large dynamic range due to the intense [110 M in protons] signal from solvent water, which severely limits the digitization of signals from compounds present at millimolar concentrations. This problem of dynamic range originates in the digital nature of the FT NMR spectrometer i.e. all analog voltages generated in the receiver coils require digitization prior to mathematical manipulation by the spectrometer computer. Therefore, the use of some type of water suppression technique is essential if good digitization and signal-to-noise levels are to be obtained for signals of metabolites in untreated biological samples. It is possible to freeze-dry samples and redissolve them in reduced volume of $^{2}H_2O$. This largely eliminates the dynamic range problem and increases the concentration of metabolites of interest. However, such action usually results in the loss of volatile or unstable compounds, and selective deuteration of metabolites with exchangeable or acidic protons, so NMR data obtained from samples prepared in this way must be interpreted with caution.

In the past we have shown that the application of a secondary irradiation field at the water resonance frequency generally results in an acceptable degree of water suppression [2]. However, in dilute solutions of biochemicals such as body fluids, this method may not be sufficient to suppress the water signal fully, and residual wings of the water peak may give rise to severe dynamic range problems. In samples containing high concentrations of macromolecules such as blood plasma, in addition to water suppression, the use of spin-echo pulse sequences are required to eliminate broad overlapping resonances which are attenuated by virtue of their short T_2 relaxation times [2, 4, 53]. Spin-echo pulse sequences are also highly effective at reducing the intensity of the water signal in biofluids after the addition of a suitable chemical agent and modification of the sample pH to selectively enhance the water relaxation rate. Such chemical exchange relaxation agents include ammonium salts, urea, and guanidinium salts and their judicious use in combination with spin-echoes allows almost complete elimination of the solvent signal and one vastly superior to simple irradiation methods [54, 57].

Proton NMR urinalysis

Urine is a complex biochemical matrix that can pose severe problems to conventional analytical procedures due to its highly variable metabolite composition and concentrations and variable ionic strength and pH. In proton NMR spectroscopy of urine the main problem is to overcome the intense water peak which can now be done routinely (see above). The wide variety of endogenous and xenobiotic metabolites that can be rapidly visualised by proton NMR of urine, makes the technique powerful in metabolic monitoring. Many examples of proton NMR urinalysis are given in Table 1. An illustration of the type of data generated by this technique is shown in Figure 1. This compared the aliphatic regions of 400 MHz proton NMR spectra from untreated urine samples (except with the addition of 10% D_2O as a spectrometer frequency lock) from a normal subject fasted for 12 and 48 hours (1A and B) and from an insulin dependent diabetic 4 hours after withdrawal from insulin therapy under controlled conditions. Each spectrum took less than 5 minutes to obtain. Note that the creatinine concentrations in each sample were approximately equal and the diabetic

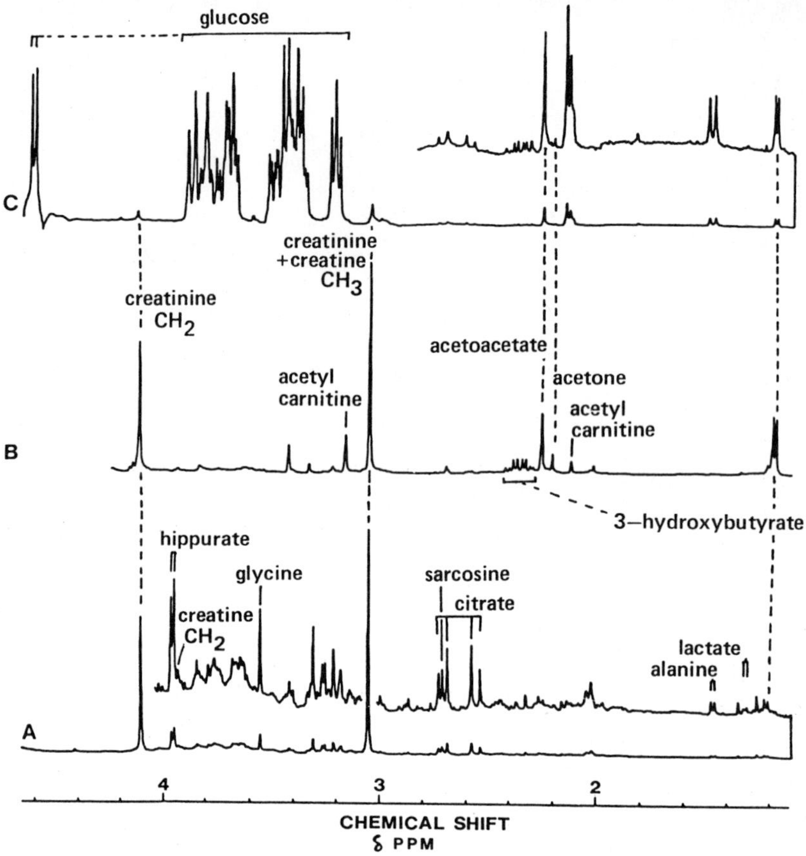

Fig. 1. Proton NMR of urine

spectrum (1C) has been plotted on a different vertical scale. Many signals from interesting and important endogenous metabolites can be assigned. In both the 48 hour fasted and diabetic urine samples signals from the ketone bodies (3-D-hydroxybutyrate, acetoacetate and acetone) and acetylcarnitine can be observed. As expected the diabetic sample also contains high concentrations of glucose. We have also applied NMR in this way to obtain quantitative information on the time course of excretion of these compounds simultaneously [4, 7]. Many other metabolic disorders can be investigated and abnormal compounds fingerprinted and identified. Inborn errors of metabolism in particular lend themselves to this type of study, as many unusual metabolites are often present at high concentration producing very specific NMR fingerprints for the particular disease state. These fingerprints can be obtained quickly with minimal biochemical effort or sample preparation [11-18]. Conventional methods for obtaining this type of data such as gas-liquid chromatography, although more sensitive than NMR (even though this sensitivity is often redundant with inborn errors of metabolism), usually require several preparation/derivatisation steps before analysis can take place. However, at present the use of NMR for characterising inborn errors of metabolism is still at the research stage, supplementing gas-chromatographic and mass spectrometric (MS) data. Many of the inborn errors of metabolism that have been studied by proton NMR (particularly by Drs Iles and Chalmers in London) are listed in Table 1.

In addition to organic acids, a variety of neutral and basic compounds can also be detected simultaneously, so that aminoacidopathies and disorders of intermediary metabolism can be studied as well as organic acidurias. The combination of metabolites that can be measured by NMR in crude biofluid samples is virtually unique, perhaps only the as yet very esoteric technique of MS-MS can detect as wide a range of metabolites in untreated biological samples. As such NMR urinalysis may be particularly suitable for population surveys of metabolic disease, as the high capital cost of the instrumentation is largely offset by the very low cost per sample and cost per analyte.

Studying the clinical biochemistry of blood plasma by proton NMR

The detection and estimation of low molecular weight compound in untreated plasma by proton NMR is hindered principally by the abundant dissolved macromolecules such as albumin and immunoglobulins [2]. These high molecular weight species give rise to broad overlapping resonances that mask the signals from the metabolites of interest. Fortunately spin-echo pulse sequences can now be employed to effectively eliminate the broad components [2]. The most widely used spin-echo experiment for this purpose is the Hahn type which utilises the pulse sequence:

$$D\text{-}(90°\text{-}\tau_2\text{-}180°\text{-}\tau_2\text{-collect FID})_n$$

where D is a delay between an successive spin-echo pulse sequences to allow T_1 relaxation and τ_2 is a short delay (typically about 60 ms) to allow magnetisation decay via spin-spin relaxation for high molecular weight species of low molecular mobility. In the Hahn spin-echo experiment, the amplitude of the observed signal (echo) at time $2\tau_2$ after the 90° pulse is dependent on the intrinsic T_2 of the observed resonance, homonuclear spin-spin coupling (which can also produce phase modulation), and a term that arises if the molecule under study diffuses through a field gradient during the second $\tau 2$ period [2]. At a τ_2 value of 60 ms (1/2J) doublet resonances with J = 8.2 Hz are phase-inverted whereas triplets and singlets remain upright [2]. Sufficient time ($2\tau_2$) then elapses for the net magnetisation associated with the broad resonances (short T_2) to decay to zero so they do not contribute to the spectrum. This effectively edits the spectrum according to molecular mobility or molecular weight hence simplifying the NMR data set. The operation of such a spin-echo experiment on heparinised blood plasma is shown in Figure 2A which is compared with the normal single pulse spectrum (Figure 2B). The spin-echo spectrum contains only resonances from low molecular weight metabolites and highly mobile lipids (for assignment details see references 2, 4, 8 and 9) that were obscured in the single pulse spectrum. By use of these techniques we were able to use NMR of plasma to monitor the onset of ketosis in fasted normal subjects and insulin dependent diabetics after insulin withdrawal, and to relate this to concomitant changes in amino acid and lipid metabolism data which were also obtained from the plasma NMR spectra [4]. Spin-echo experiments can also be used to simplify the spectra obtained from other complex biofluids such as cerebrospinal fluid [10], bile and seminal plasma [57] and in principle to study other types of metabolic disorder. It is of note that certain high molecular weight species may also contribute proton resonances in spin-echo spectra if they have local domains of high molecular mobility. Such is the case with many of the "acute phase" glycoproteins (Figure 2A) such as α_1-acid glycoprotein, which have highly mobile carbohydrate sidechains (largely N-acetylglucosamine and N-acetylneuramic acid) with long T_2 relaxation times [8]. The intensities of these signals appeared to correlate with clinical conditions in which an elevation of "acute phase" proteins is expected such as rheumatoid arthritis and monoclonal (IgG)

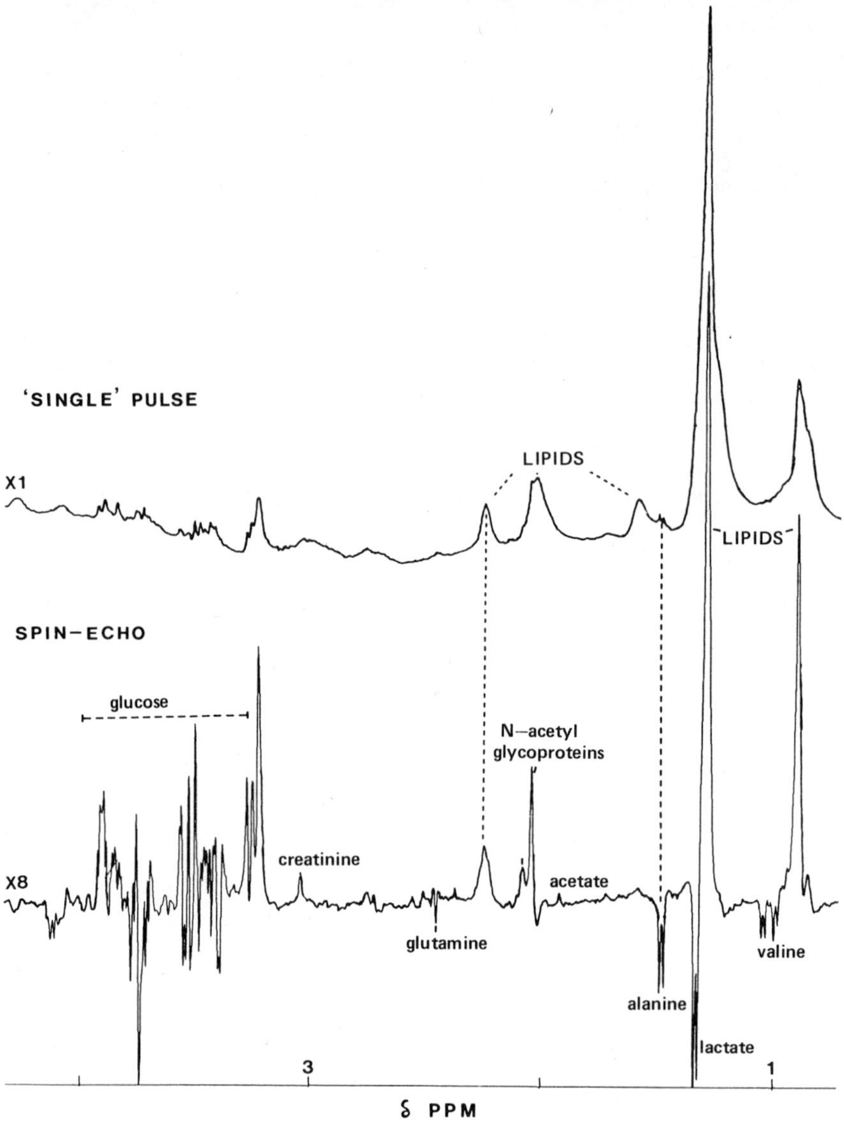

Fig. 2. Proton NMR of plasma

gammopathies. It is therefore to be expected that spin-echo spectroscopy of plasma from such subjects is likely to be of value in the study of molecular aspects of these disease states.

CAN NMR SPECTROSCOPY OF PLASMA PROVIDE A NEW TEST FOR CANCER?

A simple laboratory test for malignancy has long been sought. Recent studies suggest that a rapid proton NMR-based test on human plasma to detect malignant tumours may be feasible [23]. It has been known for some time that metastasising tumours have unusually long proton T_2 relaxation times [20], which distinguish them from malignant but non-metastatic cells. The long T_2 relaxation time is due mainly to the high level of surface fucogangliosides on the malignant cell surface. These compounds may also be closely related to the metastatic potential of the cells [19] and there appears to be a general relationship between the disorders of lipid

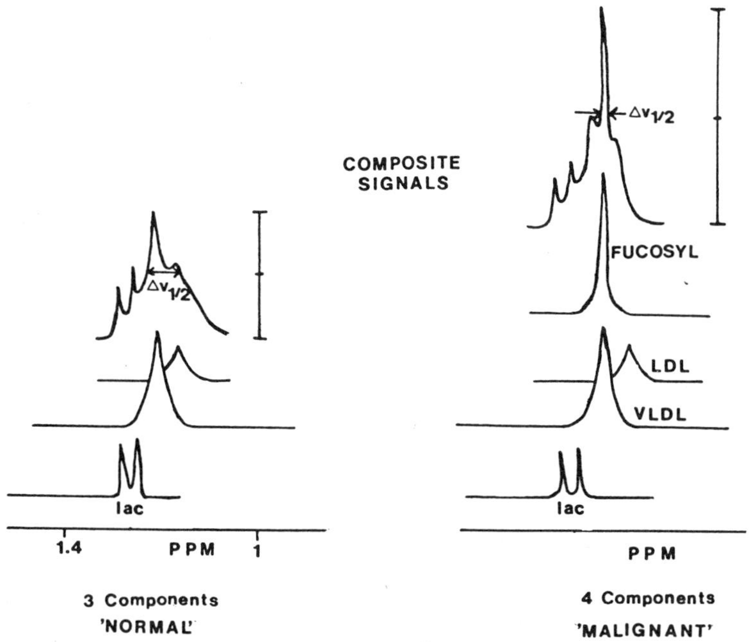

Fig. 3. ^{1}H NMR 'Cancer test'

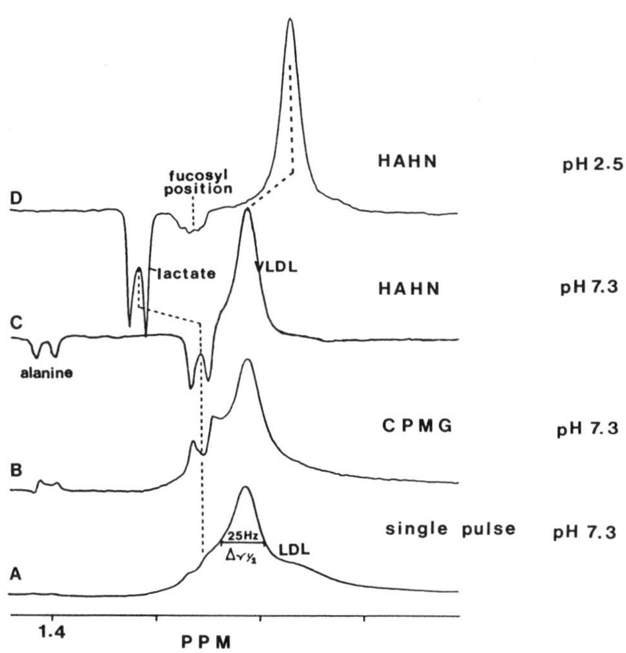

Fig. 4. For explanation see text.

metabolism as well as a link between cancer and glycosylated proteolipids [21]. It has been suggested that neutral lipid domains in the plasma membranes of malignant transformed embryonic cells resemble proteolipids and that the moiety responsible for the long T_2 relaxation time is associated with the shed proteolipid particles [22]. Proteolipid extracted from the plasma of a patient with a borderline ovarian tumour also gave proton resonances from nuclei with fast T_2's. In a detailed study of the plasma lipoproteins and proteolipid complexes from this patient [19], it was found that the latter gave rise to T_2 relaxation times of >400 ms, whereas the conventional lipoproteins had much shorter T_2's (>160 ms). The proteolipid complexes were subsequently found to have a 20% glycolipid content in which a fucosylated molecule with a long proton T_2 was identified by NMR.

In a recent report Fossel et al. [23] reported that estimation of the linewidth of the composite peak in plasma centred at about 1.2 ppm could give information on the presence of a malignant tumour in the donor. If the overall linewidth of this peak is less than ca. 28 Hz then it is reported that this is an indication of malignancy [23]; normal subjects give linewidths of about 40 Hz. The peak measured has contributions from various lipoproteins, cholesterol, free fatty acids, lactate, threonine and fucosylated moieties. A simplified but greatly exaggerated diagrammatic representation of the single pulse spectrum of this region is shown in Figure 3. In Figure 3A the "normal" situation is shown with three main components (lactate, LDL and VLDL). Although it should be noted that in non-fasted individuals chylomicrons will also contribute significantly to this peak. In Figure 3B the putative "malignant situation" is shown in which there is a further contribution from the elevated levels of fucosylated glycolipids [22]. The effect is clear, the sharp fucosyl resonance reducing the overall half height linewidth of the composite peak, in theory providing a simple means of diagnosing cancer based on the elevated levels of fucosylated glycolipid. In real situations the difference between the "normal" and "malignant" appearances of these resonances is much less distinct and the measurement of composite peak linewidth by this method is likely to give equivocal results.

Consideration of the single pulse (Figure 4A), Carr Purcell Meiboom Gill (Figure 4B) and Hahn spin-echo (Figures 4C and D) spectra of normal plasma shows the depth of the problem involved in understanding the significance of the linewidth of this composite peak. 400 MHz spectra were obtained using 0.5 ml plasma containing 10% D_2O. 64 transients were collected in each case and in spin echo spectra total spin-spin relaxation delays of 120 ms were used. Spectra in Figures 4A-C were measured at pH 7.2 and D at pH 1.6. In Figure 4A the composite signal as measured by Fossel is shown in a real sample. The very broad component (from low density lipoprotein) is eliminated in the spin-echo spectra which also show the strength of the lactate signal that is not resolved in the single pulse spectrum. Lactate has no real diagnostic value where malignancy is concerned and is a variable and major interference in this measurement. Of course it can easily be removed from samples of this type by use of lactate dehydrogenase which converts lactate to pyruvate the resonance of which is shifted well downfield of the composite peak and would not interfere with its linewidth measurement. The elimination of this type of variation is essential if meaningful linewidth values are to be obtained from the composite peak. On changing the pH of the sample to 1.6 (Figure 4D) the lactate methyl signal shifts downfield as it is protonated. This then reveals a second partially resolved doublet with a shift very similar to that of fucose which appears to be coincident with lactate at pH 7.2. Threonine, which also gives a doublet near this position has been eliminated as the source of this signal as it too shifts with lactate at low pH, whereas fucose keeps a relatively constant position. It has been suggested that it is the presence of high levels of fucosylated glycolipids (which have relatively sharp fucose methyl doublets) in the plasmas of patients that gives rise to the overall sharpness of the composite signal at 1.2 ppm. If so then it should be possible to measure the strength of this signal more directly by use of Hahn spin-echo experiments performed at low pH, as this both eliminates lactate interference and that from the bulk of the very broad lipid and lipoprotein components.

Although there appears to be some promise in the use of NMR for cancer screening or at least in the investigation of the disease process, the use of this technique in such a simplistic way to test for such serious conditions should be regarded with great suspicion until the full meaning of the measured parameters are understood. Clearly the interpretation of this NMR data will always be non-trivial. Furthermore, there is a vast amount of metabolic and biochemical information in the NMR spectrum of plasma and it would also be a pity to waste most of this by performing a single linewidth measurement on the sample in order to attempt the diagnosis of a condition.

NMR, DRUGS AND CLINICAL TOXICOLOGY

Proton NMR spectroscopy can be used to detect drugs and metabolites in a variety of biological fluids, it can also be used to obtain information on the toxic effects of the compounds under study. Indeed, such information may be obtained at the same time that the analysis of the sample for drug related material is undertaken. We have shown that proton NMR techniques can be useful in the study of the biochemical effects of a variety of different classes of metabolic poisons [44-48]. For example in a detailed study of the nephrotoxicity of mercury II chloride in the rat, the proton NMR fingerprints of urine from treated animals were grossly different from controls and significant changes in the urinary excretion profiles of 14 out of 24 endogenous metabolites were detected [46]. NMR urinalysis potentially provides a good detector for acute renal damage caused by drugs or toxins, being at least as sensitive as conventional urinary marker enzyme assays or histopathology [46-48]. However, patterns of metabolite excretion relate to functional changes in the renal tubules after exposure to the toxins, and so interpretation of these patterns can also give insight into the molecular mechanisms of toxicity [46]. Other types of toxin produce characteristic changes in metabolite profiles and NMR is proving useful in the detection of novel markers of toxicity and organ specific damage [44-48]. As yet the clinical value of this work is unclear, although proton NMR may be readily applied to the detection of many drugs and drug metabolites in biofluids after therapeutic use or in overdose situations (see Table 1). Another benefit is that in addition to data on the drugs themselves, proton NMR also gives information on endogenous metabolites which may carry prognostic information in drug overdose cases as well as diagnostic information on the nature of the overdose. We were able to illustrate this, at least in principle, in NMR studies on paracetamol (acetominophen) overdose cases, where abnormal paracetamol metabolite ratios were found in urine, plasma and cerebrospinal fluids of the patients suffering severe paracetamol-induced hepatotoxicity [10, 58].

There are many problems associated with the use of NMR methods to measure drugs in biofluids directly. Many modern drugs are used at comparatively low doses, some have very complex structures or multiple metabolic pathways. NMR spectroscopy is an inherently insensitive technique and these factors combine to limit the range of drug metabolites that are directly NMR-detectable in crude biological fluids. The sensitivity problem is of course less serious when overdose cases are being considered, however if drug mixtures or "cocktails" have been taken there may be problems of peak overlap from metabolites. Indeed, we found that when paracetamol was given at therapeutic resonances from the parent drug and its cysteinyl conjugate in urine were extensively overlapped (even at 500 Mz) defying direct quantitation [33]. Two-dimensional proton correlation methods were required to deconvolute these spectra so that all the major metabolites were NMR-detectable [34]. Consideration of the theory of two-dimensional NMR is beyond the scope of this paper (see ref. 59) but represents a "high technology" approach to solving problems posed by very complex overlapping data sets. Although two-dimensional NMR methods can be used with ease on modern high field spectrometers, they are still costly in terms of data acquisition time, data processing and computer memory and storage space. They can only be used on selected samples where particular assignment problems are presented and at present are not suitable for routine clinical samples. We have therefore investigated some simple procedures for rapid sample clean-up prior to NMR spectroscopy that will allow very complex samples to be usefully studied by NMR even at comparatively low field strengths (250 MHz).

One of the most widely used clean up procedures for biological samples prior to analysis by HPLC involves the use of small disposable solid phase extraction columns (e.g. Bond Elut™ Analytichem International, Figure 5C). These columns contain a sorbent bed consisting of silica gel with various possible bonded functionalities joined by silyl ether linkages; the bed is of similar material to that used in HPLC columns. Samples can be loaded quickly onto the top of the bed and according to the prevailing chemical conditions and bonded moiety will either allow analytes to pass through the column or be retained and possibly concentrated prior to elution with a different solvent system. The eluting solvent system can also be applied as a stepwise gradient in which retained analytes can be fractionated according to the strength of their interactions with the column; this is then a chromatographic procedure in itself. An application of this approach to the detection of a drug metabolite in urine Solid Phase Extration Chromatography with NMR detection (SPEC-NMR) is shown in Figure 5. The 250 MHz spectrum of a human urine sample collected after dosing with 600 mg of the drug oxpentifylline (Trental™) is shown in Figure 5A. The signals from the major metabolite of this drug (structure shown in figure) is difficult to detect in the background "chemical noise" [41] caused by the multiple overlapping resonances of endogenous metabolites. The sample was acidified (thus protonating the acidic function of the drug metabolite) and loaded onto a C18 (octadecyl) bonded column. The ionisation of the drug metabolite was suppressed by the lowering of the pH and together with other unionised molecules with significant non-polar interactions was then retained on the acidified column. A stepwise gradient of deuteromethanol and D_2O (e.g. 20, 40, 60, 80, and 100%) was then applied and the fractions collected and scanned by proton NMR. The spectrum of the 60% methanol fraction is shown in Figure 5B, and consists almost entirely of resonances from the drug metabolite of interest.

SPEC-NMR procedures can be applied to separate and detect a wide range of drugs and endogenous metabolites very quickly and will have widespread applications in drug metabolism studies and general toxicology [42, 43]. A further important advantage of this method is that metabolites can be concentrated from large volumes of biofluid thus aiding NMR detection. The SPEC-NMR approach has many of the attractive features of HPLC-NMR [60] without most of the disadvantages. In particular SPEC-NMR is relatively cheap and does not require a dedicated NMR instrument as is the case with HPLC-NMR which uses the spectrometer as an online detector. Furthermore, the analytical strategy involved in HPLC and indeed other chromatographic

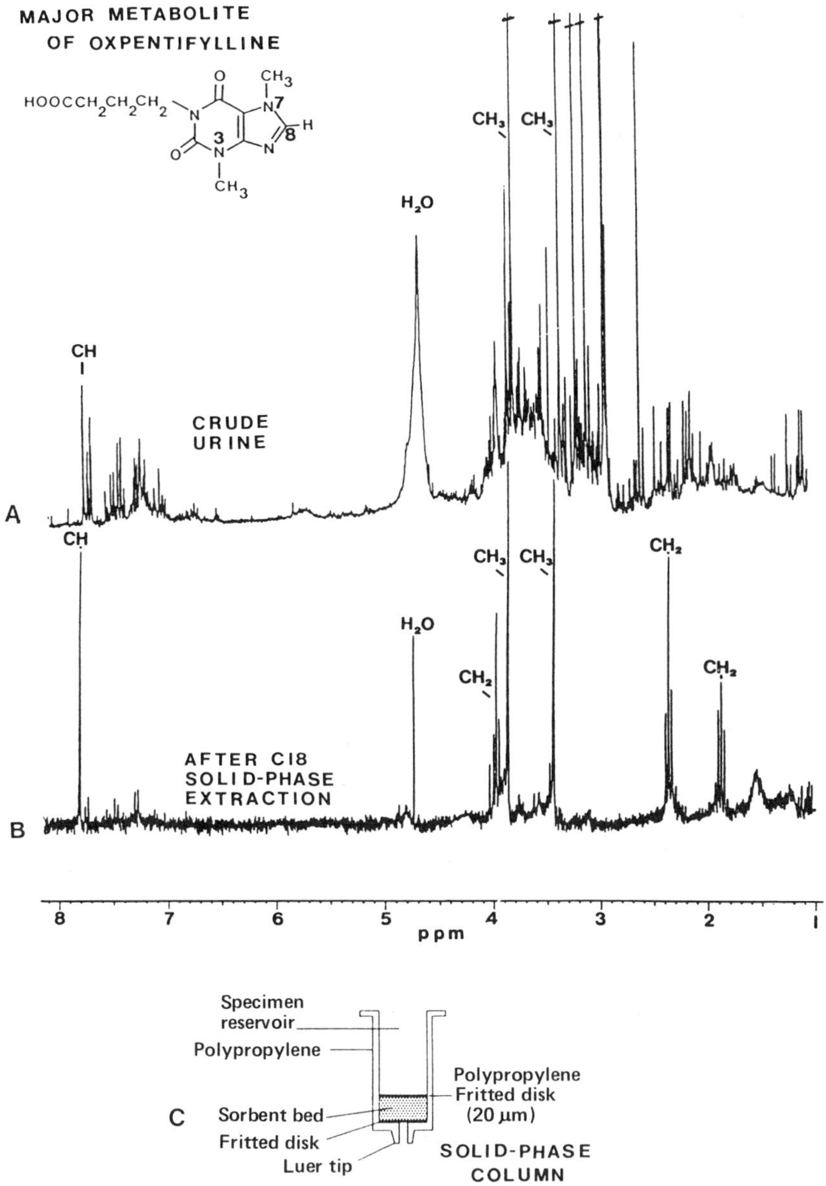

Fig. 5. SPEC-NMR of urine.

techniques always involves optimisation of the separation procedure. In the case of SPEC-NMR this is much less important as the NMR spectrometer is effectively a multicomponent detector and mixtures of compounds can be studied with relative ease, so that the SPEC-NMR technique represents a completely different analytical strategy. With relatively little method development it will be possible to devise SPEC-NMR protocols for detecting many of the commonly used and abused drugs present in biofluids. These methods with the provision of only modest NMR instrumentation may be of considerable value in future clinical and forensic toxicology studies.

DISCUSSION AND CONCLUSIONS

High field proton NMR spectroscopy can rapidly provide both qualitative and quantitative data on both endogenous and exogenous compounds present in a diverse range of biofluids. Whilst such studies are performed more rapidly using very high field/frequency NMR spectrometers (e.g. 400-600 MHz) which minimize peak overlap, and thus allow the interpretation of complex patterns of resonances, such instruments though desirable are not essential. Good results can be obtained with relatively modest instrumentation of 200 MHz (such spectrometers are widely available) although there is often the requirement for some form of sample pretreatment, such as solid phase extraction, before useful measurements can be obtained.

The range of clinical chemical problems that can be addressed by high resolution proton NMR spectroscopy is vast, including metabolic and therapeutic monitoring, diagnosis and screening for inborn errors of metabolism, possible tests for cancer, and clinical toxicology and drug overdose studies. At present the main limitations are cost and sensitivity. The capital cost of a high field NMR spectrometer is high, and skilled staff and expensive maintenance are required. However, the expected lifespan of an instrument could be 15-20 years, and when calculated, the cost per sample or per analyte is very low indeed as sample preparation is often minimal. Furthermore, NMR spectroscopy is a technique that is still developing rapidly. Instrument sensitivity is increasing steadily and much higher field strength instrumentation will become available in the next few years. NMR is a computer based technique and computer software and hardware for NMR data processing is also improving very quickly. It is widely thought that computer-based pattern recognition programmes will soon become available. This will allow spectrometers to be provided with a metabolic database so that identification of metabolites in biofluids may become, at least in part, machine-based, greatly simplifying the task of spectral assignment. Given this and the current explosion of interest in biomedical NMR spectroscopy, it is hard to believe that there is not a major role for this technique to play in the clinical chemistry laboratory of the 21st century.

ACKNOWLEDGEMENT

I am grateful to the National Kidney Research Fund for supporting this and related work.

REFERENCES

1. Bock JL. Clinical Chemistry 1982; 28: 1873-7.
2. Nicholson JK, Buckingham MJ, Sadler PJ. Biochemical Journal 1983; 211: 605-15.
3. Traube M, Bock JL, Boyer JL. Annuals of Internal Medicine 1983; 98: 171-3.
4. Nicholson JK, O'Flynn M, Sadler PJ, Macleod A, Juul SM, Sonksen PH. Biochemical Journal 1984; 217: 365-75.
5. Nicholson JK, Bales JR, Sadler PJ, Juul SM, Macleod A, Sonksen PH. The Lancet 1984; ii: 751-2.
6. Bales JR, Higham DP, Howe I, Nicholson JK, Sadler PJ. Clinical Chemistry 1984; 30: 426-32.
7. Bales JR, Bell JD, Nicholson JK, Sadler PJ. Magnetic Resonance in Medicine 1986; 3: 849-56.
8. Bell JD, Brown J, Nicholson JK, Sadler PJ. FEBS Lett 1987; 215: 311-5.

9. Bell JD, Sadler PJ, Macleod AF, Turner PR, La Ville A. FEBS Lett 1987; in press.
10. Bell JD, Brown JCC, Sadler PJ, Macleod AF, Sonksen PH, Hughes RD, Williams R. Clinical Science 1987; 72: 563-70.
11. Iles RA, Snodgrass GJ, Chalmers RA, Stacey TE. The Lancet 1984; ii: 1221-2.
12. Hind AJ, Chalmers RA, Iles RA. Biochemical Society Transactions 1985; 13: 201-2.
13. Hind AJ, Chalmers RA, Iles RA. Clinical Science 1984; 67: 53 p.
14. Lehnert W, Hunklers D. Journal of Pediatrics 1986; 145: 260-6.
15. Iles RA, Hind AJ, Chalmers RA. Clinical Chemistry 1986; 31: 1795-1801.
16. Yamaguchi S, Koda N, Eto Y, Aoki K. The Lancet 1984; ii: 284.
17. Yamaguchi S, Koda N, Eto Y, Aoki K. Journal of Paediatrics 1985; 106: 620-2.
18. Yamamoto H, Koda N, Yamaguchi S, Eto Y, Aoki K. Clinical Chemistry 1987; 33: 714-5.
19. Mountford CE, May GL, Wright LC, et al. The Lancet 1987; 829-34.
20. Mountford CE, Wright LC, Holmes KT, Mackinnon WB, Gregory P, Fox RM. Science 1984; 226: 1415-8.
21. Skipsi VP, Barclay M, Archibald FM, Stock CC. Progress in Biochemical Pharmacology 1975; 10: 112-34.
22. Wright LC, May GL, Dyne M, Mountford CE. FEBS Lett. 1986; 203: 241-60.
23. Fossel ET, Carr JM, McDonagh J. New England Journal of Medicine 1986; 315: 1369-76.
24. Parl FF, Harris TM. New England Journal of Medicine 1987; 316: 1411-2.
25. Ross BD, Barker PB, Eley CGS, Schmidt PG, Roberts JD. New England Journal of Medicine 1987; 316: 1412.
26. Regan MC, Cottrell C. New England Journal of Medicine 1987; 316: 1412.
27. Small DM, Hamilton JA. New England Journal of Medicine 1987; 316: 1412-3.
28. Campbell RA. New England Journal of Medicine 1987; 316: 1413.
29. Van Blitterswijk WJ. New England Journal of Medicine; 316: 1413.
30. Mountford CE. New England Journal of Medicine 1987; 316: 1413-4.
31. Fossel ET, Carr JM, McDonagh J. New England Journal of Medicine 1987; 316: 1414.
32. Everett JR, Jennings K, Woodnut G, Buckingham MJ. Journal of the Chemical Society Chemical Communications 1984; 894-5.
33. Bales JR, Sadler PJ, Nicholson JK, Timbrell JA. Clinical Chemistry 1984; 30: 1631-6.
34. Bales JR, Nicholson JK, Sadler PJ. Clincal Chemistry 1985; 31: 757-62.
35. Tulip K, Timbrell JA, Wilson ID, Troke J, Nicholson JK. Drug Metabolism and Disposition 1986; 14: 746-9.
36. Coleman MD, Norton RS. Xenobiotica 1986; 16: 69-77.
37. Miyata Y, Ando H. Kagaku Keisatu Kenkyusho Hokuku, Hokagaku Hen 1984; 37: 215-9.16.
38. Sanins SM, Timbrell JA, Nicholson JK. Human Toxicology 1986; 5: 123.
39. Sanins SM, Nicholson JK, Timbrell JA. Toxicology Letters 1986; 31: 242.
40. Nicholson JK, Sadler PJ, Tulip K, Timbrell JA. In: Reid E, ed. Methodological Surveys in Biochemistry and Analysis Bioactive Analytes. Plenum Press, 1986; 16: 321-36.
41. Wilson ID, Ismail IM, Fromson J, Nicholson JK. Journal of Pharmaceutical and Biomedical Analysis 1987; 5: 157-63.
42. Wilson ID, Nicholson JK. Analytical Chemistry 1987; in press.
43. Wilson ID, Nicholson JK. Journal of Pharmaceutical and Biomedical Analysis 1987; in press.
44. Nicholson JK, Drury JE, Timbrell JA, Higham DP, Sadler PJ. Human Toxicology 1986; 5: 115-6.
45. Nicholson JK, Timbrell JA, Higham DP, Sadler PJ. Human Toxicology 1984; 3: 334-5.

46. Nicholson JK, Timbrell JA, Sadler PJ. Molecular Pharmacology 1985; 27: 644-51.
47. Gartland KPR, Nicholson JK, Timbrell JA, Bonner FW. Human Toxicology 1986; 5: 122-3.
48. Gartland KPR, Bonner FW, Nicholson JK. In: Reid E, Cook GMW, Luzio JP, eds. Cells, Membranes and Disease. New York: Plenum Press, 1987: 397-408.
49. Bloch F, Hansen WW, Packard ME. Physics Review 1946; 69: 127.
50 Harris RK. Nuclear Magnetic Resonance Spectroscopy: A Physicochemical View. Pitman Books London 1983; 250 pps.
51. Gadian DG. NMR and its Application to Living Systems. Oxford University Press 1982; 197 pps.
52. Hahn EL. Physics Review 1950; 80: 580.
53. Rabenstein DL, Nakashima TT. Analytical Chemistry 1979; 51: 1465A-1474A.
54. Rabenstein DL, Fan S. Analytical Chemistry 1986; 58: 3178-84.
55. Connor SC, Everett JE, Nicholson JK. Magnetic Resonance in Medicine 1987; 4: 461-8.
56. Bell JD, Brown JCC, Sadler PJ. Biochemical Society Transactions 1986; 14: 1263-4.
57. Connor SC, Everett JE, Nicholson JK. Analytical Chemistry 1987; in press.
58. Bales JR, Bell JD, Nicholson JK, et al. Magnetic Resonance in Medicine 1987; in press.
59. Bax A, Lerner L. Science 1986; 232: 960-7.
60. Laude DA, Wilkins CL. Analytical Chemistry 1987; 59: 546-51.

DIAGNOSTIC RELATED GROUPS: THE INFLUENCE OF REFUNDING OF COST ON THE ORGANIZATION OF CLINICAL LABORATORIES

Albert L. Giles

AVATAR, Inc., Voorhees, New Jersey, USA

In 1983, with the Federal Government's adoption of diagnostic related groups as a basic reimbursement methodology the hospital industry and clinical laboratories of the United States were highly impacted. This change has created some traumatic impacts upon management philosophy within the hospital and has caused the laboratory to be viewed as a different type of operating entity. The hospital clinical laboratory has reacted by adapting and substantially altering its production and management philosophy and approach.

Key among these changes has been a much higher level of competition by the commercial laboratories for physician office referral testing. The independent laboratories of the nation have faced a number of dynamic changes. Among them was realignment of the major commercial interest owning the laboratories, a highly competitive entrance into the marketplace of hospital laboratories seeking out patient work, and finally the challenge of aggressive marketing of new technologies which created high levels of cost benefit for the physician office laboratories.

These factors have caused substantial realignment in the independent clinical laboratory field and create an entirely different world for the professionals operating these organizations.

Technological advancement, both in hard scientific techniques and information technology, an increasing appetite for economy in insurance coverages and appropriateness of medical treatment have all caused a reduction in overall demand for laboratory work, despite a technological trend in which new testing provides broad areas of more precise medical information than has ever been available before. I shall attempt then, in this presentation, to highlight the major forces causing the organizational changes and provide a snapshot of the clinical laboratory both as it exists in hospital and independent environment today and present several thoughts with regard to future trends which will dictate the strategies necessary for maintenance, expansion and general improvement of the clinical laboratory.

Diagnostic related group reimbursement was originally adopted by the Congress of the United States in 1983 as a method to drastically curtail the payment levels of the Medicare program which covers the aged of the

United States. The technique was adopted because of the achievements of similar programs in four state environments over the previous five years. It should be noted that reimbursement by diagnosis was hardly a new theory. Its foundations can be found in European systems that date back to the 1930's. In the course of medical economics, various government and private experiments have attempted to link payment with diagnosis every four of five years, somewhere in the world. Certainly, the imposition of the methodology for the reimbursement of the large number of Medicare recipients in the U.S. was the largest, direct application of the theory and it has had profound impact.

Prior to this methodology hospitals were basically reimbursed by reviewing their actual cost. This type of reimbursement was started in the 30's by the Blue Cross plans of the U.S. Though commercial coverage and private payment created a reason for the hospitals to continue to have a system of charges for each service, the expansion of the Blue Cross program, the addition of Medicare in 1966, and the growth of a number of new type HMO and PRO organizations in the U.S. all prompted more concentration on cost reporting, cost analysis and cost reimbursement. Though the American hospital system is generally dominated by a not-for-profit institution, all of the cost systems generally allowed some level of "cost plus" reimbursement.

At the time of adoption, the projected funding for the Medicare program was inadequate to actuarily support its long run existence. The cost of health care was virtually increasing at twice the rate of other costs in the national economy and Congress, as well as the private sector, was frantically seeking a solution that would halt this escalation.

As a practical matter, hospitals started to work under the yoke of three reimbursement methodologies. There still was enough commercial reimbursement and private payers to require the maintenance of the system of charges. Though the federal government moved swiftly into the DRG payment methodologies, the Blue plans and many developing HMOs essentially stayed with the older cost reimbursement methodologies. The DRG system was implemented using historic costs as they could be apportioned across hospital departments and generally related to the mix of DRG patients the institution had historically treated. It's important to understand that a good deal of the inaccuracy that developed was because DRG coding system was not standardized as a medical diagnostic system. In many cases it had overlaps and in other areas it had insufficient codes. Hence, reclassifying cases into the DRG matrix for evaluation of their previous cost to the hospital was at best a nebulous process.

Also key to interpreting what has happened in a managerial sense, to clinical laboratories, is understanding that the DRG system also had the effect of providing an incentive to administrations. It said, in essence, that any amount of money that can be saved by altering the method of delivery of laboratory services to the hospital could be kept, since the DRG payment would continue to be adjusted by a mix of national, regional and local cost factors. In short, the previous year's cost did not become the standard by which the next year's reimbursement was determined at least by the major governmental program. This provided some profound shifts in the administration's economic philosophy in a majority of the hospitals.

In retrospect, the nation's hospitals have had greater prosperity under the DRG reimbursement system than ever before recorded. The completion of their fiscal years in both 1985 and 1986 demonstrated much higher numbers of institutions with greater operating reserves and improved financial results. One must finally realize that the threat of the DRG reimbursement caused a substantial management re-thinking by the administrations, provided an incentive to economize and triggered a number of management innovations and changes in operational philosophy, which has indeed altered the operating structure and modality of the average hospital clincial laboratory.

Most profound among the management shifts for the hospital clinical laboratory was the realization that the laboratory was just another cost center of the hospital. For years laboratories have enjoyed the position of being money-making centers for the hospitals. Often they were able to achieve favorable consideration and capital expenditures as well as maximum operating staffing because of this situation. The shift away from cost reimbursement by a major program, which minimized this "privileged" status created a situation in which the hospital looked at the laboratory as just another set of increasing expenses. This, coupled with the general fear by hospitals of the reimbursement impacts, caused a tightening of budgets and a series of management constraints which required the managers of clinical laboratories to change the way they did operational planning and caused a large number of changes for the laboratory. Some of these were:

- There have been a large number of job re-evaluations. In many cases the credentials and experience required to operate highly sophisticated scientific computer oriented analyzers decreased. This happened both because of equipment simplification and because of higher speed technology in which the operator can have a lower impact. Automatic transfer of information from testing apparatus into data collection systems has also hastened this process. Hence, generally, the educational credentials, certification and level of pay of the operators of computerized analyzers decreased. Operational analysis increasingly revealed the ebb and flow on a time basis of the work demands placed on the laboratory as well as operational systems which segregated stat testing into separate units, creating a higher utilitarian approach to getting work that required a great deal of speed without interrupting more intense medical interpretive procedures. Hence we end up with a smaller number of workers doing a much higher number of the tests. They have lesser training and are totally production oriented. At the average hospital, the set of esoteric tests was redefined, with fewer tests performed at the hospital because it was not cost effective.

- Personnel Utilization - many laboratory staffing policies have changed with more part time workers being utilized. Variable hours have left employees completing two short shifts at the hospital at peak production periods, achieving a total time less than the normally mandated 8 hours, but receiving reimbursement for the full 8 hours because of their effective contributions in a highly productive mode of operation.

- Enhanced management and supervisory training. The introduction of numerous commercial factors into the independent laboratories have also caused substantial change of the last five years. The impact of this higher level of awareness is use of automated equipment with higher data transfer capacity and the rapid expansion of high technology testing. Laboratory management has also become much more economically aware that due to increased competition, they must maintain the price levels for testing at a low enough level to compete with a generation of equipment which allows physicians to do their own laboratory work within their offices and to charge the patients for it.

  This trend towards massive expansion of doctors' office laboratory testing with improved technology using smaller, cost effective equipment has not been as great as first anticipated. The trend is being slowed as newer reimbursement methodologies are taking the incentive away from physicians charging for such testing.

In short, independent laboratories have suffered heavy competitive challenges, both from each other and from expanding technology in the doctor's office, but long term trends promise a return of a majority of the physician ordered testing to the independent laboratory.

- In my previous comments I have separated the hospital and the independent laboratory. It's terribly important to realize as well, that approximately 18% of the laboratories in the U.S. have changed ownership. A majority of these have been hospitals setting up the laboratories as separate corporate entities, sometimes in clusters with several other hospitals, but always in a situation where they had the capacity to competitively compete and provide the physician-ordered outpatient laboratory work. These new laboratories have a tendency to be greater in capacity and larger providers of laboratory service within their given community.

- Increased volumes of work coupled with automated work control systems have required a virtual explosion in the clerical needs of most laboratories. In 1978 the average ratio of technical workers to clerical personnel in the laboratory was six to one. In a recent survey, in 1985, those odds have changed to one clerical per one and a half technically trained personnel. These clericals are inputting data, splitting specimens, performing the phlebotomy tasks and providing overall direction to result reporting.

- Another discrete and very important differential in the operation of the clinical laboratory is the time classes of work. For years there was stat testing, which was done immediately, and normal turn-around time. Today, a sophisticated laboratory has up to five levels of work, 1) stat, which generally means immediately done upon receipt, the outside parameter being a result phoned back to a user in two hours, 2) as soon as possible, meaning the next time a major analyzer is run, results usually turn-around in as little as an hour and as high as five hours, 3) end of day, this methodology promises work received by some given cutoff point during the day, (generally noon) will have results reported at the end of the lab's major operating day (6:00 pm in most laboratories), 4) overnight, virtually guarantees the results the following day, and, of course, 5) timed testing such as microbiology and other reference testing which requires more than a 24 hour turn-around time. This time emphasis, which has also been supported by the need to classify cases quickly for DRG purposes has had great impact upon the need to control the operational flow of the work through the laboratory.

In summary then, the following provide a view of the clinical chemistry laboratory in the United States in 1987:

1. There are less clinical chemistry laboratories. Compared to 1975 there are approximately 20% less licensed laboratories of all types.

2. The laboratory is much larger in its work volumes. The volumes have been contributed to by new technology of broader and more definitive testing. They have also been expanded by programmatic expansions of testing such as employment drug testing and Aids exposure testing. Their volumes are not as great as they might have been since there have been definitive programs controlling overutilization of inpatient

ordering and restrained diagnostic developmental programs for first treating attending physicians.

3. The increased volume, compressed number of laboratories and broadened repertoire have created more specialized divisions of the laboratory and have provided enhanced professional development opportunities in these specialization areas.

4. The modern clinical chemistry laboratory is better equipped with lower cost equipment because of vast technological development. The equipment is enhanced in technical capacities, quality control, and accuracy. But, even more impressive is the capacity of the equipment in data generation, retention and comparison.

5. In fact, these data capacities are so great that many medical planners look forward to the integration of virtually all laboratory data for inpatient and outpatient episodes for a majority of American patients by the year 1995. More of the laboratories are operated by for-profit corporations, despite the fact that a majority serve at least one hospital's needs directly. This shift, in a majority of the laboratories, to profit orientation has created greater opportunities for individual economic achievement and in general, through profit plans and broader retirement, have created a more secure economic future for the laboratory worker.

6. Laboratory professionals generally have a much greater need for supervisory and management skills. This need is being met by continuing educational opportunities as well as changes in basic training programs.

7. The number of persons working in laboratories is relatively static. The largest change is that there are many more with clerical training and no technical training occupying the current positions. These personnel are developing semi-technical skills such as phlebotomy and specimen alloquoting, et cetera, but they are not professionals as we have known and viewed them in the laboratory previously.

In short then, the clinical laboratory stands as a marvel of technological improvement, drastically improving its capacity to provide high quality and rapid testing results in an environment which is requiring highly developed management skills and is perfecting operational turn-around time and techniques.

BIBLIOGRAPHY

Altshuler CH. Use of comprehensive laboratory date as a management tool. Clin Lab Med 1985 Dec; 5(4): 673-95.

Detsky AS, et al. Do interns and residents order more tests than attending staff? Results of a house staff strike. Med Care 1986 June; 24(6): 526-34.

Giles AL. Laboratory utilization - reflective of changes in reimbursement, practice methodology, or herd instincts? 8th Beckman Conf in Clin Chem 1985; 71-81.

Iezzoni LI. The impact of reimbursement changes on doctors' office testing. Clin Lab Med 1986 June; 6(2): 329-43 (53 references).

Iezzoni LI, et al. The clinical impact of DRG-based physician reimbursement. Boston U Med Center.

Vladeck, BC. Diagnosis related group-based hospital payment: the real issues. Bull NY Acad Med 1986 Jan-Feb; 62(1): 46-54.

Williams SV. The impact of DRG-based prospective payment onclinical decision making. Med Decis Making 1985; 5(1): 23-9.

THE ROLE OF CLINICAL CHEMISTRY IN PREVENTIVE MEDICINE

K. Miyai

Department of Laboratory Medicine, Osaka University Medical School, Osaka, Japan

The recent trend in clinical medicine has been to emphasize the treatment of incurable diseases rather than their prevention. However, as the saying goes, "Prevention is the best treatment"; that is, medical care should try to prevent a disease before it occurs rather than to treat a disease after it has developed. It is, therefore, imperative that laboratory methodology develop to the extent that it can detect subclinical conditions and predispositions to diseases prior to their clinical manifestations.

It is in this area of presymptomatic diagnosis that clinical chemistry must play a prominent part. In this regard, there are several approaches and strategies that one can follow. In this lecture I will be discussing two such examples, mass screening and DNA diagnosis, with examples derived from research activities in my laboratory.

[I] MASS SCREENING FOR EARLY DIAGNOSIS OF ASYMPTOMATIC DISEASES

At present, various screening approaches can be found as part of different health examination protocols. One such approach is Automated Multiphasic Health Testing and Services (AMHTS). However, these algorithms are generally unsuccessful when applied to the healthy population because of the following complications: (1) There is a high frequency of random abnormalities not related to the specific disease state being evaluated. (2) A high incidence of false positives puts unnecessary cost on the patient and society. (3) There is a false sense of confidence in those patients with intermediate testing results.

In order to carry out mass screening effectively, it is necessary to meet the conditions listed here. (1) The frequency of the disease must be sufficiently high. (2) Early diagnosis by clinical features alone must be difficult. (3) The laboratory analysis itself must be suitable for screening; that is, it must have adequate sensitivity, ease of use and acceptably low cost per sample. (4) Presymptomatic treatment must have an overall benefit to and be accepted by the people as a whole.

Frequently methods used for routine testing or diagnosis of a disease that has already developed are inappropriate for screening programs. For instance the effectiveness of the combination of tumor markers for detecting early stage of cancer is still controversial.

In order to overcome these difficulties, it is necessary to develop new methods and/or criteria that are specific for screening objectives.

I-1 MASS SCREENING FOR PROLACTINOMA

Prolactin secreted from a prolactinoma stimulates breast function and inhibits gonadal function that cause amenorrhea-galactorrhoea syndrome, so that its diagnosis is not so difficult in women. On the other hand, in men, the signs and symptoms are so obscure that the tumors are rarely detected until they grow and compress the optic nerve causing visual disturbance. We therefore, tried to perform mass screening by measuring serum prolactin by radioimmunoassay for the early diagnosis of this disease (1).

Materials and Methods

Principle of Paired Assay: To reduce the labor required for mass screening, we devised a new method so called "Paired assay" in which we measured prolactin concentration not in each serum sample but in mixture of serum samples from two subjects. Since the normal range of serum prolactin is 7.5 to 28 μg/l, the concentration in the mixed sample of two normal subjects (normal pair) should be less than 28 μg/l. On the other hand, if serum of patient with over 50 μg/l of prolactin is mixed, the prolactin concentration in this mixed sample (abnormal pair) must exceed $(7.5 + 50)/2 = 29$ μg/l, thus, differentiating the abnormal pair from the normal pair is possible. If the paired sample is abnormal we measure again the prolactin in each sample of this pair to determine which subject is abnormal. The reliability of this method was confirmed by the fact that the prolactin values measured by the paired assay (y) are well correlated with the expected mean values (x) calculated for the values of individual assays ($y = 1.0x + 0.9$).

Protocol of Mass Screening: In practice we designed the protocol as shown in Fig. 1. When the prolactin concentration was over the cut off point in the first paired assay using 25 μl each of serum, the second and third individual assays using 50 μl serum were carried out. When the values were over the cut off point, a questionnaire was sent and a new blood sample was requested. When the cause of the hyperprolactinemia, such as pregnancy or drug, was not clear and serum prolactin was again over the cut off point, we recalled the subjects and examined them by X ray, CT scan and so on.

Results and Discussion

We carried out this mass screening testing as part of a health examination for the employees of some firms. As shown in Table 1, out of 10,550 so called normal subjects, 5 patients with pituitary prolactinoma were found. The prevalence is estimated to be 1 : 2800 in men and 1 : 1050 in women. Since the adult population of Japan is 67 million there are presumably about 44,000 cases of prolactinoma in Japan.

All of the patients with prolactinoma found by this mass screening test have no signs and symptoms, and other hormones are within normal limits except a slight decrease in testosterone in some male patients. For example, one male patient had macroprolactinoma and deformation of the pituitary fossa. The tumor was removed by surgery and he is now doing well. In another married women, microprolactinoma was demonstrated by CT scan. She had menstruation and no signs and symptoms but she had no children.

Table 1. Screening for prolactinoma in the general population

	Male	Female	Total
Total subjects screened	8450	2100	10550
Hyperprolactinemia	15	26	41
Examined	10	16	26
Prolactinoma	3	2	5
	(1 : 2800)	(1 : 1050)	
Empty sella syndrome	1	0	1
Pregnancy	-	6	6
Drugs	4	4	8
"Big" prolactinemia	1	2	3
Normalized	1	2	3

Adult population in Japan 67,000,000
Estimated number of prolactinoma 44,000
(See ref. 1)

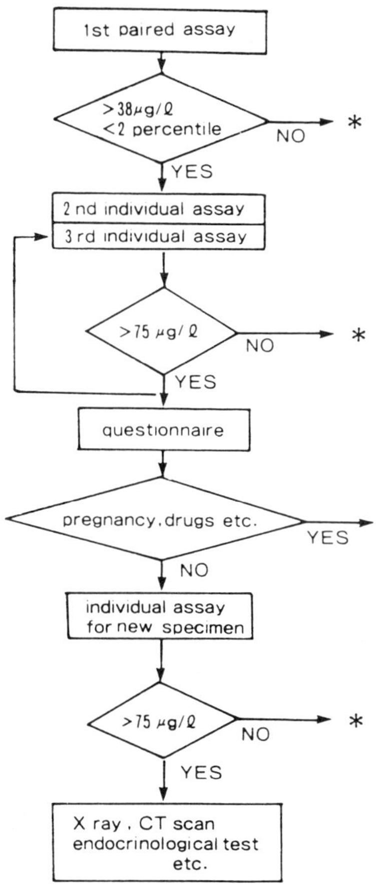

Fig. 1. Protocol of mass screening for prolactinoma

After she was treated with bromocriptin she finally succeeded in pregnancy and delivered a healthy baby.

The cost benefit of this screening program is uncertain since the prolactinoma is benign. However early diagnosis and follow up of these patients seem helpful for the selection of an adequate treatment.

I-2 MASS SCREENING FOR INBORN ERRORS OF METABOLISM

During the past 20 years neonatal screening programs for inborn errors of metabolism have been instituted throughout the world. The main diseases for neonatal screening include phenylketonuria, maple syrup urine disease, homocystinuria, histidinemia, galactosemia, hypothyroidism and so on. For this purpose of neonatal screening small amounts of blood are taken from heel pad of babies. The blood specimens are dried on the filter paper, and sent to laboratories. The most common method is the measurement of the amino acid by the bacterial inhibition assay, so called Guthrie test.

However, the most important recent development in this field is the introduction of immunoassays. For example, adrenogenital syndrome with 21-hydroxylase deficiency is screened by measuring 17- OH progesterone. Cystic fibrosis, one of the common lethal autosomal recessive disorder, can be detected by measuring trypsin. Neural tube defect of fetus also can be predicted by an elevation of α fetoprotein in mothers' serum during pregnancy. These substances can be measured by immunoassay.

I-2-1 NEONATAL SCREENING FOR HYPOTHYROIDISM

The most effective mass screening program using immunoassay is screening for neonatal hypothyroidism. Since thyroid hormone is known to be essential for the development of the brain from the late fetal to the neonatal period, congenital deficiency of this hormone results in their growth retardation and irreversible mental retardation. Such tragic disturbances can be prevented by early treatment with thyroid hormone by 3 months of age. But early diagnosis is difficult by signs and symptoms alone. Mass screening is therefore considered to be effective.

Materials and Methods

Fig. 2 shows the incidence and changes in parameters of various types of congenital hypothyroidism. A decrease in total thyroxine (T_4) is useful for detecting all types of the disease except thyroid hormone resistance but it gives false negative results for mild primary hypothyroidism and false positive results for various conditions such as thyroxine binding globulin (TBG) deficiency which does not have to be treated. Measurement of free T_4 can reduce the false positive results due to TBG deficiency but the method is still insensitive. TSH which increases by a negative feedback mechanism is useful in detecting even mild cases of primary hypothyroidism, although rare cases of tertiary and secondary hypothyroidism are missed.

Until now these parameters in dried blood samples on filter paper have been determined by radioimmunoassay as follows:
total T_4 (2), free T_4 (3), TBG (4), TSH (5, 6).

However, the severe restrictions on the use of radioisotopes is one of barriers to promoting the spread of mass screening. Thus non isotopic immunoassays such as enzyme immunoassay, fluorescence immunoassay and luminescence immunoassay have been developed as follows: total T_4 (7), free T_4 (8), TBG (9), TSH (10, 11, 12, 13).

Types	Incidence	Circulating hormone level		
		T_4	Free T_4	TSH
Primary	1/3,000 1/7,000	⇣	⇣	↑
Secondary Tertiary	<1/100,000	⇣	⇣	⇢
Peripheral	rare	↑	↑	↑
TBG deficiency	1/1,000~ 1/5,000	⇣	→	→

Fig. 2. Parameters for neonatal screening for hypothyroidism

Results and Discussion

Results of the one step semiautomated sandwich enzyme immunoassay of TSH are shown here (14). In this assay, antihuman TSH rabbit IgG coated tubes and anti TSH IgG-β-galactosidase conjugate were used. The enzyme activity of the tubes was measured using 4-methyl-umbelliferyl-β-D-galactophyranoside as substrate by an autoanalyzer which was manufactured for this purpose. The calibration curve and the distribution of the TSH values are printed out automatically as shown in Fig. 3.

Table 2 shows a summary of neonatal screening for hypothyroidism in Japan. It was 1975 when our pilot study started and in 1979 a nation wide screening program began. Since then more than 700 patients have been

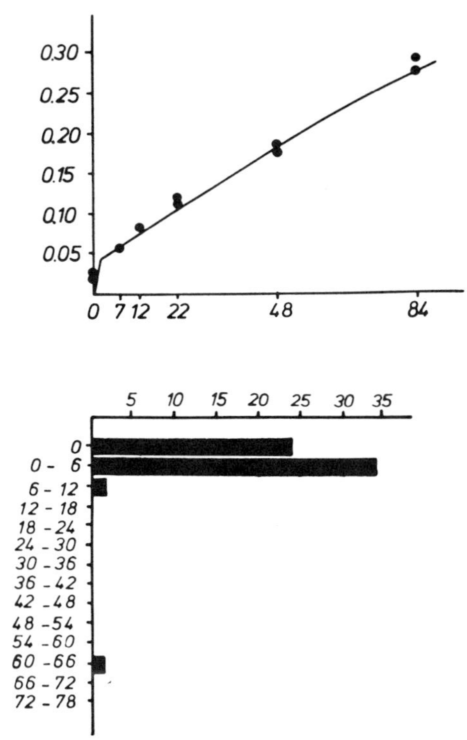

Fig. 3 Calibration curve and distribution of TSH by semiautomated enzyme immunoassay for neonatal hypothyroid (14)

Table 2. Summary of neonatal hypothyroid screening by TSH assay

	Osaka		Japan
	RIA 1975-1983	EIA 1983-1986	RIA + EIA 1979-1983
Total screened	281,468	70,280	7,417,034
Hypothy-roidism	48	14	735
Incidence	1/5800	1/5020	1/10091

detected in 7.4 million babies and almost all patients are normal after thyroid hormone therapy. At present, 1.5 million babies are born per year and 99% of them are screened.

One of the reasons that mass screening for neonatal hypothyroidism is so widespread is its high cost benefit. Assuming that one TSH test costs $7, and that live births in Japan are 1.5 million, 200 hypothyroid patients can be detected by the test, at a total cost for mass screening of $10 million. On the other hand, if the screening tests are not performed the expenditure for one case of untreated hypothyroidism is estimated as $20,000 per year the expenditure for 200 patients per year would be $4 million, which cumulates to $200 million for 50 years. Thus we can say that this mass screening for neonatal hypothyroidism yields a cost benefit ratio of 1 : 20.

Successful prevention of mental retardation due to congenital hypothyroidism by mass screening is one of good example of the role of clinical chemistry in preventive medicine.

[II] GENETIC ANALYSIS FOR PREDISPOSITION TO DISEASES

Although there are undoubtedly several approaches that can be used to clarify a subject's predisposition to disease, the genetic approach is amongst the most fundamental. Recent developments in gene technology have made this a reality. The results are that such information can now be included in genetic counseling programs.

II-1 HLA ANTIGENS AND DISEASE SUSCEPTIBILITY

The association of specific HLA antigen with disease susceptibility has been extensively investigated. The first group is the association common in any race. As an example a significant combination of HLA B27 with ankylosing spondylitis is well known. A second group is the association of HLA and some organ specific autoimmune diseases such as the association of HLA DR or DQ with insulin dependent diabetes (IDDM).

II-1-1 HLA AND CONGENITAL HYPOTHYROIDISM

Table 3 shows an example of our data on the relationship between congenital hypothyroidism due to thyroid dysgenesis and their HLA A and B

Table 3. Frequencies of HLA-A24 antigen and congenital hypothyroidism

	Frequency			
	Phenotype A24		Congenital hypothyroidism A24	
	+	-	+	-
General population	0.568	0.432	1/5,000	
Patients' mother	0.962	0.038	1/3,000	1/57,000

(See ref. 15)

antigens in Japan (15). The frequency of A24 antigens in control subjects in the general population is 0.568 in Japan. The frequency of this antigen in patients' mothers is 0.962 which is significantly higher than controls. The incidence of congenital hypothyroidism in the general population is estimated to be 1 : 5000 as mentioned previously. Thus the frequency of the disease in babies whose mother have A24 would be calculated as 1 : 3000 whereas the frequency in babies whose mother have no A24 is only 1 : 57000. If HLA antigens were measured routinely, families with a high or low risk of congenital hypothyroidism could be predicted.

II-2 DNA ANALYSIS FOR GENETIC DISEASES

Several efforts are being made to investigate predispositions to diseases on the basis of DNA. For example, restriction fragment length polymorphism analysis is effective in this purpose. As an example, Rotwein carried out restriction fragment length polymorphism analysis in the 5' flanking region of the insulin gene and found that allelic frequency of presence of 1.6 KB insertion in the region is significantly higher in non insulin dependent diabetes mellitus (NIDDM) (0.316) when compared with nondiabetic controls (0.195) and IDDM (0.192). Although it is a controversial subject, it is emphasized that this polymorphism may provide a genetic marker of this disease (16).

On the other hand many congenital malformations and genetic defects can now be directly diagnosed by DNA analysis prior to pregnancy, and genetic counseling can be performed.

II-2-1 GENETIC ANALYSIS OF CONGENITAL TSH DEFICIENCY

As an example we describe our data of congenital familial isolated TSH deficiency, which was discovered by us in 1971 (17). In this family, the proposita were 2 sisters of 3 siblings. Their parents' marriage was consanguineous with an inbreeding coefficient of 1/64 and their pituitary thyroid functions were normal. The patients had typical signs, symptoms and laboratory findings of congenital hypothyroidism. Serum TSH assessed by bioassay and radioimmunoassay using different antibodies were consistently undetectable and showed no response to TRH indicating that the patients have TSH deficiency. TSH and gonadotropin are glycoprotein hormones which

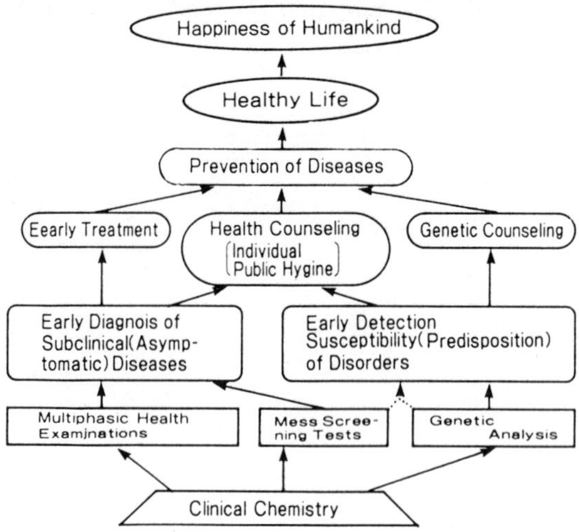

Fig. 4. Role of clinical chemistry in prevention medicine

consist two peptide chains designated α and β subunits. The α subunit (common α) is identical among these hormones. The secretion of gonadotropin is normal in these patients. The serum levels of the free α subunit were very high and showed an increase response to TRH. These findings indicate that the TSH producing cells are present and the α-subunit gene is intact. Thus we thought that abnormalities of the TSH β gene is most likely the cause of this disease. We therefore, attempted to isolate the normal human TSH β gene and succeeded in the identification of the structure (18). We then tried to clone and analyze the TSH ß gene of the patients and finally we found one point mutation. The codon GGA corresponding to Gly^{29} is substituted by AGA (Arg^{29}). In order to differentiate this abnormal mutant sequence from the normal sequence, we used a restriction enzyme Mae I. This enzyme splits for the abnormal sequence (CTAG) and produces shorter fragment but does not split normal sequence (CTGG).

By this method, we analyzed the family and we could find heterozygous carriers of this disease in some members of the same family. We could advise them of the risk of this disease before pregnancy. If they have a risk, we can diagnose the child at birth and prevent mental retardation by early treatment.

「千金方」 遜子邈

上医医未病之病　上医は未だ病まざるの病を医す
THE BEST DOCTORS PREVENT THE DISEASES

中医医欲病之病　中医は病まんと欲する病を医す
THE BETTER DOCTORS TREAT THE DISEASES
AT EARLY STAGE

下医医既病之病　下医は既に病めるの病を医す
THE MEDIOCRE DOCTORS TREAT THE
SEVERE DISEASES

Fig. 5. For explanation see text

Although TSH deficiency is rare, this is one of example that gene analysis or DNA diagnosis may play an important role in prediction and prevention genetic diseases.

III CONCLUSION

To summarize this lecture on "Role of clinical chemistry in preventive medicine", I propose a scheme (Fig. 4) although it is rather dogmatic and I introduce words of a famous ancient orient doctor (Fig. 5).
In conclusion I believe that the role of clinical chemistry in preventive medicine is rapidly expanding and will become one of the most significant aspects of clinical chemistry in near future.

ACKNOWLEDGMENT

I should like to thank our colleagues within the division for their coworks.

REFERENCES

1. Miyai K, Ichihara K, Kondo K, Mori S. Asymptomatic hyperprolactinaemia and prolactinoma in the general population - Mass screening by paired assays of serum prolactin. Clin Endocrinol 1986; 25: 549-554.
2. Dussault JH, Coulombe P, Laberge C, Letarte J, Guyda H, Khoury K. Preliminary report on a mass screening program for neonatal hypothyroidism. J Pediatr 1975; 86: 670-4.
3. Mizuta H, Miyai K, Ichihara K, et al. Radioimmunoassay of "Free thyroxin" in dried blood spots on filter paper - Preliminary observations on the effective differentiation of subjects with congenital hypothyroidism from those with subnormal thyroxin-binding globulin and normal subjects. Clin Chem 1982; 28: 505-8.
4. Robertson EF, Wilkins AC, Oldfield RK, Pollard AC. Blood spot thyroxine-binding globulin: A means to reduce recall rate in a screening strategy for neonatal hypothyroidism. Clin Chem 1980; 97: 604-7.
5. Irie M, Enomoto K. Measurement of thyroid-stimulating hormone in dried blood spot. Lancet 1975; 2: 1233-4.
6. Van den Berg JWO, Van Blankenstein M, Bosman-Jacobs EP, et al. Solid phase radioimmunoassay for determination of conjugated cholic acid in serum. Clin Chem Acta 1976; 73: 277-83.
7. Miyai K, Ishibashi K, Kawashima M. Enzyme immunoassay of thyroxine in serum and dried blood samples on filter paper. Endocrinol Japan 1980; 27: 375-80.
8. Hata N, Miyai K, Ito M, et al. Enzyme immunoassay of free thyroxin in dried blood samples on filter paper. Clin Chem 1985; 31: 750-3.
9. Hata N, Ito M, Mizuta H, Nose O, Miyai K. Enzyme immunoassay of thyroxin-binding globulin in dried blood samples of filter paper. Clin Chem 1983; 29: 1437-40.
10. Kato N, Ishii S, Naruse H, Irie M, Arakawa H, Tsuji A. Enzyme immunoassay of thyroid-stimulating hormone using dried blood samples a simple technique of screening for congenital hypothyroidism. Anal Letters 1980; 13: 1555-65.
11. Miyai K, Ishibashi K, Kawashima M. Two-site immunoenxymometric assay for thyrotropin in dried blood samples on filter paper. Clin Chem 1981; 27: 1421-3.
12. Soini E, Kojola H. Time-resolved fluorometer for lanthanide chelates - A new generation of nonisotopic immunoassays. Clin Chem 1983; 29: 65-8.

13. Weeks I, Sturgess M, Siddle K, Jones MK, Woodhead JS. A high sensitivity immunochemiluminometric assay for human thyrotropin. Clin Endocrinol 1984; 20: 489-95.
14. Miyai K, Tsuruhara T, Kusuda S, et al. Semiautomated enzyme immunoassay of thyrotropin as a mass screening test for neonatal hypothyroidism. Pediatr Res 1984; 18: 1289-92.
15. Rotwein PS, Chirgwin, Province M, et al. Polymorphism in the 5'flanking region of the human insulin gene: a genetic marker for non-insulin-dependent diabetes. New Engl J Med 1983; 308: 65-71.
16. Miyai K, Azukizawa M, Kumahara Y. Familial isolated thyrotropin deficiency with cretinism. New Eng J Med 1971; 286: 1043-8.
17. Hayashizaki Y, Miyai K, Kato K, Matsubara K. Molecular cloning of the human thyrotropon-β subunit gene. FEBS Letters 1985; 188: 394-400.

CHAPTER 2
GENES, GENE-TECHNOLOGY AND DNA-PROBES

RECOMBINANT DNA TECHNOLOGY IN THE DIAGNOSIS OF HUMAN INHERITED DISEASE

J. Schmidtke* and D.N. Cooper**

*Institut für Humangenetik der Universität Göttingen, FRG
**Haematology Department, King's College Hospital, London, UK

Over 3,000 Mendelian traits implicated in the pathology of human inherited disease, have been catalogued to date (1). Only a very small proportion can be diagnosed either antenatally or preclinically by conventional protein analysis. Moreover, many diseases are not amenable to antenatal genetic analysis since their diagnostic proteins are not present in the accessible foetal tissues. The advent of recombinant DNA technology has promised to circumvent these problems since direct investigation of the genetic material obviates the need for specific tissue samples. In addition, antenatal diagnosis and carrier detection of many genetic defects should be possible without the prerequisite of needing to identify either the primary gene product or the biochemical mechanism of the disease. Differing approaches to disease diagnosis will be described and the extent of their application to disease diagnosis to date, presented.

DIRECT ANALYSIS OF GENETIC DISEASE USING GENE PROBES

In order for the base-pair change or deletion event responsible for the disease phenotype to be detectable using restriction enzymes, two conditions must be met. First, the nature of the disease must be sufficiently well understood for the locus responsible to be identified, isolated, and used as a hybridization probe to examine the disease-associated allele(s). Second, for the aberration to be detected, a restriction site must be introduced or removed in the case of a point mutation, or the length of DNA between sites must be altered by sequence additions, or rearrangements.

Although the application of direct analytical technique to disease diagnosis is still in its infancy, Table 1a illustrates the extent to which it has already been applied. Coverage of the thalassaemias and haemoglobinopathies is by no means exhaustive here due to the disproportionate number of globin gene mutations detected to date. This is well reviewed (2).

Direct analysis of intragenic defects may also be accomplished using chemically-synthesized oligonucleotide probes thereby obviating the requirement for the mutation to lead to an altered restriction site. Discrimination between genotypes relies upon a base-pair mismatch between the oligonucleotide and a given allele being sufficient to abolish hybridization under the conditions used. This approach has been used for analysis of α_1-antitrypsin deficiency, sickle-cell anaemia, and the thalassaemias (Table 1a). However, its application to the diagnosis of diseases known to be molecularly heterogeneous requires extensive investigation of each specific mutation.

The analysis of various diseases characterized by severe chromosomal lesions has also been attempted using a variety of cloned probes. These are listed in Table 1b.

A veritable armoury of gene probes is at present available which are potentially useful in the direct analysis of disease loci (3).

INDIRECT ANALYSIS OF GENETIC DISEASE USING RFLPS DETECTED BY GENE PROBES

If a gene defect is not a gross deletion, then it may often go undetected due to the lack of a suitable restriction enzyme (Table 2). One alternative is to use DNA polymorphisms flanking the locus of interest as genetic markers. RFLPs are neutral base-pair changes which introduce or remove a restriction site, or sequence deletions, additions or rearrangements which affect the length of DNA between sites. RFLPs are not rare, occurring in the human genome approximately every 200 to 300 base-pairs (4, 5), a frequency which demonstrates the extensive variation still to be exploited in clinical medicine. Inheritance of a disease allele can thus be monitored over the generations by following the inheritance of readily detectable RFLPs linked to the gene in question. A list of diseases to which this approach has been applied is given in Table 3.

INDIRECT ANALYSIS OF GENETIC DISEASE USING RFLPS ASSOCIATED WITH LINKED DNA SEGMENTS

When for a particular disease, the relevant gene is not available as a probe, direct analysis is impossible. Instead, linkage between a cloned DNA segment and the locus of interest can be established and the inheritance of RFLPs associated with the linked DNA segment investigated. The use of linked RFLPs is merely an extension of classical linkage analysis; it provides the indirect means to detect the presence, and follow the inheritance of, genetic lesions and base-pair substitutions at a particular locus without needing to possess a cloned copy of the locus itself. The tighter the linkage between the DNA segment and the locus of interest, the smaller will be the number of recombinants as both loci tend to segregate together. A list of genetic diseases for which linkage with polymorphic DNA segments has been established is given in Table 4.

Most of the DNA polymorphisms identified to date are due to the presence or absence of particular restriction sites, while deletions, insertions, and copy number variation seem to occur less frequently. Although RFLPs seem to be abundant in the genome, most exhibit a low allele frequency. Clearly much effort must be indirected toward the further identification and localization of clinically useful RFLPs before such analysis becomes a routine procedure in medical genetics.

Disease diagnosis is an essential basis for causal therapy and presymptomatic diagnosis is in many instances a prerequisite for successful treatment or prevention. The ability to detect presymptomatically a disease which is at present not amenable to therapy may however present serious ethical problems. Furthermore, the relative simplicity of the methodology and the low cost incurred in prenatal diagnosis may well divert both interest and financial resources away from curative and social measures. We believe that in this context, the focus of research attention should be directed toward the analysis of the molecular basis of inherited disease and its cure by substitution or direct somatic gene therapy.

Note: References to table 1 - 4 may be obtained from the authors.

Table 1a. Direct analysis of a genetic disease using gene probes to detect intragenic defects. Data shown in Tables 1 - 4 include reports published up until 31st May 1987

Disease	Gene Probe
Achondroplasia	Collagen (type II)
Adenosine deaminase deficiency	Adenosine deaminase
Adrenal hyperplasia	Steroid 21-hydroxylase
Antithrombin III deficiency	Antithrombin III
Alpha-1-Antitrypsin deficiency	Synthetic oligonucleotide Alpha-1-antitrypsin
Apolipoprotein E deficiency	Apolipoprotein E
Atherosclerosis	Apolipoprotein A-1
Chorionic somatomammotropin deficiency	Chorionic somatomammotropin
Chronic granulomatous disease (CGD)	CGD 'candidate gene'
Diabetes mellitus (maturity onset diabetes of the young)	Insulin
Ehlers-Danlos Syndrome (Type II)	Alphal (1) Collagen
Elliptocytosis, hereditary	Protein 4.1
Factor X deficiency	Factor X
Growth hormone deficiency (Type A)	Growth hormone
Gaucher's disease	Glucocerebrosidase
Haemochromatosis, idiopathic	HLA-DR-beta
Haemoglobin C & SC disease	Synthetic oligonucleotide
Haemophilia A	Factor VIII Factor VIII and synthetic oligonucleotide
Haemophilia B	Factor IX Synthetic oligonucleotide
Heavy chain disease	Immunoglobulin heavy chain mu
Hereditary persistence of fetal haemoglobin	Beta-globin, Gamma-globin, Synthetic oligonucleotide
Hypercholesterolaemia	Low density lipoprotein receptor
Hyperproinsulinaemia	Insulin
Hypoxanthine-guanine phosphoribosyltransferase (HPRT) deficiency	HPRT
Immunoglobulin K-chain deficiency	Immunoglobulin C-kappa
Lesch-Nyhan syndrome	HPRT
Leukaemias and Lymphomas	T-cell receptor alpha-chain T-cell receptor beta and gamma-chain
Lymphomas	Immunoglobulin heavy chain
Marfan syndrome	Alpha-2(I) collagen
Ornithine transcarbamylase deficiency	Ornithine transcarbamylase
Osteogenesis imperfecta (Type II)	Alpha-1(I) collagen Alpha-2(I) collagen
(Type I)	Alpha-1(I) collagen
(mild atypical form)	Alpha-2(I) collagen
Phenylketonuria	Phenylalanine hydroxylase

Table 1a (cont'd)

Disease	Gene Probe
Prealbumin amyloidoses, autosomal dominant	Prealbumin
Porphyria	Uroporphyrinogen decarboxylase
Purine nucleoside phosphorylase deficiency	Purine nucleoside phosphorylase
Sandhoff disease (infantile form)	Beta-hexosaminidase
Severe combined immunodeficiency	Adenosine deaminase
Sickle cell anaemia	Beta-Globin, Synthetic oligonucleotide
Steroid sulphatase deficiency	Steroid sulphatase
Tangier disease	Apolipoprotein AI
Thalassaemias	Alpha- and beta-globin, Synthetic oligonucleotide, Beta-globin, Alpha-globin, Zeta-globin
Thrombophilia, hereditary	Protein C
Triose phosphate isomerase deficiency	Triose phosphate isomerase
Tyrosinemia II	Tyrosine aminotransferase
von Willebrand's disease (Type III)	von Willebrand factor

Table 1b. Analysis of chromosomal deletions and aneuploidies using cloned genes and DNA segments as probes

Disease	Gene Probe
Adrenal hypoplasia	Chromosome X DNA segments
Aniridia	Catalase
Beckwith-Wiedeman-Syndrome	Chromosome 11 DNA segments, Insulin, Insulin-like grotwh factor
Bladder Cancer	Insulin, C-Ha-ras 1
Cat-eye syndrome	Chromosome 22 DNA segment
Cri du Chat Syndrome	Chromosome 5 DNA segments
Choroideremia	DXYS1, DXS3
Chronic granulomatous disease	Chromosome X DNA segments
Down-syndrome	D21K9, DS21D1, SOD-1
Duchenne and Becker muscular dystrophy	Chromosome X DNA segments
Glycerol kinase deficiency	Chromosome X DNA segment (DXS84), Chromosome X DNA segments
Ichthyosis, X-linked	Chromosome X DNA segment (GMG X9)
Klinefelter syndrome	Chromosome X DNA segments
McLeod syndrome	Chromosome X DNA segments,
Norrie's disease	Chromosome X DNA segment DXS7 (L1.28)
Prader-Willi-Syndrome	Chromosome 15 DNA segments
Retinitis pigmentosa	Chromosome X DNA segments
Retinoblastoma	Chromosome 13 DNA segments Esterase D
Wilms'tumour	Chromosome 11 DNA segments
Wilms' tumour/aniridia	Follicle-stimulating hormone beta-subunit
Wolf-Hirschhorn syndrome	Chromosome 4 DNA segments
5p-deletion	Chromosome 5 DNA segments
5q-syndrome	c-fms Granulocyte-macrophage colony stimulating factor
7q-deletion	c-met, D7S8, D7S18
18p-syndrome	Chromosome 18 alphoid repeat
20q-syndrome	c-src
Xq-deletion	Chromosome X DNA segments, Factors VIII, IX[a]
Yp-deletion	Chromosome Y DNA segments

[a]Deletion of gene or DNA segment excluded

Table 2. Studies which failed to detect gene defects using cloned gene and DNA probes

Disease	Gene Probe
Abetalipoproteinaemia	Apolipoprotein B-100
Adenine phosphoribosyl transferase (APRT) deficiency	APRT
Adenosine deaminase deficiency	Adenosine deaminase
Analbuminaemia	Albumin
Angioneurotic edema	Complement 1 inhibitor
Apolipoprotein A1/C3 deficiency	Apolipoprotein A4
Apolipoprotein CII deficiency	Apolipoprotein CII
Apolipoprotein E deficiency	Apolipoprotein E
Ataxia telangiectasia	T cell antigen receptor-beta
Atopic dermatitis	Immunoglobulin C-epsilon
Bare lymphocyte syndrome	Beta-2-microglobulin, HLA genes
Charcot-Marie-Tooth disease (HMSN-1b)	Antithrombin III
Citrullinaemia	Agrininosuccinate synthetase
Complement C2 deficiency	Complement C2
Congenital afibrinogenaemia	Fibrinogen
Cystic fibrosis	Chromosome 4 DNA segments, Chromosome 19 DNA segments, Neuropeptide Y
Dihydropteridine reductase deficiency	Dihydropteridine reductase
Dysplastic Nevus syndrome	Chromosome 1 DNA segments
Familial dysautonomia	Beta-nerve growth factor
Glucosidase (alpha) deficiency	Alpha-glucosidase
Gyrate atrophy	Ornithine aminotransferase
Hereditary fructose intolerance	Aldolase B
Hunter disease	X chromosome DNA segments
Hypercholesterolaemia	Apolipoprotein CII
Hypoparathyroidism, dominant	Parathyroid hormone
Immunoglobulin A deficiency	Immunoglobulin alpha-1, alpha-2
Langer-Giedion syndrome	Thyroglobulin
Leukaemia (acute promyelocytic)	Tumour antigen p53
McArdle's disease	Glycogen phosphorylase
Malignant melanoma	c-Ha-ras-1
Manic depressive illness (bipolar-affective disorder)	Tyrosine hydroxylase, Insulin, c-Ha-ras 1
Marfan syndrome	Alpha-2(I) collagen, Alpha-I(III) collagen
Muscular dystrophy, Emery-Dreyfuss type	Chromosome X DNA segments
Multiple endocrine neoplasia (type 2A)	Various
Myotonic dystrophy	Insulin receptor
Neurofibromatosis (peripheral)	Beta-nerve growth factor, Complement C3, Apolipoprotein C2, Chromosome 1 DNA segments, Chromosome 4 DNA segments, Minisatellite
Osteogenesis imperfecta (type IA)	Alpha-1 (II) collagen
Sandhoff disease (juvenile)	Beta-hexosaminidase
Senile erythroderma	Immunoglobulin C-epsilon
Schizophrenia	Pro-opiomelanocortin

Table 2 (cont'd)

Disease	Gene Probe
Spinocerebellar ataxia (autosomal dominant)	Various
Thrombosis	Antithrombin III
Torsion dystonia	pro-opiomelanocortin, glutamic acid decarboxylase
Uroporphyrinogen decarboxylase deficiency (porphyria)	Uroporphyrinogen decarboxylase
von Willebrand's disease (Type I)	von Willebrand factor
Wilms' tumour	c-Ha-ras 1

Table 3. Indirect analysis of genetic disease using gene probes to detect closely linked polymorphisms

Disease	Gene Probe
Alpha-1-Antitrypsin deficiency	Alpha-1-Antitrypsin
Apolipoprotein CII deficiency	Apolipoprotein CII
Atherosclerosis	Apolipoprotein A-1
Carbamyl phosphate synthetase I deficiency	Carbamyl phosphate synthetase
Cataract, Coppock-like	Gamma-crystallin
Diabetes mellitus* (Type II)	Insulin
Ehlers-Danlos syndrome (Type IV)	Alpha-3 (I) collagen
Growth hormone deficiency Type I	Growth hormone
Haemophilia A	Factor VIII
Haemophilia B	Factor IX
Hypoparathyroidism	Parathyroid hormone gene
Hypothyroidism	Thyroglobulin
Hypercholesterolaemia	Low-density lipoprotein receptor gene
Hyperlipidaemia (Type III)	Apolipoprotein A-1, Apolipoprotein C2
Hypertriglyceridaemia	Apolipoprotein A-1
Lesch-Nyhan syndrome	Hypoxanthine-guanine phosphoribosyl-transferase (HPRT)
Marfan syndrome	Alpha-2 (I) collagen
Ornithine transcarbamylase deficiency	Ornithine transcarbamylase
Oestogenesis imperfecta	
Types I, IV (mild autosomal dominant)	Alpha-2 (I) collagen
(Type I)	Alpha-1 (I) collagen, Alpha-2 (I) collagen
(Type IV)	Alpha-2 (I) collagen
Phenylketonuria	Phenylalanine hydroxylase
Sickle cell anaemia	Beta globin
Thalassaemia-beta	Beta-globin, Gamma-globin
Thrombosis	Antithrombin III
von Willebrand's disease (autosomal dominant)	von Willebrand factor

* A firm association between diabetes and polymorphisms 5' to the insulin gene is questionable.

Table 4. Indirect analysis of genetic disease using linked DNA segments to examine the co-inheritance of DNA polymorphisms

Disease	Gene Probe
Adrenoleukodystrophy	DXS52 (St14)
Adult polycystic disease	Alpha-globin/hypervariable repeat region 3' to alpha-globin gene
Alport syndrome-like hereditary nephritis	DXS3, DXS1, DXS17
Alzheimer's disease	D21S16
Agammaglobulinaemia	DXS3 (p19-2), DXS17 (S21), Chromosome X DNA segments
Anhidrotic ectodermal dysplasia	DXS146 (pTAK8)
Charcot-Marie-Tooth disease (dominant, X-linked)	DXYS1
Choriodermia	DXYS1, DXS11, DXYS12
Chronic granulomatous disease	DXS164 (PERT 84), DXS84 (754)
Cleft palate/ankyloglossia (X-linked)	DXYS1 (pDP34)
Cystic Fibrosis	D7S15 (LAM4-917), c-met, D7S8 (pJ 3.11), T-cell receptor-Beta, Alpha-2(I) collagen, D7S18 (7C22)
Dyskeratosis congenita	DXS52, Factor VIII, DXS15
Emery-Dreyfuss muscular dystrophy	DXS15, Factor VIII
Fragile X-mental retardation syndrome	Factor IX, DXS52 (St14), DXS15
Haemophilia A	DXS15 (DX13), DXS52 (St14)
Haemophilia B	DXS51 (52a), DXS100 (p4SH, p4Sd)
Huntington's chorea	D4S10 (G8)
21-hydroxylase deficiency	HLA class I and II probes
Hypohydrotic ectodermal dysplasia	DXYS1, DXS14 (58-1), DXS3 (19-2)
Hypophosphataemia (dominant)	DXS41, DXS43
Hunter syndrome	DX13, various
Ichthyosis, X-linked	DXS143
Immunodeficiency, X-linked with hyperimmunoglobulinaemia	DXS42
Kennedy disease (spinal muscular atrophy)	DXYS1
Lowe oculocerebrorenal syndrome	DXS10, DXS42
Lymphoproliferative syndrome, X-linked	DXS42
Manic depression (bipolar affective disorder)	c-Ha-ras-1, Insulin
Menkes kinky hair disease	DXS7 (LI.28)
Muscular dystrophy	
Becker	DXS85 (782), DXS9 (Lambda-RC8), DXS41 (99.6), DXS43 (D2), DXS84 (754), Ornithine transcarbamylase, DXS7 (LI.28), DXS28 (C7), DXS164 (PERT 87)
Duchenne	DXS85 (782), DXS9 (Lambda-RC8), DXS41 (99.6), DXS43 (D2), DXS84 (754), Ornithine transcarbamylase, DXS7 (LI.28), DXS28 (C7), DXS67 (B24) DXS164 (PERT 87)

Table 4 (cont'd)

Disease	Gene Probe
Myotonic dystrophy	Complement C3 gene, Apolipoprotein CII, D19S7, Chromosome 19 DNA segments, D19S19, D19S16 (pJSB11), D19S15 (pJSB6), D19S19 (LDT152)
Neuropathy, X-linked	DXYS1, DXS14 (p58-1)
Norrie's disease	DXS7 (L1.28)
Ocular albinism (X-linked)	DSX85
Retinitis pigmentosa	DXS7 (L1.28)
Retinoschisis	DXS52 (Lambda-RC8), DXS85, DXS16
Spastic paraplegia	DX13, DXS52 (St14)
Skeletal dysplasia X-linked	DXS52 (St14), DXS15 (DX13)
Spinocerebellar ataxia	Chromosome 6 DNA segments
Steroid-sulphatase-X-linked ichthyosis	DXS9 (Lambda-RC8)
Wilson's disease	D13S4, D13S10

REFERENCES

1. McKusik, Mendelian Inheritance in Man, 7th ed. John Hopkins, 1986.
2. Antonarakis SE, Kazazian HH, Orkin SH. DNA polymorphism and molecular pathology of the human globin gene clusters. Hum Genetics 1985; 69: 1-14.
3. Schmidtke J, Cooper DN. Nucl Acids Res 1987; 15: 1.
4. Cooper DN, Schmidtke J. DNA restriction fragment length polymorphisms and heterozygosity in the human genome. Hum Genet 1984; 66: 1-16.
5. Hofker MH, Wapenaar MC, Goor N, Bakker E, Van Ommen G-JB, Pearson PL. Isolation of probes detecting restriction fragment length polymorphisms from X chromosome-specific libraries: potential use for diagnosis of Duchenne muscular dystrophy. Hum Genet 1985; 70: 148-56.

PATHOGENESIS AND (PRENATAL) DIAGNOSIS OF INHERITED DISEASES: GENETIC HETEROGENEITY AND CLINICAL DIVERSITY

A.J.J. Reuser

Department of Cell Biology and Genetics, Erasmus University Rotterdam

Diagnosis of human genetic diseases is performed at different levels. In many instances, the patient is first seen by the general practitioner who may refer to a specialist for detailed clinical investigation. Laboratory tests on blood, urine or tissue specimens may be required to support a differential diagnosis. These tests are often performed by the clinical chemist. Blood cells or tissue sections are analyzed by the pathologist, and a karyotype analysis by the cytogeneticist may be needed in addition. At a following level, molecular defects of proteins are studied, whereas analysis of DNA can be considered as the ultimate level of diagnosis of inherited diseases.

The new options and the limitations, the advantages and the disadvantages of DNA analysis as a diagnostic tool in comparison with other methods of investigation are discussed. Attention is specifically devoted to the phenomenon of genetic heterogeneity, a major problem for diagnosis at the gene level, and to prenatal diagnosis of inherited diseases, a major field of application.

SOME FIGURES

A total of 3,907 hereditary traits are listed in the 1986 edition of McKusick's catalogue of Mendelian inheritance in man (1). The present estimation of the total number of genes in man is between 50,000 and 70,000 (1, 2). Thus, mutations in only 4 to 8% of the genes are known to be associated with a human genetic disease. This percentage is expected to increase in future, but it may be limited by nature in that mutations in certain genes are not compatible with life or, in contrast, do not abolish an essential function.

At present, the number of cloned genes is about 400, but "new" genes are added almost daily to the list (2, 3). Many of these cloned genes are associated with inherited diseases, and have been assigned to specific human chromosomes or segments thereof. In addition, cloned DNA segments with unknown function, pseudo-genes (without protein product), and unspecified sequences homologous to known genes have been "mapped". Their total number is at present also about 400 (3), and together with the cloned genes they can serve as marker in linkage analysis studies.

The chromosomal localization of many genes has been determined by correlating the segregation of human cellular and molecular traits with human chromosomes or chromosome fragments in man-rodent somatic cell hybrids. Cloned DNA fragments can be used in this procedure as a probe to identify human genes on Southern blots. An alternative procedure is gene localization by direct in situ hybridization of DNA or RNA probes with metaphase spreads of human chromosomes. At present the assignment of 500 loci has been confirmed, an additional 250 loci have been assigned, provisionally (4). The relative positions of 160 loci associated with inherited diseases are indicated in the catalogue of McKusick (1). The function of these loci is not always known, but diagnosis of associated diseases is sometimes possible via DNA analysis.

METHODS OF DNA ANALYSIS

Linkage analysis is based on the fact that DNA sequences among different individuals are polymorphic, especially in non-coding regions. As a result, a restriction enzyme cutting the DNA at a site determined by a highly specific order of 4 to 8 nucleotides may find a restriction site in one molecular species of DNA whereas this site is absent in another. DNA fragments of different length will arise which can be separated in agarose gels and visualized after transfer to nitrocellulose filters by hybridization with a radioactivity labeled complementary DNA probe (Southern blot analysis). The method is discussed in detail by Prof. Dr. J. Schmidtke, in these Proceedings. Within one family the disease trait will be linked to a specific pattern of restriction fragments. A great advantage of linkage analysis is that it allows (prenatal) diagnosis of diseases without knowledge of the metabolic or protein defect.

Diagnosis by this method is in principle applicable to all inherited diseases provided that polymorphic sites are present around the mutant gene and probes are found which are sufficiently close to the affected gene to minimize the risk of a recombination event between the probe and gene in question during meiosis. The chances for recombination will be smaller as the distance between the probe and mutant gene is shorter. Knowledge of the chromosomal localization of the mutant gene is a prerequisite for the selection of suitable probes.

A disadvantage of linkage analysis is that an informative polymorphism is not encountered in all families. Each family needs to be investigated separately. Participation of the index patient, parents and sometimes further relatives is essential.

Some of the problems related to linkage analysis are overcome when probes are available of the mutant gene itself, allowing a more direct analysis of the mutation. The risk of recombination can be eliminated and only the index patient (and parents) need to be tested. Gene deletions are detectable by altered length of DNA fragments on Southern blots, when a suitable gene probe and proper restriction enzymes are applied. There are two methods for the detection of point mutations. When a restriction site is altered by the mutation, a restriction fragment with altered length will be obtained when the DNA is cut with the proper restriction enzyme. A more general method to diagnose point mutations is one in which oligonucleotide probes are used which specifically hybridize with either the wild type or the mutant gene. Both methods are discussed by Prof. Dr. J. Schmidtke. The major limitation of these direct procedures is that they are only applicable when the exact type and location of the mutation are known. The occurrence of genetic heterogeneity in the form of allelic mutations poses a problem and makes the method suitable for homogeneous monogenic diseases only.

Sickle cell anemia is an example of such a disease and is characterized by substitution of glutamic acid by valine in the ß chain of haemoglobin, caused by a mutation in the middle of the 6th codon. α-1-antitrypsin deficiency is another example of a disease that can be diagnosed with a probe directly directing the point mutation, but in this case the type of mutation needs first to be confirmed. In northern Europe the Z-type mutant is most common whereas the S-type mutation prevails in southern Europe.

Indeed, allelic heterogeneity can severely hamper direct diagnosis at the DNA level. For example, almost 500 genetic variants of the 4 types of globin genes have been described (5). In many instances it is more practical to perform diagnosis by linkage analysis (following the pattern of inheritance of the mutant gene via linkage to nearby polymorphic sites) than by direct analysis of the point mutation which first needs to be determined. Great reliability is ensured when DNA sequences of the gene in question are used as a probe. When the incidence of a disease is high in semi-isolated populations, such as ß-thalassemia in Cyprus and Sardinia, a genetically homogeneous disease may be anticipated and it becomes worthwhile to identify the exact type of mutation to make diagnosis even more reliable, and to develop a simple method for carrier detection.

ADVANTAGES OF DIAGNOSIS BY DNA ANALYSIS

The possibility of carrier detection either by linkage analysis or more favourably by direct analysis of the mutant gene is one of the major advantages of DNA analysis over other methods of investigation for the diagnosis and prevention of inherited diseases. Assessment of the heterozygous state by assay of metabolic function is both impractical and unreliable in recessive diseases. Establishment of heterozygosity by measurement of the deficient enzyme activity or protein level is often impeded by overlap of heterozygous and control levels. These problems are encountered by DNA analysis directly establishing the genetic status. Moreover, in carrier detection for X-linked diseases DNA analysis avoids the pitfalls of procedures relying on gene expression which are influenced by X-inactivation (6, 7).

The heterozygous state of siblings of patients may be established using linkage analysis. However, linkage studies are not informative in studying non-related individuals. As mentioned above, this restriction does not apply when probes directly detecting the mutation are used. Carrier detection by direct analysis of the mutation may become an important tool for the prevention of diseases with a relatively high incidence in a population. A number of these diseases are listed in Table 1. The applicability in population screening will depend on the homogeneity of the disease, and the availability and price of the technology.

Another advantage of DNA analysis in the diagnosis of inherited diseases is that any cell type (except erythrocytes) can be used as a source of DNA diagnose diseases that are exclusively or selectively expressed in tissues not easily accessible for diagnosis at the protein or metabolite level. This is of particular advantage in the prenatal diagnosis of inherited diseases.

DNA ANALYSIS IN PRENATAL DIAGNOSIS

Before DNA analysis became feasible, prenatal diagnosis of metabolic disorders was limited to those expressed in amniotic fluid cells (fibroblasts and epithelial cells), and to disorders as, for instance,

Table 1. Prevalent Human Inherited Diseases

Disease	incidence	mode of inheritance	gene/marker on chromosome
Adrenal hyperplasia	1/10,000	ASR	6
Cystic fibrosis (U.S. whites)	1/ 2,000	ASR	7
Duchenne muscular dystrophy	1/ 6,000	XLR	X
Haemophilia	1/10,000	XLR	X
Huntington's chorea	1/ 2,000	ASD	4
Myotonic dystrophy	1/10,000	ASD	19
Phenylketonuria	1/10,000	ASR	12
Polycystic kidneys (dominant type)	1/ 2,500	ASD	16
Sickle cell anemia (U.S. blacks)	1/ 625	ASR	11
Tay-Sachs disease (U.S. Jews)	1/ 3,000	ASR	15
Gaucher's disease 1 (U.S. Jews)	1/ 2,000	ASR	1

ASD; autosomal dominant, ASR; autosomal recessive, XLR; X-linked recessive (data adapted from Weatherall, 1985 Ref. 13, and Galjaard, 1980 Ref. 14).

cystic fibrosis which is almost invariably accompanied by decreased activity of fetal intestinal enzymes in the amniotic fluid supernate (8, 9). Amniocentesis is routinely performed in the 16th or 17th week of pregnancy with less than 1% risk for loss of the pregnancy. Fetal blood sampling by ultrasonically guided umbellicalcord puncture or under fetoscopic control enables diagnosis of disorders only detectable in blood or blood cells. However, availability of DNA analysis has reduced the number of indications for some of these diseases (haemoglobinopathies, haemophilia etc.), and extended the possibilities of prenatal diagnosis for diseases not expressed in amniotic fluid (cells) or fetal blood.

Nowadays, prenatal diagnoses are carried out on chorionic villi biopsies which are taken in the 10th week of pregnancy. Karyotype analysis can be performed without culturing cells from the villi and also most enzyme assays are carried out directly on the biopsy material. Cell culture is sometimes needed, for instance, in the diagnosis of Lesch-Nyhan syndrome when incorporation of 3H-thymidine is measured. A sufficient amount of DNA can be extracted from 40 mg of chorionic villi for analysis via restriction enzyme polymorphism or oligonucleotide hybridization. By September 1986, the world-wide experience with first trimester diagnosis was almost 17,000 procedures (data collected by Prof. Dr. H. Galjaard). Chromosome anomalies form the majority of indications, metabolic diseases are a minority at present. This may change in some areas with high incidence of disorders now detectable by DNA analysis. Diseases that have been diagnosed at the DNA level in the first trimester are listed in Table 2. In our Department of Clinical Genetics, the experience with diagnosis at the gene level includes cystic fibrosis, α-1-anti-trypsin deficiency and haemoglobinopathies. The number of tests is still small compared to the total number of approximately 1,000 first trimester diagnoses performed in Rotterdam since

Table 2. World-wide Survey of First Trimester Fetal Diagnosis by DNA Analysis (up to July 1986)

Adrenoleukodystrophy	Haemophilia A	Phenylketonuria
α-1-antitrypsin def.	Haemophilia B	Polycystic kidney dis.
Becker's musc. dystr.	21-hydroxylase def.	Retinitis pigmentosum
Chr. granulom. dis.	Myotonic distr.	Retinoblastoma
Cystic fibrosis	Norrie's dis.	Sickle cell anemia
Duchenne's mus. dystr.	Osteogenesis imp.	α-thalassemia
Huntington's chorea	Polycyst. kidney dis.	β-thalassemia

A total of 950 pregnancies at risk have been investigated and 270 affected embryo's were detected. (These data were collected by Prof. Dr. H. Galjaard, of the Department of Cell Biology and Genetics, Erasmus University, Rotterdam, The Netherlands).

1985. This will change in future considering the rate in which new disease markers are detected.

The amniocentesis program in our Center (Table 3) monitored nearly 13,000 pregnancies; the birth of over 600 severely handicapped children was prevented, but even important was the reassurance given to parents of the other 12,000 children.

PROBLEMS RELATED TO DIAGNOSIS AT THE GENE LEVEL

For every single disease it must be established whether DNA analysis is preferable to diagnosis at the metabolite or (enzyme) protein level. As mentioned above, the disease may be genetically heterogenous implying that the same metabolic disorder can be caused by allelic mutations of the same gene. This makes analysis at the DNA level for such a disease impractical.

Glycogenosis type II, a rare inborn error of metabolism, is chosen to illustrate this. This disease is caused by deficiency of the enzyme acid α-glucosidase which is localized in lysosomes and essential for the degradation of glycogen. Enzyme deficiency results in lysosomal glycogen accumulation, predominantly in skeletal and heart muscle cells. Clinical symptoms of the disease may be evident immediately from birth, with rapidly progressive muscular weakness and cardiomegaly causing death in the first or second year of life. However, clinical variants exist with onset of symptoms in the second or even third decade of life. This adult form of the disease is much less progressive and patients may live beyond the sixth decade. The primary cause of clinical diversity appears to be allelic heterogeneity, whereby each mutation has its own, rather specific effect, on the transcription of the mutant α-glucosidase gene, the biosynthesis of the mutant enzyme, or its intracellular transport to the lysosomes. In some cases of glycogenosis type II mRNA is completely or partially absent, in others only precursor forms of acid α-glucosidase are formed. Some mutant precursors do not find their way to the lysosome by lack of a recognition signal for binding to the mannose-6-phosphate receptor which mediates selective transport to these organelles (10, 11).

The data indicate that there is not a single specific mutation that leads to either the infantile, the juvenile or the adult form of this disease. Thus, the clinical phenotype will be hardly predictable on the basis of the gene mutation detected, and diagnosis at the gene level becomes unpractical. Enzyme activity measurements are more suitable for

Table 3. Experience with Prenatal Diagnosis in the Department of Clinical Genetics, University Hospital, Rotterdam (September 1986)

Indication	number of pregnancies investigated	number and percentages of fetal anomalies	
Advanced maternal age	5,258	143	2.7
Advanced paternal age	29	-	-
Previous chromosomal anomaly	1,162	14	1.2
Parental chrom. transl.	239	122	51.0
X-linked diseases	249	116	46.6
Risk for neural tube defect	4,157	60	1.4
Metabolic diseases	700	141	20.1
Other indications	834	86	10.3
Total	12,628	630	5%

diagnosis of the different clinical forms of glycogenosis type II. The residual activity of acid α-glucosidase in cultured fibroblasts and muscle cells from adult patients is significantly higher than in cells from patients with the severe infantile form of the disease (11). This directly affects the accumulation of lysosomal glycogen which is more abundant in infantile than adult forms of glycogenosis type II (12).

Enzyme or protein function is mostly generated by a series of complex processes including transcription, translation, posttranslational modification and transport to the intracellular or extracellular site of action. Therefore, the effect of a mutation at the functional level will often be unpredictable and diagnosis at the metabolite or protein level is advisable unless the disease is extremely homogeneous with respect to type of mutation and clinical expression. Even the same mutation may give rise to a variable phenotype because of differences in the genetic background of individuals and diversity of epigenetic factors.

Besides genetic heterogeneity there are problems of a different nature related to diagnosis at the gene level. Recombinant DNA technology is a relatively new methodology and not yet available in all countries where DNA analysis of inherited diseases is desired. This applies for example to the (prenatal) diagnosis of haemoglobinopathies with a high incidence in African and Asian countries.

Multifactorial diseases are less likely to become diagnosable at the DNA level, unless a "major" gene turns out to be responsible for such common malformations as spina bifida, congenital heart defects, facial clefts etc.

IMPLICATIONS OF DNA ANALYSIS IN GENETIC COUNSELING

Major improvements and new perspectives from diagnosis at the DNA level are carrier detection, and prenatal diagnosis of an increasing number of genetic disorders. Especially in autosomal dominant traits with late onset of (often neurological) symptoms, the availability of a polymorphic DNA marker may open the possibility of presymptomatic diagnosis. This may have far reaching consequences for those at risk to suffer later in life from an incurable disease. Huntington's chorea is such a disease. Through linkage analysis with a marker on chromosome 4, a 95% accurate diagnosis in informative families can be obtained. The risk of misdiagnosis due to genetic recombination is 5%. An individual identified as carrier of the dominant mutant gene needs careful support to minimize psychological and social dysfunction. Conflicts of interest may arise with spouse, children or insurance company. While diagnosis of Huntington's chorea at the DNA level does not improve the prognosis of the disease, it may contribute to prevention of the disease via genetic counseling and family planning.

DNA analysis will certainly help to unravel the molecular cause of an increasing number of inherited diseases. This may not only lead to a (more) reliable diagnosis, but also to the development of strategies for therapy.

CONCLUSIONS

DNA analysis is an important complementary tool in the diagnosis of human inherited diseases. Diagnosis at the gene level is sometimes required for the (prenatal) diagnosis of diseases that are not expressed in amniotic fluid or amniotic fluid cells, in fetal blood, or in chorionic villi cells, and more in general for the diagnosis of diseases expressed in tissues not easily accessible for studies at the (enzyme) protein or metabolite level. For diseases with an unknown molecular cause, diagnosis at the DNA level via linkage analysis is the only alternative. Carrier detection is undoubtedly most reliable at the DNA level. Diagnosis of inherited diseases by DNA analysis can be sufficient, but diagnosis at the protein and functional level is often preferable because of genetic heterogeneity and variable clinical expression. Diagnosis at the clinical or pathological level will always be essential, but is not always feasible (e.g. prenatal diagnosis and carrier detection), or sufficient. The patient will benefit most from the combined clinical, pathological and molecular (genetic) approaches for the most accurate diagnosis and the best treatment.

ACKNOWLEDGEMENTS

Many valuable suggestions for my lecture and this manuscript were given by Prof. Dr. M.F. Niermeijer from the Department of Clinical Genetics, University Hospital, Rotterdam. I like to thank him for his stimulating advice and his critical comments. I am grateful to Dr. D. Halley for helpful suggestions regarding the manuscript and to Drs. W.J. Kleijer, E.S. Sachs, M.G.J. Jahoda, M.F. Niermeijer and H. Galjaard for providing data on the experience with prenatal diagnosis in the Department of Clinical Genetics in Rotterdam. Prof. Dr. H. Galjaard is acknowledged for providing the results of the world-wide surveys. I thank Cary Meijerink for secretarial assistance. The studies on glycogenosis type II were supported by a grant from the Netherlands Organization for the Advancement of Pure Research (ZWO) under the auspices of the Netherlands Foundation for Medical and Health Research (MEDIGON), and the Princess Beatrix Fund.

REFERENCES

1. McKusick VA. Mendelian inheritance in man. 7th ed. The John Hopkins University Press, Baltimore, 1986.
2. Stanbury JB, Wijngaarden JB, Fredrickson DS, Goldstein JL, Brown MS, (eds). The metabolic basis of inherited disease. 5th ed. McGraw-Hill Book Company, New York, 1983.

3. Beaudet AL. Bibliography of cloned human and other selected DNA's. Am. J. Hum Genet 1985; 37: 386-406.
4. Human Gene Mapping 8. Helsinki Conference 1985. Cytogenet Cell Genet 1985; 40: 1-832.
5. Honig GR, Adams JG. Human hemoglobin genetics. Springer Verlag, Wien, 1986.
6. Bakker E et al. Prenatal diagnosis and carrier detection of Duchenne muscular dystrophy with closely linked RFLP's. Lancet 1985; 8439: 655-658.
7. Bakker E, Bonten EJ, De Lange LF, Veenema H, Majoor-Krakauer D, Hofker MH, Van Ommen GJB, Pearson PL. DNA probe analysis for carrier detection and prenatal diagnosis of Duchenne muscular dystrophy: a standard diagnostic procedure. J Med Genet 1986; 23: 573-580.
8. Brock DJH, Barron L, Bedgood D, Hayward C. Prospective prenatal diagnosis of cystic fibrosis. Lancet 1985; 8439: 1175-1178.
9. Claass AHW, Kleijer WJ, Van Diggelen OP, Van der Veer E, Sips HJ. Prenatal detection of cystic fibrosis. Comparative study of maltase and alkaline phosphatase activities in amniotic fluid. Prenatal Diagnosis 1985; 6: 419-427.
10. Reuser AJJ, Kroos M, Oude Elferink RPJ, Tager JM. Defects in synthesis, phosphorylation and maturation of acid α-glucosidase in glycogenosis type II. J Biol Chem 1985; 260: 8336-8341.
11. Reuser AJJ, Kroos M, Willemsen R, Swallow D, Tager JM, Galjaard H. Clinical diversity in glucogenosis type II. Biosynthesis and in situ localization of acid α-glucosidase in mutant fibroblasts. J Clin Invest 1987; 79: 1689-1699.
12. Van der Ploeg AT, Kroos M, Van Dongen JM, Visser WJ, Bolhuis PA, Loonen MCB, Reuser AJJ. Breakdown of lysosomal glycogen in cultured fibroblasts from glycogenosis type II patients after uptake of acid α-glucosidase. J Neurol Sci 1987; 79: 327-336.
13. Weatherall DJ. The new genetics and clinical practice. Oxford University Press, Oxford, 1985.
14. Galjaard H. Genetic metabolic diseases: early diagnosis and prenatal analysis. Elsevier/North Holland, Amsterdam, 1980.

MOLECULAR BIOLOGY OF THE HUMAN ALDOLASE ISOENZYME GENE FAMILY

Francesco Salvatore, Paola Izzo, Paola Costanzo and Rita Santamaria

Istituto di Scienze Biochimiche, II Facoltà di Medicina, via S. Pansini, 5, I-80131 Naples, Italy

The aldolase isoenzyme family, because of its phenotypic properties, can be taken as a model system with which to approach the study of some fundamental problems of cell biology studied at the molecular level, as well as the study of the molecular biology of an inherited disease, namely hereditary fructose intolerance (HFI) (1). The tissue-specific expression of the same gene or of a gene family, the modulation of gene expression, in relation to development and differentiation, and the cell de-differentiation, that are very often involved in neoplastic transformation, and that lead to resurgence of foetal enzyme or isoenzymes expression, are still largely unknown in their biochemical correlations.

In this paper, after a very brief description of some of the properties of the aldolase isoenzyme family at protein level, we shall concentrate on studies performed by us on the aldolase isoenzyme family, both at messenger RNA and gene level.

PHENOTYPIC PROPERTIES OF THE HUMAN ALDOLASE ISOENZYME SYSTEM

The reaction mechanism for the aldolase enzyme is an aldolic fission at a carbon bond that results in the production of two trioses, the substrate being either the fructose-bisphosphate or the fructose-monophosphate (see reviews in 2, 3). The main isoenzyme forms, A, B and C, show completely different kinetic properties with respect to substrate affinity measured by the activity ratio toward the two substrates, the bisphosphate and the monophosphate sugars (see Table 1). The activity ratio varies from fifty for the muscle aldolase A to one for the liver-specific aldolase B, being intermediate for the brain-specific aldolase C. All these main three forms are made of four identical monomers with an Mr of 40-37×10^3. The amino acid composition is different and this determines no cross-reactivity when each isoenzyme is tested against specific antisera versus the others. The hybrid tetrameric forms, made of two different monomers, are A-B in kidney and A-C in brain (see 2, 3). Table 1 also provides the chromosomal localization of the three genes that are responsible for the phenotypic expression of the isoenzyme family, i.e., the gene for the A monomer on chromosome 16 (4), the gene for the B monomer

Table 1. The aldolase isoenzyme family: molecular properties*

Type	Main localization	Mr	Sub-units	Subunit Mr	Substrate affinity	FDP/FIP ratio	Human chromosome localization
A	muscle	$160x10^3$	4 A	$40x10^3$	FDP>>FIP	50	Chromosome 16
B	liver	$158x10^3$	4 B	$39x10^3$	FDP = FIP	1	Chromosome 9
C	brain	$148x10^3$	4 C	$37x10^3$	FDP > FIP	25	Chromosome 17
A-B Hybrids	kidney	/	4A-B (A_3B,A_2B_2,AB_3)	/	/	/	
A-C Hybrids	brain	/	4A-C (A_3C,A_2C_2,AC_3)	/	/	/	

* The data are summarized from the available literature (see in 2, 3) on the protein molecules, except those of the last column, which are taken from our own recent data (6).

on chromosome 9 (5), and the gene for the C monomer on chromosome 17 (6). These results were obtained by us (6) using the Southern blot technique to analyze DNA from a series of rodent versus human somatic cell hybrids, for the presence of specific aldolase A, B or C related sequences.

The tissue specificity, already known on the basis of protein isoenzyme occurrence (see 2, 3) has been supported by various groups (see in 2, 3), including ours, using Northern blot hybridization experiments. Our results confirm that the C enzyme is expressed only in brain, while the A enzyme is expressed in muscle, kidney and brain, and as a minor spot in lung; there is no hybridization in the liver (7). The B enzyme is expressed in the liver, and also the kidney shows some expression of this type of aldolase (7). The observation that the levels of mRNA correlate with the levels of the isoenzyme protein present in various tissues suggests that aldolase expression is regulated via the control of mRNA levels.

The aldolase system is also highly modulated during development (embryogenesis and growth). In fact, the level of aldolase A mRNA decreases from the beginning of gestation to birth, and it continues to decrease during growth to the adult stage, whereas the level of aldolase B in the liver increases drastically up to birth and during development to the adult stage (8, 9).

A third feature of this system concerns gene expression during carcinogenesis and cell neoplastic transformation (10). In particular, the phenomenon of resurgence of the expression of a messenger which is not present in the adult stage, is shown in rats treated with a chemical hepatocarcinogen (11). In these animals there is a drastic increase of aldolase A messenger expression starting from the fourth week after the chemical treatment. Aldolase B expression, on the other hand, remains unchanged. We obtained similar results by hybridizing total RNA extracted from various human tissues and from human hepatoma cell lines to a human aldolase A cDNA probe (12). Interestingly, no hybridization occurred with normal human liver, whereas the hepatoma cell lines showed a strong hybridization signal (12).

Therefore, it can be said that the aldolase system shows all the phenotypical features of an ideal system with which to approach the study of the above-mentioned basic problems of cell biology.

MESSENGER RNAs OF THE HUMAN ALDOLASE SYSTEM

Table 2 reports a list of all the messenger RNAs sequenced during the last four years (7, 13-23). Here we shall describe our contribution to cloning, characterization and sequencing of cDNA for aldolase C, A and B. The aldolase C cDNA was isolated for the first time by us in a mouse brain library; however we elucidated only the NH_2-terminal part (7). More recently, another partial sequence of a rat brain aldolase C cDNA, referring to the COOH-terminal portion, has appeared (23).

Two cDNA of the human aldolase A, one from the liver sequenced by Sakakibara et al. (20), and one from the fibroblast sequenced by us (22), have been recently published. These two sequences led to the complete elucidation of the human aldolase A sequence, which was only partially known from protein chemistry analysis (24). These cDNA sequences are identical except for the 5′ non-coding region.

The importance of the detection of two mRNAs is that it demonstrates the multiplicity of human messenger RNA for aldolase A. Furthermore, the known mRNAs are completely different in the 5′ non-coding region, and therefore they will be useful in studies on this region at the gene level, particularly for tissue-specificity.

THE ALDOLASE ISOENZYME GENE FAMILY

The study of the structure and expression of aldolase genomic clones has been approached by several groups, including ours (25-29). In fact, we have fully characterized and sequenced the human aldolase A genomic clone (28), and also the human aldolase C gene (29).

By using the mouse aldolase C cDNA as probe we screened a human genomic library and have cloned, characterized and sequenced about 4500 bp of the aldolase C gene (29). The sequenced gene is split in nine exons including all the coding regions and the 3′ non-coding region and part of the 5′ non-coding region. After nucleotide sequence translation, we showed the complete amino acid sequence of this specific human brain protein, which was previously thought to be thirty amino acids shorter than the other aldolases (2): therefore, unless a post-translation partial proteolytic event occurs as a maturation process, the earlier data will have to be revised.

The structural organization of the aldolase A gene shows the presence of eight exons, which include the complete coding region and the 3′ non-coding region up to the polyadenylation signal. However, in the 5′ non-coding region, upstream from the initiation codon, there are four exons that correspond to part of the sequences found in the corresponding region of mRNAs, both in man and in other mammals (18, 20-22) (see Figure 1).

A structural analysis comparison between the sequences at the gene level and the cDNAs isolated so far and sequenced, both from man and higher mammals, allowed us to conclude for the existence of three classes of human messenger RNA. The first one includes a minor species of aldolase A messenger RNA found in human liver (20). The second class includes rabbit (18) and rat muscle aldolase A (21). Class three includes, besides the human fibroblast mRNA (22), the rat II and rat III species of messenger

Table 2. Sequenced aldolase messenger RNAs

Isoenzyme	Tissue	Length (bp) (without poly A)	References
A	Rabbit muscle	1,375	Tolan et al., 1984 (18)
	Rat muscle	1,343 (Rat I)	Mukai et al., 1986 (21)
	Mouse muscle	772 (partial)	Paolella et al., 1986 (7)
	Rat ascites hep.	1,392 (Rat II)	Mukai et al., 1986 (21)
		1.440 (Rat III)	
	Human liver	1,464	Sakakibara et al., 1985 (20)
	Human fibroblast	1,437	Izzo et al., 1987 (22)
B	Rat liver	1,547	Tsutsumi et al., 1984 (15)
	Human liver	115 (partial)	Costanzo et al., 1983 (13)
	Human liver	571 (partial)	Besmond et al., 1983 (14)
	Human liver	1,596	Paolella et al., 1984 (16)
	Human liver	1,389 (partial)	Rottman et al., 1984 (17)
	Human liver	1,652	Sakakibara et al., 1985 (19)
C	Mouse brain	741 (partial)	Paolella et al., 1986 (7)
	Rat brain	725 (partial)	Skala et al., 1987 (23)

RNA, isolated from ascites hepatoma (21). We demonstrated that the fibroblast messenger species is present in various organs and tissues, thereby providing evidence for the ubiquitary presence of this messenger in mammalian tissues. The S_1 mapping experiments, in fact, indicates that only muscle contains the specific messenger RNA which includes exon III within the 5′ non-coding region; this is not present in hepatoma or in fibroblast tissue. Therefore, the muscle species is very specific for this tissue. On the other hand, in muscle, fibroblast and hepatoma there are at least two species of messenger containing exon IV within the 5′ non-coding region. These RNAs are both ubiquitary and differ only in their length at the 5′ non-coding region.

In summary, we have a conclusive pattern of four different messengers starting from the unique 5′ non-coding region of the aldolase A gene. The interesting feature of the system is that, as has been found only in a very few other cases (30-33), the different messengers are produced by multiple promoters on the same gene sequence that initiate transcription and proceed through a differential splicing mechanism. This differential splicing is

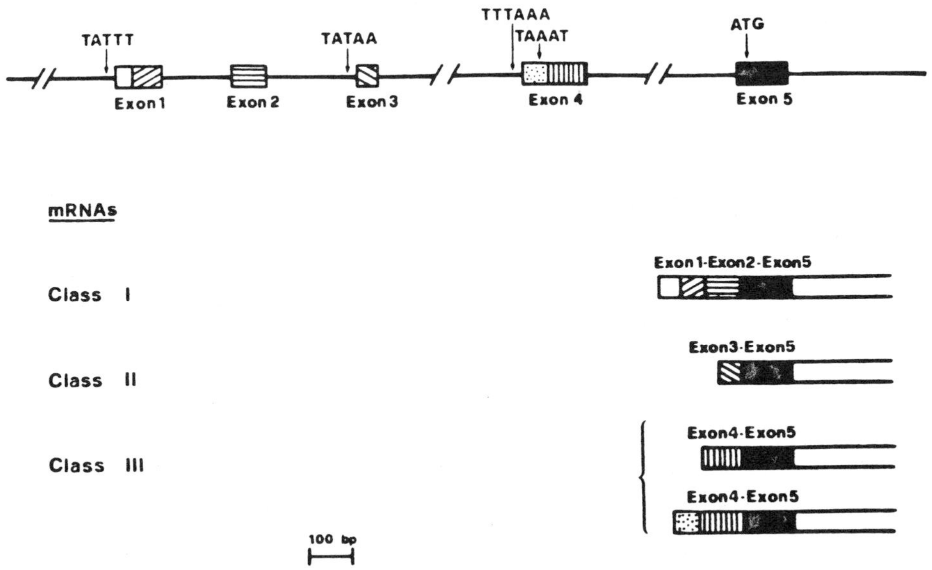

Fig. 1. Intron/exon organization of the 5' end of human aldolase A gene (upper part) and various mRNA classes (lower part).

guided to the appropriate target defined by the presence of the splicing consensus signals.

Therefore, when the first messenger is to be formed, the first two exons are cut and the splicing proceeds until the AG sequence, which is in close vicinity to the ATG initiation codon. When muscle mRNA is to be formed, its promoter proceeds, and after transcription of the exon III, splicing occurs up to the previous point. Finally, for the formation of the ubiquitary fibroblast species, a third and a fourth promoter are operative, and they transcribe the fourth exon by consecutive splicing at the same point as the previous one. In conclusion, the data allow us to conclude for a peculiar model of gene transcriptional control, which produces multiple messenger RNA from a unique gene sequence. The presence of a TATA box, or a TATA-like structure upstream from exons I, III, IV supports the envisaged mechanism of mRNA formation (see Fig. 1).

ALDOLASE B AND HEREDITARY FRUCTOSE INTOLERANCE

Aldolase B is mainly devoted to the utilization of exogenous fructose, after its phosphorylation to fructose monophosphate. Therefore, the B isoenzyme exerts a physiological function different from that of aldolase A; the latter being devoted mainly to the glycolytic and glycogenetic pathways, which are prevalent in muscle, liver, kidney and intestine (2, 3). The lack of the B enzyme is the cause of hereditary fructose intolerance (1). Hopefully, this disease will in future be diagnosed using DNA recombinant techniques, and therefore patients will no longer be subjected to cumbersome and, on occasions, painful liver needle biopsy. We were one of the first groups to unravel the complete human sequence of this human liver enzyme from the cDNA nucleotide structure (14, 16).

The aldolase B probe was hybridized to peripheral DNA after digestion with several restriction enzymes, and it revealed the presence of a few bands for each restriction enzyme (16). Furthermore, utilizing as probes

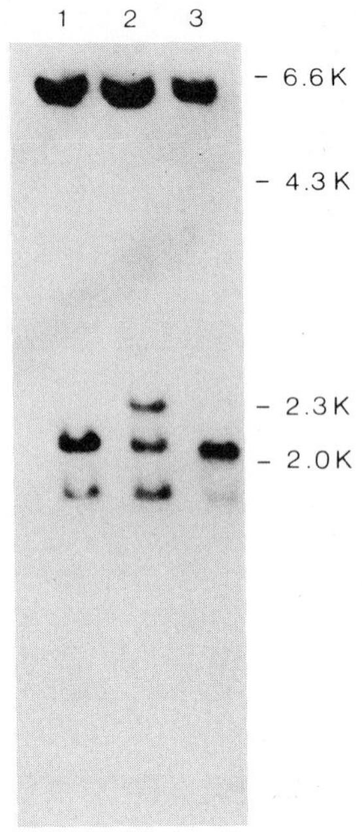

Fig. 2. Southern blot of DNA digested with Pvu II and hybridized to aldolase B cDNA from three individuals (N. 2 shows the 2.3 kb extra band).

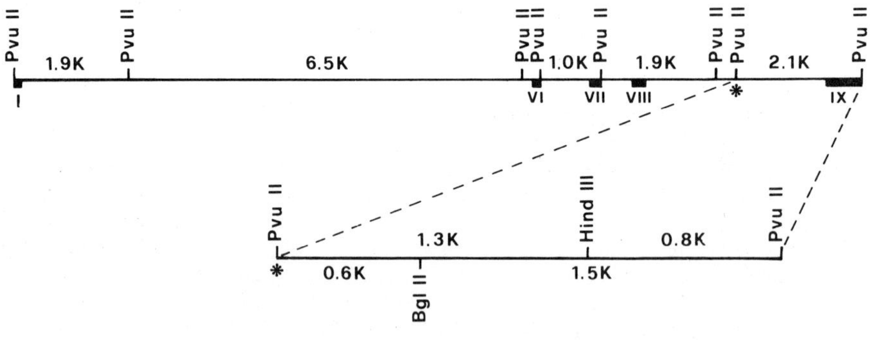

Fig. 3. Restriction map of human aldolase B gene; the Pvu II polymorphic site is indicated with the asterisk. (Taken from ref. 35, with the permission of the publisher).

Table 3. Allelic frequencies for Pvu II RFLP in the normal population

No. analyzed subjects	83
No. heterozygous subjects	19
No. homozygous subjects for the frequent allele	63
No. homozygous subjects for the rare allele	1
Allelic frequencies	
Frequent allele (2.1 kb)	0.873
Rare allele (2.3 kb)	0.127
Expected frequencies for each genotype	
Homozygote for the frequent allele	0.764 (76.4%)
Heterozygote	0.22 (22%)
Homozygote for the rare allele	0.016 (1.6%)

limited regions at the 5′ and 3′ ends, the number of bands appeared even less numerous, which indicates the likely existence of only one gene copy.

In an attempt to identify possible restriction fragment length polymorphisms (RFLP), a variety of enzymes have been used on DNA from several subjects. Only with the Pvu II enzyme, did we find a pattern that is different from the most common one, in that it showed an extra 2.3 kb band (34) (see Figure 2). We have also isolated and partially characterized a genomic clone of aldolase B. It contains the whole coding region, and the 5′ and 3′ non-coding regions. In order to map the polymorphic site, subclones have been used as probes to selectively detect the hybridization bands. The bands obtained after digestion with Pvu II enzyme were then ordered along the gene (see Fig. 3, upper part). Then, utilizing as probe a subclone corresponding to the 2.1 kb band and using a double digestion with other restriction enzymes, namely Hind III and Bgl II (see Fig. 3, lower part) that have a restriction site within the segment, it has been possible to map the polymorphic site (35). We choose two individuals with DNA that showed the extra 2.3 kb band and two individuals without the extra band. The four DNA were digested with Pvu II, electrophoresed and hybridized with a probe corresponding to the 2.1 kb fragment at the extreme 3′ of the gene. The probe hybridized with the 2.1 kb band in all individuals and to the 2.3 kb band in the individuals having the extra band. This result, together with the decreased intensity of the 2.1 kb band in the presence of the 2.3 kb band, indicates that the 2.3 kb band derives from the 2.1 kb band. To identify which of the two terminal Pvu II sites generated the polymorphism, the subclone previously described was hybridized to genomic DNA digested with Pvu II plus Hind III and with Pvu II plus Bgl II, which asymmetrically cut the 2.1 kb band. When Hind III was used, the larger band increased in size in the polymorphic subjects, thus indicating that the Pvu II site, situated on the left, is the one which is lacking in the polymorphic subjects. The smaller band, in fact, was unchanged. When Bgl II was used together with Pvu II, the larger band did not show any variation in the variant individuals, thus confirming that the Pvu II on the left site is the one involved in the polymorphism (see Fig. 3).

Out of 83 subjects analyzed, 19 were found to be polymorphic. Table 3 shows the frequency of the polymorphism: 0.127 for the rare allele and 0.873 for the frequent allele. The genotypic frequencies are 76.4% for the homozygote for the frequent allele, 22% for the heterozygote and 1.6% for the homozygote for the rare allele (see Table 3).

Now this polymorphism and its mapping are known, it is possible to conduct studies in families with one or more members affected by HFI. These studies may lead to a reliable tool for the diagnosis of the disease and of the heterozygous state.

ACKNOWLEDGEMENTS

This work was supported by grants from the C.N.R. (Rome), with special reference to the P.F. Ingegneria Genetica and the P.F. Oncologica, and from M.P.I., Rome, Italy.

REFERENCES

1. Gitzelmann R, Steinmann B, Van den Berghe G. In: Stanbury JB, et al., eds. The Metabolic Basis of Inherited Disease, 5th ed. New York: McGraw-Hill Book Company 1983: 118-140.
2. Horecker BL, Tsolas O, Lai CY. In: Boyer PB, ed. The Enzymes. New York: Academic Press 1972; Vol 7: 213-258.
3. Salvatore F, Izzo P, Paolella G. In: F. Blasi, ed. Human Genes and Diseases. Wiley & Sons. Chichester. Horiz Biochem Biophys 1986; Vol 8: 611-665.
4. Kukita A, Yoshida MC, Fukushige S, Sakakibari M, Joh K, Mukai T, Hori K. Hum Genet 1987; 76: 20-26.
5. Henry I, Gallano P, Besmond C, et al. Ann Hum Genet 1985; 49: 173-180.
6. Santamaria R, Buono P, Paolella G, et al. Ninth International Workshop on Human Gene Mapping. Paris, 6-11 September 1987; n 440; 350.
7. Paolella G, Buono P, Mancini FP, Izzo P, Salvatore F. Eur J Biochem 1986; 156: 229-235.
8. Hommes FA, Draisma MI. Biochim Biophys Acta 1970; 222: 251-252.
9. Kellen JA, Chan A, Caplan B, Malkin A. Enzyme 1980; 25: 228-235.
10. Schapira F, Hatzfeld A, Weber A. In: Markert CL ed. Isozymes New York: Academic Press 1975: 987-1003.
11. Daimon M, Tsutsumi K, Sato J, Tsutsimi R, Ishikawa K. Biochem Biophys Res Commun 1984; 124: 337-343.
12. Izzo P, Costanzo P, Lupo A, Salvatore F. In: Cisimo F et al., eds. Human Tumor Markers. Berlin: Walter de Gruyter 1987: 889-898.
13. Costanzo F, Castagnoli L, Dente L, et al. EMBO J 1983; 2: 57-61.
14. Besmond C, Dreyfus JC, Gregori C, et al. Biochem Biophys Res Commun 1983; 117: 601-609.
15. Tsutsumi K, Mukai T, Tsutsumi R, et al. J Mol Biol 1985; 181: 153-160.
16. Paolella G, Santamaria R, Izzo P, Costanzo P, Salvatore F. Nucl Acids Res 1984; 12: 7401-7410.
17. Rottman WR, Tolan DR, Penhoet EE. Proc Natl Acad Sci USA 1984; 81: 2738-2742.
18. Tolan DR, Amsden AB, Putney SD, Urdea MS, Penhoet EE. J Biol Chem 1984; 259: 1127-1131.
19. Sakakibara M, Mukai T, Yatsuki H, Hori K. Nucl Acids Res 1985; 13: 5055-5069.
20. Sakakibari M, Mukai T, Hori K. Biochem Biophys Res Commun 1985; 131: 413-420.
21. Mukai T, Joh K, Arai Y, Yatsuki H, Hori K. J Biol Chem 1986; 261: 3347-3354.
22. Izzo P, Costanzo P, Lupo A, et al. Eur J Biochem 1987; 164: 9-13.
23. Skala H, Vibert M, Lamas E, Maire P, Schweighoffer F, Kahn A. Eur J Biochem 1987; 163: 513-518.

24. Freemont PS, Dunbar B, Fothergill LA. Arch Biochem Biophys 1984; 228: 342-352.
25. Burgess, DG, Penhoet EE. J Biol Chem 1985; 260: 4604-4614.
26. Joh K, Arai Y, Mukai T, Hori K. J Mol Biol 1986; 190: 401-410.
27. Tolan DR, Penhoet EE. Mol Biol Med 1986; 3: 245-264.
28. Costanzo P, Izzo P, Lupo A, Rippa E, Paolella G, Salvatore F. It J Biochem 1987; 36, in press.
29. Buono P, Paolella G, Mancini FP, Izzo P, Salvatore F. It J Biochem 1987; 36, in press.
30. Young RA, Hagenbüchle O, Schibler U. Cell 1981; 23: 451-458.
31. Ingrahm MA, Evans GA. Mol Cell Biol 1986; 6: 2923-2931.
32. Battey J, Moulding C, Taub R, et al. Cell 1983; 34: 779-787.
33. Frunzio R, Chiarotti L, Brown AL, Graham DE, Rechler MM, Bruni CB. J Biol Chem 1986; 261: 17138-17149.
34. Paolella G, Santamaria R, Buono P, Salvatore F. Nucl Acids Res 1986; 14: 5568.
35. Paolella G, Santamaria R, Buono P, Salvatore F. Hum Genet 1987; 77: 115-117.

CHAPTER 3
THERAPEUTIC DRUG MONITORING AND TOXICOLOGY

THERAPEUTIC DRUG MONITORING AND TOXICOLOGY: RELEVANCE OF MEASURING METABOLITES

Frederik A. de Wolff and Peter M. Edelbroek

Toxicology Laboratory, University Hospital of Leiden, Leiden
The Netherlands

1. INTRODUCTION

Therapeutic Drug Monitoring and Toxicology or TDM-T is a relatively young branch of clinical chemistry. In some countries it is not considered as being part of clinical chemistry but of pharmacology or other disciplines. However, as TDM-T is based on chemical analysis of patient samples, it forms without doubt a specialty in clinical chemistry. This justifies the incorporation of a symposium on TDM-T in this international congress on chemical chemistry.

TDM-T consists of two closely related subjects:

1. Therapeutic Drug Monitoring, which is defined as the measurement of prescribed drugs in body fluids - usually serum - in order to individualize the dosage regimen and to prevent toxicity or underdosage, and

2. the toxicology component of TDM-T, which can be defined as the identification and measurement of drugs which are not necessarily prescribed to the patient, and of other potential poisons, both in body fluids and tissue samples. The aim of these toxicological analyses is not only to establish the diagnosis in poisoned patients, but also to provide a basis for specific treatment of the intoxication.

It should be stressed that the definition of toxicology given here is not generally applicable; toxicology as a biomedical science is usually defined as the study of the interaction between potentially noxious substances and living organisms. When in this article the term "toxicology" is used, the first mentioned definition applies.

2. HISTORY OF TDM-T

TDM-T came into real existence in the 1970's. Before that date, very few drug analyses were used for therapeutic purposes, e.g. sulfonamides (1) and salicylate (2). Diagnosis of poisoning in that period was mainly

Table 1: List of drugs for which TDM may be useful

1. Anticonvulsants:	phenytoin, carbamazepine, valproate
2. Cardiac drugs:	digoxin, anti-arrhymics?
3. Antibiotics:	aminoglycosides, sulfonamides
4. Antiasthmatics:	theophylline
5. Analgesics:	salicylate
6. Antidepressants:	lithium, tricyclics
7. Cytostatics:	methotrexate, cisplatinum
8. Immunosuppresives:	cyclosporin

limited to forensic post-mortem cases. At the end of the 1960's, the first more specific drug assays were designed thanks to the development of gas chromatography and the dawn of clinical pharmacology. One of the most important landmarks was the introduction of the analysis of the anticonvulsant drug phenytoin in the laboratory, and its clinical application (3).

In the "Silver Seventies" the literature on TDM-T increased tremendously. Some general characteristics in that period are, that analytical and clinical studies have been published on a very large number of prescribed drugs. Many people believed that medication with most, if not all, drugs could be improved by measuring serum levels. Technically, this was made possible by the development of new techniques such as capillary gas chromatography and high-performance liquid chromatography. In addition, many studies advocated the use of exotic matrices such as saliva, cerebrospinal fluid, hair, sweat, or even tissue biopsies instead of serum. And those who preferred to continue using serum were often tempted to measure the fraction of drug which is not bound to serum proteins even on a routine basis. Also, the measurement of drug metabolites came into being, which was strongly facilitated by the introduction of HPLC in the clinical laboratory. Many of these sophisticated procedures, however, did not survive or were not introduced in the routine clinical laboratory, mainly because their contribution to diagnosis and treatment was very small in comparison to the high cost. The present status of TDM-T is such that drug analysis is really useful for a very limited number of substances. A list of these drugs, which is by no means exhaustive, is given in Table 1.

It may also be noted that serum is in most instances the matrix of choice for drug analysis. In addition, the assay of the non-protein bound fraction of a drug in serum is only required in some special cases. The measurement of metabolites in the TDM-T laboratory, however, is of increasing importance, which will be expounded below.

As was mentioned before, TDM-T is a relatively young branch of clinical chemistry. It is therefore interesting to have an impression of the quantitative contribution of it to the whole of clinical chemical assays. Therefore, the figures which were obtained in 1986 in the University Hospital of Leiden, which is a 1000-bed general teaching hospital, were analysed. "Classical" general clinical chemistry had a production of over 1.8 million assays and endocrinology of 80,000 analyses, whereas the contribution of TDM-T amounted to 15,000 assays, which is only 0.8% of the production of the chemical laboratories excluding haematology and immunology.

If these figures are compared with the number of publications on TDM-T which appeared in the 1986 issue of "Clinical Chemistry", it is noted that 84 articles (11.5%), including three major review articles, deal with this subject. In this number of publications, those on toxic trace elements, which play an important role in TDM-T, are included. These observations may indicate that, although the number of analyses is relatively small, TDM-T is still in a process of development and draws the attention of many clinical chemists.

3. MEASUREMENT OF METABOLITES

As mentioned before, measurement of metabolites draws attention of clinical chemists working in the TDM-T field. It may be questioned whether this feature is really relevant, or that it is more or less superfluous like some other topics such as routine measurement of unbound fractions or saliva levels. An attempt is made to answer this question using 4 examples from different fields:

a. Active metabolites in TDM
b. Inactive metabolites in TDM
c. Metabolites in clinical toxicology
d. Metabolites in post-mortem toxicology.

3a. Active metabolites in TDM

Assessment on a routine basis of metabolites in addition to the parent drug is, generally speaking, only useful when this contributes significantly to the diagnosis or treatment. This is the case, e.g., with the tricyclic antidepressant amitriptyline (AT) which is demethylated to the active substance nortriptyline (NT). This compound is, in turn, hydroxylated mainly to E-10-hydroxynortriptyline which may have an antidepressant effect, and probably contributes to the cardiotoxicity of AT (4).

The interpretation of AT and NT levels is still a matter of debate. There is no consensus on the therapeutic range of AT, but probably it is between 75 and 200 μg/L for the sum total of AT and NT. Most studies do not find a significant correlation between plasma levels and the derived antidepressant effect. Only a few studies describe a positive relationship (5). Others suggest a positive relationship between the NT/AT ratio and clinical improvement (6, 7). Patients with a high demethylation capacity seem to respond better to the therapy than those with a lower capacity to form NT from AT.

A good example of a drug of which the determination of metabolites is essential is the new anticonvulsant, oxcarbazepine, which is chemically related to the classical anti-epileptic carbamazepine (CBZ).
CBZ itself is an active substance, and so is its major metabolite, CBZ-epoxide. CBZ is a strong enzyme inducer, which may interfere with the metabolism of co-medication, whereas the epoxide, as a reactive substance, is probably responsible for at least part of the side-effects which are seen in CBZ therapy. Therefore, oxcarbazepine was developed, which does not induce cytochrome P-450 liver enzymes and which is not converted into the epoxide but into an active metabolite, 10,11-dihydro-10-hydroxy-CBZ. Both substances are ultimately metabolized to the inactive substance CBZ-diol. At steady-state of a therapeutic dosage of CBZ, the parent drug is present in plasma in a concentration of approximately 7 mg/L. The level of the epoxide is approximately 15% of this concentration and as this percentage is rather constant, it is usually sufficient for TDM to measure only the parent compound. In the case of oxcarbazepine, however, the parent drug is present in plasma in very low concentrations. The active metabolite 10,11-dihydroxy-10-hydroxy-CBZ is especially responsible for the therapeutic effect of this drug. In the case of oxcarbazepine medication it is therefore more important to measure the metabolite than the parent drug.

3b. Inactive metabolites in TDM

With regard to measurement of metabolites, there is a common misbelief that only active metabolites should be determined but there is no place in TDM for the analysis of inactive metabolites. This statement is not always correct, as can be demonstrated with another anticonvulsant, phenytoin (PHT). This drug is oxidized in the liver by a cytochrome P-450 isozyme to the inactive hydroxy compound hydroxyphenylphenylhydantoin (HPPH), which, in turn, is conjugated to glucuronide prior to excretion in the urine.

PHT is an effective but very difficult drug to administer, first, because there is a strong interindividual variation in metabolism, and secondly because it is subject to saturable kinetics in the therapeutic range. This means that a small dose increase may lead to a large increase in plasma levels, and hence toxicity (8).

Measurement of the inactive hydroxy-metabolite in addition to the parent drug can be very helpful to differentiate between hypo- and hypermetabolism on the one hand, and other causes of disarranged therapy, such as noncompliance or overdosage, on the other. This can be demonstrated with the following case:

A 68-year-old retired postman was admitted with a cerebral infarction. After dismissal he developed a focal epilepsy for which PHT was prescribed in the usual starting dosage of 300 mg daily. Eleven days after onset of the medication he was admitted for the second time with a number of symptoms which lead the neurologist to differential diagnosis of either a second cerebral infarction, or a PHT intoxication. A toxic serum concentration of 37.4 mg/L was found, the upper limit of the therapeutic range being 20 mg/L. This finding explained the symptoms observed. The dose was decreased by 50%, but in spite of that two weeks later an even higher serum level of 41.3 mg/L was measured. It was then decided to discontinue medication completely, and during this drug-free period a serum half-life of 7.5 days was found, which is about 10 times longer than in the average patient. Finally, maintenance dosage of 100 mg daily was found to produce an effective serum level in the lower therapeutic range, and with this medication the patient was kept free from epileptic fits until his death 5 years after the first admission.

A number of possibilities were considered as the cause of this intoxication. Accidental or intentional overdosage were deemed unlikely by the clinician. Liver function tests were normal except for a slightly elevated gamma-GT activity which is almost always seen in patients treated with the enzyme-inducer PHT. Hypometabolism as a result of liver disease could therefore be excluded. This patient also received the anticoagulant phenprocoumon, and it could be possible that this drug, like its relative dicoumarol, interfered with phenytoin oxidation. Such interaction, however, has never been reported in patients on this particular drug combination. It was therefore considered most likely that the observed PHT intoxication in this patient was the result of a deficiency of the enzyme responsible for phenytoin oxidation.

To evaluate this hypothesis, four different tests were performed: debrisoquine hydroxylation, antipyrin plasma half-life, antipyrin metabolite formation, and the PHT to HPPH metabolic ratio in the urine. Contrary to expectations, this patient was a rapid oxidizer of debrisoquine and a rapid metabolizer of antipyrin, which could be explained by selective induction of hydroxy-antipyrin formation by PHT. The PHT to HPPH ratio in the urine of this patient, however, was increased threefold in comparison with a normal group.

This observation leads to the conclusion that the observed intoxication must be ascribed to PHT hypometabolism (9). Another important conclusion is, that this diagnosis could only be confirmed by measuring the inactive metabolite of phenytoin, HPPH. All other, established, methods to diagnose impaired drug oxidation failed to demonstrate hypometabolism.

3c. Metabolites in clinical toxicology

In acute clinical toxicology, overdosage with hypnotic drugs is very common. Over the last decade, intoxication with drugs of the benzodiazepine group has gradually replaced barbiturate overdosage almost completely. Although the morbidity and mortality from hypnotic overdosage is reduced because benzodiazepines are less toxic than barbiturates, it poses special problems for the laboratory diagnosis of drug-induced coma. The reason for that is that these drugs are present in serum in low concentrations even in overdose because of their large volume of distribution, and that they are usually extensively metabolised. This can be demonstrated with e.g. flurazepam. Even in poisoning, the parent drug is usually not detected, because it is rapidly metabolised into the mono-desethyl and N-1-desalkyl compounds. The latter metabolite, which is active, has an elimination half-life of 50-100 hours. For diagnosis of benzodiazepine overdosage it is therefore necessary to have at one's disposal a method with which the metabolites can be identified. This can be achieved for instance by capillary gas chromatography.

A special problem in the clinical drug laboratory is drug abuse, which is defined here as long-term intake of relatively low doses of a substance and is, in a way, "chronic clinical toxicology". A class of drugs which is commonly abused and poses problems to the clinician and the laboratory are the laxatives. These drugs can easily be obtained as over-the-counter drugs, and are abused mainly by people suffering from obstipation or from psychiatric eating disorders such as anorexia nervosa or bulimia. Chronic laxative intake may lead to symptoms like chronic diarrhea, catharic colon or hypokalemia. In these cases it is advisable to screen patients for laxative abuse.

A most commonly abused laxative is bisacodyl, an acetylated diphenol compound. The acetyl groups are readily hydrolysed in the stomach, and the unchanged compound is never found in the urine as such. In addition, the diphenol compound is hydroxylated in the liver, and therefore in urine extracts of patients taking bisacodyl the hydrolysed products plus the hydroxylated metabolite are found. The detection method of choice is high-performance thin-layer chromatography of urine extracts. Urine has to be pretreated with α-glucuronidase before extraction to enhance sensitivity. The metabolic pattern of spots on the chromatogram is specific for bisacodyl intake. The method is very sensitive; even a single 5 mg dose can thus be detected as long as 32 hours after intake (10).

If we look for the unchanged drug, bisacodyl abuse can never be confirmed because it never appears in the urine as such. This shows that metabolite detection in a case like this is a prerequisite.

3d. Metabolites in post-mortem toxicology

In post-mortem toxicology it may be essential to know not only which drug may be the cause of death, but also when the drug was taken or administered. Drug levels in body fluids or tissues alone can seldom, if ever, be interpreted such that it can be estimated when ingestion took place. The following case demonstrates that assay of metabolites in addition to the parent drug may be helpful to elucidate a forensic toxicological problem.

This young man was a former heroin addict who was treated with a relatively low methadone dose of 20 mg daily and the benzodiazepines chlordiazepoxide and flurazepam. He was arrested in a riot and locked in in a police cell. The next morning he was found deceased. At autopsy no signs of violence were observed. At toxicological examination the following substances were found: ethanol, amphetamine and diazepam very low levels, and chlordiazepoxide and flurazepam metabolites in concentrations which were in agreement with the prescribed doses. Methadone was also found in the urine, but since the deceased took 20 mg of it daily it could not be concluded that this drug contributed or lead to death on the basis of this qualitative test only. Therefore, it was deemed necessary to quantify methadone and its metabolite in different samples.

Methadone is mainly metabolised to 2-ethylidene-1,5-dimethyl-3,3-diphenyl-pyrrolidine (metabolite I). Both substances were measured in blood, urine, and liver by means of HPLC and GC-MS. The blood level of unchanged methadone was 0.6 mg/L. From published pharmacokinetic data (11, 12) it is concluded that the expected plasma level at 20 mg daily is in the range of 0.04-0.16 mg/L. This means that the methadone level in the deceased was 4-15 times higher than could be expected. It was therefore very likely that the victim died from methadone overdosage. However, one question could not be answered at that stage. For the judicature it was essential to know whether this intoxication was the result of acute or chronic overdosage. To answer this question, the assay of metabolite I in addition to the parent compound proved to be of utmost importance. In patients on chronic methadone, the metabolite to methadone ratio in urine greatly exceeds one, whereas in patients who died from an acute overdose this ratio is always lower than one (13, 14). In this victim, a metabolite to methadone ratio of 0.74 was found. From this result it could be concluded that he died from an acute methadone overdose. This finding proved to be in agreement with the results of the morbid anatomist, and was later confirmed by witnesses from the drug scene who told that the deceased had ingested all available pills before being locked in. A case like this could never have been solved satisfactorily if only the unchanged drug had been determined. This means that in forensic toxicology, like in clinical toxicology, it is of utmost importance to be aware of drug metabolites, and to measure them when necessary.

4. FUTURE TRENDS IN TDM-T

It is tempting to conclude this chapter on TDM-T by presenting some views on possible trends which will mark the future of this branch of clinical chemistry. When observing the literature on TDM, it may be noted that, generally speaking, the situation has stabilized over the last few years. Most experts in this field agree that TDM is useful for a relatively small number of commonly used drugs by measuring the parent compound in plasma or serum. From an analytical point of view, the development and subsequent introduction of immunochemical methods such as those using enzymatic of fluorescing labels were an important breakthrough. These methods are extremely useful for routine TDM. Only for special cases e.g. as those described in sections 3a and b, more versatile methods such as GC, HPLC and GC-MS are indispensable. It goes without saying that these techniques should be available in specialized TDM-T laboratories in addition to immunochemical techniques. For acute toxicology, a situation similar to that of TDM exists. Likewise, no big changes may be expected in the next few years.

As to chronic toxicology, it is expected that important changes in this field may impinge strongly upon the workload of the clinical laboratory. Until recently, the human toxicologist mainly dealt with the toxicology of comparatively high doses. These lead, generally speaking, to high concentrations which are comparatively easy to deal with from an analytical point of view. Over the last few years, however, there is an increasing concern about health effects of chronic exposure to low levels of chemicals, both at the workplace and from the environment.

Some groups of chemicals which may have an effect on human health in the long run are metals, organic solvents, pesticides, and mutagenic substances. Toxic metals such as Pb, Cd, Hg, Al etc. show different toxic effects which often depend on the chemical state (organic, inorganic, metallic) of the element. Organic solvents or volatiles may be neurotoxic at chronic low-level exposure. Pesticides have a wide range of toxicity, but neurotoxicity and mutagenicity may be mentioned as the most common biological effects. Mutagenic substances other than pesticides may originate from tobacco, food, polluted air, etcetera.

The major pitfall in the biological monitoring of exposure to these environmental chemicals is that, as a rule, very low amounts are absorbed in the body. This often poses special problems to the analyst, and extremely sensitive methods and hence expensive equipment are required. On the other hand, in many cases the onset of a biological effect is very slow and may occur when the noxious substance can no longer be detected in the body even with the most sensitive techniques. Much of this type of work is still in the research phase, but it is very likely that the results will find their way to the clinical laboratory in the years to come.

It may be expected that in the future the specialized clinical laboratory will play an increasing role in the biological monitoring of exposure to industrial and environmental chemicals. One method that is very promising for this purpose, is the estimation of exposure to alkylating substances by measurement of adducts to hemoglobin in blood and to nucleic acid bases in urine. Exposure to ethylene oxide, for instance, can be estimated by measuring the extent of hemoglobine alkylation (15). Another type of monitoring is the measurement of health effects, or rather biological effects, after chronic exposure (biological effect monitoring). Classic examples are the assay of urinary $ß_2$-microglobuline excretion to detect an early effect of cadmium on the kidney, and of Zn protoporphyrin to indicate lead exposure.

These examples accentuate the close relationship between human toxicology and clinical chemistry. As most of these "health effect parameters" are - though sensitive - not specific for the toxic substance, their application has some limitations. In conclusion, the developments in the human toxicology of chronic exposure deserves the attention of the clinical chemist. In this field, measurement of metabolites may perhaps be even more important than in TDM.

REFERENCES

1. Bratton AC, Marshall EK. A new coupling component for sulfonamide determination. J Biol Chem 1939; 128: 537-550.
2. Trinder P. Rapid determination of salicylates in biological materials. Biochem J 1954; 57: 301.
3. Koch-Weser J. Serum drug concentrations in clinical perspective. In: Richens A, Marks V, eds. Therapeutic Drug Monitoring. Churchill Livingstone 1981: 1.
4. Edelbroek PM. Therapeutic and toxic effects of amitriptyline in relation to its metabolism. Thesis, Leiden 1987.

5. Task force on the use of laboratory tests in psychiatry. Tricyclic antidepressants - blood level measurements and clinical outcome: An APA task force report. Am J Psychiatry 1985; 142: 155-162.
6. Jungkunz G, Kuss HJ. On the relationship of nortriptyline: amitriptyline ratio to clinical improvement of amitriptyline treated depressive patients. Pharmacopsychiatria 1980; 13: 11-116.
7. Rickels K, Weise C, Case G, Hucker H. Tricyclic plasma levels in depressed outpatients treated with amitriptyline. Psychopharmacology 1983; 80: 14-18.
8. Richens A. Clinical pharmacokinetics of phenytoin. Clin Pharmacokinet 1979; 4: 153-169.
9. De Wolff FA, Verweij P, Ferrari MD, Buruma OJS, Breimer DD. Impairment of phenytoin parahydroxylation as a cause of severe intoxication. Therap Drug Monit 1983; 5: 213-215.
10. De Wolff FA, De Haas EJM, Verweij M. A screening method for establishing laxative abuse. Clin Chem 1981; 27: 914-917.
11. Inturrisi CE, Verebely K. The levels of methadone in the plasma in methadone maintenance. Clin Pharmacol Ther 1972; 13: 633-637.
12. Vereleby K, Volavka J, Mulé S, Resnick R. Methadone in man: Pharmacokinetic and excretion studies in acute and chronic treatment. Clin Pharmacol Ther 1975; 18: 180-190.
13. Robinson AE, Williams FM. The distribution of methadone in man. J Pharm Pharmacol 1971; 23: 353-358.
14. Garriot JC, Sturmer WQ, Mason MF. Toxicological findings in six fatalities involving methadone. Clin Toxicol 1973; 6: 163-173.
15. Törnquist M, Osterman-Golkar S, Kantiainen A, Jensen S, Farmer PB, Ehrenberg L. Tissue doses of ethylene oxide in cigarette smokers determined from adduct levels in hemoglobin. Carcinogenesis 1986; 7: 1519-1521.

BIOLOGICAL MONITORING OF EXPOSURE TO TOXIC TRACE ELEMENTS

J. Savory and M.R. Wills

University of Virginia Medical Center, Departments of Pathology and Biochemistry, Box 168, Charlottesville VA 22908 USA

INTRODUCTION

The evaluation of trace metal status has usually been accomplished by the analysis of either urine or blood. Serum or plasma are the biological fluids which are most frequently analyzed with the anticipation that the trace metal concentration will reflect tissue levels; the assumption being that an equilibrium exists between the circulation and the tissues. This assumption is not always valid and in many instances either the serum or plasma trace metal concentration is not a good indicator of tissue levels.

Direct tissue analysis is considered to be the best means of assessing trace element toxicity, but the difficulties of performing this invasive procedure make it of marginal value as a routine monitoring procedure.

One technique for assessing body burden of toxic metals has been to measure plasma concentrations following infusion of a chelating agent. Pilot studies of in vivo neutron activation analysis have also yielded some promising results.

The analytical methodology is a critically important aspect of any toxic metal monitoring program. Major problems exist with establishing trace metal methods in the laboratory. Contamination of specimens during collection, storage, shipping and analysis are major difficulties because of the ubiquitous nature of many metals in the environment. Combination of reagents and equipment pose similar concerns. Sensitivity, accuracy and precision are difficult to achieve, and there are very few reference or definitive methods available. Matrix differences affect results and the way in which standardization is carried out. Among the most commonly used techniques are flame and electrothermal atomic absorption spectrometry, inductively coupled plasma emission spectrometry and neutron activation analysis. Tissue localization can be accomplished using histochemical stains, energy dispersive x-ray microanalysis or laser microprobe mass spectrometry.

Many toxic metals are measured routinely in the modern clinical chemistry laboratory. From the 1960's onwards, particular attention has been focussed on lead, arsenic, cadmium, and mercury mainly due to their relative abundance as both industrial and overall environmental contaminants. Selenium, nickel and manganese also have received considerable attention. However, at the present time, aluminum is the most widely monitored toxic metal in the hospital and clinic setting. A review of aluminum monitoring provides an excellent means of describing a monitoring program for a wide variety of toxic metals. Aluminum measurements are difficult, contamination control is a serious problem, and interpretation of results provides a real challenge to the clinical chemist.

Aluminum toxicity is only well established in individuals with chronic renal failure particularly those being treated by intermittent hemodialysis.

ALUMINUM TOXICITY

Aluminum is the most abundant metal and the third common element in the earth's crust. In nature it occurs only in the combined state and is never found in the metallic form. In the combined state it is most commonly found in minerals where it is present as aluminum silicates and slightly less commonly as aluminum oxides. Aluminum is also found in varying quantities in vegetation and in animal tissues. Aluminum as the metal is used extensively for a variety of industrial purposes and various salts of aluminum are used in a wide variety of industrial, medicinal and domestic applications. Because of its widespread occurrence in the environment and foodstuffs, it is virtually impossible for the human species to avoid exposure to aluminum; there is no evidence that it is an essential ion in any biochemical process or pathway.

The major routes of exposure to aluminum are by either the inhalation of dust orally in foods, fluids and medicines. Attention was first drawn to the potential role of aluminum as a toxic metal over fifty years ago. At that time aluminum cooking utensils were being introduced, the metal in these together with the aluminum present in city drinking-water supplies and a variety of medicines were considered to represent a potential health hazard (1). Betts in this monograph on aluminum poisoning refers to several early studies on aluminum toxicity. One of the earliest of these investigations was by Professor J.W. Mallet of the University of Virginia who in 1888 published a series of experiments on the effects of aluminum-containing baking powders upon gastric digestion. Professor Mallett reported: "From general nature of the results obtained, the conclusion may fairly be deducted that not only alum itself, but the residues which its use in baking powder leaves in bread cannot be viewed as harmless but must be ranked as objectionable and should be avoided when the object aimed at is the production of wholesome bread". Another one of these early reports, this by Cushing in 1906, is of significance in that neurotoxicity due to aluminum administration was observed in experimental animals. Although the subject has been controversial, with the exception of the health hazards associated with industrial exposure, aluminum was dismissed as a toxic metal in comprehensive reviews in 1957 and 1974 (2, 3).

Although the toxicity of aluminum in the presence of normal renal function remains to be clearly defined, it has been established, in the past decade, that aluminum accumulates in serum and tissues and has a toxic action in patients with chronic renal failure. An increased serum aluminum concentration, and toxicity, may occur in patients with chronic renal failure who are on long-term treatment with either hemodialysis or peritoneal dialysis; it may also occur in some patients who have not been dialyzed (4). The latter group of patients usually consists of children who are receiving oral treatment with aluminum hydroxide (5, 6).

Hyperaluminemia, which is associated with toxic clinical sequelae, may also occur in adults who are not on dialysis treatment (7). In patients with chronic renal failure the clinical toxic phenomena associated with an increased body burden of aluminum include a specific encephalopathy (dialysis encephalopathy), a metabolic bone disease (osteomalacic dialysis osteodystrophy), and an anemia; aluminum may also be responsible for some of the other clinical features associated with end-stage renal disease (4). In patients with chronic renal failure the source of the increased concentration of aluminum in serum and its accumulation in various tissues is derived from the intestinal absorption of aluminum from either aluminum-containing medications or tap-water and also by the passage of aluminum from the dialysate across the dialysis membrane.

ALUMINUM MONITORING

The upper limit of the reference range for serum aluminum concentration is up to 10 µg/L. Low concentrations of aluminum occur in tissues (less than 10 mg/kg), except for lung tissue where over 100 mg/kg of aluminum can be found in adults. Brain gray matter from normal subjects contains approximately 2 mg/kg. Aluminum excretion in urine has a median value of approximately 17 µg/L (8). Recommendations on aluminum monitoring in hemodialysis programs have been made in a report of an International Workshop held to study the hazards of aluminum toxicity related to renal insufficiency (9).

Reverse osmosis is the recommended method for water treatment since it provides water with a low aluminum content (less than 10 µg/L) a low content of other cations and eliminates organic contaminants which may contribute to the problems associated with hemodialysis. Deionization of the water is the next preferred mode of treatment, but is is recognized that it requires more careful monitoring since aluminum loading of the resin can occur with the danger of subsequent erratic elution. Concentrations of water aluminum not exceeding 10 to 15 µg/L are necessary to minimize significant body uptake of aluminum during dialysis. These concentrations can be achieved using reverse osmosis and are recommended.

Between-batch differences in the aluminum content of dialysate concentrate exist. The recommendation is that the final aluminum concentration of the dialysate after dilution with treated water should be less than 15 µg/L and preferably less than 10 µg/L.

There is evidence that chronic ambulatory peritoneal dialysis (C.A.P.D.) treatment can remove some of the aluminum present in the plasma. The effectiveness of transfer of aluminum species during this type of treatment is of importance and needs further investigation.

Because of these factors the concentration of aluminum in the C.A.P.D. fluid should be less than 10 µg/L.

In view of the problems with maintaining extremely low serum aluminum concentrations, especially with patients taking aluminum binders, the following criteria are recommended for the management of dialysis patients:

- Serum aluminum concentrations should never exceed 200 µg/L. Levels in excess of this concentration correlate with the development of dialysis encephalopathy or osteodystrophy.
- Concentrations over 100 µg/L should be viewed with concern and require careful monitoring.
- Concentrations from 60 to 100 µg/L appear to cause no problems to the patient in the short-term.

In industrial exposure, urinary excretion of aluminum is markedly increased but the significance of this increase is not known. Serum concentrations of aluminum are not increased in such individuals.

FREQUENCY OF MONITORING

Hospital hemodialysis: It is considered appropriate for the main water supply, the treated water, and the dialysis fluid to be monitored weekly. If a reverse osmosis system is used and has been shown to be reliable, then monthly monitoring should be adequate. The dialysis fluid to be monitored should be sampled immediately before it comes into contact with the membrane since contamination by metal parts of the system is a possibility. Serum aluminum should be monitored in all dialysis patients four to six times each year.

Home hemodialysis: It is considered appropriate for the main water supply, the treated water, the dialysis fluid, and serum aluminum to be monitored when the patient comes for check-up (usually two or three times per year). If there are important fluctuations in the aluminum content of the main water supply and if reverse osmosis is not employed, then more frequent monitoring is desirable. For peritoneal analysis and C.A.P.D. serum aluminum should be monitored every two or three months.

ALTERNATE MEANS OF ASSESSING ALUMINUM ACCUMULATION

The infusion of desferoxamine (DFO) has been used as a diagnostic test for aluminum related osteodystrophy (10, 11). In one of these reports (10) the effect was studied of a standard intravenous dose of DFO on plasma aluminum concentrations in 54 patients on hemodialysis. Stainable bone aluminum, bone histologic findings, and bone aluminum content were studied. Baseline plasma aluminum concentrations of greater than 200 μg/L were associated with aluminum-related osteodystrophy (specificity, 93%), but concentrations of less than 200 μg/L did not exclude the diagnosis (sensitivity, 43%). After administration of DFO the increase in plasma aluminum concentration was 534 $\pm$ 260 (SD) and 214 $\pm$ 92 μg/L in patients with and without aluminum-related bone disease, respectively (p less than 0.001), and correlated with the bone aluminum content ($r = 0.64$). An increment in plasma aluminum concentration of greater than 200 μg/L identified 35 of the 37 patients with aluminum-related osteodystrophy; sensitivity was 94% and specificity, 50%. These workers concluded that the DFO infusion test was noninvasive, well tolerated, and of value particularly in excluding the diagnosis of aluminum-related osteodystrophy.

Recently Kelleher et al. (12) have applied the exciting new technique of in vivo neutron activation analysis and have examined total body and partial body (hand) aluminum levels in patients with end-stage renal failure. Patients maintained on chronic hemodialysis had higher mean body burdens of aluminum than did those managed clinically without dialysis. Most of the patients examined indicated elevated levels of body or skeletal aluminum. A significant correlation was observed between the in vivo aluminum/calcium ratio obtained for the hand measurement and the increase in serum aluminum levels following a desferoxamine infusion test. The direct in vivo monitoring of hand Al/Ca values in patients may provide an alternative means of detecting aluminum intoxication.

Bone biopsy specimens provide the best means of assessing aluminum accumulation. Levels of bone aluminum are elevated in nearly all patients undergoing hemodialysis especially those with osteomalacia (13).

We have attempted to use lymphocytes as a readily accessible "tissue" as a specimen for the evaluation of both aluminum and nickel accumulation in hemodialysis patients (14). Lymphocyte aluminum levels were increased in these patients but did not correlate with serum concentrations. In other studies we have found no correlation between lymphocyte aluminum concentration and total body aluminum content using neutron activation analysis (in published data).

SPECIMEN COLLECTION AND TRANSPORTATION

The blood collection procedure that is used at the present time in the authors' laboratory involves drawing blood with a stainless steel needle into a plain glass vacuum tube (No. 6430, Becton-Dickinson and Co., Rutherford, NJ 07070). The blood is allowed to clot for approximately 20 minutes before centrifugation. Serum is then transferred to a 17 x 100 mm polypropylene tube (Falcon, Oxnard, CA) and stored for final analysis. A reference range was established using this method by collecting and analyzing blood specimens from 50 healthy individuals. The resulting aluminum range was 1-12 $\mu g/L$.

The above procedure does involve some contamination because of exposure of the blood to glass. Aluminum is rapidly leached out of glass presumably by transferrin which is a strong binder of aluminum. For more exact work especially for trace metals such as chromium, nickel and cobalt, it is best to use acid washed plastic containers. Blood specimens can be obtained by use of polyethylene I.V. cannulae.

The ideal method of collection for plasma aluminum is to use a plastic collection tube containing lithium heparinate as an anticoagulant. Ten ml polycarbonate tubes containing lithium heparinate are available from Sherwood Medical Industries, Ltd. (Crawley, West Sussex, U.K.) and as they become more widely available they should replace the glass vacuum tubes for collection of blood for aluminum measurements.

The present authors have developed procedures for collection and storage of urine and fecal specimens as follows: Twenty-four hour urine specimens are collected in plastic containers (Scientific Products, McGraw Park, IL) and the total volume is recorded. A 10 ml aliquot is transferred to a polypropylene tube (Falcon, Oxnard, CA) which is stored at 4°C.

Fecal specimens are collected directly into plastic bags which are weighted at the end of a 24-hour time period, placed in a plastic container and frozen.

Bone specimens for aluminum analysis are taken from the iliac crest at the time of biopsy or at autopsy (15, 16) and the specimen placed in an aluminum-free plastic container. Bone for histological staining is fixed in 10% buffered formalin.

As stated above, blood drawn into glass vacuum tubes should be transferred within one hour after blood is collected into a plastic tube with a tight fitting cap. If the tube is to be shipped, the tube's cap must be wrapped in such a way as to prevent leakage upon handling. There is no indication in the literature that storing the specimen at room temperature results in a change in the aluminum concentration as compared to storing the sample at 4°C. Specimens for aluminum analysis have been received in the authors' laboratory both at room temperature and on ice packs. Samples from healthy individuals when shipped at room temperature through the U.S. mail gave aluminum values all within our reference range.

Tissue samples should be placed into aluminum-free plastic containers. Brain specimens need to be frozen until analysis (17), but bone can be kept at room temperature.

Specimen collection for aluminum analysis has the potential of contributing varying amounts of the element into the specimen. Each step of the collection procedure must be scrutinized to determine if it is contributing any contaminating aluminum to the sample. Each laboratory should check all materials before collecting patient specimens. Once the collection procedure is established, the procedure should be checked regularly to verify that little or negligible contamination is being contributed by the technique. A quality control check involves drawing blood from healthy individuals once a week, which monitors the sample collection procedure and method of transport for any spurious aluminum contamination. An aluminum value within the reference range would be acceptable for this specimen. Tissue samples from healthy, non-diseased persons at autopsy can serve as controls for bone, brain and muscle aluminum determinations.

Urine and fecal specimens can be shipped frozen on dry ice. Urine specimens are then stored at 4°C, but for convenience, and aesthetic reasons, feces are sealed into 32 oz. plastic containers (Cole-Palmer, Chicago, IL) and kept frozen until processing.

SOURCES OF CONTAMINATION IN ANALYSIS

As in specimen collection, every item used during analysis should be regarded as a potential source of aluminum contamination. Glassware, pipet tips, plastic tubes, sample cups, the working environment and the water utilized must all be checked to ensure that they are adding negligible amounts of aluminum to the procedure. The room chosen for the analyses should have a limited access to ensure that dust in the working environment is being circulated as little as possible. Sample preparation should be carried out in an environmental laminar flow hood; this precaution helps to minimize contamination by dust particles.

Water utilized for standard curve preparation, rinsing of glassware and sample dilutions must be of high purity. The water should produce a resistivity of at least 18 megaOhms. All glassware should be made aluminum-free, which can be accomplished by using either acid solutions or saturated disodium EDTA.

STANDARD REFERENCE MATERIALS AND QUALITY ASSURANCE

The availability of standard reference materials is of considerable importance in the application of reliable assays for aluminum. Aqueous standards are available from commercial sources, one being Fisher Scientific Company (Fair Lawn, NJ 07410 USA). This material is in acid solution and is a stock solution of 2 g/L. Dilution to working standards must incorporate an intermediate standard if accuracy is to be maintained. The National Bureau of Standards (Washington, DC 20234 USA) also provides an aqueous standard solution which contains 10.00 mg/L of aluminum in a 10 percent acid medium. In addition, bovine serum (NBS-RM-8419) with a quoted aluminum concentration also is available from the National Bureau of Standards.

Quality assurance in aluminum measurements follows established principles used in all areas of clinical chemistry. Internal quality control materials, usually pooled plasma at two or three different concentrations, should be used on a daily basis. The concentrations of aluminum in these materials should coincide with various decision levels. Interlaboratory control materials also are an important part of any quality assurance scheme. Presently, such a program is conducted by Dr. Andrew Taylor through the Robens Institute (University of Surrey, Guildford Surrey, GU2 5XH England). The institute mails 3 water and 3 serum samples monthly to laboratories for aluminum analysis. The data and statistical information of the 38 participating laboratories is then made available for review.

One useful material to use for monitoring the performance of aluminum assays is plasma from normal individuals. Such a pool would have an aluminum concentration of approximately 5 μg/L and is useful for examining the recovery of aluminum added to normal serum.

ANALYTICAL METHODS FOR THE DETERMINATION OF ALUMINUM

The key to the success in aluminum monitoring lies in the availability of accurate and precise analytical methods together with guidelines for contamination-free specimen collection. Many problems have existed in the past with aluminum methodology and these have been detailed in the study of Versieck and Cornelis (18) on the variability in reported normal plasma concentrations. Even with reliable methods, mean normal values varied between 0.07 and 1.55 μmol/l (2 and 42 μg/l). Other reports have listed normal mean concentrations ranging from 2.67 to 54 μmol/l (72 to 1460 μg/l) although obvious interferences were present in these methods (18).

Several methods are available for the determination of total aluminum in biological materials. Chemical and physiochemical methods are insensitive and inaccurate; X-ray fluorescence is specific but lacks sensitivity; neutron activation analysis is complex and subject to interferences, although it is a sensitive technique.

The greatest degree of success of any technique for the determination of aluminum in biological specimens has been with electrothermal atomic absorption spectrometry (EAAS). In this technique the sample is placed in a graphite tube mounted in the light path of the spectrophotometer. The source of the light is a hollow cathode lamp that contains the metal being analyzed and emits characteristic wavelengths. First, the graphite tube is heated with direct current to dry the sample at a low temperature, then the sample is ashed to destroy organic matter and burn off inorganic species that may interfere, and finally the temperature is quickly raised and the metal under analysis vaporizes and absorps the light being passed through the graphite tube. There are several advantages of the graphite furnace techniques. Sample pretreatment can usually be eliminated, sample requirements are small (2-100 μl), graphite furnaces are capable of attaining the high temperature needed to form ground state atoms, and the atoms stay in the light path for a relatively long time resulting in increased sensitivity.

Several problems may be encountered in the EAAS determination of aluminum, including difficulties with untreated graphite tubes (19), matrix interferences (19, 20), and standardization procedures (20).

Most EAAS procedures have been developed with the use of auto-sampling to improve precision. Pyrolytically coated graphite tubes are recommended together with a pyrolytic graphite platform. These tubes minimize reactions between the analyte and the graphite tube. Argon is preferred over nitrogen as the purge gas since argon produces a larger and less variable signal. Furnace signals occur rapidly and in order to follow the absorbance profile, it is necessary to incorporate fast digital electronics into the instrument. The absorbance signal must be integrated to provide accurate results; peak absorbance signals will produce inaccurate quantitative measurements. Measurements also require elimination of matrix effects between samples and standards, usually accomplished by some type of background correction. The Zeeman correction system in the authors' experience provides the most sensitive and reliable results.

Matrix modifiers are an important part of the modern atomic absorption technology. Most analytes require a matrix modifier in order to stabilize the analyte until thermal conditions have become constant. For aluminum, magnesium nitrate permits the use of high char temperatures without losses.

The procedure recommended by the present authors uses a stabilized temperature platform with pyrolytically coated graphite tubes. In this procedure aqueous aluminum standards and serum samples are diluted with an equal volume of an aqueous solution containing magnesium nitrate (2 g/L). The autosampler is programmed to deliver a 10 μl aliquot of the sample or standard onto the platform for final analysis. Aqueous standards are prepared fresh daily and the linearity range is 0-125 μg/L.

The present authors have developed a procedure which minimizes any potential matrix effects by protein precipitation (21). This protein precipitation technique was originally developed for serum nickel (22) and markedly reduces matrix effects in the final atomic spectroscopic analysis. One attractive feature of this protein precipitation method is that the protein free supernatant can be used for the measurement of other trace metals such as nickel, chromium and cobalt. For urine analysis sample aliquots are diluted 1:1 with distilled water before application to the AAS stabilized temperature platform (23).

Fecal analysis requires considerably more complicated preparation steps than serum or urine analysis. The procedure developed in the authors' laboratory (24) is summarized as follows: Frozen specimens are thawed and weighed in the original plastic container used for collection. Distilled water is added (1 ml per 2 g feces) and the sample is homogenized in a sealed can on a paint shaker. A 10 ml aliquot is ashed at 550°C in a muffle furnace, dissolved in dilute HNO_3, and analyzed by EAAS.

Solid tissues must be homogenized, dried, ashed, and/or dissolved to produce a liquid sample prior to EAAS analysis. Bone samples for analysis are washed free of marrow by a strong stream of distilled water. Any existing fat or muscle is scraped with a plastic knife or other blade found to be aluminum-free. The bone sample can then be processed by any of several methods to obtain a solution for injection into the graphite furnace.

Soft tissue samples, such as brain, liver, or muscle, can be homogenized before processing, and this can be accomplished easily by pummelling the tissue in a "Stomacher" blender (Fisher Scientific). Distilled water 5 ml is added to the bag with the tissue. The sealed bag is placed in the blender and blended for 5-15 min. which completely homogenizes the sample (25). The homogenate can then be processed as described for the fecal samples.

A potentially valuable new technique involves wet ashing of biological samples in a microwave oven under pressure. A recent report describes such a technique (26). We use a similar technique for ashing bone, liver and brain for aluminum measurements. We use a teflon-lined microwave transparent acid digestion pressure bomb (Parr Instrument Co., Moline, IL 61265 USA). Pure 50% HNO_3 is used with 1-2 minutes digestion in the microwave. No contamination or losses are observed with this technique, and complete ashing is accomplished.

Aluminum can be localized in tissue by histochemical staining, electron probe X-ray microanalysis, and by laser microprobe mass spectrometry. These last two are sophisticated techniques which will be powerful tools in future studies of aluminum toxicity.

SUMMARY

Biological monitoring of exposure to toxic trace elements is a specialized but important part of clinical chemistry. Although each element has its own unique needs, there are some common approaches especially for monitoring many of the toxic metals. The present review uses aluminum monitoring as a means of addressing some of these important aspects common to many toxic metals.

Choice of specimen to be used to predict toxic element accumulation is important. Also of importance are guidelines for monitoring, specimen collection, transportation, processing, analysis and quality assurance. For aluminum and many other toxic metals, electrothermal atomic absorption spectrometry is the technique used in most laboratories. Future work will probably involve tissue localization with some of the newer microprobe analyzers.

REFERENCES

1. Betts CT. Aluminium Poisoning. Toledo: Research Publishing, Co, 1526.
2. Campbell IR, Cass JS, Cholak J, Kehoe RA. Aluminum in the environment. AMA Arch Arch Industr Health 1957; 15: 359-448.
3. Sorenson JR, Campbell IR, Tepper LB, Lingg RD. Aluminum in the environment and human health. Environ Health persp 1974; 8: 3-95.
4. Wills MR, Savory J. Aluminium Poisoning: Dialysis, encephalopathy osteomalacia and anemia. Lancet 1983; 2: 29-33.
5. Griswold WR, Reznik V, Mendoza SA, Trauner D, Alfrey AC. Accumulation of aluminum in a nondialyzed uremic child receiving aluminum hydroxide. Pediatrics 1983; 71: 56-8.
6. Andreoli SP, Bergstein JM, Sherrard DJ. Aluminum intoxication from aluminum-containing phosphate binders in children with azotemia not undergoing dialysis. New Engl J Med 1984; 310: 1079-84.
7. Kaye M. Oral aluminum toxicity in a non-dialyzed patient with renal failure. Clin Nephrol 1983; 20: 208-11.
8. Friberg L, Nordberg G.F., Vouk VB, eds. Handbook of the toxicology of metals, 2nd edition. Elsevier/North Holland Biomedical Press, Amsterdam, 1986.
9. Savory J, Berlin A, Courtoux, Yeoman B, Wills MR. Summary report of an international workshop on the role of biological monitoring in the prevention of aluminium toxicity in man: aluminium analysis in biological fluids. Ann Clin Lab Sci 1983; 13: 444-51.
10. Milliner DS, Nebeker HG, Ott SM, et al. Use of the deferoxamine infusion test in the diagnosis of aluminum-related osteodystrophy. Ann Int Med 1984; 101: 775-9.
11. Berland Y, Charhon SA, Olmer M, Meunier PJ. Predictive value of desferrioxamine infusion test for bone aluminium deposits in hemodialyzed patients. Nephron 1985; 40: 433-5.
12. Hodsman AB, Sherrard DJ, Alfrey AC, et al. Bone aluminum and histomorphometric features of renal osteodystrophy. J Clin Endocrinol Metab 1982; 54: 539-46.

13. Brown SS, Sundeman FW Jr., eds. Progress in nickel toxicology. Oxford UK: Blackwell Scientific Publications, 1986: 137-140.
14. Alfrey AC, LeGendre GR, Kaehny WD. The dialysis encephalopathy syndrome. Possible aluminum intoxication. New Engl J Med 1976; 294: 1848.
15. Maloney NA, Ott SM, Alfrey AC, Miller NL, Coburn JW, Sherrard DJ. Histological quantitation of aluminum in iliac bone patients with renal failure. J Lab Clin Med 1982; 99: 206-16.
16. Crapper DR, Krishnan, Quittkat S. Aluminium, neurofibrillary degeneration and Alzheimer's disease. Brain 1976; 99: 67-80.
17. Versieck J, Cornelis R. Normal levels of trace elements in human blood plasma or serum. Anal Chim Acta 1980; 116: 217-54.
18. Gorsky JE, Dietz AA. Determination of aluminum in biological samples by atomic absorption spectrophotometry with a graphite furnace. Clin Chem 1978; 24: 1485-90.
19. Alderman FR, Gitelman HJ. Improved electrothermal determination of aluminum in serum by atomic absorption spectroscopy. Clin Chem 1980; 26: 258-60.
20. Brown S, Bertholf RL, Wills MR, Savory J. Electrothermal atomic absorption spectrometric determination of aluminum in serum with a new technique for protein precipitation. Clin Chem 1984; 30: 1216-8.
21. Sunderman FW Jr, Crisostomo MC, Reid MC, Hopfer SM, Nomoto S. Rapid analysis of nickel in serum and whole blood by electrothermal atomic absorption spectrophotometry. Ann Clin Lab Sci 1984; 14: 232-41.
22. Leung FY, Henderson AR. Improved determination of aluminum in serum and urine with use of a stabilized temperature platform furnace. Clin Chem 1982; 28: 2139-43.
23. Brown S, Mendoza N, Bertholf RL, et al. Absorption of aluminum from aceglutamide aluminum in healthy adult males. Res Comm Chem Path Pharm 1986; 53: 105-16.
24. Sunderman FW Jr, Marzouk A, Crisostomo MC, Weatherby DR. Electrothermal atomic absorption spectrophotometry of nickel in tissue homogenates. Ann Clin Lab Sci 1985; 15: 299-307.
25. Aysola P, Anderson P, Langford CH. Wet ashing in biological samples in a microwave oven under pressure using poly(tetrafluoroethylene) vessels. Anal Chem 1987; 59: 1582-3.

PRACTICAL APPLICATIONS OF THERAPEUTIC DRUG MONITORING:

The impact of technological developments

David W. Holt* and Atholl Johnston**

*Poisons Unit, Guy's Hospital, London SE1 9RT
and **Clinical Pharmacology St Bartholomew's Hospital, London EC1 7BE

This brief review will focus on three topics related to our practice of drug measurements as a guide to therapy. It will attempt to show how technological developments are improving both our ability to measure drug concentrations and to interpret the findings.

1. IMPROVEMENTS IN SENSITIVITY AND SELECTIVITY

The demand for analytical techniques capable of measuring low drug concentrations, using small sample volumes, with resolution from either endogenous or exogenous compounds is due to a number of factors. For the investigation of new molecules and some established compounds, single-dose pharmacokinetic studies are common practice. The trend to highly potent compounds, given in low dosage, results in the need for analytical methods capable of measuring concentrations in the ug/l range. Even greater demands for sensitivity are made if free-drug concentration measurements are required, or if only very small sample volumes are available, for instance from neonatal patients.

Selectivity is of utmost importance because patients are frequently on multiple drug therapy and because many compounds are extensively metabolised to closely related molecules which may be implicated in the pharmacological or toxicological properties of the parent compound.

Over the past decade our principal approach to the development of drug assays, chiefly for the measurement of cardio-active compounds, has been to use high-performance liquid chromatography (HPLC), using unmodified silica columns together with non-aqueous eluents containing ionic promoting agents (1). Whilst ultraviolet detection has been suitable for a broad range of compounds, enhancement of sensitivity and, in some instances, selectivity has been achieved by the use of fluorescence and electrochemical detection.

The following examples serve to show some of the practical problems which have been encountered and the methods used to resolve them.

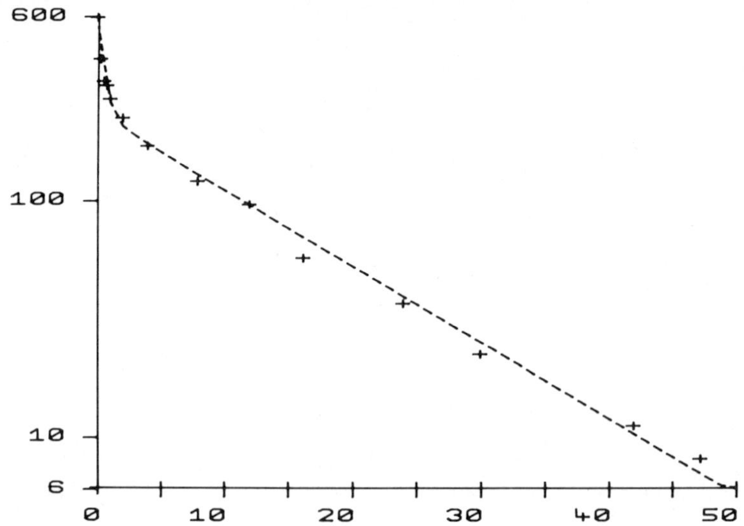

Fig. 1. i.v. Flecainide. Patient B1 (age 15). Plasma flecainide concentrations following intravenous administration of 2 mg/kg to a 15 year old child.

1.1 <u>Pediatric use of flecainide</u>

This Class 1c antiarrhythmic agent has proved useful for the control of a variety of arrthythmias in pediatric patients. Its pharmacokinetics are poorly documented in this age group and we have used a sensitive HPLC assay for its measurement following single intravenous dosage in children (2). The assay, based on a sample size of 50 μl, uses fluorescence detection and has a limit of accurate measurement which is adequate for measurements of plasma flecainide for up to 50 hours following a 2 mg/kg injection of the drug (Fig. 1).

Data from such studies have been used to establish that the elimination half-life of this drug in paediatric patients is markedly shorter than in adults. Furthermore, routine measurements as a guide to therapy have been performed in patients as young as one day old. Much higher doses than those used in adult patients (on a mg/kg basis) have been required, together with an increased frequency of dosing. The availability of this assay has been most useful, particularly when distinguishing between therapeutic failure and underdosage.

1.2 <u>Free-drug concentrations of penticainide</u>

This Class 1a antiarrhythmic drug, structurally related to disopyramide, possesses concentration-dependent binding to plasma proteins. During the early clinical evaluation of this compound we have had to develop methodology suitable for the measurement of free-drug concentrations, obtained by equilibrium dialysis or ultracentrifugation, at concentrations below 100 μg/l. Whilst a method developed using UV detection (3) had adequate sensitivity for the measurement of total plasma concentrations of the drug the sample size obtained following ultracentrifugation was insufficient to achieve the required sensitivity. However, use of electrochemical detection enabled a limit of accurate measurement of 10 μg/l to be obtained using a 50 μl sample volume. This

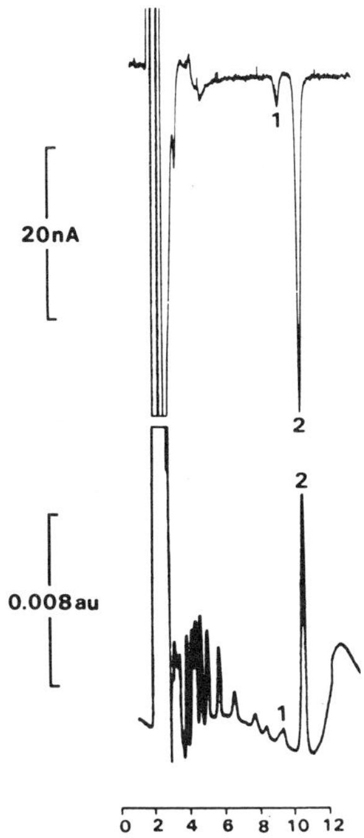

Fig. 2. Simultaneous HPLC chromatograms with UV detection (lower) and electrochemical detection (upper) of an extract of an enzymatic placental digest, following lignocaine anaesthesia. The original lignocaine (1) concentration was 0.39 mg/kg wet weight and the sample size was 79 mg.

assay is now being used to investigate the effects of other protein-bound drugs on the binding of penticainide.

1.3 Obstetric use of lignocaine

The use of drugs during pregnancy is always accompanied by concern that harm could be inflicted on the developing fetus. A recent problem referred to this laboratory involved the use of lignocaine as a local anaesthetic during trans-uterine chorionic villus sampling (CVS) and intra-uterine blood transfusion. The obstetricians wished to document whether significant uptake of lignocaine was occurring in fetal tissue during CVS, or into the fetal circulation during transfusion.

The method of analysis adopted was based on an initial enzymatic digestion of small tissue samples obtained during therapeutic abortion, exposed to the same anaesthetic procedure as was used during CVS. This was followed by HPLC of an organic extract of the digest. A simultaneous chromatogram showing both UV and electrochemical detection (Fig. 2) shows the potential for increased selectivity and sensitivity using electrochemical detection. Using tissue sample sizes of 100 mg and blood sample sizes of 100 ul we have achieved lower limits of accurate measurement of 0.25 mg/kg wet weight and 25 μg/l, respectively. The reference range for this drug when used as an antiarrhythmic agent is 2-5 mg/l.

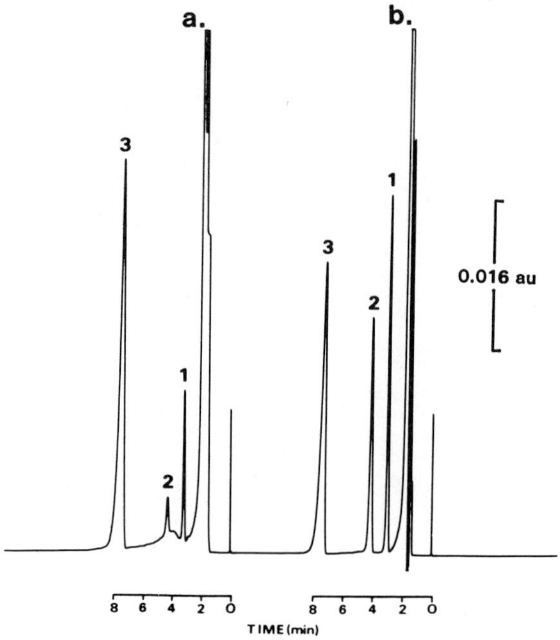

Fig. 3. HPLC chromatograms of extracts of (a) enzymatic digest of post-mortem liver and (b) plasma standard containing desethylamiodarone (DEA) (1) and amiodarone (A) (2). The original tissue concentration, before dilution of the digest in human plasma, was 44 mg/kg (A) and 170 mg/kg (DEA). The plasma standard contained 1 mg/l of both compounds.

Whilst tissue concentrations as high as 10 mg/kg and plasma concentrations as high as 17.5 mg/l have been measured in samples from this study, the clinical significance of these findings is not yet clear.

1.4 Tissue concentrations of amiodarone

The Class III antiarrhythmic agent amiodarone has an exceptionally large volume of distribution (4). We have used enzymatic digestion of small tissue samples, followed by HPLC analysis of organic extracts of these digests, to establish a data base for the tissue distribution of amiodarone and its desethyl metabolite (5). The combined effects of the organic extraction at defined pH and an absorption maximum for both compounds of 240 nm has provided sufficient selectivity to effect the measurement of both compounds in a variety of tissues, without interference from endogenous compounds (Fig. 3).

The clinical value of tissue measurements of the drug and its metabolite in the differential diagnosis of their toxic effects is illustrated by a recent case. A 64 year old woman with a history of atrial fibrillation refractory to quinidine and disopyramide received amiodarone 200 mg three times daily for seven days, then 200 mg daily. She died after a very short course of the drug, having developed jaundice, elevated hepatic enzyme concentrations and signs of hepatic encephalopathy. Plasma and liver concentrations of amiodarone/desethylamiodarone at the time of death were 0.6/0.7 mg/l and 44/170 mg/kg wet weight, respectively. These concentrations were consistent with the low dose of the drug she had been prescribed but were substantially lower than concentrations we have seen previously when amiodarone-induced hepatic toxicity has been diagnosed (6). Whilst an idiosyncratic reaction to amiodarone could not be excluded in this case, there was a low probability that it was implicated in the cause of death.

1.5 Obstetric use of verapamil

The administration of digoxin and verapamil to pregnant patients in an effort to control fetal arrhythmias is well established (7). For optimal dosing of a drug under such circumstances documentation of maternal/fetal transfer of the drug is useful to assess the potential for maternal drug measurements as a guide to fetal therapy.

However, several factors may modify the interpretation of results. In the case of drugs metabolised by the liver, such as verapamil, the pregnancy induced increase in hepatic clearance may necessitate very high maternal doses and limit the practicality of such therapy. Furthermore, there are virtually no data concerning the end-organ sensitivity of fetal tissue to the effects of such antiarrhythmic drugs.

With these points in mind we have monitored a number of cases in which verapamil has been used for the control of fetal tachyarrhythmias. A sensitive and selective HPLC method for the measurement of verapamil and a number of its metabolites has been developed (8). Fluorescence detection has been used and the limit of accurate measurement, using a 100 μl sample volume, is 5 μg/l.

Date	Verapamil (mg/d)	Verapamil (μg/l)	Norverapamil (μg/l)
03/20	240	5	10
03/23	360	5	30
03/26	480	4	22
03/27	480	10	22
Delivery Maternal		40	42
Cord		8	15

The main finding has been that, despite high maternal doses of verapamil, very low concentrations are achieved in maternal plasma and there is a low transplacental passage of the drug to the fetus. The Table illustrates the results from one case, in which there was a poor fetal response to oral verapamil given to the mother, despite an encouraging response when the mother received intravenous verapamil. The very low verapamil levels achieved in the mother (reference range for these doses 100-300 μg/l) and limited placental transfer perhaps explain the poor fetal response. However, a clinical response has been achieved in some cases with similarly low concentrations of verapamil. Whether this is due to increased fetal sensitivity to the drug or to the natural progression of the underlying condition has not been established.

1.6 Selective measurement of propafenone and its metabolites

This class 1c antiarrhythmic agent is metabolised to 5-hydroxypropafenone (5-OHP) and N-depropylpropafenone (NDPP), the former pathway being genetically linked. For the assessment of the relationship between plasma propafenone concentrations and clinical effect selective

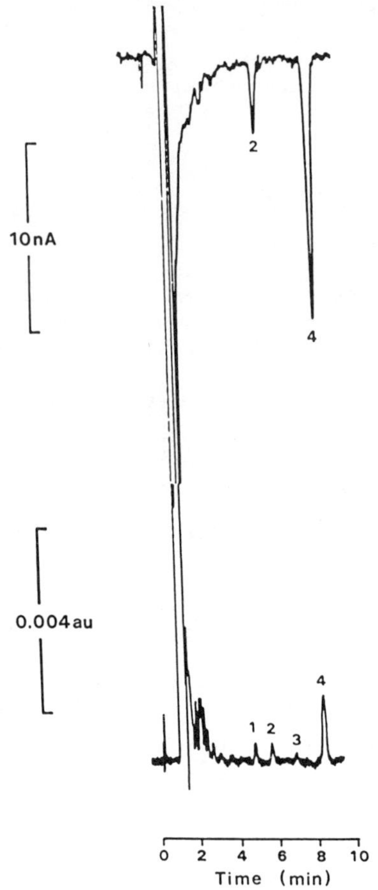

Fig. 4. Simultaneous HPLC chromatograms with UV detection (lower) and electrochemical detection (upper) of an extract of a plasma standard containing propafenone (1), 5-hydroxypropafenone (2) and N-depropyl propafenone (3). Original propafenone concentration 50 μg/l, remaining compounds 25 μg/l.

methodology, with the sensitivity necessary to measure both metabolites has proved essential. We have used an HPLC method with UV detection at 215 nm. A considerable improvement in sensitivity for 5-OHP has been achieved using electrochemical detection, since the phenolic hydroxy group of this metabolite has an excellent electrochemical response (Fig. 4).

Our preliminary clinical findings have shown a comparatively long elimination half-life for propafenone in a patient who was a poor metaboliser of debrisoquine. Plasma concentrations of 5-OHP were measurable using our methodology, at variance with one group who were unable to detect this metabolite in poor metabolisers of debrisoquine (9).

2. MONOCLONAL ANTIBODIES FOR RADIOIMMUNOASSAY

Monoclonal antibodies have been introduced for some drug immunoassays with improvements in specificity. The implications of two monoclonal antibodies developed by Sandoz AG for the measurement of cyclosporin will be discussed here.

There are a number of problems associated with the measurement of this novel immunosuppressant drug, whether HPLC or radioimmunoassay (RIA) is used (10). To date, the underlying problem of RIA has been the specificity of the antibody supplied with commercial kits. Only polyclonal, polyvalent, antisera have been available and this has resulted in discrepancies between HPLC and RIA results. Higher results are produced by RIA because the antibodies cross-react with metabolites of the parent compound. This difference has not precluded a clinical application of the RIA results (11) but, following liver and heart transplantation, particularly high results may be produced by RIA, reducing the clinical value of the measurement.

Monoclonal antibodies with high specificity for a selected portion of a molecule can be produced. Once the strain has been established the antibody can be supplied indefinitely, with a guaranteed spectrum of cross-reactivity towards metabolites of the drug, endogenous and exogenous molecules. From over 180 monoclonal anti-cyclosporin antibodies (12) Sandoz have selected two for inclusion in a new RIA kit. One antibody is specific for the parent compound, exhibiting virtually no cross-reactivity with metabolites of the drug. There is very good agreement between the results produced by the monoclonal specific RIA and HPLC, even for samples from patients who have received liver or heart transplants (Shaw L, personal communication).

The other monoclonal antibody has a broader spectrum of cross-reactivity with the metabolites of cyclosporin than the current, polycloncal, antisera. Results produced by the RIA based on this monoclonal antibody are, in general, 30-50% higher than those produced by current RIA procedures.

These monoclonal antibodies, in particular the specific, are likely to have a substantial impact on the monitoring of cyclosporin. Preliminary clinical findings in renal transplant patients suggest that measurements made by the specific antibody are more closely associated with graft rejection than measurements made using a polyclonal antibody (Taube D, personal communication). In addition, the specific antibody will be of value in defining the pharmacokinetics of the drug, especially in paediatric patients, since only small sample volumes are necessary for the RIA, with increased sensitivity compared with HPLC.

The role of the monoclonal non-specific antibody has not yet been defined. Research is in progress to establish whether an assay which includes some measure of metabolite concentrations correlates better with signs of cyclosporin toxicity than reference to a specific measurement of the drug.

3. COMPUTERISATION

Two aspects of the effects of computerisation on the application of drug measurements to therapeutics will be considered.

3.1 Computer-optimised dosing

Bayesian forecasting of drug dosing is based on a statistical approach which utilises pharmacokinetic data derived in a large patient population. From these data estimates of the dose necessary to achieve plasma concentrations within a specified range can be made. The dose is individualised by the input of specific patient data including age, sex, clinical status and plasma drug concentrations.

The population-based pharmacokinetic factors for a drug are derived from concentration data collected at a variety of time points following intravenous or oral drug dosing. As a result, the interpretation of plasma concentration data is less dependent on sampling time. Thus, it is not essential to collect samples pre-dose (trough), with obvious benefits for the use of the system in an out-patient setting.

The errors of prediction of plasma concentrations tend to diminish as more individual patient data points are added to the original equation. One group (13) has shown this trend for three drugs in common use, in addition demonstrating the superiority of Bayesian forecasting over the use of a 'simple' approach based on a nomogram.

There are some practical limitations to the use of this computerised approach. Firstly, a large data base is needed if clinically useful predictions are to be made for patients suffering a wide range of illnesses. It is worth noting, however, that one group has shown that the use of a relatively small data base of non-specific RIA cyclosporin measurements gave a good prediction of achieved concentrations (14).

Secondly, there is a need for good laboratory/clinician communication. The hardware necessary to achieve this is easily available; the will to put it into operation is often harder to find. Good communication between the two specialties may be hampered by staff change-overs and in the basic level of understanding of the value of the approach to drug monitoring. Such problems can only be overcome if there is a continuing policy of education for all staff and a fundamental commitment by the laboratory staff to involve themselves in the problems facing the clinician.

Thirdly, data relating drug concentrations to clinical effect are required if clear guidance on dosage requirements is to be offered. Some of the practical problems involved in the assessment of this relationship will be mentioned in the next section.

If Bayesian forecasting is to be put into operation as a routine it is clear that good quality commercially available software would be an advantage to most users without the resources to develop their own data base. One manufacturer (Abbott Diagnostics) has ventured into this field to produce a pharmacokinetic package. The menu-driven packages are simple to operate and run on IBM compatible equipment. The population-based statistics can be modified by the input of individual patient data, to obtain estimates of drug doses necessary to produce plasma concentrations within an optimum range. There is a clear graphical output of data which can be incorporated into patient records. The system is also useful as a teaching aid to show the effects of such factors as disease, or alterations in formulation, on the expected range of plasma concentrations within a dosage interval. Currently, the Abbott Pharmacokinetic System is available for theophylline, gentamicin and tobramycin; future releases will include digoxin and amikacin.

3.2 Fitting concentration/effect data

Establishing a relationship between clinical events and measured drug concentrations is beset by difficulties; measuring the drug is rarely the most difficult! There are a number of factors which warrant careful attention.

Firstly, there should be clear clinical end-points and it is best if these can be judged by objective, recordable, methods. Secondly, it may be necessary to take into account the effects of drug metabolites or the effects of other drug therapy. In some instances only one sterioisomer is active, limiting the value of total concentration measurements. Finally, there may be a natural variation or progression of the symptoms or signs being recorded. Hence, good control data are needed.

Antiarrhythmic drugs are ideally suited to the use of this approach since suppression of arrhythmia can be recorded either in the clinical laboratory or by ambulatory monitoring. The response to therapy can then be compared with control data and drug concentrations.

Recently, we have used a non-linear least squared regression model of time, drug concentration and drug effect (15) to examine the relationship between plasma propafenone concentrations and the suppression of ventricular extrasystoles. Very good fits have been achieved for some patients. However, it should be noted that, even with objective data in patients deemed to have a good clinical response, the between patient variation in the quality of fit and in effective concentration range was very large.

CONCLUSIONS

This review has set-out to show that sensitive and selective methodology is available for the measurement of a wide range of drugs at the concentrations attained during chronic therapy or following single dosage. These measurements can be used to investigate both the basic properties of the compounds and to assist in optimising their clinical use. In some instances the sensitivity which can be attained out-strip our ability interpret the clinical findings.

The introduction of monoclonal antisera for the RIA measurement of cyclosporin, in particular that specific for the parent compound, is likely to have a substantial impact on the routine monitoring and investigation of the drug.

Finally, computerisation greatly facilitates our ability to perform the lengthy and tedious calculations needed to utilise population derived pharmacokinetic data as a guide to dosage optimisation. In addition, the rapid transmission of the findings to clinical colleagues, together with a clear written and graphical summary should help to reduce the potential for inappropriate dosage adjustments. For the analysis of the relationship between pharmacokinetic and pharmacodynamic data easily usable programs can be devised to establish clinically useful parameters, such as the minimum effective concentration.

ACKNOWLEDGEMENTS

We thank Rupinda Bhamra, Simon Walker and Sophie Monkman (Poisons Unit), for skilled technical assistance and Dr R Miller (Abbott Diagnostic, Texas, USA), for the gift of a demonstration program of the Abbott Pharmacokinetic System.

REFERENCES

1. Flanagan RJ, Bhamra RK, Walker S, Monkman SC, Holt DW. Measurement of cardio-active drugs in biological samples by HPLC. J Liquid Chromatogr 1987; (in press).
2. Bhamra RK, Flanagan RJ, Holt DW. High-performance liquid chromatographic method for the measurement of mexiletine and flecainide in blood-plasma or serum. J Chromatogr 1984; 307: 439-444.

3. Walker S, Flanagan RJ, Holt DW. High-performance liquid chromatographic measurement of penticainide and desalkylpenticainide in biological fluids. Biomed Chromatog 1987; (in press).
4. Holt DW, Tucker GT, Jackson PR, Storey GCA. Amiodarone pharmacokinetics. Am Heart J 1983; 106: 840-847.
5. Adams PC, Holt DW, Storey GCA, Morley AR, Campbell RWF. Amiodarone and its desethyl metabolite: tissue distribution and morphological changes during long-term therapy. Circulation 1985; 72: 1064-1075.
6. Lim PK, Trewby PN, Storey GCA, Holt DW. Neuropathy and fatal hepatitis in a patient receiving amiodarone. Br Med J 1984; 288: 1638-1639.
7. Allan LD, Anderson RH, Sullivan ID, Holt DW, Tynan M. Evaluation of foetal arrhythmias by echocardiography. Br Heart J 1983; 50: 240-245.
8. Cole SCJ, Flanagan RJ, Johnston A, Holt DW. Rapid high-performance liquid chromatographic method for the measurement of verapamil and norverapamil in blood-plasma or serum. J Chromatogr 1981; 218: 621-629.
9. Siddoway LA, Thompson KA, McAllister CB et al. Polymorphism of propafenone metabolism and disposition in man: clinical and pharmacokinetic consequences. Circulation 1987; 75: 785-791.
10. Holt DW. Measurement of cyclosporin: methodological problems and clinical correlates. Prog in Transplantation Vol 3. Eds Morris P, Tilney N. Churchill Livingstone, Edinburgh 1986; 32-54.
11. Holt DW. Monitoring cyclosporine: a practical guide. Sandoz AG, Basle, Switzerland. 1987.
12. Quesniaux V, Tess R, Schreier MH, Maurer G, Van Regenmortel MHV. Potential of monoclonal antibodies to improve therapeutic monitoring of cyclosporine. Clin Chem 1987; 33: 32-37.
13. Whiting B, Kelman AW, Bryson SM et al. Clinical pharmacokinetics: a comprehensive system for therapeutic drug monitoring and prescribing. Br Med J 1984; 288: 541-545.
14. Kahan BD, Kramer WG, Williams C, Wideman CA. Application of Bayesian forecasting to predict appropriate cyclosporine dosing regimens for renal allograft recipients. Transplantation Proc 1986; 18 (Suppl 5): 200-203.
15. Holford NHG, Sheiner LB. Understanding the dose-effect relationship: clinical application of pharmacokinetic-pharmacodynamic models. Clin Pharmacokinetics 1981; 6: 429-453.

LABORATORY DIAGNOSIS OF ACUTE POISONING:

CONSEQUENCES FOR TREATMENT

Andrew Heath

Poisons Therapy Group, Department of Anaesthesia and Intensive Care, Sahlgren's Hospital, 413 45 Gothenburg Sweden

The use of analytical techniques in the diagnosis of acute poisoning is nothing new - such methods were well known to characters such as Sherlock Holmes and Hercule Poirot. The mechanisation and computerisation of clinical chemistry have to some extent left the sleuths behind, replacing individual acumen with a computer printout. Such a printout may or may not contain data relevant to the patient, often arriving or being recognized too late to contribute to the outcome of the poisoning. How then can resources best be utilized? Seen from a clinical perspective, this presentation will examine the limitations, possibilities, and impact on therapy of laboratory diagnosis in acute poisoning.

The objectives of any laboratory service to a clinical toxicology program should be clearly defined, and specific characteristics recognized. An understanding of the clinical contribution of laboratory support to diagnosis in the individual patient must be realised. Finally a forum for interchange between chemist and clinician must be established. A structured communication can be organised, maintaining mutual understanding of both parties expertise (1). This review will address these questions and more specifically look at:

* basic requirements for emergency toxicological analysis
* indications for a full toxicological screen
* indications for repeated "stat" toxicological analysis
* pitfalls in the interpretation of laboratory results
* clinical biochemistry - useful analyses.

The objectives set up for a laboratory service should parallel clinical resources and ambitions. At a hospital admitting only a small number of poisonings yearly, it may be too costly to have a full screening program as part of an on-call service. Limited resources should be centered around the important minority of cases where a stat toxicological analysis is essential in evaluating and monitoring therapy - such as hemodialysis, hemoperfusion, repeat dose activated charcoal, and hyperbaric oxygen.

Table 1 is based upon local pattern of poisoning. On a regional basis, larger hospitals with more patients and expertise in toxicology should provide a full screening service. A structured system of regional laboratories exists in many countries, for example the Federal Republic of

Table 1. Basic requirements for emergency toxicology analysis for hospitals admitting emergency toxicology cases, Sweden, 1984 (2)

Analysis	Toxin	Type of analysis	Therapy
Essential	Paracetamol	Quantitative	Antidote
	Iron	Quantitative	Antidote
	Methanol	Quantitative	Antidote
	Ethanol	Quantitative	Antidote
	Salicylate	Quantitative	Elimination
	Lithium	Quantitative	Elimination
	COHb	Quantitative	Elimination
	Ethyleneglycol	Qualitative	Elimination
Useful	Digoxin	Quantitative	Antidote
	Acetylcholinesterase	Quantitative	Antidote
	Theophylline	Quantitative	Elimination
	Isopropanol	Quantitative	Elimination
	Paraquat	Qualitative	Elimination
Possible	Amatoxin	Qualitative	Elimination
	Barbiturate	Qualitative	Elimination
	Tricyclics	Qualitative	Symptomatic
	Phenothiazines	Qualitative	Symptomatic
	Methemaglobin	Quantitative	Antidote

of Germany. Access to a tertiary laboratory with a full screening service is an essential ingredient of any toxicology research and teaching program, and for evaluating toxicological trends. Criteria should exist whereby a graded increase in resources is employed if required. For example, in a severely ill patient poisoned with an unusual toxin, a full screen may be necessary although not available at a small health care facility.

A number of characteristics must be recognized to ensure an optimal relationship between laboratory and clinician. The list above (Table 1) is adapted to Swedish conditions existing in 1984; many countries need barbiturate and paraquat analyses available at all levels. The analyses offered should match poisoning trends; since the pattern of poisoning is constantly changing (Fig. 1) any list of available analyses should be regularly reviewed.

Let us then look at the impact of laboratory resources on the diagnosis and management of acute poisoning. The diagnosis of acute poisoning is based on the case history, a physical examination of the patient, and the results of toxicological analysis. In the overwhelming majority of cases of poisoning the diagnosis is apparent from the case history, and supported by the clinical examination. A hurried, tired resident may often find it easier to order a battery of laboratory tests than take an extensive history. It is, however, the history which usually provides the diagnosis, to be confirmed or rejected by a laboratory test. The case history also provides vital information of the circumstances behind the overdose; for example if the patient has left a letter or if there were other signs of a planned suicide. Such facts are of great help when deciding on the degree and type of psychiatric intervention. In most western countries benzodiazepines, tricyclic antidepressants, and other tranquillizers are together with alcohol the major cause of admission for self-poisoning (Fig. 1). In these cases a quantitative, stat analysis is unlikely to affect the clinical course. Indeed, most of these patients improve very quickly. In tricyclic antidepressant poisoning 90% of patients are awake within 24 hours, and it is obvious that in many such situations full laboratory screening is not justified. However, there is no question that a full screen provides a better service - i.e. is superior - to a system which requests the analysis of specific substances. With both blood and urine samples and using different analytical techniques Hepler et al. (3) have shown a positive incidence of 72% compared to 49% when only blood was available; in this study two or more toxins were found in 61% of patients in whom both blood and urine were analysed as compared to 24% where only blood was available. Of 252 patients, in only 30% were the clinicians predictions 100% correct; in 36% of patients screened no history or evidence other than a "suspected" ingestion was reported, and in 16% no toxins were found at all.

When then should widespread screening be performed? Routine samples should be collected in all cases, and include blood, urine and gastric contents as well as samples of the suspected poison (4). In most of these cases samples can be discarded within 24 hours as the patient improves. If the diagnosis is unclear, or the toxin unusual, or an antidote required, the samples exist from admission, allowing if necessary a full screen to be performed. It is the responsibility of the admitting doctor to ensure that samples are taken; it is the responsibility of both chemist and clinician

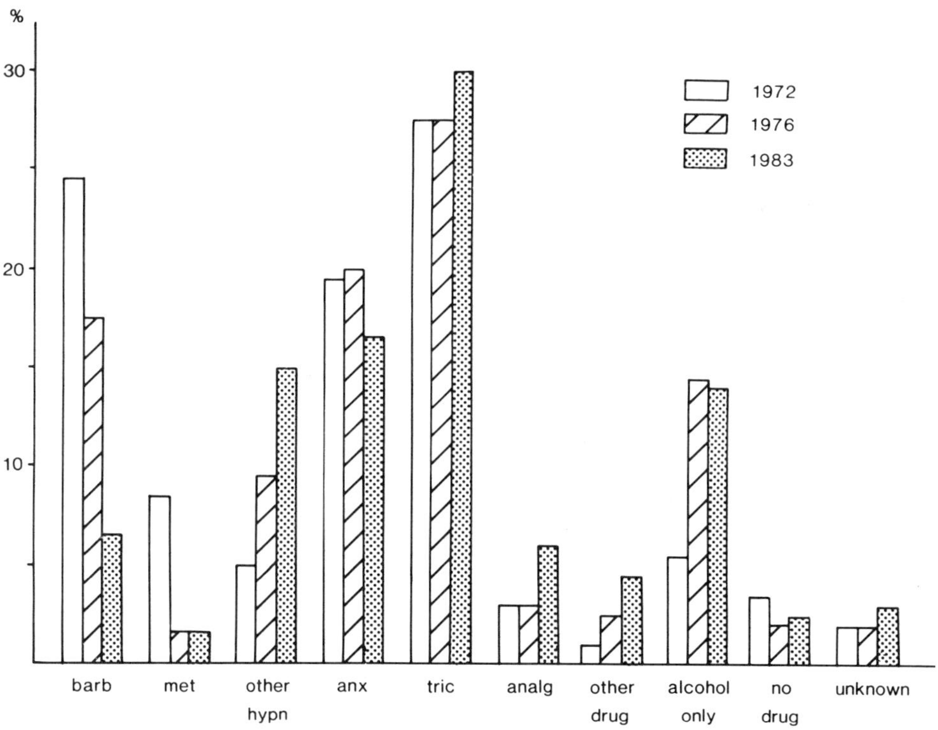

Fig. 1. Self-poisonings admitted to ICU. Admissions expressed as % of total admissions in 1972, 1976, and 1983. Bar = barbiturate, met = methaqualone, other hypn = other hypnotic, tric = tricyclic antidepressant, anx = anxiolytica, analg = analgetic. No drug includes poisonings from non-medicinal agents.

Table 2. Patient categories indicating full toxicology screening

* coma
* unexplained acidosis
* head injury
* poor case history
* seizures (particularly children)
* unclear cardiotoxicity

to decide what analysis should be performed and how quickly. Criteria indicating a full screen should be established (1). In, for example, certain groups of severely ill patients such criteria can be defined (Table 2). Some of the screen may need to be performed stat - such as methanol in unknown metabolic acidosis - whereas the remainder of a screen may be completed at a more convenient time.

In some of these groups poisoning may be an important differential diagnosis - or may exist concurrently, complicating clinical evaluation - for example a patient who is drunk with a head injury. In young adults poisoning is the most common cause of seizures among non-epileptic patients.

Another question which should be addressed is when repeated, quantitative stat analysis is indicated (Table 3). The widespread use of sustained release preparations is a common indication for a repeat level, particularly for drugs with a narrow therapeutic margin such as theophylline, where peak concentrations may develop upto or later than 12 - 24 h (5).

An assessment of the severity of the poisoning from a single plasma concentration without due consideration of the clinical picture is a common pitfall. For example, a level taken after ingestion may produce unnecessary concern if the test was taking in the distribution rather than elimination

Table 3. Some indications for repeated stat analysis in clinical toxicology

* continued absorption (sustained release preparations, formation of bezoars)
  Examples - theophylline, beta-blockers, lithium, early paracetamol, meprobamate poisoning.

* to terminate dialysis or hemoperfusion
  Examples - ethylene glycol, theophylline poisoning.

* to terminate treatment with hyperbaric oxygen
  Examples - carbon monoxide poisoning.

* to terminate therapy with repeat dose activated charcoal
  Examples - ASA, carbamazepine, theophylline, barbiturate poisoning.

* to monitor antidote therapy
  Examples - digoxin concentrations during FAB therapy, methemaglobin levels induced by cyanide antidotes, ethanol therapy in methanol poisoning.

Table 4. Pitfalls in the interpretation of single quantitative toxicological analyses

* level interpreted too early (theophylline, paracetamol)
* discrepancy between toxic levels and symptoms in acute compared to chronic overdose (lithium, ASA, theophylline)
* "toxic" level may be asymptomatic (methanol)

phase. When assessing paracetamol overdose plasma levels taken before 4 hours should not be used as a guideline for antidote therapy. On the contrary, if absorption continues an initial low level may induce false security. Another pitfall is the evaluation of toxic levels in patients in acute versus chronic overdose. In several poisonings, the clinical symptomatology differs in chronic overdose, with severe symptoms presenting at lower plasma levels. For example, in theophylline overdose the risk for seizures is much greater at a given plasma concentration in chronic overdose, with a 50% probability at 40 mg/l compared to 120 mg/l for acute overdosage patients (6).

In lithium poisoning, neurological sequelae are seen in chronic overdose at plasma levels which may be only half of those seen in patients after acute overdose and with no sequelae (7). In salicylate poisoning a non cardiogenic pulmonary oedema is a feature of chronic ingestion rather than acute overdose. The conclusion to be drawn from such observations is that treatment should be based on both toxicological analysis and a clinical assessment, where both the case history and the presence of symptoms of prognostic value are considered.

The use of hemodialysis in methanol poisoning is usually based on blood methanol concentrations. However, a number of reports suggest that chronic alcoholics drinking methylated spirits can develop high methanol concentrations when stopping drinking (8). As the blood ethanol level falls to zero methanol concentrations may be as high as 50 mg/dl. However, these patients develop no formate or lactic acidosis, and do not show any of the classical symptoms of methanol poisoning. Thus the decision to dialyse should again be based on the case history, clinical signs, and presence of a metabolic acidosis, and not on blood methanol concentrations only. The results of routine clinical biochemistry tests can be of considerable help in the diagnosis of acute poisoning, when a specific toxicological analysis may not be readily available. Furthermore, the assessment of fluid and electrolyte disturbances, acid-base changes, and renal and hepatic failure with biochemical tests is essential for prognosis and therapy. The correlation between organ function and a given biochemical test is often far greater than between organ function and a single plasma toxin level.

Thus it is my firm conviction that clinical biochemistry results may be of far greater value in the initial assessment of the overdose patient than many specific toxicological analyses. In the work up of a patient with alcohol poisoning the difference between the measured and calculated osmolality indicates the presence of alcohol or other low volatiles (9). If then ethanol, methanol and isopropranol do not make up the difference, then ethylene glycol should be suspected, and treatment instigated, even if a specific analysis is not available (10). The diagnosis of ethylene glycol poisoning is supported by a metabolic acidosis and the presence of monohydrate and dihydrate crystals in the urine. As previously mentioned,

Table 5. Elements necessary to a successful clinician - laboratory dialogue

* agreed criteria upon which toxicological evaluation is initiated
* a delineation of the type of samples necessary for the laboratory
* a programme by which the samples are transported to the laboratory
* a recognition of the limitations of the service to be performed by the laboratory
* a mutual understanding of how best to serve the patient

blood gas analyses are of great help. An anion gap metabolic acidosis (excepting salicylate, methanol and ethylene glycol) usually indicates the formation and accumulation of lactate due to hypoxia or shock, and may be used as a rough guide to the severity of the poisoning - for example, in cyanide, strychnine, camphor or carbonmonoxide poisoning.

Total creatine phosphokinase concentrations are another example of a readily available analysis of value. Although non-specific, very high values may indicate muscle cell necrosis and provide an early warning of impending acute renal failure. S-potassium concentrations are another analysis of particular value in certain poisonings. In acute digoxin poisoning there is a clear relationship between the severity of poisoning and the degree of hyperpotassemia; on the other hand in poisoning with $beta_2$-agonists, such as terbutaline, the S-potassium falls with increasing toxicity (12).

The correct utilisation of both specific, toxicological analyses and other clinical biochemistry resources requires considerable communication between clinician and chemist. My own experience is that such a dialogue cannot be taken for granted but must be worked with; unfortunately a lack of interdisciplinary exchange is common in many larger hospitals. Hepler (1) outlines a number of points which, when recognized may contribute to better medicine (Table 5).

By mutual understanding and direct communication we can hopefully obtain the best value for the resources available. Good medicine can be practiced with surprisingly few resources by combining toxicological analysis, biochemical tests and clinical acumen. However, a comprehensive analytical approach is superior to a system of requests for specific toxins. At county and regional hospital level, a wider program is essential for following toxicology trends, teaching, and as a reference.

REFERENCES

1. Hepler BR, Sutheimer CA, Sunshine I. Role of the toxicology laboratory in the treatment of acute poisoning. Medical Toxicology 1985; 1: 61-75.
2. Persson H, Hanson A, Tryding N. Behov av kliniskt kemiska undersökningar finns vid akuta förgiftningar. Läkartidningen 1984; 81: 1321-1322.
3. Hepler BR, Sutheimer CA, Sunshine I. The role of the toxicology laboratory in emergency medicine. II. Study of an integrated approach. J Toxicol Clin Toxicol 1985; 22: 503-528.

4. Volans G, Widdop B. Laboratory investigations in acute poisoning. Brit Med J 1984; 289: 426-428.
5. Heath A, Knudsen K. Role of extracorporeal drug removal in acute theophylline poisoning. Medical Toxicology 1987; 2: 294-308.
6. Olson KR, Pentel PR, Kelley MT. Physical assessment and differential diagnosis of the poisoned patient. Medical Toxicology 1987; 2: 52-81.
7. Baud FJ, Bismuth C, Galliot M, Bruneau X. Acute poisonings and therapeutic overdoses with lithium. Toxicokinetic study in plasma and red blood cell. The Proceedings of the III World Congress of the World Federation of Associations of Clinical Toxicology and Poison Control Centres, Brussels, 1986. Abstract 044. To be published.
8. Heath A. Methanol poisoning. Lancet 1983; 1: 1339-1340.
9. Pappas AA, Gadsen RH, Taylor EH. Serum osmolality in acute intoxication: A prospective clincal study. AJCP 1985; 84: 74-79.
10. Jacobsen DJ, McMartin KE. Methanol and ethylene glycol poisonings. Mechanisms of toxicity, clinical course, diagnosis and treatment. Medical Toxicology 1986; 1: 309-334.
11. Heath A, Hultén B-Å. Terbutaline concentrations in self-poisoning: A case report. Human Toxicol 6: in press 1987.

CHAPTER 4
BIOSENSORS AND BIOCHIPS

Biosensors for medical use
P. Bergveld and P.J.A. v.d. Starre

Enzyme sensors and immuno sensors
M. Aizawa

BIOSENSORS FOR MEDICAL USE

Bergveld P.* and P.J.A. van der Starre**

*University of Twente, Box 217, 7500 AE Enschede, The Netherlands **Medical Centre De Klokkenberg, Box 90108 4800 RA Breda, The Netherlands

ABSTRACT

After an introduction concerning the basic features of extra- and intracorporeal measurement of blood chemical constituents and in this context the application of an ISFET catheter-tip pH-sensor, experimental results are reported and discussed, achieved with the Sentron 6F pH catheter during post-operative monitoring following coronary artery bypass surgery.

1. INTRODUCTION

The measurement of blood chemical constituents can be performed in various ways, as illustrated in Figure 1 and listed below.

Off-line determination of chemical and biochemical quantities from blood samples in the hospital laboratory. The advantage of this technique is that, depending on the analysis equipment which is available, a large variety of blood chemical constituents can be determined from one sample, with the best possible accuracy. A disadvantage is the fact that the measurement is related only to one particular moment and that the sample should be treated carefully during the time lag between sample collection and actual measurement. For the determination of transients and moreover the short-term prediction of a trend, this technique is not appropriate.

On-line analysis with bedside laboratory equipment, in which case two systems can be distinguished, operating with blood waste or with blood recycling.
Due to the extracorporeal placement of the sensor, measuring, rinsing and recalibration procedures can intermittently be carried out in the waste system, resulting in fact also in a sampling technique, however with a much higher frequency than in the off-line case mentioned above. In this case the biocompatibility of the sensor is less stringent due to the possibility of rinsing with heparinized calibration liquids or adding heparine to the input bloodstream.

With the recycling system the methods mentioned above are less convenient. The sensor should obey nearly the same requirements as in the

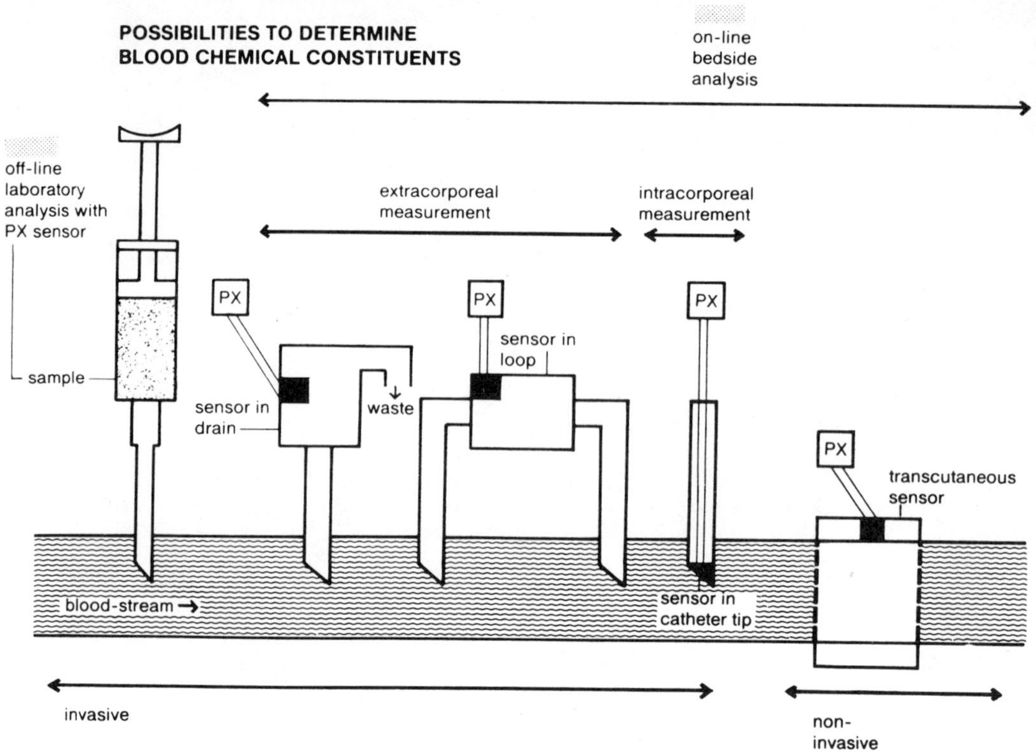

Fig. 1. Various ways to determine blood chemical constituents

case of an intracorporeal measurement. In both cases of extracorporeal bedside measurements the amount of blood, which is necessary for a chemical analysis, should be as small as possible, which requirement can be better fulfilled the smaller and faster the applied sensors are. This is the reason that Sibbald used ISFET based sensors for his four-function microanalyser, applied as a bedside monitor for pH, pNa, pK and pCa measurement (1).

A real continuous measurement can only be fulfilled with an in vivo electrode. The sensor is in this case placed directly in the bloodstream by means of a catheter-tip mounting. The best suitable sensor is in this case again an ISFET, due to its small dimensions and solid-state character as well as to the ease of a multi-sensor design. The Sentron pH catheter-tip

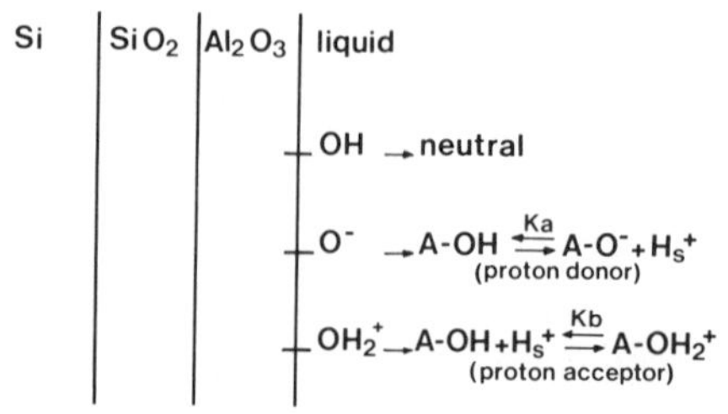

Fig. 2. Schematic representation of inorganic oxide-electrolyte interface with corresponding chemical reactions

sensor, as described below, is the most far developed example of this type of in vivo pH determination.

A serious problem with catheter-tip sensors is the impossibility to recalibrate the sensors after implantation. This means that the stability of the sensor characteristics, including the signal conditioner, should be such that a reliable long-term measurement can be guaranteed within the desired accuracy, as for instance 0.02 pH over 10 hours.

A desirable approach is of course a transcutaneous measurement which decreases the risk of infection as is the case with the methods mentioned above. Transcutaneous measurement of blood chemical constituents is however limited to blood gases which can diffuse through the skin, for instance under temperature activation. Ionic components can not be measured with this technique.

2. THE ION SENSITIVE FIELD EFFECT TRANSISTOR (ISFET)

Basically the ISFET is composed of a piece of silicon, covered with a thin layer of SiO_2, usually provided with a thin additional layer of an inorganic material such as Al_2O_3 or Ta_2O_5 (see Fig. 2).

The ion sensitivity of the device originates from the acidic and basic inorganic oxide-solution interface, resulting in a surface potential ψ_o(2).

The theory which describes the interaction between an inorganic insulator and an adjacent electrolyte is based on the existence of a discrete number of hydroxyl groups, the so-called surface sites, which are dissociated in dependence of the pH of the electrolyte. It is usually considered that only one type of site is present in the case of a chosen inorganic gate oxide, such as SiO_2, Al_2O_3 or Ta_2O_5. This means that each particular surface site (OH-group) can be neutral, act as a proton donor (acidic reaction) or as a proton acceptor (basic reaction), as schematically represented in Figure 2.

The corresponding acidic and basic reactions are characterized by their equilibrium constant K_a and K_b. Using the site-dissociation model, the resulting surface potential ψ_o can be calculated from the total number, N_s, of surface sites per unit areas and the equilibrium constant K_a and K_b:

$$\psi_o = 2.3 \frac{kT}{q} \frac{\beta}{\beta+1} (pH_{pzc} - pH) \tag{1}$$

The sensitivity parameter β can be expressed in terms of K_a, K_b, N_s and the double layer capacitance C_{DL}:

$$\beta = \frac{2q^2 N_s (K_a K_b)^{1/2}}{kT C_{DL}} \tag{2}$$

while

$$pH_{pzc} = - \log [\frac{K_a}{K_b}]^{1/2} \tag{3}$$

is the pH at the point of zero charge (pzc), thus the pH for which ψ_o = o.

So the actual ion sensitivity of an ISFET originates from the interfacial phenomenon at the gate oxide-solution interface, which is fully

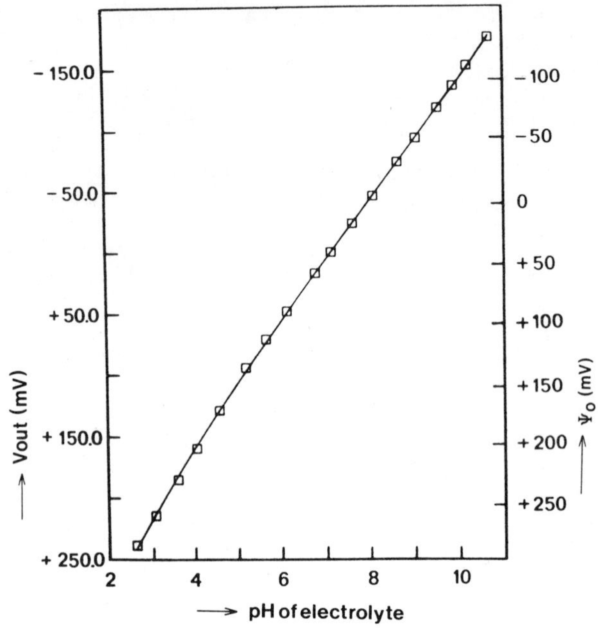

Fig. 3. Response of Al_2O_3-ISFET. ψ_o is insulator-electrolyte interface potential, V_{out} is amplifier output voltage

determined by the characteristics of the particular oxide. For measurements in blood it appears that Al_2O_3 is a very suitable oxide. It combines a relatively high and linear pH sensitivity, as can be seen from Figure 3, with an acceptable degree of biocompatibility, probably related to the value of its pH_{pzc}, which is 7.9. This means that the surface is a little positively charged around pH = 7.

Note that the surface potential ψ_o, which is the measurand of interest, exists at the surface of an insulator. This means that it can not be measured in the usual way by connecting it to an amplifier. Only the ISFET concept, makes it possible to measure insulator surface potentials. By connecting a reference electrode which is inserted in the electrolyte to the silicon substrate, the surface potential ψ_o is applied over the oxide layer, resulting in a corresponding electric field. This field controls the amount of electrons at the surface of the silicon, the so-called inversion layer.

The density of electrons can be measured by means of two diffused contacts, the so-called source and drain of the field effect transistor, as schematically shown in Figure 4. Connecting a voltage V_{ds} between the source and the drain, will result in a drain current I_d which is proportional to the density of electrons and thus to the electric field and its origin, to which the surface potential ψ_o contributes.

$$I_d = C_1 + S(\psi_o + C_2) \quad (4)$$

where C_1 and C_2 are constants depending on various solid-state parameters of the device and of the reference electrode and S is the electronic sensitivity of the ISFET, depending mainly on its geometry.

For an actual ISFET measurement, special electronic circuits have been developed (3) which convert the value of I_d into a voltage which exactly reflects the value of ψ_o, as given by V_{out} in Figure 3. So in fact an ISFET

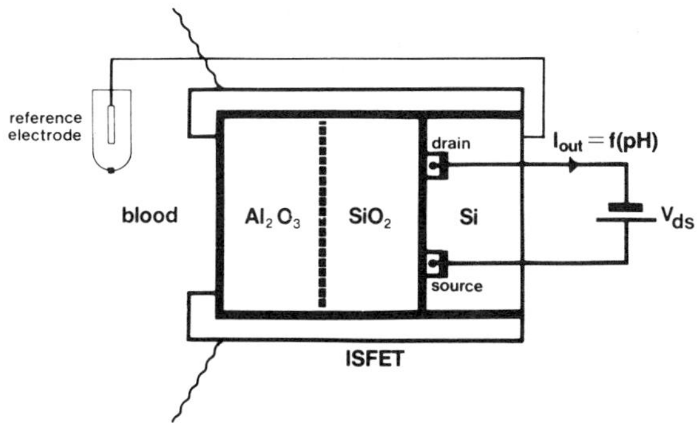

Fig. 4. Schematic representation of ISFET measuring system

including its amplifier, measures the value of ψ_0 as given by equation (1), but in such a way that there is no galvanic connection between the electronics and the electrolyte. A thin layer of insulating oxide, typically 100 nm thick, separates the "electronic world" from the "ionic world", which is however enough to prevent any undesired electrochemical reaction which would certainly occur with a galvanic connection. It will be obvious that, besides this capacitive nature of the ISFET concept, also the solid-state and rigid performance and the small dimensions are very favorable for incorporating ISFET chips in a catheter wall.

3. THE SENTRON CATHETER-TIP PH-ISFET SENSOR

Mounting an ISFET chip behind a side window of a catheter does not result directly in a realistic clinical product. A complete catheter-tip pH sensor, including a tip reference electrode and signal conditioner, has been developed by Sentron in Roden (The Netherlands) (4, 5, 6). The system consists of a floating analog amplifier, necessary for patient safety, and a digital processing system, and provides the user with automatic compensations for temperature sensitivity and long-term drift. For this reason the chip also contains a temperature sensitive resistor, while the catheter is provided with a connector containing a Programmable Read Only Memory (PROM). This memory contains all the essential parameters of the ISFET and integrated temperature resistor, such as pH sensitivity, temperature behavior, predicted drift behavior etc. In this way the system recognizes the individual ISFET properties and can calculate the actual blood pH during a certain time span, using only one initial calibration. This calibration value can be achieved by means of a sample, collected at the start of a measurement and being analysed with a conventional blood-pH analyser.

The ISFET is protected against possible electrostatic damage and electrocutery interference and can measure the actual blood pH continuously during 36 hours within an accuracy of 0.02 pH units.

4. THE CLINICAL APPLICATION

In the circumstances of major surgery, substantial changes in pH may occur in the early post operative period (7), with important consequences for the condition of the patient (8, 9). In the first clinical report (10), on the continuous intra-arterial measurement of pH in patients who had undergone coronary artery bypass surgery, interesting phenomena were discovered, particularly during weaning of mechanical ventilation. The

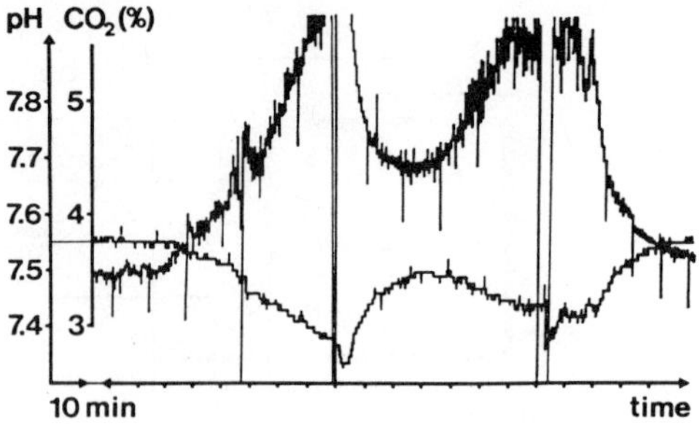

Fig. 5. Simultaneous recording of pH and end-tidal CO_2 during shivering

above described way of monitoring pH appeared to be reliable and stable, and showed a remarkably short response time.

Since pH is mainly determined by changes in ventilation in this post-operative episode, a study was undertaken to verify if the continuous intra-arterial pH-monitoring could be compared with a conventional way of monitoring ventilation, i.e. end-tidal CO_2 by means of a capnograph.

5. CLINICAL CONDITIONS

In 20 patients, who had undergone uncomplicated CABG, with a normal left ventricular function and no signs of peripheral vascular disease, a 6F pH-catheter was inserted into a femoral artery through a valved introducer system (11). All patients were mechanically ventilated (Siemens Servoventilator), aiming at a pCO_2 between 35 and 45 mm Hg and a pO_2 above 100 mm Hg.

Monitoring consisted of the conventional monitoring systems with respect to the circulation, consisting of ECG, heart rate, systemic and pulmonary artery pressures, left and right filling pressures, cardiac output, systemic and pulmonary vascular resistance. Rectal and peripheral temperatures were continuously monitored as well. Apart from minute volume and airway pressure, end-tidal CO_2-monitoring (Siemens capnograph, main stream analyser) was applied to verify the right setting of the ventilator. Both the readings of the pH-catheter and from the capnograph were simultaneously recorded on a flat-band recorder (Kipp). At specific moments during ventilation, attention was focussed on these readings, assuring that the observed changes in pH could not be explained by substantial changes in for example cardiac output.

6. THE CLINICAL RESULTS

The simultaneous readings of pH and end-tidal CO_2 appeared to correlate satisfactorily during stable mechanical ventilation. Changes in CO_2-production caused by for example shivering (Fig. 5) and bronchial suction (Fig. 6) were recorded in both the pH- and CO_2-readings. It appeared that the pH-readings were more stable, even during substantial movements of the patients, but that the capnographic recording lost it's usefullness in those conditions. In Figure 7 a typical weaning pattern is depicted, in which the patient starts to breath on his own, causing unreliable readings of CO_2, as opposed to the pH-recording, which remained very stable. After extubation continuous CO_2-monitoring is impossible, but continuous pH-monitoring is not interrupted, as is depicted in Figure 8.

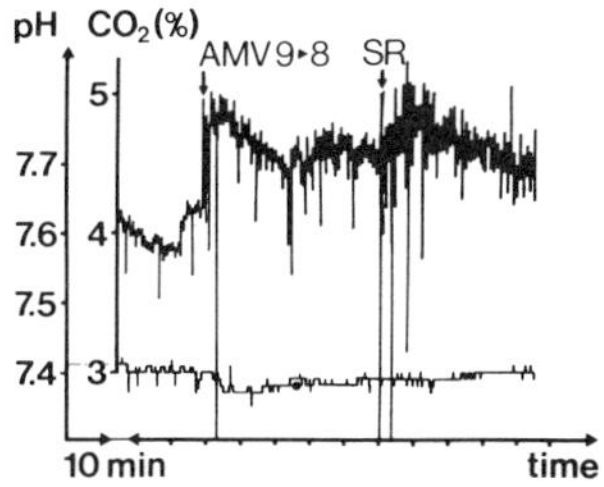

Fig. 6. Simultaneous recording of pH and end-tidal CO_2 during bronchial suction

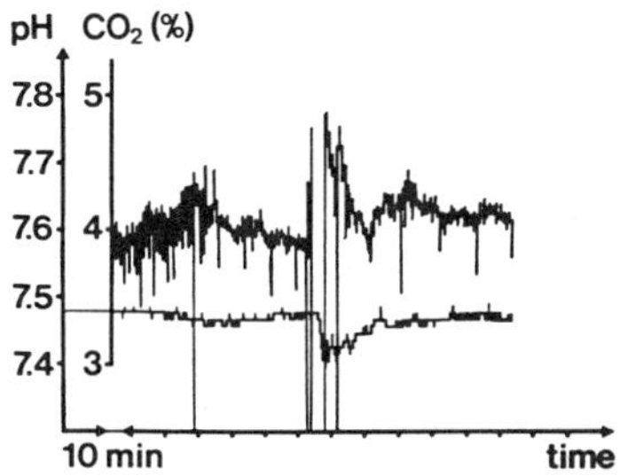

Fig. 7. Simultaneous recording of pH and end-tidal CO_2 during weaning of mechanical ventilation

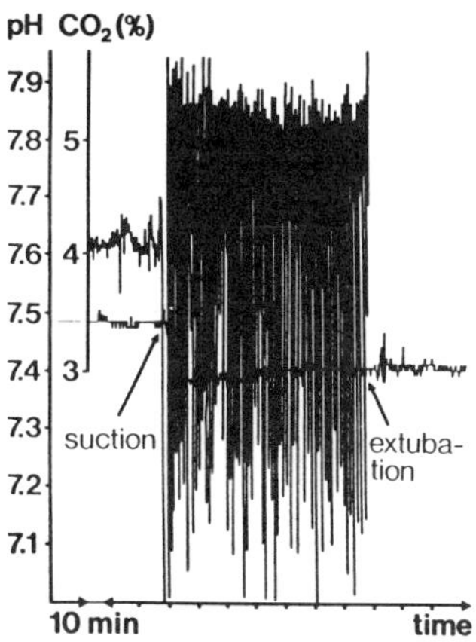

Fig. 8. Simultaneous recording of pH and end-tidal CO_2 during suction and extubation

7. DISCUSSION

The findings of the study indicate that continuous monitoring of pH and end-tidal CO_2 correlate well in patients who had undergone major surgery, assuming that these changes cannot be attributed to other factors than CO_2-production and elimination. During stable conditions both ways of monitoring are reliable, but when circumstances like movements and bronchial suction influence the stability, pH-monitoring appeared to remain constantly reliable as opposed to end-tidal CO_2-readings.

As was reported in an earlier clinical report (10), the in vivo blood pH monitoring by means of an ISFET showed also very short response times, indicating that the intermittent determination of pH by means of an off-line blood-gas analyser could not have disclosed the phenomena which were depicted. In patients with a stable circulatory condition, relatively substantial changes in the acid-base status don't cause dramatic problems, but in unstable patients they might have detrimental consequences. Continuous monitoring will not completely replace intermittent blood gas analyses, but particularly the continuous pH-monitoring by means of an ISFET gives additional information which might influence the clinical management of the patient.

REFERENCES

1. Sibbald A. A miniature flow-through cell with four-function Chemfet integrated circuit for simultaneous measurements of K^+, H^+, Ca^{2+} and Na^+ ions. Analytica Chimica Acta 1984; vol 259: 47-62.
2. Bousse L, De Rooij NF, Bergveld P. Operation of Chemically Sensitive Field-Effect Sensors as a function of the insulator-electrolyte interface. IEEE Trans on Electr Dev vol ED-30 1983; 10: 1263-1270.
3. Bergveld P. The operation of an ISFET as an electronic device. Sensors and Actuators 1982; 1: 17-29.
4. Bergveld P. The development and application of FET-based biosensors. Biosensors 1986; 2: 15-33.
5. Schepel SJ, Koning G, Oeseburg B, Langbroek HJM, Zijlstra WG. Performance of a pH monitoring system in vivo. Med & Biol Eng & Comp 1987; vol 25: 63-67.
6. Ligtenberg HCG. Solutions for some basic problems of ISFET sensor. Ph D Thesis Twente University of Technology, Enschede, The Netherlands.
7. Hinds CJ. Current management of patients after cardiopulmonary bypass. Anaesthesia 1982; 13: 170.
8. Cingolani HE, Faulkuer SL, Mattiazzi AR et al. Depression of human myocardial contractility with respiratory and metabolic acidosis. Surgery 1975; 77: 427.
9. Clowes GHA, Sabga GA, Komitaxis A et al. Effects of acidosis on cardiovascular function in surgical patients. Ann Surg 1961; 154: 524.
10. Van der Starre PJA, Harinck-de Weerd JE, Schepel SJ, Kootstra GJ. Use of an arterial pH catheter immediately after coronary artery bypass grafting. Critical Care Medicine, vol 14, 1986; 9: 812-815.
11. Seldinger SI. Catheter replacement of the needle in percutaneous arteriography: A new technique. Acta Radiol 1953; 9: 368.

ENZYME SENSORS AND IMMUNOSENSORS

Masuo Aizawa

Department of Bioengineering, Tokyo Institute of Technology
Ookayama, Meguro-ku, Tokyo 152, Japan

1. INTRODUCTION

Advances in sensor technology have opened the way for new and sophisticated measurement techniques and instrumentation for biotechnological process control, environment control, and medical diagnosis. Among various sensors, a biosensor can offer high selectivity for a specific substance, which has not been attained by synthetic sensor materials. The new breed of biosensors will present a unique combination of matrix-bound biochemical, transducer, and microelectronic components to allow almost instantaneous determination of substrate, analyte, or ligand concentrations (1, 4).

Molecular recognition of substrates by enzymes, organelles, microorganisms or tissue slices is followed by conversion into the corresponding products which are detected and recorded by the electronic device. The physico-chemical changes associated with molecular recognition caused by binding of the substance to be determined or by the enzymatic substance conversion are transduced by potentiometric or amperometric electrodes, thermistor, field effect transistors (FET), optoelectronic detectors, optical fibers, or other devices into an electrical signal. Biosensors are classified into enzyme sensors, microbial sensors, immunosensors, and others.

The molecular recognition part of enzyme sensor consists of the matrix-bound enzyme which converts a specific substance (determinant) selectively into products. The enzymatic reaction is accomplished by a change in chemicals, photons, and temperature. Such a change is transduced into electric signals at the transducer. Enzyme sensors may be classified as:

(1) Enzyme electrode
(2) Enzyme thermistor
(3) Enzyme transistor
(4) Enzyme field effect transistor (ENFET)
(5) Enzyme photodiode
(6) Optical enzyme sensor.

The interaction between antigen and antibody molecules may be extremely specific under favorable condition as a consequence of binding geometries. Immunoassay, which relies its high selectivity and sensitivity on the molecular recognition of antibody is classified into two categories: non-labeling and labeling ones. Since the pioneering work of Yalow and Berson radioisotopes have been employed as the label of choice. However, a number of nonisotopic alternatives have been developed including electron spin resonance for detecting radical labels, nephelometry, fluorescence, chemiluminescence and enzyme labels. Moreover, immunoassay with an electroactive label has become attractive due to the wide dynamic range and low detection limits of modern electroanalytical methods. On the basis of these immunoassays, there have been proposed variety of immunosensors, which are classified as follows:

(1) Non-labeled Immunosensors
 a) Transmembrane potential
 b) Electrode potential
 c) Piezoelectric
 d) Polarographic
 e) FET
 f) Surface plasmon
 g) Capacitance

(2) Labeled Immunosensors
 a) Enzyme-linked amperometry
 b) Enzyme-linked potentiometry
 c) Enzyme-linked luminescence
 d) Liposome-linked luminescence
 e) Electrochemical luminescence
 f) Other optical immunosensors
 g) Electro-active immunosensors.

2. ENZYME SENSORS

Enzyme Electrodes for Glucose: A glucose sensor installed in a glucose analyzer became commercially available in the mid-1970s. In the early 1980s, extensive development of glucose analyzers and sensors has taken place. These analyzers, which have been developed for clinical use, require 5-20 μl of serum sample for blood glucose analysis.
The sensor consists of a glucose oxidase (GOD) membrane and a Clark-type oxygen electrode. Glucose in a sample solution is oxidized with the resulting consumption of oxygen when contacted with the membrane-bound GOD. The decrease of dissolved oxygen is sensitively detected with the oxygen electrode. The output change of the sensor reflects the concentration of glucose in solution (5). Hydrogen peroxide is formed in the GOD catalyzed reaction. A GOD membrane may be coupled with a hydrogen peroxide electrode to form another type of glucose sensor. Both oxygen electrode and hydrogen peroxide electrode-based glucose sensors are now commercially available.
Aizawa et al. modified the platinum electrode surface with a GOD-entrapped polyaniline thin layer (6). The modification was accomplished by the electropolymerization of aniline in the presence of GOD in a neutral aqueous solution. The GOD-entrapped polyaniline membrane rejects the permeation of chemicals except gas molecules such as dissolved oxygen. A microenzyme sensor can be fabricated by the electrochemical polymerization.
Higgins et al. developed an alternative amperometric detection method, based on glucose oxidase, that is not dependent on oxygen as a mediator of electron transfer (7). The electrode uses a substituted ferricinium ion as a mediator of electron transfer between immobilized glucose oxidase and a graphite electrode.

In the condition diabetes mellitus, the determination of blood glucose levels rapidly, conveniently, precisely, and economically is important for its diagnosis and effective management. The minituarization of a glucose monitoring system is an essential requirement for its clinical application. Shichiri et al. have developed a needle-type glucose sensor using a platinum electrode covered with immobilized glucose oxidase (8). The current output was not significantly altered by changes in the oxygen tension of the solution in the range from 25-150 mmHg. The glucose sensor was connected with a microcomputer system to calculate the insulin infusion rate and a syringe-driving system to deliver insulin.

Biosensors have considerable advantages over conventional techniques. Serum or blood are the most common fluids tested and the first biosensor will probably be geared to the usage of these samples. A bench-top type of glucose analyzer has been commercially available and is being used in clinical chemistry laboratories. The blood or serum sample is injected onto a membrane impregnated with glucose oxidase, and the concentration of glucose is read out within 40 s. The instrument is automatically flushed and calibrated at the push of a button.

Varied Enzyme Electrode: In a similar manner to the glucose sensor, a number of enzyme electrodes have been developed for determination of various organic substrates as listed in Table 1.

Enzyme electrodes for enzyme activity: Enzyme electrodes are sometimes mistaken for electrodes which measure enzyme activities. There are electrode systems which are specifically designed to assay for enzyme activity such as the use of the NH_3 gas electrode for kinetic determination of trypsin activity based on the hydrolysis of a synthetic substrate. However, most enzyme electrodes couple an amperometric device with a specific enzyme that acts upon selected substrates. An enzyme electrode for pyruvate, which consists of membrane-bound pyruvate oxidase and an oxygen electrode, was coupled with immobilized oxaloacetate decarboxylase to determine glutamate pyruvate transaminase (GPT) and glutamate oxaloacetate transaminase (GOT) activities.

Enzyme FET Sensors: A family of new biosensors has been studied in the last decade using the MOSFET structure. The new biosensors consist of immobilized enzyme attached to ion-sensitive field effect transistor (ISFET) which is essentially an insulated gate field transistor without a metal gate. The interfacial potential at the electrolyte/insulator interface produced by the ions in solution will affect the channel conductance of the ISFET in the same way as the external gate voltage applied to the reference electrode. Several ISFETs have been developed for the determination of pH, Na^+, K^+, and Ca^{2+} (9-11).

An enzyme FET (ENFET) sensitive to penicillin was constructed by depositing a co-cross-linked penicillinase albumin layer over a pH sensitive FET (12). Immobilized penicillinase recognizes penicillin in solution with a resulting change in pH, which is detected by the FET. There have been developed so many enzyme FETs (13-15). Lundstrom et al. showed that the Pd-SiO_2-Si structure of MOSFET can be used for monitoring hydrogen dissolved in electrolytes (16). The NH_3 sensitivity is enhanced by depositing small amounts of other catalytic metals such as Ir on the Pd film. These Pd-MOSFETs were used to form a new type of biosensor, the enzyme transistor (17). The advantages of these bioelectronic sensors, including enzyme FETs and enzyme transistors, are the possibilities of minituarization and direct integration with microelectronics. It should even be possible to develop multichannel devices with several different enzymes combined with different sensitive areas on the chip.

Enzyme Photodiode Sensors: A photodiode or photo transistor is coupled with immobilized enzyme which catalyzes a luminescent reaction. The device is termed enzyme photodiode (or transistror) (18). The concentration of hydrogen peroxide is determined by counting photons at a constant concentration of luminol. Peroxidase was immobilized onto a Si

Table 1. Enzyme Sensors for Organic Molecules

Determinant		Biosensor	Biosensor Assemblies	
			Receptor	Transducer
Saccaride	Glucose	Enzyme sensor	GOD	O_2,H_2O_2 elec.
		Enzyme thermistor	GOD/Catalase	Thermistor
		Microbial sensor	P.fluorescens	O_2 electrode
Alcohol	Ethanol	Enzyme sensor	Ethanol oxidase	O_2 electrode
		Microbial sensor	T.Brassicae	O_2 electrode
Amino acid	Glutamate	Enzyme sensor	Glutamate DH	$NH^+{}_4$electrode
		Microbial sensor	E.coli	CO_2 electrode
Acid	Acetic acid	Enzyme sensor	Alcohol oxidase	O_2 electrode
		Microbial sensor	T.brassicae	O_2 electrode
	Uricate	Enzyme sensor	Uricase	O_2 electrode
		Enzyme thermistor	Uricase	Thermistor
Lipid	Choleste-rol	Enzyme sensor	Cholesterolesterase	Pt electrode
			Cholesteroloxidase	H_2O_2electrode
		Enzyme thermistor	Cholesteroloxidase	Thermistor
	Phosphaty dilcho-line	Enzyme sensor	Phospholipase	H_2O_2electrode
			Choline oxidase	H_2O_2electrode
Anti-biotics	Penicil-lin	Enzyme sensor	Penicillinase	H^+ electrode
		Enzyme thermistor	Penicillinase	Thermistor
	Cepharose-pollin	Microbial sensor	C.freundii	H^+ electrode
		Enzyme thermistor	Cepharosepollinase	Thermistor
Urea		Enzyme sensor	Urease	H^+, NH_3, CO_2 electrode
		Enzyme thermistor	Urease	Thermistor
ATP		Enzyme thermistor	Hexokinase	Thermistor

photodiode to form an enzyme photodiode sensitive to hydrogen peroxide. A glucose sensor was constructed by incorporating glucose oxidase together with peroxidase onto a Si photodiode.

Enzyme Optrode: The term optrode, formed by combining "optical" and "electrode" emphasizes that the use of optical sensors is very similar to that of electrodes (19). The devices involve a molecular recognition part on the end of a fiber optic. In operation, interactions with the determinant lead to a change in optical properties of the molecular recognition part, which is probed and detected through the fiber optic. Depending on the particular device, the optical property measured can be absorbance, reflectance, luminescense, or something else. Several enzyme optorodes have been developed.

Table 2. Electrochemical Label Immunoassays

Label	Electroactive species	Determinant	Investigator
Catalase	Oxygen	AFP, HCG IgG, IgA, IgM	Aizawa et al.
Glucose oxidase	Oxygen	Albumin	
Peroxidase	I^-	HBs	Biotieux et al.
	F	IgG	Alexander et al.
Urease	NH_3	Albumin, c-AMP	Van Weeman et al.
Alkaline phosphatase	Phenol	α_1-Acid glycoprotein	Heineman
Ferrocene	Ferrocene	Codeine Morphine	Weber et al.
		Lidocaine	Gleria et al.
Mercury	Mercury	Estriol	Heineman et al.
DTPA-In^{3+}	In^{3+}	Albumin	Heineman et al.
$-NO_2$	$-NO_2$	Estriol	Wehmeyer et al.
Erythrocyte	$TMPA^+$	Antibody	Meyerhoff et al.

3. LABELED IMMUNOSENSORS

Enzyme Immunosensors Using Oxygen Electrode: An enzyme immunosensor is an analytical device which is dependent on the immunochemical affinity for selectivity and on the chemical amplification of a labeling enzyme for sensitivity (20-29) (Table 2).

In case that catalase, which catalyzes the evolution of oxygen from hydrogen peroxide is a labeling enzyme for an antigen, the enzyme immunosensor is constructed by assembling an antibody-bound membrane and an oxygen sensing electrode. In heterogeneous enzyme immunoassay, the labeling enzyme is measured by amperometry with the oxygen sensing device. The enzyme immunosensor requires an extremely short time for measuring the labeling enzyme. Consequently, rapid and highly sensitive enzyme immunoassay may be accomplished with the enzyme immunosensor (26-29).

The enzyme immunosensor is prepared by attaching the antibody-bound membrane to a Clark-type oxygen electrode which has an oxygen permeable plastic (e.g. Teflon) membrane on the cathode surface and responds sensitively and rapidly to oxygen.

A sheet of anti-α-fetoprotein (AFP) antibody membrane, attached to the sensor was placed in contact with 260 "catalase units" of catalase-labeled AFP at pH 7 and 37°C for 1 h. After thorough washing, the sensor was placed in phosphate buffer. When it gave a steady state current from dissolved oxygen, 100 μl of 3% hydrogen peroxide was injected. The sensor responded very rapidly to the generation of oxygen (29). A steady state current was obtained within 30 s.

A calibration curve for competitive enzyme immunoassay for AFP with the sensor shows that AFP can be determined in the range $5x10^{-11}$ - $5x10^{-8}$ $g.ml^{-1}$. The standard deviation for 25 assays of 10^{-9} g of AFP was 15%. All the assays were performed with different antibody membranes. It is noted that the enzyme immunosensor provides very rapid and sensitive enzyme-linked immunosorbent assay (ELISA) as compared with photometric ELISA.

Ochratoxin A (OTA), a secondary metabolite of Aspergillus ochraceus, Penicillium viridicatum and strains of some other species of both genera, causes nephropathy in animals and most probably in man. Recently also carcinogenicity has been demonstrated in feeding experiments. Its occurence specially in grains seems to be prevailing in not only European and North American countries but also in Asian countries. OTA has most commonly been determined by chemical analysis which requires a tedious separation from a sample. Biossay and radioimmunoassay have also been conducted with problems in operation. A simplified and rapid analysis of OTA is urged to be developed. An enzyme immunosensor for OTA has recently been developed (30). The toxin sensor consists of an amperometric oxygen electrode and an OTA-bound membrane. Catalase-labeled OTA, which is added at a fixed amount to a sample solution, may competitively react with membrane-bound OTA and free OTA to be determined. The sensor is then assayed for amperometric determination of catalase activity. OTA was selectively and sensitively determined by measuring the sensor output. Competitive immunoassay of OTA was carried out using the immunosensor for OTA. OTA was determined under a constant concentration of catalase-labeled antibody. Sensor response was obtained for each concentration of free OTA, where the experiment was carried out in the PBS buffer at 35°C. The change of sensor output decreased with the addition of free OTA to be determined. The standard curve shifted to the lower concentration range with a decrease in antibody concentration. The minimum detection limited was in the order of 10^{-10} $g.ml^{-1}$, when the concentration of labeled antibody was fixed at 20 $\mu g.ml^{-1}$.

Bioaffinity Sensors with Preformed Metastable Ligand-Receptor Complex: A new biosensor has been designed on the basis of bioaffinity difference between two ligands, i.e., one a determinant and the other an analogue compound in a given binding reaction. In general an analogue compound shows lower affinity to the binding protein than a determinant does. Therefore, one can expect the following displacement reaction when a membrane-bound analogue compound complexed with its binding protein is exposed to a determinant molecule. The binding protein is displaced from the membranebound analogue molecule. The displacement is supposed to depend on the bioaffinity difference of an analogue molecule and a determinant as well as the determinant concentration. The determinant concentration may be easily measured by detecting the residual molecular complex which remains on the membrane surface. High sensitivity can be attained by chemical amplification by the usage of an enzyme catalyst as a label. A biosensor based on the above principle may be termed "Bioaffinity sensor" (31-33). A bioaffinity sensor for thyroxine (3,5,3',5'-tetraiodothyroxine: T_4), a thyroid hormone, is composed of membrane-bound T_4 and enzyme-labeled antibody, i.e., thyroxine is chemically immobilized on a membrane. The membrane-bound T_4 is undergone immunochemical reaction with enzyme-labeled anti-T_4 antibody to form an immunocomplex. Membrane-bound T_4 has less affinity to the antibody than free T_4 in solution. Attachment of the membrane where antigen-antibody complex is formed on the surface of a galvanic-type oxygen electrode results in a bioaffinity sensor for T_4. When the sensor is immersed in a solution containing free T_4 to be determined, the antigen-antibody complex will be dissociated upon exposure to T_4. The dissociation may be enhanced with an increase in the analyte concentration. The released enzyme-labeled antibody will then form a stable complex with T_4. Consequently, T_4 can be determined by measuring the enzyme activity of enzyme-labeled antibody on the T_4 membrane.

In case of a bioaffinity sensor for bovine insulin, porcine insulin is used as the analogue compound. A porcine insulin-bound plate is undergone immunoreaction with peroxidase-labeled anti-bovine insulin antibody to form an immunocomplex. When the plate on which surface the immunocomplex is formed is immersed on a solution containing free bovine insulin to be determined, the complex may be dissociated. The dissociated enzyme-labeled antibody then forms a stable complex with bovine insulin in a solution. Insulin is thus determined by measuring peroxidase retained on an immunoplate. The luminol-H_2O_2 system is employed to detect peroxidase activity. The emitted light is transferred to a photomultiplier through an optofiber.

Electrochemical Luminescence-Based Optical Immunosensor: The authors aimed at employing aromatic hydrocarbon such as pyrene as an electroactive label for immunoassay. It was found, however, that the electrochemical luminescence of the label was extremely sensitive to immunochemical reaction. The new findings have lead us to establish a novel immunosensor for homogeneous immunoassays (34). The electrochemical luminescence immunoassay is based on the following principle. The antigen (human serum Albumin) labeled with an aromatic hydrocarbon (aminopyrene) was prepared. The pyrene-labeled HSA emits luminescence by electrochemical technique, while the labeled HSA complexed with antibody generates less luminescence depending on the antibody concentration. This may probably be caused by the steric hindrance to the access of the aromatic hydrocarbon to the electrochemical luminescence (ECL) detection. In the further investigation, luminol was found to exhibit electrochemical luminescence in an aqueous solution. Luminol was also employed as an electroactive label for immunoassay. In addition, an optical fiber electrode has been developed to improve the ECL measurement. The optical fiber electrode is an optically transparent electrode which is fabricated on the top of an optical fiber. Any electrochemically induced optical changes can be sensitively detected with the optical fiber electrode. Such a device should pave the road to optical chemical sensors and biosensors.

REFERENCES

1. Updike SJ, Higgins GP. Nature 1967; 214: 986.
2. Guilbault GG. Handbook of Enzymatic Methods of Analysis. Marcel Dekker, New York, 1976.
3. Aizawa M. In: Seiyama T et al., eds. Chemical Sensors. Kodansha-Elsevier 1983; p 683.
4. Suzuki S. Biosensors. Kodansha, Tokyo, 1983.
5. Koyama M, Sato Y, Aizawa M, Suzuki S. Anal Chim Acta 1980; 116: 307.
6. Shinohara H, Chiba T, Aizawa M. Sensors and Actuators (in press).
7. Cass AEG, Davis G, Francis GD et al. Anal Chem 1984; 56: 667.
8. Shichiri M, Kawamori R, Goriya Y et al. Diabetologia 1983; 24: 179.
9. Bergveld P. IEEE Trans Biomed Eng 1970; BME-17 (1): 70.
10. Matsuo T, Wise KD, IEEE Trans Biomed Eng 1974; BME-21 (6): 485.
11. Moss SD, Janata J, Johnson CC. Anal Chem 1975; 47: 2238.
12. Caras S, Janata J. Anal Chem 1980; 52: 1935.
13. Miyahara Y, Moriizumi T, Shiokawa S, Matsuoka H, Karube I, Suzuki S. J Chem Soc Japan, 1983; 1983: 823.
14. Hanazato Y, Shiono S. In: Seiyama T et al., eds. Chemical Sensors. Kodansha-Elsevier 1983: p 513.
15. Kuriyama T, Kimura J, Kawana Y. 16th Intl Conf Solid State Devices Materials. Kobe 1984; LC-12-3.
16. Lundstrom I. Sensors and Actuators 1981; 1: 403.

17. Danielsson B, Winquist F, Mosbach K. In: Seiyama T et al., eds. Chemical Sensors. Kodansha-Elsevier 1983; p 507.
18. Aizawa M, Ikariyama Y, Kuno H. Anal Lett 1984; 17(B7): 555.
19. Seitz WR. Anal Chem 1984; 56: 16A.
20. Aizawa M, Kato S, Suzuki S. J Membrane Sci 1977; 2: 125.
21. Aizawa M, Suzuki S, Nagamura Y, Shinohara R, Ishiguro I. Chem Lett 1977; 1977: 779.
22. Aizawa M, Kato S, Suzuki S, Nagamura Y, Ishiguro I. Koubunshi Ronbunshu 1977; 34: 813.
23. Aizawa M, Suzuki S, Nagamura Y, Shinohara R, Shiguro I. J Solid-Phase Biochem 1979; 4: 25.
24. Aizawa M, Kato S, Suzuki S. J Membrane Sci 1980; 7: 1.
25. Aizawa M. In: Seiyama T et al., eds. Chemical Sensors. Kodansha-Elsevier 1983; p 683.
26. Aizawa M, Morioka A, Matsuoka H et al. J Solid-Phase Biochem 1976; 1: 319.
27. Aizawa M, Morioka A, Suzuki S. J Membrane Sci 1978; 3: 251.
28. Aizawa M, Morioka A, Suzuki S, Nagakura Y. Anal Biochem 1979; 94: 22.
29. Aizawa M, Morioka A, Suzuki S, Anal Chim Acta 1980; 115: 61.
30. Hongyo K, Uda T, Ueno H, Aizawa M, Sano S, Shinohara H. (submitted).
31. Ikariyama Y, Aizawa M. Proc 2nd Sensor Symp 1982; p 97.
32. Ikariyama Y, Aizawa M. Proc 3rd Sensor Symp 1983; p 17.
33. Ikarayama Y, Aizawa M. Anal Chem 1985; 57: 496.
34. Ikariyama Y, Kunoh H, Aizawa M. Biochem Boiphys Res Commun 1985; 128: 987.

CHAPTER 5
AUTOIMMUNE ENDOCRINOLOGICAL AND NEUROLOGICAL DISORDERS

ANTIBODIES TO THE THYROTROPIN (TSH) RECEPTOR AND AUTOIMMUNE THYROID DISEASE

L.D. Kohn, J. Chan, F. Alvarez, S. Doi, A.D. Kohn, S. Aloj, C. Rotella, O. Isozaki, K. Tahara, S. Shifrin, R. Zarrilli, and E.F. Grollman

Section on Cell Regulation, Laboratory of Biochemistry and Metabolism, National Institute of Arthritis, Diabetes, and Digestive and Kidney Diseases, National Institutes of Health, Bethesda, MD 20892, USA

INTRODUCTION

Graves' disease is an autoimmune disorder of the thyroid characterized by a) a diffusely enlarged thyroid gland (goiter); b) symptoms of hyperthyroidism; and, on occasion, c) two connective tissue complications: exophthalmos and pretibial myxedema. The weight of evidence suggests that Graves' disease is a disturbance of the immune system which results in the entrance, into the sera, of thyroid stimulating autoantibodies (TSAbs) which stimulate the thyroid and induce the hyperthyroid state. The idea evolved that TSAbs were related to the thyrotropin (TSH) receptor since IgG preparations from the sera of many Graves' patients could stimulate thyroid adenylate cyclase activity as did TSH and could inhibit TSH binding (1, 2).

The concept that all the signs and symptoms of Graves' disease reflected the action of autoantibodies to the TSH receptor developed, however, several problems (3). First, when numerous laboratories simultaneously evaluated Graves' IgG preparations for their thyroid stimulatory (TSAb) and TSH binding inhibition (TBIAb) activities, at least one-third of the preparations existed with only one activity and no correlation existed between the levels of activity when both were present simultaneously. Second, there appeared to be a distinct class of antibodies in Graves' sera which stimulated thyroid cells to grow and form goiters which were not TSAbs (4, 5). Third, evidence in experimental models which suggested that the TSH receptor might be involved in exophthalmos (6) were argued to be nonapplicable to the human situation. Last, no evidence existed that TSAbs had any relationship to pretibial myxedema (7, 8).

In sum, clinical data led to a belief that antibodies to thyroid membranes existed in the sera of Graves' patients and were important in the pathogenesis of the disease. It was, however, less clear that these antibodies were all directed against the TSH receptor. Resolution of these discrepancies required a greater understanding of the characteristics of individual antibodies within the spectra present in a single patient's serum. Resolution also required knowledge of the potential role of

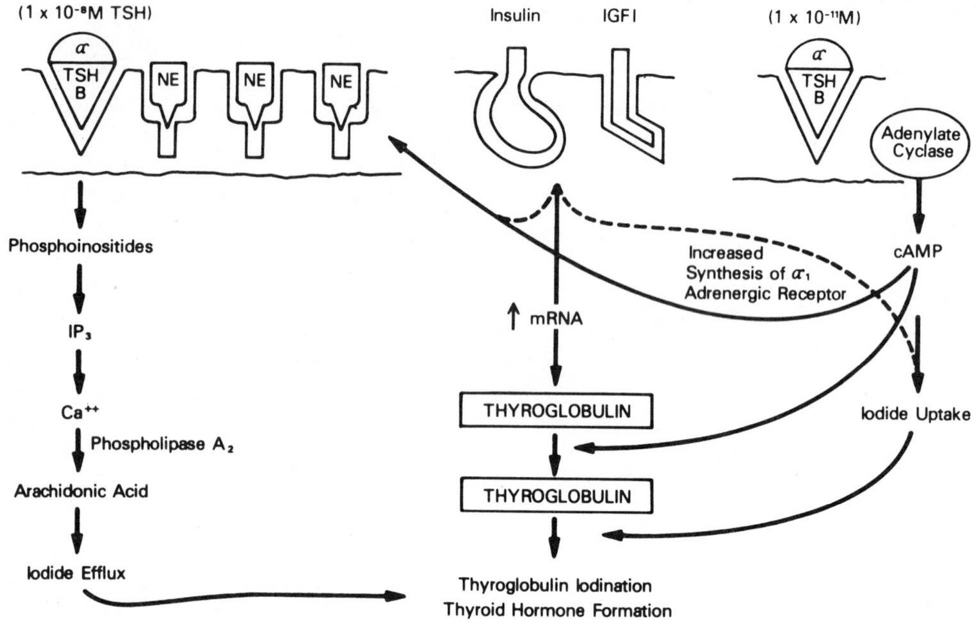

Fig. 1. Proposed model of the TSH receptor, composed of glycoprotein and ganglioside component. After the TSH β-subunit interacts with the receptor, the hormone changes its conformation and the α-subunit is brought into the bilayer where it interacts with other membrane components. The end result includes a change in organization of the membrane bilayer, a change in the transmembrane electrochemical gradient, changes in lipid turnover, and the expression of other receptors in addition to the initiation of growth and cAMP signals. The β-subunit of TSH is presumed to carry the primary determinants recognized by the glycoprotein receptor component but in no way does the model exclude an α-subunit contribution either direct or by conformational perturbations of β. The high affinity TSH binding site on the glycoprotein receptor component is also the site where autoantibodies which inhibit TSH binding (TBIAbs) interact. The stimulatory antibodies (TSAbs) interact primarily with the gangliosides. The interaction of a TSAb with the ganglioside results in stimulation of adenylate cyclase activity. The ability of a TBIAb to block TSH stimulation of adenylate cyclase action, together with the observation that there is competitive agonism of TSAbs with respect to TSH, establishes that TSH must interact with the ganglioside to stimulate adenylate cyclase activity. TBIAbs are not necessarily inactive biologically, rather they can, like TSH, stimulate functional responses linked to a phosphoinositide/Ca^{++}/phosphoinositol signal system. "Mixed" antibodies interact with both components of the TSH receptor. Growth of thyroid cells can be derive from either signal system but is maximal if both are operative simultaneously.

different determinants on the TSH receptor for TSH binding, thyroid cell growth, adenylate cyclase stimulation, pretibial myxedema, or exophthalmos.

THE THYROTROPIN RECEPTOR: A COMPLEX STRUCTURE OF MORE THAN ONE MEMBRANE COMPONENT

Perhaps the problem of the relationship of the autoantibodies to the TSH receptor might have been resolved more rapidly if there had been a clear idea of the structure of the TSH receptor. Two membrane components were identified which bound TSH with specificity (9, 10, 11) (Figure 1). The first, a membrane glycoprotein, was generally agreed to be that component of the TSH receptor which bound TSH to the cell with high affinity; its loss, for example when cells were exposed to trypsin, resulted in a loss in both TSH binding and TSH stimulated functions. The role of the second component, a membrane ganglioside, was more controversial. The evidence for its physiologic importance derived mainly from the observations that (i) higher order gangliosides, with the ability to interact with TSH, were absent in a thyroid tumor which had lost its functional TSH receptor; (ii) resynthesis or reconstitution of gangliosides in membranes from this tumor could cause both a return of TSH binding and the ability of TSH to stimulate adenylate cyclase activity; and (iii) this reconstitution was effected by a thyroid specific ganglioside. The ganglioside was suggested to modulate the apparent specificity, affinity, and capacity of the glycoprotein receptor component and induce a conformational change in the hormone believed necessary for subsequent message transmission. It was suggested to couple the high affinity binding site to the adenylate cyclase signal system by acting as an emulsifying agent to allow bound hormone to intercalate within the lipid bilayer and interact with other membrane components within the hydrophobic environment of that bilayer. Since the relationship of either binding component to autoantibodies in Graves' patients was not clear, a monoclonal antibody approach was undertaken (12, 13, 14, 15, 16, 17).

MONOCLONAL ANTIBODIES CONFIRM THE TWO COMPONENT TSH RECEPTOR MODEL AND RELATE THE MULTICOMPONENT RECEPTOR STRUCTURE TO AUTOANTIBODIES IN THE SERA OF GRAVES' PATIENTS

The monoclonal antibody approach merged two separate sets of studies. In the first, crude solubilized thyroid membrane preparations were injected into mice followed by spleen cell fusion with non-IgG producing mouse myeloma cells. In the second, lymphocytes from patients with active Graves' disease were fused with a non-IgG secreting mouse myeloma cell line, since each Graves' autoantibody could be presumed to be the product of a B cell present in these patients. In each case hybridomas secreting antibodies capable of binding to thyroid membranes in the absence but not in the presence of TSH were isolated. The following criteria were then used to identify anti-TSH receptor producing clones. (i) The antibody had to inhibit TSH binding to thyroid membranes, or, conversely, be itself prevented from binding to thyroid membranes by TSH. (ii) Binding inhibition had to be specific and had to be competitive as opposed to noncompetitive or uncompetitive. (iii) The antibody had to competitively inhibit TSH-stimulated thyroid functions, i.e., adenylate cyclase activity, iodide uptake, or thyroid hormone release, or, conversely, had to mimic TSH activity and exhibit properties of competitive agonism in all the functional assays. Over 30 antibodies satisfying the above criteria have now been identified; using these criteria, the antibodies can be broadly grouped into three classes: inhibitors, stimulators, and mixed antibodies (Table 1).

Table 1. Representative monoclonal antibodies to the TSH receptor

Clone No.	TSH receptor source	Primary classification	Growth activity (degree)	Exophthalmogenic activity	Pretibial myxedema activity
13D11	bovine	inhibitor	no	yes	no
11E8	bovine	inhibitor	no	yes	no
59C9	human	inhibitor	no	no	no
60F5	human	inhibitor	yes(++)	no	--
129H8[a]	human	inhibitor	yes(+)	weak	no
122G3[a]	human	inhibitor	no	weak	no
22A6	bovine	stimulator	yes(+)	no	no
206H3[a]	human	stimulator	yes(++)	no	--
307H6[a,b]	human	stimulator	yes(++)	yes	weak
304D3[a,b]	human	stimulator	yes(+)	no	no
308L2[a,b]	human	stimulator	yes(+)	no	no
410F9[a]	human	stimulator	yes(+)	no	no
52A8	human	mixed	yes(++++)	no	yes
208F7[a]	human	mixed	yes(+++)	no	yes

[a]Heterohybridomas.
[b]From the same patient who had exophthalmos.
[c]Inhibitor = TBIAb; Stimulator = TSAb.

The first group, termed "inhibitor" antibodies, are representative of thyrotropin binding inhibiting antibodies (TBIIs or TBIAbs) in patients. These antibodies competitively inhibit TSH binding to thyroid membrane preparations but do not stimulate adenylate cyclase activity. Rather, they competitively inhibit TSH stimulated adenylate cyclase activity; TSH stimulated iodide uptake by thyroid cells; or TSH stimulated thyroid hormone release by mice in vivo. These antibodies inhibit TSH binding to the glycoprotein component of the TSH receptor but do not interact with thyroid gangliosides. We can thus conclude that (i) the glycoprotein component of the TSH receptor (Figure 1) is the initial site of TSH binding to the thyroid cell; (ii) TBIAbs are antibodies to this component of the receptor; and (iii) a TBIAb is not a thyroid stimulating antibody (TSAb).

The second group of antibodies (Table 1), the stimulators, are representative of the thyroid stimulating antibodies (TSAbs) in patients. They stimulate adenylate cyclase activity in thyroid cells; stimulate iodide uptake; and cause release of thyroid hormones in a mouse bioassay (Table 1). They are competitive agonists with respect to TSH. Although these antibodies bind to thyroid membranes and are specifically inhibited in this binding by TSH, they are weak inhibitors of ^{125}I-TSH binding and do not significantly inhibit ^{125}I-TSH binding to the glycoprotein component of the TSH receptor. Instead they react with thyroid ganglioside preparations; this reaction is specific in that they do not react with brain ganglioside preparations. We can thus conclude the following. First, TSAbs interact with a thyroid ganglioside (Figure 1) and not the high affinity TSH binding site on the glycoprotein component of the receptor. Second, the ganglioside is not the initial binding site for TSH on the thyroid cell membrane; however, TSH must interact with a ganglioside on the thyroid membrane in order to stimulate adenylate cyclase activity. TSAbs, which interact with the gangliosides, would otherwise not exhibit competitive agonism with respect to TSH. Third, the ganglioside must be a component of the TSH receptor since TSH and TSAbs are competitive agonists. Fourth, the TBIAbs and TSAbs are both TSH receptor autoantibodies but are directed at different domains of the TSH receptor. This is confirmed in mixing experiments (18, 19) wherein a TBIAb, 11E8, can be shown to inhibit TSH activity but not the activity of the 22A6 or 307H6 TSAb monoclonals or the activity of Graves' thyroid stimulating IgG.

A third group of monoclonal antibodies to the TSH receptor are termed "mixed" antibodies (Table 1). They can be distinguished since they both stimulate in the cAMP, iodide uptake or mouse bioassays and competitively inhibit of ^{125}I-TSH binding. Their monoclonal nature is established by repetitive subcloning and typing. They interact with both the glycoprotein and ganglioside components of the TSH receptor. They are different from a TSAb in that a TBIAb, 11E8, can partially inhibit the activity of "mixed" antibody, 208F7, in mixing experiments. The existence of these antibodies initially suggested that the ganglioside and glycoprotein receptor components might exist as a tight complex. As noted below they subsequently were recognized to be important in thyroid cell growth and appear to recognize a portion of the insulin-like growth factor (IGF) as well as TSH receptor.

An important clinical implication of the mixing experiment using monoclonal antibodies is as follows. If a Graves' patient has both an "inhibiting", "stimulating", or "mixed" antibody present in his or her serum, the phenotypic expression will be a hyperthyroid state. This is evident since the TBIAb will not significantly inhibit a TSAb or mixed antibody despite its capacity to inhibit TSH.

GROWTH AUTOANTIBODIES IN GRAVES' DISEASE

As noted in the introduction, recent studies indicated that Graves' patients had antibodies in their sera which could promote growth but which did not bear a simple relationship to autoantibodies related to thyroid hyperfunction. The existence of a panel of monoclonal antibodies to the TSH receptor allowed the asking of several simple questions. What was the activity of the TSH receptor monoclonals on growth? Were there TSH receptor antibodies which could stimulate growth but not cause hyperfunction? If such antibodies existed, what was their relation to TSH receptor structure and transducing signals?

The answer to this question depended on an assay using a continuous line of functioning rat thyroid cells whose growth had been shown to be dependent on the presence of TSH and on measurements of cell number or radiolabeled thymidine incorporation (US patent 4,608,341, 1986). When the monocloncal TSH receptor antibodies were tested it was found that all TSAbs stimulated the growth activity of FRTL-5 rat thyroid cells, whether measured as cell number or as [^{3}H]-thymidine uptake (Table 1). All "mixed" antibodies also stimulated growth activity but had 3-fold or higher ratios of growth to adenylate cyclase stimulatory activity than the TSAbs, when normalized to equivalence with respect to adenylate cyclase activity. "Inhibitor" antibodies could, surprisingly, be either inhibitors of TSH stimulated growth as well as adenylate cyclase activity (122G3, 59C9) or stimulators of growth but inhibitors of TSH binding and TSH stimulated adenylate cyclase activity (129H8, 60F5). Although these data clearly state that the same structural TSH receptor is used to signal growth as well as adenylate cyclase activity, differences in signal coupling and differences in the linkage of the receptor components to the signals seemed likely.

That this was true was suggested by the following observations. TSH growth activity could be partially inhibited by indomethacin; indomethacin is a cyclooxygenase inhibitor which limits arachidonic acid processing to fibroblast assays and originated from a patient with exophthalmos who also had TSAbs without fibroblast activity. The 307H6 data clearly established that an antibody to the thyrotropin receptor could exist which also stimulated fibroblasts; separate studies using patient IgGs established that the fibroblast assay was a good measure of exophthalmos. Thus approximately 90% of Graves' patients with exophthalmos were positive in this assay whereas patients with thyroid disease but no exophthalmos were uniformly negative in the assay.

The fibroblast stimulating activity of 307H6 was not dependent on a cAMP signal, since similar, or up to 100 times higher antibody concentrations did not cause cAMP elevations in the fibroblasts and TSH itself was inactive in both assays.

Consistent with this observation, some monoclonal antibodies to the TSH receptor which were in the "inhibitory" group were found to have potent activities as stimulators of collagen biosynthesis in fibroblasts. Thus, in the same assay, 11E8 and 13D11 (Table 1), two mouse monoclonal antibodies to the bovine TSH receptor were as active in stimulating collagen incorporation of [^{3}H]proline as the most potent IgG preparations from exophthalmos patients but were active at ~1000- to 10,000-fold lower IgG concentrations. It seemed possible to conclude (i) that the fibroblast collagen biosynthesis assay was a valid means of measuring exophthalmogenic autoantibodies whether TSAb positive or not and (ii) that selected populations of autoantibodies to the TSH receptor did appear able to induce this exophthalmos-linked activity.

The antibodies clearly reflected a common thyroidal and connective tissue antigen. Thus, preadsorption of the 11E8 antibody on thyroid membranes resulted in the loss of fibroblast activity whereas preadsorption on liver or kidney membranes had no such affect. Nevertheless TSH was not only inactive in the fibroblast assay, it also did not inhibit the activity of these monoclonals in the fibroblast assay. This was of interest in several respects. First it fit with data in experimental exophthalmos studies wherein TSH and the antibody were additive activators not antagonists. Second, it indicated that more than the TSH receptor determinant was necessary for the fibroblast activity. Since exophthalmos has been linked to cellular rather than humoral immunity (26), one possible explanation was that a T-cell-like recognition phenomenon existed wherein a tissue specific antigen determinant was also involved in the recognition phenomenon.

To further ascertain the specificity of this phenomenon, we have tested IgGs prepared from the sera of a number of patients with different autoimmune diseases in which the thyroid system was not involved, for example, patients with myasthenia gravis, rheumatoid arthritis, or systemic lupus. All were inactive in the fibroblast collagen synthesis system, with one exception: IgGs prepared from two of four patients with type B insulin resistance, a form of insulin resistance due to the development of autoantibodies to the insulin receptor. The activity of the two positive antibodies was more than additive with the 11E8 and 307H6 activities they were also antibodies capable of interacting only weakly with the insulin receptor, and more potently with the IGF-I receptor. The two negative antibodies reacted only with the insulin receptor. Neither insulin nor IGF-I alone could enhance collagen biosynthesis in the fibroblasts nor could they, alone, inhibit the activity of the 11E8 monoclonal to the TSH receptor, thereby raising the issue that TSH and IGF-I, are both required for the expression of the fibroblast activity.

Pretibial Myxedema - Since the data presented above provided a new approach to understanding the pathogenesis of exophthalmos in Graves' derivatives such as prostaglandins. Indomethacin partially inhibited the growth activity of the mixed antibodies 208F7 and 52A8; it had no affect on the growth action of TSAbs such as 307H6; but it completely inhibited the growth activity of the 129H8 and 60F5 TBIAbs. In short, it appeared that the TSH modulated growth activity of the thyroid cell involved both adenylate cyclase and Ca^{++}/phospholipid modulation signals. Further, phospholipid signalling seemed to be linked more to the glycoprotein receptor component; adenylate cyclase action to the ganglioside component.

Several clinical implications emerge from these data. First, thyroid growth promoting activity and thyroid adenylate cyclase stimulatory activity should not bear a simple relationship since mixtures of these antibodies exist in any one patient. Second, it is possible, based on the monoclonal antibody studies, that thyroid stimulation in Graves' disease might result from a small goiter potently stimulated by a 307H6 type antibody to actively release thyroid hormones. Alternatively Graves' might be associated with a large goiter, caused by 129H8 antibody, whose release of excess thyroid hormone reflects the existence of an excess of responsive tissue. It is also possible that the presence of (i) a mixed antibody or (ii) a mixture of a TBIAb antibody which is a growth stimulator and a TSAb, such as 307H6, would result in two independent and potent stimulators being present simultaneously and, most likely, the most severe cases of thyroid stimulation. In the last mixture, it must be remembered (see above) that the 129H8 TBIAb would not inhibit the TSAb; the presence of the two autoantibodies would thus be complementary in their pathogenic action.

Recent studies indicate that insulin and insulin-like growth factor I (IGF-I) are necessary for TSH to exert its growth action. The action of insulin and IGF-I, with respect to growth, has been associated with the tyrosine kinase activity and/or phosphorylation of the β-subunit of the insulin and IGF-I receptors (20). This association rests, on the evidence that there is sequence homology and immune cross-reactivity between viral oncogene products and the β–subunits of the insulin and IGF-I receptors. Recent studies suggest that the mixed monoclonal antibodies directed against the TSH receptor can also immunoprecipitate the phosphorylated β-subunit of the IGF-I receptor (21). This result suggests that the interrelatedness of the growth activity induced by the insulin, IGF-I, and TSH receptors has its counterpart in a close association of the structural components of the two receptors.

MONOCLONAL ANTIBODIES TO THE TSH RECEPTOR, NONTHYROIDAL TSH RECEPTORS, AND THE CONNECTIVE TISSUE COMPLICATIONS OF GRAVES' DISEASE

Exophthalmos - Nonthyroidal TSH "receptors" have been argued to be important in the pathogenesis of exophthalmos and the connective tissue complications of Graves' disease (22, 23, 24). Thus, in the 1970's, studies of exophthalmos in experimental animal models suggested that there was a TSH receptor in retroorbital tissues and that exophthalmos involved expression of the TSH receptor in the presence of an abnormal serum gamma-globulin found in exophthalmos patients. Until the monoclonal antibody approach cited above, there was no simple approach which could be reasonably anticipated to prove or disprove this point.

Monocloncal antibodies derived from the lymphocytes of patients with Graves' disease without exophthalmos were compared with those of patients with Graves' disease plus exophthalmos. One stimulating antibody (307H6), derived from the exophthalmos group, was found to stimulate collagen biosynthesis in human skin fibroblasts (25); this antibody had all the characteristics of a stimulator antibody to the TSH receptor defined above. It was only one of over a dozen TSAb monoclonals which were active in the patients the same monoclonal approach was applied to a second connective tissue complication of Graves' patients, pretibial myxedema (27).

A glucosaminoglycan (GAG) is an acidic polysaccharide chain which consists of repeating disaccharide (acetylated hexosamine and uronic acid) units. GAGs are usually covalently bound to a protein backbone; in that state, they are termed proteoglycans (28, 29). Proteoglycans are localized in matrices between and on the surface of both connective tissue and epithelial cells. Elevated levels of proteoglycans have been reported in the affected skin of patients with pretibial myxedema. Evidence also exists that lymphocytes from patients with pretibial myxedema, but not their IgGs, can increase glycosaminoglycan production of human fibroblasts.

This last evidence was consistent with our own observations (30). Thus, IgGs from Graves' patients, either with or without exophthalmos and pretibial myxedema, were not able to significantly affect glycosaminoglycan production of skin fibroblasts. We, fortunately, however turned our attention to the continuous line of differentiated rat thyroid cells, FRTL-5, to establish if these cells were able to synthesize proteoglycans and to see if this biosynthetic activity could be stimulated by TSH, by a select group of monoclonal antibodies directed against the TSH receptor, or by polyclonal IgG preparations from Graves' patients.

FRTL-5 thyroid cells can incorporate [^{3}H]-glycosamine into proteoglycans. TSH increases the biosynthesis of proteoglycans associated with the cell layer (10-fold); the TSH effect is mediated by cAMP; however, the TSH effect on proteoglycan synthesis requires the presence of insulin or IGF-I. In the absence of TSH, and insulin and in the presence of only 0.2% serum, IGF-I was, however, noted to stimulate proteoglycan biosynthesis. Under the same conditions, fifteen of fifteen patients with Graves' disease and pretibial myxedema, independent of the presence of exophthalmos, significantly increased proteoglycan biosynthesis in FRTL-5 cells by comparison to IgG from normal subjects. The same IgG preparations had no effect on proteoglycan synthesis in fibroblasts. Polyclonal IgGs, purified from the serum of Graves' patients without pretibial myxedema, as well as from normal subjects or patients with non toxic diffuse goiter, were not able to significantly increase proteoglycan biosynthesis in FRTL-5 cells. IgGs from patients with Graves' disease but without pretibial myxedema, behaved on average as did normal subjects in this assay despite the fact that 80-90% had TSAb activity.

A number of monoclonal TSH receptor autoantibodies linked to the pathogenesis of the thyroid abnormalities of Graves' disease were tested for their ability to stimulate proteoglycan biosynthesis in FRTL-5 cells. Two antibodies were stimulatory 52A8 and 208F7 (Table 1). Many antibodies had no effect (Table 1). The active antibodies, 52A8 and 208F7, fit a single grouping: "mixed" antibodies. These are also the most potent growth stimulators of FRTL-5 thyroid cells and appear to interact with both TSH and IGF-I receptor determinants.

In sum, there is good evidence which links pretibial myxedema to the activity of a select group of monoclonal antibodies to the TSH receptor which appear to also react with the IGF-I receptor. They act by a noncAMP mechanism and are distinct antibodies from those triggering ophthalmopathy, despite the fact both are subgroups of TSH receptor autoantibodies which circulate in the sera of patients with Graves' disease.

ANTIBODIES TO THE TSH RECEPTOR AND THE FUNCTION OF THE THYROID CELL

The prime function of the thyroid cell is to produce thyroid hormones. Thyroid hormone formation requires the thyroid cell to take up iodide, synthesize thyroglobulin, and iodinate thyroglobulin. Evidence has accumulated that iodide uptake is a TSH regulated, cAMP mediated process wherein TSH induces the synthesis of key components of the iodide transport system (31, 32). In contrast, iodination of thyroglobulin involves changes in phosphoinositides, increases in Ca^{++} and arachidonic acid, and the formation of leukotriene intermediates (33, 34, 35, 36, 37, 38, 39, 40).

Just as in growth, the presence of insulin and IGF-I are key components of the functional events. Thus, in recent studies (41), insulin and IGF-I have been shown to be able to induce the synthesis of thyroglobulin in the absence of TSH and by their action at a mRNA level. The TSH-cAMP signal potentiates the IGF-I/insulin stimulation of thyroglobulin biosynthesis.

The monoclonal antibodies can be used to evaluate which components of the receptor structure and signal system are important for the regulation of function (Figure 2). In these studies, a pure TSAb can mimic the ability of TSH to induce the synthesis of the iodide transport system. A TBIAb, but not a TSAb, can cause iodination of thyroglobulin, i.e., the Ca^{++}/phos-

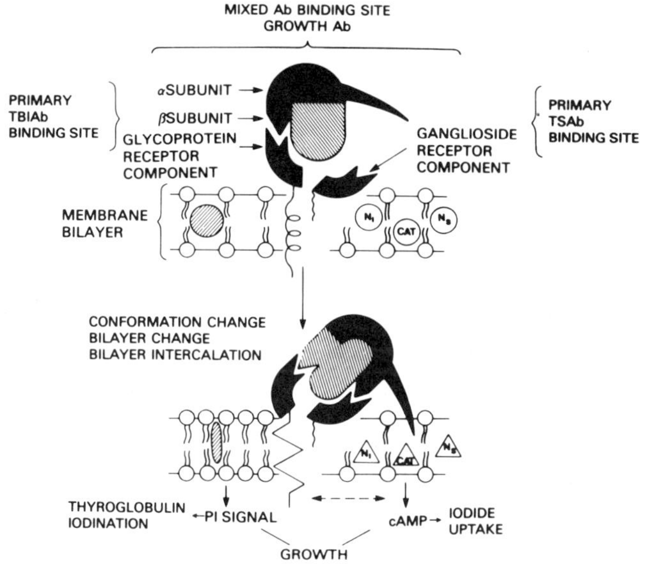

Fig. 2. Proposed model of thyroid hormone formation in rat FRTL-5 thyroid cells. TSH acts through more than one signal. Multiple other hormones are necessary for full expression of function, particularly insulin, IGF-I, and α_1-adrenergic agents. TSH, using a cAMP signal is key for concentrative iodide uptake into the cells; a phosphoinositide/Ca^{++} signal is used for iodide efflux into the follicular lumen and for iodination of thyroglobulin. α_1-Adrenergic agents can replace TSH in the latter but not the former situation. Insulin/IGF-I are prime inducers of thyroglobulin synthesis; TSH amplifies and modulates thyroglobulin synthesis and processing.

pholipid signal system seems more directly coupled to the glycoprotein receptor component. Neither a TSAb nor a TBIAb, even one of the latter which cross reacts with the insulin and IGF-I receptor β-phosphorylated subunit, can alone induce thyroglobulin biosynthesis, i.e., insulin and IGF-I are absolute requirements for this function even without a simultaneous growth effect.

In sum, as is the case for growth, a wealth of hormonal factors are necessary for full expression of thyroid hormone formation. In a population of autoantibodies, the normal complex regulation process can break down if the appropriate mixture of antibodies is not present. Further, although the antibody mechanisms are identical to TSH, each antibody elicits only a portion of the TSH action because, presumably, of its ability to interact with only a portion of the total TSH receptor. These observations suggest that the diversity of the presenting clinical syndrome in Graves' disease may reflect, at least in part, the diversity of the autoantibody population. Patients with large as opposed to small goiters and patients with greater or lesser degrees of hyperthyroidism may reflect the imbalance of the autoantibody population.

USE OF THE MONOCLONAL TSH RECEPTOR ANTIBODIES TO FURTHER DEFINE THE STRUCTURE OF THE TSH RECEPTOR

Current data suggest that the holo-TSH receptor is a ~70K protein which forms ~280K aggregates (42). TSH causes a dramatic change in receptor turnover without down regulation. Under these conditions there is protease breakdown of both the ~280K and 70K fragments. One prominent fragment is 40-50K. This has been shown in separate experiments to be the site of binding of the TSH receptor ganglioside (43).

The ganglioside from rat FRTL-5 cells is a unique minor component by comparison to the total FRTL-5 cell ganglioside pool. It separates as a disialoganglioside on DEAE, yet migrates in an area on thin layer plates which is ordinarily associated with low order gangliosides, i.e., GM_2 and GM_3. The ganglioside can inhibit the ability of a TSAb to increase cAMP levels in FRTL-5 thyroid cells; it can also reconstitute TSAb and TSH-stimulable cyclase activity in a 1-8 thyroid tumor cell previously shown to have no TSH receptor activity in association with a defect in ganglioside synthesis. The ganglioside appears to contain fucose, as previously suggested, and loses its reactivity with the TSAb after neuraminidase digestion (44).

The TSH receptor appears to have been cloned using an identification procedure involving autoantibodies to Graves' disease and the monoclonal antibodies to the TSH receptor. It has no sequence analogy to the IGF-I receptor; in humans it is on chromosome 22 but can be identified as a pseudogene on the X chromosome; and it appears to be identical in Graves' and non-Graves' thyroids. Pursuit of the cloning will hopefully explain the major unknowns of Graves' which remain. For example, why do the antibodies emerge? Why is there a female preponderance of the disease?

SUMMARY

The present report summarizes experiments with monoclonal antibodies to the TSH receptor. The data provide further insight into the TSH receptor structure and into the basis of autoimmune antibodies implicated in the pathogenesis of Graves' disease. They resolve many clinical questions and provide new approaches to enhance our understanding of autoimmune disease.

REFERENCES

1. Adams DD. Thyroid-stimulating autoantibodies. Vitam Horm 1981; 38: 119-203.
2. Orgiazzi J, Chopra IJ, Williams DE, Solomon DH. Evidence for normal thyroidal adenyl cyclase, cyclic AMP-binding and protein-kinase activities in Graves' disease. Clin Endocrinol Metab 1975; 40: 248-55.
3. Pinchera A, Fenzi GF, Macchia E, Vitti P, Monzani F, Kohn LD. Immunoglobulines thyreostimulantes. Antigens correspondants (Thyroid-stimulating immunoglobulins. The related antigens). Ann Endocrinol (Paris) 1982; 43: 520-33.
4. Drexhage HA, Botazzo GF, Bitensky L, Chayen J, Doniach D. Thyroid growth-blocking antibodies in primary myxoedema. Nature 1981; 289: 594-6.
5. Valente WA, Vitti P, Rotella CM, et al. Antibodies that promote thyroid growth. A distinct population of thyroid-stimulating autoantibodies. N Engl J Med 1983; 309: 1028-34.
6. Kohn LD, Winand RJ. Structure of an exophthalmos-producing factor derived from thyrotropin by partial pepsin digestion. J Biol Chem 1975; 250: 6503-8.
7. Cheung HS, Nicoloff JT, Kamiel MB, Spolter L, Nimni ME. Stimulation of fibroblast biosynthetic activity by serum of patients with pretibial myxedema. J Invest Dermatol 1978; 71: 12-7.
8. Jolliffe DS, Gaylarde PM, Brock AP, Sarkany I. Pretibial myxoedema: stimulation of mucopolysaccharide production of fibroblasts by serum. Brit J Dermatol 1979; 101: 557-60.
9. Kohn LD. Relationships in the structure and function of receptors for glycoprotein hormones, bacterial toxins and interferon. In: Cuatrecasas P, Greaves MF, eds. Receptors and Recognition, series A, 1978; 5: 134-212.
10. Kohn LD, Valente WA, Laccetti P, Cohen JL, Aloj SM, Grollman EF. Multicomponent structure of the thyrotropin receptor: relationship to Graves' disease. Life Sci 1983; 32: 15-30.
11. Kohn LD, Aloj SM, Tombaccini D, et al. The thyrotropin receptor. In: Litwack G. Biochemical Actions of hormones, 1985; 12: 457.
12. Yavin E, Yavin Z, Schneider MD, Kohn LD. Monoclonal antibodies to the thyrotropin receptor: implications for receptor structure and the action of autoantibodies in Graves disease. Proc Natl Acad Sci USA 1981; 78: 3180-4.
13. Valente WA, Yavin Z, Yavin E, et al. Monoclonal antibodies to the thyrotropin receptor: the identification of blocking and stimulating antibodies. J Endocrinol Invest 1982; 5: 293-301.
14. Valente WA, Vitti P, Yavin Z, et al. Monoclonal antibodies to the thyrotropin receptor: stimulating and blocking antibodies derived from the lymphocytes of patients with Graves disease. Proc Natl Acad Sci USA 1982; 79: 6680-4.
15. Kohn LD, Alvarez F, Marcocci C, et al. Monoclonal antibody studies defining the origin and properties of autoantibodies in Graves' disease. Annals NY Acad Sci 1986; 475: 157-73.
16. Eisenbarth, Haynes R. Monoclonal Antibodies: Probes for Study of Autoimmunity and Immunodeficiency. New York: Academic Press, 1983: 221.
17. Ealey PA, Kohn LD, Ekins RP, Marshall NJ. Characterization of monoclonal antibodies derived from lymphocytes from Graves' disease patients in a cytochemical bioassay for thyroid stimulators. J Clin Endrocrin Metab 1984; 58: 909-14.
18. Ealey PA, Valente WA, Ekins RP, Kohn LD, Marshall NJ. Characterization of monoclonal antibodies raised against solubilized thyrotropin receptors in a cytochemical bioassay for thyroid stimulators. Endocrinology 1985; 116: 124-31.

19. Ullrich A, Bell JR, Chen EY, et al. Human insulin receptor and its relationship to the tyrosine kinase family of oncogenes. Nature 1985; 313: 756-61.
20. Proceedings Fuji Hormone Conference 1985; 3: 159.
21. Winand RJ, Kohn LD. The binding of (^{3}H)thyrotropin and ^{3}H-labeled exophthalmogenic factor by plasma membranes of retro-orbital tissue. Proc Natl Acad Sci USA 1972; 69: 1711-5.
22. Kohn LD, Winand RJ. Experimental exophthalmos. Alterations of normal hormone-receptor interactions in the pathogenesis of a disease. Israel J Med Sci 1974; 10: 1348-63.
23. Winand RJ, Kohn LD. Stimulation of adenylate cyclase activity in retro-orbital tissue membranes by thyrotropin and an exophthalmogenic factor derived from thyrotropin. J Biol Chem 1975; 250: 6522-6.
24. Rotella CM, Zonefrati R, Toccafondi R, Valente WA, Kohn LD. Ability of monoclonal antibodies to the thyrotropin receptor to increase collagen synthesis in human fibroblasts: an assay which appears to measure exophthalmogenic immunoglobulins in Graves' sera. J Clin Endocrinol Metab 1986; 62: 357-67.
25. The Age and Orbit in Thyroid Disease. Raven Press, 1984: 103.
26. Tamai H, Hirota Y, Kasagi K, et al. The mechanism of spontaneous hypothyroidism in patients with Graves' disease after antithyroid drug treatment. J Clin Endocrinol Metab 1987; 64: 718.
27. Comper WD, Laurent TC. Physiological function of connective tissue polysaccharides. Physiologic Reviews 1978; 58: 255-315.
28. Hassel JR, Kimura JH, Hascall VC. Proteoglycan core protein families. Annu Rev Biochem 1986; 55: 539-67.
29. Weiss SJ, Philp NJ, Ambesi-Inpiombato FS, Grollman EF. Thyrotropin-stimulated iodide transport mediated by adenosine 3',5'-monophosphate and dependent on protein synthesis. Endocrinology 1984; 114: 1099-107.
30. Weiss SJ, Philp NJ, Grollman EF. Effect of thyrotropin on iodide efflux in FRTL-5 cells mediated by Ca^{2+}. Endocrinology 1984; 114: 1108-13.
31. Maayan ML, Ingbar SH. Effects of epinephrine on iodine and intermediary metabolism in isolated thyroid cells. Endocrinology 1970; 87: 588-95.
32. Corda D, Kohn LD. Thyrotropin upregulates α_1-adrenergic receptors in rat FRTL-5 thyroid cells. Proc Natl Acad Sci USA 1985; 82: 8677-80.
33. Corda D, Marcocci C, Kohn LD, Axelrod J, Luini A. Association of the changes in cytosolic Ca^{2+} and iodide efflux induced by thyrotropin and by the stimulation of alpha 1-adrenergic receptors in cultured rat thyroid cells. J Biol Chem 1985; 260: 9230-6.
34. Philp NJ, Grollman EF. Thyrotropin and norepinephrine stimulate the metabolism of phosphoinositides in FRTL-5 thyroid cells. FEBS Lett 1986; 202: 193-6.
35. Burch RM, Luini A, Mais DE, et al. α_1-adrenergic stimulation of arachidonic acid release and metabolism in a rat thyroid cell line. Mediation of cell replication by prostaglandin E2. J Biol Chem 1986; 261: 11236-41.
36. Santisteban P, Kohn LD, Di Lauro R. Thyroglobulin gene expression is regulated by insulin and insulin-like growth factor I, as well as thyrotropin, in FRTL-5 thyroid cells. J Biol Chem 1987; 262: 4048-52.
37. Bartalena L, Fenzi G, Vitti P, et al. Interaction of the thyrotropin receptor on rat FRTL-5 thyroid cells with thyrotropin and a thyrotropin-stimulating autoantibody from Graves' patients. Biochem Biophys Res Commun 1987; 143: 266-72.
38. Lacetti P, Tombaccini D, Aloj S, Grollman EF, Kohn LD. Gangliosides, the thyrotropin receptor, and autoimmune thyroid disease. In: Ledeen RW, Yu RK, Rapport MM, Suzuki K, eds. Ganglioside Structure, Function, and Biomedical Potential. Plenum Press, 1984; 355: 355-67.

PREDICTING TYPE I DIABETES

George S. Eisenbarth

Joslin Diabetes Center, Boston, Mass., USA

Accumulating data has led to the general hypothesis that Type I diabetes of the NOD mouse (non-obese-diabetic), BB rat (BioBreeding rat) and man results from chronic autoimmune beta cell destruction (1). The information supporting this hypothesis is briefly summarized in Table 1 and includes the detection of anti-islet autoantibodies prior to diabetes; presence of T lymphocytic islet infiltrates; the observation that in all three species a gene within the major histocompatibility complex is essential for the development of diabetes; the ability of multiple forms of immunotherapy to prevent diabetes in animal models and randomized placebo controlled trials of cyclosporine A in man (2) (the nephrotoxicity of cyclosporine is a major concern which prevents widespread clinical application of cyclosporine therapy of Type I diabetes); the recurrence of disease in transplanted islets which have been treated to prevent tissue rejection in animal models and similar recurrence in identical twin pancreatic transplants; and in the finding both animal models that T lymphocytes can transfer diabetes.

With the information concerning immunopathogenesis (Table 1) it is not surprising that immunologic and genetic assays are being applied to predict the development of diabetes in susceptible populations. Table 2 lists a number of immunologic assays studied to date. Populations in whom such assays are utilized include:

a. Evaluation of individuals with transient or stress related hyperglycemia (4). This occurs even in the pediatric age group and it would be useful to be able to reassure individuals not developing Type I diabetes and alert those at high risk (e.g., autoantibody positive).

b. Identification of gestational diabetics with Type I diabetes (~5 to 10% of the individuals in whom diabetes develops during pregnancy). These individuals are at risk for ketoacidosis after pregnancy even if hyperglycemia transiently resolves.

c. Evaluation of potential living related renal transplant donors to diabetic recipients. Living related donors are usually selected because of HLA identity to the family member with diabetes and thus are at increased genetic risk of diabetes though this risk is decreased by the usual long-term discordance at the time of renal failure. There is fear that if diabetes develops in an individual with a single kidney (e.g., transplant donor) there would be more rapid progression of diabetic renal disease.

Table 1. Evidence for autoimmunity in Type I diabetes

	NOD MOUSE	BB RAT	MAN
Autoantibodies	yes	yes	yes
Lymphocytic infiltrate	yes	yes	yes
MHC association	I-A beta^{NOD}	RT1-u	DQ beta 3.2 DR3
Effective immunotherapy	yes	yes	yes (? safety)
Recurrence of disease in transplants without tissue rejection	yes	yes	yes
Transfer disease with T lymphocytes	yes	yes	Not studied

d. Identification of the 5 to 10% of "clinical" Type II diabetics over age 40 who are actually Type I diabetics, will rapidly require insulin therapy and are at risk for ketoacidosis (5).
e. Population screening including relatives of Type I diabetics and eventually the general population to identify individuals with approximately an 8% risk/year of developing diabetes (6-8). Islet cell antibody positive women who are deciding when to have children may be able to complete childbearing prior to overt diabetes and its fetal risks.
f. Monitoring of immunotherapy of Type I diabetes and identification of diabetic candidates for immunotherapy.

Autoantibody identification with several available assays provide the best current predictors of the development of diabetes. In particular, the two assays which provide useful prognostic data are:

a. "Cytoplasmic islet cell" antibodies (ICA) by indirect immunofluorescence (depending on the assay format).
b. Radioassay determined insulin autoantibodies (depending on the assay format).

There are other potential important autoantigens including data that detection of autoantibodies to a 64K protein may provide useful information.

INDIRECT IMMUNOFLUORESCENCE ICA ASSAYS

Approximately 8%/year of relatives of Type I diabetics expressing "cytoplasmic" islet cell antibodies determined with several specific assays and develop overt diabetes (9). In Bottazzo and co-workers family study (Barts-Windsor) 12/24 CF-ICA positive relatives have developed diabetes in 7 years. In contrast with their standard ICA assay only one of 33 ICA positive relatives (but CF-ICA negative) developed diabetes. The latter is little different from the approximately 1/20 relatives who will develop diabetes independent of ICA knowledge (10).

Assays of islet cell antibodies which utilize frozen sections of normal pancreas vary tremendously between laboratories (11). For reasons which are not clear, results of Maclaren and co-workers standard ICA assay (FITC anti-IgG) are similar to our protein-A assay or Bottazzo's

Table 2. Immunologic assays for prediction of Type I diabetes

		Prevalence	Specific (<0.25% normals)
1.	Anti-insulin autoantibodies		
	a. Radioimmunoassay	~60%	yes
	b. ELISA assays	~40%	no
2.	"Cytoplasmic" ICA		
	a. Human pancreas		
	i. Protein A, complement fixing	~60-80%	yes
	ii. Anti-IgG ICA	70-90%	no (depending on assay)
	b. Rat pancreas		
	i. Protein A	~60%	yes
	ii. Monoclonal anti-IgG	~60%	no
	c. Anti-Surface		
	i. Rat islets	30%	no
	ii. RINm5F	30%	no
3.	Anti-64K	80%	?
4.	Anti-ganglioside	?	?

complement-fixing ICA assay. In part differences between assays are secondary to a lower sensitivity in end-point titers of the protein-A and CF-ICA assay in contrast to routine ICA assays. Thus defining a titer greater than for example 1:20 as positive for many routine FITC anti-IgG assay would probably increase the specificity of such standard ICA assays. The immunofluorescent ICA assays are limited by the supply of normal human pancreas, subjective microscopic readings and low titers of the antibodies detected. We have recently found that an assay utilizing rat pancreas can provide an alternative to human pancreas sections with modifications of our standard protein A assay (12).

In our own studies of 36 ICA+ relatives with follow-up from months to 9 years, 12 have developed overt diabetes and no one has gone a decade ICA+ without becoming diabetic. Within 3 years, 70% of ICA+ relatives have lost first phase insulin secretion on intravenous glucose tolerance testing with diabetes developing after loss of first phase insulin secretion. Overt diabetes occurs with a mean of 10 months for children and 30 months for adults after first phase insulin secretion on IVGTT is below the first percentile of normal secretion. An islet cell antibody screening program in the general population by Maclaren and co-workers indicates that similar to the family data, ICA positivity in the general population (1/400 school children are ICA positive) is predictive of Type I diabetes.

An obvious pathway is to purify and characterize the target antigen of cytoplasmic islet cell antibodies and develop either a radioassay or ELISA assay based on purified antigen. Our studies suggest that the target antigen of cytoplasmic islet cell antibodies is a glycoconjugate and most likely a ganglioside (13). Treatment of human or animal pancreas sections with periodate destroys reactivity, borohydride reduction subsequently restores reactivity similar to effects on binding of anti-ganglioside monoclonal antibody 3G5 (consistent with the oxidation and reduction of sialic acid residues). The antigen of sections is sensitive to neuraminidase (cleaves sialic acid residues), stable to acetone and sensitive to treatment with methanol or chloroform:ethanol. Chloroform:methanol extraction of whole pancreas followed by Folch partition and subsequent purification by thin layer chromatography, DEAE chromatography and C18 Sep Pak purification results in recovery of material which upon incubation with ICA positive sera blocks ICA staining as well as islet reactivity of monoclonal anti-ganglioside antibodies (but not anti-protein antibodies) (14).

Development of assays utilizing such glycolipids will probably require extensive purification and characterization of the target antigen as related glycolipids are targets for naturally occurring antibodies such as blood group antigens. To date despite considerable indirect data that the cytoplasmic islet cell antigen is a glycolipid, few assays based on direct binding of patients autoantibodies and giving a positive signal have been developed.

INSULIN AUTOANTIBODIES

Quantitation of autoantibodies to insulin provides the first radioassay providing prognostic information (15). We detect cytoplasmic islet cell antibodies in approximately 60% of individuals developing Type I diabetes and insulin autoantibodies in the same percentage of children developing diabetes. Ninety percent express one or the other autoantibody (16). There are ELISA solid phase assays for insulin autoantibodies as well as fluid phase radioimmunoassays. Standardization of these assays similar to ICA is currently a problem being addressed by the IDW committee (Immunology of Diabetes Workshop) with a recent meeting in Australia (January 1987) and a planned meeting in the United States (October 1987). Initial studies indicate that ELISA assays have relatively high false positivity (17) (50% in long-term discordant twins, 5-10% in the general population) and do not correlate with the results of radioassays suggesting that dramatically different species of antibodies are detected by the two assay formats.

Dr. Soeldner and co-workers at the Joslin have developed a competitive insulin autoantibody radioassay (17). This radioassay appears as sensitive as ELISA assays but more specific. Sixty percent of children developing diabetes but only 10% of adults developing diabetes are positive, while none of approximately 100 normal individuals were positive. Insulin autoantibodies are highest in the youngest pre-diabetic individuals and may correlate with the rate with which different patients develop diabetes.

ADDITIONAL AUTOANTIBODY ASSAYS

Antibodies to a 64K protein are present in approximately 60-80% of patients developing Type I diabetes (18). These antibodies have been detected by Baekkeskov and Lernmark utilizing S^{35}-methionine labeled human islets, followed by immunoprecipitation, polyacrylamide gel electrophoresis and autoradiography. The difficulty of the assay and source of tissue (human islets) has limited specificity testing.

A number of assays measure islet cell antibodies reacting with the "surface" of normal rat islet cells or tumor rat islet cell lines (19, 20). Many of these assays have suffered from a high false positive rate which limits their utility (with as many as 30% of first degree relatives positive). A recent study from Australia (Colman and Harrison) indicate that antibodies to bovine albumin are present in children with diabetes and may give artifactual islet surface staining with bovine albumin adsorbed to the surface of islet cells.

Other islet antigens have not been studied as autoantigens and in particular monoclonal antibodies identify a series of islet proteins shared with other tissues (22-24).

PREDICTIVE VALUE OF CURRENT ASSAYS

Clinically relevant autoantibody assays are available for insulin and by immunofluorescence for cytoplasmic ICA. Selected assays when positive identify individuals at high risk of developing Type I diabetes (in our studies, Maclaren and co-workers studies, and Bottazzo and co-workers studies, approximately 8%/year of antibody positive relatives develop diabetes). Utilizing a "specific" ICA assay (e.g., protein-A double immunofluorescence, complement fixing ICA) one can calculate predictive parameters with Bayes theorem, assuming 60% of first degree relatives who develop diabetes are ICA+ and less than 1/200 individuals in the general population are antibody positive.

As is apparent from Table 3 a positive result with a specific ICA assay, provides useful information for both relatives of Type I diabetics and the general population (for relatives, prior odds 5%, post test odds of developing diabetes = 98%) but a negative assay changes the probability of Type I diabetes very little (5% to 2.1%). For each ICA positive individual found, 400 "normal" school children need to be screened or 50 relatives of Type I diabetics.

Unanswered questions concerning anti-islet assays include how often would one screen and what factors would justify screening? Obviously if effective preventive therapy (after activation of autoimmunity) were developed, population based screening would become mandatory. Barring such therapy, screening would be designed to prevent metabolic morbidity or mortality at onset or in specialized situations (military career, consideration for the timing of pregnancy, insurance purposes). I believe it is likely that some forms of preventive therapy will be a reality within a decade and thus screening may assume major importance.
Screening at one year, at school entry and at high school and perhaps at age 30 would probably identify the great majority of individuals developing diabetes as there is a latency of years of ICA positivity preceding diabetes. The duration of this latency period apparently increases with the age at which diabetes develops. At present a combination of specific assays (cytoplasmic ICA and anti-insulin antibodies) would need to be utilized. For such generalized screening improved quantitative assays are needed to replace current indirect immunofluorescent assays.

ENDOCRINOLOGY PREDICTOR ASSOCIATED WITH THE DEVELOPMENT OF TYPE I DIABETES

An accumulating body of data from family studies (in particular from Boston, Gainesville, Denver and Barts-Windsor) indicate that the rate at which Type I diabetes develops can be tracked using a radioimmunoassay for insulin concentrations following administration of intravenous glucose (first phase insulin secretion) (8, 9). In our studies from the time insulin secretion reached the first percentile 88% (7/8) became diabetic with an average of 10 follow-up months, while amongst adults similarly defined, 44% (4/9) became diabetic within 30 months. Currently we routinely employ intravenous glucose tolerance testing in antibody positive individuals. Insulin radioimmunoassays with low "backgrounds" when there is no insulin in a human sera and sensitivity of 5 μU should prove adequate for defining risk.

SUMMARY

The information concerning the prediction of diabetes which I have summarized is of recent "vintage" and a number of significant controversies remain. In particular with some immunologic assays autoantibodies appear to be transient. Nevertheless, a concensus is developing that with the most specific immunologic assays such transient abnormalities are not detected. Larger numbers of immunologically abnormal individuals need to be followed for a longer time period to more precisely define the predictive value of current and future assays. The promise of this area of investigation coupled with development of safer immunotherapy is the prevention of Type I diabetes.

REFERENCES

1. Eisenbarth GS. Type I diabetes mellitus. A chronic autoimmune disease. N Engl J Med 1986; 314: 1360-1368.
2. Feutren G, Assan G, Karsenty G, et al. Cyclosporine increases the rate and length of remissions in insulin-dependent diabetes of recent onset. Results of a multicenter trial. Lancet July, 1986: 119-123.
3. Sutherland DE, Sibley R, Xu XZ, et al. Twin to twin pancreas transplantation: reversal and reenactment of the pathogenesis of Type I diabetes. Trans Assoc Am Physicians 1984; 97: 80-87.
4. Herskowitz RD, Ricker AT, Srikanta S, et al. Asymptomatic or transient hyperglycemia in children: Identification by IVGTT of a subgroup with rapid progression to diabetes. Society for Pediatric Research 1986.
5. Irvine WJ, Gray RS, Steel JM. Islet cell antibody as a marker for early stage Type I diabetes mellitus. In: Irvine WJ, ed. The Immunology of Diabetes. Edinburgh Teviot Publishing, 1980; 117-154.
6. Gorsuch AN, Spencer KM, Lister J, et al. Evidence for a long prediabetic period in Type I (insulin-dependent) diabetes mellitus. Lancet 1981; 2: 1363-1365.
7. Riley W, Maclaren N. Islet cell antibodies are seldom transient. Lancet 1984; 1: 1351-1352.
8. Srikanta S, Ganda OP, Soeldner JS, Eisenbarth GS. First degree relatives of patients with Type I diabetes mellitus: Islet cell antibodies and abnormal insulin secretion. N Engl J Med 1985; 313: 461-464.
9. Srikanta S, Ganda OP, Gleason RE, Jackson RA, Soeldner JS, Eisenbarth GS. Pre-Type I diabetes: linear loss of beta cell response to intravenous glucose. Diabetes 1984; 33: 717-720.
10. Gale EAM, Bottazzo GF. Can we predict Type I (insulin dependent) diabetes? In: World Book of Diabetes Practice, Elsevier, 1987; 2: 25-29.
11. Gleichmann H, Bottazzo GF. Progress toward standardization of cytoplasmic islet cell antibodies. Diabetes 1987; 36: 578-584.
12. Dib SA, Colman PG, Dotta F, Tautkus M, Rabizadeh A, Eisenbarth GS. Expression of "cytoplasmic" islet cell antigens by rat pancreas. Diabetes 1987; in press.
13. Nayak RC, Omar MAK, Rabizadeh A, Skrikanta S, Eisenbarth GS. "Cytoplasmic" islet cell antibodies: Evidence that the target antigen is a sialoglucoconjugate. Diabetes 1985; 34: 617-619.
14. Colman PG, Nayak RC, Connelly J, Rabizadeh A, Soeldner JS, Eisenbarth GS. Islet cell antibodies: Clinical utility and target antigen(s). In: Jaworski MA, Molner GD, Ragotti RV, Singh B, eds. The Immunology of Diabetes Mellitus. Excerpta Medica, Amsterdam. Elsevier Science Publishers, Netherlands, 1986, pp. 345-350.
15. Palmer JP, Asplin CM, Clemons P, et al. Insulin antibodies in insulin dependent diabetes before insulin treatment. Science 1983; 222: 1337-1339.

16. Vardi P, Dib SA, Tuttleman M, et al. Competitive insulin autoantibody RIA: Prospective evaluation of subjects at high risk for development of Type I diabetes mellitus. Diabetes 1987; in press.
17. Wilkin T, Hoskins PJ, Armitage M, et al. Value of insulin autoantibodies as serum markers for insulin dependent diabetes mellitus. Lancet 1985; 2: 480-482.
18. Baekkeskov S, Landan M, Kristensen JK, et al. Antibodies to a 64,000 Mr human islet cell antigen precede the clinical onset of insulin dependent diabetes. J Clin Invest 1987; 79: 926-937.
19. Lernmark A, Freedman ZR, Hoffman C, et al. Islet cell surface antibodies in juvenile diabetes mellitus. N Engl J Med 1978; 299: 375-380.
20. Eisenbarth GS, Morris MA, Scearce RM. Cytotoxic antibodies to cloned rat islet cells in serum of patients with diabetes mellitus. J Clin Invest 1981; 67: 403-408.
21. Colman PG, Campbell IL, Kay TW, Harrison LC. The molecular weight 64,000 autoantigen in Type I diabetes: Evidence against its surface location on human islets. Diabetes 1987; in press.
22. Eisenbarth GS, Shimizu K, Bowring MA, Wells S. Expression of receptors for tetanus toxin and monoclonal antibody A2B5 by pancreatic islet cells. Proc Natl Acad Sci (USA) 1982; 79: 5066-5070.
23. Crump A, Scearce R, Doberson M, Kortz WJ, Eisenbarth GS. Production and characterization of a cytotoxic monoclonal antibody reacting with rat islet cells. J Clin Invest 1982; 70: 659-666.
24. Srikanta S, Eisenbarth GS. Islet cell antigens. Molecular Biol Med 1986; 3: 113-127.

ANTIBODIES TO ACETYLCHOLINE RECEPTORS IN MYASTHENIA GRAVIS

Angela Vincent

Department of Neurological Science, Royal Free Hospital
London, England

INTRODUCTION

Myasthenia gravis is a condition in which there is weakness which increase with effort, and improves with the rest and treatment with anti-cholinesterase drugs. It often effects eye and facial or bulbar muscles; respiratory weakness is less commmon but can be life-threatening. In a high proportion of cases the thymus gland is hyperplastic with germinal centres, and about 10% of patients have a thymoma.

An autoimmune basis for myasthenia gravis (MG) had been suspected for some time but it was not until the late 1950s that serious attempts were made to prove the nature of the defect. Nastuk found fluctuating complement levels in MG sera (1) which lead him to suggest an immunological basis for the disease, and Strauss et al. demonstrated the presence of anti-muscle antibodies (2). Meanwhile Simpson (3) recognised the high female incidence, the increased prevalence of autoimmune diseases, and the occurrence of neonatal myasthenia in infants born to myasthenic mothers, and proposed that myasthenia was an autoimmune disease due to antibodies to an 'endplate protein'. In the 1970s this hypothesis was validated.

PHYSIOLOGY OF NEUROMUSCULAR DEFECT

Neuromuscular transmission in normal tissue is dependent on the nerve-evoked release of about 50 packets or quanta of acetylcholine (ACh) from the nerve terminal. ACh diffuses across the synaptic cleft and binds to postsynaptic ACh receptors (AChR). This opens the AChR-ion channel which allows Na^+ to enter and depolarise the muscle fibre, the endplate potential (epp). The epp activates voltage-dependent Na^+ channels, resulting in a muscle action potential. In MG the number of AChRs is reduced (4) and the epp is therefore, reduced in amplitude (5). If it falls below the critical threshold for activating the Na^+ channels the muscle fails to contract. The miniature endplate potentials (mepps), which are the result of spontaneous release of individual packets of ACh, are also reduced in amplitude (5), and measurement of mepp amplitude on intercostal biopsies is a sensitive diagnostic test for MG.

ACETYLCHOLINE RECEPTORS

The ability to measure AChRs in muscle biopsies (4) followed the demonstration that a polypeptide from the venom of Bungarus multicinctus (alpha-Bungarotoxin, a-BuTx) blocked neuromuscular transmission irreversibly, and that the radiolabelled toxin bound specifically and irreversible to the endplate. This 8000 MW toxin has now been used extensively to quantify AChRs in muscle, to assay for AChRs during purification and to label AChRs for immunoprecipitation assays. Related neurotoxins from other species have been used in affinity chromatography to purify AChR from extracts of skeletal muscle of electric organ (see 6).

The AChR is an oligomeric protein consisting of five subunits ($\alpha_2,\beta,\gamma,\delta$) arranged around the central ion channel (see 7). a-BuTx binds to the two alpha subunits and its binding is blocked by cholinergic ligands. The amino acid sequence is now known (as the result of DNA cloning techniques, see 8) and predictions based on the primary sequence suggest that each subunit is looped through the membrane four times by hydrophobic alpha helices. In addition a fifth amphipathic helix is consistent with the presence of an extracellular NH2 terminus and an intracellur COOH terminus.

ANTIBODIES TO ACETYLCHOLINE RECEPTORS

In 1973 Patrick and Lindstrom (9) found that rabbits injected with purified Torpedo electroplax AChR developed the typical weakness of MG including improvement with anti-acetylcholinesterase treatment. This condition, termed experimental autoimmune myasthenia gravis, EAMG, was associated with small mepps and could be transferred between animals by injection of serum containing anti-AChR antibodies (see 6).

Rat and mouse monoclonal antibodies have been raised against purified AChR preparations from various species. In the large studies performed by Lindstrom and his colleagues a substantial proportion of the antibodies against intact AChR bind mutally exclusively to a region on each alpha subunit called the main immunogenic region (m.i.r.) and these antibodies do not bind strongly to denatured receptors (10). Antibodies against the m.i.r. appear to bind to sequences between amino acids 37 and 85 on the alpha subunits (11).

The demonstration that EAMG could be induced by immunisation against purified AChRs stimulated the search for antibodies against AChRs in MG. Appel et al. showed that MG IgG could inhibit the binding of a-BuTx to rat AChR, and could immunoprecipitate AChR (12). Immunoprecipitation of ^{125}I-a-BuTx labelled human AChR was found to be positive in over 85% of MG patients and this is now the assay in common use (13).

In our laboratory (14) human muscle is taken from amputated limbs and extracted in Triton X 100 detergent and labelled with ^{125}I-a-BuTx. 1-5 ul MG serum is incubated overnight with 30-50 fmoles of ^{125}-I-BuTx binding sites and the patient's IgG is precipitated with a goat anti-human Ig serum. Radioactivity in the pellet shows a linear relationship with the amount of MG serum added, until all the ^{125}-I-a-BuTx-AChR is precipitated, and indicates the presence of anti-AChR antibodies after subtraction of control serum results (Fig).

Muscle from ischaemic limbs is frequently denervated and contains larger amounts of AChR due to the appearance of extrajunctional receptor. Some MG sera react preferentially with endplate or with extrajunctional AChR and we use a mixture of different extracts in the assay. Rat muscle is a more convenient source of AChR but gives a lower percentage of positive results; fetal calf AChR is probably almost as good as human (15). There have been several reports of ELISA assays for anti-AChR determination. These are generally not so satisfactory since it is difficult to purify sufficient AChR from muscle to coat the plates, and it is difficult to exclude the possibility that anti-muscle, rather than anti-AChR, antibodies are being detected. The great advantage of the immunoprecipitation assay is that it uses the unique specificity of a-BuTx to identify the protein to which the antibodies bind. On the other hand an ELISA using a m.ab against the cytoplasmic aspect of human AChR as first layer might be useful for detecting antibodies against the extracellular aspect.

Normal sera give values which can be regarded as negative and are less than 0.2 nmoles/litre if individual specificity controls are performed. However some sera from other conditions e.g. thymoma without MG, penicillamine-treated rheumatoid arthritis and primary biliary cirrhosis, do give low positive results but this may be related to an increased incidence of MG in these conditions. We found a very low incidence of anti-AChR in asymptomatic first degree relatives of MG patients or in elderly individuals (see 14).

In MG patients the titre of antibodies varies widely. About 70% are between 2.0 and >1,000 nM and form a fairly normal distribution when plotted on a semi-log scale (Fig 1). Below 2.0 nM, however, there is a subgroup of patients, often male, whose antibodies frequently show specificity for endplate AChR. Some of these are negative (<0.2 nM in our hands) on repeated testing, while others may become positive over a period of time if serial samples are taken before treatment is initiated (14).

The anti-AChR titre gives little indication of disease severity between individuals: whereas patients with purely ocular symptoms frequently have low antibody titres even patients in sustained clinical remission may have high levels. Although the absolute level in an individual is no indication of their clinical state a fall in antibody of about 50%, as happens following plasma exchange, is often sufficient to produce dramatic improvement (16).

PATHOGENIC EFFECTS OF ANTIBODIES

The pathogenic significance of anti-AChR is now hardly disputed. Passive transfer of MG IgG to mice resulted in small mepps and reduced a-BuTx binding (17). Plasma exchange produced temporary clinical improvement which followed the same time-course as changes in anti-AChR (16). The clearest evidence for a specific pathogenic action of anti-AChR came from the work of Engel and his associates. They showed that IgG and complement was present at MG endplates and their distribution was similar to the distribution of AChRs, as indicated by peroxidase-a-BuTx staining (18).

MECHANISMS OF ANTIBODY MEDIATED ACHR LOSS

The mechanisms by which anti-AChR produces a loss of functional receptor are probably diverse. Anti-AChR can induce an increased rate of degradation of endplate receptors. This was demonstrated both in tissue culture and at intact endplates of animals with passively transferred MG (19). However, an increase in AChR synthesis was also found in passive-transfer experiments (20) and the role of increased AChR degradation in human MG is difficult to establish.

Table 1. Different clinical subgroups of myastenia Gravis patients

	Young onset	Thymoma	Old onset
Age of onset	<40 yrs	30-60 yrs	>40 yrs
Thymic pathology	Hyperplastic	Tumour	Atrophy
Anti-AChR synthesis by thymic cells	+++	+,-	+,-
Most frequent HLA association	B8, Dr3	none	B7, Dr2

A functional defect of receptors due to binding of an antibody to the ACh binding site, the ion channel or some other part of the AChR molecule has been difficult to establish. Most physiological studies on muscles incubated in MG sera have shown no effect, or the effect was reversed by washing which suggests a non-immunological basis (e.g. 21).

The role of complement in MG is clearly important and can account for the marked morphological changes found at most endplates. Both C3 and C9 have been demonstrated histochemically and the latter is present mostly on the debris which is found in the synaptic cleft (see 18). This not only explains the simplification of the postsynaptic membrane and loss of normal architecture of the endplate but raises the possibility that release of

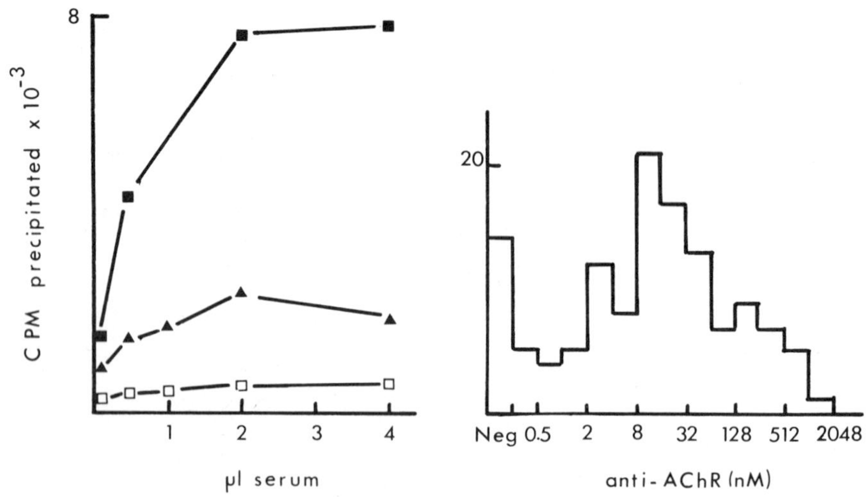

Fig. 1. Anti-AChR antibodies in MG. a. Titration of two MG sera (,Δ) and one control serum ( ) against a constant amount of ^{125}I-a-BuTx-AChR extract. One MG serum (Δ) only binds to the normal, endplate AChR which represents about 20% of this extract. b. Distribution of anti-AChR antibody results in a large group of MG patients. Note that some are negative.

AChR/anti-AChR/Complement complexes into the extracellular space may provide antigenic stimulus to regional lymphnodes.

The role of antibodies in anti-AChR negative patients has been investigated by passive transfer of Ig to mice (22). The presence of a defect in transmission was confirmed in mice treated with 10/12 anti-AChR negative preparations: there was substantial electromyographic decrement without any evidence of AChR loss or antibody bound. Thus in these patients a defect in transmission may be the result of antibodies binding to non-AChR determinants at the endplate. Further electrophysiological studies on treated mice suggest that in some of the anti-AChR negative patients a presynaptic abnormality may be present (Burges, Wray, Vincent and Newsom-Davis unpublished observations 1986).

HETEROGENEITY OF ANTI-ACHR ANTIBODIES

It is clear from a number of studies that anti-AChR antibodies are restricted to IgG class but unrestricted in their light chain or subclass (e.g. 23). Moreover they can be shown to cross-react variably with different mammalian AChR preparations although they do not react well with Torpedo receptor. Further heterogeneity is indicated by the variable reaction with the a-BuTx binding sites on the AChR, and the ability of some sera to discriminate between rat or human endplate and extrajunctional AChR.

A different approach to looking at the specificity of the anti-AChR is to use monoclonal antibodies as probes in competition experiments. There was marked variability between different sera, each of which gave a unique profile of inhibition by different m.abs. M.abs directed against the main immunogenic region (m.i.r.) inhibited a high proportion of MG antibodies suggesting that this is also the immunodominant region of human AChR in MG (24).

We have used ten m.abs directed at human AChR to characterise MG antibodies by competition. These m.abs bind to five regions on the extracellular surface of the receptor (25), probably on the alpha subunits. Antibodies binding within a region share idiotopes, whereas those binding to different regions are idiotypically distinct. Two of the regions appear to be part of the m.i.r. (26). The remaining three may be present on only one of the alpha subunits, and one of them is available only on extrajunctional AChR.

ANTI-ACHR IN DIFFERENT SUBGROUPS OF MG PATIENTS

There is good evidence for the existence of three main groups of MG patients defined by their age of onset, thymic pathology and HLA associations (27) (see Table 1).

Anti-AChR specifities, as demonstrated by inhibition by m.abs showed some significant differences between the three groups. In particular inhibition by m.ab C3 was higher in young onset patients than in the other two groups (26). This suggested the possibility that thymic anti-AChR was contributing this particular antibody specificity but anti-AChR made by thymic lymphocytes had almost identical specificity to that present in the serum of the same individual (26) and anti-AChR specificities did not change significantly after thymectomy even when total anti-AChR fell considerably. Moreover, immunofluorescent staining of thymic tissue with m.abs showed that all antigenic regions of the human AChR were present on thymic myoid cells in both control and MG thymus (28).

PROSPECTS

Our knowledge of the pathogenetic role of anti-AChR in MG should make it possible to devise specific forms of treatment rather than relying on immunosuppression for those cases which do not respond to thymectomy. Unfortunately, this is proving to be more difficult than anticipated. The idiotypes produced appear to be heterogenous and differ between patients (29) which makes the prospects of specific B cell killing based on anti-idiotype antibodies very remote. On the other hand T cell lines specific for AChR are now being raised from MG patients (30), and the exact peptide sequences which stimulate the T cells is beginning to be established. This approach does offer some hope of treatment based on eradicating the cells involved in the autoimmune response.

CONCLUSIONS

The defect in neuromuscular transmission in MG is mainly due to antibody binding to acetylcholine receptors at the neuromuscular junction of skeletal muscle. This antibody is present in about 85% of MG patients but is frequently negative in ocular MG and in a small proportion of patients with generalised disease. It is best measured by immunoprecipitation of ^{125}-I-a-BuTx-labelled human AChR which is a specific and quantitative test.

In general the antibodies do not bind to functional determinants on the AChR but cause loss of receptor by increased degradation and complement mediated damage. Anti-AChR negative sera may effect some other pre or postsynaptic function.

The antibodies in MG are very heterogeneous and each patient shows a unique profile of specificities for different regions on the human receptor. There are some differences in specificity between different groups of patients, however, and these may relate to the role of the thymus. Nevertheless antibody produced by thymic lymphocytes is similar in specificity to that made elsewhere, and the receptor found on thymic myoid cells does not differ from muscle AChR. The involvement of the thymus in young onset patients, and the nature of the association of MG with thymoma is still not clear.

Specific therapy directed at the T cells which are sensitised to AChR is a possibility when the exact peptide sequences involved have been defined.

ACKNOWLEDGEMENT

I thank the Medical Research Council of Great Britain for support.

REFERENCES

1. Nastuk WL, Plescia O, Osserman KE. Changes in serum complement activity in patients with myasthenia gravis. Proc Soc Exp Biol Med 1960; 105: 177-184.
2. Strauss AJL, Seegal BC, Hsu KC, Burkholder PM, Nastuk WL, Osserman KE. Immunofluorescence demonstration of a muscle binding, complement fixing serum globulin fraction in myasthenia gravis. Proc Soc exp Biol Med 1960; 105: 184-191.
3. Simpson JA. Myasthenia gravis: a new hypothesis. Scot Med J 1960; 5: 419-436.

4. Fambrough DM, Drachman DB, Satyamurti S. Neuromuscular junction in myasthenia gravis. Decreased acetylcholine receptors. Science 1973; 182: 293-295.
5. Elmqvist D, Hofman WW, Kugelberg J, Quastel DM. An electrophysiological investigation of neuromuscular transmission in myasthenia gravis. J Physiol 1964; 174: 417-434.
6. Vincent A. Immunology of acetylcholine receptors in relation to myasthenia gravis. Physiol Rev 1980; 60: 756-824.
7. Lindstrom J. Probing nicotinic acetylcholine receptors with monoclonal antibodies. Trends in Neurosciences 1986; 9: 401-407.
8. Claudio T. Recombinant DNA technology in the study of ion channels. Trends in Pharmacological Sciences 1986; 7: 308-312.
9. Patrick J, Lindstrom J. Autoimmune response to acetylcholine receptors. Science 1973; 180-872.
10. Tzartos S, Lindstrom J. Monoclonal antibodies used to probe acetylcholine receptor structure: localisation of the main immunogenic region and detection of similarities between subunits. Proc Nat Acad Sci 1980; 77: 755-759.
11. Barkas T, Mauron A, Roth B, Alliod C, Tzartos S, Ballivet M. Mapping the main immunogenic region and toxin binding site of the nicotinic acetylcholine receptor. Science 1987; 235: 77-80.
12. Appel SH, Almon R, Levy N. Acetylcholine receptor antibodies in myasthenia gravis. New Eng J Med 1975; 293: 760-761.
13. Lindstrom J, Seybold ME, Lennon VA, Whittingham, Duane DD. Antibody to acetylcholine receptor in myasthenia gravis. Neurology 1976; 26: 1054-1059.
14. Vincent A, Newsom-Davis J. Acetylcholine receptor antibody as a diagnostic test for myasthenia gravis: results in 153 validated cases and 2967 diagnostic assays. J Neurol Neurosurg Psychiatr 1985; 48: 1246-1252.
15. Gotti C, Mantegazza R, Clementi F. New antigen for antibody detection in myasthenia gravis. Neurology 1984; 34: 374-377.
16. Newsom-Davis J, Pinchin AJ, Vincent A, Wilson SG. Function of circulating antibody to acetylcholine receptor in myasthenia gravis investigated by plasma exchange. Neurology 1978; 28: 266-272.
17. Toyka KV, Drachman DB, Griffen DE, et al. Myasthenia gravis. Study of humoral immune mechanisms by passive transfer to mice. New Engl J Med 1977; 296: 125-131.
18. Engel AG. Myasthenia gravis and myasthenic syndromes. Ann Neurol 1984; 16: 519-534.
19. Stanley EF, Drachman DB. Effect of myasthenic immunoglobulin on acetylcholine receptors of intact neuromuscular junctions. Science 1978; 200: 1285-1286.
20. Wilson S, Vincent A, Newsom-Davis J. Acetylcholine receptor in mice with passively transferred myasthenia gravis. ii Receptor synthesis. J Neurol Neurosurg Psychiatr 1983; 46: 383-387.
21. Lerrick AJ, Wray D, Vincent A, Newsom-Davis J. Electrophysiological effects of serum factors in myasthenia gravis studied in mouse diaphragm. Ann Neurol 1983; 13: 186-191.
22. Mossman S, Vincent A, Newsom-Davis J. Myasthenia gravis without acetylcholine-receptor antibody: a distinct disease entity. Lancet 1986; i: 116-118.
23. Vincent A, Newsom-Davis J. Acetylcholine receptor antibody characteristics in myasthenia gravis. 1. Patients with generalised myasthenia or disease restricted to ocular muscles. Clin Exp Immunol 1982; 49: 257-265.
24. Tzartos S, Seybold ME, Lindstrom JM. Specificities of antibodies to acetylcholine receptors in sera from myasthenia gravis patients measured by monoclonal antibodies. Proc Natl Acad Sci 1982; 79: 188-192.

25. Whiting PJ, Vincent A, Schluep M, Newsom-Davis J. Monoclonal antibodies that distinguish between normal and denervated human acetylcholine receptor. J Neuroimmunol 1986; 11: 223-235.
26. Vincent A, Whiting P, Schluep M, et al. Antibody specificity and heterogeneity in myasthenia gravis. In: Drachman DB, ed. Myasthenia gravis. NY Acad Sci 1987; in press.
27. Compston DAS, Vincent A, Newsom-Davis J, Batchelor JR. Clinical, pathological HLA antigen and immunological evidence for disease heterogeneity in myasthenia gravis. Brain 1980; 103: 579-601.
28. Schluep M, Willcox NH, Vincent A, Dhoot GK, Newsom-Davis J. Acetylcholine receptors in human myoid cells in situ: an immunohistological study. Ann Neurol 1987; in press.
29. Lang B, Roberts AJ, Vincent A, Newsom-Davis J. Anti-acetylcholine receptor idiotypes in myasthenia gravis analysed by rabbit anti-sera. Clin Exp Immunol 1985; 60: 637-644.
30. Hohlfeld R, Toyka KV. Strategies for the modulation of neuroimmunological disease at the level of autoreactive T lymphocytes. J Neuroimmunol 1985; 193-204.

POLYGLANDULAR AUTOIMMUNE SYNDROME AND BRONCHIAL ASTHMA

Melvin Blecher

Department of Biochemistry, Georgetown University Medical Center, Washington, D.C. 20007 (USA)

INTRODUCTION

One of the basic caveats in endocrinology is that glandular abnormalities tend to occur together. As many as 25% of patients with evidence of hypofunction in one gland have evidence of other end organ-specific or non-specific autoimmune disease.

Ever since the early studies of Hashimoto and of Schmidt, which established an autoimmune etiology to polyglandular failure, the association of multiple autoimmune endocrine disorders has been termed variously as Schmidt's Syndrome, Organ Specific Autoimmunity, Polyglandular Failure Syndrome, and Polyglandular Autoimmunity Syndrome.

These names do not accurately describe the syndrome as we are now aware that concurrent multiplicity of autoimmune diseases is not limited to endocrine glands. The neuromuscular junction is also a target for lymphocytic infiltration and consequent autoimmune disease. There is also evidence that neurological disorders such as schizophrenia and Alzheimer's syndrome may be autoimmune diseases. I will present evidence later that in a subset of bronchial asthmatic patients lung smooth muscle may become autoantigenic. It is also clear that autoimmune disorders involving connective tissues may be associated with polyglandular autoimmune disorders. Thus, it is probably more accurate to describe the associations as Polyautoimmune Disease; however, for historical reasons, I will continue to refer to the associations as Polyglandular Autoimmunity (PGA).

CLASSIFICATION OF AUTOIMMUNE DISORDERS

On the basis of disease associations, Neufeld, Maclaren and Blizzard, at the Universities of Florida and Virginia, in 1980 suggested a nomenclature for dividing PGA into several discrete syndromes, based upon the incidence of associations and clinical characteristics. Although the substantial overlap of clinical entities makes categorization difficult, three broad categories have been distinguished by Neufeld and his colleagues, and named, appropriately, Types I, II and III. As will be seen later, these divisions are of great value for clinical and laboratory

Table 1. Clinical characteristics of polyglandular autoimmunity

Characteristic	Type I	Type II
Age at onset	12-15	20-60
HLA association	none	B 8/DR 3/Dw 3
Chromosome location		6 (MHL)
Genetics	autosomal recessive	multiple generations affected
Female:male	1.4	1.8

Table 2. Discordance of times of onset of pairs of autoimmune diseases

Autoimmune combination	Average interval between appearance
Addison's disease and IDDM	6 years
Graves' disease and pernicious anemia	3.3

Table 3. Frequency distribution in polyglandular autoimmune syndromes

Disorder	Incidence in Type I %	Type II
1. Addison's disease	60	100
2. Hypoparathyroidism	89	0
3. Chronic mucocutaneous candidiasis	75	0
4. 2 + 3	70	0
5. 1 + 2 + 3	33	0

diagnoses of present disorders and for predicting what autoimmune diseases may occur in the future.

Table 1 describes the clinical characteristics of the Types I and II syndromes. Generally, Type I appears in childhood, with the peak age at about 12, whereas Type II is an adult syndrome with the peak appearance at about age 30.

No consistent HLA association has been found in population studies with Type I patients, whereas HLA B-8, DR3, Dw3 loci are clearly associated with Type II syndrome.

This distinction is considered significant because it suggests different etiologies for each syndrome. Expression of HLA-DR antigens (class II major histocompatibility antigens) on cells in humans is normally restricted to B lymphocytes, activated T lymphocytes, antigen-presenting cells, capillary endothelium, and certain epithelial cells. It is thought that the principal function of these antigens is to facilitate recognition among cells of the immune system. In particular, helper T lymphocytes cannot initiate an immune response to an antigen unless the antigen is "presented" to them by a cell expressing HLA-DR. It has been proposed that the initial event in organ-specific autoimmune disease may be an aberrant expression of HLA-DR by the target cells. The presence of these class II antigens would convert the target cell into a functional antigen-presenting cell, and an autoantigen specific for the target cell could then be presented to helper T lymphocytes, which might then initiate an autoimmune reaction. In support of this hypothesis, aberrant expression of HLA-DR has been seen on thyrocytes in autoimmune thyroid diseases, on bile duct epithelium in primary biliary cirrhosis, and on insulin-containing beta cells in recent-onset Type I diabetes mellitus.

There are several exceptions to the aforementioned triad of HLA antigen associations. In Hashimoto's goitrous thyroiditis, the association is limited to DR5. The triad is also not associated with pernicious anemia (PA), except when this disease is associated with other endocrinopathies. In myasthenia gravis (MG), this HLA triad is present in young females with the disease, but not in males with associated thymomas. And Type I diabetes mellitus patients with associated Hashimoto's thyroiditis are essentially HLA B8 negative, whereas the B8 association is strong in Type I diabetes mellitus patients with associated Addison's disease, Graves' disease and atrophic thyroiditis.

Studies of families of patients with PGA have demonstrated that the haplotypes involved in transmission of disease susceptibility are often B8-DR3 negative, whereas the patients are positive. This suggests that expression of the disease requires complementation with a B8-DR3-linked gene. It has been suggested that DR3 is in linkage disequilibrium with a defective immuno suppression gene, which predisposes to the development of autoimmune disease.

Diseases of the Type I syndrome are inherited in an autosomal recessive pattern, whereas in Type II patients it is common to find multiple affected siblings, although not a parent and a child with the syndrome.

In both types, there is a pronounced female preponderance, and the explanation for this is not yet forthcoming.

Table 2 provides data that demonstrates that, although there is some correlation between the age of onset of multiple members of the PGA syndrome, many years may separate the onset of different diseases. Thus, it

Table 4. Frequency distribution (continued)

Disorder	Type I %	Type II
Addison's disease	60	100
Graves' disease	12	60
Type I diabetes mellitus	1	52

Table 5. Frequency distribution (continued)

	Type I %	Type II
Alopecia	29	< 1
Malabsorption syndrome	2	< 1
Pernicious anemia	14	< 1
Premature gonadal failure (primarily ovarian)	15	3.5
Chronic active hepatitis	13	< 1
Vitiligo	9	4.5
Myasthenia gravis		ND

Table 6. Frequency distribution of steroidal cell autoantibodies in polyglandular autoimmunity

Autoantibodies to	Type I %	Type II
Adrenal cortex	100	86
Steroidal cells (ovary, testes, placenta)	96	13

Elder M, et al. J Clin Endo Metab 1981; 52: 1137

is probably clinically wise periodically to monitor patients with one expression of a potential PGA syndrome for the appearance of others.

The next group of figures shows the disease associations that constitute Types I and II PGA.

Table 3 describes the triad of diseases that is the hallmark of the Type I syndrome, namely, Addison's disease, hypoparathyroidism and chronic mucocutaneous candidiasis (CMC). Of this triad, hypoparathyroidism and CMC occur in about 70% of the cases. The presence of any two of this triad is sufficient to make the diagnosis. A patient with only a single element of the syndrome must still be considered if any members of her family are known to have one or more of any of the other components.

Of the triad of diseases that is characteristic of Type I, only Addison's disease also appears in Type II patients. However, a different triad distinguishes Type II PGA patients from Type I. This is shown in Table 4. The Type II syndrome is distinguished by the occurrence of two or more autoimmune disorders in the same patient, selected from among Addison's disease, Graves' disease, and Type I diabetes mellitus. The incidence of the latter two diseases is quite low in the Type I patient.

Several non-endocrine diseases also occur in these patients. The incidence of evidence of alopecia, celiac disease, pernicious anemia, gonadal failure, chronic active hepatitis and vitiligo is rather substantial in Type I patients, much lower in Type II. Myasthenia gravis is occasionally seen in Type II patients, but its incidence is not known with certainty (Table 5).

Types I and II associations are also distinguished by the presence and types of steroidal cell autoantibodies (Table 6). Type I patients who present with both adrenal and gonadal failure (primarily premature ovarian failure) exhibit autoantibodies against adrenal cortical, gonadal and placental tissues. As circulating steroidal autoantibodies cross react with antigens on all steroid-producing tissues it is logical to suggest that all of these tissues have an autoantigen in common, perhaps the cell surface receptor for trophic pituitary hormones, or perhaps the enzymes of the steroid producing pathways which are known to be shared by all steroid producing cells.

Associations of autoimmune endocrine diseases without the occurrence of primary adrenal insufficiency are well known. In most such cases, thyroid autoimmune disease coexists with a second tissue-specific autoimmune disease. These are categorized as Type III PGA, and the nature of the coexisting disease provides the subclassification. This is shown in Table 7.

In subtype IIIA, the associated disease is Type I diabetes mellitus.

In subtype IIIB there is a high incidence of associated pernicious anemia. Because of the high prevalence of autoimmune thyroid disease among patients with pernicious anemia, routine evaluation of all pernicious anemia patients should include a serum T4, T3 resin uptake, and TSH. In addition, an evaluation of thyroid microsome autoantibody status may provide a parameter by which to assess the risk and future development of thyroid autoimmune disease. In patients with autoimmune thyroid disease, routine assessment of B12 status should also be considered, particularly for patients over the age of 40 in whom there is the greatest chance for such an association.

Table 7. Disorders associated with Type III polyglandular autoimmunity

Autoimmune thyroid disease without Addison's disease, plus:

Subtype	Associated disease
IIIA	Type I diabetes mellitus
IIIB	Pernicious anemia
IIIC	Connective tissue disorders (vitiligo, alopecia, etc.)

Table 8. Autoimmune connective tissue disorders

Disorder	HLA type
Acanthosis nigricans types A & B with insulin resistance	DR 4
Systemic lupus erythematosus	DR 3/DR 4
Sjogren's syndrome	Dw 3/DR 3/Dw 2/D
Rheumatoid arthritis	DR 4
Progressive systemic sclerosis (diffuse scleroderma)	-

Table 9. Autoantibody cross reactivity in multiple organ autoimmunity

Combination tested	Cross reactivity
Thyroid microsomes	
+ gastric parietal cells	0
+ thyrotropin receptors	0
+ thyroglobulin	0

Knight JG, et al. Nature 1984; 308: 318

In subtype IIIC connective tissue autoimmunity is the associated disorder.

The major autoimmune connective disease disorders are listed in Table 8. Acanthosis nigricans with associated insulin resistance is associated with HLA- DR4. Systemic lupus erythematosus (SLE) is associated with HLA loci DR3 and DR4. Sjogren's syndrome with many HLA loci, namely, Dw3, Dw2, DR3 and D. Rheumatoid arthritis (RA) is primarily associated with DR4. No HLA associations have been found for scleroderma.

The disease associations just described, along with the cross reactivity among steroidal cell autoantibodies, have given rise to the rather obvious suggestion that some patients may produce autoantibodies that recognize autoantigens on more than one involved tissue, and that this cross reactivity is responsible for the clustering of disease entities. This notion has received support from the observations of Notkins and colleagues at the National Institutes of Health (USA) that monoclonal antibodies raised against antigens in single organs extensively cross reacted with autoantigens in other tissues.

Despite this prediction, Knight et al. (Table 9) could find no evidence of cross reactivity among autoantibodies directed to different autoantigens derived from patients with autoimmune thyroid and gastric diseases. However, the hypothesis may still be valid in other cases, as Robert Schwartz has provided striking evidence that the autoantibodies produced in SLE recognized antigenic determinants occurring in several seemingly unrelated molecules.

In the next group of tables, I will consider the different autoantigens that have been recognized in autoimmune disorders, and will discuss the tests of which I am aware that are currently employed both to detect and to quantify the corresponding autoantibodies.

Table 10 lists the techniques currently employed to detect autoantibodies. Highly sensitive, but not lending themselves to routine assays in a clinical chemistry laboratory, are immunohistochemical and indirect immunofluorescence techniques, both of which are microscopic methods. Highly sensitive and lending themselves both to quantification and routine assay are ELISA methods (both solid state and in free solution), radioimmunoassay (RIA), radioimmunodisplacement (otherwise known as radioreceptor assays (RRA) or immunoradiometric assays (IRMA), and immunoblotting. 32P-labeled DNA probes are beginning to become available, and quantification should be possible by combining radioautography with instruments that measure the intensity of spots.

Less sensitive and of limited quantitative value are the serological tests and those based upon a preliminary separation of protein by SDS-PAGE, followed by immunoblotting with radiolabeled antigens or antibodies. Functional assays are generally not particularly sensitive, and thus not very suitable for routine analysis. There are exceptions, however; functional assays based upon inhibition by anti-receptor antibodies of hormone-stimulated adenylate cyclase activity in plasma membrane preparations or of glucose entry into adipocytes can be quite sensitive, precise and accurate.

In Type I diabetes mellitus various types of islet cell autoantigens, including a cytoplasmic glycolipid, have been studied, primarily by immunofluorescence microscopic techniques, for which a commercial kit is available. However, an ELISA method has been developed for islet cell cytoplasmic proteins, and work is in progress to commercialize this technique (Table 11).

Table 10. Methods to detect and quantify autoantibodies

Highly sensitive	Less sensitive
1. Immunohistochemical	1. Serological
a) Indirect immunofluorescence	a) Immunodiffusion
b) ELISA	b) Hemagglutination
2. Radioimmunoassays	c) Complement fixation
3. Radioimmunodisplacement	2. Functional assays
4. Immunoblotting	3. SDS/PAGE with immunoblotting
5. DNA probes	

Table 11. Type I diabetes mellitus

Pancreatic autoantigen	Autoantibody assays
Islet cells	Immunofluorescence*
Islet cytoplasmic protein	Immunofluorescence* (Bouin's fixed pancreas) ELISA (biotin-avidin)
Islet cytoplasmic glycolipids	Immunofluorescence (unfixed)

* Commercial kits available

Table 12. Autoimmune insulin resistant diabetes mellitus

Autoantigen	Autoantibody assays
Insulin receptor	Immunodisplacement (RRA) Functional assays: e.g., glucose uptake tyrosine kinase

In Type II diabetes mellitus of the autoimmune type, the insulin receptor is the autoantigen (Table 12). Anti-insulin receptor autoantibodies have been associated with profound insulin resistance and hyperglycemia, reflecting the insulin binding blocking action of the anti-receptor IgG antibodies. This property of the autoantibody is the basis of currently employed assays for the presence and titre of this serum protein: a competition radioreceptor assay, and two types of functional assays. In some patients, however, the anti-receptor antibody mimics insulin's actions, and can lead to profound, even intractable, hypoglycemia, and death. The same kinds of assays can be used for such autoantibodies, although the direction of the changes in the functional assays will be the opposite.

What is known and conjectured about the laboratory diagnosis of autoimmune hypoparathyroidism is shown in Table 13. What are available currently are indirect immunofluorescence assays for non-specific parathyroid cell autoantigens. Such assays are employed in those cases in which the parathyroid gland is immunocompromised and primary hypoparathyroidism is the diagnosis. However, cases have been reported in which patients suffering from PGA are resistant to endogenous and administered parathormone, just as in pseudohypoparathyroidism. In such instances, it is quite likely that the PTH receptors of bone cells and renal tubular cells have become autoantigenic and lymphocytic infiltration produces anti-receptor antibodies. By analogy to other anti-receptor autoimmune disorders, one can conceive of assays for circulating anti-PTH receptor antibodies based upon competitive radioreceptor assays or those based upon inhibition of cAMP synthesis in tissue preparations. Once the purified PTH receptor protein becomes available, it should be a simple matter to devise a solid state ELISA assay for such an autoantibody.

Addison's disease and premature gonadal failure can be considered together, as both involve steroid-producing cells (Table 14). Currently, indirect immunofluorescence methods applied to cryocut sections, for which commercial kits are available, are the methods of choice for visualizing autoantigenic structures in steroid-producing tissues, such as the adrenal cortex, ovary, testis and placenta. An example of such autoantigens are the cytoplasmic cell antigens recently found in adrenal cells. However, in view of the distinct possibility that, in at least some cases, cell surface protein receptors for the trophic hormones ACTH, FSH, and HCG/LH have become autoantigenic, blocking autoantibodies are likely to be found in the serum of such patients. In such cases, both radioimmunodisplacement and functional assays would make possible routine assays for the autoantibodies. Examples of simple functional assays would be those based upon the ability of the autoantibodies to inhibit the effects of trophic hormones as activators of adenylate cyclase in plasma membrane preparations or as stimulators of steroidogenesis in cultured steroid-producing cells. Again, with the advent of the availability of solubilized and purified trophic hormone receptors, solid assays will become available. A new assay for anti-adrenal cell surface antigen serum antibodies was described last month at the U.S. Endocrine Society meeting. The Lernmark group in Denmark together with Maclaren's group in Florida have devised a solid state assay using a mouse adrenal cell line that detected and titred adrenal cell surface antibodies in the sera of 60% of Addisonian patients.

For historical reasons assays for autoantibodies in autoimmune thyroid disease have probably been subject to more research and development and controversy than any other autoimmune disorder, and, consequently, more assays are available.

Thyroid microsomes and thyroglobulin have long been recognized as major autoantigens in autoimmune thyroid diseases, and at least five assays

Table 13. Autoimmune hypoparathyroidism

Autoantigens	Autoantibody assays
Parathyroid tissue	Indirect immunofluorescence
PTH receptors (bone cells, renal tubules)	Radioimmunodisplacement Inhibition of cAMP synthesis

Table 14. Addison's disease and premature gonadal failure

Autoantigens	Autoantibody assays
Adrenal cortical cells	Indirect immunofluorescence*
ACTH receptor protein	Inhibition of ACTH-induced steroidogenesis, cAMP production Radioimmunodisplacement
Steroidal cells (ovary, testes, placenta)	Indirect immunofluorescence* Functional assays
FSH receptor	Radioimmunodisplacement

* Commercial kit available

Table 15A. Autoimmune thyroid diseases

Autoantigens	Autoantibody assays
Microsomes	Hemagglutination* Complement fixation Radioimmunoassay ELISA Western blot
Thyroglobulin	Hemagglutination* Radioimmunoassay ELISA

* Commercial kits available

for the corresponding autoantibodies are in current use: a hemagglutination assay for which commercial kits are available; complement fixation assays; RIA; several excellent solid state ELISA assays; and, a Western blot method recently devised by Rapoport's group in San Francisco for new autoimmune thyroid disease-related autoantigens (Table 15A). In addition to what is displayed in the table, I should mention that Caryon's group in Marseille reported last month at the meeting of the U.S. Endocrine Society that they had identified a specific antigenic microsomal protein that is responsible for the appearance of the anti-microsomal autoantibodies in sera of patients with autoimmune thyroid disease; this microsomal protein is thyroid peroxidase which, you will recall, is one of the enzymes responsible for organification of iodine in this gland.

I might note parenthetically that a group from Nagasaki reported at the same meeting that they have developed an assay for serum thyroglobulin that is not subject to interference by the anti-thyroglobulin autoantibodies present in sera of patients with autoimmune thyroid disease. The assay is a sandwich assay that uses in the solid state monoclonal antibodies from a mouse hybridoma that interacts with serum thyroglobulin, but not with the autoantibody to the patients own thyroglobulin.

The major assays that are in current use for the anti-thyrotropin antibodies produced in Graves' disease and Hashimoto's thyroiditis (have been reviewed (see p. 131)), and so I will mention here only that commercial kits are already available for the radioimmunodisplacement and ELISA assays (Table 15B).

Several assays for antibodies for ill-defined thyroid autoantigens are known (Table 15C). A RIA for double stranded DNA is available in kit form. Indirect immunofluorescence has been used to detect non-specific nuclear proteins. An ELISA assay and a procedure involving SDS/PAGE coupled to immunoblotting have been reported for antibodies to autoantigens in orbital muscle. And Rapoport has used molecular cloning of autoantibodies to identify new autoantigens.

Myasthenia gravis (MG) is anomolous in that its incidence in PGA appears to be low. However, the incidence of this autoimmune disease in the general population is not low, and tests for the autoantibodies are routine in various centers, although not in the routine clinical laboratories per se. The acetylcholine receptor is an important autoantigen in MG, and three assays for its autoantibodies are in current use, the very popular double antibody immunoprecipitation assay, an immunoadsorbent RIA, and an ELISA assay (Table 16). Another test useful in the serological evaluation of a patient with MG is the anti-striational antibody tests, assayed by the technique of indirect immunofluorescence, which makes use of skeletal muscle sections or glycerinated myofibrils as substrate.

Chronic atrophic gastritis can be dealt with along with pernicious anemia because about 90% of the sera of patients with pernicious anemia contain antibodies - which are frequently cytotoxic in vitro - to gastric parietal cells. Anti-parietal cell autoantibodies are also found in atrophic gastritis. These autoantibodies are currently being detected by the classical indirect immunofluorescence tests, as well as by a cytotoxicity-based assay and a leukocyte migration test (Table 17).

Antibodies to intrinsic factor are currently being detected and quantified by a most useful assay based upon radioimmunoprecipitation with radiolabeled cobalamin. However, it should be possible to devise additional assays for the autoantibody. One can anticipate a radioimmunodisplacement assay in which labeled cobalamin and autoantibody compete for binding to intrinsic factor. Another possible assay is a variation on this theme in

Table 15B.

Autoantigens	Autoantibody assays
Thyrotropin receptor (gangliosides)	Radioimmunodisplacement* ELISA* Immunobinding Thyroid cell growth Thyroglobulin release Follicle formation Cyclic AMP synthesis

* Commercial kits available

Table 15C.

Autoantigens	Autoantibody assays
Double stranded DNA Nucleoproteins Orbital muscle New ATRA	Radioimmunoassay* Indirect immunofluorescence* ELISA SDS/PAGE with immunoblotting Molecular cloning

* Commercial kits available

Table 16. Myasthenia gravis

Autoantigens	Autoantibody assays
Muscle tissue	Indirect immunofluorescence
Contractile element	Indirect immunofluorescence
Acetylcholine receptor immunoprecipitation	Double antibody
	Immunoadsorbent RIA ELISA

which the autoantibody competes with a cobalamin-intrinsic factor complex for binding to a tissue receptor. Solid state assays are also a distinct possibility, and the prevalence of pernicious anemia in PGA should make developmental efforts in this area cost-effective.

Development of quantitatively useful assays in the autoimmune connective tissue disorders has lagged behind other areas because it has been particularly difficult to identify specific autoantigens to use in assays of autoantibodies in the circulation. nDNA, usually referred to as ANA, is present in the sera of over 99% of lupus patients, and was the autoantigen first described in this disorder. ANA autoantibodies are routinely detected clinically by indirect immunofluorescence (Table 18A). Three non-specific nuclear proteins and one cytoplasmic protein have also been described as autoantigens in SLE, and these are detected by immunoprecipitations. A reportedly specific nuclear protein kinase NII has been identified as an autoantigen, and antibodies to this protein are assayed by an RIA.

Autoantibodies to nuclear RNA polymerase I in SLE are assessed by an activity inhibition assay (Table 18B). The presence of autoantibodies to the nuclear polymer poly(ADP-ribose), a polymer that appears to be involved in DNA repair, is detected by a binding assay. Autoantibodies to neuronal tissue in SLE have been detected by indirect immunofluorescence using a neuronal cell line.

In rheumatoid arthritis, immunofluorescence and immunoprecipitation are the methods of choice to detect autoantibodies to several non-specific nuclear autoantigens. These methods are also used to detect the so-called rheumatoid factors. Rheumatoid factors are apparently autoantibodies directed against the Fc portion of IgG, which is an autoantigen in this disease (Table 19). In addition, Fc receptors on lymphocytes and macrophages may become autoantigenic in rheumatoid arthritis, and these can be assayed by a serological assay involving 51Cr uptake by coated sheep erythrocytes.

In Sjogren's syndrome, in which the functioning of salivary and tear ducts has become compromised, the autoantibodies to the non-specific tissue autoantigens are generally tested by classical indirect immunofluorescence techniques (Table 20). However, an ELISA assay has recently become available for autoantibodies to the nuclear LA(SS-B) and cytoplasmic Ro(SS-B) proteins that are so prevalent in this syndrome.

AUTOIMMUNE BRONCHIAL ASTHMA

Finally, I want to mention the newest candidate autoimmune disease, bronchial asthma. It has been hypothesized that the impaired response of some asthmatic patients to beta adrenergic agonists is a consequence of the autoimmune production of antibodies directed against the beta adrenergic receptors of lung smooth muscle. Such autoantibodies would block the binding of beta agonists to its receptors on lung smooth muscle, and this blockade would permit an exaggerated response to alpha adrenergic and cholinergic agonists. According to the hypothesis, this imbalance would lead to a hallmark of bronchial asthma, namely, bronchospasm.

In collaboration with Drs. Jocelyn Hicks and Shelby Josephs of Childrens' Hospital National Medical Center in Washington, we tested this hypothesis by applying a newly-devised radioimmunodisplacement assay for blocking serum antibodies to 376 asthmatic children and 58 non-asthmatic patients at Children's Hospital. The screening test (reaction 1) consists of a competition of serum autoantibodies against radiolabeled

Table 17. Chronic atrophic gastritis/pernicious anemia

Autoantigens	Autoantibody assays
Parietal cells	Indirect immunofluorescence
	Complement dependent cytotox.
	Leukocyte migration test
Intrinsic factor	Radioimmunoprecipitation with cobalamin
	Radioimmunodisplacement with cobalamin
	Binding of cobalamin-intrinsic factor to tissue

Table 18A. Systemic lupus erythematosus

Autoantigens	Autoantibody assays
nDNA	Indirect immunofluorescence
n Ribonucleoprotein	Immunoprecipitation
Nuclear protein Sm	Immunoprecipitation
La(SS-B), nuclear	Immunoprecipitation
Ro(SS-B), cytoplasmic	Immunoprecipitation

Table 18B.

Autoantigens	Autoantibody assays
Neuronal tissue	Indirect immunofluorescence Neuro-2a cells
Poly(ADP-ribose)	Binding assay
Nuclear protein kinase NII	Radioimmunoassay
RNA polymerase I	Activity

Table 19. Rheumatoid arthritis

Autoantigens	Autoantibody assays
Nuclear	Immunofluorescence
	Immunoprecipitation
RNA polymerase I	Activity inhibition
Fc fragment	Rheumatoid factor binding
Fc receptors	^{51}Cr uptake by coated RBC

Table 20. Sjogren's syndrome

Autoantigens	Autoantibody assays
Liver cell membranes	Indirect immunofluorescence
Salivary duct epith.	Indirect immunofluorescence
Salivary duct cyto.	Indirect immunofluorescence
Smooth muscle cells	Indirect immunofluorescence
DNA	Indirect immunofluorescence
La(SS-B) and Ro(SS-A)	Ouchterlony double diffusion
	ELISA
Circulating immune complex	Clq binding and solid state assays

Table 21. Screening test for β-adrenergic receptor autoantibodies

	Number		Relative IHYP binding	
Group	(patients)	(tests)	(% of control) (Mean ± SD_M)	P*
Normal pool serum	(-)	(251)	81 ± 7	-
Nonasthmatic patients	(58)	(58)	82 ± 10	NS
General asthmatic patients	(58)	(94)	75 ± 14	<0.0005
High-risk asthmatic patients	(318)	(589)	73 ± 14	<0.0005

* determined by students t tests

iodohydroxybenzylpindolol (a powerful beta adrenergic antagonist) for binding sites on canine lung plasma membrane preparations:

Beta Adrenergic Receptors + 1251-IHYP —> [R-1251-IHYP]
+ Ab —> [R-Ab]

The data are summarized in Table 21. We found that sera of some chronically and acutely asthmatic children inhibited the specific binding of the beta-antagonist to lung smooth muscle membranes. The inhibitions were small, but highly significant statistically.

By reference to a frequency distribution histogram (one such is shown in Fig. 1), we then identified 30 of the patients whose sera contained the highest titre of the beta blocking factor (inhibitions greater than 2 S.D. from the mean). We subjected these sera to immunodepletion using anti-human IgG (gamma chain) and anti-human IgA (alpha chain) (Table 22). Immunodepletion of these two antibodies from the sera of 22 of the 30 patients removed the beta blocking activity. For historical reasons, we assume that the blocking antibody is an IgG, but we are not yet certain of this.

Our results indicated that, in about 5% of the chronic asthmatics and in about 8% of acute asthmatics, there was present in their circulation an anti-beta-adrenergic receptor autoantibody. These results support the hypothesis that this blocking autoantibody mediates the autonomic dysfunction of at least some atopic patients.

CONCLUSIONS

Neufeld's classification of polyglandular autoimmune failure, as modified to include non-glandular autoimmune disorders, has several

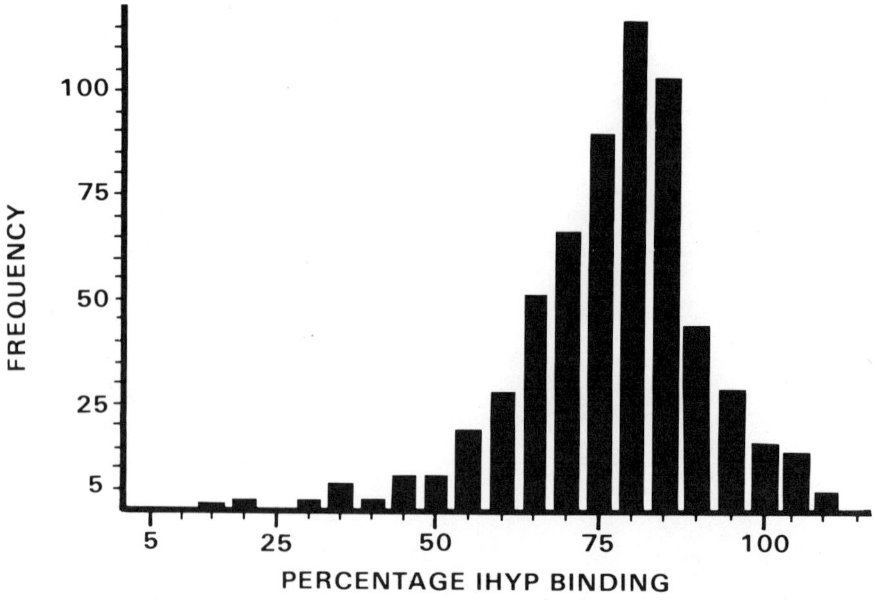

Fig. 1. Histogram of percentage IHYP Binding. n = 589, $\bar{x}$ = 73, SD = 14. The number of patients (high-risk asthmatics) is 318.

Table 22. Binding assays after immunodepletion

Group	Patients tested*	Confirmed by repeat assay**
General asthmatic patients	2	2
High-risk asthmatic patients	28	20 (23)

* Basis of selection: relative IHYP binding below 2 SD_M (98% confidence limits)

** "Confirmation": IHYP binding restored to the normal range or increased absolutely by at least 20% after removal of IgG and IgA. The value in parentheses represents three additional patients whose sera were "confirmed" in only one of two trials.

important clinical implications for disease recognition, prognosis, and health maintenance screening. Familiarity with the scheme should alert the physician and the clinical laboratory to the possible presence of additional autoimmune diseases when a single autoimmune endocrine failure or another autoimmune entity is present. The presence of two autoimmune endocrine failures should allow classification of the syndrome and thus permit some insight into the probable clinical course. For example, one may infer from the classification tables that Type II PGA patients may have a better outcome than Type I patients. This is so because Type II PGA not only has a later onset, but also has a more protracted course and lacks a significant association with chronic atrophic gastritis, pernicious anemia, or steroidal cell failure.

The heritable nature of PGA has been firmly established; therefore, careful screening of immediate relatives is advisable. As patients with PGA are at risk for additional endocrine failures, as well as for nonendocrine autoimmune disease, periodic screening would seen to be indicated for these patients. Although the appearance of new autoimmune disease is quite variable, a reasonable plan would be screening biannually, with the choice of tests being guided by the classification schemes. For example, a young person with autoimmune adrenal failure and Type I diabetes mellitus (Type II PGA) should be observed for the development of autoimmune thyroid disease, whereas the individual with chronic mucocutaneous candidiasis and hypoparathyroidism (Type I PGA) should be observed for steroidal cell failure and the other commonly associated autoimmune disorders.

At the moment, I suspect that most of the testing for autoantibodies is being carried out, not by central hospital clinical chemistry laboratories or even reference laboratories, but rather by specialized laboratories operated by clinical scientists. Although the reason for this is partly historical and partly financial, I believe that the main reason is because the tests available have not always carried with them the accuracy, precision, reliability and other attributes required by the professional clinical chemistry laboratory. However, the quantitative assays that are now being developed for many autoimmune diseases (e.g., RIA, RRA, ELISA, IRMA), plus the availability of more and more commercial kits, should make it feasible in the near future to do these diagnostic and prognostic tests on a routine basis.

CHAPTER 6
COAGULATION

Fibrinolysis
F. Bachmann

Modulation of endothelial cell coagulant properties
D.M. Stern and P.P. Nawroth

Deficiencies of protein C and protein S and thromboembolic disease
J.H. Griffin

FIBRINOLYSIS

F. Bachman

Central Hospitalier Vaudois, Lab Centrale d'Haematologie
Switzerland

INTRODUCTION

Over the last several years, there has been an explosion of new discoveries in the field of fibrinolysis. The cDNA and consequently the primary amino sequences of most of the components of the fibrinolytic system have been determined (reviewed in part in 1), the regulation of gene expression by hormones, growth factors and phorbolesters extensively studied, the enzyme kinetics of plasminogen activation worked out (reviewed in 2), two major plasminogen activator inhibitors characterized (reviewed in 3, 4), the role of plasminogen activators in cancerigenesis and the regulation of pericellular proteolysis explored (reviewed in 5, 6), correlations between a deficient fibrinolytic activity and the incidence of idiopathic deep vein thrombosis established and major progress in thrombolytic therapy achieved (reviewed in 2, 7).

The following review does not attempt to embrace the whole field of plasminogen activation but will be restricted, in a large measure, to the mechanisms inducing the lysis of fibrin. Therapeutic thrombolysis will be the subject of another review in this volume. The choice of references reflects the authors desire to cite recent articles and is influenced by his own subjective view of the field.

Plasmin, the enzyme primarily involved in the breakdown of fibrin, exists in plasma in a precursor form, the zymogen plasminogen. The latter is activated by two major pathways, the tissue-type plasminogen activator (t-PA) and the urinary-type plasminogen activator (u-PA) pathway (Fig. 1).

The enzymes of the fibrinolytic system are all serine proteases i.e. they have in their active site the amino acid serine. All consist of two chains, linked together by one or more disulfide bridges. The N-terminal A-chain confers to the enzyme substrate-binding specificity (8). It is composed of typical "building stones" which are encountered in other proteins and serine proteases, particularly those of the blood coagulation system. These comprise the kringles, folded structures approximately 80 amino acids long, held together by three disulfide bridges (Fig. 2), the regions homologous to the epidermal growth factor (EGF), the finger regions which exhibit sequence homology to the fibronectin fingers and in

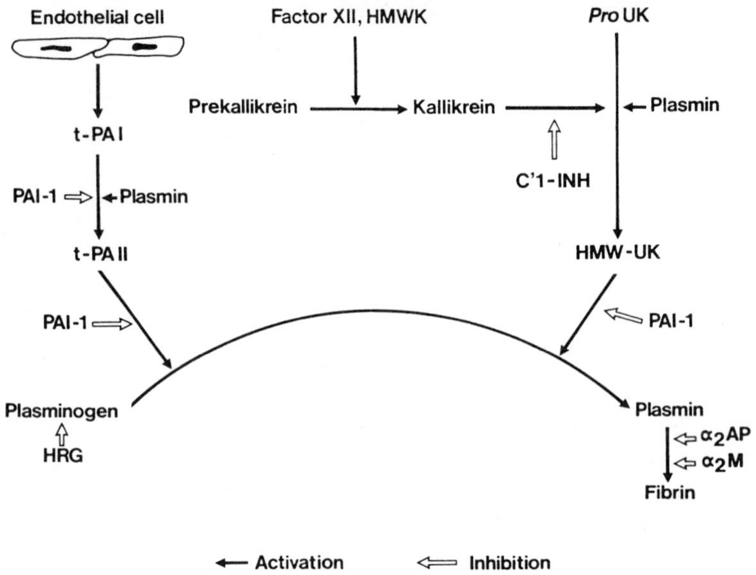

Fig. 1. The fibrinolytic system of human plasma.

t-PA I : single-chain tissue-type plasminogen activator, also called sct-PA
t-PA II: two-chain t-PA (tct-PA)
PAI-1 : plasminogen activator inhibitor-1 (endothelial type)
HMWK : high molecular weight kininogen
Pro UK : pro-urokinase, also called single chain urinary-type PA (scu-PA)
HMW-UK : high molecular weight urokinase, also called two chain u-PA (tcu-PA)
C'I-INH: Cl-inhibitor
α2-AP : $Alpha_2$-antiplasmin
α2-M : $Alpha_2$-macroglobulin

kallikrein a region of four tandem repeats of 90 to 91 amino acids which are also encountered in clotting factor XI.

The C-terminal portion of the serine proteases contains the proteolytically active B-chain. The B-chains consist of approximatively 250 amino acids and contain in their active site histidine, apartic acid and serine. They show considerable sequence homology between each other.

The natural inhibitors of the fibrinolytic system belong with few exceptions to the family of the serpins (serine protease inhibitors) of which the best known representatives are the $alpha_1$-protease inhibitor and antithrombin III. They too, show marked sequence homology.

These inhibitors are very important for the fine regulation of enzymatic pathways and are to a large extent responsible for the localisation of fibrinolytic activity on a clot surface.

Several modulators of fibrinolytic activity intervene at various phases of the fibrinolytic mechanism; these are the two competitive inhibitors histidine-rich glycoprotein and one or several competitive inhibitors forming complexes with pro-urokinase, the kringle 4 binding protein and thrombospondin.

THE COMPONENTS OF THE FIBRINOLYTIC SYSTEM

Plasminogen

The biochemistry of plasminogen, a glycoprotein of 92 kD, has been recently reviewed (9). The two major plasminogen forms, differing in carbohydrate content and composition are both synthetized in the liver. Human plasma contains approximately 200 mg/l of plasminogen (2 μM; Table 1). Its biological half-life is 2.2 days.

The complete amino acid sequence has been established in the seventies (reviewed in 10). The human cDNA corresponding to amino acid residues 272-790 has been cloned (11). The gene for plasminogen resides in chromosome 6 (12) (Table 2).

The heavy chain portion (A-chain) of the native glu-plasminogen (N-terminal: glutamic acid) contains 560 amino acids. In the presence of plasmin, glu-plasminogen is converted to lys-plasminogen by proteolytic cleavage of the peptide bond Lys 76-Lys 77 with simultaneous liberation of an activation peptide of about 8 kD (Fig. 3).

The most salient features of the A-chain are the five kringles which confer to the plasminogen its affinity for fibrin, alpha$_2$-antiplasmin and thrombospondin (Fig. 4). These kringles contain lysine binding sites (LBS). A LBS with very high affinity for lysine (K_d approx. 9 uM) is situated on kringle 1, 5 further LBS with lower affinity (k_d approx. 5 mM) are found on kringles 2 to 5. The light chain portion (B-chain) of plasminogen consists of 241 amino acid residues. The three amino acids which make up the active site are His 602, Asp 645 and Ser 740. The transformation of glu-plasminogen into lys-plasminogen and glu-plasmin and lys-plasmin has been studied in detail in vitro. The plasminogen activators t-PA and tcu-PA and perhaps even scu-PA hydrolize the peptide bond Arg 560-Val 561. This cleavage creates a two chain molecule, glu-plasmin, which is held together by two disulfide bridges (Fig. 3).

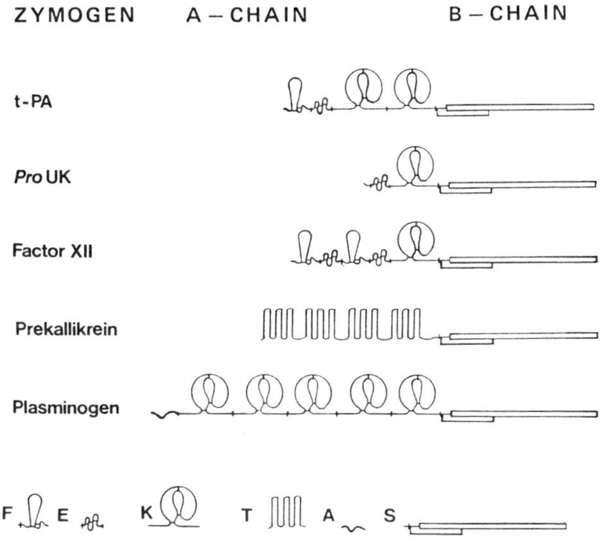

Fig. 2. Building stones of fibrinolytic serine proteases.
F: Finger domain E: epidermal growth factor domain;
K: kringle; A: activation peptide; T: tandem domain;
S: light (B-) chain, containing serine protease katalytic site

Glu-plasmin will cleave, in vitro, the peptide bond Lys 76-Lys 77, resulting in the formation of lys-plasminogen, respectively lys-plasmin. Lys-plasminogen has a higher affinity for fibrin and a short half-life of less than one day. Normal human blood contains the plasmin inhibitor alpha 2-antiplasmin which rapidly neutralizes free plasmin. Lys-plasminogen is therefore not produced under physiological conditions (for example: stimulation of t-PA release, resulting in a 100 fold higher free t-PA level in plasma).

Tissue-type plasminogen activator (T-PA)

The cDNA of t-PA has been cloned and the complete amino acid sequence established (15-20). The human t-PA gen spans over 20 kb (19, 20) consists of 14 exons and is situated on chromosome 8 (21, 22). The intron-exon organization reveals that each exon codes for a particular domain of the t-PA molecule (Fig. 5). Thus the leader sequence of the pre-pro-t-PA is encoded by the first exon, the signal peptide by exon II, the pro-sequence by exon III. The mature t-PA protein carries glycine at its N-terminal (23); 6 amino acids inwards starts a sequence which has homology to the 9 fibronectin fingers; it is encoded by exon IV. Moving further inwards, there is a domain resembling epidermal growth factor (exon V), followed by the two kringles, each one expressed by two exons (VI-IX).

The light (B) chain of t-PA is encoded by exons X to XIV and shows considerable sequence homology to the catalytic chain of other serine proteases such as trypsin, chymotrypsin, elastase, plasmin and many coagulation factors. The active site residues are His 325, Asp 374 and Ser 481.

It has been clearly established that the isolated B chain of t-PA maintains catalytic activity (24, 25) but has no affinity for fibrin (25, 26). The fibrin binding regions of the A-chain have been localized to the finger region and the second kringle (26-30). In the interaction between fibrin and the second kringle a lysine-binding site (LBS) is involved, but the affinity of the finger domain for fibrin cannot be reduced in the presence of lysine or omega-carboxylic acids (27).

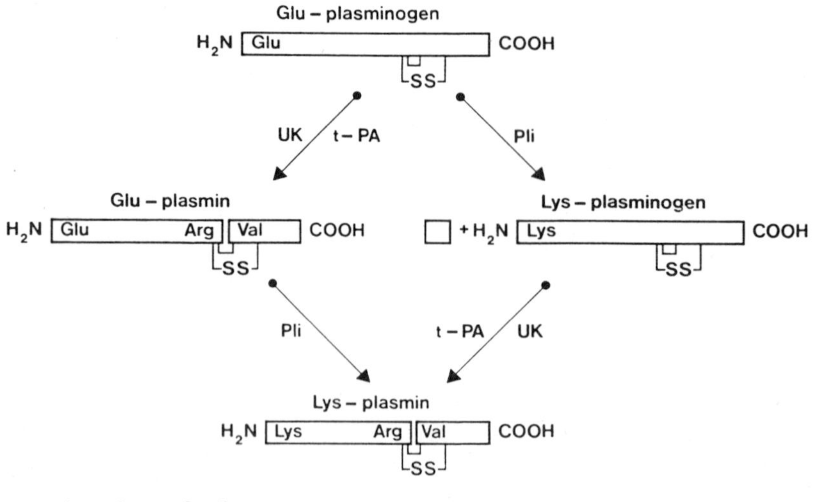

Fig. 3. Activation of plasminogen.
In the absence of plasmin inhibitors, t-PA or u-PA convert glu-plasminogen to glu-plasmin, the latter then converts glu-plasmin(ogen) to lys-plasmin(ogen).

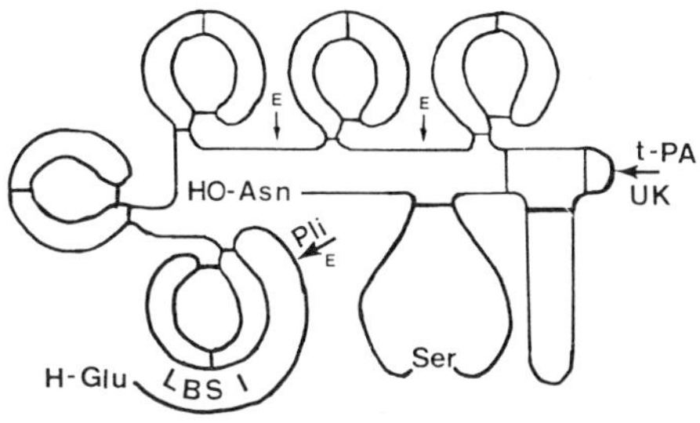

Fig. 4. Model of glu-plasminogen with the high affinity lysine binding site (LBS I). The cleavage sites for plasmin (pli) and elastase (E) are indicated (adapted from Lijnen and Collen (13)).

By lysine-Sepharose chromatography t-PA can be separated into two variants (I and II) of different carbohydrate content and molecular weight (31). Variant I contains 12.8% carbohydrate, linked to Asn 120, 187 and 451, the lighter variant II contains 7% oligosaccharides attached to Asn 120 and 451 (numbering system according to ref. 23).

In human plasma the t-PA antigen concentration is very low, approximately 5 ng/ml (Table 1). Most of it is complexed to PA inhibitor-1 and probably not more than 5% circulates in its free form (3-5). However, the local free t-PA concentration can be several times higher, since endothelial cells are capable of releasing free sct-PA (reviewed in ref. 1-5, 14, 32, 33). In the presence of small amounts of plasmin, tissue kallikrein or factor Xa, sct-PA (t-PA I) is converted to tct-PA (t-PA II). The physicochemical and enzymatic properties of these two forms have been extensively studied and have been reviewed recently (32-35). In the absence of fibrin, both sct-PA and tct-PA are poor activators of glu-plasminogen (33-35).

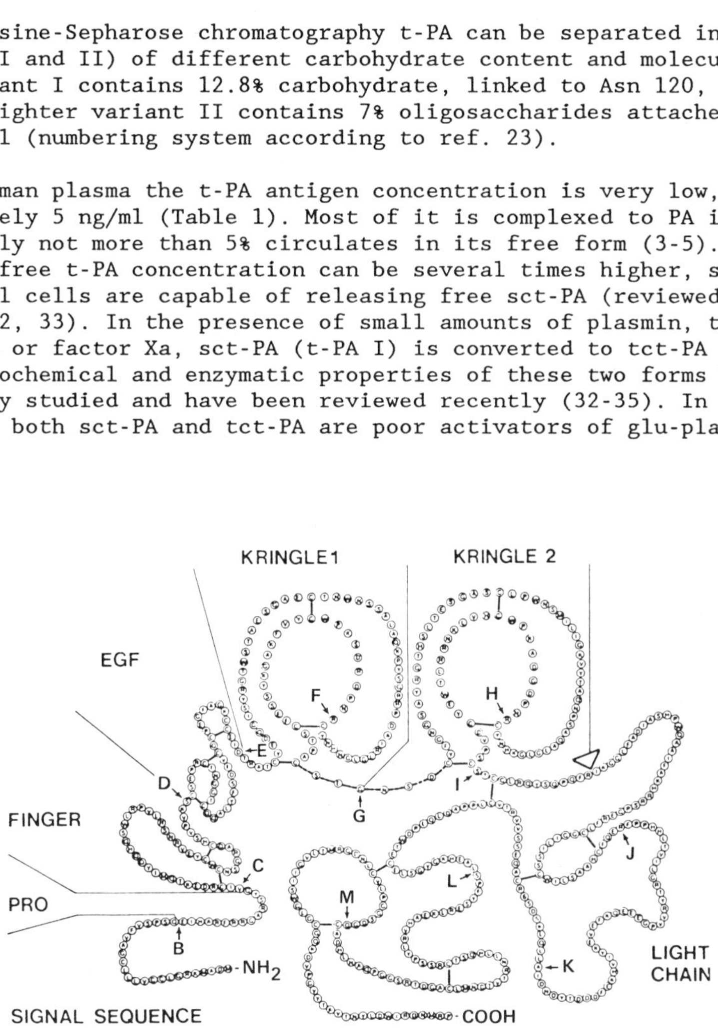

Fig. 5. A model for the secondary structure of the t-PA precursor protein, including signal peptide and prosequence. The solid black bars indicate the potential disulfide bridges. The arrows (B-M) indicate the map position of the individual introns in the protein. The cleavage site, dividing the heavy and the light chain is indicatedwith a triangle (from Blasi et al. (1), on the basis of the work of Ny et al. (20)).

This is due to the relatively high K_m of 8 to 65 μmol.l^{-1} and the low k_{cat} of approximatively 0.01 to 0.06.s^{-1} of the reaction, resulting in a very low turnover number of 0.0005.s^{-1} of glu-plasminogen. Thus, at physiological concentrations of glu-plasminogen (1 μM), one t-PA molecule takes approximately 30 min to activate one molecule of glu-plasminogen into glu-plasmin. In the presence of physiological concentrations of fibrin, t-PA and glu-plasminogen bind to fibrin, forming a ternary complex. This results in a decrease of the K_m and allows a much more efficient conversion of fibrin-bound glu-plasminogen (higher k_{cat} in the order of 0.1.s^{-1}). The turnover rates at physiological glu-plasminogen and fibrin concentrations are 200 to 400 times faster than in the absence of fibrin (Fig. 6). In this acceleration, sites within the fibrinogen sequence A alpha 148-197 appear to play a major role (36).

Kinetic studies on the activation of glu-plasminogen by t-PA particularly in the presence of fibrin are complicated by several facts: the generation of small amounts of glu-plasmin will: 1) convert sct-PA to tct-PA, 2) release an activation peptide from the N-terminal portion of glu-plasmin(ogen) resulting in the formation of lys-plasmin(ogen) which has a higher affinity for fibrin than glu-plasminogen, 3) cleave preferentially portions of the C-terminal end of the alpha chain of fibrin exposing lysine-rich regions; this creates new binding sites for plasminogen (37, 38).

To better appreciate the functional differences between sct-PA and tct-PA Tate et al. have constructed a sct-PA which cannot be cleaved into a tct-PA by plasmin. They modified, by in vitro mutagenesis, the cDNA codon for Arg 278 into a codon specifying Glu and expressed the mutant in chinese hamster ovary cells (39). From their and the studies of others it results that tct-PA is more active against a number of chromogenic substrates (Table 3) and is inhibited more rapidly by DFP and serpins (3, 4, 34, 40, 41). Sct-PA has a higher affinity to fibrin (39) than tct-PA. It activates plasminogen to plasmin in the absence of fibrin even more poorly than tct-

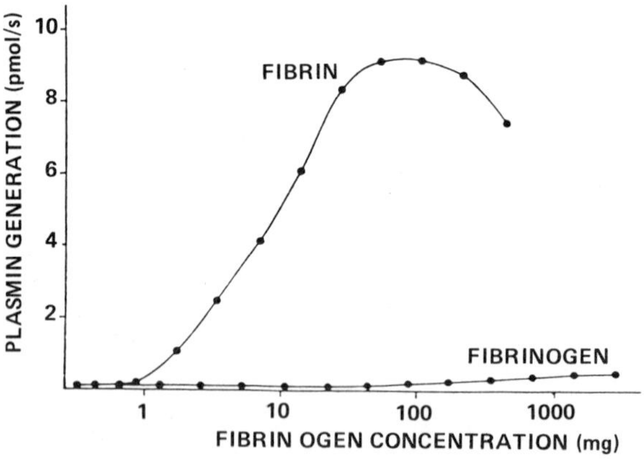

Fig. 6. Effect of increasing amounts of fibrin on the rate of plasmin generation. In this experiment 60 pM of two-chain t-PA, 0.5 μM glu-plasminogen and variable amounts of fibrinogen were mixed. The reaction was started by addition of 0.5 NIH units of thrombin per ml and the rate of plasmin generation monitored by the plasmin substrate H-D-Val-Leu-Lys-pNA which was present in the reaction mixture. In control experiments no thrombin was added resulting in the data points labelled "fibrinogen" (Taken from Ranby and Wallen (34)).

Table 3. Functional differences between sct-PA and tct-PA in the absence of fibrin

System	Difference
Amidolytic activity on H-D-Isoleu-Pro-Arg-pNA (S-2288) and D-Val-Gly-Arg (S-2322) at pH 7.3	tct-PA is 8 to 10 times more active
Activation of glu-plasminogen	tct-PA is 10 to 50 times more active
Inactivation by DFP, 25°C, pH 8	tct-PA is inactivated 3 times more rapidly
Inhibition by PAI-1	no significant difference (?)
Inhibition by PAI-2	tct-PA inactivated 100 times more efficiently than sct-PA
Binding to fibrin	sct-PA binds better

PA (33, 34, 39). In the presence of fibrin both forms activate glu-plasminogen about equally well (33, 34).

In the literature printed before 1985 a considerable confusion existed as to the specific activity of t-PA. Some investigators defined their own units, other compared t-PA activity with international units for urokinase, and incorrectly used the form IU for t-PA activity also. In 1984 the ICTH (International Committee on Thrombosis and Haemostasis) proposed and the WHO adopted a t-PA standard with a specific activity of 500,000 IU/mg. Ampouls containing 1000 IU (500 ng) can be obtained from Dr. P.J. Gaffney, National Institute for Biological Standards and Control, London (42).

In vivo, both sc and tct-PA have a short half-life of about 5 min. (43). The disappearance rate of t-PA appears to be independent of the formation of inactive complexes with plasminogen activator inhibitor-1 since in patients infused with recombinant t-PA (routinely given in an over 100 fold excess over PAI-1) the half disappearance rate is also about 5 min. (44).

Urinary-type plasminogen activator

The complete primary amino acid sequence of high molecular weight urokinase (HMW-UK), the two chain form of u-PA (tcu-PA) has been established in 1982 (45, 46). Native u-PA can be found in the urine (47, 48) and is produced in form of a single-chain (scu-PA) by various cell-cultures (49-51 and reviewed in ref. 1, 5, 6). The scu-PA concentration in human plasma has not yet been clearly established and mean values indicated in the literature vary from 2 to 20 ng/ml (reviewed in 52) corresponding to 40 to 400 pM (Table 1). The cDNA of u-PA has been isolated and the nucleotide sequence determined (53-55). The human u-PA gene is 6.4 kb long and situated on chromosome 10 (21). It contains eleven exons (55) and the intron-exon organisation of the u-PA gene closely resembles that of the t-PA gene (Fig. 7).

Some of the exons in the t-PA gene have no correspondence in the u-PA gene. Exon III of t-PA is totally and exon IV partially missing in the u-PA gene; consequently the finger domain is not expressed in u-PA. Exons VIII and IX of t-PA coding for the second kringle are also absent in the u-PA gene. Figure 8 illustrates some of the correlations between exon structure and the functional domains of (pre-pro) u-PA. Exon II contains the coding sequence for the signal peptide consisting of a stretch of 20 amino acids. Exons III and IV code for the EGF-domain, exon V and VI for the single kringle. The 5′ region of exon II coding for the peptide connecting the A and B chain of u-PA is 39 base pairs longer than the corresponding exon X of the t-PA gene. The 3′ region of exon VII and exons VIII to XI code for the B chain.

Pro-u-PA contains 411 amino acids and has a MW of 54,000. A hypothetical model of pro-u-PA is illustrated on figure 9. The primary amino acid sequence exhibits roughly 50% homology with the corresponding t-PA domains (see Figure 5). The A-chain contains the EGF region, a single kringle and a connecting peptide. The B-chain consists of 252 amino acids. The three active site residues are His 204, Asp 255 and Ser 356. Plasmin and kallikrein cleave scu-PA at position 158 (56) producing tcu-PA (HMW-UK), held together by a disulfide bridge. Commercial, highly purified UK often has phenylalanine at its C-terminal end of the A-chain (45). It is not known at present, which enzyme is capable of cleaving the Phe 157-Lys 158 bond. Thrombin and plasmin are furthermore capable to split the Arg 156-Phe 157 bond, resulting in an inactive protein. Plasmin and other enzymes present in urine can cleave the Lys 134-Lys 135 bond, producing LMW-UK (MW 33,000) consisting of a remnant of 21 amino acids of the A-chain, and the intact B-chain with preserved catalytic activity (57):

plasmin, others? ↓ (Lys 134 - Lys 135) · ? ↓ (Glu 143 - Leu 144) · thrombin ↓ (Arg 156 - Phe 157) · ? ↓ (Phe 157 - Lys 158) · plasmin, kallikr. ↓ (Lys 158 - Ile 159)

Ser ... Lys - Lys ... Glu - Leu ... Arg - Phe - Lys - Ile ....
1 134 135 143 144 156 157 158 159

rudimentary A-chain of LMW-UK (135–157)

commonly diserved A-chain in commercial HMW-UK preparations (1–157)

A-chain of "physiological" tcu-PA (1–158)

N-terminal of B-chain (159–)

LMW-pro-u-PA produced in tissue culture (144–)

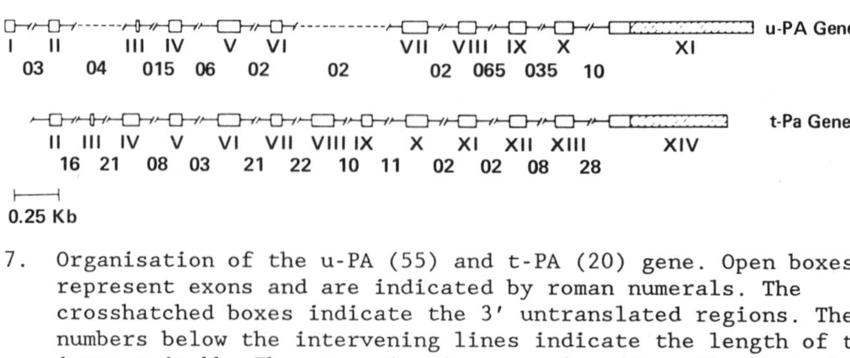

Fig. 7. Organisation of the u-PA (55) and t-PA (20) gene. Open boxes represent exons and are indicated by roman numerals. The crosshatched boxes indicate the 3′ untranslated regions. The numbers below the intervening lines indicate the length of the introns in kb. The exon size is on scale. (From Blasi et al. (1)).

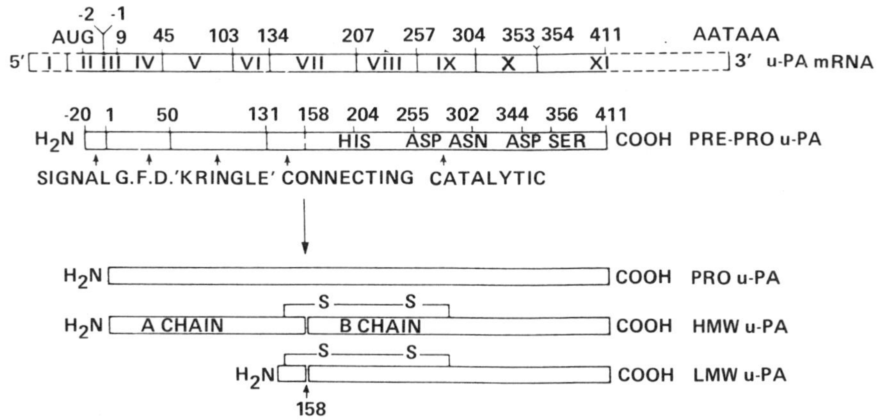

Fig. 8. Correlation of transcribed u-PA mRNA to domains of the mature pre-pro u-PA protein and conversion of pro-u-PA (scu-PA) to HMW-UK (tcu-PA) and LMW-UK. Exons are indicated by roman numerals, amino acid positions by arabic numerals. G.F.D.: growth factor domain (From Blasi et al. (1)).

Still another cleavage product can be observed in the conditioned medium of CALU-3, a human lung adenocarcinoma cell line and of monkey kidney cell cultures (58, 59). The cleavage of the Glu 143-Leu 144 bond by an unidentified enzyme results in a low molecular weight pro-UK with intact Lys 158-Ile 159 bond. This molecule has a MW of 32,000 and similar enzymatic properties as the 54,000 MW scu-PA.

The single chain u-PA has many properties of a zymogen. It incorporates DFP much more slowly (60, 61) than tcu-PA, has a poor activity on chromogenic substrates such as pyro- Glu-Gly-Arg-pNA (S-2444) and on fibrin plates (47, 61). It does not form complexes with its natural inhibitors PAI-1 and PAI-2. In a purified in vitro system employing scu-PA, glu-plasminogen, 500 KIU/ml of aprotinin and dansyl-Glu-Gly-Arg-chloromethyl-ketone to block plasmin and tcu-PA mediated secondary reactions, there was hardly any (0.4%) activation of glu-plasminogen to glu-plasmin (61). These observations lead several authors to postulate that pro-UK is a true zymogen with little katalytic activity (47, 60-62).

Lijnen et al. and Collen et al., on the other hand, observed that pro-u-PA brings about the conversion of plasminogen to plasmin even in the presence of aprotinin, physiological concentrations of alpha 2-antiplasmin, and/or saturating concentrations of the chromogenic substrate S-2251 (63, 64). For this reason, they proposed that the term pro-UK or pro-u-PA be replaced by the more descriptive term scu-PA (40). Since it is not certain, that the pro-u-PA preparations of Lijnen et al. and of Collen et al. were absolutely free of tcu-PA and that the methodology used to quench forming plasmin was 100% effective, the question as to whether scu-PA is or is not a true zymogen has not been settled yet.

Although pro-u-PA has fibrin-specific thrombolytic activity (fibrinolysis >> fibrinogenolysis) it does not bind to any appreciable degree to fibrin (65). Experiments with ^{125}I-pro-UK added to fibrinogen and thrombin resulted in over 95% recovery of ^{125}I in the supernatant after centrifugation of the clot. ^{125}I-pro-UK had been converted to a two-chain molecule and in the process, lost most of its amidolytic activity (65). It therefore appears that cleavage of pro-UK at the position Arg 156-Phe 157 results in an inactive molecule and binding studies which are based on the

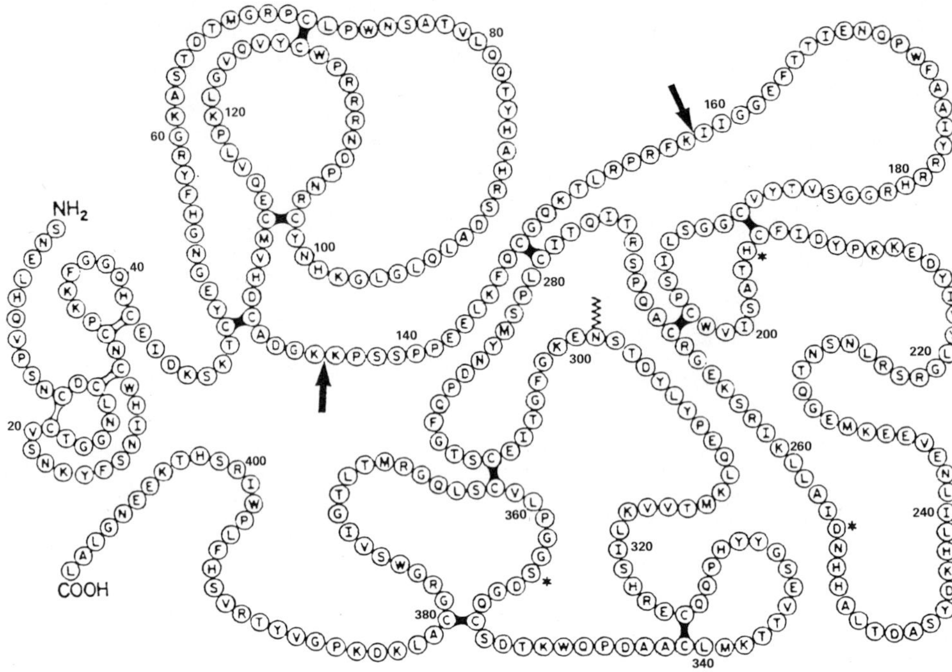

Fig. 9. Schematic diagram of the secondary structure of u-PA. The solid black bars indicate the potential disulfide bridges based on homology with other serine proteases and kringles. The assignment of disulfide bridges in the domain which resembles EGF is less certain and are indicated with open bars. The arrows indicate the cleavage sites between lysine and isoleucine that generates the two chain molecule and between the two lysines, which converts the high molecular weight form to the low molecular weight form. The zig-zag line at Asp 302 indicates the unique glycosylation site.

determination of the disappearance of u-PA activity in the clot supernatant (60) do not, in fact, reflect the binding of pro-UK to fibrin.

On a molar basis, pro-UK has a 3 to 4 times better thrombolytic effect than tcu-PA while causing less fibrinogen degradation (66-75). Thus it appears that the presence of fibrin in the circulation induces pro-u-PA by some as yet poorly understood mechanism, to take on thrombolytic activity without inducing fibrinogenolysis. Lijnen et al. have found that human plasma competitively inhibits the activation of plasminogen by pro-u-PA and attributed, at least in part, the inhibitory activity to Gc-globulin (63). To these authors, the presence of competitive inhibitors in human plasma explains the relative stability of pro-u-PA in plasma and the lack of activation of the plasma fibrinolytic system in the absence of fibrin (Fig. 10).

It is also possible, that there is always some plasmin and/or kallikrein present on the surface of a clot and that one or both of these enzymes convert locally clot bound glu-plasminogen to plasmin which then in turn attacks fibrin (see also below, contact activation of the fibrinolytic system). The therapeutic intravenous infusion of 70 mg of recombinant pro-UK into patients results in: 1) circulating u-PA antigen levels which are roughly 1000 times higher than the preinfusion levels, 2) the lysis of coronary thrombi comparable to that obtained with equivalent concentrations of t-PA, 3) a proteolytic state characterized by low fibrinogen, plasminogen and alpha$_2$-antiplasmin levels (75). The half-disappearance rate of recombinant pro-UK, measured after termination of the infusion was 8 min. (75).

The urokinase-receptor

Whereas t-PA synthesis occurs mainly in endothelial cells (76), immunocytochemical staining of tissues with anti-UK antibodies reveals that many cells of different origin, such as fibroblasts, epithelial cells, pneumocytes, decidual cells of the placenta and others, produce (pro)u-PA (77). It is generally believed that t-PA plays its major role in the prevention of the extension of intravascular thrombi, while u-PA probably is important for a host of pericellular proteolytic mechanisms, such as tissue remodelling, tumor invasion, embryogenesis, fertilization and inflammatory processes (5, 6). Recently it has been discovered that many cells contain a receptor which is specific for pro-UK and HMW-UK but does not bind LMW-UK (78-81).

The sequence which is critical for the binding has been located to the EGF domain of human scu-PA and tcu-PA, and resides probably in the amino acid sequence 18-30 (82). Cell bound scu-PA can be converted to tcu-PA, without dissociation from the membrane (83). K_d for binding is of the order of 10^{-9} to 10^{-10} M and there are about 50,000 binding sites in the monocyte-like U937 cell. No internalization of the bound u-PA can be detected after several hours incubation of cells at 37°C. There exists an absolute specificity for the binding of u-PA since related molecules containing the EGF domain such as t-PA and clotting factors IX and X do not bind to the UK-receptor (reviewed in ref. 82). Since membrane bound u-PA maintains its ability to activate plasminogen to plasmin and is protected to some extent from extracellular PA-inhibitors, it is likely that the presence of u-PA receptors on many normal and malignant cells provides an efficient mechanism for extracellular matrix breakdown.

Factor XII

Since factor XII does not seem to be necessary for maintaining normal hemostasis, but is involved in the contact activation of the fibrinolytic system it is included here in the list of fibrinolytic components. Its physicochemical and functional properties have been recently reviewed (74). Human factor XII is a single-chain glycoprotein of MW 76,000, containing 17% carbohydrate. Its plasma concentration is about 30 mg/l (Table 1). The complete primary amino-sequence has been established (85, 86). The cDNA has been isolated and the nucleotide sequence determined (87, 88) (Table 2). Factor XII contains several of the domains encountered in other components of the fibrinolytic system. The N-terminal is composed of a type II finger

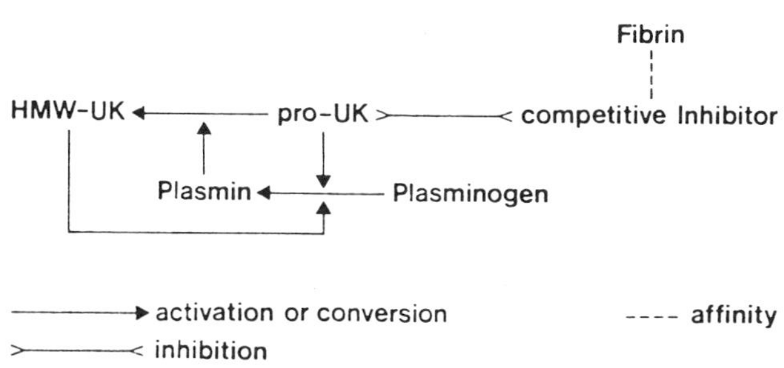

Fig. 10. Hypothetical model to explain the fibrin specificity of pro-UK.

Table 1. Physiochemical properties of components of the fibrinolytic system

	MW[1] kD	Chain(s)	Plasma concentration mg/l	M	Carbodydrate percent	MW on basis of aa sequence kD
Plasminogen	92	1	200	2 μM	2	88
t-PA	68	1	0.005	70 pM	7/13	59
u-PA	54	1	0.008	150 pM	7	46
Prekallikrein	88	1	40	450 nM	15	69
Factor XII	80	1	30	375 nM	17	66
Alpha 2-antiplasmin	70	1	70	1 μM	13	51
Alpha 2-macroglobulin	725	4	2,500	3 μM	8	4 x 161
PAI-I	52	1	0.05	1 nM		43
PAI-2	46/70	1	< 0.005	< 100 pM		44
Cl-inhibitor	105	1	180	1.7 μM	35	53
HRG	75	2	100	1.5 μM	14	58

1 Usually as determined by SDS-PAGE; these MWs after substraction of carbohydrate content often do not fit with MW calculated from the primary amino acid sequence.

t-PA: Tissue-type plasminogen activator; u-PA: urinary-type PA. PAI-1: PA inhibitor; PAI-2: PA inhibitor-2.
HRG: histidine-rich glycoprotein.

Table 2. Genom and synthesis of some components of the fibrinolytic system

	Gene length kb	Chromosome	mRNA kb	Pre-Pro protein nr. aa	Protein nr. aa
Plasminogen	> 21	6	sequence incomplete		790
t-PA	29	8	2.7	562	530
u-PA	6.4	10	2.4	431	411
Factor XII	> 20		2.6		596
Prekallikrein			2.4	638	619
Alpha 2-antiplasmin			2.2	464 ?	452
Alpha 2-macroglobulin, unit		12	4.6	1,474	1,451
PAI-1		7	3.0	402	379
PAI-2			1.9	415	393
C1-inhibitor	> 16	11	1.8	500	478
HRG			2.0	525	507

kb: kilobases; aa: amino acids. Other abbreviations as in Table 1.

domain. Moving toward the C-terminal we find an EGF region, a type I finger domain, another EGF region, and a kringle followed by the serine protease B-chain (Fig. 2). Factor XII plays a key role in mediating via kallikrein the activation of scu-PA to tcu-PA (see below, contract activation).

High molecular weight kininogen (HMWK)

HMWK is a cofactor of the contact activation system. It is a glycoprotein of MW of 110,000. The cDNA has been cloned recently (89). Its physicochemical and functional properties have been recently reviewed (90).

Prekallikrein

Human plasma prekallikrein is a glycoprotein of MW 88,000 containing 15% carbohydrate. It is present in plasma at a concentration of 35 to 45 mg/l (Table 1) and circulates as a noncovalent complex with HMWK (reviewed in ref. 91). It is converted to the serine protease kallikrein by factor XIIa through cleavage of the peptide bond Arg 371-Ile 372. The complete primary amino acid sequence has been determined by a combination of Edman degradation and cDNA sequencing techniques (92) (Table 2). The A-chain of prekallikrein consists of 371 amino acids and exhibits 4 unusual tandem repeats of 90 to 91 residues. These structures have hitherto only been found in factor XI (92). The B-chain (248 amino acid residues) contains the typical serine protease catalytic region (Fig. 2). Kallikrein converts scu-PA to tcu-PA with high efficiency (56).

Alpha 2-antiplasmin

The serpin alpha$_2$-antiplasmin is the primary inhibitor of plasmin. It is a single-chain molecule of MW of 70,000 and contains approximatively 13% carbohydrates. Its plasma concentration is 70 mg/l, i.e. 1 μM which is about half of the concentration of plasminogen (2 μM) on a molar basis. It is synthetized by the liver and correspondingly reduced in patients with advanced impairment of hepatic function. Its biological half-life is 3.3 days.

The cDNA of alpha$_2$-antiplasmin has been cloned recently (93, 94) and the primary amino acid sequence derived. Alpha$_2$-antiplasmin has three major functional properties: 1) it inhibits plasmin very rapidly, 2) interfers with the adsorption of plasminogen to fibrin and, 3) undergoes crosslinking with the alpha-chains of fibrin during clotting. Alpha$_2$-antiplasmin has specific binding sites for plasmin and fibrin (95) (Fig. 11).

The former, situated in the C-terminal region of alpha$_2$-antiplasmin will bind reversibly to the LBS I of plasmin and might be called "LBS binding site" (96). This reaction is extremely rapid (t/2 of approx. 0.1 s); the K_d of 2.10^{-10}M illustrates the very high affinity of alpha$_2$-antiplasmin to plasmin. In the presence of lysine or 6-amino-hexanoic acid (EACA), two components which fix to the LBS I of plasmin, the interaction between alpha$_2$-antiplasmin and plasmin is greatly inhibited. Alpha$_2$-antiplasmin from which a 2,000 MW C-terminal sequence had been cleaved also reacted quite slowly with plasmin (reviewed in (95)).

In a second step, a serine residue in the active site of plasmin reacts with the reactive site residues Arg 364-Met 365 (94) of alpha$_2$-antiplasmin, with subsequent cleavage of this peptide bond. An 11,000 MW C-terminal sequence is liberated but remains noncovalently bound to the LBS I of plg (96). The resulting MW 150,000 covalent complex probably is held together by the newly formed Ser_{plg}-Argalpha$_2$-antiplasmin bond (94).

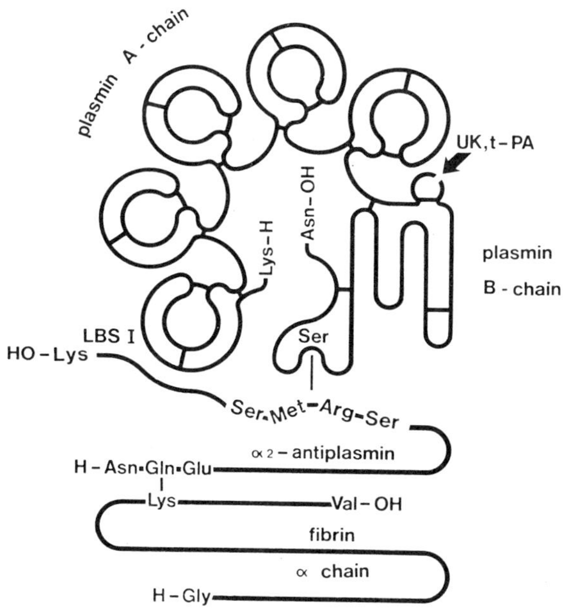

Fig. 11. Interaction of alpha$_2$-antiplasmin with plasminogen and the alpha-chain of fibrin. For lys-plasmin(ogen) only the high affinity lysine binding site is indicated (LBS I). H designates N-terminal, OH- C-terminal of a protein. Figure constructed on the basis of references (14) and (95).

The cross-linking site of alpha$_2$-antiplasmin with fibrin is Gln 2, very close to the N-terminal (Fig. 11). Crosslinking is established via F XIIIa which catalyses a glutamyl-lysine bond between the Glu 2 of alpha$_2$-antiplasmin and a lysine in the C-terminal portion of the alpha-chain of fibrin. Thus, alpha$_2$-antiplasmin has 3 functional sites: the "LBS binding site", the reactive site and the crosslinking site.

Lys-plasminogen has a higher affinity for native fibrin than glu-plasminogen. This is probably due to the fact that the N-terminal peptide of native glu-plasminogen covers the high affinity LBS I (Fig. 4) or to different configurations of these 2 forms of plasminogen. Alpha$_2$-antiplasmin competes with the binding of plasminogen to fibrin as do lysine, epsilon-amino-caproic and tranexamic acid, two antifibrinolytic agents. On the other hand, the crosslinking of alpha$_2$-antiplasmin to fibrin also enhances the indirect binding of plasminogen to fibrin as illustrated in Fig. 11. During coagulation approximately 20% of the alpha$_2$-antiplasmin crosslinks to the C-terminal portion of the alpha-chains of fibrin, resulting in a clot-bound concentration of alpha$_2$-antiplasmin of 0.2 uM. This concentration equals that of fibrin-bound plasminogen. Because of this molar balance between alpha$_2$-antiplasmin and plasminogen, a normal blood clot does not lyse spontaneously despite the fixation of some t-PA to the clot. However, a clot prepared from plasma deficient in alpha$_2$-antiplasmin, submerged in normal plasma will lyse in a few hours. This well illustrates that the 20% crosslinked alpha$_2$-antiplasmin is more efficient in the prevention of spontaneous clot lysis than the remaining 80% free alpha$_2$-antiplasmin in the plasma milieu.

Fibrin plays a very important role in the interplay between t-PA, plasminogen and alpha$_2$-antiplasmin. The plasma of patients congenitally deficient in alpha$_2$-antiplasmin does not exhibit signs of increased fibrinolytic activity: no free plasmin can be detected and fibrin(ogen) degradation products (FDP's) are absent. After clotting of the plasma, there is progressive formation of plasmin and of FDP's. If one adds ^{125}I-labelled fibrin monomers to unclotted alpha$_2$-antiplasmin deficient plasma ^{125}I is released from the monomers. Addition of alpha$_2$-antiplasmin abolishes the generation of free ^{125}I (97). These experiments clearly show that under these conditions fibrinolysis can be induced by fibrin itself.

<u>Alpha$_2$-macroglobulin</u>

Alpha$_2$-macroglobulin is a major proteinase inhibitor of wide specificity. It is a glycoprotein of MW 725,000, consisting of 4 identical chains of MW 161,000 and containing 8 to 11% carbohydrate (Table 1). Its plasma concentration is 2.5 g/l (3 μM). The tetramer is arranged as a pair of dimers, each consisting of two monomers linked together via a disulfide bond. The complete sequence has been established and the gene locus attributed to chromosome 12 (98, 99) (Table 2). Alpha$_2$-macroglobulin contains two reactive sites:

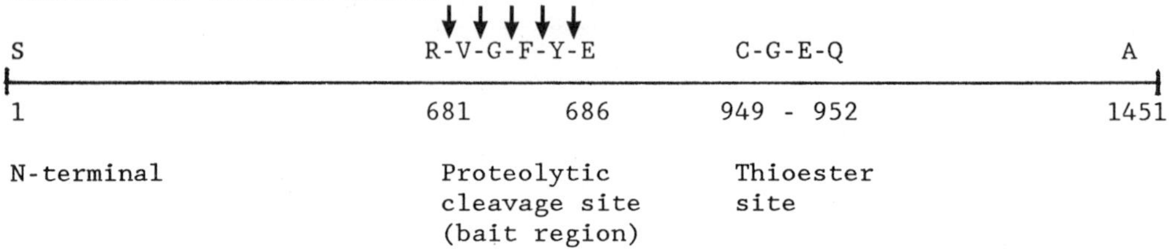

the sequence Arg 681-Val-Gly-Phe-Tyr-Glu 686 represents a "multipurpose" proteolytic cleavage site. The bait region offers substrate specificity to many different proteases. These cleave peptide bonds, indicated by arrows, according to their catalytic specificity (reviewed in (100)). Cleavage leads to a conformational change of the alpha$_2$-macroglobulin molecule and activates the thiol ester bond leading to its hydrolysis

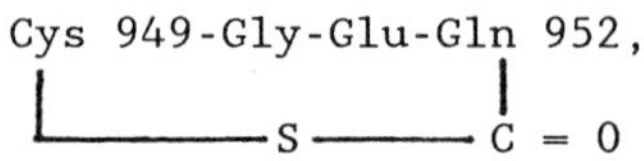

and, as a consequence, to covalent binding of the proteinase to the reactive glutamate in position 952 (99).

Alpha$_2$-macroglobulin is a "second defense line" inhibitor of many components of the fibrinolytic system and inactivates, at relatively low rates, plasmin, kallikrein, tcu-PA, t-PA and the streptokinase-plasmin(ogen) complex. Following extensive activation of the fibrinolytic system such as occurs during acute disseminated coagulation and therapy with thrombolytic agents, the generated plasmin will first be neutralized by alpha$_2$-antiplasmin. However, since plasminogen exists in an about 2 fold molar excess over alpha$_2$-antiplasmin, the latter will be consumed before it can inactivate all circulating plasmin. In this situation alpha$_2$-macroglobulin plays the role of a scavenger protease inhibitor and neutralizes excess plasmin by the formation of covalent plasmin/alpha$_2$-macroglobulin complexes. The latter are recognized by specific receptors on the surface of fibroblasts, macrophages and cells of the reticuloendothelial system, bound and neutralized.

Alpha$_2$-macroglobulin does not belong to the antithrombin III/alpha$_1$-antiprotease family of serpins. It exhibits sequence homology with the complement components C3 and C4 (99).

Plasminogen activator inhibitor 1 (PAI-1)*

Since the sixties, several authors reported on the existence of specific PAI's in human blood, but the lack of appropriate kinetic data did not allow the identification of a specific PAI in human plasma (reviewed in (3, 4)). At the Lausanne International Fibrinolysis Congress in 1982, Kruithof et al. demonstrated that human plasma contains small amounts of a specific PAI which rapidly inactivates physiological concentrations of t-PA and tcu-PA. This inactivation was accompanied by the formation of a PA/PAI complex of MW of 110,000 after the addition of t-PA, and of 95,000 after the addition of tcu-PA to inhibitor-rich plasma (101). PAI-1 is identical with the endothelial type PAI (102-108) and is also produced by hepatocytes, smooth muscle cells, fibroblasts and malignant cell lines (reviewed in (3)). In the following years the physicochemical and functional properties of plasmatic PAI-1 were rapidly elucidated by several research groups and the inhibitor purified to homogeneity (108, 114). Between June 5 and October 13, 1986 seven different, partly related-partly unrelated, research groups submitted a total of 4 manuscripts dealing with the cDNA sequence of PAI-1 (115-118). The race to elucidate the structure-function relationships of PAI-1 well illustrates the physiological importance attributed to this inhibitor.

PAI-1 is a glycoprotein of MW of about 52,000. It consists of 379 aminoacids. The gene for PAI-1 is located on chromosome 7. PAI-1 is present in human plasma at a molar concentration which is approximately 1000 times lower than that of alpha$_2$-antiplasmin (Tables 1 and 2), but 2 to 3 times higher than that of the two PA's it inhibits. PAI-1 belongs to the family of the serpins. In the amino acid stretch 328-359 (alpha$_1$-protease inhibitor numbering 340-371), flanking the reactive site P$_1$-P'$_1$ it has 55% homology with antithrombin III, 41% with PAI-2, 39% with alpha$_2$-antiplasmin, 32% with C1-inhibitor and 21% with alpha$_1$-protease inhibitor (Table 4).

PAI-1 reacts equally well with sct-PA, tct-PA, tcu-PA, but not with scu-PA or SK (110-113, 120). The second order rate constants are of the order of 10^7 $M^{-1}s^{-1}$. The interaction of the PA with PAI-1 probably results first in a reversible complex, which in a second step becomes covalent after the cleavage of the Arg 346-Met 347 bond has taken place. Finally the 33 amino acid stretch Met 347-Pro 379 is released from the complex resulting in a final molecular weigth which is approximately 4,200 less than the composite MW of the activator and the inhibitor (14).

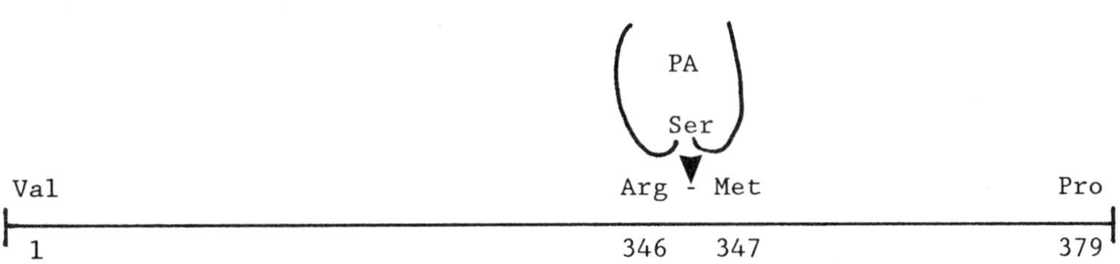

In human blood, the plasma compartment accounts for only about three forth of the total PAI-1 content; a larger compartment is constitued by the platelets where PAI-1 is probably stored in the alpha-granules (122-124) and rapidly released upon activation of platelets by a variety of agonists (123).

* Nomenclature as proposed by the International Committee on Thrombosis and Fibrinolysis (40).

Table 4. Homologies of serpins in the region flanking the reactive site

	340									350							
Alpha$_2$-AP	L	S	E	V	G	V	E	A	A	A	A	T	S	-	I	A	M
PAI-1	V	N	E	S	G	T	V	A	S	S	S	T	A	V	I	V	S
PAI-2	V	N	E	E	G	T	E	A	A	A	G	T	G	G	V	M	T
C1-inh.	L	T	E	T	G	V	E	A	A	A	A	S	A	-	I	S	V
AT III	V	N	E	E	G	S	E	A	A	A	S	T	A	V	V	I	A
Alpha$_1$-PI	I	D	E	K	G	T	E	A	A	G	A	M	F	L	E	I	P
			*		*		+	*	+	●		●					

				360													370	
	$P_2P_1P_1P_2P_3$																	
Alpha$_2$-AP	S	R	M	S	L	S	S	-	-	-	F	S	V	N	R	P	F	L
PAI-1	A	R	M	A	P	E	E	-	-	-	I	I	M	D	R	P	F	L
PAI-2	G	R	T	G	H	G	G	P	Q	-	F	V	A	D	H	P	F	L
C1-inh.	A	R	T	L	L	-	-	-	-	V	F	E	V	Q	Q	P	F	L
AT III	G	R	S	L	N	P	N	R	V	T	F	K	A	V	R	P	F	L
Alpha$_1$-PI	-	M	S	I	P	-	-	-	P	E	V	K	F	N	P	P	F	V
		+									●					*	*	+

The amino acid numbering system for alpha$_1$-protease inhibitor is used (119). Alpha$_2$-AP: alpha$_2$-antiplasmin, PAI-1 and PAI-2: plasminogen activator inhibitor 1 and 2, C1-inh.: C1-inhibitor, AT III: antithrombin III, Alpha$_1$-PI: alpha$_1$-protease inhibitor. The number of homologies exceeding 3 are indicated with ● (4), + (5) and * (6) homologies.

It has not yet been clearly determined if the PAI-1 stocked in the platelets is produced by megakaryocytes. The plasma concentration of PAI-1 is not correlated to the beta-thromboglobulin level; it therefore appears unlikely that plasma PAI-1 is merely a platelet-release product (123). PAI-1 is produced by the endothelial cells (102-106), but also by cultured hepatocytes (125). At the present time, it is not known if vessels or the liver are the primary synthetic site.

PAI-1 isolated from the conditioned medium of endothelial cell cultures is very labile (126-128). Its activity in the medium decays with a half-life of 20 min., probably due to a large measure to oxydation of methionine and cysteine residues (129). In chloramine T inactivated PAI-1, activity could be restored after treatment with sulfoxide peptide reductase in the presence of dithiothreitol indicating that at least one critical methionine is responsible for the loss of PAI-activity upon iodination with chloramine T. Inactivated "latent" PAI-1 in tissue culture fluid and in plasma can be partly "reactivated" by treatment with denaturing agents such as urea, guanidine hydrochloride and SDS (126, 127).

A number of stimuli increase production and/or release from a variety of cells. Thrombin concentrations of 1 U/ml stimulated PAI-1 activity in the serum-free medium several fold. Although t-PA antigen production and release were also increased, the net effect of thrombin was to increase the inhibitory potential of the culture medium (130-132). Dexamethasone stimulates PAI-1 production in HTC hepatoma cells (133, 134). Transforming growth factor-β induces PAI-1 production in normal cultured human lung fibroblasts (135), and endotoxin leads to marked stimulation of production and to release of PAI-1 from endothelial cells (132, 136, 137). Many disease states are associated with elevated plasma PAI-1 levels (see below, fibrinolysis in health and disease).

Plasminogen activator inhibitor 2 (PAI-2)

In 1968 Kawano et al. discovered that extracts of human placenta strongly inhibited urokinase (138). It has now become clear that this inhibitor is different from PAI-1 and identical with that isolated from monocytes/macrophages, granulocytes, human buffy coats and the human macrophage-like U-937 cell line (139-142). PAI-2 has been purified to homogeneity from human placenta (143) and from U-937 cells (41). It is to be expected that the complete cDNA sequence will be published in 1987 (144, 145). PAI-2 exists in 2 forms; an intracellular non-glycosylated form has a MW of 46,000; the secreted PAI-2 is glycosylated and has a MW of 70,000. It contains 393 amino acids.

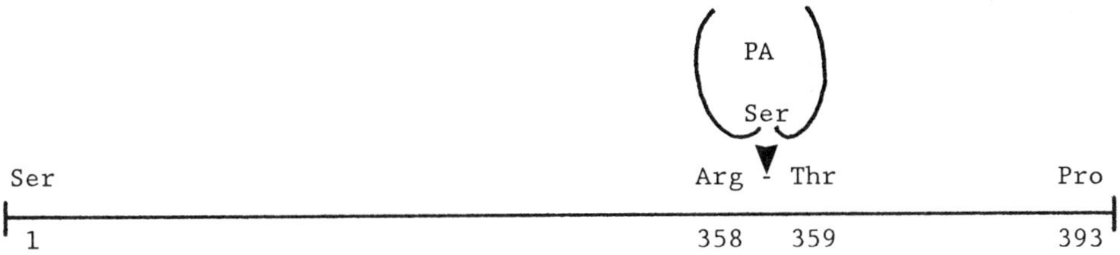

By analogy with other serpins (Table 4) the reactive site probably is constitued by the peptide bond Arg 358-Thr 359. In the amino acid stretch 341-373 (alpha$_1$-protease inhibitor numbering 340 to 371), flanking the reactive site, PAI-2 has 63% homology with antithrombin III, 43% with alpha$_2$-antiplasmin, 41% with PAI-1, 39% with C1-inhibitor and only 25% with alpha$_1$-protease inhibitor (Table 4). It is astonishing that the sequence homology is considerably better with antithrombin III than with PAI-1, since the second order rate constants of PAI-1 and PAI-2 are 2 to 3 orders of magnitude greater for the interaction with tct-PA and tcu-PA than with thrombin. Apparently there must exist other sites on the PAI-1 and PAI-2 molecule which mediate binding to PA's. PAI-2 is not measurable with presently available assays in normal human plasma. Its plasma level rises steeply in pregnancy to mean values of 250 ng/ml in the third trimester (146, 147). It is not known at the present time if the PAI-2 found in pregnancy plasma is of placental origin, where PAI-2 can be detected by immunohistochemical methods in the trophoblastic epithelium (148). PAI-2 reacts well with tct-PA and tcu-PA but poorly with sct-PA and scu-PA (41, 149; Table 5).

Two other inhibitors, PAI-3 and protease nexin, exhibit second order rate constants with PA's which are quite a bit lower than those of PAI-1 and PAI-2 (Table 5). PAI-3 has been found in urine and plasma (40) but is not yet thoroughly characterized. Protease-nexin is synthesized by fibroblasts, heart muscle cells and kidney epithelial cells. It is not found in human plasma (reviewed in (3)).

C1-inhibitor

C1-inhibitor is a highly glycosylated plasma protease inhibitor of MW 105,000. It consists of 478 amino acids (Tables 1 and 2) and inhibits activated subcomponents of the first component of complement, factors XIIa and XIa, plasma kallikrein and plasmin. Its primary amino acid and cDNA structures reveal that it belongs to the family of serpins (148-150; Table 4). The gene has a length of at least 16 kb and is situated on chromosome 11 (150). By virtue of its inhibitory effect on components of the contact

Table 5. Physico-chemical and functional characteristics of PA-inhibitors

Inhibitor	PAI-1	PAI-2	PAI-3	Protease nexin
Sources	plasma platelets endothelial cells hepatoma cells fibrosarcoma cells melanoma cells	pregnancy placenta leucocytes monocytes histiocytic lymphoma cells	urine plasma	fibroblasts fibrosarcoma cells epithelial celss
Designation used	endothelial cell type PAI t-PA inhibitor antiactivator plasma PAI platelet PAI PA-inhibitor 1	placental PAI u-PA-inhibitor PA-inhibitor 2	urokinase inhibitor	protease nexin
MW	54,000	47,000 (60,000)**	50,000	51,000
pI	4.5-5.0	5.0 (4.4)**	--	7.5-7.8
Glycosylated	yes	no (yes)**	yes	--
Heparin binding	no	no	yes	yes
Rate constant of inhibition (in $M^{-1}s^{-1}$)			8.10^3 (-H)*	
tcu-PA (urokinase)	$>10^7$	9.10^5	9.10^3 (+H)*	2.10^5
tct-PA	$>10^7$	2.10^5	10^3	3.10^4
sct-PA	$>10^7$	9.10^3	10^3	1.10^5
thrombin	--	1.10^2	4.10^4 (-H)*	6.10^5 (-H)*
plasmin	--	1.10^2	--	1.10^5

** (between brackets the characteristics of the glycosylated form of PAI-2). * ((-H) and (+H): kinetics of inhibition in the absence and presence of heparin). (Taken from (40)).

activation system and on plasmin it probably participates in the inhibition of contact dependant fibrinolysis, i.e. the conversion of scu-PA to tcu-PA.

Histidine-rich glycoprotein (HRG)

Molecular cloning has recently revealed the primary structure of HRG. It is a glycoprotein of MW of 75,000, consisting of 507 amino acids. It has a histidine-rich region in the amino acid stretch 330-389, which shows homology with histidine-rich regions in human and bovine kininogen (153). HRG is a competitive inhibitor of plasminogen as is epsilon-aminocoproic acid and has a fairly high affinity for the LBS I of plasminogen (K_d approx. 1 μM). On the basis of the molar concentrations of plasminogen (2 μM) and of HRG (1,5 μM), Lijnen et al. postulate that approximately 50% of the circulating plasminogen exists in the form of a reversible complex with HRG (154); this evidently reduces the proportion of plasminogen available for binding to fibrin during and after coagulation. In studies comparing the effect of HRG and of alpha$_2$-antiplasmin on the binding of plasminogen to fibrin Ichinose et al. arrived at the conclusion that alpha$_2$-antiplasmin interferes more effectively with this reaction than HRG (155).

Modulators of fibrinolytic activity

Thrombospondin, a large glycoprotein of MW of 450,000 is found in the alpha granules of platelets and is released upon thrombin-stimulation of platelets (reviewed in 156). Thrombospondin binds to HRG and to plasminogen. It is a non competitive inhibitor of the activation of plasminogen by t-PA in the presence of fibrin, suggesting that thrombospondin may form a ternary complex with fibrin and t-PA. In the absence of fibrin thrombospondin accelerates plasminogen activation by t-PA by a factor of 30 (K_{cat}/Km approx. 0.006 μM^{-1} sec^{-1}). This intermediate efficiency might be well suited for matrix breakdown related to cell migration and tissue remodeling in the extravascular spaces (Fig. 12).

Recently a novel component of the fibrinolytic system has been identified. This MW 68,000 protein, tentatively called tetranectin, consists of four subunits of MW of 17,000 and binds to kringle 4 of plasminogen. Its plasma concentration is 15 mg/l. Kringle 4 binding protein accelerates the conversion of plasminogen to plasmin by t-PA in the presence of poly-D-lysine and possibly of fibrine (157).

PHYSIOLOGY OF CLOT LYSIS

After stimulation of t-PA release from endothelial cells by exercise, stasis, or the injection of DDAVP, fibrinolytic activity rapidly disappears from the circulation (43). Despite the short half-life of t-PA, the biologic effect extends over a considerably longer period. Agnelli and co-workers have injected recombinant t-PA into rabbits in which they previously produced a ^{125}I-fibrin-labeled thrombus in a jugular vein (158). There was no detectable circulating activity 30 minutes after a bolus dose of 75,000 IU of t-PA/kg, but the thrombolytic effect persisted up to 2

These observations support the findings of Tran et al. who demonstrated that t-PA-poor clots immersed in post-DDAVP plasma (rich in free t-PA) took up t-PA from the surrounding milieu in a time-dependent fashion (159); the final clot-bound t-PA concentration reached values several fold above those of the surrounding milieu (Fig. 13).

In the absence of fibrin t-PA has little effect on plasminogen due to the unfavorable katalytic rate constant. Even a hundred-fold increase of

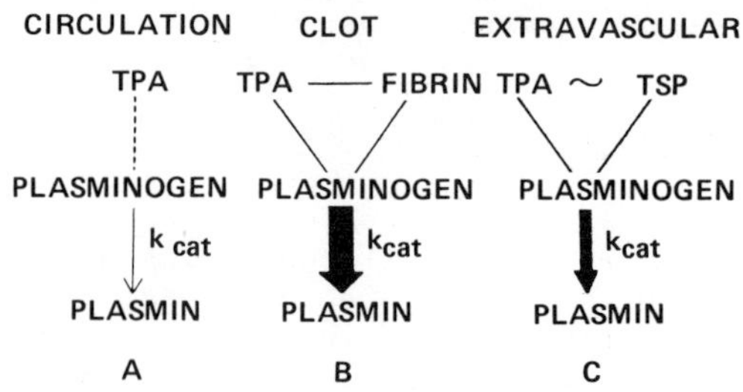

Fig. 12. Kinetics of plasminogen (Plg) activation by t-PA under three experimental conditions. In the fluid phase (A) activation of Plg to plasmin by t-PA is kinetically at a disadvantage, represented in this model as a narrow line (K_{cat}/Km). On a fibrin clot (B) activation is maximal, represented as a heavy arrow. On a thrombospondin (TSP) surface (C), activation is of intermediate efficiency (Taken from Silverstein et al. (156)). hours, as evidenced by the sustained release of radiolabel from the jugular thrombi.

free t-PA such as occurs in good responders after strenuous exercise or the injection of DDAVP does not produce plasminemia; should a small amount of plasmin be formed it will be very rapidly inhibited by alpha$_2$-antiplasmin.

On a fibrin surface these interactions proceed quite differently. A ternary complex between t-PA, glu-plasminogen and fibrin forms, facilitating the conversion of clot bound plasminogen to plasmin. Clot-bound plasmin is protected somewhat from the action of alpha$_2$-antiplasmin (Fig. 14).

During coagulation, approximately one tenth (0.2 μM) of the glu-plasminogen present in human plasma binds to fibrin. Simultaneously, an equimolar quantity of alpha$_2$-antiplasmin binds covalently to the lysine 303 residues of the alpha chains of fibrin (160). If the clot has adsorbed moderate amounts of t-PA, a fraction of the clot-bound plasminogen will be converted into plasmin. Despite the slower inhibition of clot-bound plasmin by alpha$_2$-antiplasmin, most of the generated plasmin will be neutralized by clot-bound alpha$_2$-antiplasmin. Effective fibrinolysis therefore necessitates additional fixation of plasminogen to the clot. Several authors have shown that t-PA triggers such a mechanism. It is dependent upon partial digestion of fibrin, which then binds larger amounts of plasminogen than does the native fibrin (37, 38, 161, 162). Although this process is plasmin dependent, it does not involve the conversion of glu-plasminogen into the lyse form (162).

These observations suggest that t-PA plays an important role in the dynamic equilibrium between clotting and lysis. The presence of CA^{2+} in the plasma allows the crosslinking of alpha$_2$-antiplasmin to fibrin, mediated by Factor XIIIa; this ensures the initial stability of the hemostatic plug. Clot lysis is initiated only if additional t-PA can be taken up from the surroundings; stasis upstream from the occluded vessel is one of the stimuli leading to release of t-PA from the endothelium. The additional uptake of t-PA by the clot results in conversion of small amounts of plasminogen to plasmin, which in turn partially digests fibrin. The latter exposes additional binding sites for the fixation of Glu-plasminogen to fibrin. Only when this additional clot-bound plasminogen has been converted to plasmin can the inhibitory effect of crosslinked alpha$_2$-antiplasmin be overcome and clot lysis proceed at an accelerated rate.

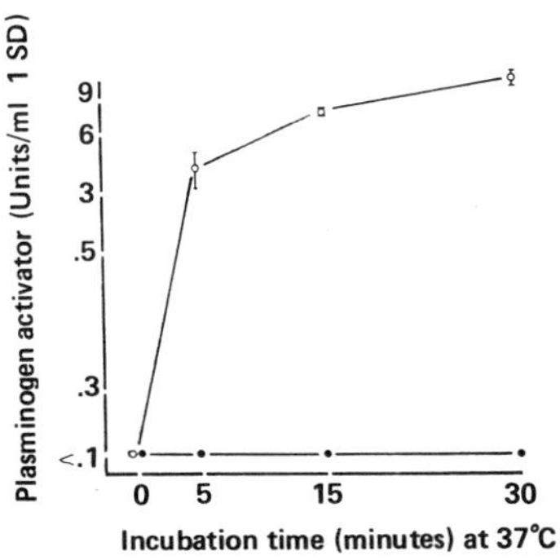

Fig. 13. Uptake of t-PA from the surrounding milieu onto clots. Aliquots of 0.2 ml of noncrosslinked plasma clots were incubated in 1.5 ml of normal (•——•) or of post-DDAVP plasma containing 4 IU of t-PA/ml (o——o). After the indicated incubation times, clots were washed, depolymerized in 0.017-M acetic acid, and t-PA assayed on fibrin plates (adapted from Tran-Thang et al. (159)).

This clot-localized process is subject to many modulators. PAI-1 released upon platelet activation may have an effect on the stability of venous thrombi through inactivation of free sct-PA. Its role probably is much greater in the stabilization of arterial thrombi which contain a large proportion of platelets and might adsorb less t-PA due to the fact that endothelial cells of the arterial bed produce less t-PA than those lining veins. The protein C pathway probably constitutes another modulating factor of fibrinolytic activity. Several authors have demonstrated that activated protein C forms complexes with PAI-1, thereby reducing the inhibitory effect of PAI-1 on fibrinolysis (163-165). For maximal activation of protein C by thrombin, thrombomodulin, phospholipids and the cofactor protein S are needed (165-166). The observation that patients with homozygens protein C deficiency do not exhibit a defect of t-PA release after exercise (167) does not rule out an imbalance of global fibrinolytic activity on a local level. Indeed, it appears that in vitro studies of fibrinolytic activity in protein C or protein S deficient plasma do not adequately reflect physiological mechanisms taking place on a damaged endothelial lining. It is well possible that, in heterozygous protein C deficiency, PAI-1 is destroyed at a lower rate than normal on the surface of platelet containing clots leading to persistently increased inhibitory activity of fibrinolysis (166, 167).

According to Lijnen et al. fibrin might displace a competitive inhibitor of scu-PA (63). Scu-PA might then directly activate clot bound plasminogen (Figure 10). Small amounts of clot bound plasmin certainly are capable of converting scu-PA to tcu-PA which then will activate clot bound plasminogen to plasmin. The u-PA pathway appears to be much less efficient to bring about clot lysis in the absence of contact activation than the t-PA pathway, since one needs approximately 70 to 100 IU of scu-PA/ml of plasma (15 nM) to achieve clot lysis within 2 hours whereas only 20 IU of free t-PA (0.6 nM) suffice for this effect. After activation of the contact phase, the urokinase pathway becomes much more efficient.

Thus if plasma is treated with dextran sulfate or ellagic acid and phospholipids in the presence of scu-PA, followed by clotting, 7 IU of scu-PA suffice to bring about 50% clot lysis within 90 min. (168).

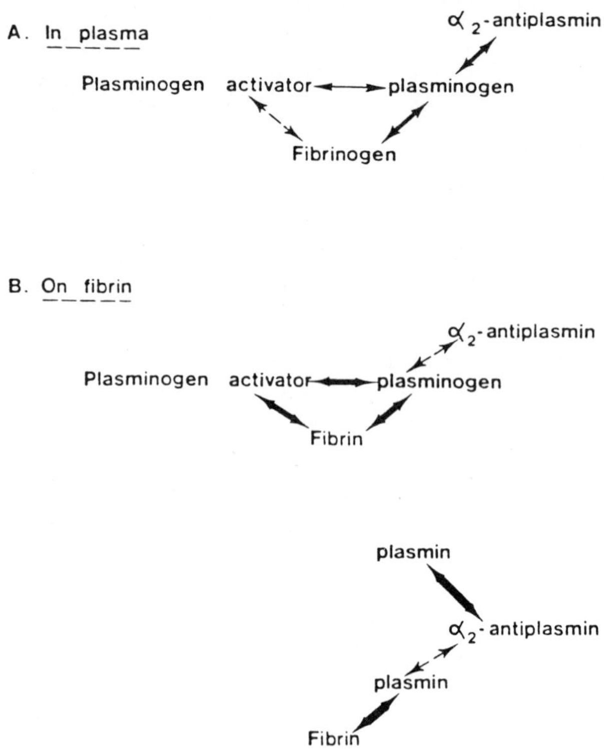

Fig. 14. Physiological fibrinolysis. Schematic representation of the interactions between fibrinogen, plasmin(ogen), alpha$_2$-antiplasmin and plasminogen activator. The size of the arrows is roughly proportional to the affinity between the different components (Taken from Lijnen et al. (19)).

THE CONTACT DEPENDANT PATHWAY OF FIBRINOLYSIS

Considerable controversy exists as to the components of the contact system involved in the stimulation of fibrinolytic activity after treatment of plasma with kaolin, ellagic acid or dextran sulfate (reviewed in (169)).

Kluft et al. postulate that there exists, in human plasma, a separate third pathway for the activation of plasminogen to plasmin through an as yet unidentified factor XII dependant proactivator which is converted to a plasminogen activator after contact phase activation (169). Our results indicate that contact activation and the generation of kallikrein fully account for the stimulation of fibrinolytic activity by Factor XII activating agents, such as dextran sulfate (Fig. 1). During contact phase activation kallikrein converts scu-PA into tcu-PA (56) and stimulates the fibrinolytic potential of scu-PA about twenty fold; conversion of scu-PA into the two chain form does not take place in plasma deficient in kallikrein (170) and is extremely slowed down in plasma deficient in Factor XII or HMWK. This conversion is independant of plasmin (Hauert et al. unpublished data). These observations suggest an important physiological role for scu-PA, conteracting excessive clot formation in clinical situations associated with contact activation of the coagulation system.

Figure 15 represents a rather simplistic model of the balance between profibrinolytic and fibrinolysis inhibiting proteins. Patients with deficiency of alpha$_2$-antiplasmin often have a bleeding tendancy (literature references in (171)). On the other hand, a heterozygous deficiency of plasminogen or dysplasminogenemia (literature references in (172)), of factor XII or of kallikrein (reviewed in (173)) may lead to thrombophilia. Protein C (and S) has been indicated at the bottom of figure 15, since it cannot be determined at the present time if the wellknown tendancy of protein C or S deficient patients for thromboembolic disorders is primarily due to lack of inactivation of factors Va and VIIIa or to diminished destruction of PAI-1.

Of all constitutive factors described (deficiencies or functional abnormalities of antithrombin III, protein C, protein S, heparin cofactor II, plasminogen, factor XII, prekallikrein, and of some particular fibrinogen variants) a defect of fibrinolytic activity represents by far the most commonly disserved abnormality of the hemostasis system associated with a tendancy to develop thrombosis. However, while the first mentioned conditions are, as a rule, genetically transmitted, this is not the case in the majority of patients with a fibrinolytic defect (174).

Most authors determined fibrinolytic activity in plasma euglobulins prepared before and after 10 to 20 min. of venous occlusion or after an intravenous injection of 4 ng/kg of DDAVP, using the classical fibrin plate method. A few authors found lower mean baseline fibrinolytic activity in patients with thrombophilia; nearly all investigators observed a decreased fibrinolytic response to a venous stasis test or to the infusion of DDAVP in about 25 to 40% of patients with a history of idiopathic thrombosis; i.e. occurring in the absence of any of the known risk factors (surgery, cancer, advanced age, presence of varices, heart failure, obesity, oral contraception) (174-178).

It has recently become clear that an inadequate response to venous occlusion or to DDVAP is due, in the large majority of cases, to an increased PAI-1 plasma level. In about 10 to 20% of patients with a poor response to venous stasis there seems to exist a defect of t-PA production,

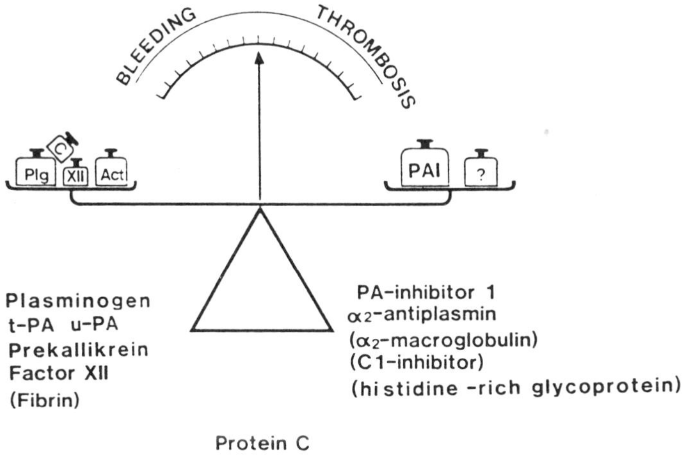

Fig. 15. Homeostasis of the fibrinolytic system. If the precarious balance between profibrinolytic and antifibrinolytic factors is disturbed, bleeding or thrombosis may result.

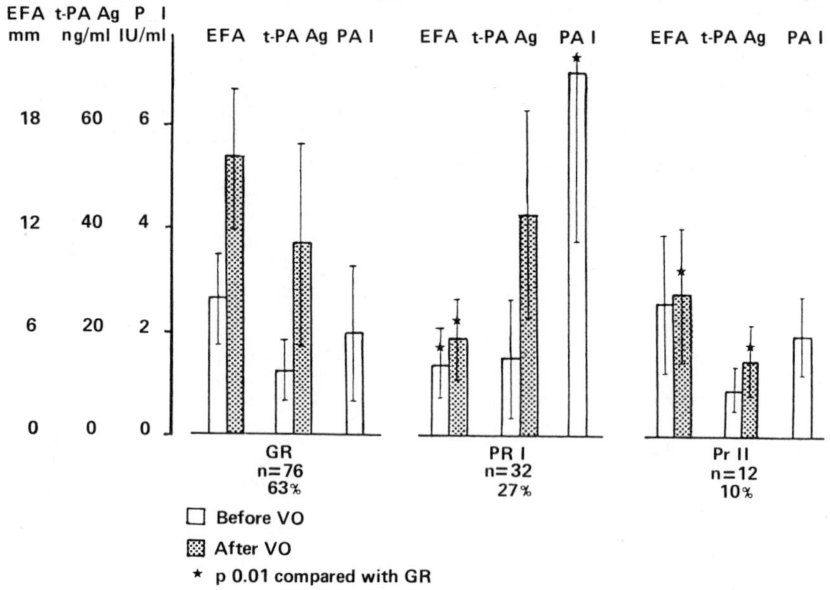

Fig. 16. Fibrinolytic parameters in 120 patients with idiopathic deep vein thrombosis. GR: good responders; PR: poor responders; EFA: euglobulin fibrinolytic activity; VO: venous occlusion (Taken from Juhan-Vague et al. (174)).

or of release from the endothelium. The largest study so far was conducted by Juhan-Vague et al. who studied a collective of 120 patients with idiopathic deep venous thrombosis. These authors found a normal fibrinolytic response after venous stasis in 63% of their patients. There were 37% poor responders which could be divided into 2 groups. Approximately three forth exhibited increased PAI-1 levels at rest, the remainder showed a poor release of t-PA antigen in response to venous occlusion (Figure 16).

The only prospective study reported so far examined euglobulin lysis times in 121 patients with a proven episode of venous thrombosis and/or pulmonary embolism. A post venous stasis euglobulin lysis time test (ELT) was performed at earliest 3 months after the last thromboembolic episode and patients followed for a mean period of 56 months. The recurrency rate in patients having had ELT's of less than 60 minutes was significantly lower (4.8%/year) than in those with ELT's over 60 min. (179).

Increased PAI-1 levels have also been found in patients with coronary artery disease (180, 181), after surgery (182, 183), in old (184) and in obese subjects (185), and in many other pathological conditions (186). PAI-1 appears to be an acute-phase reactant and may therefore be increased in many processes accompanied by inflammation, wound healing and infection.

CONCLUSION

A large amount of new knowledge of hitherto unknown components of the fibrinolytic system has vastly increased our understanding of the interactions of these proteins participating in controlled breakdown of fibrin. Much work remains to be done, however, before we understand fibrinolysis as well as blood coagulation. We should devote particular attention to phenomena which take place on endothelial and clot surfaces to receptors of plasminogen activators and to messenger systems which control the expression of fibrinolytic and inhibitory protein on the DNA level.

A thorough understanding of the regulation of fibrinolysis on the gene level might allow in the future to influence the activity of the fibrinolytic system to the benefit of our patients.

ACKNOWLEDGEMENT

Part of the studies reported here were supported by grants Nr. 3.461-0.83, 3.334-0.86 and 3.387-0.86 of the Swiss National Fondation for Scientific Research. I thank my collaborators Drs. E.K.O. Kruithof, W.-D. Schleuning and J. Hauert for many helpful discussion and for letting me report some of the unpublished results. The secretarial assistance of Mrs Josette Guillemin and Suzanne Portner is gratefully acknowledged.

REFERENCES

1. Blasi F, Riccio A, Sebastio G. Human plasminogen activators. Genes and proteins structure. In: Blasi F, ed. Human Genes and Diseases. New York: John Wiley & Sons Ltd. 1986: 377-414.
2. Collen D. Molecular mechanisms of fibrinolysis and their application to fibrin-specific thrombolytic therapy. J Cell Biochem 1987; 33: 77-94.
3. Sprengers ED, Kluft C. Plasminogen activator inhibitors. Blood, 1987; 69: 381-387.
4. Kruithof EKO. Inhibitors of plasminogen activators. In: Kluft C, ed. Tissue-type plasminogen activator: Physiological and Clinical Aspects. Boca Raton: CRC Press 1987; in press.
5. Danø K, Andreasen PA, Grøndahl-Hansen J, Kristensen P, Nielsen LS, Skriver L. Plasminogen activators, tissue degradation, and cancer. Adv Cancer Res 1985; 44: 139-266.
6. Saksela O. Plasminogen activation and regulation of pericellular proteolysis. Biochim Biophys Acta 1985; 823: 35-65.
7. Verstraete M, Collen D. Thrombolytic therapy in the eighties. Blood 1986; 67: 1529-1541.
8. Muellertz S. Fibrinolysis. General aspects, characteristic features and perspectives. Fibrinolysis 1987; 1: 3-12.
9. Castellino FJ. Biochemistry of human plasminogen. Semin Thromb Hemost 1984; 10: 18-23.
10. Young CL, Barker WC, Tomaselli CM, Dayhoft MO. Serine proteases. In: Atlas of Protein Sequence and Structure. Silver Spring: The National Biomedical Research Foundation 1978: 73-93.
11. Malinowski DP, Sadler JE, Davie EW. Characterization of a complementary deoxyribonucleic acid coding for human and bovine plasminogen. Biochemistry 1984; 23: 4243-4250.
12. Swisshelm K, Dyer K, Sadler E, Disteche C. Localisation of the plasminogen gene (plg) to the distal portion of the long arm of chromosome 6 by in situ hybridization. Cytogenet Cell Genet 1985; 40: 756.
13. Lijnen HR, Collen D. Interaction of plasminogen activators and inhibitors with plasminogen and fibrin. Semin Thromb Hemost 1982; 8: 2-10.
14. Collen D. On the regulation and control of fibrinolysis. Thromb Haemost 1980; 43: 77-89.
15. Lucas MA, Fretto LJ, McKee PA. The binding of human plasminogen to fibrin and fibrinogen. J Biol Chem 1983; 258: 4249-4256.

16. Varadi A, Patthy L. Beta(Leu_{121}-Lys_{122}) segment of fibrinogen is in a region essential for plasminogen binding by fibrin fragment E. Biochemistry 1984; 23: 2108-2112.
17. Pennica D, Holmes WE, Kohr WJ, et al. Cloning and expression of human tissue-type plasminogen activator cDNA in E. coli. Nature 1983; 301: 214-221.
18. Edlund T, Ny T, Rånby M, et al. Isolation of cDNA sequences coding for a part of human tissue plasminogen activator. Proc Natl Acad Sci USA 1983; 80: 349-352.
19. Fisher R, Waller EK, Grossi G, Thompson D, Tizard R, Schleuning W.-D. Isolation and characterization of the human tissue-type plasminogen activator structural gene including its 5' flanking region. J Biol Chem 1985; 260: 11223-11230.
20. Ny T, Elgh F, Lund B. The structure of the human tissue-type plasminogen activator gene: Correlation of intron and exon structures to functional and structural domains. Proc Natl Acad Sci USA 1984; 81: 5355-5359.
21. Rajput B, Degen SF, Reich E, et al. Chromosomal locations of human tissue plasminogen activator and urokinase genes. Science 1985; 230: 672-674.
22. Verheijen JH, Visse R, Wijnen JTh, Chang GTG, Kluft C, Meera Khan P. Assignment of the human tissue-type plasminogen activator gene (PLAT) to chromosome 8. Hum Genet 1986; 72: 153-156.
23. Wallen P, Pohl G, Bergsdorf N, Rånby M, Ny T, Jornvall H. Structural characterization of tissue plasminogen activator purified by immunosorbent chromatography. In: Davidson JF, Bachman F, Bouvier CA, Kruithof EKO, eds. Progress in Fibrinolysis. Edinburgh: Churchill Livingstone, 1983; Vol 6: 338-343.
24. Dodd I, Fears R, Robinson JH. Isolation and preliminary characterisation of active B-chain of recombinant tissue-type plasminogen activator. Thromb Haemost 1986; 55: 94-97.
25. Rijken DC, Groeneveld E. Isolation and functional characterization of the heavy and light chains of human tissue-type plasminogen activator. J Biol Chem 1986; 261: 3098-3102.
26. Van Zonneveld A-J, Veerman H, Pannekoek H. Autonomous functions of structural domains on human tissue-type plasminogen activator. Proc Natl Acad Sci USA 1986; 83: 4670-4674.
27. Zonneveld AJ, Veerman H, Pannekoek H. On the interaction of the finger and the kringle-2 domain of tissue-type plasminogen activator with fibrin. J Biol Chem 1986; 261: 14214-14218.
28. Verheijen JH, Caspers MPM, Chang GTG, De Munk GAW, Pouwels PH, Enger-Valk BE. Involvement of finger domain and kringle 2 domain of tissue-type plasminogen activator in fibrin binding and stimulation of activity by fibrin. Embo J 1986; 5: 3525-3530.
29. Kagitani H, Tagawa M, Hatanaka K, et al. Expression in E. coli of finger-domain lacking tissue-type plasminogen activator with high fibrin affinity. Febs Lett 1985; 189: 145-149.
30. Ehrlich HJ, Bang NU, Little SP, et al. Biological properties of a kringleless tissue plasminogen activator (t-PA) mutant. Fibrinolysis 1987; 1: 75-81.
31. Rijken DC, Emeis JJ, Gerwig GJ. On the composition and function of the carbohydrate moiety of tissue-type plasminogen activator from human melanoma cells. Thromb Haemost 1985; 54: 788-791.
32. Bachmann F, Kruithof EKO. Tissue plasminogen activator: Chemical and physiological aspects. Semin Thromb Hemost 1984; 10: 6-17.
33. Rijken DC, Wijngaards G, Collen D. Tissue-type plasminogen activator from human tissue and cell cultures and its occurrence in plasma. In: Collen D, Lijnen HR, Verstraete M., eds. Thrombolysis, Biological and Therapeutic Properties of New Thrombolytic Agents. Edinburgh: Churchill Livingstone 1985: 15-30.

34. Rånby M, Wallén P. Enzymatic properties of tissue-type plasminogen activator. In: Collen D, Lijnen HR, Verstraete M., eds. Thrombolysis, Biological and Therapeutic Properties of New Thrombolytic Agents. Edinburgh: Churchill Livingstone 1985: 31-48.
35. Norrman B, Wallén P, Rånby M. Fibrinolysis mediated by tissue plasminogen activator. Disclosure of a kinetic transition. Eur J Biochem 1985; 149: 193-200.
36. Nieuwenhuizen W, Vermond A, Voskuilen M, Traas DW, Verheijen JH. Identification of a site in fibrin(ogen) which is involved in the acceleration of plasminogen activation by tissue-type plasminogen activator. Biochim Biophys Acta 1983; 748: 86-92.
37. Suenson E, Petersen LC. Fibrin and plasminogen structures essential to stimulation of plasmin formation by tissue-type plasminogen activator. Biochim Biophys Acta 1986; 870: 510-519.
38. Tran-Thang C, Kruithof EKO, Atkinson J, Bachmann F. High-affinity binding sites for human Glu-plasminogen unveiled by limited plasmic degradation of human fibrin. Eur J Biochem 1986; 160: 599-604.
39. Tate KM, Higgins DL, Holmes WE, Winkler ME, Heyneker HL, Vehar GA. Functional role of proteolytic cleavage at arginine-275 of human tissue plasminogen activator as assessed by site-directed mutagenesis. Biochemistry 1987; 26: 338-343.
40. Collen D. Report of the meeting of the subcommittee on fibrinolysis, (ICTH), Jerusalem, Israel. Thromb Haemost 1986; 56: 415-416.
41. Kruithof EKO, Vassalli J-D, Schleuning W-D, Mattaliano RJ, Bachmann F. Purification and characterization of a plasminogen activator inhibitor from the histiocytic lymphoma cell line U-937. J Biol Chem 1986; 261: 11207-11213.
42. Gaffney PJ, Curtis AD. A collaborative study of a proposed international standard for tissue plasminogen activator (t-PA). Thromb Haemost 1985; 53: 134-136.
43. Verstraete M, Su CAPF, Tanswell P, Feuerer W, Collen D. Pharmacokinetics and effects on fibrinolytic and coagulation parameters of two doses of recombinant tissue-type plasminogen activator in healthy volunteers. Thromb Haemost 1986; 56: 1-5.
44. Verstraete M, Bounameaux H, De Cock F, Van de Werf F, Collen D. Pharmacokinetics and systemic fibrinogenolytic effects of recombinant human tissue-type plasminogen activator (rt-PA) in humans. J Pharmacol Exp Ther 1985; 235: 506-512.
45. Günzler WA, Steffens GJ, Ötting F, Kim SM, Frankus E, Flohe L. The primary structure of high molecular mass urokinase from human urine. The complete amino acid sequence of the A chain. Hoppe Seylers Z Physiol Chem 1982; 363: 1155-1165.
46. Steffens GJ, Günzler WA, Ötting F, Frankus E, Flohe L. The complete amino acid sequence of low molecular mass urokinase from human urine. Hoppe Seylers Z Physiol Chem 1982; 363: 1043-1058.
47. Husain SS, Gurewich V, Lipinski B. Purification and partial characterization of a single-chain high-molecular-weight form of urokinase from human urine. Arch Biochem Biophys 1983; 220: 31-38.
48. Stump DC, Thienpont M, Collen D. Urokinase-related proteins in human urine. Isolation and characterization of single-chain urokinase (pro-urokinase) and urokinase-inhibitor complex. J Biol Chem 1986; 261: 1267-1273.
49. Cajot J-F, Kruithoff EKO, Schleuning W-D, Sordat B, Bachmann F. Plasminogen activators, plasminogen activator inhibitors and procoagulant analyzed in twenty human tumor cell lines. Int J Cancer 1986; 38: 719-727.
50. Corti A, Nolli ML, Soffientini A, Cassani G. Purification and characterization of single-chain urokinase-type plasminogen activator (pro-urokinase) from human A431 cells. Thromb Haemost 1986; 56: 219-224.

51. Stump DC, Lijnen HR, Collen D. Purification and characterization of single-chain urokinase-type plasminogen activator from human cell cultures. J Biol Chem 1986; 261: 1274-1278.
52. Darras V, Thienpont M, Stump DC, Collen C. Measurement of urokinase-type plasminogen activator (u-PA) with an enzyme-linked immunosorbent assay (ELISA) based on three murine monoclonal antibodies. Thromb Haemost 1986; 56: 411-414.
53. Holmes WE, Pennica D, Blaber M, et al. Cloning and expression of the gene for pro-urokinase in Escherichia coli. Biotechnol 1985; 3: 923-929.
54. Verde P, Stoppelli MP, Galeffi P, Di Nocera P, Blasi F. Identification and primary sequence of an unspliced human urokinase poly(A)$^+$ RNA. Proc Natl Acad Sci USA 1984; 81: 4727-4731.
55. Riccio A, Grimaldi G, Verde P, Sebastio G, Boast S, Blasi F. The human urokinase plasminogen activator gene and its promoter. Nucleic Acids Res 1985; 13: 2759-2771.
56. Ichinose A, Fujikawa K, Suyama T. The activation of pro-urokinase by plasma kallikrein and its inactivation by thrombin. J Biol Chem, 1986; 261: 3486-3489.
57. Schaller J, Nick H, Rickli E, Gillessen D, Lergier W, Studer RO. The two-chain structure of human low molecular weight urinary urokinase, evidenced by sequence analysis. In: Davidson JF, Bachmann F, Bouvier CA, Kruithof EKO, eds. Progress in Fibrinolysis. Edinburgh: Churchill Livingstone 1983; Vol VI: 348-352.
58. Stump DC, Lijnen HR, Collen D. Purification and characterization of a novel low molecular weight form of single-chain urokinase-type plasminogen activator. J Biol Chem 1986; 261: 17120-17126.
59. Rijken DC, Binnema JD, Los P. Specific fibrinolytic properties of different molecular forms of pro-urokinase from a monkey kidney cell culture. Thromb Res 1986; 42: 761-768.
60. Kasai S, Arimura H, Nishida M, Suyama T. Proteolytic cleavage of single-chain pro-urokinase induces conformational change which follows activation of the zymogen and reduction of its high affinity for fibrin. J Biol Chem 1985; 260: 12377-12381.
61. Pannell R, Gurewich V. Activation of plasminogen by single-chain urokinase or by two-chain urokinase - A demonstration that single-chain urokinase has a low catalytic activity (pro-urokinase). Blood 1987; 69: 22-26.
62. Nielsen LS, Hansen JG, Skriver L, et al. Purification of zymogen to plasminogen activator from human glioblastoma cells by affinity chromatography with monoclonal antibody. Biochemistry 1982; 21: 6410-6415.
63. Lijnen HR, Zamarron C, Blaber M, Winkler ME, Collen D. Activation of plasminogen by pro-urokinase. I. Mechanism. J Biol Chem 1986; 261: 1253-1258.
64. Collen D, Zamarron C, Lijnen HR, Hoylaerts M. Activation of plasminogen by pro-urokinase. II. Kinetics. J Biol Chem 1986; 261: 1259-1266.
65. Gurewich V, Pannell R. Inactivation of single-chain urokinase (pro-urokinase) by thrombin and thrombin-like enzymes: Relevance of the findings to the interpretation of fibrin-binding experiments. Blood 1987; 69: 769-772.
66. Pannell R, Gurewich V. Pro-urokinase: A study of its stability in plasma and of a mechanism for its selective fibrinolytic effect. Blood 1986; 67: 1215-1223.
67. Kasai S, Arimura H, Nishida M, Suyama T. Thrombolytic properties of an inactive proenzyme form of human urokinase secreted from human kidney cells. Cell Struct Funct 1985; 10: 151-159.
68. Gurewich V, Pannell R, Louie S, Kelley P, Suddith RL, Greenlee R. Effective and fibrin-specific clot lysis by a zymogen precursor form

of urokinase (pro-urokinase). A study in vitro and in two animal species. J Clin Invest 1984; 73: 1731-1739.
69. Zamarron C, Lijnen HR, Van Hoef B, Collen D. Biological and thrombolytic properties of proenzyme and active forms of human urokinase - I. Fibrinolytic and fibrinogenolytic properties in human plasma in vitro of urokinases obtained from human urine or by recombinant DNA technology. Thromb Haemost 1984; 52: 19-23.
70. Collen D, Stassen JM, Blaber M, Winkler N, Verstraete M. Biological and thrombolytic properties of proenzyme and active forms of human urokinase - III. Thrombolytic properties of natural and recombinant urokinase in rabbits with experimental jugular vein thrombosis. Thromb Haemost 1984; 52: 27-30.
71. Matsuo O, Bando H, Okada K, et al. Thrombolytic effect of single-chain pro-urokinase in a rabbit jugular vein thrombosis model. Thromb Res 1986; 42: 187-194.
72. Lijnen HR, Van Hoef B, Collen D. Comparative kinetic analysis of the activation of human plasminogen by natural and recombinant single-chain urokinase-type plasminogen activator. Biochim Biophys Acta 1986; 884: 402-408.
73. Stump DC, Stassen JM, Demarsin E, Collen D. Comparative thrombolytic properties of single-chain forms of urokinase-type plasminogen activator. Blood 1987; 69: 592-596.
74. Van de Werf F, Jang IK, Collen D. Thrombolysis with recombinant human single-chain urokinase-type plasminogen activator (rscu-PA): dose-response in dogs with coronary artery thrombosis. J Cardiovasc Pharmacol 1987; 9: 91-93.
75. Van de Werf F, Vanhaecke J, De Geest H, Verstraete M, Collen D. Coronary thrombolysis with recombinant single-chain urokinase-type plasminogen activator in patients with acute myocardial infarction. Circulation 1986; 74: 1066-1070.
76. Kristensen P, Larsson L-I, Nielsen LS, Grøndahl-Hansen J, Andreasen PA, Danø K. Human endothelial cells contain one type of plasminogen activator. FEBS Lett 1984; 168: 33-37.
77. Larsson L-I, Skriver L, Nielsen LS, Grøndahl-Hansen J, Kristensen P, Danø K. Distribution of urokinase-type plasminogen activator immunoreactivity in the mouse. J Cell Biol 1984; 98: 894-903.
78. Vassalli JD, Baccino D, Belin D. A cellular binding site for the M_r 55,000 form of the human plasminogen activator, urokinase. J Cell Biol 1985; 100: 86-92.
79. Del Rosso M, Dini G, Fibbi G. Receptors for plasminogen activator, urokinase, in normal and rous sarcoma virus-transformed mouse fibroblasts. Cancer Res 1985; 45: 630-636.
80. Stoppelli MP, Corti A, Soffientini A, Cassani G, Blasi F, Assoian RK. Differentiation-enhanced binding of the amino-terminal fragment of human urokinase plasminogen activator to a specific receptor on U937 monocytes. Proc Natl Acad Sci USA 1985; 82: 4939-4943.
81. Bajpai A, Baker JB. Urokinase binding sites on human foreskin cells. Evidence for occupancy with endogenous urokinase. Biochem Biophys Res Commun 1985; 133: 994-1000.
82. Blasi F, Stoppelli MP, Cubellis MV. The receptor for urokinase-plasminogen activator. J Cell Biochem 1986; 32: 179-186.
83. Cubellis MV, Nolli ML, Cassani G, Blasi F. Binding of single-chain prourokinase to the urokinase receptor of human U937 cells. J Biol Chem 1986; 261: 15819-15822.
84. Tans G, Rosing J. Structural and functional characterization of factor XII. Semin Thromb Hemost 1987; 13: 1-4.
85. Fujikawa K, McMullen BA. Amino acid sequence of human β-factor XIIa. J Biol Chem 1983; 258: 10924-10933.
86. McMullen BA, Fujikawa K. Amino acid sequence of the heavy chain of human α-factor XIIa (activated Hageman factor). J Biol Chem 1985; 260: 5328-5341.

87. Que BG, Davie EW. Characterization of a cDNA coding for human factor XII (Hageman factor). Biochemistry 1986; 25: 1525-1528.
88. Cool DE, Edgell C-JS, Louie GV, Zoller MJ, Brayer GD, MacGillivray RTA. Characterization of human blood coagulation factor XII cDNA. Prediction of the primary structure of factor XII and the tertiary structure of β-factor XIIa. J Biol Chem 1985; 260: 13666-13676.
89. Takagaki Y, Kitamura N, Nakanishi S. Cloning and sequence analysis of cDNA's for human high molecular weight and low molecular weight prekininogens. Primary structures of two human prekininogens. J Biol Chem 1985; 260: 8601-8609.
90. Van Iwarden F, Bouma BN. Role of high molecular weight kininogen in contact activation. Semin Thromb Hemost 1987; 13: 15-24.
91. Mannhalter ChH. Biochemical and functional properties of factor XI and prekallikrein. Semin Thromb Hemost 1987; 13: 25-35.
92. Chung DW, Fujikawa K, McMullen BA, Davie EW. Human plasma prekallikrein, a zymogen to a serine protease that contains four tandem repeats. Biochemistry 1986; 25: 2410-2417.
93. Sumi Y, Nakamura Y, Aoki N, Sakai M, Muramatsu M. Structure of the carboxyl-terminal half of human α 2-plasmin inhibitor deduced from that of cDNA. J Biochem 1986; 100: 1399-1402.
94. Holmes WE, Nelles L, Lijnen HR, Collen D. Primary structure of human α 2-antiplasmin, a serine protease inhibitor (serpin). J Biol Chem 1987; 262: 1659-1664.
95. Aoki N, Harpel PC. Inhibitors of the fibrinolytic enzyme system. Semin Thromb Hemost 1984; 10: 24-41.
96. Mimuro J, Koike Y, Sumi Y, Aoki N. Monoclonal antibodies to discrete regions in α 2-plasmin inhibitor. Blood 1987; 69: 446-453.
97. Ichinose A, Aoki N. The initiation of fibrinolysis in α 2-plasmin inhibitor deficient plasma. Role of fibrin. Thromb Res 1986; 41: 847-854.
98. Sottrup-Jensen L, Stepanik TM, Kristensen T, et al. Primary structure of human α 2-macroglobulin. V. The complete structure. J Biol Chem 1984; 259: 8318-8327.
99. Kan CC, Solomon E, Belt KT, Chain AC, Hiorns LR, Fey G. Nucleotide sequence of cDNA encoding human α_2-macroglobulin and assignment of the chromosomal locus. Proc Natl Acad Sci USA 1985; 82: 2282-2286.
100. Travis J, Salvesen GS. Human plasma proteinase inhibitors. Ann Rev Biochem 1983; 52: 655-709.
101. Kruithof EKO, Ransijn A, Bachmann F. Inhibition of tissue plasminogen activator by human plasma. In: Davidson JF, Bachmann F, Bouvier CA, Kruithof EKO, eds. Progress in Fibrinolysis. Edinburgh: Churchill Livingstone 1983; Vol VI: 362-366.
102. Dosne AM, Dupuy E, Bodevin E. Production of a fibrinolytic inhibitor by cultured endothelial cells derived from human umbilical vein. Thromb Res 1978; 12: 377-387.
103. Esnard F, Dupuy E, Dosne AM, Bodevin E. Partial characterization of a fibrinolytic inhibitor produced by cultured endothelial cells derived from human umbilical vein. Thromb Haemost 1982; 47: 128-131.
104. Loskutoff DJ, Van Mourik JA, Erickson LA, Lawrence D. Detection of an unusually stable fibrinolytic inhibitor produced by bovine endothelial cells. Proc Natl Acad Sci USA 1983; 80: 2956-2960.
105. Levin EG. Latent tissue plasminogen activator produced by human endothelial cells in culture: Evidence for an enzyme-inhibitor complex. Proc Natl Acad Sci USA 1983; 80: 6804-6808.
106. Philips M, Juul AG, Thorsen S. Human endothelial cells produce a plasminogen activator inhibitor and a tissue-type plasminogen activator-inhibitor complex. Biochim Biophys Acta 1984; 802: 99-110.
107. Sprengers ED, Verheijen JH, Van Hinsbergh VWM, Emeis JJ. Evidence for the presence of two different fibrinolytic inhibitors in human endothelial cell conditioned medium. Biochim Biophys Acta 1984; 801: 163-170.

108. Van Mourik JA, Lawrence DA, Loskutoff DJ. Purification of an inhibitor of plasminogen activator (antiactivator) synthesized by endothelial cells. J Biol Chem 1984; 259: 14914-14921.
109. Chmielewska J, Rånby M, Wiman B. Evidence for a rapid inhibitor to tissue plasminogen activator in plasma. Thromb Res 1983; 31: 427-436.
110. Thorsen S, Philips M. Isolation of tissue-type plasminogen activator-inhibitor complexes from human plasma. Evidence for a rapid plasminogen activator inhibitor. Biochim Biophys Acta 1984; 802: 111-118.
111. Kruithof EKO, Tran-Thang C, Ransijn A, Bachmann F. Demonstration of a fast-acting inhibitor of plasminogen activators in human plasma. Blood 1984; 64: 907-913.
112. Verheijen JH, Chang GT, Kluft C. Evidence for the occurrence of a fast-acting inhibitor for tissue-type plasminogen activator in human plasma. Thromb Haemost 1984; 51: 392-395.
113. Wiman B, Chmielewska J, Rånby M. Inactivation of tissue plasminogen activator in plasma. Demonstration of a complex with a new rapid inhibitor. J Biol Chem 1984; 259: 3644-3647.
114. Wagner OF, Binder BR. Purification of an active plasminogen activator inhibitor immunologically related to the endothelial type plasminogen activator inhibitor from the conditioned media of a human melanoma cell line. J Biol Chem 1986; 261: 14474-14481.
115. Ny T, Sawdey M, Lawrence D, Millan JL, Loskutoff DJ. Cloning and sequence of a cDNA coding for the human β-migrating endothelial-cell-type plasminogen activator inhibitor. Proc Natl Acad Sci USA 1986; 83: 6776-6780.
116. Pannekoek H, Veerman H, Lambers H, et al. Endothelial plasminogen activator inhibitor (PAI): a new member of the Serpin gene family. EMBO J 1986; 5: 2539-2544.
117. Ginsburg D, Zeheb R, Yang AY, et al. cDNA cloning of human plasminogen activator-inhibitor from endothelial cells. J Clin Invest 1986; 78: 1673-1680.
118. Wun TC, Kretzmer KK. cDNA cloning and expression in E. coli of a plasminogen activator inhibitor (PAI) related to a PAI produced by Hep G2 hepatoma cell. FEBS Lett 1987; 210: 11-16.
119. Carrell RW, Jeppsson J-O, Laurell C-B, et al. Structure and variation of human α_1-antitrypsin. Nature 1982; 298: 329-334.
120. Kruithof EKO, Tran-Thang C, Bachmann F. The fast-acting inhibitor of tissue-type plasminogen activator in plasma is also the primary plasma inhibitor of urokinase. Thromb Haemost 1986; 55: 65-69.
121. Andreasen PA, Riccio A, Welinder KG, et al. Plasminogen activator inhibitor type-1: reactive center and amino-terminal heterogeneity determined by protein and cDNA sequencing. FEBS Lett 1986; 209: 213-218.
122. Erickson LA, Ginsberg MH, Loskutoff DJ. Detection and partial characterization of an inhibitor of plasminogen activator in human platelets. J Clin Invest 1984; 74: 1465-1472.
123. Kruithof EKO, Tran-Thang C, Bachmann F. Studies on the release of a plasminogen activator inhibitor by human platelets. Thromb Haemost 1986; 55: 201-205.
124. Sprengers ED, Akkerman JWN, Jansen BG. Blood platelet plasminogen activator inhibitor: Two different pools of endothelial cell type plasminogen activator inhibitor in human blood. Thromb Haemost 1986; 55: 325-329.
125. Sprengers ED, Princen HMG, Kooistra T, Van Hinsbergh VWM. Inhibition of plasminogen activators by conditioned medium of human hepatocytes and hepatoma cell line Hep G2. J Lab Clin Med 1985; 105: 751-758.
126. Hekman CM, Loskutoff DJ. Endothelial cells produce a latent inhibitor of plasminogen activators that can be activated by denaturants. J Biol Chem 1985; 260: 11581-11587.

127. Levin EG. Quantitation and properties of the active and latent plasminogen activator inhibitors in cultures of human endothelial cells. Blood 1986; 67: 1309-1313.
128. Kooistra T, Sprengers ED, Van Hinsbergh VWM. Rapid inactivation of the plasminogen-activator inhibitor upon secretion from cultured human endothelial cells. Biochem J 1986; 239: 497-503.
129. Lawrence DA, Loskutoff DJ. Inactivation of plasminogen activator inhibitor by oxidants. Biochemistry 1986; 25: 6351-6355.
130. Gelehrter TD, Sznycer-Laszuk R. Thrombin induction of plasminogen activator-inhibitor in cultured human endothelial cells. J Clin Invest 1986; 77: 165-169.
131. Levin EG, Stern DM, Nawroth PP, et al. Specificity of the thrombin-induced release of tissue plasminogen activator from cultured human endothelial cells. Thromb Haemost 1986; 56: 115-119.
132. Hanss M, Collen D. Secretion of tissue-type plasminogen activator and plasminogen activator inhibitor by cultured human endothelial cells: modulation by thrombin, endotoxin, and histamine. J Lab Clin Med 1987; 109: 97-104.
133. Loskutoff DJ, Roegner K, Erickson LA, et al. The dexamethasone-induced inhibitor of plasminogen activator in hepatoma cells is antigenically-related to an inhibitor produced by bovine aortic endothelial cells. Thromb Haemost 1986; 55: 8-11.
134. Coleman PL, Patel PD, Cwikel BJ, Rafferty UM, Sznycer-Laszuk R, Gelehrter TD. Characterization of the dexamethasone-induced inhibitor of plasminogen activator in HTC hepatoma cells. J Biol Chem 1986; 9: 4352-4357.
135. Laiho M. Saksela O, Andreasen PA, Keski-Oja J. Enhanced production and extracellular deposition of the endothelial-type plasminogen activator inhibitor in cultured human lung fibroblasts by transforming growth factor-β. J Cell Biol 1986; 103: 2403-2410.
136. Crutchley DJ, Conanan LB. Endotoxin induction of an inhibitor of plasminogen activator in bovine pulmonary artery endothelial cells. J Biol Chem 1986; 261: 154-159.
137. Colucci M, Paramo JA, Collen D. Generation in plasma of a fast-acting inhibitor of plasminogen activator in response to endotoxin stimulation. J Clin Invest 1985; 75:818-824.
138. Kawano T, Morimoto K, Uemura Y. Urokinase inhibitor in human placenta. Nature 1968; 217: 253-254.
139. Vassalli J-D, Dayer JM, Wohlwend A, Belin D. Concomitant secretion of prourokinase and of a plasminogen activator-specific inhibitor by cultured human monocytes-macrophages. J Exp Med 1984; 159: 1653-1668.
140. Kopitar M, Rozman B, Babnik J, Turk V, Mullins DE, Wun T-C. Human leucocyte urokinase inhibitor. Purification, characterization and comparative studies against different plasminogen activators. Thromb Haemost 1985; 54: 750-755.
141. Chapman HA, Stone OL. Characterization of a macrophage-derived plasminogen-activator inhibitor. Similarities with placental urokinase inhibitor. Biochem J 1985; 230: 109-116.
142. Wohlwend A, Belin D, Vassalli JD. Plasminogen activator-specific inhibitors produced by human monocytes/macrophages. J Exp Med 1987; 165: 320-339.
143. Åstedt B, Lecander I, Brodin T, Lundblad A, Loew K. Purification of a specific placental plasminogen activator inhibitor by monoclonal antibody and its complex formation with plasminogen activator. Thromb Haemost 1985; 53: 122-125.
144. Ye RD, Wun T-CJ, Sadler JE. Characterization of cDNA for a plasminogen activator inhibitor from human placenta. Circulation 1986; 1: 988.
145. Schleuning WD, Medcalf RL, Hession K, Rothenbuehler R, Kruithof EKO. Plasminogen activator inhibitor 2: molecular cloning and regulation of gene transcription during phorbol ester mediated differentiation of U-937 human histiocytic lymphoma cells (submitted).

146. Kruithof EKO, Tran-Thang C, Gudinchet A, et al. Fibrinolysis in pregnancy: A study of plasminogen activator inhibitors. Blood 1987; 69: 460-466.
147. Lecander I, Åstedt B. Isolation of a new specific plasminogen activator inhibitor from pregnancy plasma. Br J Haematol 1986; 62: 221-228.
148. Åstedt B, Hägerstrand I, Lecander I. Cellular localisation in placenta of placental type plasminogen activator inhibitor. Thromb Hemost 1986; 56: 63-65.
149. Åstedt B, Bladh B, Christensen U, Lecander I. Different inhibition of one and two chain tissue plasminogen activator by a placental inhibitor studied with two tripeptide-p-nitroanilide substrates. Scand J Clin Lab Invest 1985; 45: 429-435.
150. Salvesen GS, Catanese JJ, Kress LF, Travis J. Primary structure of the reactive site of human C1-inhibitor. J Biol Chem 1985; 260: 2432-2436.
151. Tosi M, Duponchel C, Bourgarel P, Colomb M, Meo T. Molecular cloning of human C1 inhibitor: Sequence homologies with α 1-antitrypsin and other members of the serpins superfamily. Gene 1986; 42: 265-272.
152. Bock, SC, Skriver K, Nielsen E, et al. Human C1 inhibitor: Primary structure, cDNA cloning, and chromosomal localization. Biochemistry 1986; 25: 4292-4301.
153. Koide T, Foster D, Yoshitake S, Davie EW. Amino acid sequence of human histidine-rich glycoprotein derived from the nucleotide sequence of its cDNA. Biochemistry 1986; 25: 2220-2225.
154. Lijnen HR, Hoylaerts M, Collen D. Isolation and characterization of a human plasma protein with affinity for the lysine binding sites in plasminogen. J Biol Chem 1980; 255: 10214-10222.
155. Ichinose A, Mimuro J, Koide T, Aoki N. Histidine-rich glycoprotein and α 2-plasmin inhibitor in inhibition of plasminogen binding to fibrin. Thromb Res 1984; 33: 401-407.
156. Silverstein RL, Leung LLK, Nachman RL. Thrombospondin: A versatile multifunctional glycoprotein. Arteriosclerosis 1986; 6: 245-253.
157. Clemmensen I, Petersen LC, Kluft C. Purification and characterization of a novel, oligomeric, plasminogen kringle 4 binding protein from human plasma: Tetranectin. Eur J Biochem 1986; 156: 327-333.
158. Agnelli G, Buchanan MR, Fernandez F, Van Ryn J, Hirsh J. Sustained thrombolysis with DNA-recombinant tissue type plasminogen activator in rabbits. Blood 1985; 66:399-401.
159. Tran-Thang C, Kruithof EKO, Bachmann F. The mechanism of in vitro clot lysis induced by vascular plasminogen activator. Blood 1984; 63: 1331-1337.
160. Kimura S, Aoki N. Cross-linking site in fibrinogen for α 2-plasmin inhibitor. J Biol Chem 1986; 261: 15591-15595.
161. Harpel PC, Chang T-S, Verderber E. Tissue plasminogen activator and urokinase mediate the binding of Glu-plasminogen to plasma fibrin I. Evidence for new binding sites in plasmin-degraded fibrin I. J Biol Chem 1985; 260: 4432-4440.
162. Tran-Thang C, Kruithof EKO, Bachmann F. Tissue-type plasminogen activator increases the binding of Glu-plasminogen to clots. J Clin Invest 1984; 74: 2009-2016.
163. Sakata Y, Curriden S, Lawrence D, Griffin JH, Loskutoff DJ. Activated protein C stimulates the fibrinolytic activity of cultured endothelial cells and decreases antiactivator activity. Proc Natl Sci USA 1985; 82: 1121-1125.
164. Van Hinsbergh VWM, Bertina RM, Van Wijngaarden A, Van Tilburg NH, Emeis JJ, Haverkate F. Activated protein C decreases plasminogen activator-inhibitor activity in endothelial cell-conditioned medium. Blood 1985; 65: 444-451.
165. De Fouw NJ, Haverkate F, Bertina RM, Koopman J, Van Wijngaarden A, Van Hinsbergh VWM. The cofactor role of protein S in the acceleration of

whole clot lysis by activated protein C in vitro. Blood 1986; 67: 1189-1192.
166. D'Angelo A, Lockhart MS, D'Angelo SV, Taylor FB. Protein S is a cofactor for activated protein C neutralization of an inhibitor of plasminogen activation released from platelets. Blood 1987; 69: 231-237.
167. Aznar J, Dasi A, Espana F, Estellés A. Fibrinolytic study in a homozygous protein C deficient patient. Thromb Res 1986; 42: 313-322.
168. Hauert J, Bachmann F. In vitro clot lysis induced by single-chain urokinase (scu-PA) and activation of the contact phase system. Thromb Haemost 57, abstract, XI Int Congr Thromb Haemost, Bruxelles, 1987.
169. Kluft C, Dooijewaard G, Emeis JJ. Role of the contact system in fibrinolysis. Semin Thromb Hemost 1987; 13: 50-68.
170. Hauert J, Bachmann F. Prourokinase in the euglobulin fractions. Thromb Hemost 1985; 54: 122.
171. Knot EAR, Ten Cate JW, Lamping RJ, Gie LK. α 2-Antiplasmin: Functional characterization and metabolism in a heterozygote deficient patient. Thromb Haemost 1986; 55: 375-378.
172. Scharrer IM, Wohl RC, Hach V, Sinio L, Boreisha I, Robbins KC. Investigation of a congenital abnormal plasminogen, Frankfurt I, and its relationship to thrombosis. Thromb Haemost 1986; 55: 396-401.
173. Goodnough LT, Saito H, Ratnoff OD. Thrombosis or myocardial infarction in congenital clotting factor abnormalities and chronic thrombocytopenias: A report of 21 patients and a review of 50 previously reported cases. Medicine 1983; 62: 248-255.
174. Juhan-Vague I, Valadier J, Alessi MC, et al. Deficient t-PA release and elevated PA inhibitor levels in patients with spontaneous or recurrent deep venous thrombosis. Thromb Haemost 1987; 57: 67-72.
175. Nilsson IM, Ljungner H, Tengborn L. Two different mechanisms in patients with venous thrombosis and defective fibrinolysis: Low concentration of plasminogen activator or increased concentration of plasminogen activator inhibitor. Br Med J 1985; 290: 1453-1456.
176. Stalder M, Hauert J, Kruithof EKO, Bachmann F. Release of vascular plasminogen activator (v-PA) after venous stasis: Electrophoretic-zymographic analysis of free and complexed v-PA. Br J Haematol 1985; 61: 169-176.
177. Marbet GA, Duckert F, Biland L, Widmer LK. Quantitative und qualitative Hämostaseveränderungen bei 373 Patienten mit rezidivierenden Venenthrombosen. Schweiz med Wschr 1985; 115: 1505-1506.
178. Jørgensen M, Bonnevie-Nielsen V. Increased concentration of the fast-acting plasminogen activator inhibitor in plasma associated with familial venous thrombosis. Br J Haematol 1987; 65: 175-180.
179. Korninger C, Lechner K, Niessner H, Gössinger H, Kundi M. Impaired fibrinolytic capacity predisposes for recurrence of venous thrombosis. Thromb Haemost 1984; 52: 127-130.
180. Paramo JA, Colucci M, Collen D, Van de Werf F. Plasminogen activator inhibitor in the blood of patients with coronary artery disease. Br Med J 1985; 291: 573-574.
181. Hamsten A, Wiman B, De Faire U, Blombäck M. Increased plasma levels of a rapid inhibitor of tissue plasminogen activator in young survivors of myocardial infarction. N Engl J Med 1985; 313: 1557-1563.
182. D'Angelo A, Kluft C, Verheijen JH, Rijken DC, Mozzi E, Mannucci PM. Fibrinolytic shut-down after surgery: Impairment of the balance between tissue-type plasminogen activator and its specific inhibitor. Eur J Clin Invest 1985; 15: 308-312.
183. Kluft C, Verheijen JH, Jie AFH, et al. The postoperative fibrinolytic shutdown: A rapidly reverting acute phase pattern for the fast-acting inhibitor of tissue-type plasminogen activator after trauma. Scand J Clin Lab Invest 1985; 45: 605-610.

184. Rånby M, Bergsdorf N, Nilsson T, Mellbring G, Winblad B, Bucht G. Age dependence of tissue plasminogen activator concentrations in plasma, as studied by an improved enzyme-linked immunosorbent assay. Clin Chem 1986; 32: 2160-2165.
185. Vague P, Juhan-Vague I, Aillaud MF, et al. Correlation between blood fibrinolytic activity, plasminogen activator inhibitor level, plasma insulin level, and relative body weight in normal and obese subjects. Metabolism 1986; 35: 250-253.
186. Juhan-Vague I, Moerman B, De Cock F, Aillaud MF, Collen D. Plasma levels of a specific inhibitor of tissue-type plasminogen activator (and urokinase) in normal and pathological conditions. Thromb Res 1984; 33: 523-530.

MODULATION OF ENDOTHELIAL CELL COAGULANT PROPERTIES

D.M. Stern and P.P. Nawroth

Thrombosis/Hematology Research, Oklahoma Medical Research Foundation, Oklahoma City, Oklahoma, USA

Endothelium has traditionally been thought of as an inert container responsible for confining the blood to the intravascular space. Recent studies, however, have indicated an active role for endothelium in the regulation of procoagulant and anticoagulant reactions. These hemostatic properties are mediated by specific receptors on the cell surface and released products.

The anticoagulant protein C/protein S pathway is closely linked to the vessel wall (1). Endothelium provides a receptor, thrombomodulin, which binds thrombin and promotes activation of protein C. Bovine endothelium also expresses binding sites for protein S which facilitate formation of the activated protein C/protein S complex (2). Assembly of this anticoagulant complex on the endothelial cell surface occurs in response to low concentrations of protein S and activated protein C suggesting this may be a physiologic site at which these reactions take place. In addition to providing binding sites, endothelium also synthesizes the vitamin K-dependent cofactor, protein S. Synthesis and release of protein S occurs in a constitutive manner (3, 4). Release of intracellular protein S is facilitated by γ-carboxylation as indicated by its intracellular accumulation in the presence of warfarin. Morphologic studies have shown protein S to be present in endothelium, but not vascular fibroblasts or smooth muscle cells. At the electron microscopic level, protein S has been visualized in cisternae of rough endoplasmic reticulum, the trans face of the Golgi, and a population of intracellular vesicles located near the cellular periphery. The existence of the latter pool of protein S-containing vesicles indicates that there is an intracellular storage pool. By analogy with the von Willebrand Factor, we have hypothesized that there must be mechanisms for release. In this context, the adrenergic agonist norepinephrine (5) has been shown to stimulate endothelium to induce protein S release and degradation, leading to diminished anticoagulant activity and to down-regulation of protein S-cell surface binding sites. Norepinephrine-induced release of intracellular protein S results from the interaction of this vasoactive amine with a class of α-1 adrenergic receptors not previously observed on endothelium. Attenuation of norepinephrine-induced release of protein S by pertussis toxin in association with the ADP-ribosylation of a 41,000-dalton membrane protein indicates that this intracellular transduction pathway involves a regulatory G protein. Other studies have indicated that protein S release occurs in response to maneuvers which elevate cytosolic calcium (4) or activate protein kinase C, suggesting that hydrolysis of phosphoinositides may mediate this response. In addition to protein S release, norepinephrine also induces loss of cell surface binding sites for protein S. This blocks effective formation of the activated protein C/protein S complex on the cell surface. The net result of these changes is to down-regulate the protein C/protein S anticoagulant mechanism on the vessel wall. These observations define a relationship between a vessel wall anticoagulant property and the autonomic nervous system, and indicate an intracellular mechanism by which coagulation can respond to environmental stimuli.

Since thrombin is required to initiate the protein C pathway, we have considered whether thrombin formation could occur on the endothelial cell surface. In the quiescent state, formation of low amounts of thrombin in the vicinity of receptors for the protein C/protein S mechanism would serve an anticoagulant function, priming this system. In a perturbed state, larger amounts of thrombin could potentially participate in the pathophysiology of localized thrombosis. A schematic depiction of this pathway is shown in Figure 1 (6). Our work has focussed on the interaction of Factors IX and IXa with endothelium. The results of work by our laboratory and other investigators (7-11) has indicated that Factors IX and IXa bind to endothelium in a specific, high affinity manner. In the presence of Factors VIII and X, this site becomes relatively selective for the enzyme, allowing the Factor X activation complex to assemble in response to low concentrations of Factor IXa. Interaction of Factor IX/IXa with endothelium involves association with a cell surface protein, M_r~140,000, which metabolic labelling studies have shown to be of endothelial cell origin. The potential physiologic importance of this receptor is suggested by the clinical syndrome observed in two brothers with an abnormal Factor IX molecule, Factor IX_{ala}, which has decreased affinity for its endothelial cell sites.

On the surface of quiescent endothelium, multiple anticoagulant mechanisms predominate, and probably only small amounts of thrombin can form. Stimulation of endothelium, however, with agents such as monokines can fundamentally change this situation. Treatment of endothelium in vitro with tumor necrosis factor/cachectin (TNF) can lead to the induction of the procoagulant cofactor tissue factor (Figure 2A-B) (12, 13). Induction of tissue factor occurs in a time-dependent and dose-dependent manner. In order for tissue factor to be induced, de novo protein synthesis followed by cell surface expression of tissue factor must occur. Thus, the interaction of TNF with endothelium does not directly lead to cytotoxicity, but re-directs the expression of cellular coagulant activities. Since tissue factor is a potent procoagulant which can initiate activation of coagulation, it was important to understand if tissue factor induction could occur in vivo. This was tested by infusing rabbits with Interleukin-1 (IL-1), subsequently sacrificing the animals and assaying the native aortic endothelium for tissue factor (Figure 2C-D) (14). Tissue factor induction is observed in the native aortic endothelium.

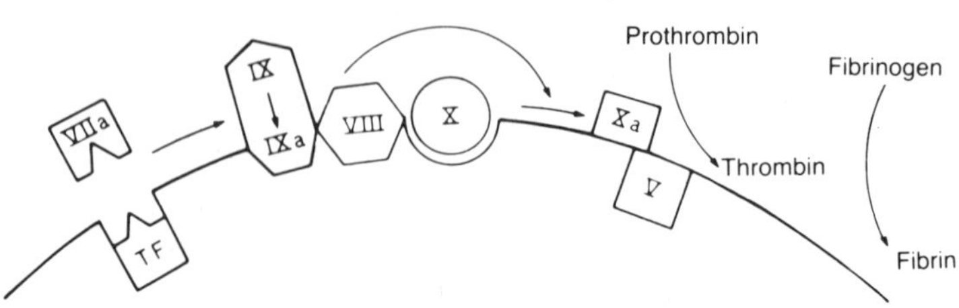

Fig. 1. Schematic depiction of a procoagulant pathway on the endothelial cell surface. TF = tissue factor.

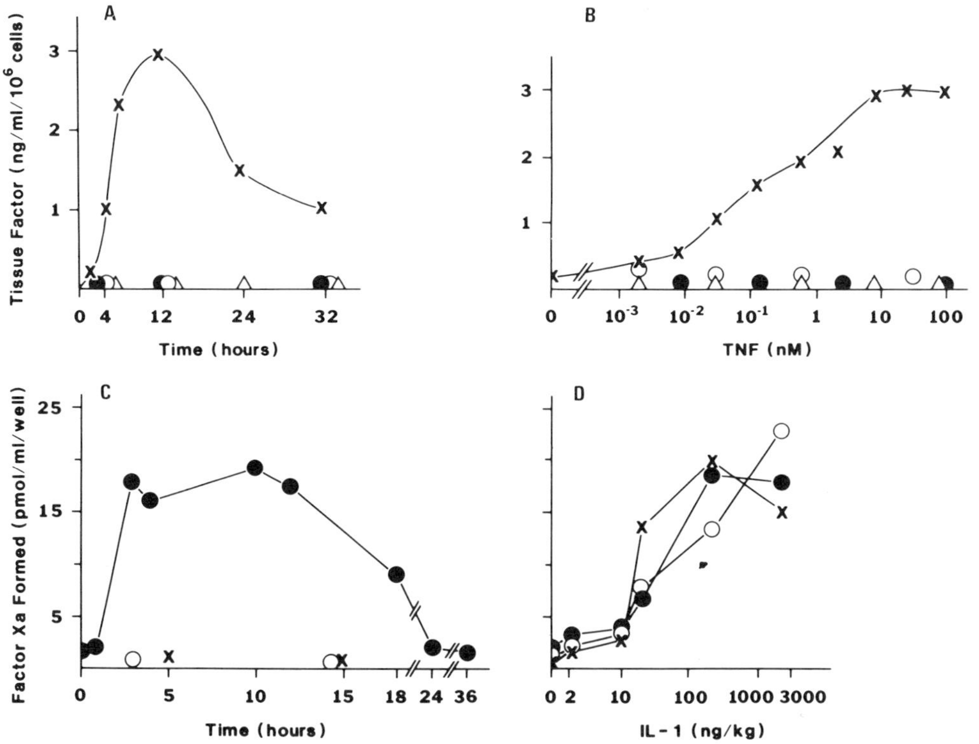

Fig. 2. Effect of TNF and IL-1 on endothelial cell tissue factor. A. Time course of tissue factor induction by TNF in cultured endothelium. Monolayers were incubated in serum-free medium alone (Δ) or in the presence of TNF (10 nM) (x). Cycloheximide or heat-treated TNF (o) was added where indicated. B. Dependence of tissue factor induction on the dose of TNF. Monolayers were incubated with the indicated concentration of TNF (x) for 12 hours. Heat-treated TNF (•) or cycloheximide (o) was added to certain cultures. C. Time course of tissue factor induction in aortic segments following IL-1 infusion into rabbits. Rabbits were infused with recombinant IL-1 (0.5 μg/kg body weight, •), heat-treated IL-1 (0.9 μg/kg body weight, o), or a control protein (0.8 μg/kg body weight, x). Animals were sacrificed at the indicated times and the tissue factor activity of native aortic endothelium assayed by monitoring Factor VIIa-dependent Factor X activation. D. Dependence of tissue factor induction on the dose of IL-1. Eight hours after infusion of the indicated dose of IL-1 rabbits were sacrificed, aortas were isolated and segments from proximal (o), mid (x) and distal (o) portions were assayed for tissue factor.

In addition to the induction of the procoagulant tissue factor, both TNF and IL-1 have been shown to suppress endothelial cell anticoagulant pathways, including the protein C/protein S and fibrinolytic systems (14, 15). Furthermore, fibrin formation was observed on the endothelium of rabbits infused with IL-1 (14).

From these data, it is clear that stimulation of endothelium can shift the balance of cellular coagulant properties favoring activation of the clotting system. Modulation of endothelial cell coagulant properties in response to mediators of the host response contributes to an emerging picture of the active role of endothelium in the regulation of the hemostatic system.

REFERENCES

1. Esmon D. Science 1987; 235: 1248-1353.
2. Stern D, Nawroth P, Harris K, Esmon C. J Biol Chem 1986; 261: 713-718.
3. Fair D, Marlar R, Levin E. Blood 1986; 67: 1168-1171.
4. Stern D, Brett J, Harris K, Nawroth P. J Cell Biol 1986; 102: 1971-1978.
5. Brett J, Godman G, Nawroth P, Stern D. Fed Proc 1986; 45: 1073A.
6. Nawroth P, Stern D. J Cell Biochem 1985; 28: 253-264.
7. Heimark R, Schwartz S. Biochem Biophys Res Commun 1983; 111: 723-731.
8. Stern D, Drillings M, Nossel H, Hurlet-Jensen A, LaGamma K, Owen J. PNAS (USA) 1983; 80: 4119-4123.
9. Varadi and Elodi. Blood 1987; 69: 442-446.
10. Rimon S, Melamed R, Savion N, Nawroth P, Stern D. J Biol Chem 1987; 262: 6023-6031.
11. Stern D, Nawroth P, Kisiel W, Vehar G, Esmon C. J Biol Chem 1985; 260: 6717-6722.
12. Nawroth P, Stern D. J Exp Med 1986; 163: 740-745.
13. Bevilacqua M, Pober J, Majeau G, Fiers W, Cotran R, Gimbrone M. PNAS (USA) 1986; 83: 4533-4538.
14. Nawroth P, Handley D, Esmon C, Stern D. PNAS (USA) 1986; 83: 3460-3464.
15. Bevilacqua M, Schleef R, Gimbrone M, Loskutoff D. J Clin Invest 1986; 78: 587-596.

DEFICIENCIES OF PROTEIN C AND PROTEIN S AND THROMOEMBOLIC DISEASE

John H. Griffin

Department of Immunology, Research Institute of Scripps Clinic La Jolla, California, USA

Blood clotting pathways (Fig. 1) involve sequential enzymatic activations of serine protease zymogens. Protein C and protein S are vitamin K-dependent plasma proteins central to the protein C pathway. The protein C pathway is an antithrombotic regulatory system that inhibits the blood coagulation pathways and also stimulates fibrinolysis (1, 2, 3, 4). This presentation concerns a description of the functions of protein C and protein S, of the methods to measure protein C and protein S in plasma, and of the clinical presentation of protein C and protein S deficiencies associated with venous thrombotic disease.

In recent years human protein C and protein S have been purified and biochemically characterized and the genes for human protein C and protein S have been sequenced. These proteins are homologous to other plasma vitamin K-dependent proteins in their amino acid sequence and contain ~10 γ-carboxyglutamic acid residues in the amino terminal region. This unusual amino acid is formed by carboxylation of glutamic acid in a reaction requiring vitamin K. Oral anticoagulants are vitamin K antagonists that interfere with normal carboxylation of the vitamin K-dependent factors. Protein C is a serine protease zymogen that is converted to an active enzyme by thrombin by limited proteolysis. The activation of protein C by thrombin is greatly accelerated by thrombomodulin, a non-enzymatic lipoprotein cofactor found on the surface of the endothelium in the microcirculation (5, 6). Thrombin associates with thrombomodulin with high affinity, and in the thrombin:thrombomodulin complex, the enzymatic properties of thrombin are altered such that protein C activation is favored whereas the procoagulant properties of thrombin, i.e. fibrinogen cleavage, platelet activation, and activation of Factors V and VIII, are inhibited. Hence, low levels of thrombin in the microcirculation bind to thrombomodulin and activate protein C, thereby generating the anticoagulant enzyme, activated protein C. Factor Va also accelerates protein C activation in vitro by thrombin, though the physiologic significance of this reaction is unknown (6).

Activated protein C is anticoagulant because it is a potent inactivator of Factors Va and VIIIa (3). The biochemical structures of Factors V and VIII and Va and VIIIa have been recently elucidated. Thrombin converts Factors V and VIII to highly active cofactors, Factors Va and VIIIa, by limited proteolysis. Further limited proteolysis of these

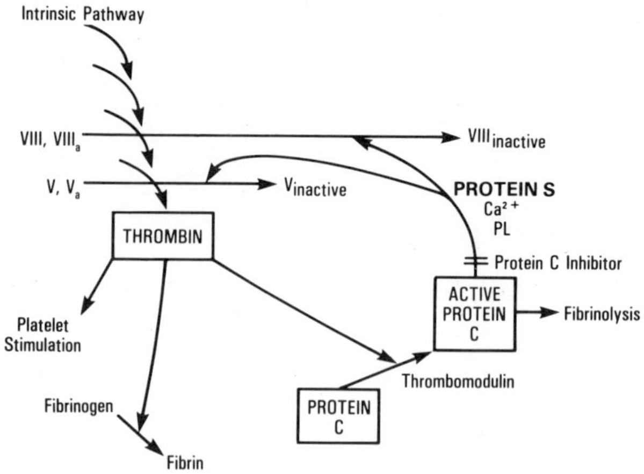

Fig. 1. Protein C pathway scheme. This scheme depicts the role of the protein C pathway as an anticoagulant mechanism to inhibit blood coagulation. A series of arrows represents the generation of thrombin by the coagulation pathways. Thrombin combines with the endothelial cell surface protein, thrombomodulin, to form a complex that neutralized the procoagulant properties of thrombin while enhancing its ability to activate protein C. Activated protein C with protein S as a cofactor inactivates Factors Va and VIIIa by limited proteolysis. This reaction is accelerated on a phospholipid surface (PL) in the presence of calcium ions. Activated protein C also stimulates fibrinolysis, at least in part by neutralizing the inhibitor of tissue plasminogen activator. Activated protein C can be neutralized by the plasma protein, protein C inhibitor. Inherited deficiencies of protein C or protein S are associated with recurrent venous thrombotic disease.

proteins by activated protein C inactivates Factors Va and VIIIa. Protein S functions as a non-enzymatic cofactor in the inactivation of Factors Va and VIIIa by activated protein C (7). These important reactions probably occur in vivo on cell surfaces, such as platelets or endothelial cells, as both of these cells contain protein S (4, 8, 9). Protein S reversibly associates noncovalently with C4b-binding protein, a regulatory protein of the complement pathway, and in plasma ~ one third of protein S is free and two thirds of protein S is bound to this complement protein. It appears that only free protein S is active as an anticoagulant cofactor in clotting assays in vitro. Recently another protein, named protein S binding protein, was described in bovine plasma and it enhances the cofactor activity of protein S. Since no similar human protein S binding protein is known, no inference about the role of a human protein S binding protein can be yet made.

Immunologic and functional assays have been developed to measure protein C and protein S in plasma. The most convenient and widely available assays for measuring protein C and protein S utilize Laurell rocket immunoelectrophoresis to determine antigen levels (SmithKline Bio-Science). The normal level of protein C in plasma is 4 μg/ml. The normal level of protein S antigen in plasma is 34 μg/ml in one report and 24 μg/ml in another report. Some technical difficulties may arise in measuring protein S antigen because protein S exists in two distinct forms in plasma: free protein S and protein S associated non-covalently with the complement protein, C4b-binding protein. Functional assays for protein C or protein S are more difficult to perform than immunologic assays. The anticoagulant activity of protein C and the cofactor activity of protein S can be measured using clotting assays and deficient plasmas. The amidolytic activity of activated protein C provides the basis for other functional assays. Reagents for such assays are becoming available commercially (Diagnostica Stago, American Bioproducts, American Diagnostica), because of the increasing interest in laboratory measurements of protein C and protein S in patients with venous thrombosis.

Deficiencies of protein C and protein S can be associated with inherited venous thrombotic disease (10, 11, 12, 13, 14, 15). Protein C deficiency was first reported in 1981 and protein S deficiency in 1984. The clinical presentation of heterozygous protein C or protein S deficiency involves recurring venous thrombotic disease that often involves thrombophlebitis, pulmonary emboli, or deep vein thrombosis. Plasma levels of protein C or protein S antigen in heterozygous patients range from 30% to 70% of normal levels, with the average near 50%. The extremely rare homozygous deficiency of protein C has been described in newborn infants who present with purpura fulminans or extensive venous thrombotic disease (16). Clinical manifestations of autosomal dominant forms of protein C or protein S heterozygous deficiency usually appear between the ages of 15 and 40 years with over 50% of affected patients presenting before the age of 30 years (10). Some protein C and protein S heterozygous deficient patients are known in whom no clinical manifestations of thrombotic disease are apparent. The current impression is that heterozygous patients in families with strong histories of thrombotic disease are at much greater risk than heterozygous deficient patients in families that have no remarkable history of thrombotic disease. In families with homozygous protein C deficient infants with purpura fulminans, the heterozygous protein C deficient adults are usually asymptomatic (16). Therefore, at this time it appears there are two clinically distinct forms of protein C deficiency: autosomal dominant heterozygous deficiency with partial penetrance and autosomal recessive. Protein C heterozygous deficiency in pediatric patients has also been associated with stroke in three cases and venous thrombosis induced by asparaginase therapy in one case.

Identification of heterozygous protein C or protein S deficiency in thrombotic patients is complicated by the fact that many patients with thrombotic disease receive oral anticoagulant therapy. Such therapy decreases all of the vitamin K-dependent proteins to some extent, with an average depression of antigenic levels of vitamin K-dependent proteins on the order of 50%. Therefore, for patients receiving oral anticoagulant therapy, it is important to measure protein C, protein S, and two other vitamin K-dependent proteins. Since oral anticoagulant therapy depresses the levels of all vitamin K-dependent proteins in a similar manner, the ratios of protein C and protein S to other factors such as prothrombin and Factor X are estimated, and when values fall below the control range, heterozygous protein C or protein S deficiency is suspected (10, 11, 14, 15). When protein C or protein S deficiency is suspected in patients taking oral anticoagulants, it is advisable to confirm the diagnosis by family studies or by analyzing plasma samples collected when the patient is not taken oral anticoagulants.

The efficacious treatment of protein C or protein S deficiency involves heparin therapy and/or oral anticoagulant therapy, especially the latter. Therapy with vitamin K antagonists seems somewhat ironic because oral anticoagulant therapy decreases protein C and protein S as well as the procoagulant vitamin K-dependent factors prothrombin, Factors VII, IX, and X. Nonetheless, it appears that the net effect of decreasing two anticoagulant factors and four procoagulant factors is to achieve a less procoagulant state. Initiation of oral anticoagulant therapy causes a rapid decrease in Factor VII and protein C since their half-lives are approximately 6 to 8 hours whereas the half-lives of the other vitamin K-dependent proteins are much longer. This may result in a transient apparent acquired deficiency of protein C at the onset of oral anticoagulant therapy. This phenomenon is thought to explain the reported tendency to warfarin-induced skin necrosis in a small percentage of protein C deficient heterozygous patients.

Other plasma proteins whose deficiency is associated with inherited venous thrombotic disease include antithrombin III, heparin cofactor II, plasminogen, and fibrinogen. Deficiencies of these proteins account for approximately 4% of families with inherited thrombotic disease whereas it is estimated, based on limited current data, that protein C and protein S deficiency may account for approximately 10% to 12% of venous thrombotic disease in patients who presented with symptoms under age 45 years. The incidence of protein C and protein S deficiency appears even higher for young patients with strongly family histories of venous thrombosis. Therefore, it is useful to assay protein C, protein S, antithrombin III, heparin cofactor II, plasminogen, and fibrinogen in patients with recurrent venous thrombotic disease.

REFERENCES

1. Clouse LH, Comp PC. The regulation of hemostasis: The protein C system. N Engl J Med 1986; 314: 1298-1304.
2. Comp PC. Hereditary disorders predisposing to thrombosis. Prog Hemostasis and Thrombosis 1986: 71-102.
3. Gardiner JE, Griffin JH. Human protein C and thromboembolic disease. In: Brown EB, ed. Progress in Hematology. New York: Gruce & Stratton, Inc, 1983; Vol XIII: 265-278.
4. Stern D, Brett J, Harris K, Nawroth P. Participation of endothelial cells in the protein C - protein S anticoagulant pathway: The synthesis and release of protein S. J Cell Biol 1986; 102: 1970-1978.
5. Esmon NL, Owen WG, Esmon CT. Isolation of membranebound cofactor for thrombin-catalyzed activation of protein C. J Biol Chem 1982; 257: 859-864.
6. Ishii H, Majerus PW. Thrombomodulin is present in human plasma and urine. J Clin Invest 1985; 76: 2178-2181.
7. Suzuki K, Nishioka J, Matsuda M, Maurayama H, Hashimoto S. Protein S is essential for the activated protein C-catalyzed inactivation of platelet-associated Factor Va. J Biochem 1984; 96: 455-460.
8. Harris KW, Esmon CT. Protein S is required for bovine platelets to support activated protein C binding and activity. J Biol Chem 1985; 260: 2007-2009.
9. Schwarz HP, Heeb MJ, Wencel-Drake J, Griffin JH. Identification and characterization of protein S in human platelets. Blood 1985; 66: 1452-1455.
10. Broekmans AW, Veltkamp JJ, Bertina RM. Congenital protein C deficiency and venous thromboembolism. N Engl J Med 1983; 309: 304-344.
11. Broekmans AW, Bertina RM, Reinalda-Poot J, et al. Hereditary protein S deficiency and venous thromboembolism. A study in three Dutch families. Thromb Haemostas 1985; 53: 273-277.
12. Comp PC, Esmon CT. Recurrent venous thromboembolism in patients with a partial deficiency of protein S. N Engl J Med 1984; 311: 1525-1528.
13. Comp PC, Nixon RR, Cooper MR, Esmon CT. Familial protein S deficiency is associated with recurrent thrombosis. J Clin Invest 1984; 74: 2082-2088.
14. Griffin JH, Evatt B, Zimmerman TS, Kleiss AJ, Wideman C. Deficiency of protein C in congenital thrombotic disease. J Clin Invest 1981; 68: 1370-1373.

15. Schwarz HP, Fischer M, Hopmeier P, Batard MA, Griffin JH. Plasma protein S deficiency in familial thrombotic disease. Blood 1984; 64: 1297-1300.
16. Seligsohn U, Berger A, Abead M, et al. Homozygous protein C deficiency manifested by massive venous thrombosis in the newborn. N Engl J Med 1984; 307: 1191-1195.

CHAPTER 7
RHEUMATIC DISEASES

Biochemical events in the destruction of cartilage in rheumatic diseases
T.E. Hardingham

New immunological aspects in the pathogenesis and diagnosis of rheumatic diseases
J.R. Kalden

New biochemical parameters in diagnosis of joint diseases
K. Kleesiek, R. Reinards and H. Greiling

Clinical biochemistry of gout
M.M. Müller

BIOCHEMICAL EVENTS IN THE DESTRUCTION OF CARTILAGE IN RHEUMATIC DISEASES

T.E. Hardingham

Kennedy Institute of Rheumatology, Hammersmith
London W6 7DW, U.K.

ARTICULAR CARTILAGE

Articular cartilage is an active tissue consisting of a large highly organised extracellular matrix in which a sparse population of specialised cells, the chondrocytes, is embedded (1). The function of the chondrocytes is to maintain the cartilage matrix by the co-ordinate synthesis and turnover of matrix components. It is the composition of the matrix which endows the tissue with its mechanical properties, and these can be understood largely in terms of the relative contributions made by the two major components, collagen and proteoglycan (2). The collagen in cartilage is of several different genetically defined types, primarily type II, but also some type VI, IX and XI, and in hypertrophic cartilage there is also production of type X collagen (3). These 'minor' collagens have quite distinct molecular features compared with type II, but their precise distribution and function in cartilage is not yet fully understood. The collagen forms a dense network of fine fibres that give structural shape and form to the tissues and the proteoglycans, which form large aggregates, are immobilised within the interfibrillar space of the network. The proteoglycans are predominantly of a high molecular weight (Mr 1-4x10^6) aggregating type (4), but smaller amounts of non-aggregating proteoglycan are also present (5, 6). Those that aggregate with hyaluronate form highly ordered supramolecular structures which occupy the space between fibrils, whereas the smaller species may be associated with the fibrillar network itself. The high concentration (up to 100 mg/ml) of proteoglycans and their polyanionic characteristics create a high osmotic swelling pressure that draws water into the tissue. The collagen network is thus under continuous tension and it is the balance between the osmotic swelling pressure of the proteoglycans and the tension in the collagen network that is the basis of the compressive properties of the tissue, which are so important for its biomechanical function (7). Maintenance of the proteoglycan content and the macromolecular organization is thus essential in normal cartilage.

CAPACITY FOR CARTILAGE REPAIR

The articular joint is made up of many different tissue components; the articular cartilage, subchondral bone, ligaments and the joint capsule, together with the muscles that act upon the joint (8). All these tissues respond and adapt to the mechanical forces put upon them during normal use, and changes in the stresses put upon any one component, or in the individual properties of one component, will cause a response and some compensatory change in other tissues (9). The healthy joint may thus be considered to reflect an equilibirum of interaction responses amongst the component tissues and the development of pathological changes may arise when adaptation to change is insufficient to sustain normal function.

Of the different tissues involved, cartilage may be most prone to failure because of its particular structure and limited capacity for repair. Cartilage is an unusual tissue as the matrix is avascular and not innervated and the chondrocytes are therefore not influenced by direct neural contact or by exposure to normal concentrations of serum components, other than by diffusion through the tissue matrix. The chondrocytes are thus isolated and in the adult they form a relatively static population of cells. New cells cannot be recruited from the circulation to replace damaged matrix and the response of the tissue to injury is limited. A number of aetiological factors acting on various components of the joint may thus eventually result in cartilage damage and some common mechanisms of degeneration in cartilage may thus be involved in clinically different joint diseases.

PROTEOGLYCAN TURNOVER IN CARTILAGE

The mechanical properties of cartilage are dependent on the integrity of its matrix. Of the major components present, the fibrillar collagen network is less metabolically active and more resistant to proteolytic attack than the proteoglycans, which are continuously being degraded and replaced by new synthesis (10, 6), (Fig. 1). A number of observations would suggest that an important step in the early stage of cartilage degeneration is some loss of proteoglycan. This would decrease the compressive stiffness of the tissue and may thereby damage the collagen network by exposing it to greater stresses. However, it does not appear that major damage to the

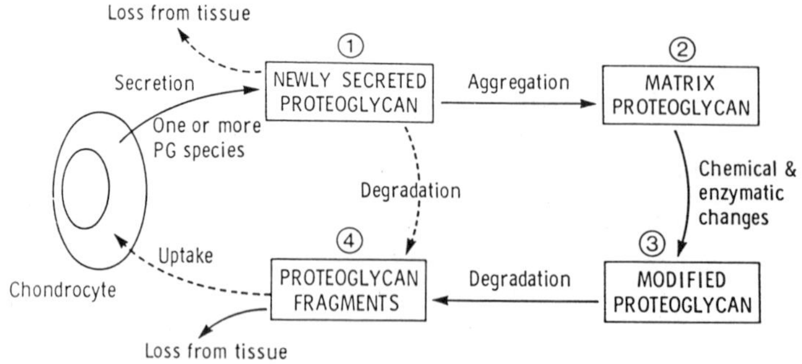

Fig. 1. Proteoglycan turnover in the extracellular matrix of Cartilage. Proteoglycans are synthesised by chondrocytes and secreted into the matrix (pool 1) from where they may be incorporated into proteoglycan aggregates with hyaluronate and link protein (pool 2) or part may be rapidly degraded and lost from the tissue. The major fraction (pool 2) is of slow turnover in mature cartilage and may undergo some structural modification by slow enzyme action or free radical attack (pool 3). Final degradation by proteolytic enzymes yields fragments (pool 4) that mainly diffuse from the tissue and appear in synovial fluid (6).

collagen network occurs until after the loss of proteoglycan is extensive. The proteoglycans appears to have some protective function in the tissue and control of their turnover has an important role in cartilage maintenance.

It is of great importance that proteoglycan degradation during normal turnover should be a conservative process, which permits a steady renewal of the proteoglycans without weakening the tissue (Fig. 1). As turnover involves the mobilisation of proteoglycans within the matrix the general characteristics have been investigated by studying the release of proteoglycans from cartilage explants maintained in culture. Initial studies showed that the chondrocytes themselves were responsible for the processes involved in turnover in the matrix surrounding them and that large proteoglycan fragments were released into the culture medium, which on testing were found to be unable to aggregate (12, 13, 14). Further investigation of the structural differences in the released proteoglycans (11) showed that they had been cleaved to separate the major part of the proteoglycan bearing all the glycosaminoglycan chain from the region involved in aggregation (Fig. 2). Cleavage within the major glycosaminoglycan part of the molecule was quite limited. The most favoured site of cleavage was thus optimally located to free the proteoglycan from aggregate structures and to permit its diffusion out of the tissue. The mechanism was therefore ideally suited to promote turnover and yet to conserve the properties of the matrix.

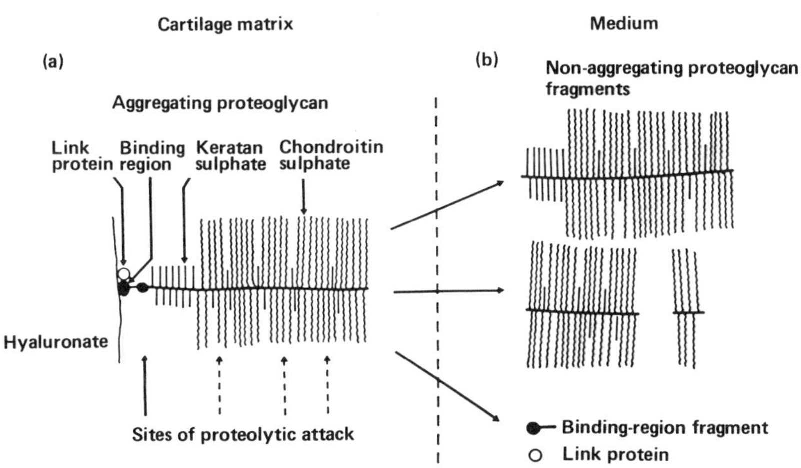

Fig. 2. Mechanisms involved in normal and accelerated proteoglycan turnover in vitro (11)

a) Schematic model of the structure of aggregating proteoglycan (4).
b) Fragments of proteoglycan released into the medium of cartilage explant cultures. The major site of proteolytic attack is close to the binding region involved in aggregation. Further limited cleavage occurs within the major chondroitin sulphate and keratan sulphate containing regions. The effects of interleukin-1 on the explants is to increase the rate of this mechanism of proteoglycan turnover.

INTERLEUKIN-1 EFFECTS ON PROTEOGLYCAN TURNOVER

These results can be contrasted with the effects of interleukin-1 on the same cartilage explant system, where it not only increased the rate of proteoglycan loss from the matrix, but also inhibited its biosynthesis by the chondrocytes (14). Comparison of the type of proteoglycan fragments released showed that although there was some reduction of their average size, the general pattern of cleavage was very similar (11), (Fig. 2). The IL-1 thus appeared to enhance the rate of mechanisms involved in normal turnover rather than initiating a new mode of attack and the matrix content of proteoglycan rapidly became depleted.

Intra-articular injection of IL-1 into rabbit joints was also shown to produce a major loss of proteoglycan from the cartilage within 24 h and this was accompanied by accumulation of polymorphonuclear and mononuclear leucocytes in the joint space (15). However, cartilage damage was not directly attributed to the leucocytes as injection of endotoxin caused leucocyte accumulation without cartilage depletion. It was also shown (16) that in animals that had been made leucopenic the intra-articular injection of IL-1 promoted as much, if not more, proteoglycan loss inspite of minimal cellular infiltration. The in vivo effect of IL-1 therefore appeared to be by direct stimulation of chondrocyte-mediated matrix degradation rather than acting on polymorphs or monocytes to stimulate the release of proteinases of further cytokines. The effects of intra-articular IL-1 injection were similar to the effects of antigen-induced inflammatory arthritis in the rabbits (Dumonde-Glynn model), which elicited leucocyte infiltration and matrix depletion, although over a slightly longer time scale (1-7 days) (15). In vitro tumour necrosis factor (TNF α) also appears to trigger matrix depletion of proteoglycans by chondrocytes (17), and the local release of IL-1 or TNF-like cytokines may be a major factor in the initial attack on cartilage in inflammatory joint disease and also during the acute episodes in more chronic disease.

TURNOVER IN A NON-INFLAMMATORY MODEL OF OSTEOARTHRITIS

Studies on animal models of osteoarthritis show changes in cartilage behaviour that are different from those in the acute inflammatory models. In the canine model of experimental osteoarthritis in which joint laxity is created by section of the cruciate ligament there are progressive changes in the cartilage over several months that resemble the early stages of natural osteoarthritis (18). There is increased water content of the cartilage matrix (it swells); sustained increase in the biosynthesis of cartilage matrix components and their turnover; induction of some chondrocyte cell division; but there is no net loss of cartilage proteoglycan (18, 19, 20, 21, 22). There is also some articular surface fibrillation, evidence of subchondral bone remodelling and extensive osteophyte development (23). There are thus a range of responses to the altered loading pattern that leads to some joint remodelling. The response of the cartilage involves increased activity in the chondrocytes, which might be broadly interpreted as an attempt at repair and regeneration. However, this activity leads to abnormal matrix turnover and renewal and this may result in progressive changes in matrix organisation that are not conducive to effective tissue repair. As the increased matrix turnover was accompanied by increased matrix synthesis, rather than by inhibition, this did not appear to be a simple response to IL-1 production. Furthermore, the differences in rates of turnover in the cartilage were maintained in explant culture over several days (21) showing that the chondrocyte response did not require continuous exposure to systemic factors for it to be sustained. This altered pattern of behaviour of the chondrocytes may thus reflect a response to changes in the matrix surrounding the cell.

CLINICAL MARKERS OF CARTILAGE DAMAGE

The ability to detect the products of proteoglycan turnover as they are released from cartilage and appear in synovial fluid or serum has raised the prospect of clinical detection of cartilage degeneration in joint disease (24). In the Dumonde-Glynn model of inflammatory arthritis the synovial fluid was shown to contain an increased amount of proteoglycan which was detected with a dye binding assay and by a radioimmunoassay for keratan sulphate (a proteoglycan component), (Ratcliffe, Pettifer, Henderson, Hardingham, unpublished work). Raised levels occurred over several days whilst rapid depletion was taking place, but returned to low levels when depletion progressed no further. In the canine model of osteoarthritis where there was increased synthesis and turnover, but no net proteoglycan depletion of the cartilage and there was also raised levels of proteoglycan components in the synovial fluids, (Ratcliffe, Saed-Nejad, Billingham & Hardingham, unpublished work). The experimental models thus show that raised levels in synovial fluid may reflect increased biosynthesis and turnover as well as active depletion of the cartilage matrix.

Investigation of clinical samples of synovial fluid have shown that only acute arthropathies appear to contain consistently high levels of proteoglycan (25). These include acute gout, acute pseudogout and Reiters syndrome, whereas the majority of RA and OA patients with more chronic knee problems do not show significantly raised levels. Analyses have been correlated (26) with the radiographic score and showed that those patients with most severe OA or RA, i.e., with greatest joint space narrowing, showed the lowest synovial fluid content of proteoglycan. Since joint space narrowing reflects the loss of cartilage, the result was correlating least cartilage with least proteoglycan in the synovial fluid. It is therefore important to understand that the detection of raised levels of proteoglycan is not an indication of disease severity, but may either reflect an acute loss of proteoglycan from cartilage or its enhanced synthesis and turnover. Raised proteoglycan is not thus a disease marker, but may be a process marker for intense activity in cartilage that could be anabolic or catabolic. It is interesting that those patients with acute arthropathies mainly have a good prognosis and it will require further development with retrospective or prospective studies to establish if measuring proteoglycan components in synovial fluid will be useful for predicting disease outcome or for monitoring treatment.

ACKNOWLEDGEMENTS

This work was carried out with the support of the Arthritis and Rheumatism Council (UK) and Medical Research Council (UK).

REFERENCES

1. Stockwell RA. In: Biology of Chondrocytes. Cambridge University Press 1979: 1-316.
2. Muir H, Hardingham TE. In: Scott JT, ed. Copemans Textbook of the Rheumatic Diseases. Edinburgh: Churchill Livingstone 1986; Vol. 1: 177-198.
3. Miller EJ. Ann NY Acad Sci 1985; 460: 1-13.
4. Hardingham TE, Beardmore-Gray M, Dunham DG, Ratcliffe A. Ciba Found Symp 1986; 124: 30-46.
5. Heinegård D, Paulsson N. In: Piez K, Reddi AH, eds. Extracellular Matrix Biochemistry. Amsterdam: Elsevier 1984: 277-328.
6. Hardingham TE. Rheumatology 1986; 10: 143-183.

7. Mow VC, Lai WM. Siam Rev 1980; 22: 275-317.
8. Gardner DL. Br Med J 1983; 286: 418-424.
9. Noble J, Alexander K. J Bone Jt Surg 1985; 67: 295-302.
10. Hardinghan T, Burditt L, Ratcliffe A. Prog Clin Biol Res 1984; 151: 17-29.
11. Ratcliffe A, Tyler JA, Hardingham TE. Biochem J 1986; 238: 571-580.
12. Sandy JD, Brown H, Lowther DA. Biochem Biophys Acta 1978; 543: 536-544.
13. Tyler JA. Biochem J 1985; 225: 493-507.
14. Tyler JA. Biochem J 1985; 227: 869-878.
15. Pettifer ER, Higgs GA, Henderson B. Proc Natl Acad Sci 1986; 83: 8749-8753.
16. Pettifer ER, Henderson B. Brit J Exp Path 1987; in press.
17. Saklatvala J. Nature 1986; 322: 547-549.
18. McDevitt CA, Gilbertson EMM, Muir H. J Bone Jt Surg (Br) 1977; 59B: 24-35.
19. Eyre DR, McDevitt CA, Billingham MEJ, Muir H. Biochem J 1980; 188: 823-837.
20. Sandy JD, Adams ME, Billingham MEJ, Plaas A, Muir H. Arthritis Rheum 1984; 27: 388-397.
21. Carney SL, Billingham MEJ, Muir H, Sandy JD. J Orthop Res 1984; 2: 201-206.
22. Carney SL, Billingham MEJ, Muir H, Sandy JD. J Orthop Res 1985; 3: 140-147.
23. Gilbertson EMM. Ann Rheum Dis 1975; 34: 12-25.
24. Thonar EJM, Lenz ME, Klintworth GK, et al. Arth Rheum 1985; 28: 1367-1376.
25. Ratcliffe A, Maini RN, Doherty M, Hardingham TE. Br J Rheum 1987; 26, Suppl 1: 66.
26. Saxne T, Heinegård D, Wollheim FA, Pettersson H. Lancet 1985; ii: 127-128.

NEW IMMUNOLOGICAL ASPECTS IN THE PATHOGENESIS AND DIAGNOSIS OF RHEUMATIC DISEASES

J.R. Kalden

Institute for Clinical Immunology and Rheumatology of the University of Erlangen-Nürnberg, 8520 Erlangen, F.R.G.

INTRODUCTION

Although research over the past three decades has considerably enlarged our knowledge in terms of tissue destructive autoimmune mechanisms it is only fair to say that the question: "What triggers autoimmune reactivity?" is still unanswered. For example, in rheumatoid arthritis, we know that cytokines such as Interferon Gamma, Interleukin-1 (IL-1), and Tumornecrosis factor (TNF) are directly involved in events leading to cartilage destruction. Mechanisms triggering the secretion of these cytokines, however, are still ill understood. We know that TNF and IL-1 will be produced and secreted by activated monocytes and macrophages or even dendritic cells upon stimulation, and, furthermore, that there is direct evidence that IL-1 and TNF do not only activate synovial lining cells but also exhibit a direct deleterious effect on chondrocytes. Interferon-Gamma, TNF or IL-1-activated chondrocytes release considerable amounts of endoproteinases which destroy the extracellular matrix. Mechanisms initiating an increased production of cytokine secretion by the different mononuclear cells have only partly been defined. One mechanism is represented by the phagocytosis of locally formed immune complexes consisting of aggregated IgG rheumatoid factors. Secondly, it is well known that in the synovial fluid as well as in the inflamed synovium the number of activated T lymphocytes which express class II (Ia) antigens or Interleukin-2 receptors (1) is significantly increased. Activated T lymphocytes secrete Interferon-Gamma which stimulates monocytic cell populations as well as macrophages to produce and secrete amongst others molecules with cartilage destructive activity, IL-1 and TNF.

Events leading to the local production of IgG rheumatoid factors which selfaggregate to immune complexes or events leading to the activation of T lymphocytes in the synovial fluid or the synovium of RA patients are still unknown. For the latter situation, a possible stimulation of autoreactive T cells by autoantigens such as proteoglycans or collagen II are under discussion. In this context, it is of interest that recent experiments from our laboratory demonstrated that chondrocytes express - upon stimulation with Interferon-Gamma - class II antigens which could enable them to act as antigen-presenting cells (2).

Since in systemic lupus erythematosus (SLE), our understanding of disease underlying mechanisms has especially progressed within the last years, new developments in diagnostic procedures as well as new etiopathogeneic aspects will be discussed in more detail as follows:

NEW ASPECTS OF THE PATHOGENESIS OF SYSTEMIC LUPUS ERYTHEMATOSUS

Systemic lupus erythematosus was initially recognized as a systemic disease based upon pathological evidence of underlying vasculitis (3). Since the description of the SLE cell phenomenon in 1948, which later turned out to be an autoantibody of the IgG class directed against nuclear antigen, this disease entity was considered as an autoimmune disease, and multiple studies were performed on the specificities of antinuclear antibody (ANA) reactivity and the formation of DNA anti-DNA immune complexes (4, 5, 6).

Presently, a large variation of different antibody-nuclear-antigen-systems are known, and attempts have repeatedly been made to associate different ANA specificities or antinuclear antibody clusters to certain forms of rheumatic diseases. ANA's can be classified in antibodies directed to single stranded (ss) or double stranded (ds) DNA, to non-histone nuclear antigens or to histones and nucleolar antigens respectively (7). A summary of relevant ANA's and their association to various connective tissue diseases is given in Table 1.

Antibodies to dsDNA have not only turned out to be a disease specific marker for SLE but they also serve as a parameter for the assessment of the disease activity. Antibodies to the Sm antigen that can be traced in up to 25% of patients with SLE are an additional disease-specific marker, and antibodies to the Ro antigen, when demonstrated in pregnant SLE patients, are frequently associated with a complete congenital heart block of the new-born child.

From the other associations of connective tissue diseases and ANA's - as listed in Table I - autoantibodies against the NRMPs should be mentioned, which are found in high titers in patients suffering from mixed connective tissue diseases, and antibodies against the LA or RO antigen, which are frequently associated with the Sjögren syndrome. In summary, there are some - although few - autoantibodies against nuclear antigens which can be used as a disease-specific marker and in addition, in certain cases, as a parameter for the assessment of the disease activity. With increasingly improved technologies, it seems to be reasonable to assume that in future even more disease-specific autoantibody phenomena or clusters of autoantibodies against nuclear antigens will help to define a certain rheumatic disease entity.

Besides different ANA's and especially antibodies against dsDNA, free nucleic acids including dsDNA have been traced in the sera of patients. In a recent communication (8), it was shown that the increase of serum nucleic acids was to correlate to a decrease in deoxyribonuclease 1 activity, a finding which prompted the discussion that a deficiency in that enzyme could represent at least one of multiple factors involved in the pathogenesis of SLE.

In spite of ongoing discussions with regard to the ultimate technology for the detection of serum immune complexes, there is overwhelming evidence that in SLE immune complexes are responsible for the disease underlying vasculitis. Serum immune complexes as demonstrated by the polyethylene glycol method correlate with the disease activity including the effect of plasmaseparation in acute disease exacerbations (9). More recent data

Table 1. Profiles of autoantibodies in connective diseases

Disease	Specificities	Occurrence % of patients varies with methods
SLE	Double-stranded DNA	50 - 70
	(Sm) antigen	21 - 30
	Deoxyribonucleoprotein (DNP)	70
	Histones	60 - 50 (idiopathic SLE) > 80 (drug-induced SLE)
	Ribonucleoprotein (RNP)	30 - 40
	SSA/Ro	30 - 40
	SSB/La	15 - 20
	Ribosomes ribosomal RNP (rRNP)	< 10
	Single-stranded DNA	variable
Sjögren's syndrome	Ro SSA	60
	La SSB	60
Scleroderma - diffuse	Nucleolar RNA, others	50
	Scleroderma-70 (Scl-70)	20
	Centromere	< 10
	RNP, Ro, La, others	variable
Scleroderma/ CREST	Centromere	50 - 95
	RNP, Ro, La, others	variable
Polymyositis/ dermatomyositis	Polymyositis-1	
	Jo-1	20 - 30
	Ml-1	variable
RA	RA-associated nuclear antigen (RANA)	80 - 90
	Histones	< 20
Overlap syndromes	RNP	< 90
	Others (scleroderma, lupus, or myositis-related)	variable

CREST = limited form of scleroderma, including calcinosis, Raynaud's phenomenon, esophageal dysmotility, sclerodactyly, and telangiectasia

indicate that immune complexes are not only formed in blood circulation but also locally, for example in kidneys, causing an immune complex-induced glomerulonephritis. The crossreactivity of anti-DNA antibodies with heparinsulfate, a constituent of the glomerulum basement membrane (10), or an unspecific binding of DNA or anti-DNA immune complexes due to charge differences to collagen IV or V, also parts of the glomerulum basement membrane, might lead to a local complex formation.

In summary, there is clear evidence that in SLE anti-dsDNA antibodies are of pathogenic importance and, furthermore, that complexes formed of DNA and their respective antibodies play a central role with regard to organ-specific and systemic disease manifestations. This is additionally

Table 2. Factors favouring IC mediated tissue destruction

1.	Defective DNA-clearance?
2.	Defective red blood clearance?
3.	Defective RES-clearance
4.	Features of DNA-anti-DNA-AB complexes

supported by defective clearance mechanisms for immune complexes as listed in Table 2. Own thus far published studies have shown a specific red cell membrane component which clears free nucleic acids from the serum and which appears to be defective in patients with SLE. Furthermore, it was demonstrated that monocytes of SLE patients express Fc-8 receptors but are unable to degrade membrane-bound immune complexes (11). In addition, a clear-cut CR1-receptor deficiency of erythrocytes has repeatedly been described in SLE (12), and finally, as taken from plasmapheresis studies, there seems to be an immune-complex-induced blockade of the RES in SLE patients, which can be unblocked by plasmaseparation (13).

As already mentioned, there is ample evidence that anti-DNA-antibodies are of central importance in the pathogenesis of SLE; the question, however, remains: "What triggers the formation of anti-dsDNA antibodies?", and this particularly since dsDNA by itself is not immunogenic. Based upon the observation that B cell clones being able to produce anti-DNA antibodies are part of the normal B cell repertoire (14), one of two major hypotheses favours the endogenous origin of anti-dsDNA antibodies. Autoreactive B cell clones could either unspecifically be activated by polyclonal B cell activators, or, more specifically, by bacterial wall antigens which share common epitopes with the dsDNA molecule, including anti-dsDNA antibody production by a mechanism called antigenic mimicry. In both situations, a defective control mechanism for B cell proliferation, possibly genetically determined, should be present.

A spontaneous hyperreactivity of B cells in SLE patients has repeatedly been demonstrated (15), the finding that murine monoclonal anti-DNA autoantibodies bind to endogenous bacteria suggests that bacterial phospholipids might provide an immunogenic stimulus for the production of antibodies that crossreact with DNA (16). In this context, in the human situation, it has been shown recently that in patients suffering from tuberculosis, not only, in 60%, a common anti-DNA idiotype (16.6) was present but also an increased activity of antibodies against a variety of antigens with which SLE-specific autoantibodies are known to crossreact, e.g. ssDNA, dsDNA, polynucleotides and cardiolipin was obvious (17); finally, a defective T suppressor cell function was described which partly correlated with the disease activity (18, 19).

With regard to the second hypothesis, postulating a primary exogenous trigger mechanism for anti-DNA antibody production, recent experiments from our laboratory are of interest, demonstrating that, from plasmapheresis fluid of patients suffering from acute SLE, a complex of RNA and DNA was to separate. As shown by different methods including cesium chloride gradient centrifugation, hyperchromic shift studies and HPLC analysis, evidence was obtained that indeed DNA and RNAse-resistant and thus double or triple helically formed RNA was present in the extracted material. Further characterization of the DNA revealed a molecular size of roughly 20 kb, purely methylated and rich in GC-pairs as contrasted to normal DNA as isolated from hepatocytes. With regard to a possible exogenous origin of antinuclear antibody, it is of interest that a significant homology of parts of the cloned and sequenced DNA from SLE patients with parts of the

POL and NF gene region of the retrovirus HIV I was found. Furthermore, when animals were immunized with the nucleid acids extracted from SLE sera, antibodies were produced that reacted with native DNA, which was not the case when animals were immunized with dsDNA. These preliminary data indicate that foreign RNA might be an antigen triggering antibody formation which not only reacts with the RNA but also crossreacts with dsDNA, and thus leading to the manifestation of systemic lupus erythematosus.

Further studies will have to prove if the pathogenesis of SLE is due to a defect in immunomodulation permitting autoreactive B cell clones to produce antibodies to dsDNA or if the disease could rather be attributed to an exogenous agent. However, a combination of both pathways seems to me to be the most likely explanation for the development of SLE, provided the respective genetic predisposition is present.

REFERENCES

1. Burmester GR, Yu DT, Irani AM, Kunkel HG, Winchester RJ. Ia + T cells in synovial fluid and tissues of patients with rheumatoid arthritis. Arthritis Rheum 1981; 24: 1370-6.
2. Jahn B, Burmester GR, Schmid H, Weselon G, Rohwer P, Kalden JR. Changes in cell surface antigen expression on human articular chondrocytes induces by gamma-interferon. Induction of Ia antigens. Arthritis Rheum 1987; 30: 64-74.
3. Baehr G, Klemperer P, Schifrin A. A diffuse disease of the peripheral circulation (usually associated with Lupus erythematosis and endocarditis). Trans Assoc Am Physicians 1935; 50: 139.
4. Eilat D. Anti-DNA antibodies: problems in their study and interpretation. Clin Exp Immunol 1986; 65: 215-22.
5. Emlen W, Pisetsky DS, Taylor RP. Antibodies to DNA. A perspective. Arthritis Rheum 1986; 29: 1417-26.
6. Schattner A, Miller KB, Kaburaki Y, Schwartz RS. Suppressor cell function and anti-DNA antibody idiotypes in the serum of SLE patients and their first-degree relatives. Clin Immunol Immunopathol 1986; 41: 417-26.
7. Hood AF. Cutaneous side effects of cancer chemotherapy. Medical Clinics of North America 1986; 70: 187-209.
8. Chitrabamrung S, Rubin RL, Tan EM. Serum deoxyribonuclease I and clinical activity in systemic lupus erythematosus. Rheumatol Int 1981; 1: 55-60.
9. Immunology of Rheumatic Diseases. Pergamon Press, 1986: 741.
10. Faaber P, Rijke TP, Van de Putte LB, Capel PJ, Berden JH. Cross-reactivity of human and murine anti DNA antibodies with heparan sulfate. The major glycosaminoglycan in glomerular basement membranes. J Clin Invest 1986; 77: 1824-30.
11. Kavai M, Csipoe I, Sonkoly I, Csongor J, Szegedi GY. Defective immune complex degradation by monocytes in patients with systemic lupus erythematosus. Scand J Immunol 1986; 24: 527-32.
12. Cook JM, Kazetchkine MD, Bourgeois P, Mignon F, Mery JP, Kann MF. Anti-C3b-receptor (CR1) antibodies in patients with systemic lupus erythematosus. Clin Immunol Immunopathol 1986; 38: 135-8.
13. Hoch S, Schwaber J. Specificity analysis of human anti-DNA antibodies. J Immunol 1986; 136: 892-7.
14. Delfraissy JF, Wallon C, Vazques A, Dugas B, Dormont J, Galanaud P. B cell hyperactivity in systemic lupus erythematosus: selectively enhanced responsiveness to a high molecular weight B cell growth factor. Eur J Immunol 1986; 16: 1251-6.
15. Carroll P, Stafford D, Schwartz RS, Stollar BD. Murine monoclonal anti-DNA autoantibodies bind to endogenous bacteria. J Immunol 1985; 135: 1086-90.

16. Sela O, el-Roeiy A, Isenberg DA, et al. A common anti-DNA idiotype in sera of patients with active pulmonary tuberculosis. Arthritis Rheum 1987; 30: 50-6.
17. Bresnihan B, Jasin HE. Suppressor function of peripheral blood mononuclear cells in normal individuals and in patients with systemic lupus erythematosus. J Clin Invest 1977; 59: 106-16.
18. Schattner A. The origin of autoantibodies. Immunology Letters 1987; 14: 143-53.

NEW BIOCHEMICAL PARAMETERS IN THE DIAGNOSIS OF JOINT DISEASES

K. Kleesiek*, R. Reinards** and H. Greiling**

*Institute of Laboratory and Transfusion Medicine, Heart Center Northrhine-Westfalia, D-4970 Bad Oeynhausen (FRG)
**Department of Clinical Chemistry and Pathobiochemistry, Aachen University of Technology, D-5100 Aachen (FRG)

An important aim in the diagnostic strategy of rheumatic diseases is the differentiation between chronic inflammatory and degenerative joint diseases. In the early phase of chronic joint diseases the separation between rheumatoid arthritis and osteoarthrosis, especially polyarticular osteoarthrosis, might be difficult. In principle osteoarthrosis is characterized by an isolated defect of the cartilage metabolism. Therefore biochemical and immunochemical methods which allow the recognition of pathobiochemical alterations of the osteoarthrotic metabolism in blood are lacking. Also during acute inflammatory phases of osteoarthrosis, no changes of biochemical blood parameters are observed.

NON-SPECIFIC SERUM PARAMETERS IN CHRONIC INFLAMMATORY JOINT DISEASES

Alterations in biochemical parameters, which are observed in blood from patients with chronic inflammatory joint diseases (e.g. increase of the concentration of copper and acute phase proteins, such as haptoglobin, ceruloplasmin, α_1-proteinase inhibitor and C-reactive protein, decrease of the iron concentration in serum, and of the hemoglobin content of erythrocytes), are indicative for an inflammatory process in general, without specific characteristics, and are also observed, when malignant tumors or acute and chronic inflammation processes of different origin are present. Particularly hyposiderinemia, caused by a shift of body iron into the macrophages of the reticuloendothelial system, is frequently observed in the initial phase of rheumatoid arthritis.

The determination of leukocyte count in blood offers only little specific diagnostic information. High leukocyte counts are found particularly in juvenile rheumatoid arthritis and septic arthritis. However, the leukocyte count does not correlate with the activity of the inflammatory processes occurring in the joint. In the case of an extremely active rheumatoid arthritis and systemic lupus erythematodes, even leukopenia is not seldom found.

A considerable advantage in an organ specific diagnosis of joint diseases can be achieved, when the occurrence of a joint effusion makes an investigation of the synovial fluid possible. The synovial fluid forms the extracellular space of the synovial system, in which metabolic processes of the adjacent connective tissue are reflected directly.

DIAGNOSTIC SIGNIFICANCE OF BIOCHEMICAL PARAMETERS IN THE SYNOVIAL FLUID

It is generally excepted that biochemical synovial analysis makes a symptomatic differentiation between inflammatory and non-inflammatory joint diseases possible. Our previous investigations lead to the development of a multivariable linear discriminant analysis in order to determine the biochemical parameters or the combination of parameters, which provide a maximal discrimination of defined joint diseases (inflammatory and non-inflammatory joint diseases, osteoarthrosis, rheumatoid arthritis, arthritis urica, traumatic arthropathy (1).

Up to 20 parameters in more than 2000 synovial fluids were determined simultaneously, including total synovia cell count, granulocyte count, relative viscosity, total protein, uric acid, glucose, calcium, inorganic phosphorus, bilirubin, creatinine, urea, LDH, LDH-1/2, rheumatoid factor, monosodium urate- and calcium pyrophosphate dihydrate crystals in synovia cell pellet.

The multivariate linear discriminant analysis (2) was employed to answer the following questions:

1. To what extent can chronic joint diseases be differentiated by biochemical parameters of the synovial fluid?
2. Which parameters or parameter combinations have the highest discriminative potency?
3. How many parameters are necessary to achieve a maximal separation between the diagnostic groups?

The classification of biochemical parameters in a diagnostic group was carried out by calculation on the linear discriminant function. The best parameter combination was searched in a stepwise procedure, starting with the best single parameter and adding one further parameter at each step. The ability of the parameters to discriminate among the joint diseases was checked by reclassifying the original diagnosis. The results of the discriminant analysis were compared with those of a diagnosis-independent cluster analysis (2).

Of all biochemical parameters in the synovial fluid, granulocyte count, LDH, cell count, protein, and in the case of arthritis urica also monosodium urate crystals and uric acid proved to be the parameters with the highest discriminative potency. The differentiation between inflammatory and non-inflammatory joint diseases can be achieved by granulocyte count, relative viscosity and LDH with only 5.7% of misclassifications and cannot be improved by including additional parameters like cell count and protein concentration. More misclassifications (25%) are observed, if the granulocytes are not taken into account.

A separation of the rheumatoid arthritis from degenerative joint diseases and traumatic arthropathy is possible with 5.9% and 5.4% misclassifications, respectively. Only a small contribution to better separation is provided by the rheumatoid factor.

It must be pointed out that the classification into a diagnostic group does not mean the diagnosis itself, but the exclusion of that diagnostic group in which no classification occurs.

However, the detection of monosodium urate- and calcium pyrophosphate dihydrate crystals in the synovial cell pellet makes an etiological diagnosis of arthritis urica and chondrocalcinosis possible. A discrimination of gouty arthritis from osteoarthrosis, rheumatoid arthritis and traumatic arthropathy, with only few misclassifications (0 - 2.5%) can be achieved when monosodium urate crystals are identified and elevated uric acid concentrations are determined.

Bilirubin is shown as a parameter, which separates degenerative joint diseases from traumatic arthropathies with misclassifications of 8.2%.

Only the development of new methods and the search for better biochemical parameters on the bases of new pathobiochemical findings will lead to an improvement of diagnostic discrimination in chronic joint diseases. In the following we present two new biochemical parameters: elastase from granulocytes (EC 3.4.21.37), which is enhanced in the plasma of patients with rheumatoid arthritis (3), and UDP-D-xylose: proteoglycan core protein β-D-xylosyltransferase (EC 2.4.2.26) from proteochondroitin sulfate synthesizing cartilage cells, which is found in high activity in the synovial fluid after destruction of the articular cartilage (4).

GRANULOCYTE ELASTASE IN PLASMA AS A MARKER OF RHEUMATOID ARTHRITIS

In chronic inflammatory joint diseases the involvement of the proteinases elastase, cathepsin G, and collagenase in the destruction of matrix components of articular cartilage, such as proteoglycans and collagen, is generally accepted (5-7). In particular the neutral proteinase elastase, which is released from granulocytes or other sources, like

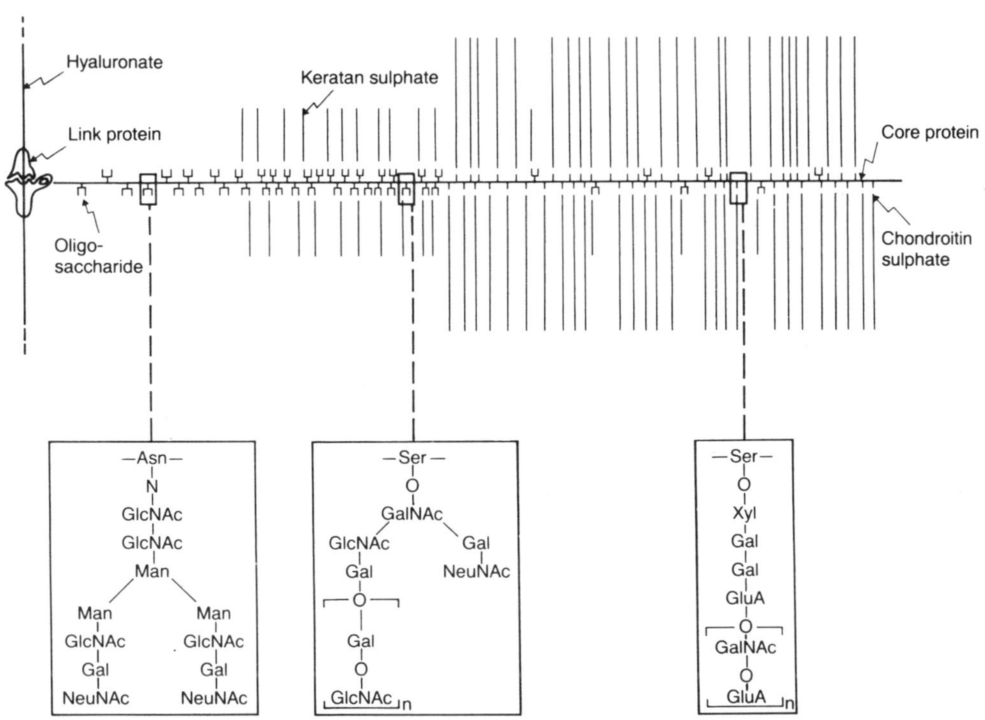

Fig. 1. Present idea of the structure of proteoglycan-hyaluronate complexes in the intercellular matrix of the articular cartilage.

macrophages, synovial fibroblasts, and chondrocytes (8, 9), plays an important role in the degradation of articular cartilage in chronic inflammatory joint diseases.

In vitro experiments have shown that granulocyte elastase degrades the chondroitin sulfate attachment region of proteoglycans into fragments with one or two polysaccharide chains and also liberates a chondroitin sulfate-free fragment (molecular weight about 20,000 - 30,000) from the hyaluronate-binding region of the protein core (10) (a schematic representation of the proteoglycan structure is given in Fig. 1). Moreover, elastase seems to be involved in the cleavage of the link protein of proteoglycan aggregates (11), and is known to be responsible for the degradation of the non-helical terminal peptides of cartilage collagen molecules. This leads to the disruption of the main intermolecular crosslinks in collagen fibres (12) and to deterioration in the tensile strength of articular cartilage (13).

In Fig. 2 several mechanisms are shown, which may cause an enzymatic destruction of the articular cartilage. An extracellular proteolytic activity of proteinases in the synovial system is still a point of discussion. The synovial fluid contains an excess of very potent proteinase inhibitors, α_1-proteinase inhibitor and α_2-macroglobulin, derived from the plasma. Thus, there seems to be little evidence for the occurrence of free proteinases.

A proteinase/proteinase inhibitor imbalance seems to lead to cartilage destruction. This may occur after inactivation of the α_1-proteinase inhibitor by phagocyte-derived oxidants (14, 15). Also oxygen-derived radicals itself are directly involved in the degradation of cartilage components, especially of hyaluronate, which is mainly degraded by hydroxyl radicals (16, 17).

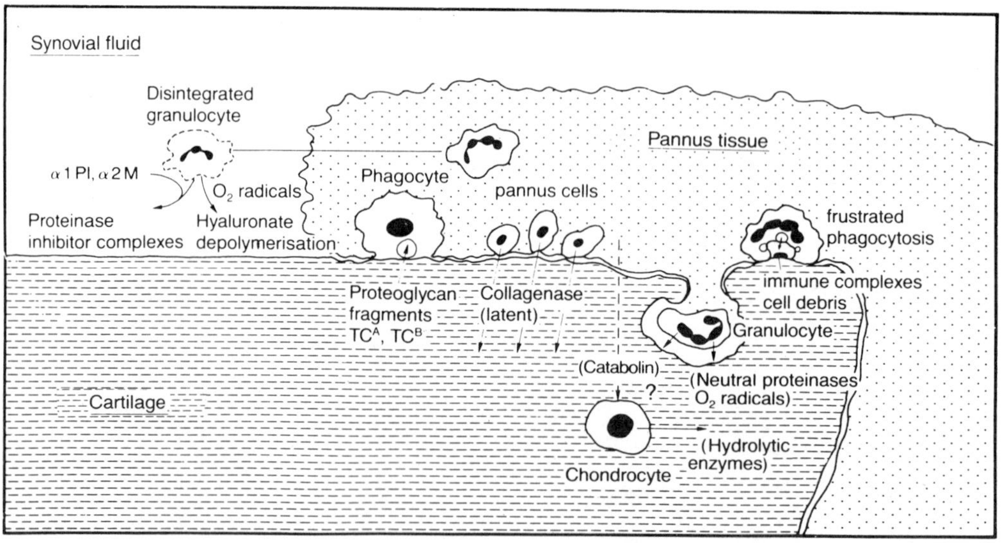

Fig. 2. Destruction of connective tissue components in rheumatoid arthritis. Schematic representation of some important pathogenic mechanisms. (α1PI = α_1-proteinase inhibitor; α2M = α_2-macroglobulin; TC^A, TC^B = tropocollagen fragments A and B obtained after cleavage by collagenase).

Another way in which the proteinase may evade the inhibitors is by direct release onto the cartilage surface by granulocytes adherent to it. This process of frustrated phagocytosis might be promoted by the presence of immun complexes at the cartilage surface. Histological studies show absorption lacunae containing granulocytes in close contact with the connective tissue (18). In those microcavities between granulocyte and matrix, proteinase may be in excess of their inhibitors. Moreover, the penetration of elastase into the polyanionic cartilage matrix is facilitated by its positive charge, whereas the higher molecular weight and anionic charge hinder α_1-proteinase inhibitor and α_2-macroglobulin from invading.

Recent observations suggest that factors, which are released from the inflamed synovium stimulate chondrocytes to excrete hydrolytic enzymes. Catabolin is such a factor, which is described by Dingle (19), and which seems to be identical with interleukin I.

In inflammatory joint diseases large numbers of granulocytes pass through the inflamed synovial membrane entering the synovial fluid, where they disintegrate very fast. Since in synovial fluid, like in plasma (20), elastase is predominantly bound to α_1-proteinase inhibitor, the concentration of elastase α_1-proteinase inhibitor complex in the synovial fluid represents the degree of granulocyte disintegration and activation (phagocytosis of particals, stimulation by soluble factors like C_{5a} and interleukin I) (21, 22).

In order to examine the diagnostic significance of granulocyte elastase as a biochemical parameter, we determined the concentration of elastase in complex with α_1-proteinase inhibitor by an enzymimmunoassay (Fig. 3) (23) in synovial fluids and plasma of patients with chronic joint diseases (3).

The median value of about 10,000 μg/l for the concentration of elastase in the inflammatory synovial fluid is significantly higher in comparison with the concentration of 41 μg/l in non-inflammatory synovial fluids (Table 1). In purulent synovial fluids extremely high elastase levels of more than 100,000 μg/l are found. In this case even the presence of free, uncomplexed elastase, obviously due to an insufficient concentration of proteinase inhibitors in the synovial fluid, could be observed.

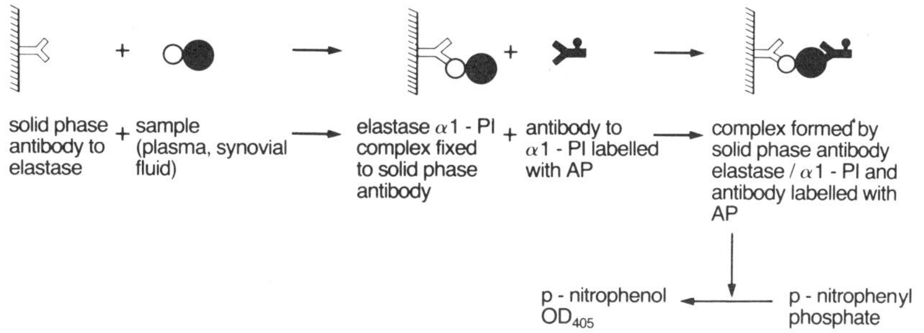

Fig. 3. Determination of the elastase α_1-proteinase inhibitor complex by an enzyme immunoassay. Principle of the method. (α_1-PI = α_1-proteinase inhibitor, AP = alkaline phosphatase, OD_{405} = optical density at 405 nm).

Table 1. Concentration of elastase in complex with α_1-proteinase inhibitor (enzyme immunoassay) and elastase activity (hydrolysis of MeO-Suc-Ala-Ala-Pro-Val-4-nitroanilide) in the synovial fluid. Additionally the concentration of the proteinase inhibitors and granulocyte count are given. For calculation of the molar concentrations the following molecular weights are used: elastase 28,000; α_1-proteinase inhibitor 54,000; α_2-macroglobulin 725,000.

		Joint Diseases	
		Non-inflammatory (n = 30)	Inflammatory (n = 39)
Elastase Median (range)	(μg/l) (nmol/l)	41.1 (0-935) 1.47 (0-33,4)	9,947 (878-101,800) 355 (31.4- 3,636)
Elastase activity (range)	(U/l,37°C)	not detectable	0.20-130 (n = 17)
α_1-Proteinase inhibitor $\bar{x}$ (s)	(nmol/l)	21,500 (12,100)	38,600 (24,200)
α_2-Macroglobulin $\bar{x}$ (s)	(nmol/l)	468 (557)	944 (1,010)
Granulocytes Median (range)	(N/μl)	3.3 (0-125)	4,520 (770-32,180)

In normal plasma, elastase concentrations of about 100 μg/l are found. Enhanced concentrations of elastase were observed in plasma of patients with inflammatory chronic joint diseases and were due to the strong gradient which exists between the concentration of elastase in the joint cavity and that in the plasma.

Fig. 4 shows the elastase concentration in synovial fluid and plasma of patients with clinically defined joint diseases. Synovial fluid and plasma were taken simultaneously from the same patient. In rheumatoid arthritis, ankylosing spondylitis, and septic arthritis, elastase levels in the synovial fluid, which exceeded in all cases the corresponding values of the plasma, are significantly higher than those in osteoarthrosis, whereas in osteoarthrosis elastase concentrations in the synovial fluid ranged from values below the normal plasma concentration up to values of about 10,000 μg/l during a phase of inflammatory activity.

More than 90% of the patients with rheumatoid arthritis demonstrated significantly enhanced concentrations of complexed elastase in the plasma, whereas in all patients suffering from osteoarthrosis normal plasma levels were found.

The definite diagnosis of rheumatoid arthritis can not be concluded solely from the elevated elastase concentration in the plasma, since increased elastase levels are also found in other inflammatory joint diseases, i.e. septic arthritis (Fig. 4). However, normal plasma concentrations of elastase exclude the diagnosis of rheumatoid arthritis with high probability.

The simultaneous determination of elastase concentrations in both synovial fluid and plasma of the same patient makes a further classification of the present chronic joint disease possible (Fig. 5). All patients with rheumatoid arthritis showed elastase levels of at least 175 μg/l and 600 μg/l in plasma and synovial fluid, respectively. Thus, the determination of elastase levels in the plasma provides a useful parameter, which reflects the chronic inflammatory process in the joint, and is helpful in the nosological differentiation of chronic joint diseases. Furthermore, the plasma concentration of this enzyme may also serve as an indicator of disease activity, as can be derived from longitudinal studies with patients, who showed a decrease of elastase concentration in the plasma after antirheumatic therapy.

UDP-D-XYLOSE: PROTEOGLYCAN CORE PROTEIN β-D-XYLOSYLTRANSFERASE AS A MARKER OF CARTILAGE DESTRUCTION

In the biosynthesis of chondroitin sulfate proteoglycan (Fig. 1) several glycosyltransferases and sulphotransferases are involved (24). Xylosyltransferase catalyzes the initial step in the formation of chondroitin sulfate side chains of proteoglycans by transferring D-xylose from UDP-D-xylose to serine residues in the nascent core protein (Fig. 6) (25, 26).

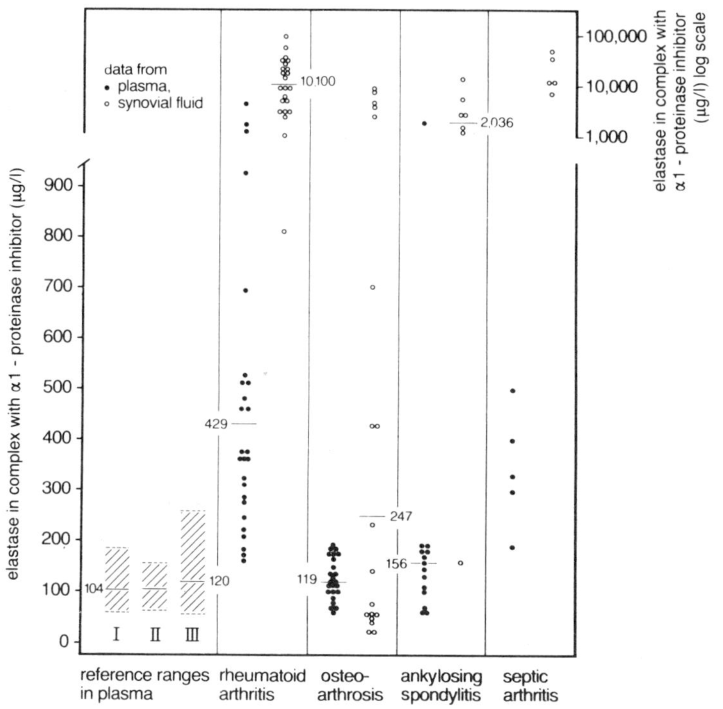

Fig. 4. Concentration of elastase in complex with α_1-proteinase inhibitor in synovial fluid and plasma of patients with different joint diseases. In addition the geometric means and 95% ranges of three reference groups are given. Reference group I: blood donors; reference group II: healthy factory employees; reference group III: patients of the hospital without rheumatic diseases and with normal haemograms.

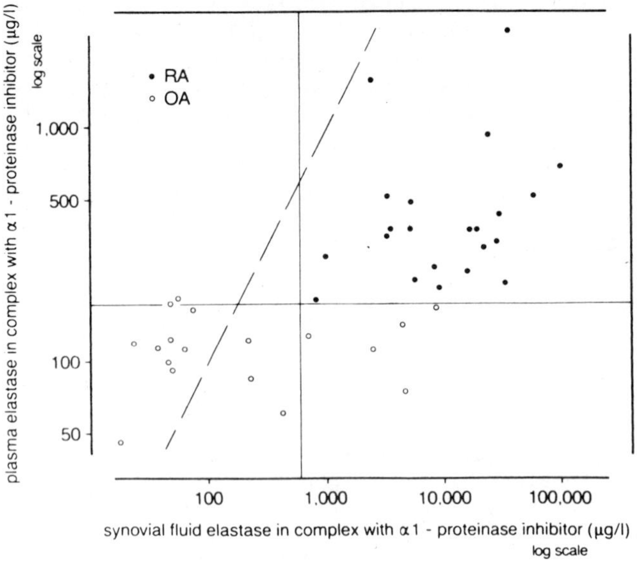

Fig. 5. Concentrations of elastase in complex with α_1-proteinase inhibitor in synovial fluid and the corresponding plasma of patients with rheumatoid arthritis (RA) and osteoarthrosis (OA). The intersecting points of the horizontal and vertical bars on the y- and x-axis are at 175 and 600 μg/l, respectively. The dashed line in the diagram represents the points of identical concentration in both synovial fluid and plasma.

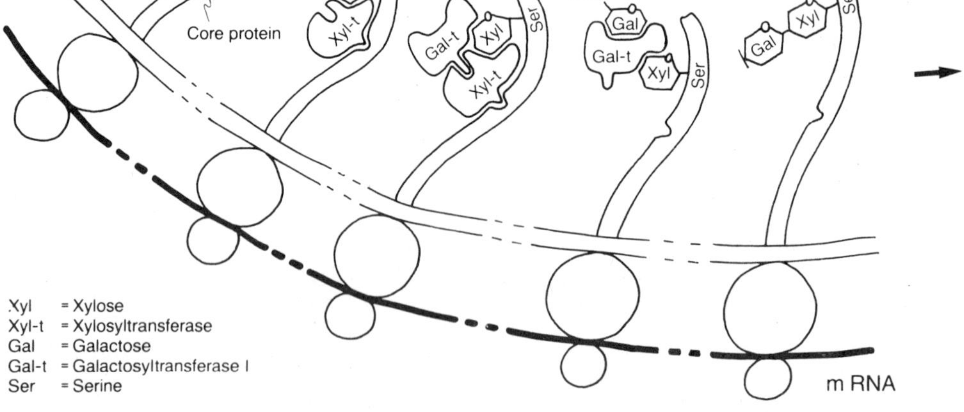

Fig. 6. Initiation of the biosynthesis of chondroitin sulfate in chondrocytes by UDP-D-xylose: proteoglycan core protein β-D-xylosyltransferase (Xyl-t = xylosyltransferase, Gal-t = galactosyltransferase I).

Table 2. ^{14}C-xylose transfer to various exogenous acceptors in the xylosyltransferase assay. Enzyme and acceptor activities were calculated in the presence of the following acceptors: core protein I (from HF-degraded proteoglycan), core protein II (from trifluoromethane sulfonic acid-degraded proteoglycan), silk from Bombyx mori, and two synthetic peptides.

Acceptor	Enzyme activity (^{14}C-xylose incorporated) [nmol.min^{-1}.l^{-1}]	Acceptor activity [nmol.min^{-1}.mmol serine^{-1}]	K_m [mg/l]
Coreprotein I (HF-degraded)	19.5	1.27	143-240
Coreprotein II (TFMS-degraded)	39.7	2.57	
Silk	72.6	2.63	182
Leu-Pro-Ser-Gly	< 0.2		
Pro-Ser-Gly	< 0.2		

HF: Hydrogen fluoride - pyridine (70%)
TFMS: Trifluoromethane sulfonic acid

We investigated the diagnostic significance of xylosyltransferase as a biochemical parameter in chronic joint diseases, since the enzyme is highly specific to cells, which are able to synthesize proteoglycans, especially to chondrocytes in articular cartilage. Moreover, xylosyltransferase is unique among the glycosyltransferases in being more loosely bound to the membranes of the rough endoplasmic reticulum of chondrocytes (27), so that it is released easily after disruption of the cells. Since destruction of articular cartilage is a common pathogenic process of all chronic joint diseases, the activity of xylosyltransferase in synovial fluids of patients with chronic joint diseases will indicate the extent of cartilage destruction.

We determined the xylosyltransferase activity in synovial fluid and serum by a radiochemical method, based on the incorporation of ^{14}C-xylose from UDP-^{14}C-xylose into an exogenous acceptor protein (28). Compared with core protein acceptors, which often show the disadvantage of considerably varying acceptor activities between different preparations, the use of silk from the silk worm Bombyx mori (29) as an acceptor offers several advantages. It has shown to be the substrate with the highest incorporation rates in the xylosyltransferase assay, compared to bovine core protein preparations from HF- and trifluoromethane sulfonic acid-degraded proteoglycans (30, 31), and the synthetic peptides Pro-Ser-Gly and Leu-Pro-Ser-Gly (Table 2), and is more invariable available. Therefore, silk was employed in our investigations as the acceptor protein to determine xylosyltransferase activities in synovial fluids and sera of patients with chronic joint diseases.

Since proteoglycan fragments, which may serve as a substrate for xylosyltransferase, are released into synovial fluid during inflammatory joint processes, we studied the influence of chondroitin sulfate peptidoglycosaminoglycans on the xylosyltransferase activity in vitro.

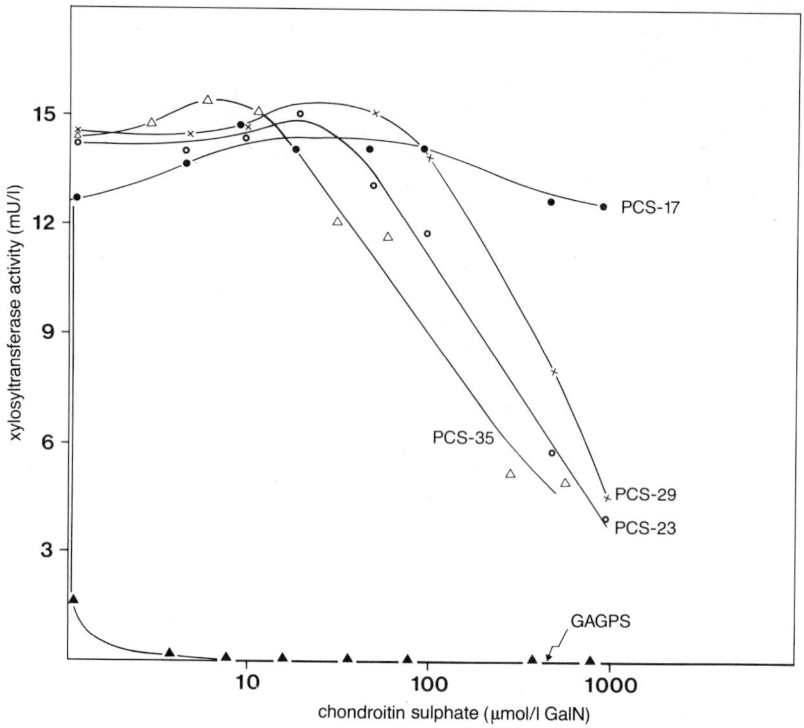

Fig. 7. Effect of peptidochondroitin sulphates of different degree of sulfation on xylosyltransferase activity in tissue extracts of human costal cartilage (PCS = peptidochondroitin sulphate; GAGPS = persulfated peptidoglycosaminoglycan). The content of sulphate (mol SO_4^{2-}/mol disaccharide) was as follows: PCS-17 = 0.31; PCS-23 = 0.52; PCS-29 = 0.71; PCS-35 = 0,96; GAGPS = 4.0.

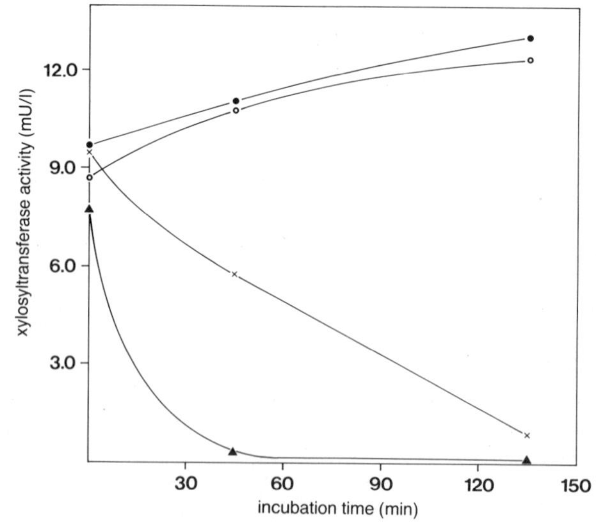

Fig. 8. In vitro effect of human granulocyte elastase on xylosyltransferase activity in tissue extracts of human costal cartilage. Elastase was added to give following final concentrations: 1.86 mg/l (O), 18.6 mg/l (x), 186 mg/l (▲), 0 mg/l (●).

Although our studies have shown that the xylosyltransferase activity is inhibited by chondroitin sulfate peptidoglycosaminoglycans as a function of concentration and degree of sulfation (Fig. 7), and furthermore that granulocyte elastase degrades xylosyltransferase activity (Fig. 8), an erroneous effect of these factors on the enzyme activities measured in synovial fluid of patients with chronic joint diseases seems to be unlikely, since xylosyltransferase activities, which have been added to synovial fluid in defined quantities, were recovered in a high rate.

In Fig. 9 the xylosyltransferase activities in synovial fluids are plotted against those in sera taken from the same patients, which were suffering from different joint diseases. In more than 80% of the synovial fluids xylosyltransferase activities were higher than those of the corresponding sera. However, no significant correlation between xylosyltransferase activity in synovial fluids and that in sera was observed.

Xylosyltransferase activities in the synovial fluid of patients with chronic-inflammatory and degenerative joint diseases ranged from 0.5 to 22.0 mU/l (4). In the synovial system articular cartilage represents the main source of these enzyme activities. In cultured synovial fibroblasts, which synthesize chondroitin sulfate apart from hyaluronate as the main glycosaminoglycan, little xylosyltransferase activity is detectable. No xylosyltransferase activity was found in extracts of granulocytes and in synovial cell pellets, even not in highly inflammatory processes.

In Fig. 10 the median and single values of xylosyltransferase activities in synovial fluid and serum taken from patients with different chronic joint diseases are given. The enzyme activities showed considerable differences between clinically defined joint diseases. The highest enzyme activities

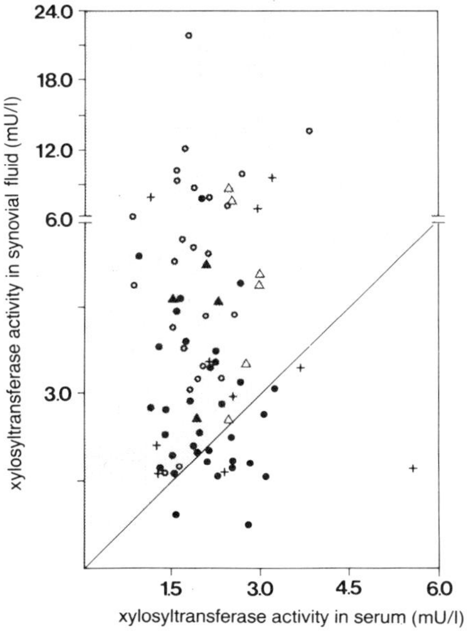

Fig. 9. Comparison of xylosyltransferase activities in synovial fluids and sera taken from the same patients, which were suffering from the following joint diseases: osteoarthrosis (●; n = 33), rheumatoid arthritis (O; n = 24), ankylosing spondylitis (+; n = 10), Reiter's syndrome (▲; n = 4), and septic arthritis (Δ; n = 6).

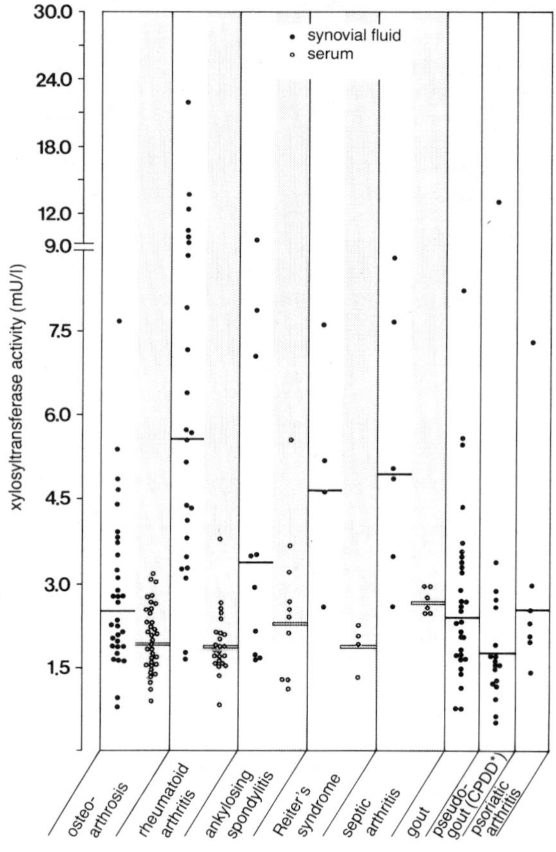

Fig. 10. Xylosyltransferase activities in synoval fluid and serum from patients with different chronic joint diseases. The median and single values are given. CPDD = calcium pyrophosphate deposition disease.

were found in synovial fluids of patient suffering from rheumatoid arthritis.

It is an interesting finding that the xylosyltransferase activities in synovial fluids of patients with arthritis urica were much lower and in the same range of those in osteoarthrosis, although the granulocyte count and elastase concentration, reflecting the degree of inflammation, corresponded with those in synovial fluids of patients with rheumatoid arthritis. Obviously xylosyltransferase is no parameter of inflammation in the precise sense, although in inflammatory processes the increased destruction of chondrocytes may lead to an enhanced release of xylosyltransferase. No correlation between xylosyltransferase activity in the synovial fluid and granulocyte number or elastase concentration was observed. The higher xylosyltransferase activities in synovial fluids of patients with rheumatoid arthritis, compared with those of patients with arthritis urica and osteoarthrosis, is probably attributed to an increased cartilage destruction during the course of this disease.

High xylosyltransferase activities in synovial fluids of patient with chronic joint diseases did not result in a parallel enhancement of enzyme activity in the corresponding sera (Fig. 9). This might be due to the high

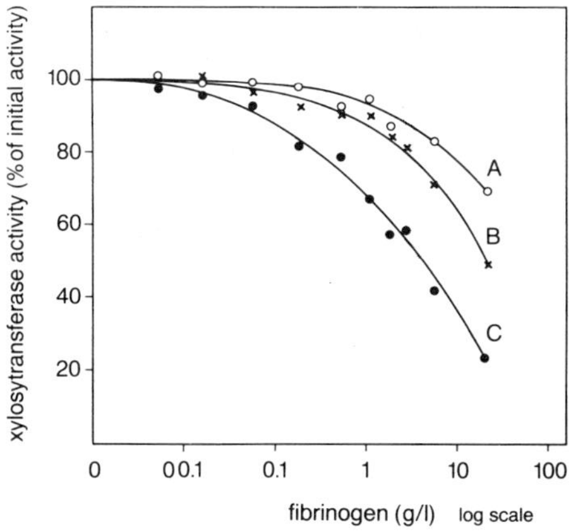

Fig. 11. Inhibitory effect of fibrinogen on xylosyltransferase activity in tissue extracts of human costal cartilage (A), synovial fluid (B), and serum (C). The hatched area indicates the physiological concentration range of fibrinogen.

degree of dilution, when the enzyme is released from the joint cavity into the blood. Data about the half life of the enzyme in blood are not yet known. Moreover, different tissues, e.g. liver and skin, must be taken into account as sources of xylosyltransferase activity in serum. The limited organ specificity of this parameter in serum is also indicated by the inhibition studies with fibrinogen, suggesting that various xylosyltransferase isoenzymes are present in blood. Addition of human fibrinogen to serum in physiological concentrations caused an about 50% inhibition of xylosyltransferase activity (Fig. 11).

A similar degree of inhibition was observed, when plasma enzyme activities were compared with those in the corresponding sera. Although the mechanism of inhibition by fibrinogen is still unclear so far, the degree of inhibition seems to be dependent on the origin of the xylosyltransferase isoenzyme, as seen from Figure 11. The inhibition of xylosyltransferase activity in blood is more pronounced than that derived from human costal cartilage. The intermediate degree of inhibition of xylosyltransferase activity in synovial fluids from patients with chronic joint diseases may be due to the presence of xylosyltransferase isoenzymes from sources outside the joint, which are able to enter the synovial system as a consequence of the enhanced permeability of the blood-synovial-barrier in inflammatory joint processes.

Our results demonstrate that xylosyltransferase activity in the synovial fluid represents an appropriate clinico-chemical parameter, indicating the degree of cartilage destruction in chronic joint diseases. The significance of xylosyltransferase activity in serum as a marker of cartilage degradation, however, seems to be limited, since various xylosyltransferase isoenzymes from different tissue sources seem to be present in blood. For that reason, the development of an immunological assay seems to be indispensable for the quantification of cartilage-specific xylosyltransferase in serum.

ACKNOWLEDGEMENTS

Parts of these investigations were supported by the Deutsche Forschungsgemeinschaft (KL 388/3-3) and by the Bundesministerium für Forschung und Technologie (01 ZU 850 3/2).

REFERENCES

1. Kleesiek K, Michels P, Greiling H. Clinical chemical synovial fluid analysis: a comparative study of the discriminative value of 20 parameters. J Clin Chem Clin Biochem 1982; 20: 688-689.
2. Brown D. ed. Biomedical computer program BMDP-79. University California Press, Berkely, 1979.
3. Kleesiek K, Reinards R, Brackertz D, Neumann S, Lang H, Greiling H. Granulocyte elastase as a new biochemical marker in the diagnosis of chronic joint diseases. Rheumatol Int 1986; 6: 161-169.
4. Kleesiek K, Reinards R, Wolf B, Okusi J, Greiling H. UDP-D-xylose: proteoglycan core protein β-D-xylosyltransferase: a new marker of cartilage destruction in chronic joint diseases. J Clin Chem Biochem 1987; in press.
5. Barret AJ. The possible role of neutrophilic proteinases in damage to articular cartilage. Agents Actions 1978; 8: 11-18.
6. Velvart M, Fehr K, Baici A, et al. Degradation in vivo of articular cartilage in rheumatoid arthritis by leucocyte elastase from polymorphonuclear leucocytes. Rheumatol Int 1981; 1: 121-130.
7. Bartholomew JS, Lowther DA, Handley CJ. Changes in proteoglycan biosynthesis following leukocyte elastase treatment of bovine articular cartilage in culture. Arthritis Rheum 1984; 27: 905-912.
8. Barrett AJ. The enzymic degradation of cartilage matrix. In: Burleigh PMC, Poole PR, eds. Dynamics of connective tissue macromolecules. Amsterdam: Elsevier-North Holland, 1975; 189-226.
9. Martel-Pelletier J, Cloutier J-M, Howell D, Pelletier J-P. Human rheumatoid arthritic cartilage and its neutral proteoglycan-degrading proteases - The effect of antirheumatic drugs. Arthritis Rheum 1985; 28: 405-412.
10. Roughley PJ. The degradation of cartilage proteoglycans by tissue proteinases. Proteoglycan heterogeneity and the pathway of proteolytic degradation. Biochem J 1977; 167: 639-646.
11. Keiser H, Greenwald RA, Feinstein G, Janoff A. Degradation of cartilage proteoglycan by human leukocyte granule neutral proteases: a model of joint injury. I. Penetration of enzyme into rabbit articular cartilage and release of $^{35}SO_4^{=}$ -labeled material from the tissue. J Clin Invest 1976; 57: 615-624.
12. Bader DL, Kempson GE, Barrett AJ, Webb W. The effects of leucocyte elastase on the mechanical properties of adult human articular cartilage in tension. Biochim Biophys Acta 1981; 677: 103-108.
13. Kempson GE, Tuke MA, Dingle JT, Barrett AJ, Horsfield PH. The effects of proteolytic enzymes on the mechanical properties of adult human articular cartilage. Biochim Biophys Acta 1976; 428: 741-760.
14. Carp H, Janoff A. Potential mediators of inflammation. Phagocyte-derived oxidants suppress the elastase-inhibitory capacity of alpha$_1$-proteinase inhibitor in vitro. J Clin Invest 1980; 66: 987-995.
15. Zaslow MC, Clark RA, Stone PJ, Calore JD, Snider GL, Franzblau C. Human neutrophil elastase does not bind to alpha$_1$-protease inhibitor that has been exposed to activated human neutrophils. Am Rev Resp Dis 1983; 128: 434-439.
16. Wong S, Halliwell B, Richmond R, Skowroneck W. The role of superoxide and hydroxyl radicals in the degradation of hyaluronic acid induced by metal ions and by ascorbic acid. J Inorg Biochem 1981; 14: 127-134.

17. Greenwald RA, Moy WW. Effect of oxygen-derived free radicals on hyaluronic acid. Arthritis Rheum 1980; 23: 455-463.
18. Shiozawa S, Shiozawa K, Fujita T. Morphologic observations in the early phase of the cartilage-pannus junction. Light and electron microscopy studies of active cellular pannus. Arthritis Rheum 1983; 26: 472-478.
19. Dingle JT, Saklatvala J, Hembrey R, Tyler J, Fell HB, Jubb R. A cartilage catabolic factor from synovium. Biochem J 1979; 184: 177-180.
20. Ohlsson K, Olsson I. Neutral proteases of human granulocytes. III. Interaction between human granulocyte elastase and plasma protease inhibitors. Scand J Clin Lab Invest 1974; 34: 349-355.
21. Goldstein JRA, Hoffstein S, Gallin J, Weissmann G. Mechanisms of lysosomal enzyme release from human leukocytes. Microtubule assembly and membrane fusion induced by a component of complement. Proc Natl Acad Sci USA 1973; 70: 2916-2920.
22. Klempner MS, Dinarello CA, Gallin JI. Human leukocytic pyrogen induces release of specific granule contents from human neutrophils. J Clin Invest 1978; 61: 1330-1336.
23. Neumann S, Gunzer G, Hennrich N, Lang H. "PMN-Elastase Assay": Enzyme immunoassay for human polymorphonuclear elastase complexed with α_1-proteinase inhibitor. J Clin Chem Clin Biochem 1984; 22: 693-697.
24. Rodén L. Structure and metabolism of connective tissue proteoglycans. In: Lennarz WJ, ed. The biochemistry of glycoproteins and proteoglycans. New York: Plenum Press, 1980; 267-371.
25. Morriss FH Jr, Fitzgerald B, Riddle LM. Fetal cartilage xylosyltransferase activity and skeletal growth in sheep. Pediat Res 1985; 19: 1240-1243.
26. Rodén L, Koerner T, Olson C, Schwartz NB. Mechanisms of chain initiation in the biosynthesis of connective tissue polysaccharides. Fed Proc 1985; 44: 373-380.
27. Hoffmann HP, Schwartz NB, Rodén L, Prockop DJ. Location of xylosyltransferase in the cisternae of the rough endoplasmic reticulum of embryonic cartilage cells. Connect Tissue Res 1984; 12: 151-163.
28. Stoolmiller AC, Horwitz AL, Dorfman A. Biosynthesis of the chondroitin sulfate proteoglycan. Purification and properties of xylosyltransferase. J Biol Chem 1972; 247: 3525-3532.
29. Campbell P, Jacobsson I, Benzing-Purdie L, Rodén L, Fessler JH. Silk - a new substrate for UDP-D-xylose: proteoglycan core protein β-D-xylosyltransferase. Anal Biochem 1984; 137: 505-516.
30. Olson CA, Krueger R, Schwartz NB. Deglycosylation of chondroitin sulfate proteoglycan by hydrogen fluoride in pyridine. Anal Biochem 1985; 146: 232-237.
31. Edge ASB, Faltynek CR, Hof L, Reichert LE Jr, Weber P. Deglycosylation of glycoproteins by trifluoromethanesulfonic acid. Anal Biochem 1981; 118: 131-137.

CLINICAL BIOCHEMISTRY OF GOUT

Mathias M. Müller

2nd Dept. of Surgery, Div. of Clinical Biochemistry
University of Vienna, Austria

In ancient documents of medicine Hippokrates, Seneca, Araetius and Galen described the distinctive picture of an acute monoarticular arthritis in the big toe, knee or hand which corresponds to our clinical description of gout today. In 1848 Garrod defined first the relationship between hyperuricemia and gout based on a congenital metabolic defect (1). However, the first specific enzyme defect responsible for a subtype of gout, hypoxanthine guanine phosphoribosyltransferase (HGPRT) deficiency, was discovered by Seegmiller and coworkers (2).

EPIDEMIOLOGY OF HYPERURICEMIA AND GOUT

The biochemical basis of each gouty arthritis are elevated blood uric acid levels. According to the physicochemical definition based on the solubility limit of urate in serum (pH 7.4, 37°C) hyperuricemia is defined as concentrations >420 µmol/l (7 mg/dl). In a variety of epidemiological studies this reference limit agreed very well with reference values established by population statistics (3-10).

Table 2 shows the prevalence of hyperuricemia in different countries for females and males using 420 µmol/l as the upper reference limit for serum uric acid concentration. From this international comparison it is obvious that the average serum uric acid level for males and females with 375 µmol/l (6.25 mg/dl) and 280 µmol/l (4.66 mg/dl) in Austria (Vienna) is rather high. These results were obtained in a study including 2,626 subjects (3), of which 2.7% showed clinical manifestations of gout. Taking the reference limit of 420 µmol/l and the percentages of hyperuricemic probands obtained in this study into consideration approximately 130,000 males and 20,000 females of Vienna would be hyperuricemic.

ETIOLOGY OF HYPERURICEMIA AND GOUT

The incidence of gout and nephrolithiasis is closely correlated with the extent and duration of hyperuricemia. The reasons for hyperuricemia and gout are multifactorial and include both genetic and environmental or acquired factors (Table 3). Hyperuricemia develops when there is an imbalance between production due to enzyme mutations and excretion of urate

Table 1. History of gout

1500 BC	Ebers papyrus	Description of colchicine
400 BC	Hippocrates	Monoarticular arthritis
200	Galen	Description of a tophus
1679	Leeuwenhoek	Crystals in a gouty tophus
1683	Sydenham	Description of gout
1763	Von Storch	Colchicum autumnale extract
1776	Scheele	Uric acid in kidney stones
1798	Wollastone	Uric acid crystals in gouty tophus
1814	Want	Colchicine identified
1848	Garrod	Hyperuricemia in gouty patients
1898	Fischer	Uric acid a purine compound
1899	Feudwiler	Inflammation due to urate cristals
1913	Folin and Denis	Determination of uric acid
1950	Buchanan et al.	De novo purine synthesis
1967	Seegmiller et al.	HGPRT deficiency (Lesch Nyhan)
1972	Sperling	PRPP synthetase superactivity

Table 2. Prevalence of hyperuricemia

Country	Serum Uric Acid (μmol/l)	Prevalence of Hyperuricemia (%)
	Males	
Australia (Maoris)	424 ± 92	47.1
(Sydney)	377 ± 74	23.0
Austria	375 ± 76	24.4
France	353 ± 71	17.6
UK	330 ± 68	7.2
USA (Framingham)	307 ± 67	4.8
(Tecumseh)	295 ± 84	7.4
	Females	
Australia (Maoris)	346 ± 93	38.7
Austria	280 ± 67	3.0
USA (Tecumseh)	251 ± 70	7.2
(Framingham)	249 ± 56	3.3
UK	234 ± 65	0.4

Table 3. Etiology of hyperuricemia and gout

Increased urate production	Decreased urate excretion
Genetic Causes	

Enzyme defects and mutations
  Glucose-6-phosphatase
  Glutathione reductase mutant
  HGPRT deficiency
  PRPP amidotransferase mutant
  PRPP synthetase mutant
	Renal underexcretion
	(normal renal function)
	  decreased urate
	  clearance

Acquired Causes	

Haemopoietic diseases
and factors
  myeloproliferative
  polycythaemia
  mononucleosis
  cytotoxic agents
Nutritional factors
  purine intake
  fructose administration
  obesity
  hypertriglyceridemia
Systemic diseases
  myocardial infarction
  psoriasis
	Renal factors
	(impaired renal function)
	  renal diseases
	  decreased urine flow
	  hypertension
	
	Drugs and metabolites
	  diuretics
	  aspirin
	  lactic acidosis
	  ketosis
	Systemic diseases
	  respiratory acidosis
	  toxaemia of pregnancy

leading to an increase of the urate pool and subsequent to urate deposits. Only in some cases of purine overproduction specific enzyme defects have been found. However, 80% of patients with genetic causes (primary) of hyperuricemia are based on a poor excretory capacity in spite of normal renal function. In most of these patients the urate clearance was shifted towards a lower range indicating that their ability to excrete urate at normal amounts at a particular serum urate concentration is less than in normal subjects. In these patients an increased serum uric acid of approximately 120 μmol/l was needed to reach normal urate excretion velocity (11). In addition hyperuricemia may be acquired during myeloproliferative disorders, due to dietary factors, to drugs, and to impairment of renal function (12) and concomitant with a variety of systemic diseases. These conditions are not inherited and are summarized as acquired causes (secondary) of hyperuricemia and gout in Table 2.

ASSESSMENT OF URATE PRODUCTION

For the exact detection of uric acid overproducers the diagnostic tools have to include isotopic in vivo methods for the determination of the miscible urate pool and the amount of purine synthesis de novo (Fig. 1), since simple laboratory methods such as serum urate and urinary urate excretion give only a hint to the amount of urate in the body (urate pool) and some indication of urate production. The miscible urate pool can be determined by intravenous administration of labelled urate and measurement of the degree of dilution of the labelled with the unlabelled urate in the body after equilibration. The isotope concentration of urate in urine (or in serum) will reflect the degree of dilution. Since the body continues to produce uric acid the rate of reduction in the concentration of radioactive urate will be a reflection of the overall rate (turnover) of new uric acid production. Following intravenous administration the isotope concentration of urinary uric acid is determined during several days and plotted semilogarithmic versus time. From the intercept of this plot the pool size and from the slope the turnover rate can be calculated. The urate pool size varies in normal men and women from 5.2 to 9.8 mmol and 3.2 to 4.1 mmol respectively (13). From this pool 45 to 85% are replaced each day. In gouty patients without an enzyme defect sizes of 20.5 ± 12.5 mmol were reported (14).

Labelled glycine as a precursor of de novo purine synthesis administered to a patient under standardized conditions will be diluted with the unlabelled glycine pool of the body and then incorporated into purines partially degraded to uric acid. Measurement of label in urinary urate will thus provide an index of purine synthesis de novo. Usually the isotope concentration of isolated urinary uric acid is followed over a period of 7 days. Approximately 0.17% of glycine are incorporated into urinary urate in normal adult subjects, 0.28% in gouty patients (15).

BIOCHEMISTRY OF PURINE METABOLISM

Before consideration of congenital enzyme defects associated with hyperuricemia and gout whose common feature uric acid overproduction is, some biochemical and regulatory aspects of purine metabolism will be discussed. Purine metabolism consists of 4 different pathways which interact with each other (Fig. 2):

1. <u>De novo synthesis</u>

The biosynthesis starts with the formation of phosphoribosylamine from the high-energy intermediate of ribose-5-phosphate phosphoribosyl

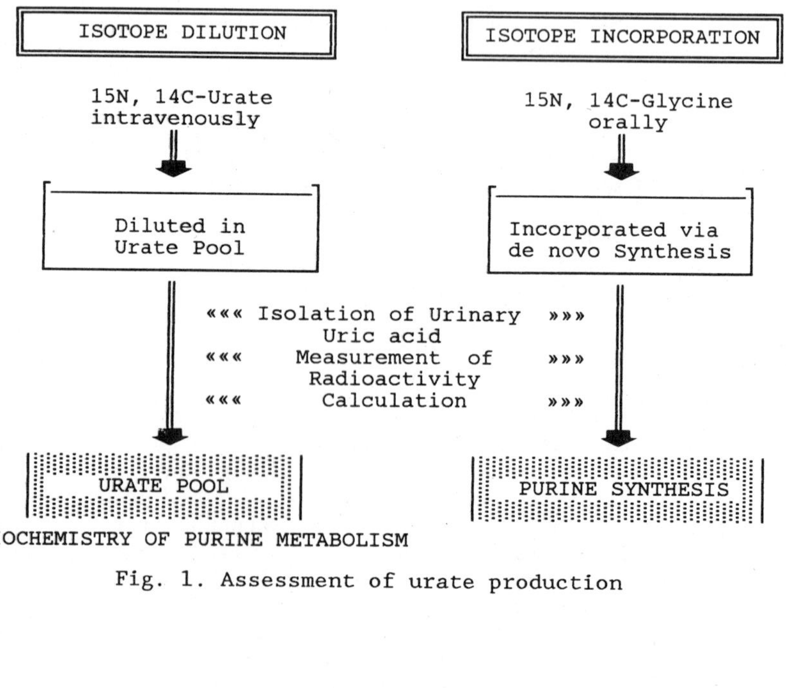

BIOCHEMISTRY OF PURINE METABOLISM

Fig. 1. Assessment of urate production

pyrophosphate (PRPP) and glutamine catalyzed by PRPP amidotransferase (PRPP-AT). This reaction is considered to be rate limiting for de novo synthesis. Subsequently the complex structure of the purine ring is synthesized by successive reactions with glycine, formate, glutamine, bicarbonate and aspartate. For the formation of inosine monophosphate (IMP) at least 5 ATP are required.

2. Interconversion

IMP is the central metabolite of purine interrelationships, since from it adenosine monophosphate (AMP) and guanosine monophosphate (GMP) are formed. Furthermore AMP and GMP may be redesaminated to IMP. Purine interconversion may be described best by two cycles with IMP as their branchpoint.

3. Degradation

Purine nucleotides are catabolized to their nucleosides by phosphatases and 5'-nucleotidase. Adenosine is desaminated to inosine. Inosine and guanosine are converted to hypoxanthine and guanine and then degraded to xanthine which is oxidized by xanthine oxidase (XOX) to uric acid.

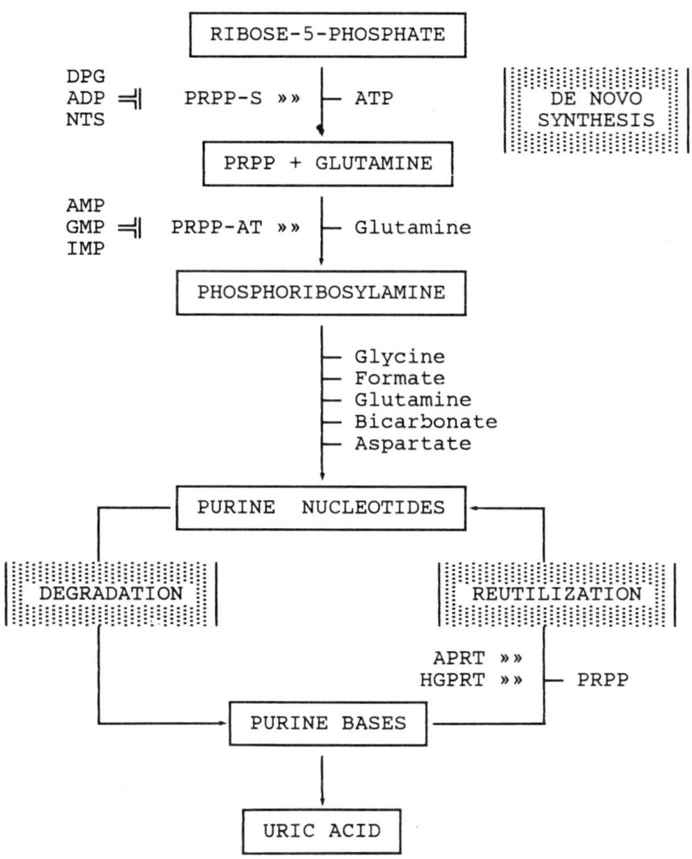

Fig. 2. Purine metabolism

4. Reutilisation (salvage pathway)

Purine nucleotides may be synthesized from preformed bases of endogenous or dietary origin. Adenine phosphoribosyltransferase (APRT) catalyzes the formation of AMP from adenine and PRPP, whereas hypoxanthine guanine phosphoribosyltransferase (HGPRT) reconverts hypoxanthine and guanine to IMP and GMP respectively.

For the reutilization of purine bases only 1 ATP is needed. Thus, from the point of expenditure of high-energy compounds purine salvage is a much more economic pathway for the body than de novo synthesis. Furthermore, the kinetic datas of enzymes involved in both pathways suggest that under physiological conditions purine salvage predominates in most tissues purine synthesis de novo.

5. Regulation

Besides, enzyme turnover as demonstrated in bacterial systems (16), the major regulatory mechanisms involve control by substrate concentrations and feedback control. The reaction catalized by PRPP-AT is considered the rate-limiting step of de novo synthesis. It is controlled both by the intracellular concentration of PRPP which is well below PRPP-AT's Km-value and therefore rate-limiting, and by the amount of purine 5'-nucleotides acting as allosteric inhibitors. Nucleotides bind to allosteric sites on the enzyme surface and thus promote aggregation of two small active subunits to an inactive large form (17). The intracellular concentrations of nucleotides indicate that PRPP-AT is normally under considerable inhibition (18). This inhibition can be overcome by high amounts of PRPP leading to the splitting of the dimer and activation of the enzyme (17). Therefore an increased availability of PRPP due to its increased formation or decreased utilization will result in an increase of de novo synthesis and uric acid formation. This mechanism is the most common biochemical basis of enzyme defects described later.

The formation of PRPP from ribose-5-phosphate and ATP by PRPP synthetase (PRPP-S) might be another rate-limiting site of de novo synthesis. ADP. 2,3-diphosphoglycerate (2,3-DPG) and PRPP inhibit PRPP-S in a competitive manner, whereas a variety of pyrimidine-, purine- and pyridine-nucleotides exhibit noncompetitive (feedback) inhibition (19, 20, 21). Furthermore, 2,3-DPG suppresses activity by disaggregating the enzyme to small inactive subunits.

ENZYME DEFECTS ASSOCIATED WITH GOUT

1. Glucose-6-phosphatase deficiency

Enzyme defects and mutants listed in Table 3 are associated with excessive uric acid production. Besides hyperlactic acidemia and suppression of tubular secretion of uric acid (22), the disturbance of the oxidative limb of the pentose phosphate pathway in glycogen storage disease type 1 (glucose-6-phosphatase deficiency) is expected to lead to overproduction of ribose-5-phosphate followed by an increased formation of PRPP. In fact, in 5 patients with glucose-6-phosphatase deficiency de novo synthesis of uric acid as determined by glycine-incorporation was 3-4 times the normal amount (Table 4). Furthermore urate pool and turnover rate size were significantly increased (23, 24). The same mechanism should be true for those gouty patients with the glutathione reductase mutant which is 28% more active than the normal enzyme (25).

Table 4. Glucose-6-phosphatase deficiency (Uric acid metabolism)

	Patients	Controls
Serum urate (μmol/l)	888 ± 143	174 - 330
Urinary uric acid (mmol/d)	2.52 ± 0.69	2.13 - 3.19
Pool size (mmol)	12.3 ± 3.19	3.22 - 7.68
% Recovery of glycine in urinary uric acid	1.18 ± 0.69	0.27 - 0.37

Jakovcic 1967, Kelley 1968

2. PRPP synthetase mutant

The superactive PRPP-S mutant is an x-linked gouty arthritis with hyperuricemia and hyperuricosuria. PRPP-S catalyzes the formation of PRPP from ribose-5-phosphate and ATP. In patients with PRPP-S overactivity 2-3 times elevated concentrations and generations of PRPP were detected in fibroblasts and erythrocytes (Table 5) (26). The reasons for the overactivity are either diminished feed-back inhibition by adenine nucleotides (27) or increased Vmax of an altered enzymeprotein (29). As a result of PRPP-S overactivity and increased formation of PRPP besides hyperuricemia and hyperuricosuria a 3-times enhanced incorporation of glycine into urinary uric acid was demonstrated (Table 6) (29).

3. HGPRT deficiency

The most impressive enzyme defect associated with hyperuricemia and gout is the complete deficiency of HGPRT, the x-linked Lesch-Nyhan syndrome, characterized also by its neurological symptoms (choreoathetosis, mental retardation, selfmutilation). In some patients with gout and marked overproduction of uric acid partial deficiency of HGPRT has been demonstrated. HGPRT salvages hypoxanthine and guanine under PRPP consumption to IMP and GMP.

Table 5. PRPP synthetase mutant (Biochemistry of enzyme mutation)

	Patients	Controls
Erythrocytes		
PRPP-S (nmol/h.mg protein)	204, 177	62 ± 18
PRPP Concentration (nmol/ml)	5.8, 5.7	2.9 ± 0.6
PRPP Generation (nmol/h.ml)	266, 217	119 - 146
Fibroblasts		
PRPP-S (nmol/h.mg protein)	1116, 1187	237 - 451
PRPP Concentration (nmol/10^6)	0.77, 0.81	0.19 - 0.41
PRPP Generation (nmol/h.10^6)	7.4, 7.9	2.3 - 2.9

Becker 1982

Table 6. PRPP synthetase mutant (Uric acid metabolism)

	Patients	Controls
Serum urate (μmol/l)	534	< 420
Urinary uric acid (mmol/d)	5.06	< 3.00
% Recovery of glycine in urinary uric acid	0.54	0.13 - 0.24

Müller 1974

Table 7. HGPRT deficiency (Biochemistry of enzyme defect in erythrocytes)

	Lesch-Nyhan	Partial Deficiency	Controls
HGPRT (nmol/h.mg prot)	< 0.01	6.85 ± 4.47	103 ± 18
PRPP (nmol/ml)	38.8 ± 4.0	4.6 ± 1.3	3.1 ± 0.5

Greene 1969, Fox 1971

Table 8. HGPRT deficiency (Uric acid metabolism)

	Lesch-Nyhan	Partial Deficiency	Controls (children)
Serum urate (μmol/l)	654 ± 174	751 ± 131	216 ± 72
Urinary uric acid (mmol/d)	4.50 ± 0.65	6.87 ± 2.34	4.00 ± 1.28
Pool size (mmol)	5.92 ± 2.64	15.3 ± 2.75	
% Recovery of glycine in urinary uric acid	2.75 ± 0.65	1.46 ± 1.13	< 0.15

Lesch 1964, Kelley 1969

Therefore as a result of HGPRT deficiency PRPP should be elevated. In fact, in erythrocytes of Lesch-Nyhan patients 6-10 times and in those of adults with partial HGPRT deficiency 2-3 times elevated PRPP concentrations were demonstrated (Table 7) (30, 31). Again this enhanced PRPP availability favours purine de novo synthesis. This could be demonstrated in Lesch-Nyhan children as well as in adults with residual HGPRT activity by increased glycine incorporation into urinary uric acid (Table 8) leading to enlarged urate pools (32, 33).

In summary, the common biochemical feature of all these enzyme defects is the elevated intracellular PRPP due to its increased formation or underutilization. Because PRPP levels are normally in the low range of the substrate-velocity curve of PRPP-AT even slightly increases of PRPP will accelerate purine synthesis. In addition in HGPRT deficiency and in the two cases of PRPP-AT mutant (34) reduced feed-back inhibition by purine nucleotides may also be responsible for overproduction of uric acid. The purpose of this presentation was to demonstrate that results of biochemical research are useful for rational explanations of clinical syndromes. However, in spite of extensive research in this field of clinical biochemistry only in a minority of patients with gout a primary defect could be demonstrated up to now. It should also be considered that environmental and nutritional factors contribute to gout and that for prevention one should "live on six pence a day and earn it".

REFERENCES

1. Garrod AB. Trans Med Chir Soc Edinburgh 1848; 31: 83.
2. Seegmiller JE, Rosenbloom FM, Kelley WN. Science 1967; 155: 1682-1684.
3. Müller MM, Kuzmits R, Frassine I, Lorant P. Wr klin Wschr 1981; 93: 492-500.
4. Susic D, Bäumer A. Z Rheumaforsch 1969; 28: 323-328.
5. Zalokar J, Lellouch J, Claude JR, Kuntz D. J chron Dis 1972; 25: 305-312.
6. Scott JT, Sturge RA, Kennedy AC, Hart DP, Buchanan WW. Adv Exp Med Biol 1977; 76B: 214-222.
7. Mikkelsen WM, Dodge HJ, Valkenburg H. Amer J Med 1965; 39: 242.
8. Hall AP, Barry PE, Dawber TR, McNamara PM. Amer J Med 1967; 42: 27-37.
9. Jeremy R, Towson J. Med J Austr 1971; 1: 1116-1118.
10. Rose BS, Prior JAM, Davidson F. In: Bennett PH, Wood PHN, eds. Population Studies of the Rheumatic Diseases. Excerpta Medica 1968: 358-362.
11. Simkin PA. Adv Exp Med Biol 1977; 76B: 41-45
12. Leumann EP, Wegmann W. Nephron 1983; 34: 51-57.
13. Scott J-T, Holloway VP, Glass HI, Arnot RN. Ann Rheum Dis 1969; 28: 366-372.
14. Müller MM, Fuchs H, Pischek G, Bresnik W. Therapiewoche 1975; 25: 514-521.
15. Wyngaarden JB, Kelley WN. In: Stanbury JB, Wyngaarden JB, Frederickson DS, Goldstein JL, Brown MS, eds. The Metabolic Basis of Inherited Disease. McGraw Hill 1983; 5th ed.: 1041-1114.
16. Jacob F, Monod H. J Mol Biol 1961; 3: 318-356.
17. Holmes EW, McDonald JA, McCord JM, Wyngaarden JB, Kelley WN. J Biol Chem 1973; 248: 144-150.
18. Nelson DJ, Bugge CJL, Krasny HC, Elion GB. Biochem Pharmacol 1973; 22: 2003-2022.
19. Fox IH, Kelley WN. J Biol Chem 1972; 247: 2126-2131.
20. Switzer RL, Sogin DC. J Biol Chem 1973; 248: 1063-1073.
21. Olszowy J, Switzer RL. J Bacteriol 1972; 110: 450-451.
22. Howell RR, Ashton DM, Wyngaarden JB. Pediatrics 1962; 29: 553.
23. Jakovcic S, Sorenson LB. Arthritis Rheum 1967; 10: 129.
24. Kelley WN, Rosenbloom FM, Seegmiller JE, Howell RR. J Pedriatr 1968; 72: 488.
25. Long WK. Science 1967; 155: 712-713.
26. Becker MA, Losman MJ, Itkin P, Simkin PA. J Lab Clin Med 1982; 99: 495-511.
27. Sperling O, Eilam G, Pershky-Brosh S, De Vries A. Biochem Med 1972; 6: 310-316.
28. Becker MA, Kostel PJ, Meyer L, Seegmiller JE. Proc Natl Acad Sci USA 1973; 70: 2749-2752.
29. Müller MM, Frank O. Adv Exp Med Biol 1974; 41B: 509-516.
30. Greene ML, Seegmiller JE. J Clin Invest 1968; 48: 32.
31. Fox IH, Kelley WN. Ann Intern Med 1971; 74: 424.

32. Lesch M, Nyhan WL. Am J Med 1964; 36: 561-570.
33. Kelley W, Greene ML, Rosenbloom FM, Henderson JF, Seegmiller JE. Ann Intern Med 1969; 70: 155-206.
34. Henderson JF, Rosenbloom FM, Kelley WN, Seegmiller JE. Clin Invest 1968; 47: 1511-1516.

CHAPTER 8
ION SELECTIVE ELECTRODES

Ion selective electrodes (ISE's) in clinical chemistry
U. Oesch, D. Amman, U. Schefer, E. Malinowska, and W. Simon

Proposed recommendations on ion selective electrode determinations of the substance concentration of sodium, potassium and ionized calcium in serum plasma or whole blood
A.H.J. Maas

The clinical application of ionized calcium
G.N. Bowers

Sodium and potassium by ion selective electrodes. Clinical interpretation
O. Siggaard-Andersen, N. Fogh-Andersen, and P.D. Wimberley

ION SELECTIVE ELECTRODES (ISE'S) IN CLINICAL CHEMISTRY

U. Oesch*, D. Amman*, U. Schefer*, E. Malinowska** and W. Simon**

*Department of Organic Chemistry, Swiss Federal Institute of Technology (ETH), CH-8092 Zürich, Switzerland
**Department of Analytical Chemistry, Technical University Ul. Noakowskiego 3, PL-00-664 Warsaw, Poland

The assay of K^+ is claimed to be the most frequently performed clinical analysis (1). In industrialized countries an average of about 0.2 determinations per capita and year is performed (1). It has been estimated that in 1986 in some countries (e.g. USA) about 70% of these electrolyte assays are done by potentiometric procedures, using K^+-selective membrane electrodes.

Under a variety of membrane types (e.g. glass electrodes for H_3O^+ and Na^+ measurements), solvent polymeric membranes have proved to be especially suited for clinical analysis since they can easily be manufactured in different sizes and shapes and are less affected by the presence of biological substrates (e.g. proteins). Most of the competitive ion-selective electrodes based on solvent polymeric membranes utilize a neutral carrier as the ion-selective component. At present more than 30 manufacturers of clinical laboratory instrumentation produce each at least one clinical analyser with potentiometric sensors for alkali and/or alkaline earth metal cations (2). Probably with one exception all of the potentiometric sensors for K^+ are based on the neutral carrier valinomycin. For a summary of the development of such electrodes see (3). Again with probably one exception neutral carriers are used throughout in sensors for Ca^{2+} (4). The potentiometric sensors for the direct assay of Cl^- in use in clinical chemistry are however all based on classical anion exchangers (5).

For clinical ion measurements in blood and urine a well-balanced optimization of all the electrode characteristics relevant for the application should be discussed rather than the superiority of a single property. Such relevant properties are selectivity, stability, response time and lifetime of the sensor (6).

SELECTIVITY

Using the Nikolskii-Eisenman formalism and assuming representative physiological concentration ranges, required log K(Pot)IJ-values (7) can be calculated for a maximally tolerable error of 1% due to interfering alkali

Table 1. Selectivity factors $\log K_{IJ}^{Pot}$ for Li^+-selective electrodes

ion	required selectivity (1% interference, worst case)[a]	observed selectivity $\log K_{LiJ}^{Pot}$	
J	$\log K_{LiJ}^{Pot}$	ETH 1810 (SSM) (10)	14-crown-4-derivative (FIM) (9)
H^+	< 2.1	1.0	-3.4
Li^+	0	0	0
Na^+	<-4.3	-2.3[b]	-2.2
K^+	<-2.8	-2.6	-2.0
Mg^{2+}	<-3.5	-4.0	-4.7
Ca^{2+}	<-3.6	-2.7	-4.4

a) Whole blood, plasma and serum. Clinical Li^+-concentration range (0.7 - 1.5 mM)

b) -2.45 (FIM)

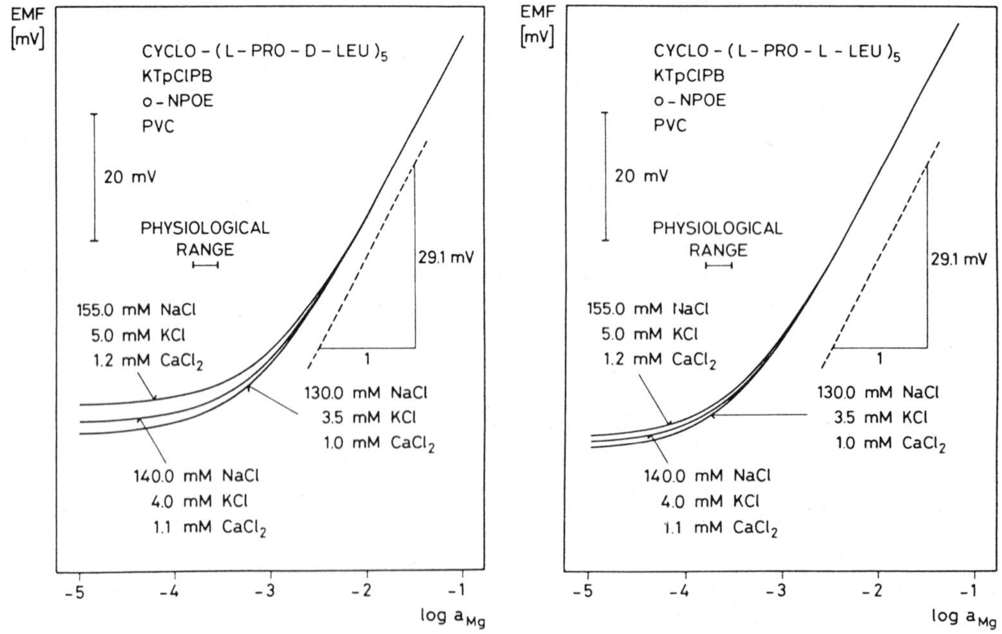

Fig. 1. Calculated EMF-response for Mg^{2+}-selective cell assemblies based on cyclodecapeptides at typical, high and low physiological cation backgrounds. Selectivity factors of Table 2 were used.

and alkaline earth metal cations or interfering anions (worst case, see (6, 8)). For blood serum and whole blood there are neutral carrier based sensors for H^+, Na^+, K^+ and Ca^{2+} available, which meet this requirement (6). Tables 1 and 2 indicate that even the most promising sensors described so far for Li^+ (9, 10) and Mg^{2+} (11) are unfortunately quite far beyond their required selectivities. Nevertheless, for Li^+ a calibration utilizing an adequate ion background or a correction for the interference induced by the presence of Na^+ leads to clinically relevant assays (12). Even less selective membrane systems exhibiting very high selectivity stability may be advantageous for such purposes (13). On the other side, a clinically relevant assay for Mg^{2+} performed with the sensors based on cyclodecapeptides (Table 2) could only be done by an additional simultaneous determination of Na^+ and Ca^{2+} in the ion background of the sample of interest (Figure 1). These determinations would then be used for a computational correction of the calibration. Such a chemometric procedure (14) represents still a severe drawback for a clinical application of these sensors because they rely on a sufficiently high stability of the selectivity factors K^{Pot}_{MgNa} and K^{Pot}_{MgCa} with time.

Furthermore there is evidence that at least some of these sensors exhibit high selectivities for monovalent magnesium complexes (11). This requires a calibration which takes care of all the relevant sample anions (e.g. Cl^-, OH^-) that undergo complex formation with Mg^{2+}. Recently, bridged 3-oxaglutaric acid diamides have been synthesized (15) that yield Mg^{2+}-selective membranes with a rejection of Na^+, K^+ and Ca^{2+} by a factor of 4900, 3 and 200, respectively (Table 2). Using these ion selectivities (obtained by the separate solution technique (7)), calculated EMF response curves for a typical extracellular ion background (see Figure 2) show a detection limit of about $7 \cdot 10^{-6}$ M Mg^{2+} activity. Although these membranes fulfil the selectivity requirements almost perfectly, the long-term reproducibilities of the electrode characteristics are still very poor and demand further membrane technological interventions in view of a successful clinical application. The poor stability is also reflected when comparing the selectivities obtained by the separate solution technique and fixed interference methods in aqueous solutions.

Table 2. Selectivity Factor log K^{Pot}_{IJ} for Mg^{2+}-selective electrodes

ion	required selectivity (1% interference, worst case)[a]	observed selectivity log K^{Pot}_{MgJ} (SSM)		
J	log K^{Pot}_{MgJ}	cyclo(-L-Pro-D-Leu-)$_5$ (11)	cyclo(-L-Pro-L-Leu-)$_5$ (11)	bridged oxaglutaric acid diamide (V233) (15)
H^+	< 8.9	0.7	0.5	1.0
Li^+	<-2.3	-1.8	-2.6	-4.2
Na^+	<-3.9	-1.2	-2.3	-3.7
K^+	<-0.9	1.1	-1.0	-0.5
Mg^{2+}	0	0	0	0
Ca^{2+}	<-2.4	-2.2	-0.1	-2.3

a) Whole blood, plasma and serum

RESPONSE TIME

The required response time of the membrane electrode cell assembly has to be compatible with the analysis time, which is in the order of ~30 s for commercial analysers. There are many different processes that contribute to the dynamic response behaviour (16). Since some of these processes (e.g. diffusion of sample ions through the stagnant layer adhering to the membrane surface) are not intrinsic to the membrane itself, it is difficult to properly compare reported response times as characteristics of membrane systems. The mathematical expression for the time constant of the response of neutral carrier based sensors however indicates that the intrinsic response time of the membrane may be reduced by adequate membrane technology (16).

STABILITY OF THE EMF OF THE CELL ASSEMBLY

In view of clinically meaningful results the performance of the electrode cell assembly should allow an adequate subdivision of the small EMF ranges corresponding to the physiological normal concentration ranges. These ranges are 6.8 mV (H^+), 19.3 mV (Li^+, therapeutical range), 2.4 mV (Na^+), 9.1 mV (K^+), 7.4 mV (Mg^{2+}), 2.4 mV (Ca^{2+}), 3.8 mV (Cl^-), and 5.6 mV (HCO_3^-) (6, 8). For a five-fold subdivision (95% confidence limit) of the physiological normal range the required standard deviations are 0.35 mV (H^+), 0.96 mV (Li^+), 0.12 mV (Na^+), 0.46 mV (K^+), 0.37 mV (Mg^{2+}), 0.12 mV (Ca^{2+}), 0.19 mV (Cl^-), and 0.28 mV (HCO_3^-) (6, 8). Certain neutral carrier

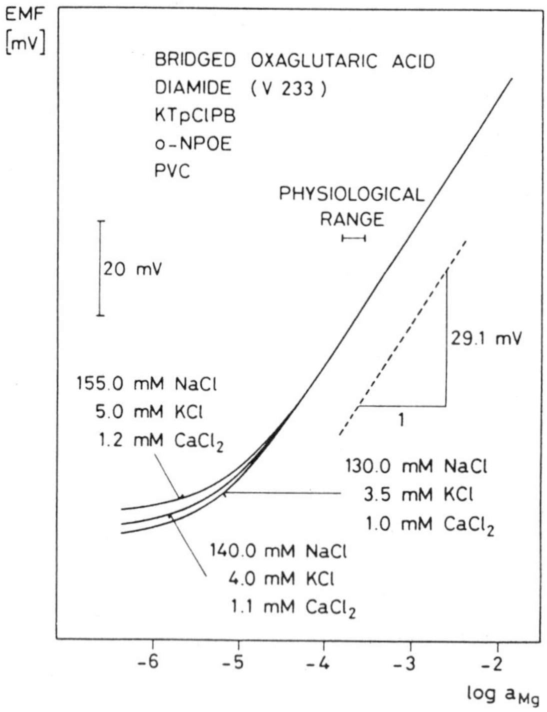

Fig. 2. Calculated EMF-response for a Mg^{2+}-selective cell assembly based on a bridged oxaglutaric acid diamide at typical, high and low physiological cation backgrounds. Selectivity factors of Table 2 were used.

based electrodes for H^+, Na^+, K^+ and Ca^{2+} which meet the selectivity requirement also exhibit the required stability (17).

LIFETIME OF THE SENSOR

The lifetime of neutral carrier based solvent polymeric membrane electrodes is to a large extent dictated by the loss of membrane components (ion carrier, plasticizer, additives) into the sample solution (6, 18, 19). Because such a loss affects the membrane characteristics (selectivity, membrane resistance), it is reasonable to expect that the lifetime and the EMF stability of the cell assembly are related (18). Through adequate membrane technology the required concentrations of membrane components may be maintained in the membrane phase for at least a desired time period. The lipophilic character of the sample (whole blood, serum, plasma) favours a substantial and fast extraction of the membrane components. Therefore, for prevention an extremely high lipophilicity P of the membrane components should be realized. The correlation between the lipophilicity P_{TLC} of a component (lipophilicity P as determined by thin layer chromatography (20)) and its partition coefficient K in the membrane/sample system can be assessed by Equations (1) and (2) (see (19)).

blood:

$$\log K = 0.48 + 0.33 \log P_{TLC} \quad \left\{ \begin{array}{ll} \text{membrane:} & \text{70\% dioctyl sebacate} \\ & \text{30\% PVC} \\ \text{sample:} & \text{serum} \end{array} \right. \tag{1}$$

urine:

$$\log K = 1.42 + 0.80 \log P_{TLC} \quad \left\{ \begin{array}{ll} \text{membrane:} & \text{70\% o-nitrophenyl octyl ether} \\ & \text{30\% PVC} \\ \text{sample:} & \text{water} \end{array} \right. \tag{2}$$

To guarantee a continuous use lifetime of the solvent polymeric membrane in contact with undiluted serum or whole blood of at least one month in a typical commercial flow analyser (6) or of at least 10 days in a typical ISFET arrangement (21) the lipophilicities $\log P_{TLC}$ given in Table 3 are required. The values obtained in this way represent a worst case situation since the assumption of a continuous use is only fulfilled in few applications like an in situ monitoring (with e.g. ISFETs) in intensive care units. In conventional commercial analysers the stream of samples is usually interrupted after each sample by calibration and/or washing solution which on one hand do not require high values for $\log P_{TLC}$. On the other hand they allow to recover an eventual depletion of components in the vicinity of the membrane surface by a resupply from the interior of the membrane. The lifetime will obviously be extended in these cases.

Only few reported plasticizers (e.g. ETH 469: $\log P_{TLC} = 10.8$ (6, 17) and ETH 2112: $\log P_{TLC} = 23.8$ (22)) and the ionophores for K^+ (valinomycin: $\log P_{TLC} = 8.6$), for Na^+ (ETH 227: $\log P_{TLC} = 7.8$; ETH 2120: $\log P_{TLC} = 8.3$ (23)), for Ca^{2+} (ETH 1001: $\log P_{TLC} = 7.5$), for Li^+ (ETH 1810: $\log P_{TLC} = 7.2$) and for H^+ (tridodecylamine: $\log P_{TLC} = 11.6$) are close to or may even meet the required lipophilicities for classical applications (Table 3).

Some other widely used plasticizers (dibutyl phthalate: $\log P_{TLC} = 4.6$; o-nitrophenyl octyl ether: $\log P_{TLC} = 5.9$ and dioctyl sebacate: $\log P_{TLC} = 10.1$) are poorer in this respect (6, 24). This also holds for many reported ion carriers, e.g. the ionophores for Na^+ (ETH 157: $\log P_{TLC} = 4.6$; a bis 12-crown-4 (23, 25): $\log P_{TLC} = 5.6$) and for K^+ (a bis 15-crown -5 (26): $\log P_{TLC} = 2.4$; BME 15 (27): $\log P_{TLC} = 2.5$) (6). For the

Table 3. Required lipophilicity log P_{TLC}[a] for a clinical use of ion sensors

analytical application[b]	required lipophilicity (log P_{TLC})			
	ionophore		plasticizer	
	whole blood or blood serum	aqueous sample (e.g. urine)	whole blood or blood serum	aqueous sample (e.g. urine)
ISFET (use over 10 days)	15.4	5.2	17.8	6.1
classical flow-through (use over 30 days)	8.4	2.3	12.8	4.1

a) The value log P_{TLC} is a substitute for the partition coefficient between water and octane-1-ol (19, 20)
b) Continuous use in streaming fluid

application of ISFETs (see Table 3) no ionophore and only the plasticizers described recently (21, 22) exhibit sufficient lipophilicities.

Valinomycin based electrodes are almost ideal in respect to the requirements a) - d) in classical applications. This is certainly one reason for the world wide acceptance in biomedical and clinical applications of this electrode. Certainly a search for a better substitute may still be an academic challenge.

Neutral carriers for anions have been reported (28). There is some evidence that this type of ionophore might be useful for a clinically relevant assay of Cl^- and possibly also of HCO_3^-. Another type of HCO_3^--selective electrodes on the basis of gas-permeable, H^+-selective neutral carrier-based polymeric membranes has been proposed (29, 30). An electrode based on this approach utilizing multilayers of thin (~10-25 μm) membranes has recently been introduced (31) exhibiting a detection limit for HCO_3^- of ~0.2 mM at a physiological anion background. The response time ($t_{95\%}$) of the electrode is in the order of 30 sec which meets the requirement for an application in a clinical analyser. In employing electrodes of this kind attention has to be given to interferences of electrically neutral, acidic components in the samples such as e.g. acetic acid, lactic acid and salicylic acid.

The reported electrode is free of interferences of these acids at physiological or therapeutical concentrations (31). A preliminary correlation study with blood serum samples between this electrode and the bicarbonate determination by a commercial blood gas analyser shows a residual standard deviation of the correlation of 0.81 mM (n = 27). Although this value does not yet meet the required stability of 0.21 mM (see above) for a bicarbonate electrode, it is nevertheless promising since the residual standard deviation of the self correlation of the blood gas analyser over the same sample set is only 0.37 mM.

This work was partly supported by the Swiss National Science Foundation, by E.I. du Pont de Nemours & Co., (Inc.), by Orion Research, Inc. and by Nova Biomedical Corp.

REFERENCES

1. Vonderschmitt DJ. Private communication.
2. Möller W. Private communication.
3. Oggenfuss P, Morf WE, Oesch U, Ammann D, Pretsch E, Simon W. Anal Chim Acta 1986; 180: 299.
4. Bowers GN Jr, Brassard C, Sena SF. Clin Chem 1986; 32: 1437.
5. Oesch U, Ammann D, Simon W., Clin Chem 1986; 32: 1448.
6. Oesch U, Anker P, Ammann D, Simon W. Proceedings of the 4th Symposium on Ion-Selective Electrodes. Pungor E, Buzas I, eds. Budapest: Akadémai Kiadó, 1985: 81.
7. Guilbault GG, Durst RA, Frant MS, et al. Pure Appl Chem 1976; 48: 127.
8. Oesch U, Ammann D, Pham HV, Wuthier U, Zünd R, Simon W. J Chem Soc, Faraday Trans 1986; 82: 1179.
9. Kitazawa S, Kimura K, Yano K, Shono T. Analyst 1985; 110: 295.
10. Metzger E, Ammann D, Schefer U, Pretsch E, Simon W. Chimia 1984; 38: 440.
11. Behm F, Ammann D, Simon W, Brunfeldt K, Halstrøm J. Helv Chim Acta 1985; 68: 110.
12. Metzger E, Ammann D, Asper R, Simon W. Anal Chem 1986; 58: 132.
13. Metzger E, Dohner R, Simon W. Anal Chem 1987; in press.
14. Otto M, Thomas JDR. Anal Chem 1985; 57: 2647.
15. Schefer U, Bongardt F, Vögtle F, Simon W. Publication in preparation.
16. Morf WE, Lindner E, Simon W. Anal Chem 1975; 47: 1596.
17. Ammann D, Anker P, Metzger E, Oesch U, Simon W. Proceedings of the International Symposium on the Theory and Application of Ion-Selective Electrodes in Physiology and Medicine, Burg Rabenstein, BRD, September 12-14, 1983. In: Kessler M, Harrison DK, Höper J, eds. Ion Measurements in Physiology and Medicine. Berlin, Heidelberg: Springer Verlag, 1985: 102.
18. Oesch U, Simon W. Anal Chem 1980; 52: 692.
19. Oesch U, Dinten O, Ammann D, Simon W. Proceedings of the International Symposium on the Theory and Application of Ion-Selective Electrodes in Physiology and Medicine, Burg Rabenstein, BRD, September 12-14, 1983. In: Kessler M, Harrison DK, Höper J, eds. Ion Measurements in Physiology and Medicine. Berlin, Heidelberg: Springer Verlag, 1985: 42.
20. Simon W, Pretsch E, Morf WE, Ammann D, Oesch U, Dinten O. Analyst 1984; 109: 207.
21. Oesch U, Simon W. Proceedings of the 4th International Conference on Solid-State Sensors and Actuators, Transducers '87, Tokyo, Japan, June 2-5, 1987. Printed by the Institute of Electrical Engineers of Japan, Tokyo, 1987: 755.
22. Oesch U, Xu A, Brzózka Z, Suter G, Simon W. Chimia 1986; 40: 351.
23. Maruizumi T, Wegmann D, Suter G, Ammann D, Simon W. Mikrochim Acta 1986; I: 331.
24. Oesch U, Ammann D, Simon W. Proceedings of the 4th Meeting of the European Working Group on Ion Selective Electrodes, Helsinki, Finland, June 25-28, 1985. In: Maas AHJ, Boink FBTJ, Saris N-EL, Spronkholt R, Wimberley RD, eds. Methodology and Clinical Applications on Ion Selective Electrodes. Copenhagen: Interprint A/S, 1986: 273.
25. Shono T, Okahara M, Ikeda I, Kimura K. J Electroanal Chem 1982; 132: 99.
26. Kimura K, Maeda T, Tamura H, Shono T. J Electroanal Chem 1979; 95: 91.
27. Toeke L, Agai B, Bitter I, et al. PCT Int Appl 1983; WO 8300149.
28. Wuthier U, Pham HV, Zünd R, et al. Anal Chem 1984; 56: 535.

29. Coon RL, Lai NCJ, Kampine JP. J Appl Physiol 1976; 40: 625.
30. Funck RJJ, Morf WE, Schulthess P, Ammann D, Simon W. Anal Chem 1982; 54: 423.
31. Oesch U, Malinowska E, Simon W. Anal Chem 1987; in press.

PROPOSED RECOMMENDATIONS ON ION-SELECTIVE ELECTRODE DETERMINATIONS OF THE SUBSTANCE CONCENTRATION OF SODIUM, POTASSIUM AND IONIZED CALCIUM IN SERUM PLASMA OR WHOLE BLOOD

Anton H.J. Maas

Heart Lung Institute, University Hospital, Utrecht, The Netherlands

1. INTRODUCTION

Ion-selective electrodes (ISEs) make possible the measurement of a wide range of analytes, which cover such diverse substances as ions, dissolved gases, enzymes etc. in serum, plasma or whole blood (1). The aims of this lecture are to outline some of the problems that have arisen with the introduction of sodium, potassium and calcium ISEs in clinical chemistry. In this context I want to discuss provisionally proposed recommendations which are prepared by a European Working Group on Ion Selective Electrodes (EWGISE)[1] of the International Federation of Clinical Chemistry (IFCC) consisting of clinical chemists and manufacturers with a common interest in exchanging ideas and information on ISEs and trying to resolve some of the problems created by the increasing use of this new technology. The group started in 1982 and has held meetings in Oslo (1983), Oxford (1984), Helsinki (1985) and Graz (1986) in which European, American, Australian and Japanese invited scientists both from profession and industry participated.

2. DEFINITIONS AND PRINCIPLES

2.1 Ion-activity

The relationship between activity and concentration of an ion B is given by the following equations (2).
The relative molal activity ($a_{m,B}$) is the molal activity coefficient (γ_B) times the molality of (m_B) divided by the unit mol/kg.:

$$a_{m,B} = \frac{\gamma_B \cdot m_B}{(\mathrm{mol/kg})} \qquad \text{Eq. 1}$$

---

[1]European Working Group Members: A.B.T.J. Boink*, P.M.G. Broughton, B.M. Buckley*, T.F. Christiansen*, A.K. Covington*, N. Fogh-Andersen, K. Saarinen, J. Kofstad, L. Larsson, A.H.J. Maas*, H. Marsoner, O. Müller-Plathe*, L. Russell, Ch. Sachs*, O. Siggaard-Andersen*, W. Simon and R. Sprokholt.
*Members of the EWGISE subcommittee.

What we call the active molality ($\tilde{m}_B$) is the activity coefficient times the molality.

$$\tilde{m} = \gamma_B \cdot m_B \qquad \text{Eq. 2}$$

The two quantities are identical except that the relative activity has the unit 1 but the active molality has the same unit as the molality (mol/kg). The substance concentration (c_B) is the molality (m_B) times the mass concentration of water of the solution (ρH_2O):

$$c_B = m_B \cdot \rho_{H_2O} \qquad \text{Eq. 3}$$

The molality has the dimension mol/kg; the mass concentration of water kg/L; hence the concentration has the unit mol/L.
The mass concentration of water in any plasma sample may be obtained by drying and weighing or by calculation from the protein and lipid mass concentration according to Waugh formula (3):

$$\rho_{H_2O}(P) = 0.991 - 0.73\,\rho_{protein} - 1.03\,\rho_{lipid} \qquad \text{Eq. 4}$$

The mass concentration of water in normal plasma is about 0.933 kg/L, decreasing with increasing concentrations of protein or lipid (range 0.80 to 0.96 kg/L).
This is the most important problem when you will express the results in concentration. All the total plasma concentration values are suppressed by the mass of lipid and protein. If we take the full consequence of this we should convert all concentrations to molality (mol/kg plasma water) which is impractical in general routine work.
The molal activity coefficient of an ion B (γ_B) may be calculated by means of the Debye-Hückel equation, which shows the relationship of the logarithm of the activity coefficient of an ion and the root of the ionic strength (I) of the solution:

$$\log \gamma_B = \frac{-A \cdot z^2_B \cdot I^{\frac{1}{2}}}{1 + \mathring{a}_B \cdot B \cdot I^{\frac{1}{2}}} \qquad \text{Eq. 5}$$

where z_B is the charge number of ion B, $\mathring{a}_B$ is the ion-size parameter, A and B are temperature-dependent constants, and I is the ionic strength. The latter is defined as the half of the sum of the products of the square valence and the molality of each ion B_i in the solution:

$$I = 0.5 \sum_{i=1}^{n} z^2_{B_i} \cdot m_{B_i} \qquad \text{Eq. 6}$$

The calculated molal activity coefficients for a monovalent ion (e.g. γ_{Na^+}) and a divalent ion (e.g. $\gamma_{Ca^{2+}}$) are shown as a function of ionic strength for pure aqueous solutions in Figure 1. It should be noticed that the value of $\gamma_{Ca^{2+}}$ is much smaller than the γ_{Na^+} and that the variation in the physiological range (I = 0.1 mol/kg up to 0.2 mol/kg) is small, however more for $\gamma_{Ca^{2+}}$ than for γ_{Na^+}.

2.2 ISE measurement

The ISE measurement is based on the use of a cell consisting of an external reference electrode (R_1) with a concentrated potassium chloride liquid/liquid junction in combination with an ion-selective electrode with an inner reference electrode (R_2) according to the scheme:

$$\overbrace{R_1 \mid KCl\ (conc.)}^{\text{Reference electrode}} \,\vdots\vdots\, \text{solution X} \mid \overbrace{M \mid \text{inner reference sol.} \mid R_2}^{\text{Ion Selective Electrode}} \qquad (a)$$

Electrodes R_1 and R_2 are connected to a millivolt meter e.g. ISE analyzer. On each side of the ion selective membrane (M) an electrical potential difference E develops across the membrane/solution boundary. The inner side of the ion-selective membrane is in contact with an inner reference solution of constant activity of ion species and hence develops a constant potential.

At the other side, the external surface of the membrane, the electrical potential difference varies linearly with the logarithm of the activity of an ion species B of sample solution X only, as described by the Nicolsky equation:

$$E = E^\circ + \frac{R \cdot T}{z_B \cdot F} \ln \left(a_B + \sum_{j=1}^{n} k^{Pot}_{B,B_j} \cdot a_{B_j}^{\;z_B/z_{B_j}} \right) \qquad \text{Eq. 7}$$

where B is the measured ion and B_j the interfering ions.

The first part is equal to the Nernstian equation and the second part contains a term for the selectivity and activity of the interfering ions.

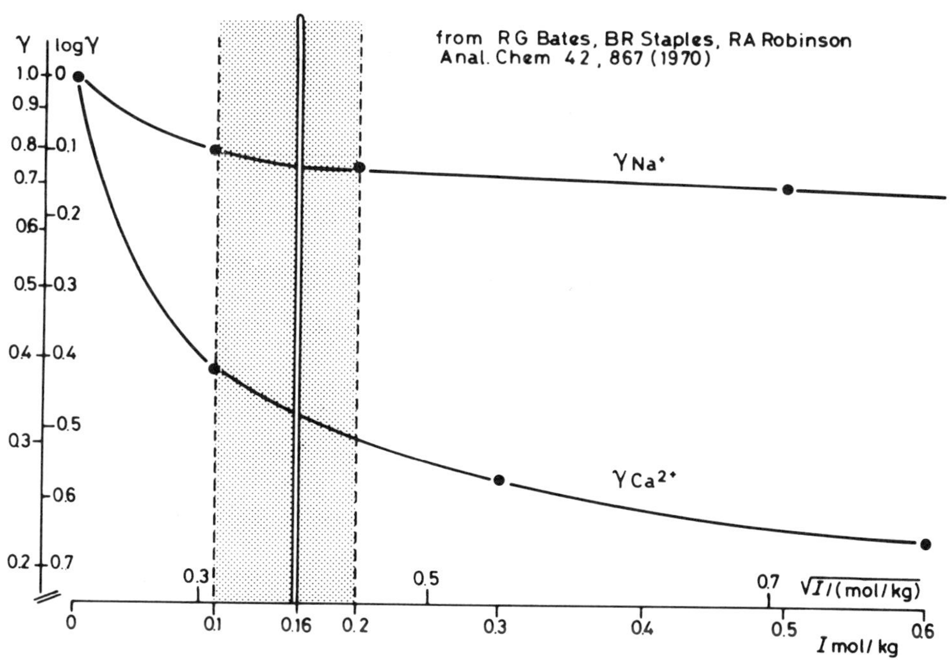

Fig. 1. The activity coefficients for a monovalent ion (γ_{Na^+}) and a divalent ion ($\gamma_{Ca^{2+}}$) as a function of ionic strength, I, for pure aqueous solutions.

Ideally the latter should be zero, resulting the Nicolsky equation in the Nernst equation. The cells should be calibrated by means of two calibrators one for adjustment of the zero point of the ISE analyzer and one for adjustment of the slope control.

3. COMPOSITION OF SODIUM, POTASSIUM AND CALCIUM IN PLASMA

3.1 Sodium and potassium

Sodium and potassium in plasma exists mostly as free hydrated ion. Estimates of the extent of binding of protein, carbonate and bicarbonate to sodium are less than 2% of total sodium and to potassium less than 3% of total potassium. Figure 2 illustrates the sodium fractions of normal blood plasma (4). Total sodium (140 mmol/L or 150 mmol/kg) is divided into protein-bound (1.4 mmol/L) and ultrafiltrable sodium (138.6 mmol/L). The latter is subdivided into "complex bound" sodium (0.6 mmol/L) and "free" sodium ions (138 mmol/L). The "free" sodium ions may again be subdivided into electrostatically "bound" sodium ions (35.9 mmol/L) and "active" sodium ions (102.1 mmol/L).

For the potassium fractions of normal blood plasma the figure is nearly identical except the scale, which is about 30 times smaller.

3.2 Calcium

About 50% of total calcium in plasma is bound and the remaining is ionized. The different calcium fractions in normal blood plasma are shown in Figure 3 (5). The concentration of total calcium in normal plasma or serum is about 2.4 mmol/L. The main fractions are protein bound (0.90 mmol/L), complex bound (0.30 mmol/L) and free calcium ions (1.20 mmol/L) of which about a third "active" calcium ions (0.36 mmol/L). The binding between protein and calcium ions is pH dependent because hydrogen ion and calcium ion compete for binding sites. With a high pH more calcium ions are

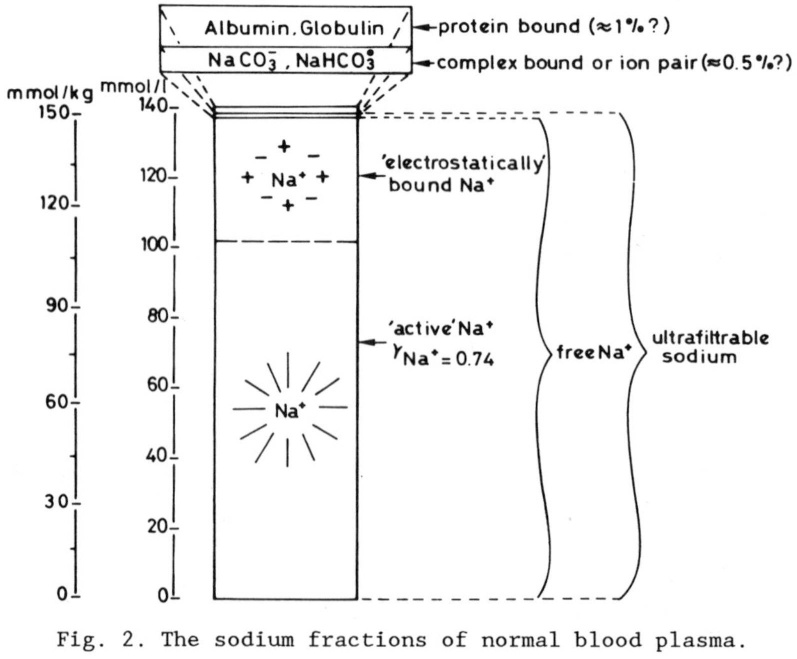

Fig. 2. The sodium fractions of normal blood plasma.

bound and the fraction ionized calcium therefore falls; opposite is the case with a low pH. Calcium ions are complexed by anions with two or more negative charges e.g. citrate but also by monovalent anions like lactate and bicarbonate. Due to the much higher bicarbonate concentration in the plasma than the others, the bicarbonate bound calcium ($CaHCO_3^+$) accounts for the major part of the complex bound fraction, i.e. about 0.25 mmol/L. The major part of the concentration of free calcium ions in the plasma is inactivated by electrostatical forces mainly due to sodium and chloride ions. The "active" fraction i.e. that fraction which an electrode can "see", constitutes as little as 0.36 mmol/L.

4. PROBLEMS OF CLINICAL CHEMICAL APPLICATION

ISEs for sodium and potassium ions may be used on two different ways namely in undiluted and diluted samples. For this the denomination direct and indirect potentiometry has been used what in fact is incorrect: in both cases the measurement is performed directly in different samples. When the sample has been diluted highly the total concentration expressed in mmol per liter serum or plasma should be measured at the end. This concentration is comparable with the concentrations measured with for example a flame emission spectrometer. The medium which the ISEs has been offered is relatively constant concerning ionic strength, pH and further composition and thus independent of the sample. This means that also the activity coefficient of the measuring ion and the mass concentration of the

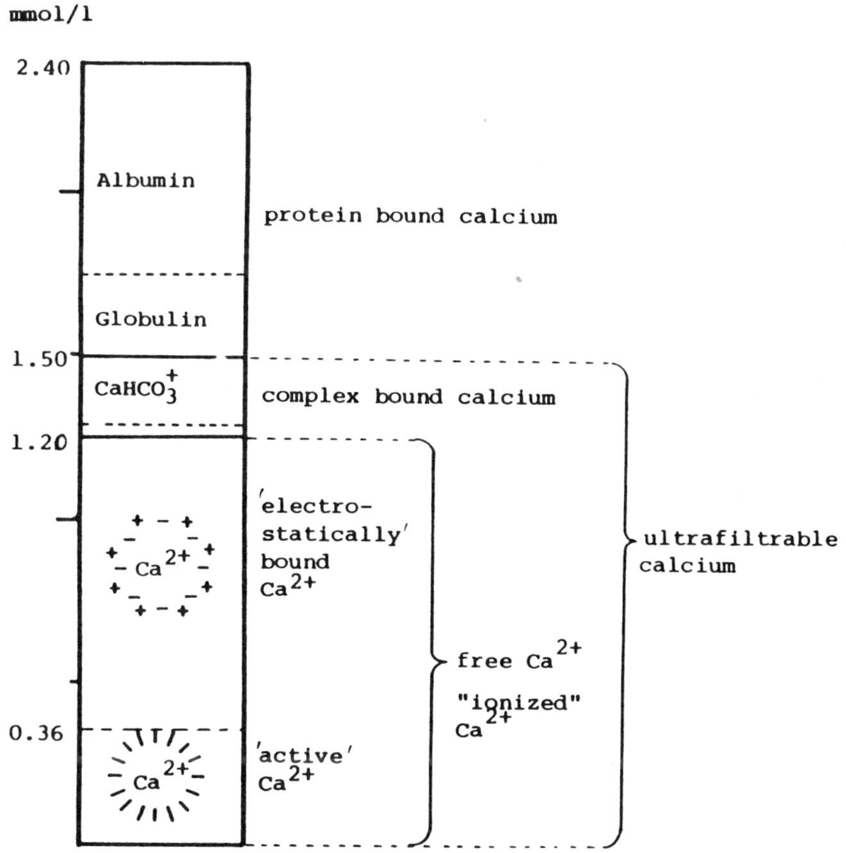

Fig. 3. The calcium fractions of normal blood plasma.

water in the diluted medium will differ hardly from sample to sample (and calibration solution) by which the measured active molality is proportional to the concentration and even in a given case equal to the concentration. Herewith it may be noticed, that some instruments should be calibrated not only with aqueous solutions but at the same time with a serum calibrator to compensate possible matrix effects which may appear in the (insufficiently) diluted sample.

The problems of measurement in undiluted specimens which are arising with the introduction of ISEs for sodium, potassium and also calcium ions in the clinical laboratories are (4):

a. What should be reported?
History has largely determined that ions such as sodium and potassium are usually measured by methods on prediluted specimens using flame atomic emission spectrometry. In principle flame atomic emission spectrometry measured the total substance concentration of the ion and ISE measures the activity of the free ion in the water phase. Hence differences between the results obtained by flame atomic emission spectrometry and ISE-analyzers have been reported by several authors. So we must decide whether we want to report activity or substance concentration in either the total volume of plasma (mol/L) or only in the waterphase of plasma (mol/L) i.e. molality (mol/kg)?

b. Differences in measuring technique.
Different instruments on the market for the measurement of sodium, potassium and ionized calcium by ISE on undiluted specimens give different results indicating a problem of measurement technique i.e. calibration solutions, selectivity of the electrodes, measuring time, liquid junction potential (suspension effect of the red cells) and temperature (6, 7, 8). In order for the routine measurement process to be with meaning it requires standardisation of methods of measurement, reference methods and reference materials to correct for systematic biases inherent in the techniques.

c. Differences in reference values.
It may be clear, that above mentioned problems at the same time lead to differences in reference values, namely for ionized calcium, for which till now no reference method is available and that due to variable preanalytical factors also sample handling should be standardized.

5. WHAT SHOULD BE REPORTED?

The answer on this question depends on:

a. the actual measured quantity.
The electrode responds to changes in the ion-activity and from puristic point of view we should report the measured activities. For normal plasma with substance concentrations of Na^+ and K^+ of 140 mmol/L and 4.5 mmol/L, respectively, as measured by flame atomic-emission spectrometry, we would find relative millimolal ion-activities of about 112 and 3.6, respectively, and for ionized calcium 0.36. This indicates, that we get entirely different values when we report activity instead of concentration and we should need a new set of reference intervals.

b. the physiological relevance.
All thermodynamic equilibria are governed by the activities of the components involved in chemical equilibria and equilibria across

biological membranes are no exception. It is reasonable therefore to suppose that the physiological effects of ions in their natural biological environment relate more to their ionic activity then to their total concentration. This concept has long been accepted with pH which is an expression of hydrogen ion activity.
So from this we should report activity.

6. RECOMMENDATIONS ON THE EXPRESSION OF RESULTS OF SODIUM AND POTASSIUM ISE MEASUREMENTS (9)

For many years, however, results of determinations of sodium and potassium ions in physiological fluids have been expressed in terms of their stoichiometric amount of substance concentration, i.e. the concentration of total sodium or potassium ion (mmol/L). Hence, the use both of substance concentrations of sodium and potassium and of their reference ranges is firmly established in clinical interpretation and practice. Furthermore, it can be envisaged that analytical systems which measure substance concentration, such as flame atomic emission spectrometry, will continue to be used alongside ion-selective electrode determinations in undiluted plasma for the foreseeable future. In consequence, the convention proposed represents a pragmatic compromise which attempts to facilitate the introduction of ion-selective electrode determinations of sodium and potassium ion concentrations in whole blood or undiluted plasma into routine clinical practice while minimising the risk of clinical misinterpretation.

6.1 Convention: reporting results of ion-selective electrode determinations of sodium and potassium

A convention is proposed whereby, for routine clinical purposes:

A. Results of ion-selective electrode determinations of sodium and potassium in whole blood and undiluted plasma should be reported in terms of concentration (mmol/L).

B. Results of measurements on standard normal specimens should conform exactly with those obtained by flame atomic emission spectrometry on the same specimens.

C. Standard plasma specimens are herein defined as having mass concentration of plasma water of 0.93 ± 0.005 kg/L, plasma bicarbonate concentration of 24 ± 2 mmol/L, plasma pH of 7.40 ± 0.05, and concentrations of albumin, total protein, cholesterol and triglycerides within the reference range for healthy subjects.

When this convention is used results reported by ion-selective electrodes in normal plasma are equivalent to the substance concentration of the ion. However, in samples with abnormal plasma water mass concentration or with abnormal concentrations of complexes of the ion, results are numerically different from true substance concentration.

For a better understanding I will discuss the quantities activity and concentration and the relationship between the activity measured by ISE and the substance concentration measured by flame atomic emission spectrometry in more detail.

6.2 The relationship between ion activity and total substance concentration of ion in plasma

The relationship between the activity measured by ISE in undiluted specimens and the total substance concentration measured by flame atomic emission spectrometry may be expressed in a mathematical form as exemplified for Na^+ in the flow chart (Figure 4). The substance concentration of total sodium (c_{tNa}) may be converted to the molality of total sodium (m_{tNa}) by dividing by the mass concentration of water ρ_{H2O}:

$$m_{tNa} = c_{tM}/\rho_{H_2O} \qquad \text{Eq. 8}$$

To obtain the molality of free Na^+ (m_{Na^+}), the sum of the molalities of a number (j = 1, 2, 3, ...) of species (Bj) bound to Na^+ e.g. protein bound (NaPr), carbonate bound ($NaCO_3^-$), bicarbonate bound ($NaHCO_3$) etc. (NaB_j) must be substracted.

$$m_{Na^+} = m_{tNa} - \sum_{j=1}^{n} m_{NaB_j} \qquad \text{Eq. 9}$$

or, introducing the equilibrium constants of these complexes:

$$Na^+ + B_j^- \underset{}{\overset{K_{NaB_j}}{\rightleftharpoons}} NaB_j \qquad \text{Eq. 10}$$

$$\sum_{j=1}^{n} m_{NaB_j} = \sum_{j=1}^{n} K'_{NaB_j} \cdot m_{Na^+} \cdot m_{B_j^-} \qquad \text{Eq. 11}$$

The equation may be written:

$$m_{Na^+} = m_{tNa}/(1 + \sum_{j=1}^{n} K'_{NaB_j} \cdot m_{B_j^-}) \qquad \text{Eq. 12}$$

The molality of free Na^+ may be converted to the active molality $\tilde{m}_{Na}{}^+$ by multiplying by the (molal) activity coefficient ($\gamma_{Na}{}^+$):

$$\tilde{m}_{Na^+} = \gamma_{Na^+} \cdot m_{Na^+} \qquad \text{Eq. 13}$$

The active molality may be converted to the relative (molal) activity by dividing by the unit of molality (mol/kg):

$$a_{Na^+} = \tilde{m}_{Na^+}/(\text{mol/kg}) \qquad \text{Eq. 14}$$

Combining and rearranging Eqs. 8, 12, 13 and 14 result in the relationship:

$$c_{tNa} = a_{Na^+} \cdot \rho_{H_2O} \cdot (1 + \sum_{j=1}^{n} K'_{NaB_j} \cdot m_{B_j^-}) \cdot (\text{mol/kg})/\gamma_{Na^+} \qquad \text{Eq. 15}$$

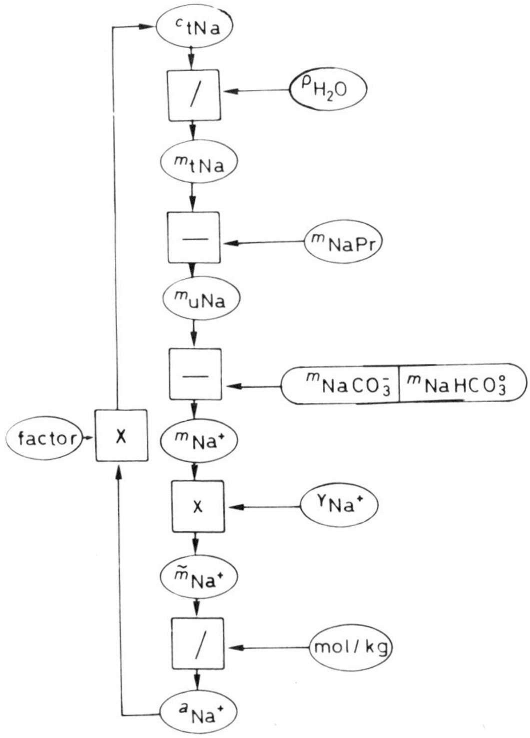

Fig. 4. Relationship between the flame atomic emission spectrometer result c_{tNa} and the ISE result (a_{Na^+}) for sodium measurement in plasma.

with which one can calculate the substance concentration of total Na on the basis of the activity of Na^+ or vice versa.

The factor

$$\rho_{H_2O} \cdot (1 + \sum_{j=1}^{n} K'_{NaB_j} \cdot m_{B_j^-}) \cdot (mol/kg)/\gamma_{Na^+} \qquad \text{Eq. 16}$$

by which the ISE measured a_{Na^+} is related to c_{tNa} may be called the ISE adjustment factor and the result the adjusted activity instead of total substance concentration. For sodium in normal plasma we may apply the following values: mass concentration of water = 0.933 kg/L; bound Na is 2% which means

$$\sum_{j=1}^{n} K'_{NaB_j} \cdot m_{B_j^-} = 0.02; \text{ and } \gamma_{Na^+} = 0.747$$

So we obtain an ISE adjustment factor of 1.27 mol/L (0.933* 1.02/0.747) for converting the measured relative molal activity to the adjusted activity ("c_{tNa}"), which value is close to the total substance concentration of sodium.

6.3 A calculated example

A calculated example of the differences between the substance concentration obtained by flame atomic emission spectrometry and the results obtained with ISE when the ionic strength or the mass concentration of water varies is shown in Table 1. When the appropriate adjustment factor (in this case 1.27 mol/L) is applied, the ISE reading, "c_{tNa}" is the same as the result obtained by flame atomic emission spectrometry for normal as the result obtained by flame atomic emission spectrometry for normal plasma. With the same factor, small discrepancies are observed when the ionic strength of the plasma varies because the activity coefficient of Na^+ is dependent on the ionic strength. However, the discrepancies are very small, even with extreme pathological variations of ionic strength, and the discrepancies do not have any practical consequences. Large differences arise only when the water concentration of the plasma changes, for example, because of severe hyper- or hypoproteinemia or severe hyperlipemia. In the example where the mass concentration of water has decreased by 15% to 0.80 kg/L, the difference between the flame atomic emission spectrometer and the ISE results is 20 mmol/L.

6.4 Standardization

For practical purposes, an accurate activity adjustment factor is not necessary as long as the main goal is achieved; i.e. values for normal plasma are identical with ISE and flame atomic emission spectrometer. This may be achieved in three ways.

a. Normal plasma sample method (single sample).
The ISE system has to be calibrated on a concentration basis with at least two aqueous electrolyte solutions of constant ionic strength (constant activity coefficient) spanning at least 5 mV thus obtaining a nernstian slope. The advantage of achieving a nernstian slope is that it enables us an easy assessment of the relative sensitivity which should generally not deviate more than plus two percent to minus five percent from the theoretical value in the physiological concentration ranges.
The aqueous calibration solutions are then assigned an appropriate value to obtain agreement between the ISE reading and flame atomic emission spectrometer result for normal plasma.

b. Pooled plasma sample method (few spiked samples).
It has been suggested to calibrate the ISE with a standard serum pool and furthermore to obtain the slope with standard serum samples of which the concentration of sodium or potassium has been increased by adding known amounts of sodium chloride or potassium chloride and decreased by partial ultrafiltration.

c. Patient plasma sample method (multi samples).
It has been suggested that the slope should be obtained from a correlation study between ISE and flame atomic emission spectrometry 100-200 sera with electrolyte concentration covering the clinical range and fulfilling the normal conditions. Results are analysed by regression analysis after which the output of the ISE electrode analyser is adjusted giving the same results as the flame atomic emission spectrometry.

With both last procedures the slope deviates from the nernstian slope, but the difference from the first approach (based on the nernstian slope) is without clinical significance as shown in Table 1 where the influence of the variations in ionic strength are shown to be very small. In addition a variation in ionic strength due to variations in sodium concentration causes a change in the liquid junction potential which tends to cancel out the change in activity coefficients. Therefore for practical purposes the calibration solutions which vary in ionic strength due to the large variation in sodium concentrations encountered in clinical practice are acceptable for sodium but for potassium requires the ionic strength not to be far out of the normal range.

Table 1. Calculated examples

Specimen	Flame c_{tNa}, mmol/L	ISE a_{Na^+} x 10^{-3}	"c"$_{tNa}$,* mmol/L	Difference c_{tNa} - "c"$_{tNa}$ mmol/L
Normal plasma $\gamma Na^+ = 0.747$ $\rho = 0.933$ kg/L I = 0.160 mol/kg	140	110.2	140	0.0
Hypernatremia $\gamma Na^+ = 0.737$ $\rho = 0.933$ kg/L I = 0.190 mol/kg	170	132.0	167.7	+2.3
Hyponatremia $\gamma Na^+ = 0.760$ $\rho = 0.933$ kg/L I = 0.130 mol/kg	110	88.1	111.9	-1.9
Hyperlipemia $\gamma Na^+ = 0.747$ $\rho = 0.800$ kg/L I = 0.160 mol/kg	120	110.2	140	-20.0

*Calculated concentrations obtained by multiplying ISE a_{Na^+} values x 1.27.

Physiologically and biochemically, results obtained by ISEs are to be preferred because they more accurately reflect the pathophysiological status of these ions in plasma water and are thus more relevant clinically than those reported by flame atomic emission spectrometry. Actually, therefore, this is a problem of the flame atomic emission spectrometer and not of the ISE. It is recommended that results obtained with severely hyperlipemic sera always be accompanied by an explanatory remark, i.e., that the flame atomic emission spectrometry result is spuriously low ("pseudohyponatremia") but that the ISE result is not affected.

7. RECOMMENDATIONS ON THE REFERENCE METHOD FOR THE DETERMINATION OF IONIZED CALCIUM IN SERUM, PLASMA OR WHOLE BLOOD (10)

7.1 Convention: reference method

By the proposed reference method for the measurement of ionized calcium in serum, plasma or whole blood, the amount of substance concentration of ionized calcium in the water phase of plasma may be reliably determined on the basis of primary reference materials. These are aqueous solutions whose compositions are established by convention to contain known amount of substance concentrations of ionized calcium and which have a constant ionic strength I of 0.160 mol/kg which value is commonly used for normal plasma.

Ionized calcium measurement in a solution may be performed by several methods based on different analytical principles: the biological frog heart method, spectrophotometry with calcium indicators and potentiometry with calcium ion-selective electrodes that respond to changes in thermodynamic activity of calcium ions. For ionized calcium measurement in plasma, potentiometry is at present the method of choice for the reference method.

The proposed IFCC reference method for ionized calcium measurement in plasma is based on the use of cell (a).

Although the electrochemical cell responds to changes in the activity of calcium ions the cell is calibrated in terms of concentration. The cell is calibrated by means of primary calibration solutions of which the composition is chosen such that the activity coefficient of the calcium ions is assumed to be identical to both calibration solutions and normal plasma.

7.2 Convention: reporting results

Ideally ionized calcium measurements should be reported as molality (mol/kg) but in principal could also be reported as substance concentration (mol/L) or as activity. To avoid a proliferation of units, the convention is hereby adopted to report ionized calcium measurements as substance concentration (mol/L). Reporting of results of ISE measurements in terms of substance concentration entails a bias when the ionic strength of the sample differs from that of the calibration solution, since ISEs response to activity rather than substance concentration. It is emphasized that with the present choice of calibration solutions the term ionized calcium concentration refers to ionized calcium in the plasma water and not in the entire volume of plasma. The concentration of ionized calcium in the plasma will be lower by a factor of 0.933.

7.3 Instrumentation and equipment

The sensor system comprises a calcium ion-selective electrode, a reference electrode and a liquid/liquid junction. The whole system is at 37°C. The associated measuring instrument provides the means of measuring the output from the sensor system.

The calcium ion-selective electrode. The calcium ion-selective electrode consists of an assembly in which a membrane, containing a calcium ion electroactive substance, encloses an inner reference solution held within a stem of plastic or glass. An internal reference electrode is immersed in the inner reference solution. As electroactive substance either a charged carrier or a neutral carrier may be used.

The following requirements for the electrode and cell are specified:

1. The relative sensitivity (S) of the Ca^{2+}-ISE should be between 95-102% of the theoretical value.

2. The interference in the physiological range should be less than 0.13 mV (corresponding to 1% error on the measured activity).

3. The response time of the cell with alternating measurements on both extreme calibration solutions should be less than 30s for 95% of the final value presuming that the sample is preheated to 37°C to avoid thermal effects.

4. The stability should be such that the drift of the electrochemical cell does not exceed 1% error in the measured activity.

5. Geometry: The measuring cell and the specimen volume should be optimized such that the carry-over from flush solution to sample is less than 1% at the calcium ion-selectivity electrode in the measuring chamber. The measuring chamber should be gas-tight because of the influence of the pH on the ionized calcium value of the sample.

Table 2. Composition of the primary calibration solutions

Solution No.	$c_{Ca^{2+}}$	$m_{Ca^{2+}}$	m_{Na^+}	m_{Cl^-}
1	1,250 ±0,006	1,266	156,25	158,75
2	0,250 ±0,001	0,253	159,25	159,75
3	2,500 ±0,01	2,526	152,50	157,50

c in mol/L; m in mol/kg; ionic strength 160.0 ± 0.5 mmol/kg. pH should be adjusted between 7 to 8.

Liquid/liquid junction. The following requirements for the liquid/liquid junction are specified:

1. The bridge solution is a potassium chloride saturated at 37°C.

2. Presence of erythrocytes at the boundary: salt bridge solution|sample causes a positive bias of +4% in the measured calcium ion activity when the measurement is performed on normal whole blood. For the reference method the effect of erythrocytes should be eliminated by avoiding the presence of erythrocytes at the liquid junction by interposing a bridge of plasma originating from the blood sample being measured.

3. The geometry of the liquid/liquid junction also influences the liquid junction potential. A stable and reproducible liquid junction is obtained with the free diffusion type of junction when it is arranged in such a way that the higher density salt bridge solution is below the test solution.

Reference electrode. The following requirements for the reference electrode are specified:

1. As internal reference electrode the silver/silver chloride electrode should be used.
2. The mercury/calomel electrode is recommended as external reference electrode.

Temperature control. The whole measuring cell including the liquid/liquid junction should be maintained at the same temperature i.e. 37.0 ± 0.05°C.

The measuring instrument. A high impedance mV meter with properties as specified in the IFCC pH document (11) should be recommended.

7.4 Calibration solutions

Primary reference materials, are issued by the NBS and the user is required to prepare the calibration solutions according to a prescribed procedure.

All calibration solutions have an ionic strength of 160 mol/kg, which actually refers to the ultrafiltered water phase of plasma, a pH between 7 and 8, and have the following composition: (see Table 2)

Solutions 1, 2 and 3, of which Solution 2 and Solution 3 differ a decade, are required to establish the relative sensitivity and should be used to establish secondary calibration solutions.
In principle the primary reference solutions should not contain buffer. For certain purposes HEPES (1 mmol/kg) may be included in each of these solutions to stabilise pH against dissolution of the glass of the containing vessel. This concentration of HEPES shows negligible binding of calcium ions.

7.5 Specimen collection

Whole blood is collected in a glass syringe (5 mL) containing a flat stainless steel plate for mixing. The syringe contains dry calcium titrated sodium heparin sufficient to achieve a final heparin concentration of 50 IU per mL of blood. The contents of the syringe has to be mixed after sampling and immediately before measuring.

If it is certain that the final heparin concentration is below 15 IU/mL blood, the use of calcium titrated heparin is not obligatory.

7.6 Measurement procedure

Concerning the measurement procedure, a typical protocol including the sample sequence of calibration and measurement has been prescribed.

7.7 Analytical variability

The true concentration in the water phase of plasma sample is theoretically given by the equation:

$$c_{Ca^{2+},i} = (\gamma_{Ca^{2+},1}/\gamma_{Ca^{2+},i}) \cdot c_{Ca^{2+}} \cdot 10^{\frac{E_i - E_1}{S \cdot 0,03077}} \qquad \text{Eq. 17}$$

where S equals the relative sensitivity, E_i and E_1 are the measured potentials in sample i and calibration solution 1 with the concentrations of calcium ion $c_{Ca^{2+},i}$ and $c_{Ca^{2+},1}$ and the relative molal activity coefficients $\gamma_{Ca^{2+},i}$ and $\gamma_{Ca^{2+},1}$, respectively.

The inaccuracy is affected by the following.

a. The residual liquid/liquid junction potential included in the difference $E_i - E_1$ introduces an uncertainty in $a_{Ca^{2+}}$. The magnitude of this uncertainty cannot be estimated using the Henderson equation in this system for plasma, for instance because this system does not fulfil the geometrical requirements of the equation and because of the high protein concentration of plasma.

b. The ratio $\gamma_{Ca^{2+},1}/\gamma_{Ca^{2+},i}$, the true value of which cannot be determined; $\gamma_{Ca^{2+},1}$ depends on the ionic strength of the calibration solution 1 and $\gamma_{Ca^{2+},i}$, varies with the ionic strength I in plasma sample i. If results are reported as $a_{Ca^{2+}}$ the variations in I will not affect the accuracy of $a_{Ca^{2+}}$ determination.

The effects of differences in ionic strength between samples and calibration solutions on the residual liquid/liquid junction potential (a) tends to be opposite in sign to their effects in (b).

An imprecision of 0.00-0.02 mmol/L should be obtained with the reference method when measuring the concentration of ionized calcium $c_{Ca}{}^{2+}$ in the biological range.

7.8 The use of the reference method

The reference method should be used to evaluate routine methods for measurement of ionized calcium in plasma. Routine methods are based on a cell similar to that of the reference method and are usually calibrated with secondary calibration solutions. These are aqueous buffer solutions to which values of ionized calcium concentration have been assigned by means of the routine instrument using the primary calibration solutions as reference material. Routine methods often produce highly precise, but not necessarily accurate ionized calcium data because of variations in electrode systems, liquid/liquid junctions, calibration and measurement procedures. In order to monitor the inaccuracy of routine ionized calcium measurements, it is necessary to employ quality control solutions e.g. plasma samples for which the ionized calcium value has been established with the reference method. The ideal solution for quality control is plasma with a known pH and concentration of ionized calcium since this simulates patients specimens. The plasma may be equilibrated in a tonometer with any desired gas mixture before measurement of pH and concentration of ionized calcium measurement by their reference methods (11). However, there are many problems involved in making plasma a stable quality control material for ionized calcium because the instability of pH and the relationship between the concentration of ionized calcium and pH in plasma. For this reason quality control materials simulating plasma with assigned values for pH and the concentration of ionized calcium have been developed.

1) pooled plasma (frozen)
2) aqueous preparation of lyophilized plasma
3) buffered aqueous protein-containing solutions of $CaCl_2$ and NaCl
4) buffered aqueous solutions of $CaCl_2$ and NaCl (and $MgCl_2$).

Each of these materials has its own particular advantage and limitations. For example, many problems in ion-selective electrode measurements arise from the liquid-junction. These errors are often not disclosed by protein-free control solutions. It is expected that further development in this area will occur.

8. SUMMARY

8.1 What should be reported

The activities of Na^+ and K^+ obtained with ISEs in undiluted serum, plasma or whole blood should be multiplied by an appropriate factor - the ISE adjustment factor - to obtain the same values as the substance concentration obtained by flame atomic emission spectrometry. The factor is primarily dependent on the mass concentration of water in normal plasma divided by the relative molal activity coefficient of Na^+ (or K^+) of normal plasma. Large differences arise only when the water concentration of the plasma changes from normal.

Clinically we must consider the result obtained with the ISE to be the relevant value.

In the case of ionized calcium the factor should convert the measured activity to substance concentration of free calcium ions in the plasma water phase and not whole plasma or to the concentration of total calcium in whole plasma. This factor is primarily dependent on the reciprocal molal activity coefficient of calcium ion of normal plasma i.e. at ionic strength I = 0.16 mol/kg.

In practice to obtain the substance concentration of ionized calcium in the sample the measuring cell should be calibrated with calibrators containing known amount of substance concentration of ionized calcium and having an ionic strength I of 0.160 mol/kg.

8.2 Differences in measuring technique

To get among the themselves the same results with sodium or potassium ISE devices the technique of measurement should be standardized. Standardization of these devices however is not necessary as long as the main goal is achieved i.e. values for normal plasma are identical with ISE and flame atomic emission spectrometer. Therefore three methods of normalization of sodium or potassium ISE results with flame atomic emission spectrometer were suggested.

A reference method for the determination of the substance concentration of free calcium ions in serum, plasma or whole blood has been purposed using calibrators with ionic strength I = 0.16 mol/kg by convention.

8.3 Difference in reference values

Recommendations for optimal conditions for the collection and processing of blood specimens measurement both for routine purposes and to establish reference ranges are in preparation (12).

9. TOPICS UNDER INVESTIGATION BY EWGISE

For users of ISEs who wish to express results of ion measurements in terms of activity, values of activity coefficients will be recommended in the near future providing calculated activity values of sodium, potassium and calcium in a number of specified aqueous solutions which are suitable for instrument calibration.

REFERENCES

1. Oesch U, Amman D, Simon W. Ion-selective membrane electrodes for clinical use. Clin Chem 1986; 32: 1448-1459.
2. Siggaard-Andersen O, Durst RA, Maas AHJ. Physico-chemical quantities and units in clinical chemistry with special emphasis on activities and activity coefficients. IUPAC/IFCC recommendation. Pure Appl Chem 1984; 56: 567-594.
3. Waugh WH. Utility of expressing serum sodium per unit of water in assessing hyponatremia. Metabolism 1969; 18: 706-712.
4. Maas AHJ, Siggaard-Andersen O, Weisberg HF, Zijlstra WG. Ion-selective electrodes for sodium and potassium: A new problem of what is measured and what should be reported. Clin Chem 1985; 31: 482-485.
5. Siggaard-Andersen O, Thode J, Wandrup J. The concentration of free calcium ions in the blood plasma, "ionized calcium". In: Siggaard-Andersen O ed. Blood pH, carbon dioxide, oxygen and calcium-ion. Proceedings of the third IFCC-EPPHBG Workshop 1980, Copenhagen 1981: 163-190.

6. Bijster P, Vink KLJ. Direct potentiometric determination of sodium in blood. I. Potentiometric response in simple sodium chloride solutions. Clin Chem 1984; 30: 6-10.
7. Bijster P, Vink KLJ. Direct potentiometric determination of sodium ion in blood. II. Influence of cations. Clin Chem 1984; 30: 865-870.
8. Boink ABTJ, Bijster P, Vink KLJ, Maas AHJ. Direct potentiometric determination of sodium in blood. III. Influence of (bi)carbonate. Clin Chem 1985; 31: 523-526.
9. Boink ABTJ, Buckley BM, Christiansen TF, Coventon AK, et al. Recommendations on the expression of results of ion-selective electrode measurement of sodium and potassium ion activities in undiluted serum, plasma or whole blood in clinical practice. In: Maas AHJ, Buckley BM, Marsoner J, Saris NE, Sprokholt R, eds. Methodology and Clinical Applications of Ion-Selective Electrodes, Graz 1987; Vol 8: 137-150.
10. Boink ABTJ, Buckley BM, Christiansen TF, Coventon AK, et al. Reference method for the determination of the substance concentration of ionized calcium in serum, plasma or whole blood. In: Maas AHJ, Buckley BM, Marsoner H, Saris NE, Sprokholt, eds. Methodology and Clinical Applications of Ion-Selective Electrodes, Graz 1987; Vol 8: 39-62.
11. Maas AHJ, Weisberg HF, Burnett RW, Müller-Plate O, et al. Approved IFCC methods. Reference method (1986) for pH measurement of blood. Clin Chim Acta 1987; 165: 97-109. J Clin Chem Clin Biochem 1987; 25: 281-289.
12. Boink ABTJ, Buckley BM, Christiansen TF, Coverson AK, et al. Recommendations on blood sampling, transport and storage for the determination of the substance concentration of ionized calcium. In: Maas AHJ, Buckley BM, Marsoner H, Saris NE, Sprokholt R, eds. Methodology and Clinical Applications of Ion-Selective Electrodes, Graz, 1987; Vol 8: 81-88.

THE CLINICAL APPLICATION OF IONIZED CALCIUM

George N. Bowers, Jr.

Clinical Chemistry Laboratory, Department of Pathology
Hartford Hospital, Hartford, CT 06115-0729, USA

HISTORICAL DEVELOPMENTS

Some of the important milestones in the development of our ability to rapidly measure ionized calcium (Ca^{2+}) in the whole blood, plasma, and serum to patients are listed in Table 1. The text also gives key references that can provide greater in-depth coverage for those wishing more detailed information.

Although the concept of different fractions of the total calcium (Ca_T) was clearly established in 1911 by the work of Rana and Takahashi (1), the classical studies of McLean & Hastings in the mid-1930s using the frog heart-muscle bioassay method to measure Ca^{2+} established beyond any doubt that Ca^{2+} and not Ca_T was the fraction of primary physiological importance in the calcium metabolism of man (2, 3). McLean & Hastings' fastidious experimental work and the application of the bioassay method of Ca^{2+} to patient studies in association with clinicians provided unequivocal evidence that Ca^{2+} was regulated by the actions of parathyroid hormone to achieve tight Ca^{2+} homeostasis in healthy persons and was abnormal in patients with parathyroid disorders.

KEY CA^{2+}/ISE DEVELOPMENTS

Prior to the early 1970's, despite the publication of many physical and chemical methods for the measurement of Ca^{2+}, no simple and reliable method was available for use in the clinical service laboratory (4). After the publication by Ross (5) in 1967 of an ion-selective electrode (ISE) sensitive mainly to calcium ions, this lack of a practical means by which to quantitate Ca^{2+} began to change rather quickly. Using Ross's new calcium ISE in a properly designed electrochemical cell, it became possible to make a rapid and reproducible potentiometric measurement of Ca^{2+} directly on very small samples of heparinized whole blood and plasma or serum. From 1965-1970 Moore (6) studied the analytical problems associated with the use of the Ross Ca^{2+}/ISE. He also explored the many clinical applications of Ca^{2+} and concluded that Ca^{2+} measurements by ISE were more relevant than Ca_T measurements in medicine.

Table 1. A brief historical overview of ionized calcium in medicine

Year(s)	Worker/Organization	Contribution
1911	Rana and Takahashi	Diffusible and non-diffusible fractions
1934-35	McLean and Hastings	Bioassay showed Ca^{2+} to be the fraction of primary biological importance
1967	Ross	ISE- liquid membrane of dialkylphosphate
1965-70	Moore	Clinical utility Ca^{2+} by ISE shown
1968-78	Orion	First commercial Ca^{2+}/ISE analyzer
1970	Moody et al	PVC matrix for ISE liquid membrane
1972	Ruzicka et al	New organophosphate calcium ISE sensor
1972-75	Simon et al	Neutral carrier calcium ISE sensor
1980	Radiometer ICA-1	Simultaneous Ca^{2+} and pH with Ca^{2+} at 7.4
1978-1980	Siggaard-Andersen et al	Extensive exploration of physicochemical factors governing Ca^{2+} measurements and intensive study of reference intervals
1982-on	EPpH/IFCC-EWGISEs	European and US Working Groups/ meetings with publication of proceedings
1984-on	EPpH/IFCC-EWGISEs Maas et al	Draft document of Ca^{2+} Reference Method
1985-on	Bowers et al	Bias of buffer binding and residual E_j
1987-?	IFCC	Accuracy and A Reference System for Ca^{2+}

Calcium ISE technology is now 20^+ years of age and is relatively mature (7). The one manufacturer of the late 1960's has been joined by at least eight other original equipment manufacturers in Austria, Denmark, England, Finland, Japan, and the USA. The original exchanger soaked nitrocellulose liquid membranes that were assembled into the ISE with considerable difficulty by each user have now been replaced by small exchangeable Ca^{2+} sensors in a PVC matrix that have markedly superior physical and chemical characteristics (8, 9). All presently available commercial Ca^{2+}/ISE analyzers except for the Radiometer ICA-1 (10) employ the neutral carrier ISE which gives a Nernstian response with a very high selectivity for calcium.

Matching these rapid advances in the ISE technology, there has been a steady growth in the numbers of knowledgeable clinical laboratory scientists who have become strong advocates for the use of Ca^{2+} in daily patient care. As a direct consequence of their interactions with the manufacturers, the analytical performance of the second-generation Ca^{2+}/ISE analyzers has improved and there is substantial agreement about the performance specifications expected of new instruments. The numerous publications on instrument evaluations and clinical applications of Ca^{2+} measurements has created a large volume of current information, especially as a result of the proceedings of special ISE meetings sponsored by the Expert Panel on pH & Blood Gases of the International Federation for

Table 2. Ca^{2+} versus Ca_T (Hartford Hospital)

Year	Ca^{2+}	Ca_T
1980	145 dialyzable	28,295
1981	172 dialyzable	28,742
1982	9,800 ionized	22,601
1983	22,752 ionized	14,997
1984	28,218 ionized	11,666
1985	31,640 ionized	7,825
1986	30,474 ionized	4,272
1987	(33,000) ionized	(4,000)

Clinical Chemistry, the European Working Group of ISEs, and various American organizations (11-18).

EXPERIENCE WITH CA^{2+} AT HARTFORD HOSPITAL

Our goal for Ca^{2+} within the chemistry laboratory was to solve the technical problems of providing a reliable uninterrupted service so that a practicing physician might fully evaluate the role of Ca^{2+} in daily patient care. To date, we have examined over a dozen different prototype and production models of Ca^{2+}/ISE analyzers produced by six different manufacturers. In contrast to our experience in the early 1970's (19), we have had only minor operational problems with todays second-generation Ca^{2+}/ISE devices (7). Since no two analyzers give exactly the same numerical value on patient specimens, we have chosen to make the output of all instruments identical to the Radiometer ICA-1 output simply because our reference intervals were determined and published using values corrected to pH 7.4 on that device (7, 20, 21). As shown in Table 2 by the volume growth of Ca^{2+} at Hartford Hospital since its introduction to replace dialyzable calcium in late 1981, our physicians quickly developed a preference for Ca^{2+} over Ca_T.

I am always surprised to note that each year about 40 ± 10% of all Ca^{2+} results on patients are beyond our stated reference intervals! 30% of our Ca^{2+} volume is from ambulatory outpatients. For the outpatients who are free of renal disorders only 5 ± 2% are abnormal. However, the majority of our outpatients obtain repeated determinations at monthly intervals to follow their course of either chronic renal dialysis (50% of these patients have an abnormal low value) or after a kidney transplant (45% of these patients have an abnormal high value) as listed below in Table 3. The importance of concurrent PTH assays in both of these groups of renal disease is well established (22) and the role of deficient vitamin $1,25\text{-}(OH)_2D_3$ production in the hypocalcemia of renal diseases has been reemphasized by new findings on the regulation of PTH synthesis by the number and function of $1,25\text{-}(OH)_2D_3$ receptors in the parathyroid gland (23).

40% of inpatients also show abnormal values but removal of repeat determinations reduces the number of patients with a disturbance in calcium metabolism to nearer 15 ± 5%. Markedly elevated Ca^{2+} in association with malignancy and hyperparathyroidism occurs frequently in hospitalized patients (24, 25). These elevations are best differentiated by use of a

Table 3. Ionized calcium testing on outpatients (May 1-14, 1987)

Group (No.)	Actual Ca^{2+} (% Lo/Hi)	pH	$Ca^{2+}/7.4$ (% Lo/Hi)	% Abnormal tests Change/No change
Control (121)	1.26 (3/1)	7.42	1.27 (1/4)	7 2/5
Dialysis (133)	1.17 (49/1)	7.46	1.19 (40/4)	54 24/30
Renal Trans (162)	1.33 (2/45)	7.38	1.32 (2/35)	50 16/34

Low = 1.15 mmol/L or less
High = 1.34 mmol/L or more

concurrent PTH assay. In common with others (26), we note that Ca^{2+} is elevated twice as often as Ca_T in patients with neoplastic disease (27).

Hypocalcemia by Ca^{2+} measurements has multiple causes as listed below in Table 4. A recently review of hypocalcemia in critically ill patients by Zaloga and Chernow (28) is recommended to the reader for its discussion of the differential diagnosis and treatment of the associated disease or deficiency states. Over the years, we have noted that the number of patients with low Ca^{2+} is about one-half those with low Ca_T since the hypoalbuminemia which directly lowers Ca_T has little or no effect on Ca^{2+} measurement. Furthermore, Ca_T measurements do not reflect the low Ca^{2+} due to chelation by citrate which is infused with repeated use of blood products. This cause of hypocalcemia is a common finding in the recovery room and surgical intensive care wards in our tertiary care hospital in parallel to the reports of others (28-30). As an extreme example, a transplant recipient of two livers who was just been discharged from our hospital after a four month stay had been treated with about 1000 separate units of blood components and many of these units contained citrate! This patient had only mild transient hypocalcemia during her entire prolonged course because Ca^{2+} was continuously monitored and any calcium deficit was promptly replaced. However, in contrast, one can readily find a number of severely ill patients on any acute care ward who have had a course of prolonged severe ionized hypocalcemia which often remains untreated.

BIAS AND A REFERENCE SYSTEM FOR IONIZED CALCIUM

All commercially available Ca^{2+}/ISE analyzers have been intercompared and it is clear from the data on both reference intervals and patient specimens that there are systematic differences in results (4, 7, 13, 19, 31-37). Each instrument's unique biases are due to arbitrary choices about calibrators, buffers, design of the liquid-liquid junction and the bridge solution used in the reference electrode, and the scheme for data reduction and output that each manufacturer has chosen in the complete absence of any agreements on standardization. Fortunately, the underlying electro-analytical reasons for these biases are been studied in several laboratories (38-40) and may soon provide the scientific information on which to develop a rational system that can unequivocally define the accuracy of Ca^{2+} measurements within the international clinical laboratory community (41-43).

Table 4. Hypocalcemia as measured with ionized calcium

1. Chelators- albumin, bicarbonate, citrate, EDTA FFA, heparin, lactate, phosphate
2. Hypoparathyroidism - 1° and 2°
3. Vitamin D and magnesium deficiency
4. Pancreatitis, sepsis, and fat emboli
5. Neoplasia and severe chronic illness

Our studies at Hartford show that all presently available Ca^{2+}/ISE analyzers have uncompensated residual liquid-liquid junction potentials (dE_j) in respect to serum that make patient results 5 ± 3% low (these estimates are based on calculation of E_j for serum and the different calibrators by use of the Henderson equation and involve assumptions such as protein having no effect on the E_j of serum). Another source of bias that is now recognized in calibration solutions is the binding of free ionized calcium to the buffers that are used to hold the pH at 7.40 and thus lowering the effective $[Ca^{2+}]$. The calcium binding to MOPS buffer at pH 7.40 at 37°C is a constant which has been experimentally determined in our and two other laboratories to be 3 ± 1.5% (39, 40). We now have evidence that a reference solution of 85 mmol/L MOPS at pH 7.4 with an ionic strength of 160 mmol/L will match the E_j of serum. The addition of 3% more calcium to a total of 1.288 mmol/L provides a reference solution with 1.25 mmol/L of $[Ca^{2+}]$ and no residual E_j to serum. The numerical output of four instruments (Radiometer ICA-1, Nova 8, AVL 984, CC 634) which were systematically different with their original calibrators have been pulled together by use of this common reference solution. On our daily QC serum Pool #32, these new outputs are 2 to 8% higher with the new reference solution than the outputs found with the original calibrators as shown in Table 5.

It is expected that agreements about a reference method and primary reference solutions will soon be reached and be widely accepted. This international reference system will clearly define the accuracy of Ca^{2+} measurements. Calibrated serum certified as to its mmol/L $[Ca^{2+}]$ value to the reference system can then be used in the development and standardization of past, present, and future Ca^{2+}/ISE analyzers. When systematic bias is reduced before the analyzer ever comes to the marketplace and each service laboratory has the ability to more closely approach the "defined true-value", the utility of Ca^{+} in medicine that

Table 5. Effect of the new reference solution

Analyzer	Original Calibrators			Serum pool #32		
	old mmol/L	new mmol/L	old bias %	old mmol/L	new mmol/L	old bias %
Nova 8*	1.25	1.15	-8.0	1.19	1.29	-7.8
Rad ICA-1	1.25	1.18	-5.6	1.22	1.29	-5.4
AVL 984	1.25	1.23	-1.6	1.26	1.28	-2.3

*CC 634 gave similar results as Nova 8.

McLean and Hastings first described a half century ago will indeed be fully realized in daily patient care.

REFERENCES

1. Rona P, Takahashi D. Über das Verhalten des Calciums in Serum und über der Blutkörperchen an Calcium. Biochem Z 1911; 31: 336-344.
2. McLean FC, Hastings AB. A biological method for the estimation of calcium ion concentration. J Biol Chem 1934; 107: 337-350.
3. McLean FC, Hastings AB. Clinical estimation and significance of calcium-ion concentrations in the blood. Am J Med Sci 1935; 189: 601-613.
4. Robertson WC, Marshall RW. Ionized calcium in body fluids. Critical Reviews in Clinical Laboratory Sciences 1981; 15: 85-125.
5. Ross JW Jr. Calcium selective electrode with liquid ion-exchanger. Science 1967; 156: 1378.
6. Moore EW. Ionized Calcium in Normal Serum, Ultrafiltrates, and Whole Blood Determined by Ion-Exchange Electrodes. J Clin Invest 1970; 49: 318-334.
7. Bowers GN Jr., Brassard C, Sena SF. Opinion Article: The measurement of ionized calcium in serum with ion-selective electrodes. A mature technology that can meet the daily service needs. Clin Chem 1986; 32: 1437-1447.
8. Moody GJ, Oke RB, Thomas JDR. A calcium-sensitive electrode based on a liquid ion-exchanger in a poly(vinyl chloride) matrix. Analyst 1970; 95: 910-912.
9. Oesch U, Ammann D, Simon W. Ion-Selective Membrane Electrodes for Clinical Use. Clin Chem 1986; 32: 1448-1459.
10. Ruzicka J, Hansen EH, Tjell JC. Selectrode - the universal ion-selective electrode. Part IV. The calcium(II) Selectrode employing a new ion-exchanger in a nonporous membrane and solid state reference system. Anal Chim Acta 1973; 67: 155-178.
11. Siggaard-Andersen O, Thode J, Wandrup J. The concentration of free calcium ions in the blood plasma: "ionized calcium". In: Siggaard-Andersen O, ed. Blood pH, carbon dioxide, oxygen and calcium ion. Proceedings of the 1980 Meeting of the IFCC Expert Panel on pH and Blood Gases. Copenhagen: Private Press, 1981: 163-190.
12. Ionized Calcium Workshop No. 1. (Sponsored by Radiometer A/S and held on May 16-18, 1982 in Copenhagen, Denmark). Engel K, Pedersen KO, Nielsen SP, Siggaard-Andersen O, eds. Scand J Clin Lab Invest 1983; 43-Suppl 165: 1-126.
13. Graham G, Burritt MF. Preliminary Report, AACC Ionized Working Group on Reference Values. Clin Chem 1983; 29: 1187.
14. Proceedings of the Workshop on Direct Potentiometric Measurements in Blood. Ladenson J, Martin R, Koch W, eds. Sponsored by the National Committee Clinical Laboratory Standards and the National Bureau of Standards on May 12-13, 1983 at Gaithersburg, MD. National Committee Clinical Laboratory Standards, Villanova, PA 19085, 1985.
15. Ionized calcium, sodium, and potassium by ion selective electrodes. Vol 5: Proceedings of the 1983 Meeting of the European Working Group on Ion Selective Electrodes on June 13-17 at Oslo, Norway. Maas AHJ, Kofstad J, Siggaard-Andersen O, Kokholm G, eds. Copenhagen: Private Press, 1984.
16. Proceedings of the Meeting of the Expert Panel on pH and Blood Gases (10th EPpHBG Meeting) and European Working Group on Ion Selective Electrodes (3rd EWGISE Meeting). Broughton PMG, Maas AHJ, eds. Held on June 24-27, 1984 at Wolfson College, Oxford, England. Copenhagen: Private Press, 1985.

17. Proceedings of the Meeting of the Expert Panel on pH and Blood Gases (11th EPpHBG Meeting) and European Working Group on Ion Selective Electrodes (4th EWGISE Meeting). Maas AHJ, ed. Held on June 25-28, 1985 at Helsinki, Finland. Copenhagen: Private Press, 1986.
18. Proceedings of the International Symposium on "Ionized Calcium: Its Determination and Clinical Usefulness" (Organized by Graham GA and Burritt MF for the Electrolyte/Blood Gas Division of the American Association for Clinical Chemistry). Moran RF, ed. Held on February 27-28, 1986 at Galveston, TX. Galveston: Univ. Texas, 1986.
19. Ladenson JH, Bowers GN Jr. Free calcium in serum. I. Determination with ion-specific electrodes and factors affecting results. Clin Chem 1973; 19: 565-574.
20. Fogh-Andersen N, Christiansen TF, Komarmy L, Siggaard-Andersen O. Measurement of free calcium ion in capillary blood and serum. Clin Chem 1978; 24: 1545-1552.
21. Fogh-Andersen N. Ionized calcium analyzer with a built-in pH correction. Clin Chem 1981; 27: 1264-1267.
22. Arnaud CD. Hyperparathyroidism and renal failure. Kidney Int 1973; 4: 89-95.
23. Korkor AK. Reduced binding of [^{3}H]1,25-dihydroxyvitamin D_3 in the parathyroid glands of patients with renal disease. NEJM 1987; 316: 1573-1577.
24. Ladenson JH, Bowers GN Jr. Free calcium in serum. II. Rigor of Homeostatic Control, Correlations with Total Serum Calcium and Review of Data on Patients with Disturbed Calcium Metabolism. Clin Chem 1973; 19: 575-582.
25. Ladenson JH, Lewis JW, McDonald JM, Slatopolsky E, Boyd JC. Relationship of free and total calcium in hypercalcemic conditions. J Clin Endocrin Metab 1979; 48: 393.
26. Lindemann K, Mueller-Plathe O. Ionized calcium for the detection and treatment of osteolytic processes. Op. cit. ref. (15): 63-70.
27. Shainfarber LH. Hypercalcemia and malignancy. Research paper submitted as a Hartford Hospital Summer Student on Aug 12, 1983 (available on request of Bowers GN Jr).
28. Zaloga GP, Chernow B. Hypocalcemia in critical illness. JAMA 1986; 256: 1924-1929.
29. Wu AH, Smith SCH. Ionized calcium in acute care medicine: Focus on liver transplantation. Op. cit. ref. (18).
30. Drop LJ, Laver MB. Low plasma ionized calcium and responses to calcium therapy in critically ill man. Anesthesiol 1975; 43: 300.
31. Larsson L, Ohman S. Higher values for ionized calcium with a new type of electrode for Orion SS-20. Clin Chem 1978; 24: 2070.
32. Larsson L, Ohman S. Nova 2 ionized calcium analyzer compared with the Orion SS-20. Clin Chem 1980; 26: 1761-1763.
33. Johnson L, Graham G, Brueningsen DA. Ionized calcium: Clinical significance and application. Am Clin Prod Rev 1985; July: 24-34.
34. Boink ABTJ, Gimpel JA, Maas AHJ. A comparison of four Ca^{2+} analyzers. ibid: 17-19.
35. Toffaletti J. Aqueous and serum-based materials compared for use as simulated calibrators for three ionized calcium analyzers. Op. cit., ref (21): 103-110.
36. Boink ABTJ, Sprokholt R. Quality control of the measurements of ionized calcium with ISE's. Op. cit., ref (21): 111-122.
37. Sachs C. Reference intervals for ionized calcium in blood. Op. cit., ref (21): 131-135.
38. Crowell JA, Bowers GN Jr. Apparent binding by buffers of ionized calcium. Clin Chem 1985; 31: 267-270.
39. Covington AK, Kataky R. Complexation of the Good buffers, HEPES & MOPS, by calcium ions. Op. cit., ref (21): 128-130.

40. Kroneis H, Ritter CHR, Marsoner HJ. Simultaneous calibration of calcium ion-selective electrodes together with sodium, potassium and pH. Paper submitted for the proceedings of the EWGISE 1986 meeting in Graz.
41. Bowers GN Jr. A reference system for ionized calcium. Op. cit. ref (12): 49-56.
42. Maas AHJ. Reference method for the determination of ionized calcium in serum, plasma or whole blood. Op. cit. ref (18): 53-79.
43. Bowers GN Jr. A need for a reference system for ionized calcium. Op. cit. ref. (18): 48-50.

SODIUM AND POTASSIUM BY ION SELECTIVE ELECTRODES

CLINICAL INTERPRETATION

O. Siggaard-Andersen, N. Fogh-Andersen, and P.D. Wimberley

Univ. of Copenhagen, Herlev Hospital, DK-2730 Herlev, Denmark

Measurements of sodium and potassium are among the most frequent analyses in the clinical laboratory. In our 700 bed hospital sodium and potassium analyses account for 18% of all analyses and 28% of these are "stat" analyses. 90% are analyses of serum (or plasma or whole blood), 9% urine, and 1% other body fluids.

FLAME PHOTOMETRY VERSUS ION SELECTIVE ELECTRODES:

For many years flame photometry was the method of choice. With the IL^R flame photometer we achieve a very high precision with a day-to-day coefficient of variation of 0.7% for Na^+ and 1.6% for K^+. During the past decade ion-selective electrodes have taken over with slightly less precision: with the SMAC-I^R we get day-to-day coefficients of variation of 1.2% for Na^+ (glass electrode) and 2.0% for K^+ (valinomycin). This instrument is based on indirect potentiometry (predilution of the sample). With the KNA-1^R measuring directly in the undiluted specimen we obtain a similar precision, which is fully adequate for clinical purposes.

The advantages of ISE's for sodium and potassium are the avoidance of gases and flames, the whole blood capability, and the simplicity of design. It is a mixed blessing that they measure activity, not stoichiometric concentration (concentration of total ion) which we are accustomed to from the flame photometer. The cell potential (E) is related to the activity of the ion (a) as follows:

$$\lg a = \lg a_1 + (E - E_1 - \Delta_1 E_j) * (\lg(a_2/a_1)) / (E_2 - E_1 - \Delta_1{}^2 E_j) \qquad (1)$$

where subscripts 1 and 2 refer to two calibration solutions with known activity of the ion (established by convention), and E_j is the residual liquid junction potential (which should be minimized by proper selection of the composition of calibration and bridge solutions). The relationship between the activity of the ion (e.g. a_{Na+}) and the stoichiometric concentration (c_{Na}) is given by:

$$a_{Na} = c_{Na} * (f_{Na+} * \rho_{H2O}{}^{-1} * \gamma_{Na+}) \qquad (2)$$

f_{Na+} is the fraction of free unbound Na^+. For normal plasma the value is about 0.98 - 0.99 and the same (or slightly lower) applies to f_{K+}. About 1% of Na^+ and K^+ are bound to HCO_3^- and protein binding may account for another 1%. ρ_{H2O} is the mass concentration of water in the sample, 0.93 kg/L for normal plasma. The value is primarily dependent on the protein and lipid concentration and with a very lipemic specimen the value may fall to about 0.8 kg/L. α_{Na+} is the activity coefficient, 0.75 for normal plasma (0.74 for K^+). The value is primarily dependent on the sodium concentration which determines the ionic strength. With a rise in the sodium concentration from 110 to 170 mmol/L, α_{Na+} falls modestly (0.760 to 0.737). We may therefore conclude that the conversion factor (the value of the parenthesis in Eqn. 2) is relatively constant, varying significantly only when ρ_{H2O} varies due to massive alterations in protein or lipid concentration.

From a biological point of view the activity is the quantity of interest because chemical equilibria, receptor binding, diffusion, etc. are dependent on the activity, not the stoichiometric concentration. The latter is relevant only in connection with balance studies, where the total input and output of the ions are to be measured.

From a practical clinical point of view the stoichiometric concentration is the quantity of interest because this is the familiar quantity and because the stoichiometric concentrations of Na and K sufficiently accurately reflect the activities. The only situation where this is not the case is when ρ_{H2O} is significantly decreased by hyperproteinemia or hyperlipemia. This is clinically well known and designated "pseudo-hyponatremia" or spurious hyponatremia. This depression of the sodium concentration by the volume effects of proteins and lipids also affects potassium and all other serum concentrations but it is generally only mentioned in medical textbooks in connection with sodium and chloride.

The problem with ISE's for sodium and potassium therefore is to convert the measured cell potentials to the stoichiometric concentration. This is based on a linear relationship between the cell potential and the logarithm of the substance concentration. This presents no problem with indirect potentiometry where the sample is diluted to minimize variations in ion binding, water concentration, and ionic strength so that the conversion factor (parenthesis of Eqn. 2) becomes identical for samples and calibration solutions and residual liquid junction potentials are minimized. With direct potentiometry, where the sample is undiluted, the conversion factor of Eqn. 2 and the residual liquid junction potentials cannot be identical for all samples. A small bias depending upon the composition of the sample is therefore unavoidable. However, the only samples where the bias is clinically important are samples with severe hyperproteinemia or hyperlipemia. With such samples the flame photometer may for example show 120 mmol/L for Na with an ISE value of 140. In other words, the ISE does not respond to "pseudo-hyponatremia". Paradoxically, from a clinical point of view it is the flame photometer value which is biased, not the ISE value. In any case, results obtained with severely hyperproteinemic or hyperlipemic specimens, whether measured by flame photometry or by direct or indirect ISE should be accompanied by an explanatory remark to avoid misinterpretation of the results.

CLINICAL INTERPRETATION OF THE SERUM SODIUM AND POTASSIUM CONCENTRATIONS

Electrolyte disturbances involving Na^+ and K^+ are common in a hospital population and the correct clinical interpretation of the serum sodium and potassium results is often complicated. We have constructed a "double chart" (Fig. 1) which illustrates this complexity and may be an aid in the correct classification of salt-and-water disturbances. The chart was first shown at an IFCC workshop on methodology and clinical applications of ion selective electrodes, Helsinki 1985 (ISBN 87-88138-10-0).

The serum sodium concentration (extracellular sodium concentration) is shown on a logarithmic scale on the ordinate of the left half of the double chart. The normal reference interval is indicated by dotted lines. Values above the reference interval indicate hypernatremia or extracellular hypertonicity. Values below indicate hyponatremia or extracellular hypotonicity. The extracellular sodium concentration (c_{Na}) is defined as the amount of extracellular sodium (n_{Na}) divided by the extracellular volume (V):

$$c_{Na} = n_{Na}/V, \tag{3}$$

hence changes in the extracellular sodium concentration may be due to changes in n_{Na} (extracellular sodium excess or deficit) or to a change in V (excess or deficit of extracellular water).

The relative extracellular water excess or deficit is shown on a logarithmic scale on the abscissa of the left half of the double chart. The normal reference interval is taken to be ± 5%. The scale is drawn from right to left, i.e. a point to the left indicates overhydration (oedema), a point to the right dehydration. For practical clinical purposes the extracellular water excess or deficit is estimated clinically (oedema, skin turgor, moisture of mucous membranes, venous filling, pulse, blood pressure, diuresis, change in body mass). The plasma protein and albumin concentrations and the hematocrit may also be employed as indicators of the plasma volume. However, in most clinical situations, for example extracellular oedema, the relative change in plasma volume is less than the relative change in total extracellular volume. If a is the relative increase in plasma protein or albumin concentration then the relative increase in plasma volume is

$$b = -a/(1+a). \tag{4}$$

If g is the relative increase in hematocrit then the relative increase in plasma volume is

$$b = -g/((1+g)*(1-h^\circ)), \tag{5}$$

where h° is the normal hematocrit for the patient. This shows that the hematocrit is about half as sensitive as an indicator of hemodilution or hemoconcentration than the plasma protein or albumin concentration. Estimation of the relative extracellular water excess or deficit is generally very imprecise, yet it is very important in order to classify the water-electrolyte disturbance correctly.

The relative extracellular sodium excess or deficit is indicated on the vertical axis in the middle of the chart. Divisions on the scale are slanting with a 45° slope to emphasize that constant amount of extracellular sodium is represented by straight lines with a 45° slope. This is achieved by using logarithmic scales on the abscissa and ordinate.

A given point in the chart represents the extracellular salt-water status of a given patient. The ordinate of the point is the serum sodium concentration, the abscissa is the relative extracellular water excess or deficit, and a 45° projection to the scale in the middle gives the relative extracellular sodium excess or deficit. The position of the point in relation to the various areas of the chart classifies the disturbance.

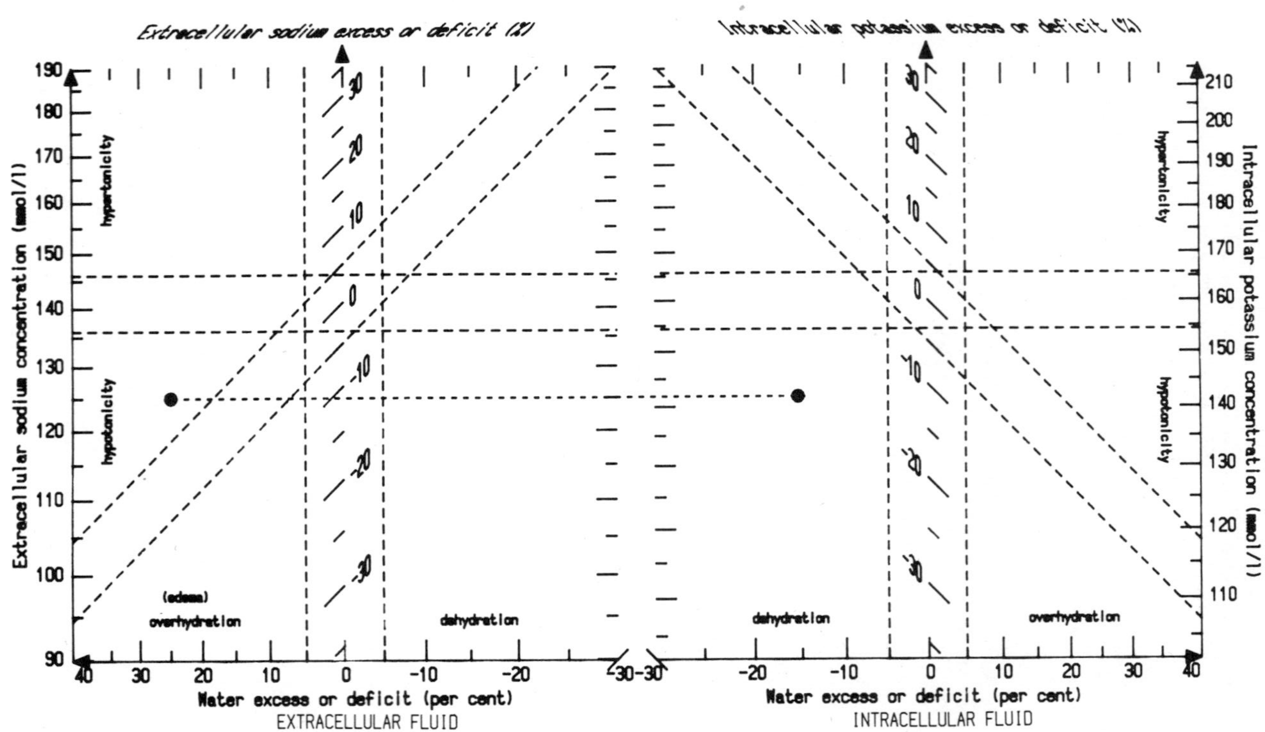

Fig. 1. Sodium-potassium chart for classification of salt-water disturbances.

There are 12 major pathological classes radiating from the oval normal area in the middle. For example area 2 (at 2 o'clock) indicates pure water deficit; area 3 indicates isotonic dehydration; area 6 is hypotonic sodium excess.

A detailed classification of the extracellular salt-water status is in itself of limited value because each extracellular disturbance may be combined with a variety of intracellular salt-water disturbances. The situation is entirely different when a hypotonic overhydration due to pure extracellular water excess is combined with an intracellular overhydration versus an intracellular dehydration. However, one thing is known: the intracellular tonicity (osmolality) is the same as the extracellular tonicity because, with few exceptions, the cell membranes are permeable to water. Osmolality ($\hat{m}$) is defined as:

$$\hat{m} = (-\ln a_{H2O})/M_{H2O}, \tag{6}$$

where a_{H2O} is the activity of water and M_{H2O} is the molar mass of water (0.018 kg/mol). If the water activities are equal in the two phases then the osmolalities (tonicities) must also be equal. If the tonicity was different water would move from the lower to the higher tonicity until the tonicity finally would be the same in the two compartments. The intracellular tonicity is mainly governed by the intracellular potassium concentration. Therefore, measuring the extracellular sodium concentration gives a good estimate of the average intracellular potassium concentration. The right half of the double chart illustrates the intracellular salt-water status.

The average intracellular potassium concentration (as indicator of the intracellular tonicity) is shown on a logarithmic scale on the ordinate to the right. The average intracellular potassium concentration is slightly higher (ca. 15%) than the extracellular sodium concentration because there are fewer small intracellular anions to contribute to the tonicity. The reference interval is taken to be 153 to 167 mmol/L.

The relative intracellular water excess or deficit is shown on the abscissa of the right half of the double chart. The scale is drawn from left to right, i.e. a point to the left indicates intracellular dehydration, a point to the right intracellular overhydration. For practical clinical purposes estimation of the intracellular hydration is very difficult (evaluation of the turgor of the thigh muscles has been suggested) and it is usually necessary to take anamnestic information into consideration (for example water intake, or water losses through the skin, gastrointestinal tract, or kidneys). Often it may be easier to estimate the following oblique coordinate.

The relative excess or deficit of intracellular potassium is shown on the vertical scale in the middle. Divisions on the scale are slanting with slope -45° to emphasize that constant amount of intracellular potassium is represented by a straight line with a slope of -45°. For practical clinical purposes the plasma potassium concentration is used as an indicator of a potassium depletion or excess. As a very rough rule of thumb the relative change in plasma potassium concentration may be assumed to equal the relative change in amount of intracellular potassium. However, it is necessary to take into account that changes in the plasma potassium concentration may be due to a redistribution of K^+ between cells and extracellular fluid, for example with acute pH changes $c_K(P)$ changes as $\Delta c_K(P) = -6 * \Delta pH$. Often anamnestic information must be taken into consideration, e.g. prolonged treatment with diuretics with or without potassium supplementation. In any case, the clinical estimation of a potassium excess or deficit is often better than the estimate of relative intracellular water excess or deficit, and, together with the tonicity (the ordinate) it provides the coordinates for the point representing the intracellular salt-water status.

Figure 1 illustrates a complicated case with hyponatremia (125 mmol/L) and hypokalemia (3.0 mmol/L). The patient had an incompensated cardiac failure with oedema and the extracellular overhydration was estimated to be +25%. According to the chart, the extracellular sodium excess is then +10%. The intracellular tonicity must be as low as the extracellular and from the chart we read that the average intracellular potassium concentration is decreased to 142 mmol/L (normally 160). A plasma potassium concentration of 3.0 mmol/L indicates a -25% potassium depletion according to our rule of thumb. This allows us to plot the intracellular point, and reading the abscissa to the point, we find an intracellular dehydration with a water deficit of -17%. This illustrates a complex situation with hyponatremia in spite of a sodium retention and with extracellular oedema but intracellular dehydration.

The chart is primarily a teaching aid, illustrating the various types of water and electrolyte disturbances. For example, it allows a calculation of the effects of a 30% pure K^+ (phosphate) depletion, versus a pure 30% Na^+ (Cl^-) depletion. With a 30% K+ depletion the intracellular point would tend to slide down vertically to the oblique -30 mark. With zero excess or deficit of intracellular water the intracellular potassium concentration would fall to 112 mmol/L. Intracellular tonicity would then be lower than extracellular tonicity and water would pass from the intracellular to the extracellular space. The intracellular point would slide upwards along the oblique -30% potassium excess line while the extracellular point would slide downwards from the normal area along the oblique 0% sodium excess line. Equilibrium would be obtained when the intracellular dehydration is about -12.5% and the extracellular overhydration is about +25%, which would represent equal volumes. The intracellular potassium concentration would then be about 127 mmol/L and the extracellular sodium concentration about 113 mmol/L. A pure 30% Na^+ depletion would similarly result in a situation with an extracellular dehydration of -22%, an intracellular overhydration of +11%, extracellular sodium concentration of about 126 mmol/L, and intracellular potassium concentration of about 143 mmol/L. It is worth noticing that the hyponatremia is more pronounced with a pure 30% K^+ depletion than with a pure 30% Na^+ depletion, at first a surprising result, which illustrates the complexity of the clinical interpretation of the plasma sodium concentration.

In summary, ISE's have provided a valuable tool suitable for mass production of Na-K results for serum and urine as well as rapid bed-side measurements on whole blood. A minor problem is the positive bias encountered in cases of severe hyperproteinemia or hyperlipidemia when Na^+ and K^+ are measured by direct ISE, which does not detect the pseudo-hyponatremia (or pseudo-hypokalemia) measured by flame photometry or indirect potentiometry. However, from a clinical point of view the result of the flame photometer is more likely to be misinterpreted than the result of direct ISE measurement. The most difficult problem in the interpretation of serum sodium and potassium values is unrelated to the use of ISE's: it is the difficulty of estimating the extracellular and intracellular water status, which is essential for a complete classification of sodium and potassium disturbances. A new "double chart" representing the extracellular and the intracellular salt-water status may be an aid in such classification.

CHAPTER 9
BEDSIDE CHEMISTRY AND SELFTESTING

CLINICAL BIOCHEMISTRY - AN INTEGRATED SERVICE

Vincent Marks

Biochemistry Department, University of Surrey, Guildford, Surrey

Clinical biochemistry as we understand it today had its origins in the testing of urine for the presence of protein as a means of distinguishing dropsy due to kidney disease, which would not respond to digitalis therapy, from that due to heart disease which would. Other urine tests, all of which could be performed with the simplest of reagents and minimum of facilities were gradually introduced into clinical practice. They were generally performed in the side room of a hospital ward or in the patient's own home and used mainly as an aid to diagnosis.

HISTORICAL DEVELOPMENTS

The growth of chemical pathology, or pathological chemistry, during the latter part of the nineteenth century and its distinction from physiological chemistry - or biochemistry - led to the establishment, within the medical schools, of chemical laboratories whose primary purpose was the creation and pursuit of knowledge relevant to the origin and nature of disease. With the passage of time the teaching hospital laboratories, in addition to carrying out research and teaching, acquired clinical or service commitments.

With the introduction of insulin therapy in 1922 the measurement of blood glucose became de rigeur and the treatment of diabetes without access to a clinical laboratory became unthinkable.

A number of other substances which could be measured in blood were characterised; some were seemingly related specifically to certain diseases and could, therefore, be used as an aid to diagnosis or monitor progress. The number of such tests grew comparatively slowly and, by 1950, only some 30 substances were generally held to be useful and available in the comparatively few hospital laboratories that existed at that time (1).

In Britain, as in most other countries, clinical laboratories grew up under the auspices of the hospital morbid anatomy/histology departments as they alone had the space, organisation and manpower to carry out the still technically demanding and cumbersome analytical procedures necessary to produce clinically valuable data (2).

In some teaching hospitals clinical biochemistry departments developed within academic departments of medicine (and surgery) and only much later, or not at all, became associated with the other laboratory disciplines now commonly grouped together as clinical pathology.

From their very origins the clinical biochemistry laboratories were concerned mainly with analysis of blood. Only those urine analyses that could not, because of their complexity, be performed at the bedside or in the ward side room were undertaken in laboratories. This tradition has persisted in most hospitals in Britain but not necessarily elsewhere in the world.

Analysis of blood and faeces - with the notable exception of occult blood - were almost invariably performed in a central hospital laboratory. This tradition too has survived even though as a consequence of technological developments it is no longer necessarily the most efficient.

FUNCTION OF CLINICAL BIOCHEMISTRY

From its historical roots and present position the main function of a hospital clinical biochemistry laboratory can be seen to fall conveniently into four main divisions (3, 4): consultative and clinical; investigation and research; education and finally, analysis. Because the last named is unique to clinical biochemistry - and the one upon which the other three depend, it has become accepted by many health workers - including hospital administrators, clinicians and even clinical biochemists themselves - as synonymous with the subject as a whole.

This concept of clinical biochemistry has been resisted by a small but growing group of clinical biochemists mainly, though not exclusively, drawn from amongst those who are both medically and scientifically trained (3-6). It has, however, been encouraged by clinicians who are unaware of the benefits to them and their patients of a truly comprehensive clinical biochemistry service and who consequently prefer to look upon clinical biochemistry as a service to them rather than a parallel medical speciality.

The custom in some countries of charging patients, through their physicians, for each item of service i.e. analysis, led in some cases to clinical laboratories becoming profit rather than cost centres (7) and their directors, who were paid on a similar basis, amongst the most highly remunerated individuals in the health care industry. Markel and Rycus (8), however, found no evidence to support the suggestion that the method of remuneration i.e. salary versus fee-for-service, of the director had any influence upon laboratory usage by clinicians. There is, however, clear evidence from other quarters that there are financial incentives to perform medical procedures and laboratory tests which carry higher remuneration than consultation alone (9).

TECHNOLOGICAL AND SOCIO-ECONOMICAL CHANGES IN THE MID-1980s

A combination of technological and politico-economical events, which began in most countries around 1980, have made changes in the established philosophy and practice of clinical biochemistry inevitable (10). These events have been most apparent in the USA but discernable, to a greater or lesser extent, elsewhere. Briefly they involve an increasing awareness that growth in medical knowledge and technology has enabled death from certain previously fatal illnesses to be postponed or even prevented providing enough resources are devoted to it. This, coupled with an increasing expectation of good health from patients, potential patients and their families can, and eventually will, impose an impossible burden upon those responsible for funding health care provision.

In the USA this had led to the concept of DRG with its potential for converting clinical laboratories from profit to cost centres almost overnight; in the UK it has led to the initiation of a thorough review of clinical laboratory management and funding (11) - the results of which are still unknown but expected to be compatible with privatisation of clinical laboratory services. Both organisational actions tacitly accept the view long held by many clinical biochemists and other practitioners that much of the analytical work currently performed in hospital and similar laboratories is of limited clinical value (12-15).

The other innovation that has made re-examination of laboratory organisation timely is the possibility of performing laboratory type analyses much nearer to the patient and more quickly than has been possible in the past.

NEAR-PATIENT TESTING

Most of the so called stat clinical chemistry tests i.e. those analytical procedures considered essential for the practice of emergency or acute medicine and surgery, and which alone justify the continued existence of the average hospital laboratory, can now be performed rapidly, reliably and economically where and when they are needed without recourse to a formal laboratory. The technological changes that have made this revolution possible have been discussed elsewhere by many different authors (18, 19) but their implications for the practice of non-acute medicine and surgery and their possible effect upon the organisation of clinical biochemistry services have, in the main, received less attention.

HOSPITAL LABORATORIES

The main justification for each major hospital having its own central laboratory has been the need to provide a rapid analytical service for acutely ill patients and, to a lesser extent, for out-patients so as to remove their need to re-attend just to learn the results of laboratory tests. Now that this justification has largely been removed - by the availability of most urgently required tests at near-patient testing sites -the continued need, indeed even the desirability, of a large central hospital laboratory designed mainly to carry out non-urgent procedures - many of which are now regarded as being of limited value anyway - becomes debatable.

Few hospital laboratories - even in the biggest centres - are large enough to have a throughput of truly diagnostic test procedures sufficient to warrant many of them being performed daily. Consequently their menu of analyses is often restricted to analytes whose frequency of requesting is sufficient to justify the purchase of the appropriate apparatus and reagents, not to mention skill, necessary for their performance. Tests that are requested but are not on the local menu are dispatched to another laboratory providing a more comprehensive service. Results obtained in this way are generally just as highly regarded by clinicians as those produced on site; in other words analytical results are no more than the tools of the trade of clinical biochemists - especially the growing number of them who consider their chief function to be the provision of advice on what tests to perform in a given clinical situation and to interpret the results in context - in exactly the same way as a knife is for a surgeon.

CENTRAL (REGIONAL) LABORATORY

It is really of no great importance just where an analytical result is produced; - much more important is the certainty that it is appropriate i.e. can be interpreted and is accurate, specific and timely. The last three conditions can best be achieved by using centralised facilities capable of providing a near comprehensive menu of analytical tests to a population of say 1-2 million people.

Such a laboratory would have a sufficiently large throughput of all but the most esoteric tests to guarantee a 24-hour turn round time. It would not necessarily be attached to a hospital, though this would probably be desirable for the motivating influence it might have upon the staff who should be encouraged to look upon themselves as health care workers rather than as industrial employees.

Specimens for analysis would be sent from the collection point i.e. hospitals, doctor's offices, industrial health clinics etc. to the laboratory at frequent intervals. The exact mode of transport would be determined by geographical considerations. In most urban and semi-urban communities dedicated road transport, provided and organised by the laboratory, would be used; in rural communities, small townships etc., a combination of road, rail and air transport would be used, as appropriate.

Within each large hospital a collection, separation and preparation unit would be required. This would be associated organisationally, and administratively, with a small specialist laboratory where the consultant clinical biochemists - whose presence on site would be required so as to fulfill the consultative, clinical and educational functions of a comprehensive clinical biochemistry - would work.

Ideally this collection/separation facility - which would, in effect, be a satellite laboratory of the central laboratory - would be adjacent to a clinical investigation unit (20) where patients undergoing diagnostic procedures can be dealt with by staff i.e. nurses and technicians with a proper understanding of the need for careful specimen collection, handling and labelling that is necessary if truly meaningful results are to be obtained. All too often, carefully performed analyses carried out on specimens received in the laboratory from dynamic function tests are rendered worthless because of poor patient preparation, sample collection or labelling prior to their receipt by the laboratory (21). These vital steps in the production of clinical biochemistry data are seldom accorded the same detailed scrutiny and quality control as the analytical procedures themselves yet they are every bit as important for proper interpretation of the results.

DOCUMENTATION AND DATA HANDLING

Concentration, in a central laboratory computer, of all the clinical laboratory data handling for the community, rather than in a number of smaller hospital computers, permits more comprehensive facilities to be installed than could otherwise be economically justified. The central computer can be linked by a view data system to all of its users e.g. doctors offices, occupational health centres and, in the case of hospitals, to the individual wards, units and outpatients departments as well as the satellite laboratories. The electronic link can function in both directions permitting results obtained by the decentralised testing facilities to be integrated in the computer with results obtained in the central laboratory and elsewhere. This enables clinicians and others interested in obtaining the latest - as well as the cumulative - results to obtain them as and when they are wanted with the minimum of delay. A hard copy can always be made available at a later date if needed.

Using information technology methods the central laboratory computer can be made to alert recipients of analytical interference from any drugs the patients might be taking and which are recorded in the pharmacy databank for that patient. Drug-test interactions are a well recognised cause of interpretive errors but seldom receive the attention they deserve either because the laboratory is not provided with the necessary information and/or does not have access to a drug-test interaction data bank (22).

ORGANISATION

Commercial models for the organisation of a central laboratory facility have existed for many years. They were, in the main, established before the technology of near-patient testing had reached its present level of development. It is, however, this development, above all others, that makes what has long been recognised as clinically desirable a practicable proposition. Thus the technology to convert current hospital based clinical biochemistry services into a more effective and efficient multi-user, multi-tiered services already exist - it will, however, require a radical change in the attitude of clinicians as well as by laboratory workers and administrators to turn the possibility into a reality.

CLINICAL PRACTICE

The habit of requesting large numbers of supposedly useful, but often irrelevant, tests either as admission "profiles" or as an almost daily "routine" during the patient's stay in hospital is now generally recognised as wasteful (23-28) but is seemingly deeply entrenched (29-31). Its replacement by recognition that the practice of medicine depends more upon clinical skill than upon the availability of numerous laboratory results is gradually dawning (32); the process could be hastened by encouragement from clinical biochemists who have much to gain from it in terms of their own job satisfaction. Clinicians also might find it more interesting to practice medicine on the basis of thoughtfulness and value judgement rather than rote.

ADMINISTRATION

An integrated laboratory i.e. the central laboratory and its hospital satellites, would constitute a single financial and organisational entity. All levels of staff engaged in the practice of clinical biochemistry would be on the laboratory payroll rather than on those of the individual hospitals served. This would make for managerial flexibility as well as facilitating training and the provision of cover during annual leave and other absences. An arrangement somewhat similar to this has been successfully operated by the Public Health Laboratory Service (PHLS) in Britain for many years.

In the integrated clinical biochemistry service proposed the vital consultative, interpretive, and educational components would be delivered locally by clinical biochemists who had established the close working relationships with clinical colleagues that are so necessary for any consultative system to work efficiently. These consultants along with the other clinical biochemistry staff working within the hospital would operate the sub-laboratories serving those specialist units e.g. accident and emergency, coronary care, intensive care and outpatients, where the volume of urgent work justified their existence. They would also train the junior doctors and ancillary staff who would be authorised to use the near-patient testing facilities located elsewhere in the hospital - on the wards for example - as well as supervising the quality control and operating the quality assurance programme.

The central laboratory and its various satellites would be managed by a board of directors on which at least one consultant clinical biochemist from each hospital served by it would sit. This would ensure some measure of uniformity throughout the service as well as bringing the benefits of scale in negotiating the price of equipment and reagents with manufacturers and suppliers. It would also make specialisation in the various branches of clinical biochemistry possible and achievable. The specialist clinical biochemists would function locally as generalists and regionally as specialists. They would use the "hospital" laboratories mentioned above to carry out the work necessary for them to develop their subject, maintain their expertise and provide a clinical service.

FINANCES

Although detailed costing have not been made of the proposals contained within this paper, preliminary investigations suggest that they would cost no more than is currently being spent on clinical biochemistry services (at least, in Britain) but would provide a more clinically useful service. This would result at least as much from the more efficient use of physician time as from as any internal laboratory efficiency savings.

REFERENCES

1. Bodansky M, Bodansky O. Biochemistry of Disease. New York: The Macmillan Company, 1952.
2. Foster WD. Pathology as a profession in Great Britain and The early history of the Royal College of Pathologists. London: The Royal College of Pathologists.
3. Astrup P. Clinical Chemistry at the Crossroads. Z Med Labor Diagn 1981; 22: 243-245.
4. Marks V. Clinical biochemistry: trade or profession. University of Surrey, Inaugural Lecture, 1972: 1-22.
5. Carter PM, Davison AJ, Wickings HI, Zilva JF. Quality and quantity in chemical pathology. Lancet 1974; ii: 1555-1557.
6. Astrup P. Clinical chemistry - a challenging discipline. Clin Chem 1975; 21: 1709-1715.
7. Conn RB. Clinical laboratories: Profit center, production industry or patient-care resource? N Engl J Med 1978; 298: 422-427.
8. Markel SF, Rycus MJ. Modes of compensation of pathologists and the charges for and use of laboratory services. JAMA 1980; 243: 2053-2055.
9. Schroeder SA, Showstack JA. Financial incentives to perform medical procedures and laboratory tests: illustrative models of office practice. Medical Care 1978; 16: 289-298.
10. Marks V. Impact of new technology on the future of clinical chemistry. Pure and Appl Chem 1985; 57: 565-570.
11. Sherman J. £ 30m saving study in NHS. London: The Times, May 13 1987.
12. Hampton JR, Harrison MJG, Mitchell JRA, Prichard JS, Seymour C. Relative contributions of history-taking, physical examination, and laboratory investigation to diagnosis and management of medical outpatients. Br Med J 1975; 2: 486-489.
13. Sandler G. Costs of unnecessary tests. Br Med J 1979; 2: 21-24.

14. Griner PF. Use of laboratory tests in a teaching hospital: long-term trends: reductions in use and relative costs. Ann Intern Med 1979; 90: 243-248.
15. Marks V. The choice between discretionary and profile testing. In: Price CP, Spencer K, eds. Centrifugal Analysers in Clinical Chemistry. Eastbourne: Praeger 1980: 259-270.
16. Fowkes FGR. Containing the use of diagnostic tests. Br Med J 1985; 290: 488-490.
17. Wiener K. Pathology measurements closer to the patient? J Clin Pathol 1980; 33: 857-863.
18. Marks V, Alberti KGMM, eds. Clinical Biochemistry Nearer the Patient. Edinburgh: Churchill Livingstone, 1985.
19. Marks V, Alberti KGMM, eds. Clinical Biochemistry Nearer the Patient II. London: Bailliere Tindall, 1986.
20. Marks V. Who should investigate patients? Bull Roy Coll Pathol 1983; 42: 2-4.
21. Zilva J. Collection and preservation of specimens for chemical pathology. Br J Hos Med 1970; Dec: 845-852.
22. Salway J. Interference from drugs. In: Marks V, Alberti KGMM, eds. Clinical Biochemistry Nearer the Patient. Edinburgh: Churchill Livingstone, 1985: 150-156.
23. Greene R. Questionable Dogma: irrelevant results. World Medicine 1975; September 10: 21.
24. Carmalt MHB, Freeman P, Stephens AJH, Whitehead TP. Value of routine multiple blood tests in patients attending the general practitioner. Br Med J 1970; 1: 620-623.
25. Kaplan EB, Sheiner LB, Boeckmann AJ, et al. The usefulness of preoperative laboratory screening. JAMA 1985; 253: 3576-3581.
26. Blery C, Charpak Y, Szatan M, et al. Evaluation of a protocol for selective ordering of preoperative tests. Lancet 1986; i: 139-141.
27. Griner PF, Glaser RJ. Misuse of laboratory tests and diagnostic procedures. N Engl J Med 1982; 307: 1336-1337.
28. Durbridge TC, Edwards F, Edwards RG, Atkinson M. Evaluation of benefits of screening tests done immediately on admission to hospital. Clin Chem 1976; 22: 968-971.
29. Sandler G. Do emergency tests help in the management of acute medical admissions? Br Med J 1984; 289: 973-977.
30. Mitchell MW, Fowkes FGR. Audit reviewed: does feedback on performance change clinical behaviour? Journal of the Royal College of Physicians of London 1985; 19: 251-254.
31. Fraser CG, Woodword FP. Strategies to modify the test-requesting patterns of clinicians. Ann Clin Biochem 1987; 24: 223-231.
32. Vaisrub S. Laboratory Testing - routine or on demand? JAMA 1980; 244: 592.

TECHNOLOGY ASSESSMENT OF NEAR-PATIENT TESTING: EXPERIENCES FROM PRIMARY HEALTH CARE IN THE NORDIC COUNTRIES

M. Hørder

Odense University Hospital, Dept. Clin. Chem., 5000 Odense, Denmark

The role of laboratory medicine in primary health care is undergoing rapid changes. Two recent developments have generated new perspectives for clinical chemistry in particular. One of those, the production and marketing by industry of apparently robust equipment with low analytical capacity, could seem to be the most important. Much more significant is probably another trend in the development of the overall health care system, i.e. the upgrading of the role of primary health care. The way these two developments may influence the role of laboratory medicine in primary health care is not the same in all parts of the industrialized world. In the Nordic countries general medicine is an integrated part of a non-profit health care system. The opportunities to be obtained in such system from the new patient-near technology cannot be assessed from its analytical performance only. An assessment should also consider the true medical needs in primary health care, the organization and economy of health care, quality requirements, and the skill and education of personnel.

The medical needs of laboratory information in primary health care is usually very different from those of the specialized hospital clinic. Less than 1% of the test results are true urgent requests. More than 50% of the results are used for monitoring of disease and therapy.

In a recent investigation in the county of Fyn in Denmark the clinical need for laboratory information in primary health care has been described (LE Matzen, P Grinsted, M Hørder, in press). The county has approx 500,000 inhabitants served by approx 280 general practitioners. Over a four weeks period the physicians registered a) if laboratory information was needed b) whether such information was for diagnostic purposes, i.e. disease of the patient was not yet identified, or for control of a known disease.

Figure 1 shows examples of the distribution of six medical problems for which the physician wanted laboratory information. It is confirmed that control of disease is as frequent as diagnosis. The kind and number of tests for each problem depend on the clinical problem. Examples are given in Table 1 which also shows that the number of tests needed for control are less than for diagnosis.

Table 1. Average number of tests by general practitioners in six clinical situations

	Diagnosis	Control
Thyroid disease	8.3	4.6
Liver disease	8.3	7.8
Hypertension	7.9	5.8
Renal disease	5.0	4.2
Allergy	4.6	2.1
Diabetes	1.9	2.1

Test results that are used for diagnostic work-up are applied as early information more than as late confirmatory diagnostic tests, Figure 2. By clinical investigation and few simple tests the physicians gain for more than half of the patients a sufficient knowledge for dealing with the problems of the patient. Availability of the broad selection of tests in the clinical laboratory will further reduce the uncertainty of diagnosis with great power. However, the fraction of patients that need high technology diagnostic support in primary health care is low (Fig 3).

It is however important that the general practitioner is well informed about the possible contributions from laboratory information in diagnosis and control of disease. The clinical value of "high technology" tests available in laboratories are not identical in primary health care and in the specialized hospital clinic due to great differences in the prevalences of disease in these two clinical settings. The sequence of applying tests described on the x-axis of Fig. 2 and 3 should therefore be discussed by

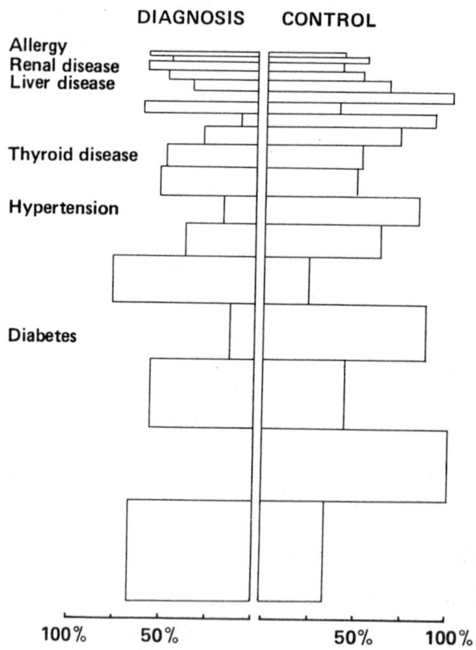

Fig 1. Purpose of laboratory information in the clinical management disease. Six of eighteen problems are shown.

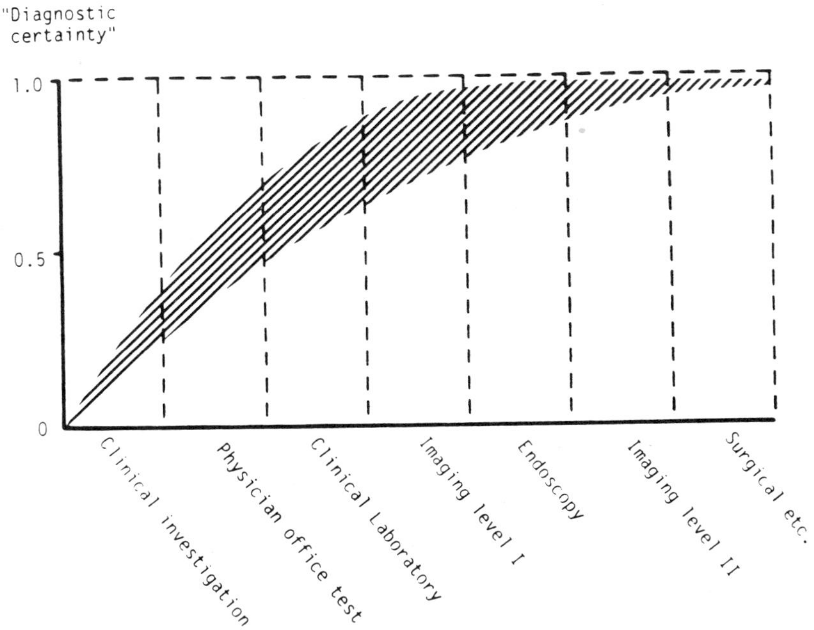

Fig 2. Diagram showing the role of diagnostic information in the diagnostic work-up of patients by primary health care physicians.

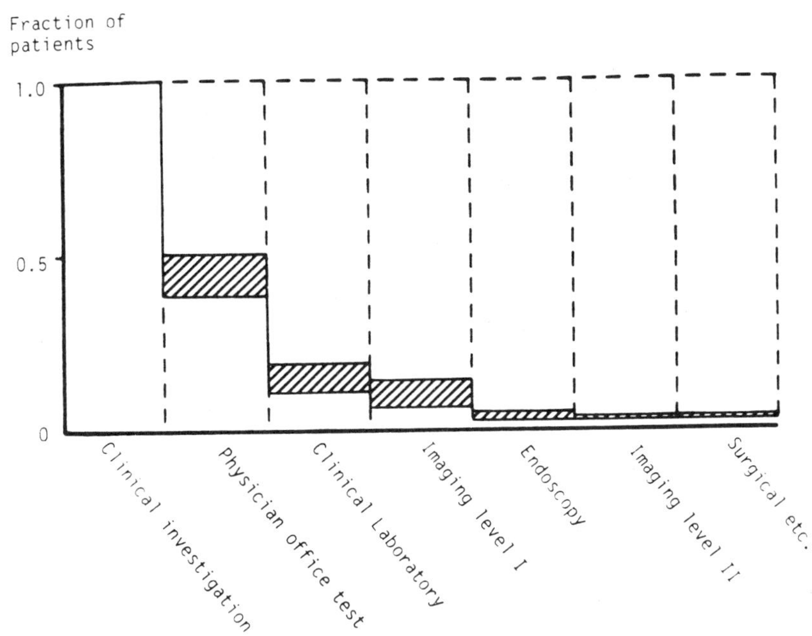

Fig 3. Sequence and frequency of diagnostic procedures in the management of patients seen by primary health care physicians.

Table 2. Need for local analyses in primary health care

B-Hemoglobin	B-Coagulation factors
B-Erythrocyte sedimentation rate	S-Potassium
U-Albumin, -glucose, -hemoglobin	S-C-reactive protein
B-Glucose	S-Creatinine
U-Cells, Cost, (microscopy)	U-Density
U-Human Chorionic gonadotropin	F-Hemoglobin
B-Leucocytes	B-Thrombocytes
B-Cells, (microscopy)	

(Opinion of doctors in Kristianstad county, Sweden, N. Tryding, 1986)

general practitioners and clinical laboratories of a region. The kind of tests that might be value as physician office tests may vary depending on the clinical problems of the region. An example appear in Table 2.

The laboratory information may be obtained in different ways. Whether small laboratories should be established in physicians offices cannot be decided only from the technical availability of instruments and reagents. From a technical point of view it is now possible to perform the main part of the tests that are necessary for primary health care. If laboratory work is performed the physician must take responsibility of quality assurance and education of the personnel. Experiences from the Nordic countries has shown that even simple tests performed in physician offices must be supported by laboratory medicine to ensure the analytical quality. In the county of Fyn in Denmark the quality of B-Hemoglobin and B-Glucose analyses were found of an unacceptable standard for 30% and 70%, respectively, when registered in 1981. Since then laboratory medicine has supported general practitioners as consultants. The experiences from the period up to 1985 has been that the personnel of physician office has lacking knowledge about proper performance of tests and of the need for quality assurance. The result of the consultant function from laboratory medicine has been an improvement in the analytical performance for both tests (Table 3). The investment in time for education and organization for quality assurance has

Table 3. Improvement in quality of tests performed in physician offices through a consultant support from clinical chemistry laboratories.

B-Hemoglobin			B-Glucose		
Concentration (mmol/l)	1981	1985	Concentration (mmol/l)	1981	1985
~ 5	67%	88%	~ 3-4	33%	31%
~7,5	70%	93%	~12-14	39%	68%
~ 10	70%	91%	variable	-	67%
Number of participants	33	56	Number of participants	33	56

been of considerable size. It is however necessary to ensure the quality of tests performed decentralized.

The tables indicate the percentage of the results from the external quality assessment scheme which are judged as acceptable (Dept. of clin. chem. + 10%, for the low glucose + 15%). B-Hemoglobin improved from 70 to 90% acceptable results. B-Glucose improved from 40 to about 70%, except for the low glucose values, which cannot be measured satisfactory with most of the methods available for physician office tests.

Technology assessment of laboratory medicine in primary health care involves a broad spectrum of considerations as illustrated in Figure 4.

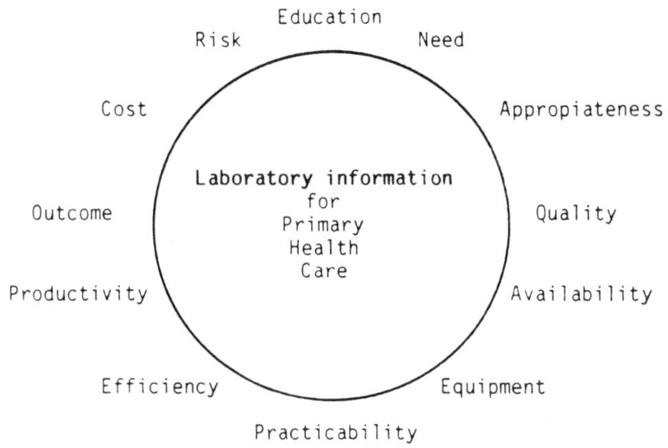

Fig. 4. Elements of technology assessment of laboratory medicine in primary health care.

As an alternative the physician may decide not to establish laboratory functions but rather concentrate on the proper clinical application of test results that are obtained from hospital or private laboratories. If these laboratories are to play a role in the growing primary health care they must change their ways of communication with primary health care doctors.

In the Nordic countries the "new" roles of laboratory medicine in primary health care is currently being evaluated by NORDKEM. The goals for the collaboration of laboratory medicine with primary health care have been published by NORDKEM in 1985 as a result of a working party by general practitioners, clinical chemists and health administrators. A synopsis of the guidelines describing the six fields of joint interests indicates how the elements of technology assessment of Figure 4 can be transformed to easily identifiable projects for developments in local regions.

GUIDELINES FOR THE COLLABORATION OF PRIMARY HEALTH CARE AND CLINICAL CHEMISTRY IN THE NORDIC COUNTRIES

General practitioners and clinical chemists within a region together should describe and publish - as rules of thumb - selected laboratory information that are valuable in dealing with major clinical problems in general practice. By regularly intervals revision should be made considering changes in medical as well as analytical prerequisites. Such guidelines for appropriate laboratory information in primary health care should be coordinated with those that are used in specialized hospital clinics; the object being to ensure that they fit a common strategy for care of health problems by primary and hospital clinics. Clinical chemistry should be involved in this coordination of application of laboratory medicine on various levels of the health care system.

The access to information from hospital departments of clinical chemistry for primary health care should not be restricted. The kind and number of laboratory data that are necessary for the general practitioners may differ due to variation of the frequency and severity of diseases in geographical regions. 24 hours service should be provided if necessary. The delivery of blood and other materials for investigation from primary health care centers to the hospital laboratories should be organized to ensure stability of the components and minimal time for transport. The way answers are referred from the laboratory to the general practitioner should fulfill the needs for easy and unequivocal interpretation.

General practitioners and other doctors of primary health care centers should ask clinical chemists of their region for assistance before the establishment of analytical facilities in their offices. The kind of advice provided by the clinical chemists should form the basis for selection of the kind of analysis and the necessary equipment and reagents. Training of personnel and the safety of laboratory work should also be considered. Once established the collaboration on analytical work between the staff of primary health care centers and the clinical chemistry of the region should be continued.

As for analytical work in hospital department of clinical chemistry quality assurance must be part of laboratory medicine of primary health care. It should be organized in close collaboration with clinical chemistry of the region. The quality assurance should consist of external quality assessment to ensure comparability of data. An appropriate system for internal quality control of analytical work in the primary health care office must also be arranged.

Training of personnel of primary health care in those procedures they perform must take place. Clinical chemistry of the region may provide most of this education. The level and content of the training programme depends on the amount of laboratory procedures performed, e.g. blood drawing and specimen procurement only or also analytical work with quality assurance. On the other hand the staff of clinical chemistry should learn about the working conditions in primary health care to be able to provide the appropriate kind of service.

Improvement in the utilization of laboratory medicine in primary health care according to these guidelines should be stimulated and ensured by laws and regulations on the way primary health care and laboratory medicine play their roles in the health system.

(A full report is available from NORDKEM, Kirelä Hospital, SF-00260 Helsinki 26, Finland).

It is noteworthy that initiatives like that of NORDKEM at present are on their way from organizations like WHO, NCCLS and ECCLS. Together they will form the framework for future collaboration of laboratory medicine with primary health care. To be of value for the practicing physician such guidelines must be followed of local arrangements involving personnel from laboratory medicine and general medicine. The appearance of patient near technology has stressed the important role of laboratory medicine for primary health care.

BEDSIDE CHEMISTRY

R. Galimany

Hospital Germans Trias i Pujol, Badalona, Barcelona, Spain

Clinical laboratories have undergone a remarkable change in the last two decades. One of the aspects which has helped to cause their dramatic expansion is the development that has occurred in the last few years in the conception and designs of instruments for bedside use. As always happens with any change in medical practice, one finds certain people enthusiastic about this new concept, whilst other people highlight its drawbacks or accept it with numerous reservations. Bedside usages run from intensive care units in many hospitals, to primitive rural medicine set-ups in developing countries. Within this broad spectrum a wide variety of situations are to be found, and to give adequate attention to them all, it is important to have sufficient information regarding their circumstances.

However, the initial application of this type of laboratory instruments has witnessed a divergence or diversification due to various factors. One of these factors has been the competition and the opening of new commercial fields for the manufacturers of these instruments, who have played a vital role in pointing out new applications of their instruments (the monitoring of blood glucose levels in the patient's home, clinical analyses in the physician's office, etc.) which, in some countries, is leading to interference between the realms of different medical specialities.

The carefully controlled production of current clinical chemistry analysers has made it possible to reach levels of quality and practical application unthinkable a few years ago. However, all the guarantees and advances offered by these automatic systems can, paradoxically, be potentially dangerous in the hands of inexperienced operators.

In the seventies, there was a huge upsurge in the appearance of new instrument technology and methodology relating to clinical laboratories. This dramatic change did not occur in the same way as with the number of analytes, which received careful study as their number increased. The design of equipment and automatic instruments for bedside use, or with bedside applications, was a small but interesting component in this dramatic increase in the technology available for clinical laboratories.

We will mention the performance of a few bedside instruments corresponding to this technology by looking at the blood components measured and the methods used: clinical chemistry, toxicology, haematology, microbiology and immunology. Among the monoparametric analysers most associated with bedside usage are those designed to measure blood glucose levels. We have looked at various blood glucose analysers, falling into two generations; the first generation being somewhat less used due to the appearance of a new, technically superior, generation: Accu-check, Glucometer, Glucoscan, Reflolux. Their operating characteristics are all very similar. The reading is by reflectometry, read from reactive strips, adapted to each analyser with its characteristics either wash-out or wipe-off. The measuring range is similar for the different models, noticeably improved in the second generation. The size of the sample required depends on the characteristics of the strips on which it must be placed; it does not depend directly on the analyser used. Sample size can be a factor leading to false results if the manufacturer's instructions for specific strips are not adhered to. Various critics have put forward significant differences in the results. For the Chemstrip bG strips, an overlarge of insufficient sample will produce a statistically significant effect only with samples with a glucose concentration over 15 mmol/l; for the Glucostix the effect can be observed in samples with a glucose concentration over 10 mmol/l. With the Glucoscan strips, significant effects are not observed because of a variation in sample size.

The incubation periods and the time necessary for colour development are different for each type of strip, as indicated by the manufacturers themselves. Some of these times are very strict and must be adhered to as this is a vital factor in order to achieve good quality results. Blood is incubated on the test pad for 60 s (Chemstrip bG), 30 s (Glucostix), or 40 s (Glucoscan) before being removed. Incubating with blood for periods less than or exceeding these recommended times had an analytically insignificant effect on the Chemstrip bG but had profound effects on the other two strips. For example, removing blood (containing 12.9 mmol/l of glucose) from the Glucoscan strip 5 s earlier or later than the prescribed 40-s interval resulted in errors of -1.7 and +1.4 mmol/l, respectively.

As regards the differences in the time required for colour development, times given by the manufacturers should be kept, as this is vital for Chemstrip and Glucostix when it comes to measuring high glucose levels. The Chemstrip bG strip shows little variation in the final colour for up to 20 s before or after the correct reading time of 60 s. On the Glucostix, colour-development times only became significant (-4 mmol/l) for blood-glucose concentrations $>$ 17.5 mmol/l; otherwise, the colour remained stable for a further 20 s after the correct reading time. By contrast, the Glucoscan strip developed a stable colour 10 s before the correct reading time, and the colour remained stable for at least a further 20 s.

As regards the system for calibration these analysers, significant advances have been made. Accu-check II memorises the calibration by reading it from a bar code specific to each lot of 50 strips. Glucometer, which in its original version required calibration using two calibrating solutions (Dextrocheck cal), in its new version requires calibration for each new lot of the reagent by inputting the calibration number. Whilst with Reflocheck, each strip has a control bar code on the back, with Reflolux, each box of strips includes its own calibration strip. Although a zero-reading is not a requirement, Refloflux has a zero adjustment for a test range reading prior to applying the sample. All the monitors, except for Accu-check, are prepared by the manufacturer to read only in units established beforehand by the user.

All the strips can work with both capillary blood and with plasma. For obvious reasons, it is interesting to know the correlation between the glucose level measurements shown by these instruments and the measurements with plasma after analysis in the hospital central laboratory. Godine et al., obtained a good correlation from 205 measurements in capillary blood and plasma with the hexokinase method in a Centrifichem analyser. This is similar to other evaluators' results. He calculated the range of the deviation of the blood glucose value in relation with the plasma glucose for varying glucose levels.

In the inaccuracy study, the majority of the observed results are good, especially in the second generation analysers. Among these, Glucoscan 2000 and Accu-check II show greater precision than Glucometer II in the 5 to 20 mmol/l range of blood glucose levels. When compared with other methods, as studied by various evaluators, the correlation is good, both with the hexokinase method, accepted as the recommended method, and with the automated Beckman glucose oxidase method (Astra-8) widely used in USA.

In a colour stability study carried out, some variations in the glucose levels were observed in the following half hour. It was also observed that after 48 hours both Chemstrip and Glucoscan showed good colour stability, whilst Glucostix's colour faded more quickly. Nonetheless, the colour was sufficiently stable in all the strips one hour after applying the sample so as to allow them to be used as a quality control colour reference in all the lectors using the same brand of strips, after due recalibration for the corresponding lot of strips. This quality control produced better results as regards accuracy than the conventional method carried out by different operators (CV%: very variable 8.3 to 9.8 for 7.8 mmol/l glucose).

It is interesting to highlight some of the points of agreement of these evaluators when they give their judgement on the practicability of these systems. As a possible improvement, they suggest the inclusion of an automatic thermostat system for automatic temperature correction, since all the tests are based on enzymatic reactions. Many evaluators have coincided in their observations on anomalies in the results obtained with serous and aqueous quality control products; they ascribe this fact to the presence of stabilisers and preservatives in aqueous materials, producing an atypical response from the chemical reaction. For example, aqueous calibrating solutions designed for one system of strips, cannot be used with another system. In the case of serous materials, the effect may be due to the viscosity of the sample.

It is noted that satisfactory results depend a great deal on the structure of the strips, as not all of them adapt adequately to the instruments, and also on the ability of the operator to correctly place the sample on the lector. One of them relates how one user, in order to economize, cut the strip in two, in such a way that its size insufficient to fill the reading field of the reflectometer.

An insufficient amount of blood placed on the strip causes an unequal adsorption over the reactive surface and insufficient colour development. Obviously the wash-out test strips have greater problems. The incubation and reading time is a vital factor for most lectors; therefore, it is necessary to strictly adhere to the times established by the manufacturer. Adequate calibration is also a vital factor. For some analysers, such as Eyetone, calibration is recommended for each operator before beginning measurement of the samples. They consider that the new versions of these systems are easier to use, and whilst many aspects of the early models have been improved, the analytical performance has not shown great improvements.

Among the causes of possible errors about which the majority of the evaluators agree, we should like to indicate the following:

The main cause of error, as indicated in all the works consulted, is faulty handling of the strip; also faulty handling of the lector device. All the authors studied the correlation of results obtained with these analysis systems comparing operators with laboratory experience and operators with no laboratory experience. They indicate that invariably a certain level of bad execution was observed in operators without experience in laboratory work or insufficient training. They observed systematic deviations not attributable to the strips. However, with trained operators, they have observed significant differences between different series of strips. All this is understandable if we take into account all the varying factors mentioned above. For an operator whose speciality is clinical laboratory work, controlling these effects will be a common feature of his work, whilst for a person with no laboratory knowledge, they may not catch his attention or be given the importance that they warrant. Evaluators also all indicates as a frequent cause of error the washing of the blood in the reaction area in those lectors that require it. The viscosity of the sample is a factor that should be borne in mind as it can be a cause of variations and will therefore influence the quality of the results.

Taking into account the various error factors, some of them uncontrollable, it is vital to carry out adequate quality control work. This should be established for these tests, as we noted earlier, and be based on a material suitable for the strips and also suitable for comparison, in order to ensure that central laboratory and bedside analyses give results that are in agreement; otherwise, doubts will only increase concerning the true value of the tests.

Between the dry chemist multitest analysers accessible to bedside we selected: The Seralyzer (Ames), Reflotron (Boehringer) and Kodak-EKTACHEM DT-60. The Seralyzer can use serum or plasma; the volume required is 30 μl. The results that appear on the display can be obtained in IS or conventional units. The length of time for each test can vary between 30 s and 3 minutes. The Reflotron analyzer can use whole blood or plasma or serum. The volume sample required is 30 μl, and it can work at a temperature of 25, 30 or 37°C; the results that appear on the display can be obtained in IS and conventional units. The length of time for each test can vary between 2 and 4 minutes. It is made up of three modules: the photometer, the reagent carrier and the pipette. Once the sample has been pipetted onto the reagent carrying strip, this is inserted in an adaptor in order to fit properly in the reading zone; without practice, its correct positioning may cause the operator some difficulties.

The Ektachem DT-60 is composed of three interconnected modules: The main module for testing substrates, including the calculator, touch-sensitive key-board, screen for visual representation and thermal print-out. The special module for enzyme determinations and the special module for measuring sodium and potassium by selective electrodes.

The reagent strip for Seralyzer is a plastic support with paper matrix. For Reflotron is a reagent carrier; for whole blood tests, the separating out of the erythrocytes is achieved by means of glass fibres transferred to the paper of the strip, and the plasma spreads towards the plasma reservoir until it reaches the layers carrying the chemical substances. Reagents can be applied to this glass fibre fabric, for example to chemically transform the ascorbic acid in the sample using ascorbate oxidase, and thus avoid this interference. Each reagent carrying strip has a magnetic band on the back; with 500 bits, this band stores all the test information. For DT-60 the reagents are individual hermetically sealed multilayered slides. Each slide is identified by a bar code which includes the analyte to be tested for and the reagent lot number. The slide is inserted in the analyser and the information is memorised, with the name of the test appearing on the screen. A sample amount is placed in a 250 μl reservoir from which the required volume is taken by means of a mechanised pipette at controlled speed. The tip of the pipette is disposable for each sample. The sample goes into a tube and its arrival at the surface of the reactive slide is announced by an acoustic signal. The slide has an incubation time of 5 minutes. Slides can be placed on top of each other, allowing the operator to begin another test during the incubation period. The order of the tests is optional.

The Seralyzer is calibrated with 2-point calibration using 2 different calibration solutions each time new lot used. The Reflotron is self-calibrating thanks to a reference detector lit by the wall of the sphere. A correction factor is incorporated which is used when a reaction is used in order to avoid interferences, in the first stage of the analytical process. Three calibrators (DT 1, 2, 3) are required for calibration of the DT-60; each of 3 ml and held stable at 4°C for 24 h. Three calibration levels are required for the substrates, programmed by the operator, and with from 1 to 5 doses of each calibrator able to be made. Calibration should be carried out every 3 months and when the reagent lot changes. Evaluators recommend checking the calibration via checks at three levels to detect possible linear drift, although the manufacturers only recommend using one level. For the ions, calibration is carried out at two levels.

For Seralyzer, individual test modules plug into the instrument and contain program memory and filters. An integrating sphere contained within the instrument is illuminated by a series of flashes from a xenon flashtube. The reagent test strip, after being reacted with sample, is placed at an opening in the sphere. Light striking the reagent test strip is reflected vertically through a collimator. The test filter placed above the collimator allows the correct wavelength of light to pass to the sample detector. Other reflected light within the sphere is passed through the reference detector. The electrical signals from both detectors are analyzed by a microcomputer, compared with stored calibration data, and the resulting concentration value is displayed. For Reflotron automatic measuring reaches the reactive field of the Ulbricht sphere. According to the type of test, three different wave lengths can be used for measuring; it has three light emitting diodes fitted, whose basic emitting wavelengths are set at 565, 642 and 950 nm. The photosensitive receiver measures the fraction of the light reflected through the indicating field after partial absorption.

DT-60 has two references (zero reflection and 100% reflection) which are measured prior to each reading in order to calibrate the reflectometer. Reflectometer with optic fibres: three optic fibre beams are connected to three diodes emitting on wavelengths of 555, 605 and 660 nm. The fourth beam is connected to a photo-diode detector. The selection of the appropriate wavelength and three successive measurements are ordered by the micro-processor that carries out a check between these three measurements and rejects the result in the event of any anomaly. After the reading, the slide is removed to a disposal compartment. For ions only one measurement (sodium or potassium) can be carried out per cycle. The volume measured out is 10 μl deposited at the same time as the reference liquid with the aid of a double pipette with disposable tips. The volume of the sample that needs to be placed in the reservoir is 100 μl. Incubation of the slide takes place at 25°C, an electrometer covers it to provide electrical contact. A bridge is established by means of a filter paper. Direct potential measurements are taken. Measuring time is three minutes.

Some evaluators bemoan the fact that the 10 μl of the sample cannot be measured out directly, because it has to pass beforehand through a 250

μl reservoir. With the ions, if short series are processed, it is annoying to have to renew the reference liquid each time as it only takes 10 μl. The operator has the possibility of identifying the parameter studied as well as inserting a sample identification number. Most evaluations agree in indicating that there is no drift for thirty days without calibration.

For their examination of the linearity, they recommend a revision of the high linearity limits for urea and cholesterol and check on glucose at three concentration levels. In their examination of accuracy, they all find values very close to the real ones; in the electrolyte study they recommend using the DT control or serum pool without additives.

During the process various checks can be carried out:

- check on the optical systems
- check on the print-out
- check on the parameters of the calculating functions in the memory
- different mechanical tests
- checks on the measurements carried out.

The Seralyzer displays a flashing number if the concentration is above or below the manufacturer's stated limits of linearity of the system. Samples that gave values less than the lower limits were reported as "less than" the flashing values. Samples that exceeded the upper limits were diluted a further threefold. Special test module for each parameter is a step necessary for to change of parameter. Humidifying of the measuring chamber is recommended for optimal results. It is important that the serum be pipetted directly onto the center of the reagent pad.

In Reflotron system a series of fiber matrix tabs performs blood lysis, filtration, fluid transport and chemical analysis; however, adequate pipetting of whole blood is a critical factor for most tests, except hemoglobin. Samples from onco-haematologics and dehydrates patients could produce untrue results of hemoglobin. In the enzyme quality control a controls from fuman serum is recommended, since the substrates are optimised for human enzymes.

INTERFERENCES

Several drugs can interfere with same Reflotron's methods at therapeutics doses:

Methyldope	: Glucose, Creatinine, GOT, GPT
Paracetamol	: GGT
Bezafibrate	: GOT, GPT, Creatinine
Dobesilate-Ca	: GOT, GPT
Noramidopirine	: Glucose.

SERALYZER: Hemolysis has a greater effect on total bilirubin values in the normal range than in samples with an above-normal concentration. For a bilirubin value of 9 mg/l, slight hemolysis interfered positively, adding 4 mg/l to the apparent bilirubin concentration. For a bilirubin value of 13 mg/l, moderate hemolysis added an apparent 8 mg/l. Hemolyzed specimens should not be assayed for total bilirubin with the Seralyzer and DT-60.

DT-60: The effect of hemoglobin in the bilirubin assays: 2 g/l increases a bilirubin value of 6.0 mg/l by 33%; hemoglobin, 10 mg/l, increases it six fold.

A novel Dry Electro-Chemical Technology (DECT) suitable for satellite and bedside testing is developed. The basic employs electrochemical sensor (ECS) as detector and utilizes paper as carrier matrix for dry reagents.

The electrochemical sensor is designed so as to make measurements by contacting the surface of a one centimeter diameter paper disc to which a small sample volume (15-50 μl) is applied. The electrochemical surface is cleaned after each measurement by simple blotting with an absorbing material and is reusable for as many as one thousand measurements. The reagent containing paper disc is discarded after single use.

Assay for several species in blood are developed by this technology. In the case of Potassium assay the paper disc does not contain any reagent, whereas for BUN the disc contains buffered urease and amonium ion produced is measured with the appropriate electrochemical sensor. Glucose and Uric Acid assays are accomplished by measuring hydrogen peroxide produced by reacting with respective buffered enzyme incorporated in the paper matrix. In all assays undiluted serum or plasma samples are used. The method characteristics is summarized in the following table:

- Detector is an ISE or a polarographic electrode (8 mm diameter)
- Dry reagents impregnated on filter paper (10 mm diameter)
- Apply calibrant (15-50 μl) on reagent paper disc
- Place electrode surface on top to measure signal
- Electrode surface is blotted with absorbing material
- Repeat measurement cycle using undiluted serum or plasma on fresh reagent paper disc
- Clean electrode surface by blotting with absorbing material.

In view of all that has been said, and the precautions outlined for the correct control and functioning of these analyzers, it would be wrong to think that these instruments do not require careful oversight, or that due to the form in which reagents are presented and used, they are accessible to untrained staff with no specialist skills in clinical laboratory work.

BIBLIOGRAPHY

Free AH. Instrumentation for bedside information. Progress and challenges. Pure and Applied Chem 1982; 54: 2063-2073.

Free AH, Free HH. Selftesting - An important contribution of clinical chemistry to health delivery. Ann Biol Clin 1985; 43: 419-438.

Brooks KE, et al. Laboratory assessment of three monitors of blood glucose: Accu-check II, Glucometer II and Glucoscan 2000. Clin Chem 1986; 32: 2195-2200.

Godine JE, et al. Bedside capillary glucose measurement by staff nurses in a general hospital. Am J Med 1986; 80: 803-806.

Jagenburg R. Determination of blood glucose by reflectance spectroscopy. Upsala J Med Sci 1986; 91: 173-174.

Von Schenck H. Laboratory Evaluation of four reflectometers for blood glucose. Upsala J Med Sci 1986; 91: 169-172.

Whipple DR, et al. Evaluation of the Accu-Check bG for rapid quantitative whole blood glucose determinations. 35th Nat Meeting of the AACC. July 21-29, 1983.

Tao I, et al. Evaluation of the Accu-Check II blood glucose monitor. Joint Meeting of the AACC and the Can Soc Clin Chem. July 13-18, 1986.

Sherwood M. Determination of blood glucose with glucostix reagent strips. Joint Meeting of the AACC and the Can Soc Clin Chem. July 13-18, 1986.

Ryan PA, et al. Evaluation of the Reflotron in three physician office laboratories. Joint Meeting of the AACC and the Can Soc Clin Chem. July 13-18, 1986.

Silverstein J. Evaluation of the Accu-Check bG blood glucose meter for accuracy. Nat Meeting of the AACC. July 21-26, 1985.

Hay WW, et al. The Eyetone blood glucose reflectance colorimeter evaluated for in vitro and in vivo accuracy and clinical efficacy. Clin Chem 1983; 29: 558-560.

Onal HNG. Evaluation of the Chempro-1000, a compact "Dry Chemistry" analyzer designed for physician's office and bedside testing. Meeting of the ASCP/CAP, November 2-8, 1985.

Binisti J, et al. Evaluation de l'analyseur KODAK DT-60. Lab Biochemie, Centre Hospitalier, Meaux, France.

Hicks J, et al. Hemoglobin, electrolytes and other major clinical laboratory analytes as measured with a physician's office analyzer KODAK DT 60. Clin Chem 1986; 32: 2201-2203.

CHAPTER 10
CLINICAL CHEMICAL ASPECTS OF TRANSPLANTATION

RELEVANT ISSUES IN CLINICAL ORGAN TRANSPLANTATION

R. van Schilfgaarde

Department of Surgery, University Hospital Groningen
Groningen, The Netherlands

INTRODUCTION

During the past three decades, clinical organ transplantation has developed from the status of highly creative pioneering to a well-established and generally-well organized clinical activity performed in a large number of transplantation centers throughout the world. At present, many thousands of organ transplantations are performed each year. Kidney transplantation has been recognized as the treatment of choice in most patients with end-stage renal disease. Heart and liver transplantation are being performed with fastly increasing frequency since their value in the adequate treatment of patients with well-defined forms of life threatening heart and liver failure has been clearly demonstrated. Sound arguments for performing pancreas transplantation in selected patients with insulin dependent diabetes mellitus are readily available. The current experience, insight and scientific knowledge in the vast field of clinical organ transplantation is based on the tremendous effort of a variety of biomedical disciplines, ranging from (immuno)genetics, immunohaematology, immunobiology, (immuno)pathology, cellular biology, biochemistry and pharmacology to several cooperating clinical disciplines such as surgery, nephrology, cardiology, hepatology and endocrinology. Many are the mutual interrelationships between these disciplines, and many other disciplines are contributing in one way or another. Together, they have not only established the current clinical practice of organ transplantation, but they also pursue a continuous effort to further improve the results achieved and to extend the feasibility of clinical transplantation to new area's of public health care.

Obviously, it is impossible to adequately review all topics relevant to clinical organ transplantation in the compact format of this paper. Therefore, I shall try to restrict myself to briefly mentioning those items which in a general fashion may be taken to illustrate the present status and current practice of clinical organ transplantation. This, I hope, may serve the purpose of presenting a general introduction to the intended readers who, being clinical chemists, as a rule can be taken to have not more than a remote relationship with the daily practice of clinical organ transplantation.

INDICATIONS FOR ORGAN TRANSPLANTATION

Replacement of a failing organ by an adequately functioning organ from another individual is the basic purpose of organ transplantation. The indication should be based on a well-balanced interpretation of available alternatives on the one hand and chances of success on the other. Obviously, such a well-balanced interpretation is dependent upon the sort of failing organ, since alternatives to transplantation are available with some but not with other organs, It is also dependent upon the underlying cause of organ failure, since different causes are associated with different individual patient conditions and with different chances of success.

The function of a failing heart or a failing liver cannot be replaced for longer periods of time by other means than heart or liver transplantation. For these patients, death is the only alternative to organ transplantation. This consideration does not imply that each patient is a suitable candidate for organ transplantation. In regard to heart transplantation, excellent results can be achieved both with cardiomyopathy and with ischemic heart disease as the original cause of heart failure. However, many patients with ischemic heart disease have generalized arteriosclerosis which renders them less appropriate candidates for transplantation both in terms of medical as well as socio-economic considerations. Similarly, other factors such as age should be taken into account. Although good results can be achieved in older recipients, the general efficacy may be higher with younger recipients which consideration is pertinent in view of the ongoing shortage of available donors. In regard to liver transplantation, the reasoning is quite similar to that with heart transplantation. Excellent results can be obtained in patients with liver failure on the basis of primary biliary cirrhosis and several other fatal liver conditions, but the efficacy of liver transplantation for primary or secondary cancer or for acute hepatic failure with fulminant hepatitis is still doubtful.

The situation with kidney transplantation is quite different when compared to heart or liver. End-stage renal failure can be adequately treated with several forms of dialysis, which theoretically implies that the indication for renal transplantation is a relative rather than an absolute one. The choice in individual cases depends upon many factors, among which the local availability of either of both options for renal function replacement represents a very practical consideration. Generally speaking it can be said that, when performed in an experienced center, renal transplantation is to be preferred over dialysis, since it is associated with a superior quality of life and with lower public health care costs. In this context, it should be mentioned that the availability of dialysis as an alternative for replacing the failing kidney is of extreme importance in the execution of a clinical kidney transplant program. Not only does it offer a prolonged period of time during which the potential kidney graft recipient can wait for his best chance of receiving an optimal transplant, but it also offers the opportunity of sustaining life during transient episodes of impaired graft function as well as after irreversible graft failure while waiting for a second kidney transplant.

The indication for pancreas transplantation in insulin dependent diabetics is, at this time, still of a very conceptual nature. Insulin dependent diabetes is associated with a substantial risk of so-called late complications, which develop in the course of several decades. These late complications include retinopathy and eventual blindness, nephropathy and eventual end-stage renal disease, invalidating neuropathy and accelerated development of macrovascular disease with cardiovascular complications. Thus, in the long run, diabetes have a high chance of severe morbidity and increased mortality, which, as a commonly accepted concept, should be primarily explained as originating from insufficient blood-glucose regulation by conventional insulin treatment. Pancreas transplantation, therefore, aims at providing the insulin dependent diabetic with a permanent source of endogenous insulin which can be taken to yield a better chance for maintaining adequate glucose homeostasis than conventional (exogenous) insulin treatment. At present, not more than circumstantial evidence is available that pancreas transplantation will actually be capable of preventing the occurrence of late complications or at least reducing their severity. This implies that the indication for pancreas transplantation should still be restricted to selected individuals, and that it should be performed according to strictly defined clinical protocols in order to collect sufficient information for reliably answering the question as to its efficacy in actually preventing late complications and thus improving the quality of life for diabetics.

ORGAN DONATION

Organs to be used for transplantation can be derived either from living donor or from cadaver donors.

In case of cadaver donors, strict criteria should be applied for uniquivocally assessing death. Although there is an ongoing debate among medical, ethical, judicial and political factions regarding the question as to the most appropriate method for assessing death, most countries have regulations set or affirmed by law in regard to this matter. Of main importance is to discern between so-called "heart-beating" and "non-heart-beating" cadaver donors which, in other words, regards the question whether "brain-death" is an adequate criterium or whether not only the brain-function must have been irreversibly and completely eliminated but also the heart must have ceased to function. Since the heart action and consequently the circulation in brain-dead cadavers is often sufficient, organs derived from these donors are usually of a better quality for the purpose of transplantation than those derived from non-heart-beating cadaver donors, in whom the organs have already been exposed to the deleterious effects of insufficient perfusion during varying periods of time.

Cadaver donors, and especially heart-beating cadaver donors, can often be multiple organ donors. An adequate surgical procedure can serve the purpose of recovering the heart, the liver and both kidneys from one donor, and even the pancreas can be removed when special technical provisions are applied in regard to the anatomically shared arterial blood supply of the liver and the pancreas. The option of multiple organ donation is of extreme importance in view of the need for sufficient numbers of organ transplants to reduce the increasingly long waiting lists of potential recipients.

Living donors can be used for kidney transplantation to related recipients. This is generally accepted to be a safe procedure, and its clinical feasibility is supported by the excellent results which are superior to those obtained with kidneys derived from cadaver donors. The use of living donors for kidney transplantation to non-related recipients is a procedure applied by very few centers, which procedure should be considered only with the utmost reluctancy if not rejected.

PRESERVATION OF ORGANS

Once an organ has been removed from the body and thus has been deprived of its blood supply, specific measures should be taken to preserve its functional integrity until transplantation and thus reperfusion in the recipient has been established. Basically, two approaches are available for organ preservation. One is flush perfusion with a preservation fluid and subsequent cold storage by immersion at about 0°C. The other is machine preservation by continuous perfusion with a preservation fluid at 5-10°C. Flush perfusion and cold storage is a simple and adequate procedure which is widely applied in the majority of transplantation centers. Continuous machine perfusion is more complicated, but longer periods of preservation are achievable and parameters for interpreting the quality of the preserved organ can be more readily obtained. Limits of preservation times differ considerably among different organs. Kidneys can be kept in good shape during at least 48 hours with cold storage after flush perfusion, although shorter periods of cold storage appear to be associated with relatively better graft function. But preservation times of hearts, livers and pancreases should be kept to a minimum. At present, the limits are still 4-6 hours with hearts, 6-8 hours with livers, and 10-12 hours with pancreases although incidentically longer periods of cold ischemia have been tolerated. Obviously, these limits have profound effects in regard to the logistics of clinical organ transplantation. Eventual organ sharing programs between centers are complex organizations by nature, but they can be more readily achieved for kidneys than for other organs, since their limited tolerance of cold ischemia necessitates an extreme grade of concordance between the donor and recipient procedures.

TRANSPLANTATION - THE SURGICAL PROCEDURE

When considering surgical aspects of the transplantation procedure as such, it appears that there are three general questions of importance. One is: where to place the transplant? The second is: will all or only part of the organ be transplanted? The third regards the surgical specifics once these decisions have been made.

Where to place the transplant?

Organ transplants can be placed either in an orthotopic or heterotopic position. Orthotopic transplantation holds that the native and failing organ is exchanged for the graft. This orthotopic technique is commonly applied with heart and liver transplantation. The failing heart is excised and the donor heart is implanted while the circulation and gas exchange are maintained by extracorporeal techniques as with other forms of open heart surgery. With orthotopic liver transplantation, there is no need for extracorporeal circulation. However, the so-called "anhepatic phase" after removal of the native liver may be associated with a variety of specific problems. One of these derives from the temporarily obstructed mesenteric venous (portal venous) outflow tract, for which several centers use an extracorporeal veno-venous (portal-systemic) bypass until the portal flow through the liver graft has been re-established. Both heart and liver can also be heterotopically transplanted. With heterotopic heart transplantation, the graft is anastomosed side to side with the diseased heart on its right side. With heterotopic liver transplantation, the graft is placed intraabdominally and caudally to the diseased liver. Since the liver is a very large organ, the heterotopic technique necessitates to reduce the size of the liver transplant because of lack of intraabdominal space. Therefore heterotopic liver transplantation is performed with a partial rather than with a whole organ liver graft. A major goal of heterotopic transplantation of either heart or liver is to provide for functional organ replacement on the one hand and for an opportunity of the failing organ to recover from its disease on the other. Therefore, "auxiliary" may be a more appropriate term than "heterotopic" to identify the surgical procedure.

Kidney transplantation is always performed heterotopically and extraperitoneally in the iliac fossa, since orthotopic transplantation has no advantage. The choice for the right or left side depends upon individual recipient factors such as previous operations and the condition of the iliac vascular tract. When renal transplantation is performed in children, a transabdominal approach may be required in recipients of small size. Removal of one or both native kidneys is not necessary in most instances, but some conditions (such as complicated urological disease, persistent infection, massive cystic disease, or persistent hypertension in spite of adequate medical treatment) may render uni- or bilateral nephrectomy advisable.

With pancreas transplantation for insulin dependent diabetes, there is no reason to remove the native pancreas which (apart from the absence of beta cell function) is not diseased. Most centers perform pancreas transplantation in the heterotopic position, quite similarly to kidney transplantation. There is a modified approach called "paratopic" pancreas transplantation, with which technique the venous connection is not made with the iliac vein (systemic circulation) but with the splenic vein (portal circulation). This, unmistakably, is the more physiologic route for delivering insulin, but the clinical efficacy of the paratopic as compared to the heterotopic technique has still to be established.

Partial or whole organ transplantation?

For heart and kidney transplantation, whole organ transplantation is the only realistic approach. But the liver, with its segmental anatomy, offers the opportunity of partial transplantation. This is of importance in two circumstances, both of which require a liver transplant of reduced size. One is orthotopic liver transplantation in small individuals in general and in children in particular. Here, only a liver graft derived from a donor of similar or smaller size can be applied, but such donors are not always readily available. Therefore, livers derived from larger donors can be applied by reducing them to a partial liver graft of smaller size, thus extending the opportunity for small individuals with hepatic failure to receive their transplant. The other condition requiring a reduced size liver transplant is with auxiliary (heterotopic) liver transplantation. Although eventual indications for this method of liver transplantation are still the subject of debate, it presently appears that at least a small number of patients may benefit from auxiliary partial liver transplantation.

The pancreas, too, can be transplanted as a partial graft. Here the commonly applied term is "segmental" pancreas transplantation, since the body and tail of the pancreas retrieve their blood supply from the splenic vessels which offers the opportunity to transplant the body and tail together as a pancreatic segment. Whole pancreas transplantation includes the head and, for anatomic reasons, also that part of the duodenum that is attached to the head. The pancreas as a whole receives it blood supply from the combination of celiac trunc (including the splenic artery) and superior mesenteric artery, and it delivers its venous effluent through the portal vein. Both whole and segmental pancreas transplantation have been shown to be clinically feasible, but whole pancreas transplants appear to be preferred since they contain a much larger beta cell mass than segmental grafts which may offer better chances of long-term success.

Specifics of the surgical procedure

Technical aspects of the surgical procedure in organ transplantation can only be reviewed in very general terms. Basic and similar for each organ is to reconstitute adequate perfusion by connecting the grafts' arterial entry and venous outflow to the recipients' arterial and venous circuits, respectively. For the kidney and the pancreas this implies in principle one arterial and one venous anastomosis, which usually are made with the iliac artery and vein. For the orthotopically transplanted liver, the venous connection is made by two anastomoses since the graft contains a segment of the caval vein which is interposed in the recipient by means of both a proximal and a distal venous vascular connection. In addition, the liver receives the portal venous blood flow, which necessitates a third venous anastomosis between the donor and recipient portal vein. The orthotopically transplanted heart needs in principle four vascular anastomoses because of its anatomy with two arterial and two ventricular compartments.

Apart from establishing vascular connections, the liver, the kidney and the pancreas require additional measures since they produce fluids (bile, urine and exocrine secretion, respectively). With liver transplantation, the bile should be conducted to the small bowel for which purpose an anastomosis between the donor and recipient choledochal duct may serve or a choledocho-jejunostomy with a Roux-en-Y loop. With kidney transplantation, the urine should be conducted to the bladder for which purpose a uretero-cystostomy is the widely applied technique. With pancreas transplantation, one has to choose between either abolishing or maintaining the exocrine secretion. Abrogation of exocrine secretion can be brought about by injecting the exocrine ductal system with a latex-like substance which quickly solidifies after injection. This technique of "ductinjection" is effective in deleting exocrine secretions since it induces a complete atrophy of exocrine tissue. However, it also interferes with the integrity of the endocrine tissue which is an obvious drawback in spite of the well-established reliability for obviating complications from the side of exocrine secretion. Maintenance of exocrine secretion can be brought about by anastomosing the pancreatic ductal system (or the duodenum to which it is connected) to either the small bowel (pancreatico-jejunostomy by means of Roux-en-Y loop) or the bladder (pancreatico-duodeno-cystostomy). Although such techniques require an extended surgical procedure and, consequently, may be associated with an increased chance of surgical complications, they yield a better chance for maintaining the integrity of both the exocrine and endocrine pancreatic tissue and thus for long-term success.

METHODS FOR CIRCUMVENTING REJECTION

Rejection is the major cause of graft failure. Its occurrence is suppressed by life-long immunosuppressive medication. Conventional immunosuppression includes azathioprine and steroids, and since 1977 cyclosporine has gained wide application in organ transplantation. Many regimens and combinations have been tested in clinical trials, and ongoing and future trials are expected to yield further information as to the specifics of optimal immunosuppressive treatment. Cyclosporine has not only excellent immunosuppressive qualities when compared to conventional treatment, but it also has the additional advantage that the steroid dosage can be substantially reduced with a concomitant reduction of the often invalidating and sometimes life-threatening complications associated with long-term steroid medication in high dosages. However, cyclosporine as such has toxic side effects as well, among which nephrotoxicity is an extremely disturbing one. Additional measures for immunosuppression can be taken by applying anti-lymphocyte or anti-thymoyte globulin, and the use of monoclonal antibodies is currently under investigation in several centers. For adequately treating eventual rejection crises, methods of monitoring rejection are of utmost importance. Considerations concerning this topic go beyond the scope of this paper.

The second approach to reduce the chance of rejection is to minimize the immunogenetic disparities between donor and recipient. Prospective application of tissue typing and HLA-matching is feasible only with kidney transplantation since, as a rule, other organs tolerate insufficient periods of ischemia for the typing and matching procedures to be completed. Matching for HLA-A and B, but especially for HLA-DR, has been shown to favourably influence the outcome of clinical kidney transplantation. Retrospective analyses with other organ transplants have not yet yielded sufficient information as to reach a conclusion.

The third approach aims at modulating the immunologic status of the recipient so as to run a lesser chance of rejecting an organ graft. This modulation is brought about by giving the prospective recipient of a cadaver kidney graft a number of blood transfusions from random donors. The efficacy of this policy is well-documented, but the extent of its effect seems lately to have been surpassed by the strong immunosuppressive action of cyclosporine. With living related kidney transplantation, excellent results have been achieved with blood transfusions from the prospective kidney donor ("donor specific transfusion", DST). The drawback of the transfusion policy is that it carries the risk of unintended sensitization in 5-10% of potential recipients.

The fourth approach aims at modulating the immunogenicity of the graft itself. This has been tried by means of several so-called "donor pretreatment" regimens, but this method has never been quite effective. Reducing graft immunogenicity by ex vivo manipulation of the graft prior to transplantation is the subject of experimental research but, at present, such techniques are not clinically applied.

RESULTS OF ORGAN TRANSPLANTATION

Results of organ transplantation can be expressed in many ways. The patient survival is an extremely important parameter for each type of organ transplant. For heart and liver transplantation, patient survival rates are almost identical to graft survival rates, since alternative modes of treatment are not available. But for kidney and pancreas transplantation, patient and graft survival rates differ widely. With pancreas transplantation, in addition, graft survival rates give only a partial impression of success since it aims primarily at deleting the late complications of diabetes and not at obviating the need of insulin injections as such. Other aspects should also be taken into account when considering results of organ transplantation. Such aspects include the quality of graft function and the incidence of transplantation associated complications and morbidity or, in other words, the quality of life. They also include the physiological effects on organ graft recipients, their social rehabilitation and, more in general, the socio-economic effects of organ transplantation programs.

At present, well over 4,000 heart transplants have been performed, of which 1,415 in 1986. The 5 year survival rate is in the order of 65%, with a 30 day mortality of 11.5%. The number of liver transplants currently performed is approximately 3,000 and the number of liver transplant centers has increased from 4 in 1983 to about 40 in Europe and 50 in North America. The 2 year survival rate can be estimated in the order of 50%.
In 1984, 17,000 kidney transplants were performed. In Europe, 1 out of every 5 patients with end-stage renal disease is carrying a functioning kidney graft, which implies that there were 23,000 individuals with a functioning kidney graft in 1984. Graft survival varies between 75 and 90% after the first year depending upon the center, the country, selection criteria and whether cadaver or living related kidneys were used. Patient

survival is almost 95% with kidney transplantation. Both graft and patient survival rates gradually reduce over the years after transplantation. Pancreas transplantation has initially yielded very disappointing results. At present, however, the graft survival rate at 1 year is in the order of 40-50%, and several of the more experienced centers report percentages in the order of 70. The number of pancreas transplants performed is well over 1,200 and, when analyzed according to time era's, results in terms of graft survival are unmistakably and rapidly improving. Information as to the effects of pancreas transplantation on long-term diabetic complications is only sparsely available.

FINAL REMARKS

The foregoing text is by no means intended to give a complete or detailed overview. It was tried to present an inventory of those items which, together, give a fair but rather superficial view not only on the current status of clinical organ transplantation but also on the way of thinking in that field and on the methods applied in daily practice. An accurate reference list would have been extremely lengthy and was therefore deleted. The interested reader is rather advised to use one of the more specific handbooks available in order to find his way to more detailed and, in its detail, possibly more accurate information.

BIOCHEMICAL ASPECTS OF TRANSPLANTATION

E. Schmidt*, F.W. Schmidt*, U. Lautz*, G. Gubernatis**, W Lauchart**, R. Pichlmayer**, S. Raygrotzky**, B. Ringe**, K. Wonigeit**, J. Bernauer***, P.L. Reichertz***, G. Tusch*** and J. Kemnitz****

*Abteilung Gastroenterologie und Hepatologie, **Klinik für Abdominal- und Transplantationschirugie, ***Medizinische Informatik, ****Pathologisches Institut, Medizinische Hochschule Hannover, FRG

The report is mainly based on own experience and, thus, virtually confined to liver transplantation (LTX). Two crucial problems are selected to demonstrate the possibilities and the limitations of clinical chemistry for clinical decision-making. The first is the indication for LTX, the other is the detection and differential diagnosis of post-operative complications.

Cirrhosis of the liver, the most common indication for LTX, is the last stage of a chronic inflammatory process. Sometimes, the course is rapid with permanent signs and symptoms of chronic active hepatitis, sometimes, it is slow: asymptomatic and, thus, unrecognized, portal hypertension, liver insufficiency and finally hepatocellular carcinoma develop gradually over years and decades. Therefore, the question, if a patient with liver cirrhosis is a candidate for LTX, has to be considered on a temporal background.

Surgery as such, and the subsequent lifelong immunosuppression bear various unavoidable risks of their own. So, it must be asked: When does the improvement of life expectancy through LTX compensate for these risks? - It can be deduced that this must be rather late in slowly progressive cirrhoses. Thus, it is questionable, if these patients with scarcely any signs of hepatic dysfunction and/or chronic inflammation are transplant candidates at all.

In the group of progressive cirrhoses the question of candidacy has two sides. The first: Does the patient need liver transplantation because his survival time and his quality of life would thereby be markedly improved, that means, would be virtually cured? The second: Is the patient in such a condition that he has a fair chance to stand the transplantation and the period afterwards? The more the cirrhosis is advanced, the more LTX becomes the only rational treatment, however, the higher is the risk that it will not succeed.
The so-called "therapeutical window" must be found for each individual patient, weighing his spontaneous life expectancy against the restriction of his operability as a consequence of his advanced liver disease.

For both purposes a series of indices exists, which comprise, apart from physiological and morphological data, clinico-chemical information on liver function and on the state of other organ systems. It could have been expected that neither the prognosis indices like the CCLI according to Orrego (1) or the Index of the Copenhagen Study Group for Liver Diseases (2) which have been designed for the evaluation of conservative therapeutical effects, nor the severity of disease indices for intensive care use, like APACHE (3) alone proved to be helpful in the indication for LTX (4). It came as a surprise, however, that upon retrograde evaluation of the fate of transplanted patients, also the combined scores did not correlate well with the final outcome (4).

In the attempt to improve the predictive estimates for candidate election and particularly for the timing of LTX, we investigated retrospectively out of 1,145 patients with liver cirrhosis, who had presented at our Medical School, 658 cirrhotics with known date of death backwards up to more than 30 years (5). The last five years were considered to be the decisive period with respect to the timing of LTX. In this interval we studied besides 10 clinical signs, 12 laboratory parameters. Six of the latter appeared interesting under the prognostic aspect and were chosen for closer inspection. Three of them are traditional components of prognostic indices: Prothrombin time (Quick test), albumin and bilirubin. The other three have proven useful in our hands for the assessment of individual progression: Cholinesterase (CHE), alanine aminotransferase (ALAT) and gamma-glutamyltransferase (GGT). It soon turned out that the courses of viral, toxic and primary biliary cirrhosis must be considered separately. From each year two values at the dates of the highest and of the lowest activity of aspartate aminotransferase in serum registered, have been documented. For clarity only the medians of the two values are plotted against time on the abscissa in Figure 1. The medians of the respective parameter, as determined in our transplanted patients with the respective type of cirrhosis immediately prior to LTX, are given as "pre-LTX value".

It can be seen that albumin and ALAT have no recognizable prognostic value in toxic and biliary cirrhoses. In viral cirrhosis the conspicious drop of ALAT activity with the consequential conversion of the DeRitis ratio occurs early, as a rule between the 5th and the 4th year prior to death. The fall of albumin is not marked, but a late and ominous sign. According to ALAT activity and albumin concentration, in all groups LTX has been performed in a late disease stage.

The courses of two other indicators of hepatic protein synthesis, prothrombin time, on Figure 1 expressed as Quick value, and CHE activity are rather similar. The decline of CHE appears steadier, whereas the Quick value exhibits plateaus, particularly in the viral and the toxic cirrhoses. Again the pre-transplantation values fit well into the medians of the last year with the exception of CHE in the viral cirrhoses, which corresponds to the median activity 2 years prior to death of the retrospectively examined collective.

The courses of the two indicators of cholestasis, bilirubin concentration and GGT activity, are rather different, easy to conceive on the ground of the different mechanisms, which lie behind their elevation in the extracellular space: Excretory failure of bilirubin on the one hand and induced synthesis of the enzyme concomitant with enhanced solubilization on the other. The causes for the latter are multifarious, and so, it is not surprising that GGT has much less prognostic value statistically than in individual courses, when these causes are known, although in Figure 1 the more or less marked decline of GGT in the last years of life is obvious.

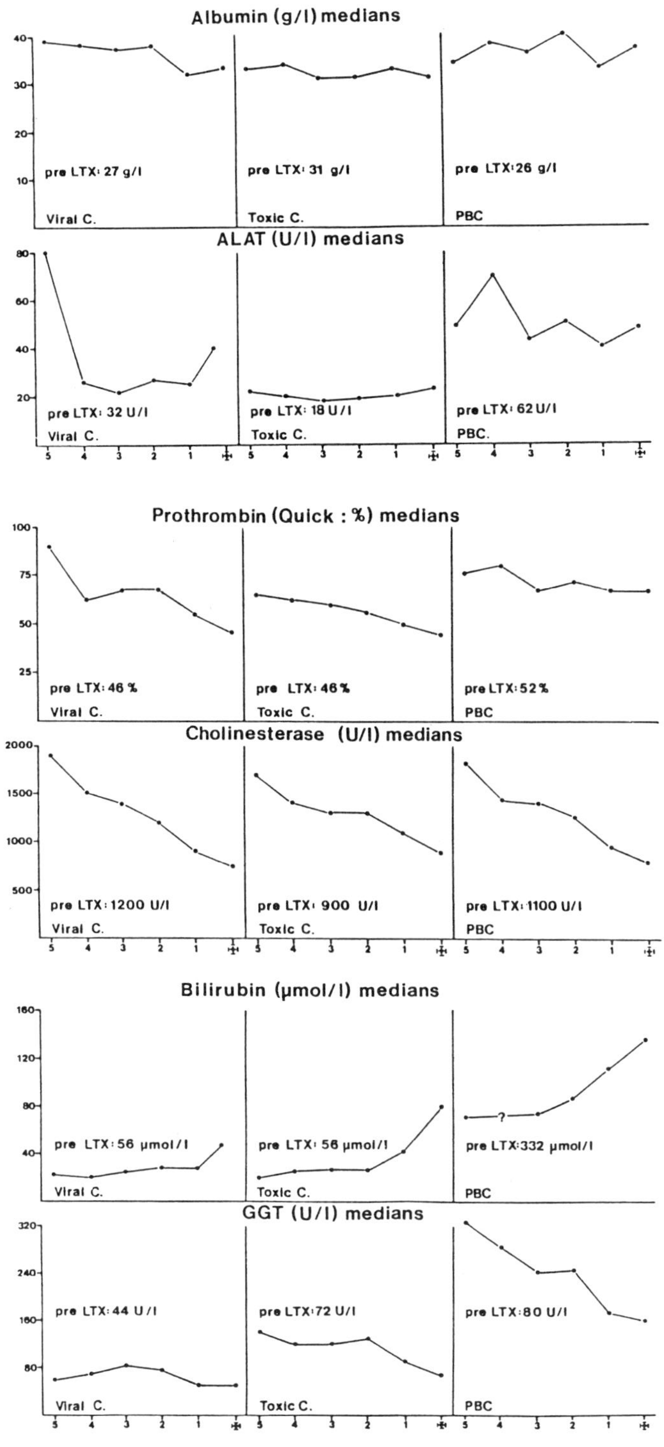

Fig. 1. For explanation see text.

This is, of course, not due to less cholestasis but to impaired synthesis, not only for secreted, but also for cellular enzymes. Both indicators of cholestasis show considerably higher values in primary biliary cirrhosis, as was to be expected. Unexplained, though, is up to now, why the severity of cholestasis in our transplant candidates exceeded so much that of the patients in the retrospective study. One reason may be that primary biliary cirrhosis, which amounts only to 3% of all patients in the retrospective study, is grossly overrepresented among liver graft recipients, with deep jaundice and unbearable itching being important components of the indication for LTX.

The satisfactory appearance of the medians of CHE, prothrombin, bilirubin and even GGT as prognostic indicators, as shown in Figure 1, is deceptive. This becomes evident, when the medians are replaced by means and ranges. The persistence of normal liver functions in part of the patients until the end underlines the fact that death from liver failure or any other liver-related event in general is only one of the possible causes of death in cirrhotics. Is does not devaluate these parameters as components of scoring systems for the timing of LTX, but stresses the need of complex profiles of the diverse liver functions on one hand and an etiology-related analysis on the other (6).

For the second problem in timing for LTX, that of the therapeutic risk, the contribution of laboratory data should be considerable. In the reduced APACHE II Index (7), besides age, so-called chronic health points, the 12-point-acute-physiology score consists of 5 clinico-physical and 7 laboratory parameters. As transplantation aptitude test APACHE II, like others, has not been successful in our experience. So, in parallel to and in interaction with the retrospective study a prospective trial has been started for all patients who present at our Medical School with liver cirrhosis under the aspect of LTX. Contingent on the type of cirrhosis

LABORATORY COMPONENTS OF THE MHH PRE-LTX SCORING SYSTEM (hypothetical)

Etiology-independent scores:

1. Stage, activity and complications:
   Hemoglobin - WBC - platelets
   ICG clearance - ammonia
2. Nutritional state (in addition to various physical data):
   CK activity/Quételet index (muscle index III)
3. Risk of infections:
   Gamma-globulins - lymphocyte count and differentiation
   - Mérieux-test

Etiology-dependent scores:

	viral	toxic	biliary
Bilirubin	1	0.25	1 x years
Prothrombin time	1	1	0.50
Cholinesterase	1	0.65	0.50
DeRitis ratio	1	0.65	0.65
Bile acids	1	0.65	0.40
Alkaline phosphatase	1	0.80	0.40
Alpha-1-fetoprotein	1	1	∅
GGT	∅	1	∅
MCV	∅	1	∅

prognosis scores on the one hand and operability scores on the other are fixed in regular intervals at each presentation. These tentative scores are based on clinical data (8), but largely also on clinico-chemical measurements.

The latter comprise for all etiologies the basal pattern and additional tentative scores for the nutritional state and risk of infection. The etiology-dependent scores are graded for toxic and biliary cirrhoses as fractions of those for viral cirrhoses, or they are omitted or added for one type or another. For the direct peri-operative risk, muscular, cardio-pulmonal and renal functions in addition to the nutritional and immunological state are of high significance. Their estimates form an important part of the risk score, together with the history of previous bleeding episodes, the recent alcohol intake and the neurological state. The reliability of pre-operatively elevated plasma renin activity as a sign predictive of impaired post-operative renal function, which has been shown for a small group of patients (9), is still under investigation.

The second part of the report is devoted to the contribution of clinical biochemistry for decision-making early after liver transplantation. In the post-operative period, beyond the usual intensive care profiles for the evaluation of vital functions, of fluid and electrolyte balance, cardiovascular, pulmonary and renal functions, coagulation and infection, a special liver-related daily protocol is required, which allows to judge the function of the graft, the quality of immunosuppression and the incidence of complications. Clinical differential diagnosis is particularly difficult then for obvious reasons. One is the unstable condition of the patient, who has frequently received massive transfusion in compensation for the intra-operative blood losses, and who may be on respiration and parenteral nutrition and often requires renewed substitution with blood, plasma or plasma fractions. These substitutions complicate diagnosis by laboratory means: Some parameters, e.g. plasma proteins become so therapy-dependent that they have no more any value as indicators of liver function. Another difficulty is the interrelation between the usual complications, mediated by the central metabolic role of the liver on one hand and the sequelae of therapeutic interventions on the other, which lead to very complex disturbances: The true nature of about one third of all episodes of hepatic dysfunction with enzyme elevations in serum, which occur after the first post-operative peak, remains unclear (10).

The nearly regular initial peak at the first post-operative day has an extremely wide range. Aminotransferase activities up to 300 fold and glutamate dehydrogenase (GLDH) elevations up to 750 fold their respective upper reference limits occur. The height of this first peak bears neither a significant correlation to the assessed pre-operative state of graft, nor to the duration and difficulty of the surgical procedure, the biochemical signs of which are frequently neutralized by the dilution of the patients plasma with transfused blood, nor appears the initial enzyme elevation to have any prognostic significance. Moreover, its prolongation, transiently considered by us to indicate primary non-function of the graft, was found to be unreliable: Enzyme levels in plasma due to release from damaged cells are so perfusion-dependent that primary non-function can show opposite pictures contingent on the degree of circulatory disturbance. If re-transplantation is not possible in time, the patient may die with extremely high enzyme elevations or with a nearly normal enzyme pattern in serum.

Later, re-elevation of cell enzymes in plasma and, less regularly and distorted by substitution, decrease of short-lived plasma-specific enzymes are, as a rule, the first practical signs of impending deterioration of liver function or of complications. They serve as alert for heightened attention to clinical symptoms, closer monitoring of the immunosuppressive therapy and specific diagnostic measures, which may lead to a correct diagnosis in time for successful therapy. In the role as alarm signals, serial enzyme determinations are indispensable.

The differential diagnostic value of enzyme patterns is more difficult to assess. Their potential and their limitation are shown in two post-operative situations, where decisions between different or even contrary treatments must be made at once. Early acute rejection, the specific complication, seems to be unavoidable to a certain degree in spite of controlled immunosuppression. Its severity is very variable. The main differential diagnostic problem is to distinguish between rejection, which needs medical treatment by enhanced immunosuppression, on one hand and vascular occlusion and biliary obstructions, which require both surgical therapy, on the other.

Figure 2 shows the pertinent enzyme patterns at the day of maximum changes. The differences between the extremely high activities of cell enzymes in vascular occlusion and those in acute rejection are significant despite the wide ranges of variation, except with GGT (and CHE). The pattern of biliary obstruction is more similar to that of rejection, although ASAT, LDH and HBDH are significantly lower and ALP is significantly higher in the predominantly cholestatic condition. Corresponding to morphology, early and slight rejection resembles more biliary obstruction, whereas in severe rejection the characteristics of circulatory failure may develop and finally prevail (10).

The second differential diagnostic problem arises from immunosuppression and - again - the therapeutical decision has to be made between opposite treatments. Insufficient immunosuppression causes rejection: A higher dosage and bulk corticosteroids become necessary. Overdosed immunosuppressive drugs, however, are hepatotoxic: Dose reduction or change of the drug is the remedy of choice. Finally, immunocompromized patients are liable to contract generalized infections: Antibiotic and antimycotic drugs together with reduced immunosuppression will be applied.

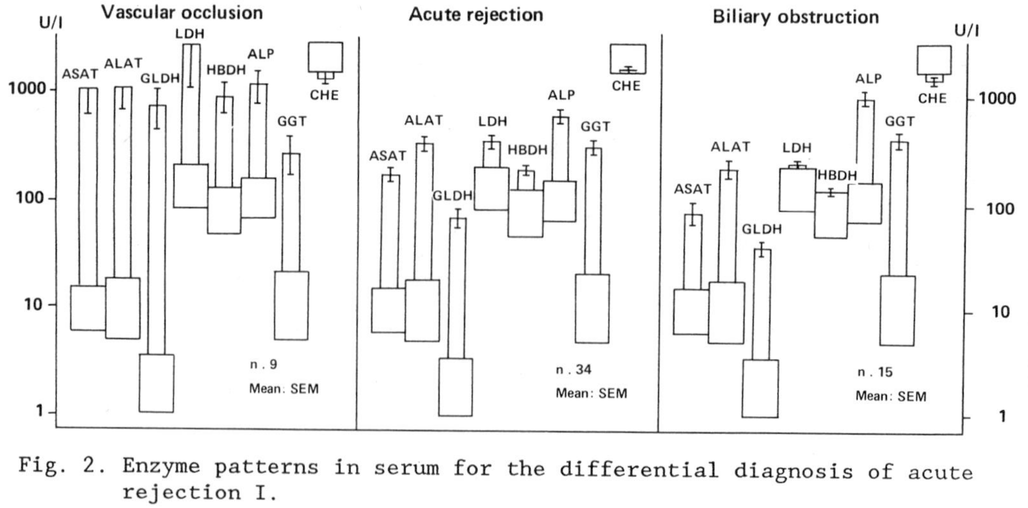

Fig. 2. Enzyme patterns in serum for the differential diagnosis of acute rejection I.

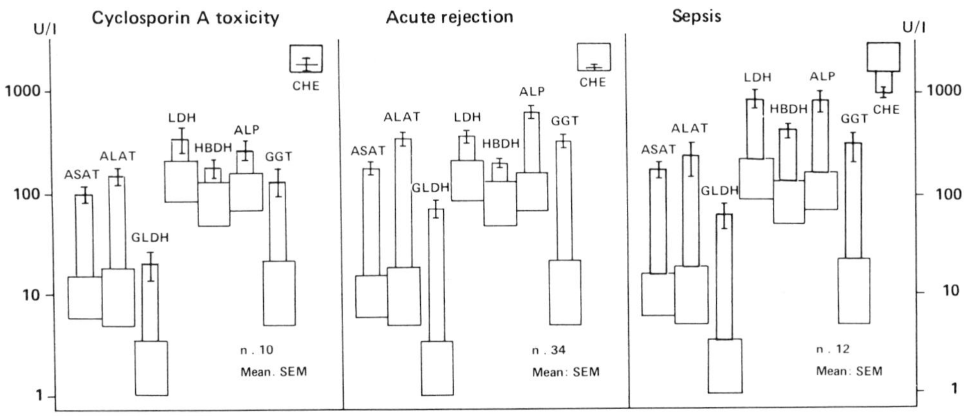

Fig. 3. Enzyme patterns in serum for the differential diagnosis of acute rejection II

Figure 3 shows the enzyme pattern of acute rejection framed by that of septicemia and that of Cyclosporin hepatotoxicity. Although Cyclosporin A is much less hepatotoxic than nephrotoxic, and despite individual dosage, overtreatment occurs, due to impairment of graft function or sometimes early after the replacement of parenteral by oral administration. Unknown factors, related to intestinal absorption, liver function and drug interactions seem to distort the relationship between dose and blood level of Cyclosporin and between blood level and overt hepatotoxicity. Therefore, serial determinations of Cyclosporin in blood are necessary for monitoring the immunosuppressive effect, but insufficient to predict hepatotoxicity (10, 11).

The mean enzyme changes in Cyclosporin hepatotoxicity, less marked than in acute rejection, are significantly different with regard to ALAT, ALP and GGT. The enzyme pattern in septicemia stands out for its high LDH. The normal ratio HBDH (LDH_{4H})/LDH indicates its non-hepatocellular origin. The other characteristic, the low CHE, is, in the early post-operative period, frequently not discernible due to superimposition by exogenous CHE.

Multivariate discriminant analysis of the enzyme patterns in the 5 different complications shows that Cyclosporin hepatotoxicity is correctly reclassified in 80%, acute vascular failure in 78%, biliary obstruction in 73%, septicemia in 64% and acute rejection only in 61% (11). The overall correct classification amounts to 70%, which is not satisfactory, but actually helpful in practice, although sometimes too late (11).

With regard to the aim to differentiate between possible complications at their very beginning, the early indicators of imminent hepatic dysfunction must be reconsidered under the aspect of candidacy for timely differential diagnosis (Table 1).

On the table seven parameters are listed in falling succession, preliminary findings, derived from 49 episodes of hepatic dysfunction, where daily registration after the first post-operative week made the evaluation possible. Prevalence, the amplitude of the first change, which must surpass daily variations to attract attention, and of course, the earliness of these first changes must be considered together, the more so, as the sequence of the individual parameters changes from one aspect to the

Table 1. Early events in the diagnosis of hepatic dysfunction I

Prevalence up to 2 weeks before maximum		Amplitude of 1st change Median (range) in %	Earliness of 1st change Median (range) in days
ASAT activity	0.92	GGT + 128 (33-532)	GLDH 4 (1-10)
Bilirubin concentraton	0.90	GLDH + 108 (13-520)	ALAT 3.5 (1- 7)
ALP activity	0.82	ALAT + 74 (10-656)	Bili. 3 (1-10)
GLDH activity	0.76	ALP + 60 (10-170)	ALP 3 (1- 7)
WBC number	0.65	ASAT + 54 (17-282)	ASAT 2 (1-79)
GGT activity	0.63	Bili.+ 30 (11-144)	GGT 2 (1- 6)
ALAT activity	0.62	WBC + 30 (8- 98)	WBC 2 (1- 4)

"Prevalence x amplitude x earliness" (medians)
GLDH: 328
GGT and ALAT: 161 - ALP: 148
ASAT: 99 - Bilirubin: 81 - WBC: 39

other. The product of all three qualities might be called a "utility score", in which glutamate dehydrogenase turns out to be best, followed by GGT, ALAT and ALP, whereas ASAT, bilirubin and WBC bring up the rear. Experience has shown this order to be true in many individual cases under the aspect as non-specific alerts.

The reason, why early differential diagnosis is still unsatisfactory, even if numerous other laboratory parameters are included in the profile, is that in reality the clinical pictures reveal themselves mostly not in a pure form, but are complex from the beginning or merge into each other with time. Thus, vicious cycles develop, in which erraneous, too late or too strong therapeutical measures play a not unimportant role. This is depicted in Figure 4.

The problem to be solved is to recognize with enough probability the site, where the cycle starts, which must not be rejection, but can just as well be hepatotoxicity or a viral, bacterial or fungal infection.

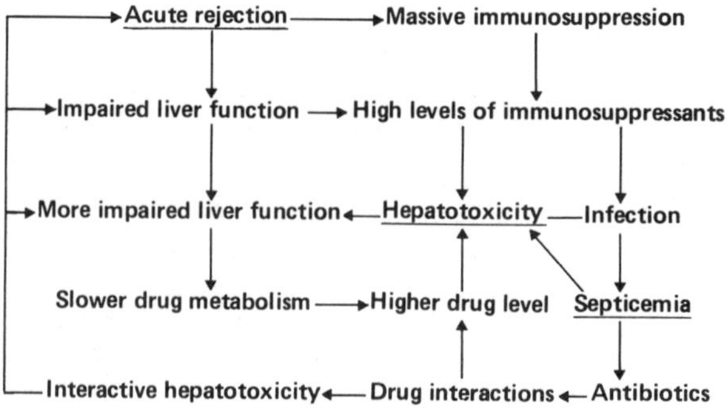

Fig. 4. For explanation see text.

REFERENCES

1. Orrego H, Israel Y, Blake JE, Medline A. Hepatology 1983; 3: 896-905.
2. Christensen E, Schlichting P, Fauerholdt L, Juhl E, Poulsen H, Tygstrup N and the Copenhagen Study Group for Liver Diseases. Hepatology 1985; 5: 843-853.
3. Knaus WA, Zimmermann JE, Wagner DP, Draper EA, Lawrence DE. Crit Care Med 1984; 9: 591-597.
4. Lauchart W. Habilitationsschrift. Med. Hochschule Hannover 1986.
5. Burghard M, Henze A, Russmann U, Lobers J, Schmidt E, Schmidt FW. In: Brölsch CE, Zelder O, eds. Experimental and Clinical Hepatology. Lancaster: MTP Press Ltd., 1986; Vol 5: 80-88.
6. Neuberger J, Altman DG, Christensen E, Tygstrup N, Williams R. Transplantation 1986; 41: 713-716.
7. Knaus WA, Draper EA, Wagner DP, Zimmermann JE. Crit Care Med 1985; 13: 818-829.
8. Pichlmayr R, Neuhaus P. Chirurg 1985; 56: 211-215.
9. Armbrecht A, Brunkhorst R, Brölsch CE, et al. In: Brölsch CE, Zelder O, eds. Experimental and Clinical Hepatology. Lancaster: MTP Press Ltd., 1986; Vol 5: 31-35.
10. Schmidt E, Schmidt FW, Ohlendorf S, et al. Clin Enzymol Symposia 1986; 5: 143-156.
11. Schmidt E, Schmidt FW, Wonigeit K, et al. In: Brölsch CE, Zelder O, eds. Experimental and Clinical Hepatology. Lancaster: MTP Press Ltd., 1986; Vol 5: 23-30.

HLA CLASS II ANTIGENS - SIGNIFICANCE IN ORGAN TRANSPLANTATION

Kathryn J. Wood

Nuffield Department of Surgery, University of Oxford, John Radcliffe Hospital, Oxford OX3 9DU, U.K.

The major and minor histocompatibility systems are primarily responsible for determining the fate of an organ graft when it is transplanted into an unmodified recipient. Incompatibility for major and/or minor histocompatibility antigens can result in graft rejection. The major histocompatibility antigens are coded by a gene complex - the major histocompatibility complex (MHC). The MHC genes and molecules are the best characterised of these two histocompatibility systems. A discussion of the MHC and its significance in transplantation will form the major part of this article, however it is important to point out that minor histocompatibility antigens also play a role in determining the survival of an allograft. Incompatibility for minor histocompatibility antigens only, in other words the situation that occurs when the donor and the recipient are identical or matched for all the MHC antigens, can result in graft rejection. The importance of minor antigens has been demonstrated using experimental systems, particularly in the mouse where these antigens have been best characterised (1, 2), but it has also been shown in clinical bone marrow transplantation. In the latter case immune responses against minor histocompatibility have been used to develop reagents for characterising minor antigens in man (3). These data show that minor histocompatibility antigens can act both as target molecules for the effector mechanisms that mediate graft destruction, as well as in the genetic control of the immune response to an organ graft (2). Many different minor histocompatibility systems exist and much more work is required to fully characterise them. At present there is insufficient information available on the molecular nature of minor histocompatibility antigens and no way of detecting these antigens routinely in man.

The genes of the major histocompatibility complex and their products have been the subject of intense investigation over many years. A MHC has been identified in all the vertebrate species examined so far, in man it is known as the HLA complex. The MHC codes for three families of polymorphic glycoproteins, the class I, II and III antigens or molecules. The class I and II antigens are expressed at the cell surface, class I antigens by all somatic nucleated cells and class II antigens by a more limited number of cell types, including antigen presenting cells and B lymphocytes. The class III molecules comprise some of the complement proteins, namely C4, C2 and factor B.

Each class I molecule is composed of two polypeptide chains, a heavy chain (Mr45k) coded by the MHC that expresses the polymorphic determinant(s) and spans the plasma membrane of the cells. In man the class I heavy chains are coded by the A, B and C loci of the HLA complex. Each class I heavy chain noncovalently associates with B_2 microglobulin (B_2M) (Mr12k). B_2M is relatively invariant in the population and the B_2M gene is not located in the MHC. The genes for class II antigens are located in the HLA-D region of the complex. Class II molecules are also heterodimeric, composed of an alpha chain and a beta chain, both of these chains are encoded by genes within the MHC. The HLA-D region was originally thought to code for only one family class II products, the HLA-DR antigens. These antigens were first characterised at the HLA Workshop in 1977, when 7 different specificities were identified (4). Since 1977 a variety of techniques have been developed that can identify HLA class II polymorphisms. These include: serology (4), the mixed lymphocyte reaction (MLR) (5), primed lymphocyte typing (PLT) (6), T cell clones (7), 2-dimensional gel electrophoresis (8, 9), DNA cloning (10) and restriction fragment analysis (11). As the analysis of the HLA-D region progressed it became clear that it coded for more than one product. Two other series of HLA-D antigens have now been characterised, HLA-DQ and HLA-DP.

The genes of the HLA-D region have recently been mapped using the technique of pure field gel electrophoresis (12). The organisation of the HLA-D region is shown in Figure 1. Many more alpha and beta genes exist than the number of protein products that have been identified at the cell surface.

The genes that are transcribed are indicated (Fig. 1). Three beta chain genes and 1 alpha chain gene have been identified in the DR subregion. In the majority of haplotypes the beta 1 gene is thought to code for the polymorphic determinants that can be identified as HLA-DR 1 to 14 using conventional serology. The beta 3 gene is thought to code for the supertypic determinants DRw52 and DRw53. In the DQ and DP subregions two alpha and two beta chain genes have been identified. Two additional loci, DZ alpha and DO beta have located, however the significance of these chains is unknown; they are not thought to form an alpha beta pair at the cell surface.

All three, HLA-DR, HLA-DQ and DP of the class II products are polymorphic. The number of different forms of each antigen that can be identified in the population is still increasing as new techniques and reagents for detecting class II molecules and their genes are developed. In all cases, the beta chain of each of the class II products exhibits most polymorphism. However, the DQ alpha chain has also been shown to be polymorphic, thus the polymorphism of HLA-DQ antigens is a function of both the alpha and beta chains of the molecules.

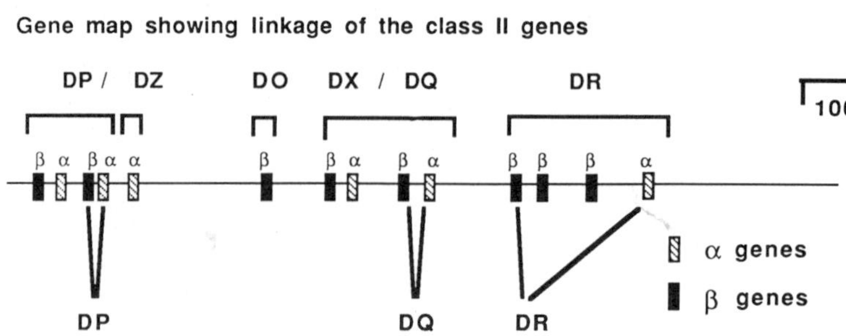

Fig. 1. Gene maps showing linkage of the class II genes

DETECTION OF HLA CLASS II ANTIGENS

The technique that can be used to detect polymorphisms in HLA class II antigens include: (i) serology using allosera or monoclonal antibodies; (ii) the mixed lymphocyte reaction, primed lymphocyte typed and T cell clones; (iii) 2-dimensional gel electrophoresis and (iv) restriction fragment analysis and DNA cloning. The serological and cellular techniques (i) and (ii) identify HLA class II antigens expressed at the cell surface, the third can detect both cell surface and intracellular class II proteins while the fourth detects polymorphisms in the genes that code for class II molecules. A combination of these various techniques is required to identify all the HLA-DR, DQ and DP antigens that have been characterised.

(i) Serology using conventional allosera can be used to type individuals for HLA-DR 1-14 and HLA DQw 1-3 (7). Allosera specific for DP antigens are not available. Monoclonal antibodies that recognise the polymorphic determinants of some of HLA-DR, DQ and DP antigens have been developed (7), however a complete set of reagents defining every HLA-D region specificity is not available. Monoclonal antibodies have been useful in defining new HLA-D determinants, for example DQw3 can be divided into two specificities, DQ 3.1 and 3.2, on the basis of reactivity with the monoclonal antibody, A-10-83 (13).

(ii) The mixed lymphocyte reaction has been used to define HLA-Dw types 1-19. Considerable controversy surrounds the nature of the determinants that activate the MLR. The contribution of HLA-DQ antigens has been shown recently by Karr et al. (14). Six HLA-DP specificities have been identified using PLT (7). HLA-DP antigens cannot be typed using conventional serology as no allosera reactive with DP products exist. A limited number of monoclonal antibodies reactive with DP antigens are available, however these do not define all DP specificities. Alloreactive T cell clones are also being used to a limited extent to define new HLA-D region specificities in combination with other techniques (7, 15). Identification of new class II specificities using T cell clones may offer advantages over techniques employing antibodies as T cells may recognise the polymorphic determinants of HLA molecules in a different way.

(iii) Characterization of HLA-D region antigens by 2-dimensional gel electrophoresis was first described in 1980 (8, 9). Both monomorphic, locus specific as well as polymorphic monoclonal antibodies can be used to immunoprecipitate class II molecules. The alpha and beta chains of the class II products can be examined by using different conditions for isoelectrofocusing in the first dimension. This technique is capable of resolving families of antigens within a single HLA DR specificity. For example a correlation has been noted between the HLA-Dw type and diversity of the beta chains on 2-D gel electrophoresis from cells that all type serologically as DR4 (16).

(iv) Polymorphism in HLA class II antigens can also be detected at the DNA level using restriction enzymes and DNA probes specific for HLA-DR, DQ or DP genes. We and others have shown that there is a positive correlation between HLA class II restriction fragment length polymorphism and DR and DQ serology (11, 17). In other words each of the HLA-DR and DQ specificities that can be defined serologically can also be identified at the DNA level by restriction fragment analysis. We have shown that only two restriction enzymes are required to type homozygous individuals for HLA-DR and DQ specificities (17). Figure 2 shows a Southern blot of DNA prepared from homozygous typing cells covering the specificities of DR 1-8, including the two splits of DR5, DRw11 and w12, and the 2 splits of DR6, DRw13 and w14, digested with the enzyme HindIII and hybridised with a DR beta chain probe.

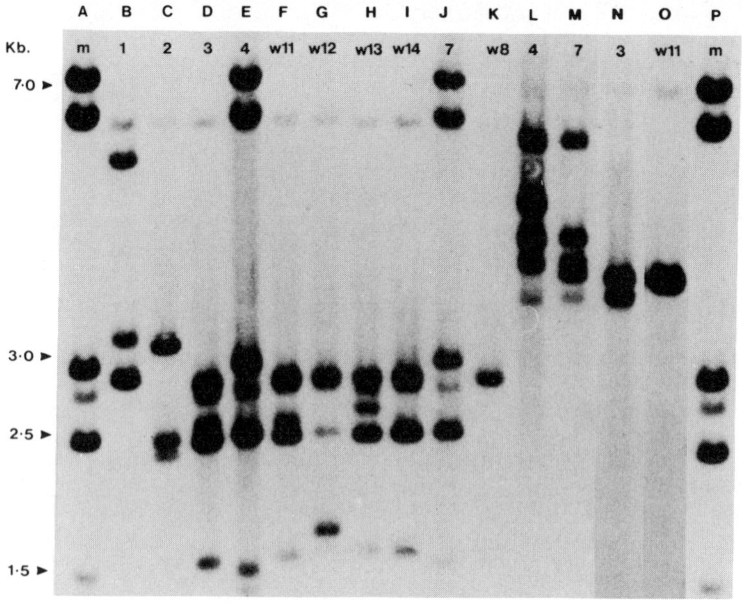

Fig. 2. For explanation see text.

The enzyme HindIII yields unique restriction fragment patterns for each DR type, except DR 4 and 7, tracks E and J. DR 4 and 7 can easily be resolved using a second enzyme, BamHl (tracks L and M). Restriction enzyme analysis has also been useful in the identification of new HLA class II specificities, for example two subtypes of DQw3 have been identified, Figure 3, (13).

Fig. 3. Identification of two subtypes of DQw3 using restriction fragment analysis. The specificities DR4 and 5 (w11, w12) are strongly associated with the serologically defined antigen DQw3. Restriction fragment analysis shows that the DQw3 antigen associated with DR4 and 5 are different. Designated DR4, DQw3.1 (track E) and DR5, DQw3.2 (tracks F & G) (14, 18).

In addition antigens that cannot be identified by serology, so called blank alleles, can be detected at the DNA level. Thus the DQ blank allele associated with DRw8 can be shown to be one of the subtypes of DQw3, DQw3.2 (17). Null DR specificities can also be identified, using restriction fragment analysis, for example the DRBr specificity has been shown to be indistinguishable from DR1 (18). By combining information obtained using different techniques typing for HLA class II antigens will be improved and new HLA-D specificities identified.

MATCHING FOR HLA

The excellent survival of renal allografts between HLA identical siblings is the best proof that the MHC influences graft survival. The importance of matching HLA antigens between donor and recipient in transplantation is controversial, as many factors play a role in determining graft outcome. These include HLA matching, preoperative blood transfusions, immunosuppressive therapy, sex, sensitization, blood group and the transplant centre. In addition matching cadaver donors and recipients for all HLA antigens would be impossible because of the extreme polymorphism of HLA products. However partial matching of donors and recipients for HLA antigens has been shown by many centres to be sufficient to achieve high graft survival rates.

The relative effects of matching for the various HLA class I and class II locus products has been investigated by many centres. The data on HLA-A,B will not be discussed here (for review see 19).

The Oxford unit were the first to report the beneficial effects of matching for HLA-DR antigens alone in cadaveric renal transplantation (20) and this finding has since been confirmed by many other centres (21). Our data show that HLA-DR matched kidneys have a one year survival rate that is about 20% greater than those kidneys that are mismatched for 2 DR antigens (Table 1).

As HLA-DR alleles appear to be less polymorphic than HLA-A,B alleles, the probability of finding DR matched kidneys is greater than finding A,B compatible kidneys, particularly when cadaver donors are used. A beneficial effect of DR matching has also been shown in heart transplantation (22).

Multicentre studies have been compared the benefit of HLA-DR and HLA-A,B matching. HLA-DR matching alone was found to significantly improve graft outcome, however if these DR matched grafts were also A,B matched the survival rate was also improved (23). This confirms that both HLA-DR and A,B matching are important.

The use of different immunosuppressive agents also influences graft survival, in particular the introduction of cyclosporine has resulted in a significant increase in the survival rate. As a result the importance of many factors, including HLA matching have been re-examined. At present the data are conflicting with some groups reporting that HLA-DR matching no longer has any effect (24). However other studies, including our own clearly show that HLA-DR matching does influence the survival of renal allografts in patients treated with cyclosporine (25).

Since the original observations showing beneficial effect of HLA-DR matching on renal allograft survival (20), other products of the HLA-D region, HLA-DQ and DP have been identified. As outlined earlier, techniques in addition to serology are required to detect these other class II specificities. As matching for HLA-DR has been shown to be beneficial, the influence of matching for these other antigens on graft survival should

Table 1. The influence of HLA-DR matching on first cadaveric renal allografts in Oxford

No. DR mismatches	No. patients	% graft survival (months)			
		3	12	36	60
0	121	82	80	74	65
1	171	72	64	57	49
2	93	65	57	52	49

also be considered. Studies examining MLR responses between donors and recipients have shown good correlations between graft survival and the proliferative response in MLR (26). As all the DR region products are thought to contribute to the MLR and this is one way of indirectly determining the role of matching for DQ and DP. As yet few studies have addressed these questions directly. One report has shown that there is no significant correlation between DQ matching and graft outcome (27), however this parameter needs to be examined by more centres. We are currently investigating this question by analysing (i) DQ serology, (ii) DR, DQ and DP restriction fragment patterns and (iii) 2-D gel patterns in donors and

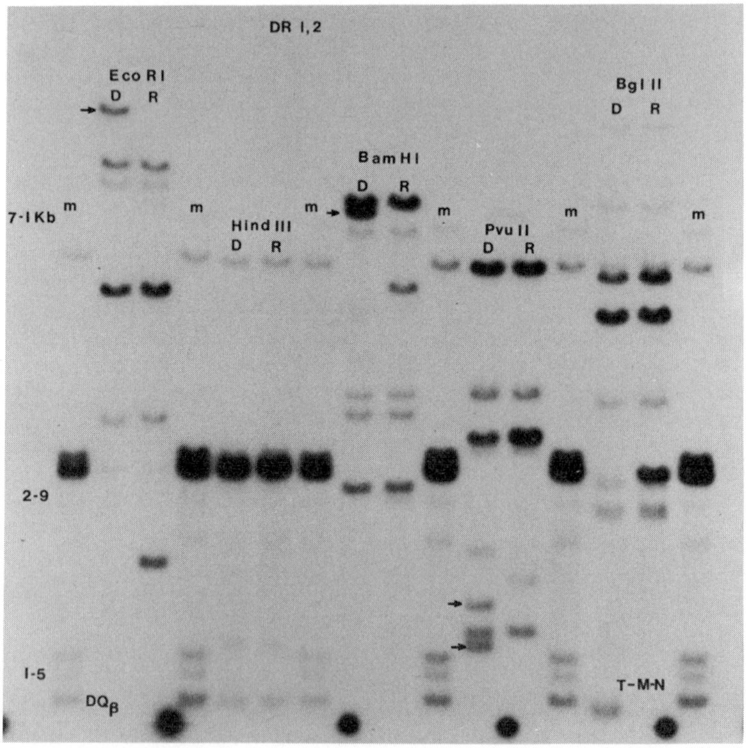

Fig. 4. DNA was isolated from the donor (D) and the recipient (R), digested with the enzymes BamHl, PvuII, EcoRI, HindIII and BglII and probed for DQ beta. M-designates marker tracks.

recipients transplanted in Oxford. Figure 4 shows the restriction fragment analysis of DNA isolated from a donor and recipient digested with 5 different enzymes and probed for HLA-DQ beta. The results show that in this case the donor and recipient were mismatched for HLA-DQ beta chain. A fully analysis of data from patterns transplanted in Oxford is currently in progress (28).

EXPRESSION OF HLA CLASS II ANTIGENS ON TRANSPLANTED KIDNEYS

The induction of donor MHC antigens on transplanted organs has been suggested to be of primary importance in the rejection process determining both the way in which the graft induces the immunological response, as well as the effector mechanisms that are responsible for graft destruction. The induction of class II antigens following transplantation during graft rejection has stimulated enormous interest. Induction of class II antigens has been reported on rejected skin (29), cardiac and renal allografts (30) in rodents and in association with renal allograft rejection in man (31). In Oxford we have studied the expression of HLA class II antigens, DR, DQ and DP on normal human kidney (32, 34) and using biopsies taken at regular intervals after transplantation (33). In normal human kidneys class II antigens are expressed on glomerular endothelium and mesnagium, and on intertubular structures - capillary endothelium and interstitial leucocytes with dendritic morphology. Distal tubules do not express class II, while cytoplasmic staining for class II antigens can be detected in the proximal tubules of some kidneys. The expression of class II antigens on large vessels varied both within and between kidneys and raised from being weakly positive to negative (32). The distribution of class II antigens on normal kidney has been re-examined using both locus specific and polymorphic monoclonal antibodies (34). HLA-DR, DQ and DP antigens are all expressed by normal kidney. The expression of HLA-DR antigens is identical to that described previously using broadly reactive antibodies (32). HLA-DQ and DP antigens are weakly expressed, but they are present on glomeruli and intertubular structures. HLA-DQ and DP antigens were not detected on any of the renal tubules.

The expression of class II antigens on renal allografts was monitored at regular intervals after transplantation using a broadly reactive monomorphic monoclonal antibody that detects DR, DQ and DP antigens, as well as using antibodies that detect only a single locus product (33, 34). Of 113 biopsies examined, 53 showed a massive and generalised induction of class II antigens after transplantation, in 14 there was a focal induction of class II expression and the remaining 46 biopsies showed normal distribution of class II expression. The form of immunosuppressive therapy was found to influence class II induction. Sixteen of the 17 patients receiving azathioprine and prednisolone had increased expression of class II antigens after transplantation, whereas class II expression was induced in only 6 of the 12 patients receiving cyclosporine. Ninety days after transplantation only 9% of the cyclosporine treated patients had increased class II expression, compared with 71% of the biopsies from patients receiving azathioprine and prednisolone. Cyclosporine has been shown to inhibit the induction of class II MHC antigens in experimental transplantation models (30) and it has been suggested that cyclosporine prevents gamma interferon release from the infiltrating leucocytes and this reduces the induction of class II antigens.

Attempts have been made to correlate the induction of class II antigens with rejection episodes. Hall et al. (31) showed a direct association between episodes of severe cellular rejection and class II expression. However in our study we found an indirect rather than a direct association with clinical rejection. 26% of rejection episodes were not associated with increased class II expression. Induction of class II expression has been noted in rat renal allografts that survive indefinitely (35) thus induction of class II MHC antigens may not always correlate in the graft rejection.

CONCLUSION

New HLA-D region specificities can be detected using a variety of assay systems, including serology, MLR, PLT, T cell clones, 2-D gel electrophoresis and restriction fragment analysis. Incompatibility for HLA between the donor and the recipient is a significant factor in determining the survival of renal allografts. Matching for HLA-DR antigens does influence the survival of renal allografts. The importance of matching for HLA-DQ and DP antigens remains to be evaluated. Class II antigens, HLA-DR, DQ and DP, are expressed by normal human kidney. Expression of class II antigens can be induced after transplantation.

ACKNOWLEDGEMENT

This work was performed by Kathryn Wood, Andrew Bushell, Susan Fuggle, Alan Ting and Peter Morris in the Nuffield Department of Surgery, University of Oxford and was supported by grants from the Medical Research Council and the National Kidney Research Fund, U.K.

REFERENCES

1. Counce S, Smith P, Barth R, Snell GD. Ann Surg 1956; 144: 198.
2. Peugh WN, Superina RA, Wood KJ, Morris PJ. Immunogenetics 1986; 23: 30.
3. Goulmy E. Prog Allergy 1985; 36: 44.
4. Bodmer WF et al. (eds). Histocompatibility Testing 1977. Munksgaard, Copenhagen.
5. Bach FH, Hirschon K. Science 1964; 143: 813.
6. Shaw S, Kavathas P, Pollack MS, Charmcot D, Mawas C. Nature 1981; 293: 745.
7. Albert ED, Baur MP, Mayr WR (eds). Histocompatibility Testing 1984. Springer Verlag, Heidelberg.
8. Charron DJ, McDevitt HO. J Exp Med 1980; 152: 185.
9. Shackelford DA, Strominger JL. J Exp Med 1980; 151: 144.
10. Korman AJ, Auffray C, Schamboec A, Shrominger JL. Proc Natl Acad Sci 1982; 79: 6013.
11. Anderson M, Bohme J, Andersson G. Hum Immunol 1985; 11: 57.
12. Hardy DA, Bell JI, Long EO, Lindsten T, McDevitt HO. Nature 1986; 323: 453.
13. Kim SJ, Holbeck SL, Nisperos, et al. Proc Natl Acad Sci 1985; 82: 8139.
14. Karr RW, Hsu SH, Bias WB. Hum Immunol 1986; 15: 198.
15. Paulsen G, Qvigstad E, Gaudernack G. J Exp Med 1985; 161: 1569.
16. Nepom BS, Nepom GT, Mickelson, et al. Proc Natl Acad Sci 1983; 80: 6962.
17. Bushell AR, Wood KJ, Morris PJ. Hum Immunol 1987; (submitted).
18. Bidwell JL, Jarrold EA, Laundy CJ. Tissue Antigens 1986; 27: 99.
19. Ting A, Morris PJ. Tissue Antigens 1985; 25: 225.
20. Ting A, Morris PJ. Lancet 1978; 1: 575.
21. Persijn GG, Gabb BW, Van Leeuwen A. Lancet 1978; 1: 1278.
22. Yacoub M, Festenstein P, Doyle M, et al. Transplant Proc 1987; 19: 2487.

23. Van Rood JJ, Hendrick GFJ, d'Amaro J, et al. Transplant Proc 1985; 17: 681.
24. Klintmalm G, Brynger H, Flatmark A, et al. Transplant Proc 1985; 17: 1026.
25. Opelz G. For the Collaborative Transplant Study. Transplantation 1985; 40: 240.
26. Langhoff E, Jakobsen BK, Platz K, et al. Transplantation 1985; 39: 18.
27. Festenstein H, Doyle P, Holmes J. N Eng J Med 1986; 314: 7.
28. Bushell AR, Wood KJ, Morris PJ. Transplant Proc 1987; 19: 834.
29. Dallman MJ, Mason DW. Transplantation 1983; 36: 222.
30. Milton AD, Spencer SC, Fabre JW. Transplantation 1986; 42: 337.
31. Hall BM, Duggin GG, Phillips J, et al. Lancet 1984; 2 : 247.
32. Fuggle SV, Errasti P, Daar AS. Transplantation 1983; 35: 385.
33. Fuggle SV, McWhinnie DL, Chapman JR, et al. Transplantation 1986; 42: 144.
34. Fuggle SV, McWhinnie DL, Morris PJ. Transplantation 1987; 4 (in press).
35. Wood KJ, Dallman MJ, Morris PJ. Transplant Proc 1987; (in press).

CYTOMEGALOVIRUS INFECTIONS AND TRANSPLANTATION REJECTION: PROGNOSTIC OUTLOOK OF ORGAN TRANSPLANT RECIPIENTS THROUGH AN IMMUNOLOGICAL VIEW

T.H. The and W. van der Bij

University Hospital Groningen, Groningen, The Netherlands

PROGNOSTIC FACTORS

In the past decennia results of organ transplantation have improved considerably by the clinical introduction of more advanced immunological methods for the prevention of graft rejections. These prognostic factors include HLA-matching of donor and recipient especially with respect to HLA class II antigens and the use of cyclosporin A as an immunosuppressive drug. The clinical success accompanying their application is illustrated in Figure 1 by the results of the Renal Transplant Unit Groningen.

In 1979 the availability of serological methods for HLA-DR typing opened the way to donor recipient matching for HLA class II antigens. The application of this innovative matching criterium is accompanied by a significant improvement of the transplantation results.

A second marked increase in cumulative graft survival data is seen after the introduction of cyclosporine as immunosuppresive agent. It proves to be more potent in controlled graft rejection than the conventional drugs like imuran and prednisolon.

Although very promising to us this prognostic outlook of organ transplant recipients is quite limited because it focusses our attention to graft survival only during a relatively short period after transplantation.

The problems which are still present are illustrated by the case history of our patient with a heart transplantation because of a severe dilatating cardiomyopathy. In such cases HLA typing and matching of donor and recipient cannot be performed because the organ preservation time is limited to six hours. There was initially a good functioning donor hart until three weeks after transplantation clinical symptoms of rejection became apparent, which obviously occurred in spite of his immunosuppressive treatment with cyclosporine and prednisolon. Antilymphocyte globulin was than given to the patient as an antirejection treatment. However, this had to be stopped because he developed clinical signs of cytomegalovirus infection.

Attention will be given to cytomegalovirus (CMV) infections as an additional prognostic factor of organ transplant recipients.

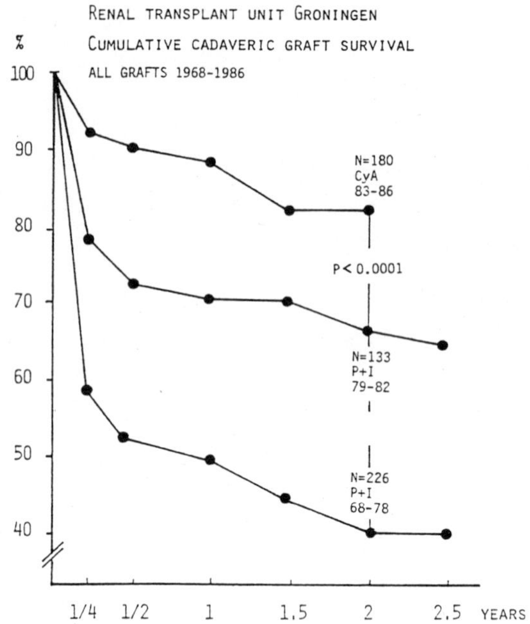

Fig. 1. Cumulative graft survival rates from the Renal Transplant Unit Groningen (Dr. A.M. Tegzess).
P+I = Immunosuppressive treatment with Prednison and Imuran; CyA = cyclosporine A; 68-78 = years; N = numbers of transplanted patients. During 79-82 the middle curve, typing and matching for HLA class II was performed and further continued in 83-86.

IMMUNOSUPPRESSIVE DILEMMA

---

graft rejection ⟷ acceptance

host immunity ⟷ cytomegalovirus infection

Fig. 2. The immunosuppressive dilemma constitutes an important prognostic factor for the graft function and quality of life of its recipient. A correct clinical diagnosis of graft rejection and cytomegalovirus infection is necessary for giving the right decision about increasing immunosuppressive treatment in the case of graft rejection of lowering it for facilitating the hosts' immunity to recover from the cytomegalovirus infection.

Emphasis will be laid on the relationship between transplantation rejection processes and host immunity against CMV infections. The introduction of alloantigens of the transplanted organ induces activation of rejection mechanisms in the host against HLA antigens, which are suppressed by immunosuppressive drugs to ensure the acceptance of donor organ (Figure 2 top line).

However, such a stimulation of the immune system and also the administration of immunosuppressive drugs appear to induce an activation of CMV infections (Figure 2, second line). The relationship between immunological mechanisms of rejections and CMV infections is reflected in the difficulty in the differential diagnosis between these two. This constitutes the immunosuppressive dilemma to which we are faced in the clinical management of transplantation recipients. It is hoped that by the application of diagnostic methods a contribution will be made for the improvement of the prognostic outlook of organ transplant recipients.

This review article consists of three parts. The first deals with some fundamental immunological aspects of immunity against CMV infections in relation with HLA antigens. The second part focusses on CMV infections and their clinical relevance in organ transplantation. The third part is directed to methods for early and rapid diagnosis of transplantation rejections and CMV infections.

PART 1. IMMUNITY AGAINST CMV INFECTIONS AND AGAINST TRANSPLANTATION ANTIGENS

The immune system plays an important role against invasions with microbial organisms and cancer growth. This immunological surveillance is based on the recognition of SELF and NON-SELF.

Non self or foreign antigens are recognized by an afferent and an efferent response which constitute an immunological circle (Figure 3). The afferent arch is mediated by T helper cells whereas the efferent one is mediated by cytotoxic T lymphocytes. Their corresponding specific antigens, in our case transplantation antigens or cytomegalovirus antigens, induce activation of the T helper cells and appearance of cytotoxic cells which can specifically kill target tissues with the corresponding antigens of transplanted organs or virus infected cells. How do these two types of cells recognize the virus-infected targets? Recently the structure of the T cell receptor has been elucidated. Figure 4 shows that three different molecules constitute the receptor complex.

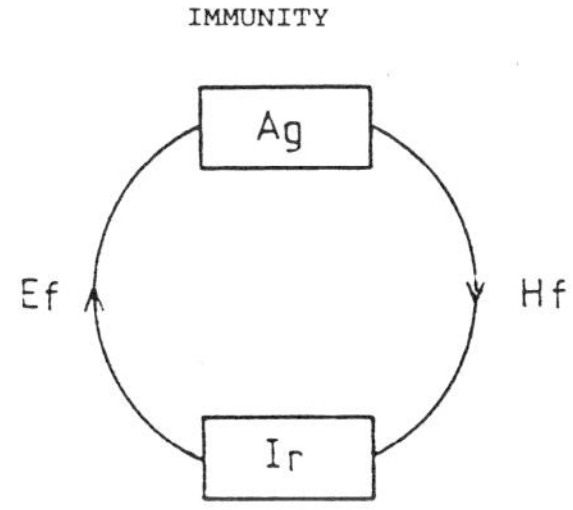

Fig. 3. The immunological circle. Mechanisms of transplantation rejection and immunity against CMV infections are based on the same principles, recognition and destruction of foreign or non-self tissues.

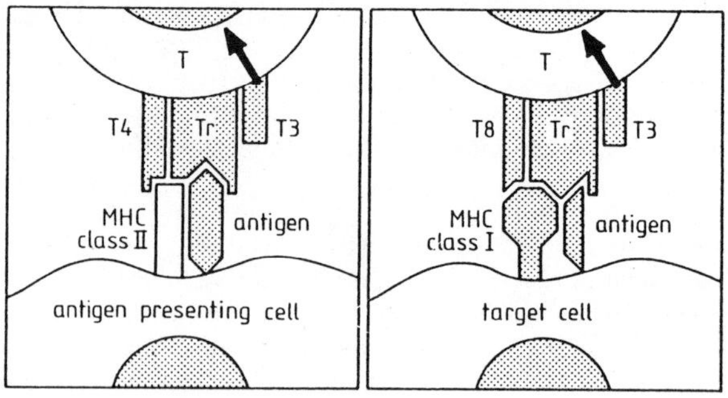

Fig. 4. The T cell recognition structure or receptor. The T helper lymphocyte (left). In the afferent arch of the immunological circle (Fig. 3) foreign antigen in the context of MHC class II antigens whereas cytotoxic T cell (right) in the efferent arch recognized antigen in relation to class I.

The T4 and T8 protein molecules are involved in the actual recognition of respectively HLA class II and HLA class I molecules from the target structure. The Tr molecule bearing the antigen binding part, consists of two protein chains which are linked to one another by a disulfide bridge. The T3 protein consists of three different subunits which are non-covalently bound to one another and to the Tr molecule. The T3 molecule is needed for the transmembrane signal transduction for cellular activation after a positive recognition event. The T cell receptor structure clearly shows that HLA antigens have an important biological significance as recognition structures for the distinction of self and non-self elements. According to these fundamental immunological mechanisms one can imagine that immunity to CMV infection and allograft rejection processes are functionally related to one another.

PART II. CYTOMEGALOVIRUS INFECTIONS AND THEIR CLINICAL RELEVANCE IN ORGAN TRANSPLANTATION

The functional relationship between transplantation rejection mechanisms and CMV infections are clinically reflected by the association of CMV infections with rejection episodes after organ transplantation. CMV infections are frequent causes of morbidity and mortality of organ transplant recipients. It is of importance to emphasize that CMV infections may cause clinical symptoms which are difficult to distinguish from transplant rejection episodes. Especially deteriorations in the physiological functions of transplanted organs may also be a result of CMV infections. Therefore there is an urgent need for diagnostic tests for the discrimination between graft rejection and CMV infection.

CMV infections counterpart of transplant rejection; epidemiology and pathology of CMV infections

Cytomegalovirus, a member of the family of Herpes viridea, is commonly spread in man. Infectious CMV is excreted in the saliva, urine, mothermilk, cervix and semen, also by apparently healthy individuals. This is the main reason for its ubiquitous nature.

CMV infections in experimental animals and natural infections in man have clearly demonstrated an intricate and close relationship between CMV and immunological response in the host. While a virus specific immune response is mounted to limit the virus spread CMV infection is also likely to cause a serious depression of hosts' immune response. The resultant effect in healthy immunocompetent individuals is usually an asymptomatic seroconversion or a mild mononucleosis. However, active CMV infection may be associated with significant morbidity and mortality in patients with a deficient (cellular) immune system, such as immunosuppressed organ transplant and bone marrow transplant recipients. The essential role of cell mediated immunity has been well recognised in bone marrow transplant recipients who developed CMV infection. Patients with CMV infections who developed CMV specific cytotoxicity lymphocyte responses survived, but patients who failed to generate these responses generally died of CMV infection. In addition, patients with fatal infection had depressed levels of natural killer cell and antibody-dependent killer cell activity before and during their CMV infection, while patients who survived did not. Thus, a correct evaluation of the crucial phase of the intricate relationship between CMV induced immunosuppression and host immune response is most important in the management of patients at risk. It is important to note that a significant number of these infections is transmitted by blood transfusions and organ transplantations, especially when seronegative recipients receive organs from seropositive donors. These individuals acquire a so called primary CMV infection and because a pre-existing immunity is not yet present the clinical symptoms in this group are more pronounced than in recipients who are already CMV-seropositive and have built up virus-specific memory T cells. However, this statement is relative because secondary CMV infections, caused by reinfections with other CMV strains or reactivation of latent CMV infections can also cause serious clinical problems in patients depending on their cellular immunocompetence under the given immunosuppressive therapy.

CMV-replication and viral antigens

The human CMV is a highly species-specific, double-stranded DNA virus with a large genome of 235 kB with a coding capacity of at least 70 proteins. Infection of fetal fibroblasts with AD169, one of the laboratory strains of CMV, is a widely used system to have the CMV specific antigens at one's disposal for diagnostic and immunological studies. After infection several CMV encoded antigens and induced host cell proteins appear intracellularly and on the surface membrane. Based on blocking of host cell metabolism and on patterns and localization of immunofluorescence with human sera three distinct phases of the infectious cycle can be defined, marking the emergence of Immediate Early, Early and Late Antigens (IEA, EA and LA) (Figure 5). The final result of the "teamwork" of these subsequent phases is the release of new infectious viral particles and death of the host cell.

There are two types of CMV infection, primary and secondary ones. Primary infections do occur in recipients who were initially CMV seronegative before transplantation. The exogenous CMV source of the infection appears to be mostly the infectiously virus in the donor organ. In the case of secondary CMV infections the recipients were already CMV seropositive before the surgical implantation of the donor organ. The source of infectious CMV appears to be endogenous by a process of (re)activation of latent CMV infection. In this respect it is important to note that more than half of the normal adult population have such a latent infection because they are CMV seropositive.

In renal transplantation more than half of the CMV seronegative recipients acquire a primary CMV infection after organ transplantation and up to ninety percent of the seropositive recipients have a secondary CMV infection.

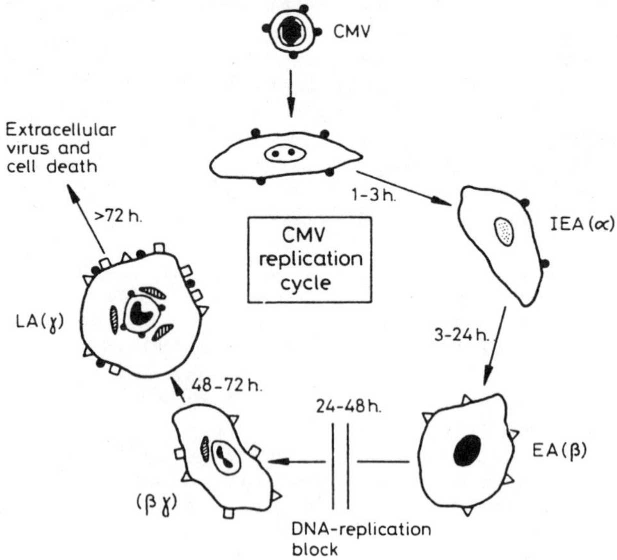

Fig. 5. The CMV replication cycle. Different stages of virus maturation are reflected by distinct virus-specific antigens.

Interestingly there exists a relationship between the incidence and severity of these infections at the one side and the immunogenicity of different grafted tissues at the other. For example immunogenicity of bone marrow grafts known to be the highest are associated with relatively high morbidity and mortality amongst the recipients caused by CMV diseases.

CYTOMEGALOVIRUS INFECTION

EXOGENOUS CMV
(donor)

CMV SERONEG. ↓ PRIMARY INFECTION

- Blood
- Organ

↓

Recipient ⟶ Infected recipient

↑

- Immunosuppression
- Allogeneic reaction

CMV SEROPOS. ↑ SECONDARY INFECTION

ENDOGENOUS CMV
(recipient)

Fig. 6. Mechanisms and sources of both primary and secondary CMV infections in organ transplantation recipients.

PART III. METHODS FOR EARLY AND RAPID DIAGNOSIS OF TRANSPLANTATION REJECTIONS AND CMV

Diagnosis of graft rejection episodes

Immunological effector mechanisms elicited by the introduction of alloantigens of allografts are produced in lymphoid organs. Specific antibodies and cytotoxic lymphocytes will infiltrate the implanted target organ by the way of the blood circulation. The immunologic methods for the diagnosis of rejections are based on the detection of these effector components especially the immune cells. The first category consists of immunohistological staining techniques performed on sections of organ biopsies. The diagnosis is made by the in situ detection of cellular infiltrates around blood vessels and interstitial tissues. A phenomenon of target cell lysis is an important hall-mark for rejection next to the amount and composition of lymphocyte infiltrates. Distinct T cell subsets (CD4 or the helper cell phenotype, and CD8 the suppressor and cytotoxic phenotype) are visualized with specific monoclonal antibodies. Although these methods are the golden standards for diagnosis of rejections there is a need for non-invasive techniques. These are provided by the second category of tests which are based on cytoimmunomonitoring of blood leucocytes. Especially after cardiac transplantation cytoimmunological monitoring provides a method to reduce biopsy frequency after transplantation. Endomyocardial biopsy is a safe, reliable and reproducible technique to diagnose acute rejection after cardiac transplantation. It is a disadvantage of the biopsy technique that daily use is not feasible and this highly invasive method is inconvenient for the patient. With the help of cytoimmunological monitoring, various parameters for rejection episodes were obtained daily. In addition, cytoimmunomonitoring predicts acute rejection even earlier than percutaneous biopsy and allows evaluation of changes in the leucocyte subpopulations and their precursor cells, diagnosing a recipient's inflammatory event. The drawback of invasive and non-invasive methods however, is that they lack specificity for transplantation rejection since microbial infections, especially CMV infections, are giving the same results. Therefore the combination of these methods together with newly developed CMV diagnostic test as will be discussed below, are important for the clinical management.

Detection of CMV in blood, a new method for the rapid and early diagnosis of CMV infections

Cytomegalovirus is transmittable by blood transfusions and especially leucocyte transfusions. These observations support our notion that we are dealing with a cell-associated virus and therefore the question arises whether CMV could be detected in certain cell types. Although for many years it is known that the virus can be transmitted by transfusions with blood from CMV-seropositive donors, viral cultures from blood cells derived from buffy coat from these donors were negative. It is important to note that infectious virus could only be isolated from leucocytes from patients with an active CMV infection.

Interestingly, Schrier and Oldstone (3) reported that HCMV specific messenger RNA is detectable in the CD4+ subpopulation of T lymphocytes in atently infected healthy subjects. The foregoing results indicate that CMV might be present in an incomplete form as CMV-DNA in certain lymphocytes from healthy individuals with a latent infection. The results of studies in patients after kidney transplantation have shown that positive virus isolation from the buffy coat (CMV-viremia) is related to those acute CMV infections with overt symptoms of disease. Using CMV specific monoclonal antibodies against immediate early antigens (CMV-IEA), prepared in our laboratory, we recently succeeded in showing these antigens in blood

Table 1. Method of CMV antigen detection in blood leucocytes

- isolation of peripheral blood leucocytes
- cytocentrifuge-preparations (100 x 10^3 leucocytes)
- indirect immunoperoxidase staining with:
- (three) monoclonal antibodies against CMV-IEA
- detection and (semi)quantification of IEA pos. leucocytes

leucocytes from patients with an active CMV infection (4). These CMV-IEA positive cells were only detected during the onset of disease in relation with the period of clinical symptoms of CMV infection and that these cells have the morphology of polymorphonuclear leucocytes and monocytes. Very rarely lymphocytes were positive which were T cells and not B lymphocytes. In a prospective longitudinal study in patients after organ transplantation the emergence of CMV-antigen positive blood leucocytes reached a peak, decreased and disappeared thereafter at the moment of the appearance or rise of virus-specific circulating antibodies. CMV antigens appeared, on average, nine days before a significant rise of antibody levels. The test was positive in patient groups with a primary as well as a secondary CMV infection. The test is virus-specific and of clincal relevance for the early and rapid diagnosis (within a few hours) of an active CMV infection and its possible therapeutic consequences. The essential steps of its methods are presented in Table 1.

First, serotyping of recipients with a sensitive ELISA method for IgG anti-CMV-LA antibodies is needed to recognize CMV-seronegative from the positive individuals. Especially the seronegative ones need blood or blood cells and, if possible, also organs from CMV-seronegative donors. Prevention of secondary CMV infection may be partially achieved by using leucocyte-free blood in stead of whole blood for donations to CMV-seropositive recipients. In addition, the use of leucocyte depleted and stored blood in stead of fresh blood also reduces CMV infectivity.

Prevention of CMV infection

In patients belonging to the above mentioned high risk groups for serious CMV infections prevention and treatment of CMV infections is essential, and such a strategy is shown in Table 2.

SUMMARY AND CONCLUSIONS

The prognostic outlook of organ transplant recipients is determined by two closely related items in casu rejection and CMV infection. The functional relationship is reflected in the clinical association. From the immunological point of view there exists a basic dilemma because immunosuppressive drugs available to control graft rejection mechanisms to ensure acceptance of transplanted organs, will facilitate persistent activation of CMV infections. The price paid for the functional insurance of the optimal function of transplanted organs is the built-in danger effecting host quality of life by persisting infections.

Improvement in the clinical management is based on prevention of primary CMV infections by CMV host and donor typing. Besides frequent monitoring of blood samples after transplantation for parameters of rejection by phenotyping effector cells and parameters for CMV infection by the direct detection of CMV antigen markers in blood leucocytes is needed

Table 2. Prevention and treatment

- serological detection
- selective use of "CMV-negative" blood products
- acceleration immune reconsitution, i.e. tapering of immunesuppression
- CMV immunoglobulins
- antiviral chemotherapy

for the optimal immunosuppressive drug regimen. The rapid and early diagnosis of CMV infections versus graft rejections have important clinical implications. In case of CMV infections lowering or even stopping of immunosuppressive drugs will favour host immunological mediated recovery from CMV infection without increased risk of organ rejection. In case, however, of rejection episodes, antirejection treatment by increasing immunosuppression is needed for saving donor organ function. The development of new strategies which favour the acceptance of transplanted organs while maintaining host immunity against infections will further increase the prognostic outlook of organ transplant recipients.

REFERENCES

1. Quinnan GV, Kirmani N, Esber E, et al. HLA-restricted cytotoxic T lymphocyte and nonthymic cytotoxic lymphocyte responses to cytomegalovirus infection of bone marrow transplantation. J Immunol 1981; 126: 2036-41.
2. Middeldorp JM. Diagnostic and immunological aspects of the antibody response to human cytomegalovirus infection. Academic Thesis., University of Groningen, 1985.
3. Schrier RD, Nelson JA, Oldstone MBA. Detection of human cytomegalovirus in peripheral blood lymphocytes in a natural infection. Science 1985; 230: 1048-51.
4. Van der Bij W, Torensma R, Van Son WJ, et al. Expression of cytomegalovirus antigens in blood leucocytes as an early diagnostic marker of active infection. Submitted.
5. Griffiths PD, Stirk PR, Ganczakowski M, et al. Rapid diagnosis of cytomegalovirus infection in immuncompromised patients by detection of early antigen fluorescent foci. Lancet 1984; ii: 1242-5.
6. Winston DJ, Ho WG, Lin CH, Budinger MD, Champlin RE, Gale RP. Intravenous immunoglobulin for modification of Cytomegalovirus infections associated with bone marrow transplantation. Am J Med 1984; 30: 128-33.
7. Condie RM, O'Reilly RJ. Prevention of cytomegalovirus infection by prophylaxis with an intravenous, hyperimmune, native, unmodified cytomegalovirus globulin. Am J Med 1984; 30:134-41.

CHAPTER 11
ENZYMOLOGY

New perspectives in clinical enzymology
D.W. Moss

Structure and function of disease-related allelozymes
L. Luzatto

The clinical enzymology of the complement system
M. Loos and J. Alsenz

NEW PERSPECTIVES IN CLINICAL ENZYMOLOGY

Donald W. Moss

Royal Postgraduate Medical School, Hammersmith Hospital
London W12 OHS, U.K.

Living processes are made up of a multiplicity of integrated chemical reactions, almost every one of which is catalysed by a specific enzyme. It is evident, therefore, that knowledge of the structure and properties of enzymes and of their biosynthesis, functions and disposal is fundamental to an understanding of normal and pathological metabolism. Clinical enzymology attempts to infer the occurrence, site, nature and extent of pathological processes from studies of the activities and properties of individual enzymes in readily obtainable samples. Classically, this approach has been virtually synonymous with the measurement of levels of enzymes in blood plasma or serum, in which the in vitro catalytic activity of enzyme molecules has been used for their specific and sensitive identification and measurement, without implying changes in function in vivo. Although the normal and abnormal functionings of secreted enzymes have long been of interest to clinical enzymologists, complex integrated series of extracellular reactions such as complement-activation or blood-coagulation have only recently begun to be analysed in enzymological rather than phenomenological terms.

The selection and interpretation of many of the established tests in diagnostic enzymology have been placed on a scientific basis by experiments on the factors which affect the transfer of enzymes from cells to plasma; first with respect to cytoplasmic enzymes and more recently for membrane-bound enzymes.

Extra- or intra-cellular enzyme function, and any leakage of intracellular enzymes that may take place, are late stages in the life-cycle of an enzyme (Figure 1). The development of molecular biology has greatly increased the ability to study the earlier stages of enzyme production. Genetic disease has in the past provided an opportunity to compare the properties and, to some extent, the structures of genetically-altered enzymes. The ability to do so has been greatly extended by gene cloning and sequencing, speeding up access to primary structures of enzymes. Furthermore, these techniques have shown that the causes of altered enzyme expression extend beyond changes in bases encoding amino acids within the enzyme molecule. These advances should also provide better understanding of the selective normal expression of enzymes and isoenzymes, and of the changes in enzyme production which take place in some acquired diseases and which are relevant to the interpretation of diagnostic enzyme measurements.

As well as acquiring its active conformation, which seems to be implicit in its primary structure, an enzyme typically undergoes post-translational modification and transport to its final functional destination. These processes and their abnormalities in disease are also within the enlarged perspective of clinical enzymology.

ESCAPE OF ENZYMES FROM THEIR CELLS OF ORIGIN

The inverse relationship between the level of cellular energy generation and the rates of leakage of cytoplasmic enzymes has been firmly established by experiments with isolated perfused liver preparations. It accounts for the sensitivity of plasma levels of enzymes such as aminotransferases and creatine kinase as indicators of reversible or irreversible tissue damage in a very wide range of conditions in which perfusion of organs with oxygenated blood and oxidizable metabolites is impaired. However, certain enzymes of diagnostic value, e.g. alkaline phosphatase and gammaglutamyl transferase, are incorporated into external or internal cellular membranes and their release into the circulation is governed by factors other than reduced energy generation.

Both alkaline phosphatase and gammaglutamyl transferase are inserted into membranes by hydrophobic domains which can be removed by proteolytic enzymes in vitro. The enzymes may appear in serum in different multiple forms: attached to membrane-fragments; with or without their hydrophobic membrane-binding domains; or in the case of gammaglutamyl transferase, in

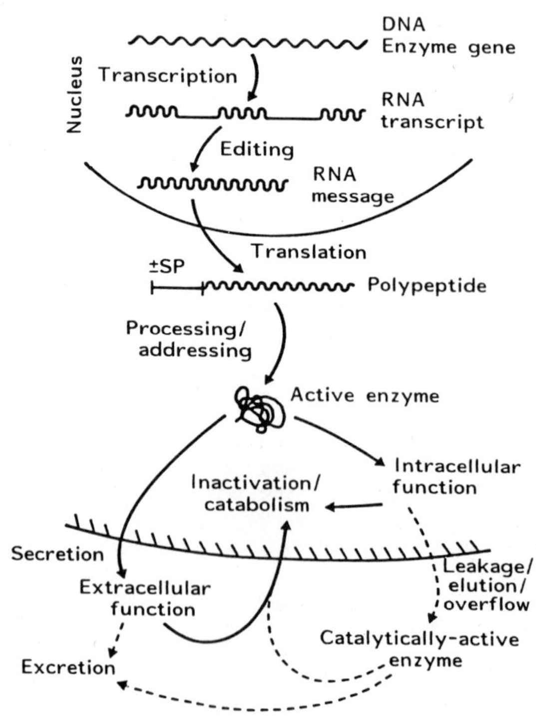

Fig. 1 Stages in the life-cycle of an enzyme.

aggregated forms which probably involve interactions between the hydrophobic domain and other components of plasma. The multiple forms can be reproduced experimentally and thus allow the physiological and pathological mechanisms of release of membrane enzymes to be inferred.

Alkaline phosphatases solubilized from specimens of human tissues such as liver or placenta with, e.g., n-butanol contain molecular segments ranging from 8 to 20 kDa molecular weight which can be cleaved by the action of bromelain without otherwise altering the molecular or catalytic properties (1). Alkaline phosphatases with molecular sizes corresponding to those of the butanol-extracted enzymes that have been treated with bromelain can be solubilized by the direct action of bromelain on the tissues, confirming that the removable moiety is involved in binding to the tissues. However, a bromelain-cleavable segment, of approximately 20 kDa, can only be demonstrated in alkaline phosphatase extracted immediately from adult intestinal tissue removed at operation, since in autopsy material the enzyme has already undergone partial proteolysis.

Alkaline phosphatases of bone or liver origin recovered by gel-filtration from sera of patients with diseases of these tissues retain their bromelain-removable segments. However, when adult-intestinal alkaline phosphatase is recovered from serum it is found not to possess the segment removable by bromelain which is present in the isoenzyme solubilized with butanol from fresh tissue (1). These observations underline the different routes by which intestinal and non-intestinal alkaline phosphatase isoenzymes gain access to the circulation (Fig. 2).

Bone and liver alkaline phosphatases, generated in increased amounts in osteoblastic and hepatobiliary disease, are released complete from their cells of origin: increased production may lead to saturation of binding-sites in membranes and increased overflow of enzyme, while increased detergent action of the plasma due to retained bile salts probably also plays a part in hepatobiliary disease. The carbohydrate side-chains of these alkaline phosphatases are terminated by sialic acid residues. Their half-lives in the plasma are of the order of 1-2 days, clearance presumably being effected by tissue macrophages. Intestinal alkaline phosphatase appears to undergo partial proteolysis in the lumen of the small intestine, before or after solubilization. The partially-proteolysed isoenzyme, lacking its hydrophobic domain, is reabsorbed in the thoracic-duct lymph and delivered to the circulation. Unlike liver and bone phosphatase, adult-intestinal alkaline phosphatase is an asialoglycoprotein and it is rapidly cleared from the plasma by hepatocyte galactosyl receptors.

The raised levels of bone and liver alkaline phosphatase in serum that are so useful in the diagnosis and monitoring of diseases of the respective tissues and organs are thus largely determined by increased rates of entry of the enzymes into the circulation as a result of increased synthesis in their cells of origin. When increased levels of intestinal phosphatases are observed, however, they can usually be attributed to reduced rates of clearance of the isoenzyme from the circulation: consequently, raised levels of intestinal alkaline phosphatase are more often observed in chronic liver disease than in diseases primarily affecting the intestine.

Different molecular-weight forms of gammaglutamyl transferase and their experimentally-formed analogues can also help in the interpretation of pathological changes that result in increased levels of this enzyme in plasma (2). Gammaglutamyl transferase that has lost its hydrophobic domain by proteolytic action in vivo is found only rarely in serum in disease. However, in certain pathological conditions such as malignant infiltration of the liver, very high molecular-weight forms of gammaglutamyl transferase, as well as of alkaline phosphatase, are frequently present in serum. These can be shown to consist of enzyme molecules attached to fragments of cell membranes. They thus indicate the occurrence of a particular type of pathological change involving cellular destruction, and have been studied as potential markers of the occurrence of hepatic metastases, although the diagnostic efficiency of measuring the high molecular-weight fractions is not markedly greater than those of measurements of total enzyme activities in this respect (Table 1).

The increased activity of gammaglutamyl transferase in serum in hepatobiliary disease is mainly due to one or more fractions of about 300 kDa molecular weight. These appear to correspond to the whole molecule, including its hydrophobic domain, through which the enzyme is able to associate with other plasma constituents, such as lipoproteins. Thus, the occurrence and nature of these multiple forms appear to depend both on enzyme release and on other factors which affect plasma composition; however, their significance has not yet been fully elucidated. An interesting aspect of the interactions between enzymes and other plasma constituents is the not infrequent participation of enzymes in immune complexes, sometimes without manifestations of autoimmune disease. Are these phenomena due to failures of self-recognition, or are immune-complex formation followed by rapid removal stages in the normal clearance of enzymes from the circulation, which in these cases have become impaired?

POST-TRANSLATIONAL MODIFICATION OF ENZYMES

Many enzymes undergo post-translational modifications of the gene-product which, in some cases, can lead to the production of tissue-specific forms. Examples of this phenomenon are the bone, liver and kidney variants of alkaline phosphatase (Table 2). Much evidence supports the view that these three enzymes are the products of a single "tissue-unspecific" structural gene. Nevertheless, they are recognizably different in such properties as electrophoretic mobility and resistance to inactivation by heat (3). These tissue-specific differences, which especially in the case of bone and liver phosphatases are extremely useful in diagnostic enzymology, can be attributed to differences in the degree of glycosylation of the single gene-product, since they can be reduced or abolished by removal of attached carbohydrate residues.

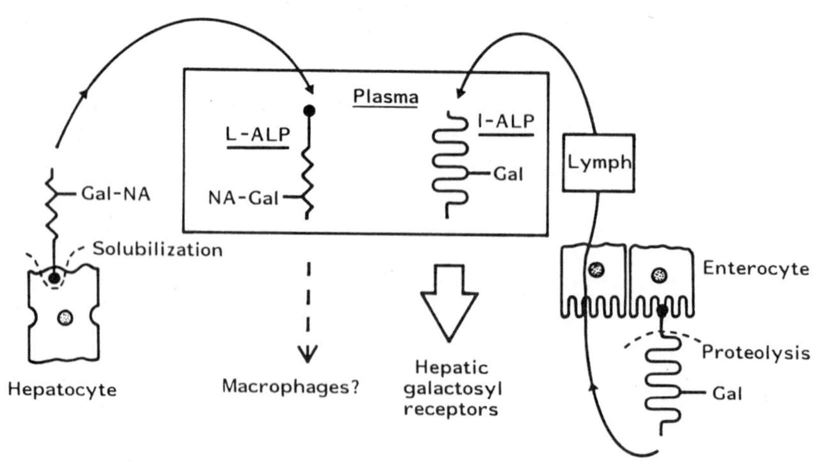

Fig. 2 Different modes of entry of intestinal (I-ALP) and non-intestinal (e.g. hepatic; L-ALP) alkaline phosphatases into the circulation (see text).

Table 1. Molecular-weight variants of gammaglutamyl transferase

Approx. size (kDa)	Occurrence in vivo	Formation in vitro	Probable nature
Greater than 10^3	Serum; especially with liver metastases	Incubation of liver tissue in serum	Enzyme-bearing cellular membrane fragments
300	Serum: normal and hepatobiliary disease	Incubation of liver tissue in serum Extraction of $>10^3$ fractions with lipid solvents or deoxycholate	Complete enzyme molecule, associated to varying extents with lipoprotein
160 increasing to 550 on storage	--------	Extraction of liver tissue with deoxycholate	Complete enzyme molecule, aggregating through hydrophobic domain
82	Serum (rarely) Bile	Incubation of liver in saline or serum Treatment of all larger-sized fractions with trypsin	Enzyme from which hydrophobic domain has been removed by proteolysis
Less than 20	--------	In small amounts by deoxycholate-treatment of 82 or 300 kDa forms	Catalytic subunit

It is not clear why these enzyme variants should differ so consistently in their glycosylated forms. However, the process can be altered in disease, particularly in cancer, giving rise to abnormal variants which, besides being diagnostically useful, may offer fresh insights into pathological processes.

ALTERATIONS IN PATTERNS OF ENZYME SYNTHESIS

Genetically-determined decreases, or more rarely increases, in enzyme production have been well documented, and include both alterations in the structural gene encoding the enzyme protein, resulting in a change in specific activity, and in expression of the gene product. Changes in enzyme production may also occur in acquired disease, while some enzyme- or-isoenzyme-determining genes are selectively expressed during normal development and differentiation.

Acquired changes in production of enzymes of diagnostic importance include the increased alkaline phosphatase in serum that has already been mentioned as a feature of obstructive liver disease. The cause of this was

Table 2. Similarities and differences between human alkaline phosphatases from liver and bone

	Liver	Bone
Genetic control	both deficient in congenital hypophosphatasia	
Peptide maps	- identical -	
Substrate specificity and response to inhibitors	- identical -	
Reaction with antibodies	complete cross-reaction with polyclonal antisera; small differences with mAbs	
Relative electrophoretic mobilities		
- native	100	91
- after brief treatment with neuraminidase	91	67
- after extensive digestion with neuraminidase	51	53
- in presence of wheat-germ lectin	-	retarded
Relative stability at 56°C		
- native, in serum	100	25
- after digestion with glycosidases	47	27

for long a subject of controversy, but it is now known to result from an increase in enzyme production in the hepatocytes, induced by biliary stasis. Hepatic gammaglutamyl transferase can also be induced by certain drugs such as barbiturates and probably ethanol, though not by biliary stasis. In these situations the increased enzyme levels in serum may be due to the rate of synthesis exceeding the capacity of the cells to incorporate the enzyme molecules into their normal intracellular locations.

Selective normal expression of specific isoenzymes can be seen among the human non-specific acid phosphatases, as it can amongst the alkaline phosphatases. The tartrate-resistant acid phosphatase has a limited, and at first sight random, distribution in a few normal and pathological cells. The isoenzyme is normally present in osteoclasts, and pathologically in Gaucher cells in the eponymous disease of spleen and in the "hairy cells" of leukaemic reticuloendotheliosis. However, the tartrate-resistant isoenzyme has now been shown to be present also in human alveolar macrophages and in rat Kupffer cells (4). These are cells of the mononuclear phagocyte system, to which osteoclasts also almost certainly belong. The isoenzyme is not present in the monocytes of blood, the precursors of the tissue macrophages. Thus, the expression of the tartrate-resistant acid phosphatase accompanies differentiation of certain macrophages. Gaucher cells are abnormal macrophages; however, the relationship, if any, of hairy cells to the macrophage system is not established.

Tartrate-resistant acid phosphatase is a member of a widely-distributed class of iron-containing proteins with acid phosphatase activity, such as porcine uteroferrin, with which the human isoenzyme displays considerable immunological cross-reactivity. Antibodies raised against porcine uteroferrin can be used to identify and measure the human tartrate-resistant acid phosphatase isoenzyme (5), while the selective expression of the human isoenzyme affords opportunities to study, and perhaps to generalize from, those factors which result in the expression of a specific isoenzyme during differentiation.

These various examples - by no means exhaustive - demonstrate the considerable progress that clinical enzymology has already made towards fulfilling its objectives, which are not confined to its established role in the diagnosis of tissue-damage, but include an understanding of the nature of pathological deviations from normal metabolic pathways.

REFERENCES

1. Kalimanovska V, Whitaker KB, Moss DW. Effect of bromelain on alkaline phosphatases of intestinal and non-intestinal tissues and serum. Clin Chem Acta 1987; 170: 219-26.
2. Echetebu ZO, Moss DW. Multiple forms of human gamma-glutamyl-transferase: modification and possible structures of different molecular weight fractions. Enzyme 1982; 27: 9-18.
3. Moss DW. Multiple forms of acid and alkaline phosphatases: genetics, expression and tissue specific modification. Clin Chim Acta 1986; 161: 123-35.
4. Efratiadis T, Moss DW. Tartrate resistant acid phosphatase of human lung: apparent identity with osteoclastic acid phosphatase. Enzyme 1985; 33: 34-40.
5. Echetebu ZO, Cox TM, Moss DW. Antibodies to porcine uteroferrin used in the measurement of human tartrate resistant acid phosphatase. Clin Chem 1987; 33: 1832-6.

STRUCTURE AND FUNCTION OF DISEASE-RELATED ALLELOZYMES

L. Luzatto

Department of Haematology, Royal Postgraduate Medical School
London, W12 OHS

Genetic variation at structural loci encoding enzymes (or polypeptide subunits of oligomeric enzymes) can cause pathology (i.e. enzymopathies) by producing in the gene product either a quantitative change, or a qualitative change or both. Purely quantitative changes might result from a deletion, or from a mutation in a regulatory region of DNA, causing decreased transcription. Qualitative changes usually result from point mutations within the coding region. These may bring about impaired catalytic activity, decreased stability or altered glycosylation of the protein, thus producing enzyme deficiency by a translational or post-translational mechanism. The clinical manifestations of an enzymopathy will depend in the first place on whether the expression of the respective gene is tissue-specific (e.g. the gene encoding phenylalanine hydroxylase is expressed only or mainly in the liver), or universal, as is the case with so-called housekeeping genes (e.g. triose phosphate isomerase is distributed ubiquitously and its deficiency causes a multi-system disease). Secondly, the clinical phenotype will depend on the precise nature of the enzyme abnormality: for instance, on whether substrate affinity is modified or not. It is not yet entirely clear why genetic variations seem to be much more common at certain enzyme loci than others. The largest numbers of pathological structural variants have been reported for pyruvate kinase (PK: gene on chromosome 15q22) (1) and for glucose 6-phosphate dehydrogenase (G6PD: gene on Xq28). For both, the predominant clinical expression is seen in red blood cells in the form of haemolytic anaemia. An extensive analysis of genotype-phenotype correlations has been carried out especially in G6PD deficiency (2). Some enzyme variants cause haemolysis only following the action of a superimposed exogenous trigger whereas others cause more severe, chronic haemolysis. In many cases the biochemical basis for one or the other phenotype has been characterized in terms of the degree of enzyme instability or the extent of unfavorable changes in enzyme kinetics. The full sequence of the enzyme has been recently elucidated (3) and the changes leading to G6PD deficiency are beginning to emerge. In due course the reason for the exceptionally large number of variants at the G6PD locus should become apparent (4).

REFERENCES

1. Valentine WN, Tanaka RK, Paglia DE. Pyruvate kinase and other enzyme deficiency disorders of the erythrocyte. In: Stanbury JB, Wyngaarden

JB, Frederickson DS, Goldstein JL, Brown MS, eds. The metabolic basis of inherited disease, fifth edition. New York: McGraw-Hill Book Company, 1978.
2. Beutler E. Glucose-6-phosphate dehydrogenase deficiency. In: Stanbury JB, Wyngaarden JB, Frederickson DS, Goldstein JL, Brown MS, eds. The metabolic basis of inherited disease, fifth edition. New York: McGraw-Hill Book Company, 1978.
3. Luzzatto L, Testa U. Human erythrocyte glucose 6-phosphate dehydrogenase: structure and function in normal and mutant subjects. Current Topics in Hematol 1978; 1: 1-70.
4. Persico MG, Vigletto G, Martini G, et al. Isolation of human glucose-6-phosphate dehydrogenase (G6PD) cDNA clones; primary structure of the protein and unusual 5′ non-coding region. Nucleic Acids Research 1986; 14: 2511-2522.

THE CLINICAL ENZYMOLOGY OF THE COMPLEMENT SYSTEM

Michael Loos and Jochem Alsenz

Institute of Medical Microbiology
Johannes Gutenberg University, Mainz, FRG

INTRODUCTION

The complement (C) system is one of the most important humoral systems mediating many activities that contribute to inflammation and host defense, e.g. various anaphylatoxin activities, chemotaxis and opsonization for phagocytosis. The C system is similar to other humoral systems, such as coagulation, fibrinolysis and the kinin system, a multifactoral system whose activation represents sequentially occurring multi-step activation cascades of the "classical" as well as the "alternative" pathway.

The C system is composed of eleven proteins of the "classical" pathway*: C1q, C1r, C1s, C4, C2, C3, C5, C6, C7, C8, and C9. There are three proteins of the "alternative" pathway** B, D, and P. Finally, there are four control proteins: $C\overline{1}$ inhibitor ($C\overline{1}$ INH) and C4b binding protein (C4b-bp) for the classical pathway, I (C3b inactivator or C3b INA) and H (β_1H or C3b INA accelerator) for the alternative pathway, and anaphylatoxin inactivator. These 19 distinct serum proteins have been highly purified, physiochemically characterized and for some of these proteins the amino acid compositions and sequences as well as the DNA sequences are known (For further details see references 1, 2, 3, 4, 5, 6, 7).

The third component, C3, which probably is the most important C component in terms of host defense, can be activated by two different pathways: the classical pathway starting with C1, C4, and C2, and the alternative pathway with C3b, a cleavage product of native C3 together with B, D, and P (Figure 1).

---

* The nomenclature used for the proteins of the classical complement pathway conforms with that agreed upon by the World Health Organization (Bull Wld Hlth Org 1968; 39: 935-938): A component in the activated state, such as C1, is designated with a bar over the number, $C\overline{1}$; the activated forms of two subcomponents of native C1, C1r, and C1s, are written with a bar over the letter, $C1\overline{r}$ and $C1\overline{s}$.

** The "Nomenclature of the alternative activating pathway of complement" was elaborated by a IUIS-WHO Nomenclature Committee: J Immunol 1981; 127: 1261-1262.

THE ENZYMES OF THE CLASSICAL PATHWAY

The classical pathway (8) can be activated by antigen-antibody complexes and involves the C-components C1, C4 and C2 (Figure 1). C1 by itself is a macromolecule composed of C1q and doublets of the proenzymes C1r and C1s which are linked by Ca^{++}ions ($C1r_2$-$C1s_2$). Only the antibodies IgM and IgG have a binding site for the C1q subcomponent of macromolecular C1. The binding sites are located in the Fc portion of the antibody molecules (IgM: C14 (Fc)5μ; IgG: C_H2) and are becoming available when the antibodies are bound to the corresponding antigen forming immune complexes (IC). In addition to IgM and IgG, a variety of other substances interact directly with C1 and C1q (9): Bacterial derived lipopolysaccharides (LPS) and lipid A; polyanionic substances such as polyiniosinic acid, DNA carageenan, heparin, dextran sulfate, polyvinyl sulfate, polyanethol sulfonate (Liquoid^R), and chondroitin sulfate; the C-reactive protein (CRP); envelopes of some RNA viruses, including Maloney leukemia virus and vesicular stomatitis virus.

Several steps are involved in the activation of C1 by IC leading to the formation of the C1 esterase, C1$\overline{s}$: Binding of C1 to IC via C1q, the recognition unit of C1, induces a conformational change that occurs within the C1q molecule forming C1q* as shown by using monoclonal anti-C1q antibodies. The appearance of C1q* induces a further conformational change in C1r leading reversibly to the exposure of the enzymatic site of one C1r monomer, designated C1r*. C1r* cleaves the second monomer of C1r to C1$\overline{r}$. C1$\overline{r}$ cleaves now the single chain C1r* molecule which activated it initially and a second two chain C1$\overline{r}$ molecule is generated. This process is called "autocatalytic activation" of C1r and it occurs in serum even in the absence of IC under certain conditions, e.g. when C$\overline{1}$ INH concentrations are below 30% of normal. The last step of the internal activation of C1 is the cleavage of the two single chain zymogen C1s by C1$\overline{r}$ leading to the formation of the C1 esterase, C1$\overline{s}$, and converting C1 to (C$\overline{1}$).

There are a number of similarities between the two subcomponents C1$\overline{r}$ and C1$\overline{s}$: both are serine esterases which are inhibited by C$\overline{1}$-INH, DFP, and

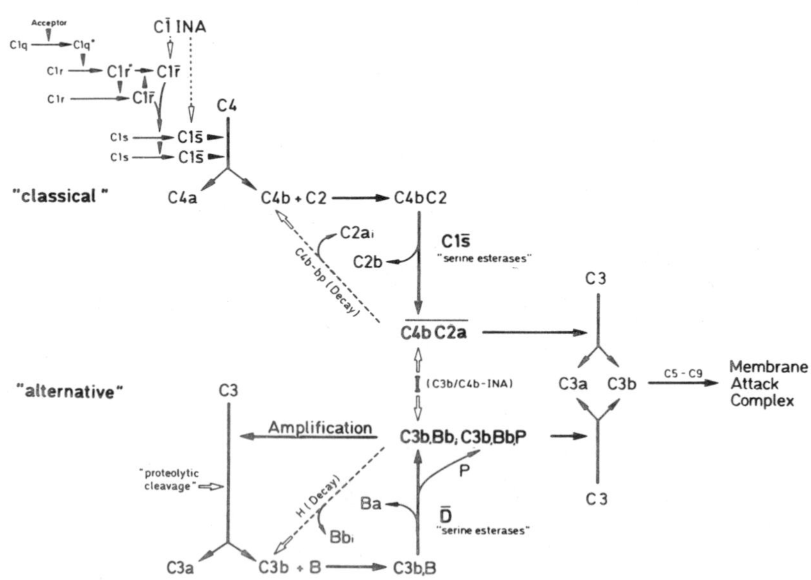

Fig. 1. The classical and the alternative pathways of the complement system.

NPGB. Both precursor forms are single chain molecules (85,000 D) which are cleaved upon activation into a larger "a" (56,000 D) and a smaller "b" (27,000 D) chain which are linked by disulfide bonds. The identities of the molecular weights of precursor chains and chain fragments were confirmed by showing a high similarity in their amino acid composition, especially of the esterolytic center. However, the two subcomponents differ in their substrate specificities: C$\overline{1r}$ can activate only C1s, but not C4 and C2; C$\overline{1s}$ can not activate C1r, but is does cleave C4 and C2.

The natural substrates of C$\overline{1s}$ in serum are C4 and C2, the next two components in the C cascade of the classical pathway. Cleavage of C4 and C2 on cell surfaces by C$\overline{1s}$ leads to the formation of the classical pathway C3 convertase, C4b2a (10). The component C4 is composed of three polypeptide chains, α, β, γ with molecular weights of 90,000, 70,000, and 33,000 daltons. When subjected to the action of C$\overline{1s}$ a 10,000 dalton peptide (C4a) is cleaved off from the α-chain and on the larger fragment of C4 (C4b) a highly labile binding site is exposed. Only when the cleavage of C4 takes place in the presence of acceptor sites for C4b, such as IC-C$\overline{1}$, C4b sites are generated, IC-C$\overline{14}$b, for the further participation of C4 in the cascade. In the absence of acceptor sites for C4b the labile binding site on C4b is no longer available, and C4 becomes iC4b incapable of furthering the C cascade.

The formation of a haemolytically active C4b site represents the binding site for C2, the second natural substrate of C$\overline{1s}$. The generation of the classical C3 convertase (C4b2a), the third enzyme besides C$\overline{1r}$ and C$\overline{1s}$ of the classical pathway, is at least a two step reaction. First, binding of native C2 bound C4b, i.e. IC-C$\overline{14}$b, is dependent on the presence of Mg^{++}ions even at 0°C and is completed within 5 to 10 min. The conversion of the C4b2 complex into the C3 convertase is dependent on the enzymatic cleavage of C2 into C2b and C2a by C$\overline{1s}$ or other proteolytic enzymes like trypsin, chymotrypsin etc. The larger fragment C2a remains bound to C4b and the formed C4b2a complex now becomes capable of cleaving the next component of the cascade, C3. In contrast to the C4b2 complex the newly formed C4b2a is no longer dependent upon Mg^{++} indicating a Mg^{++}-independent arrangement of C2a with C4b. Thus, C4b2a is no longer sensitive to chelating agents like EDTA or EGTA. The enzymatic site of the C3 convertase is located in the C2a molecule and has substrate specificity for C3 and after reassociation with the major split product C3b (C4b2a3b) enzymatic activity against C5. Therefore, C2 is the zymogen carrying the proteolytic site for C3 and C5 in the classical pathway C3 and C5 convertases. C4b2a is unstable and undergoes a time- and temperature-dependent decay unless there is a sufficient quantity of C3 in the vicinity of the cellbound complex to mediate the next site in the sequence. During this decay the C2a fragment is released in a functionally inactivate form, iC2a. The remaining C4b site is now able to take up new native C2 and a new C4b2a enzyme can be generated.

THE ENZYMES OF THE ALTERNATIVE PATHWAY

The alternative pathway of complement leads (Figure 1) under the participation of C3b, the major split product of C3, factor D and B as well as factor P (properdin) to the generation of C3 convertases (11, 12). A variety of activators of this pathway have been described such as bacterial (LPS), yeast (zymosan), or plant (inulin) polysaccharides, polyanionic substrates like dextran sulfate, a cobra venom factor (CVF), the Fab portions of immunoglobulins e.g. IgA or IgE etc.

Since the activation of this pathway requires C3b as acceptor for factor B one postulates that the binding site for B on C3 is generated by a nucleophilic modification and scission of the thioester bond in C3 resulting in formation of a molecule C3(H_2O) possessing all of the functional properties of C3b. The generated C3b like C3 molecule, serves as the receptor for factor B which in turn provides the enzymatic site of the C3 convertase of the alternative pathway. Usually, the fluid phase interaction of C3b, B, D, and P is prevented by the action of the naturally occurring regulating proteins, C3b-inactivator (I) and factor H, the C3b inactivator accelerator. However, binding of the generated C3b to microbial and mammalian cell surfaces that activate the alternative pathway, protect C3b from inactivation by I in the presence of H. Now factor B binds reversibly in the presence of Mg^{++} to C3b.

Upon cleavage of bound by $\bar{D}$ a C3 converting activity, C3b,Bb, is formed. This C3 convertase generates more C3b, the acceptor for B and amplifies the generation of alternative pathway C3 convertase. The bimolecular amplification convertase C3b,Bb undergoes an intrinsic decay whereby Bb is released in an inactive form, Bbi. The residual C3b is capable of reforming a C3b,B complex from which under the participation of D a new C3 convertase C3b,Bb is generated. The dissociation of Bb as Bbi from C3b,Bb complexes can be retarded by binding of properdin (P). This is associated with a non-enzymatic conformational change in the C3b,Bb complex leading to the formation of the P-stabilized C3 convertase, C3b,Bb,P. By affording a protected microenvironment for C3b and the convertases C3b,Bb and C3b,Bb,P an activating surface advances the alternative complement pathway slow fluid phase turnover to intense particle associated amplified cleavage of C3 into C3a and C3b. The deposition of C3b on the surfaces alters the fate of those particles by facilitating phagocytosis and cell lysis. Both enzymes of the alternative pathway, factor $\bar{D}$ and Bb associated with C3b, are DFP-sensitive serine esterases. Factor $\bar{D}$ exists in serum in very low concentrations in its activated form. In contrast to the other proteases of the complement system, factor $\bar{D}$ is not incorporated into the C3b,Bb complex. Therefore, it is concluded that factor $\bar{D}$ plays only a catalytic role cleaving the C3b associated factor B into Ba and Bb, what also may be consistent with its low serum concentration. Factor $\bar{D}$ has a trypsin-like specificity since it splits factor B at a single Arg-Lys bond into two fragments Ba (30,000 D) and Bb (63,000 D). Therefore, it is not surprising that the action of $\bar{D}$ on B can also be mimicked by other serine esterases like trypsin, pronase, and plasmin. Bb is the catalytic subunit of the alternative-pathway C3 convertases C3b,Bb and C3,Bb,P. There are striking similarities between C2 and factor B: Both are enzymatically active in the complement cascade when they are complexed to C4b or to C3b, respectively; both are serine esterases and the active subunits in the C3b convertases of classical or alternative pathway of complement; both represent in the presence of excess C3b also the active site of the C5 convertases C4b2a3b and C3b,Bb,C3b; both are very unstable enzymes that decay spontaneously with a half-life of several minutes.

THE TERMINAL COMPLEMENT SEQUENCE

Cleavage of C3 into C3a and C3b by the C3 convertase of the classical as well as of the alternative complement pathway and the association of C3b with these convertases leads to the formation of the C5 convertase which cleaves C5 in two fragments C5a and C5b. After cleavage of C5, the reaction C5b with C6, C7, C8, and C9 proceeds by protein- protein interactions without further proteolysis. The formed C5b-C9 complex, also called membrane attack complex, MAC, is inserted into membranes, to create complete transmembrane channels leading to osmotic lysis of the cell.

Table 1. Biological functions of activated complement components

C2a	Kinin-like activity. Increase of vascular permeability and contraction of smooth muscles
C3a, C4a, C5a	Anaphylatoxic peptides bind to receptors on granulocytes, macrophages, mast cells and thrombocytes
	Release of vasoactive amines
	Enhanced vascular permeability
	Contraction of smooth muscle
	Induced release of lysosomal enzymes
C5a	Chemotaxis: Induction of migration of leukocytes into an area of C activation
	Granulocyte aggregation
	Activation of intracellular process such as release of oxygen metabolites and SRS-A
C3b, C4b	Immune adherence and opsonization. Bridging between a complex or target cell bearing C3b or C4b and the responding cell having a receptor for C3b or C4b:
	Phagocytic cells, macrophages, monocytes, polymorphonuclear leukocytes, B-lymphocytes, primate erythrocytes, platelets
C5b-C9	Membrane damage: Lysis of cells, e.g. red cells and gram-negative bacteria

THE BIOLOGICAL ROLE OF THE COMPLEMENT ACTIVATION

The complement system is one of the most important humoral systems mediating many reactivities that contribute to inflammation and host defense even in the preimmune phase where specific antibodies and lymphocytes are not available. Therefore, it is not surprising that the complement cascades can be initiated by multiple ways in addition to antigen-antibody reactions. Activation of the complement cascade leads to the fragmentation of C3, C4, and C5 into low-molecular weight hormone like peptides - C3a, C4a, and C5a (Table 2) (13). These molecules induce smooth muscle contraction and enhance vascular permeability. They bind to specific receptors and induce the release of vasoactive amines such as histamine from mast cells and basophils and lysosomal enzyme release from granulocytes. C5a functions also as chemoattractant inducing the migration of leukocytes into an area of complement activation. The activities of C3a and C5a are abolished by anaphylatoxin inactivator (AI) which removes the C-terminal arginine from both molecules. The high-molecular-weight fragments of C3 and C4, C3b and C4b, bound to membranes of cells or bacteria are recognized by various cells having receptors for C3b or C4b such as phagocytic cells (macrophages, monocytes, and polymorphonuclear leukocytes), B-lymphocytes, neutrophils, erythrocytes, and platelets (14). Therefore, C3b and C4b molecules serve as bridge between a complex or target cell bearing C3b or C4b and the corresponding cell having a receptor for C3b or C4b. The consequence of C3b- or C4b-mediated bridging depend on the responding cell type: Binding to phagocytic cells triggers phagocytosis. Adherence to nonprimate platelets induces specific release of vasoactive amines and nucleotides from the platelet. Since B-lymphocytes have receptors for C3b and C4b, it is postulated that bridging brings antigens in direct contact with antibody-forming cells and that, therefore, bound complement components may play a role in the induction of an immune response. Induced by activation of both complement pathways, the formation of the multimolecular C5b-C9 complex on a cell membrane results in an impairment of osmotic regulation which may cause cytolysis. The C5b-C9 complex is incorporated into the lipid layer of the membrane leading to the formation of an transmembrane channel allowing passage of small molecules and initiating osmotic lysis of the cell. Complement-mediated lysis has been shown for many kinds of cells: erythrocytes, platelets, bacteria, viruses processing a lipoprotein envelope, and lymphocytes.

The biological importance of the complement system for the maintenance of a functional host defense is impressive illustrated by the markedly increased susceptibility to infection and the predisposition to diseases observed in some congenital or acquired deficiencies of complement components or complement regulatory proteins (Table 2).

REGULATORY FACTORS OF THE COMPLEMENT SYSTEM

A number of regulatory systems prevent unrestricted activation of the reaction sequence or bring about the rapid destruction of the activated factors at several stages of the complement system (15, 5, 7). Control is exerted by either specific control proteins or the short half-life of activated components and of some of the complexes. Examples for the latter are the spontaneous dissociation of the classical pathway (C4b2a) and alternative pathway (C3b,Bb) convertases, the rapid inactivation of the labile cell binding site in C3b and C4b by water and the short life of the C6 binding site in C5b. The main regulatory functions, however, are taken over by specific control proteins, namely C1-esterase inhibitor (C$\overline{1}$-INH), C4b-binding (C4bp), decay-accelerating factor (DAF), factor H, factor I, C3b-receptor (CRI), S-protein and C8-binding protein (C8bp).

In the classical pathway C$\overline{1}$-INH tightly controls activated C$\overline{1}$ and C1-activation (16). C$\overline{1}$-INH is a glycoprotein of 105,000 daltons containing 35% carbohydrate, that blocks C1 activity by covalent binding to the catalytic site on the light chains of C1r and C1s and by subsequent disassembly of the C$\overline{1}$ complex into two C1rC1s-$(C\overline{1}\text{-INH})_2$ complexes and C1q, which remains bound to the activator. In addition, C1-INH has been reported to inhibit autoactivation of plasma C1. The classical C3 convertase (C4b2a) is controlled by the decay-accelerating factor DAF, a 70,000 dalton molecule present within the membranes of many different cell types and by C4bp, a 570,000 dalton protein with a spidery structure. Both regulatory proteins inactivate the C4b2a-complex by dissociating the enzymatically active C2a from the convertase. In addition, DAF accelerates the decay of the alternative pathway C3 convertase and C4bp works as a cofactor for the inactivation of C4b by factor I. Factor I, also known as C3b/C4b inactivator, is a 80,000 dalton serine-esterase that cleaves C4b bound to surfaces or to C4bp in a two-step reaction to the hemolytically inactive C4c and C4b.

Furthermore, factor I is also of great importance in the control of the feedback loop of the alternative pathway. Here, the role of factor I is to cleave C3b to iC3b in order to prevent formation of C3 convertase. Required cofactors for the permanent inactivation of C3b are factor H, a 150,000 dalton glycoprotein and CR1. Binding of one of these cofactors to C3b alters the C3b molecule thus making it susceptible to cleavage by factor I. In addition, both factor H and CR1 accelerate the rate of dissociation of Bb from the C3 convertase (C3b, Bb) and inhibit C5 and factor B binding to C3b.

The terminal component lytic efficiency is restricted by S-protein and by C8bp. In the fluid phase, S-protein, a 88,000 dalton glycoprotein, is capable of stable, presumably hydrophobic interactions with the membrane binding site of the C5b67 complex, thus inhibiting the attachment of the complex to the membrane. In contrast, C8bp is a species-specific membrane protein, that inhibits the efficiency of hemolytic lesions by both C8 and C9.

Table 2. Diseases associated with complement deficiencies in

Clq	Vasculitis, nephritis, systemic lupus erythematosus or similar syndrome, hypogammaglobulinemia, and chronic bacterial infections
Clr	Recurrent infections, renal disease, systemic lupus erythematosus or similar syndrome, rheumatoid disease
Cls	Systemic lupus erythematosus, Raynaud's phenomenon
C4	Systemic lupus erythematosus
C2	Systemic lupus erythematosus or similar syndromes, glomerulonephritis, vasculitis, arthralgia, susceptibility to infections
C3	Recurrent infections with pyogenic bacteria
C5	Systemic lupus erythematosus, recurrent infections (gram-negative bacteria), recurrent gonococcal infections, defect in chemotaxis
C6, C7 or C8	Recurrent meningococcal and gonococcal infections, Raynaud's syndrome, systemic lupus erythematosus, glomerulonephritis
C$\bar{1}$INH	Hereditary angioedema (HAE), systemic lupus erythematosus-like disease
H (C3INH)	Recurrent infections with pyogenic bacteria

The importance of regulatory proteins becomes evident when inhibitor deficiencies are examined. As an example, one of the best studied complement deficiency states, the C$\bar{1}$-INH deficiency will be discussed in more detail.

C$\bar{1}$-INH DEFICIENCY

Deficiency of C$\bar{1}$-INH results in a syndrome designated as "angioedema with C$\bar{1}$-INH deficiency". This syndrome is characterized by recurrent, self-limited edema of the face, extremities and upper airways and by episodes of abdominal pain (17, 16). The present classification of C$\bar{1}$-INH deficiency bases on the underlaying biochemical abnormalities leading to a decreased plasma C$\bar{1}$-INH functional activity (Table 3).

Low levels of functionally active C$\bar{1}$-INH in HAE are caused by either a half normal rate of C$\bar{1}$-INH synthesis and an increased fractional catabolic rate of C$\bar{1}$-INH (HAE type I) or synthesis of a dysfunctional C$\bar{1}$-INH (HAE type II) or complex formation of C$\bar{1}$-INH with albumin (HAE type III). In contrast, the rare form of AAE usually results from diseases such as B-cell abnormalities with autoantibody production or lympho-proliferative disorders (AAE type I), however, can also be associated with autoantibodies to C$\bar{1}$-INH (AAE type II) as reported previously (18, 19).

In addition to the low levels of functional C$\bar{1}$-INH protein, CH50, C4 and C2 found in both HAE and AAE, AAE is further characterized by decreased Clq and Cl levels and by an onset of symptoms in the middle age.

The unique pathomechanism found in patients with AAE type II provides an excellent example for the understanding of the reactions resulting from decreased or absent functional levels of C$\bar{1}$-INH (Figure 2):

Patients with AAE type II synthesize normal amounts of functionally active 105,000 dalton C$\bar{1}$-INH molecules. Due to an unknown event, these patients develop autoantibodies to C$\bar{1}$-INH that inhibit C$\bar{1}$-INH binding to C$\bar{1}$s and/or

Table 3. Angioedema with C$\bar{1}$-INH deficiency

Hereditary angioedema (HAE)	
Type I	decreased synthesis of C$\bar{1}$-INH
Type II	synthesis of dysfunctional C$\bar{1}$-INH
Type III	complex formation of C$\bar{1}$-INH with albumin
Acquired angioedema (AAE)	
Type I	associated with diseases, e.g. B-cell abnormalities
Type II	associated with autoantibodies to C$\bar{1}$-INH

the control of C1 esterase activity. When C$\bar{1}$-INH functional activity in plasma drops below about 35% of normal, C1 is activated autocatalytically, as shown for other types of angioedema, or is activated by the C$\bar{1}$-INH-anti-C$\bar{1}$-INH antibody complexes. The uncontrolled C$\bar{1}$ subsequently consumes C4 and C2. In addition, C$\bar{1}$-INH complexes with the autoantibody is cleaved at the susceptible Arg-Thr bond in the active site of C$\bar{1}$-INH by one of its target proteases, presumably by C1s, to the inactive 96,000 dalton C$\bar{1}$-INH molecule found in patients' plasma. In contrast to normal C$\bar{1}$-INH-target protease interactions, where such a cleavage results in the formation of an irreversible, covalent C$\bar{1}$-INH-enzyme complex, the complex in these patients dissociates again, into the serine esterase and the modified 96,000 dalton C$\bar{1}$-INH.

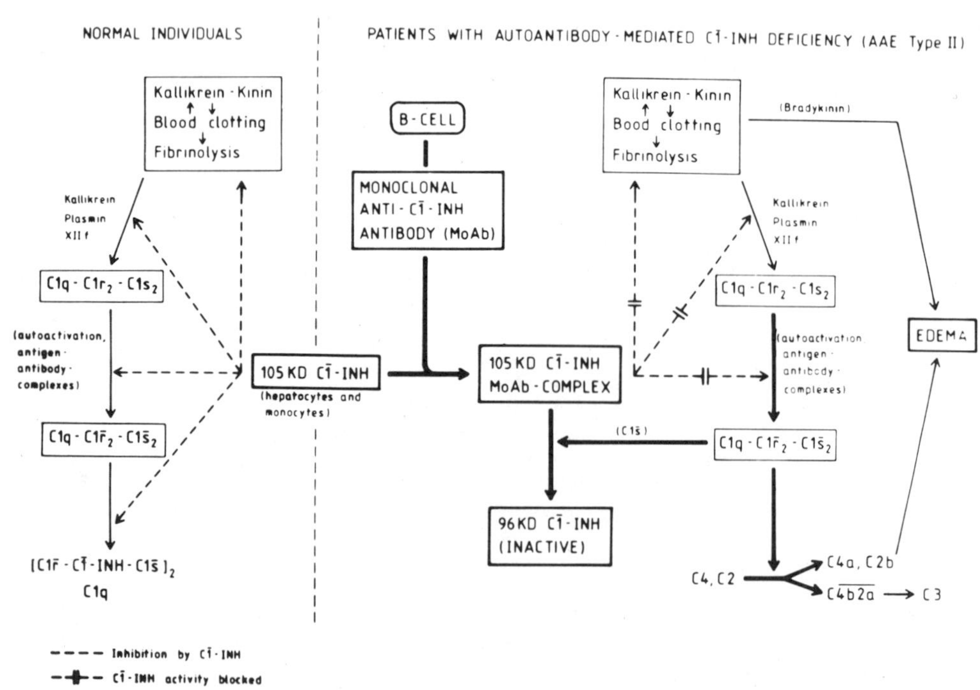

Fig. 2. Proposed pathomechanism for autoantibody-mediated C$\bar{1}$-INH deficiency (AAE Type II).

The cause of the angioedema is still unknown. Since C1̄-INH inactivates not only C1̄, but is also an important regulator of the kallikrein-kinin (kallikrein), blood-clotting (Hageman factor and factor XIa) and fibrinolytic (plasmin) system it still remains to be determined which system or combination of systems is responsible for the clinical symptoms.

REFERENCES

1. Cooper NR. The classical complement pathway: Activation and regulation of the first complement component. Adv Immunol 1985; in press.
2. Fothergill JE, Anderson WHK. A molecular approach to the complement system. Curr Top cell Regul 1980; 13: 259-311.
3. Müller-Eberhard HJ. Complement. A Rev Biochem 1975; 44: 697-724.
4. Müller-Eberhard HJ, Schreiber RD. Molecular biology and chemistry of the alternative pathway of complement. Adv Immunol 1980; 29: 1-53.
5. Müller-Eberhard HJ. Chemistry and function of the complement system. In: Dixon FJ, Fisher DW, eds. The Biology of immunologic disease. Sinauer Associates, Inc., Sunderland, MA, USA 1983.
6. Porter RR, Reid KBM. Activation of the complement system by antibody-antigen complexes: the classical pathway. Adv Protein Chem 1979; 33: 1-71.
7. Ross GD. Immunobiology of the complement system. Academic press, Inc. 1986.
8. Loos M. The classical complement pathway: Mechanism of activation of the first component by antigen-antibody complexes. Prog Allergy 1982a; 30: 135-192.
9. Loos M. Antibody-independent activation of C1, the first component of complement. Ann Immunol Paris 1982b; 133C: 165-179.
10. Loos M, Heinz HP. Generation of the classical pathway C3 convertase (EAC4b2a) by proteolytic enzymes. Acta path microbiol immunol scand Sect C 1984; Suppl 284, 92: 67-74.
11. Pangburn MK, Müller-Eberhard HJ. The alternative pathway of complement. Springer Semin Immunopathol 1984; 7: 163-192.
12. Müller-Eberhard HJ. The membrane attack complex. Springer Semin Immunopath 1984; 7: 193-219.
13. Hugli TE. Structure and Function of the anaphylatoxins. Springer Semin Immunopath 1984; 7: 193-219.
14. Schreiber RD. The chemistry and biology of complement receptors. Springer Semin Immunopath 1984; 7: 221-249.
15. Fries III LF, Frank MM. Molecular mechanisms of complement action. In: Stamatoyannopoulos G, Nienhuis AW, Leder P, Majerus PW, eds. The molecular basis of blood diseases. Saunders Company 1987.
16. Schapira M, De Agostini A, Schifferli JA, Colman RW. Biochemistry and pathophysiology of human C1̄ inhibitor: current issues. Complement 1986; 2: 111-126.
17. Cullmann W, Opferkuch W. C1̄ inhibitor in Progress Allergy. In: Rother K, Rother U, Karger AC, eds. Heriditary and acquired complement deficiencies in animals and man. 1986; 39: 311.
18. Alsenz J, Bork K, Loos M. Autoantibody-mediated deficiency of C1 inhibitor. N Engl J Med 1987; 316: 1360-1366.
19. Jackson J, Sim RB, Whelan A, Feighery C. An IgG autoantibody which inactivates C1̄ inhibitor. Nature 1986; 323: 722-724.

CHAPTER 12
CHEMI- AND BIOLUMINESCENCE

LUMINESCENCE IMMUNOASSAYS

William Graham Wood* and Huis-Zhen Yuan**

Medizinische Universität zu Lübeck, Ratzeburger Allee 160, D-2400 Lübeck* and Shanghai Childrens' Hospital, Kang Ding Lu Shanghai, P.R. China**

INTRODUCTION

The first known documentation of bioluminescence was made some 3,500 years ago, when a Chinese scholar described the pleasing light of glow-worms.

Over the following centuries many well-known historical personages such as Aristoteles, Francis Bacon and Leonard da Vinci, have taken an interest in luminescence. It was Robert Boyle, who around 1670, first described the scientific basis of bioluminescence, which still hold today, some 300 years later (1).

Chemiluminescence, or the production of light in the test-tube using chemical reagents, is a much newer field and started in the organic chemistry laboratories in the last quarter of the 19th century when Radziszewski described the luminescence arising from the oxidation of lophine in 1877 (2). Further advances came from Albrecht, who in 1928 published the Luminescence of Luminol (3) and from Gleu and Petsch, who described lucigenin luminescence in 1935 (4). In the last three decades the names Gundermann, Rauhut and Schroeder must be mentioned in the further development of chemiluminescent compounds suitable for analytical purposes.

LUMINESCENCE IMMUNOASSAYS

a. Assay labels

Luminescence Immunoassays can be subdivided into several classes as shown in Table 1.

The most robust luminescence immunoassays use chemiluminescent labels or detection systems, which despite the low quantum efficiency in aqueous solution (less than 0.05 Einstein/mol), offer a reagent stability which cannot be obtained in bioluminescent systems.

Table 1. Types of Luminescence Immunoassays

a) Using Bioluminescence -	(1) Antibody or antigen labelled with an enzyme e.g. luciferase, kinase or dehydrogenase
	(2) Antibody or antigen labelled with a substrate or cofactor e.g. NAD derivative or luciferin
b) Using Chemiluminescence	(1) Antibody or antigen labelled with luminogen - e.g. 1,2 diacyl-hydrazide or acridinium ester
	(2) Antibody or antigen labelled with an enzyme - e.g. peroxidase -
	(3) Antibody or antigen labelled with an indirect marker - e.g. avidin or biotin

b. Assay design

All successful luminescence immunoassays are based on solid phase techniques - coated tubes, balls or magnetic microparticles being the most popular supports (5, 6). Where possible, immunometric assays (ILMA) should be performed (7, 8).

The possibility of using solid phase antigen techniques (SPALT) has been exploited with success both commercially and on a laboratory scale (9).

Often, assay design includes the separation of the sample incubation and the incubation with the luminogen to remove the possibility of prior oxidation of the label. This is not always necessary, as one-step immunoluminometric assays with excellent precision and accuracy are commercially available for measuring serum samples (Henning Lumitest TSH). The addition of NaN_3 to the incubation buffer inhibits many serum enzymes capable of producing active oxygen species, which could lead to label oxidation during the incubation.

The abbreviations in luminescence immunoassays are numerous and non-standardized. Table 2 shows the assay types, using nomenclature analogous to radioimmunological methods, where relevant.

Table 2. Luminescence immunoassay types

LIA	- Luminescence Immunoassay (cf RIA)
CELIA	- Chemiluminescence Immunoassay
ILMA	- Immunoluminometric Assay (cf IRMA)
SPALT	- Solid Phase Antigen Luminescence Technique
LUCIA	- Luminescent Cofactor Immunoassay
LEIA	- Luminescence Enhanced Enzyme Immunoassay
EELIA	- Enzyme Enhanced Luminescence Immunoassay

Table 3. TSH ILMA for dry blood spots

1 Blood spot (4.5 mm diameter)
200 μl Assay buffer
50 μl Anti-TSH-ABEN (monoclonal)
1 Anti-TSH coated tube (monoclonal)
Incubate 4h on horizontal rotator (170-190 rpm)
Discard blood spot and wash with 2 x 5 ml 0.25 ml/l Tween 20 and add 300 μl catalyst solution.
Load luminometer, initiate light reaction with 300 μl alkaline peroxide and integrate signal over 4 s.

ILMA and SPALT use luminescent-labelled antibodies, LUCIA uses labelled cofactor (e.g. Streptavidin-ABEI) whereas LEIA and EELIA can used luminescent labelled antigen or antibody. CELIA uses luminescent labelled antigen.

The choice of assay depends on the analyte and antibodies available. Where possible, ILMA should be used for analytes with two or more independent antibody-binding sites and SPALT for haptens or peptides with only one antibody binding site (or where only 1 Antibody is available, but the antigen/analyte is sufficiently cheap and is available for coating the solid-phase).

A "universal label" is possible, when LUCIA is used with the avidin-biotin system. If the antigen/antibody is coupled to a biotin analogue (e.g. the 8-aminooctanoic acid derivative) only one label - e.g. streptavidin or avidin - luminogen is needed for all such assays. It is interesting to know, that by varying the avidin/streptavidin - signal carrier, that the "universality" is extended to EIA FIA and RIA techniques using one basic set of components for a given analyte! This is not just theoretical but works in practice as has been demonstrated in this laboratory.

RESULTS

The proof of any assay lies in the practicability and results obtained. The assay must be as simple as the "competitors" to have a chance of commercial acceptance. A few assays are presented here to show, that luminescence immunoassays can "hold their own" in all respects. Table 3 shows the scheme of one-step luminescence immunoassay for TSH neonatal screening, which allows the results to be given out on the same day.

Table 4. Lactoferrin ILMA flow sheet

10 μl sample/standard (plasma or BAL)
200 μl anti-lactoferrin-ABEI (rabbit)
1 anti-lactoferrin-coated ball (goat)
Incubate for 90 min.
Wash with 2 x 5 ml Tween 20 (0.25 ml/l)
Transfer ball to cuvette
Add 300 μl catalyst solution and measure as in Table 3

Table 5. Lactoferrin ranges in plasma from different control and experimental groups

Percentile	Blood Donors (Serum)	Blood Donors (Plasma)	Pre-dialysis	Post dialysis
2.5	0.1 mg/l	0.1 mg/l	0.1 mg/l	0.1 mg/l
16	0.17 mg/l	0.11 mg/l	0.18 mg/l	0.36 mg/l
50	0.32 mg/l	0.17 mg/l	0.27 mg/l	0.56 mg/l
84	0.68 mg/l	0.31 mg/l	0.45 mg/l	1.06 mg/l
97.5	1.65 mg/l	0.42 mg/l	1.13 mg/l	2.25 mg/l
no. in group	125	52	37	37

This assay eliminates the need for separate removal of blood spot and the "tube transfer" step needed in the coated ball technique. The lower detection limit of the assay is under 2 mU/l TSH (WHO 80/558) and is suited not only for neonatal screening, but for screening children in remote areas of endemic goitre where samples cannot be processed immediately (e.g. Southern Tanzania - where this assay has been used on "airmail-samples").

The use of luminescence immunoassays is either in the field of endocrinology where low detection limits are required, or in perinatology where the amount of sample available is limited. It is possible to measure several important parameters in as little as 10 μl of plasma in cases of neonatal sepsis (10).

An example is the lactoferrin assay, which can be performed in conjunction with alpha 1-antitrypsin-elastase for leucocyte-involved illnesses such as pneumonia due to bacterial causes. It is interesting to know, that this assay can either be performed on 10 μl undiluted bronchioalveolar lavage (BAL) or plasma. The lactoferrin flow scheme is shown in Table 4, relevant reference and experimental ranges in different patient groups in Table 5.

Here the importance of chosing the right material for analysis can be demonstrated in the comparison between blood-donors plasma (intact leucocytes) and serum (destroyed leucocytes). As lactoferrin in blood normally originates from leucocytes, any damage to these cells will result in elevated levels (see pre- and post dialysis, where the dialysis membranes are known to destroy leucocytes).

CONCLUSION

This short article describes how luminescence immunoassays can be set-up and used.

The choice of the area in which the extreme sensitivity can be used lies either in the low analyte concentration present (endocrinology) or small sample volume available (paediatrics).

The entry of such assays into the routine laboratory dependant on commercial kits, depends upon the ability of kit and measuring equipment producers to supply simple and robust methods and luminometers to meet routine needs.

ACKNOWLEDGEMENTS

The author wishes to thank the following for technical assistance: Carsten Boger, Astrid Jacobs, Christel Rodust and Claudia Rohde.

REFERENCES

1. Phil Trans Roy Soc (London) 1667; 2: 605-12.
2. Radziszewski BR. Untersuchungen über Hydrobenzamid, Amarin und Lophin. Berichte D Chem Ges 1877; 10: 70-5.
3. Albrecht HO. Über die Chemiluminescenz des Aminophthalzäurehydrazids. Z Phys Chem 1928; 136: 321-30.
4. Gleu K, Petsch W. Die Chemiluminescenz der Dimethyl-diacridyliumsalze. Angew Chem 1935; 48: 57-9.
5. Wood WG, Gadow A. Immobilisation of antibodies and antigens on macro solid phases - A comparison between adsorptive and covalent binding. J Clin Chem Clin Biochem 1983; 21: 789-97.
6. Sturgess ML, Weeks I, Mposko CN, Laing I, Woodhead JS. Chemiluminescent labeled-antibody assay for thyroxin in serum, with magnetic separation of the solid-phase. Clin Chem 1986; 32: 532-5.
7. Weeks I, Campbell AK, Woodhead JS. Two-site immunochemiluminometric assay for human α_1-fetoprotein. Clin Chem 1983; 29: 1480-2.
8. Wood WG, Fricke H, Haritz J, et al. An evaluation of four different luminescence immunoassay systems: CELIA (chemiluminescent immunoassay), SPALT (solid-phase antigen luminescence technique), ILMA (immunoluminometric assay) and ILSA (immunoluminometric labelled second antibody). J Clin Chem Clin Biochem 1984; 22: 349-56.
9. Wood WG, Fricke H. Solid phase antigen luminescent immunoassays (SPALT) for the determination of insulin, insulin antibodies and gentamicin levels in human serum. J Clin Chem Clin Biochem 1982; 20: 835-32.
10. Braun J, Schultek T, Tegtmeler KF, Florenz A, Rohde C, Wood WG. Luminometric assays of seven acute-phase proteins in minimal volumes of serum, plasma, sputum and bronchioalveolar lavage. Clin Chem 1986; 32: 743-7.

LUMINESCENCE METHODS FOR SUBSTRATE AND ENZYME DETERMINATIONS

W. Miska*, R. Hauber and R. Geiger

Abtg. Klin. Chem. Klin. Biochem. Chir. Klinik, Nuszbaumstr 20, D-8000 München 2, FRG; Dermatol. Klinik Universität München, Frauenlobstr. 9, D-8000 München 2, FRG*

Bioluminescence is a natural phenomenon found in many lower forms of life. Naturally occurring bioluminescent systems differ with regard to the structure and function of enzymes and cofactors, as well as in the mechanism of the light emitting reactions (1, 2). More recently, highly sensitive bioluminescent and luminescent methods have become available for the determination of ATP and many different analytes (3, 4).

In this communication we describe and demonstrate highly sensitive bioluminogenic enzyme substrates and their application in enzyme activity tests, enzyme immunoassays and protein blotting techniques on the basis of the bioluminescence of the firefly Photinus pyralis (Figure 1). The scheme of the bioluminogenic enzyme activity measurements are shown in Figure 2.

Luciferin (Photinus pyralis) derivatives have been developed recently. The compounds represent a new type of bioluminogenic enzyme substrates allowing an ultrasensitive determination of enzymes in the femtogram range (5). This is about 10^2 to 10^4 times more sensitive than enzyme determinations performed with chromogenic or fluorogenic substrates (5).

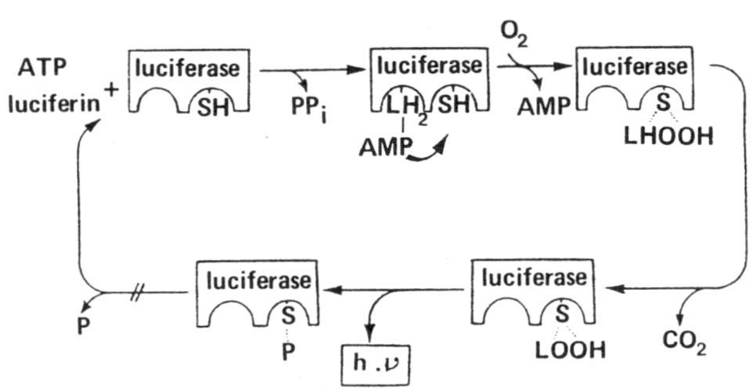

LH_2–D(–)-luciferin; P–Oxyluciferin; EDTA, Mg^{2+}, DTT, HEPES, pH 7.75, 25° C

Fig. 1. Scheme of firefly bioluminescence reaction.

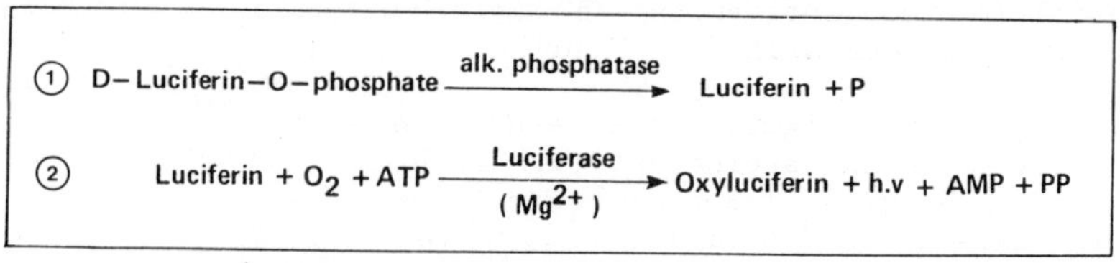

Fig. 2. Test principle of the bioluminogenic enzyme substrates.

Enzyme substrates on the basis of D-luciferin (Photinus pyralis) as a leaving group have been developed for carboxypeptidases A and B, for arylsulfatase, carboxylic esterase and for alkaline phosphatase.

Before use of the synthesized luciferin substrates in enzymatic tests the following questions had to be answered: 1) can different amounts of luciferin be quantified in the luminometric assay?, 2) are the derivatives themselves substrates for luciferase?; is a light emitting reaction induced?, 3) are the derivatives inhibitors of luciferase?; are the luminometric assays disturbed by these compounds?, 4) can the derivatives be cleaved by enzymes?, and 5) is D-luciferin released?

As to the first question luciferin can easily be quantitated. As shown in Figure 3 there is a linear relation between light output and used luciferin concentrations.

Enzymatic studies clearly demonstrate that luciferin derivatives applied in the luminometric assay up to 1 mmol/l cause no light emission. In other words, the luciferin derivatives are not substrates for firefly luciferase and free of luciferin.

Depending on the concentrations of luciferin derivatives added to the luminometric assay (see Table 2), luciferin derivatives can inhibit the light emitting reaction. To prevent this interference, which is essential for the sensitivity of the detection systems of the immunoassays, concentrations of luciferin derivatives were later used in the detection systems which did not interfere in the luminometric assay (6).

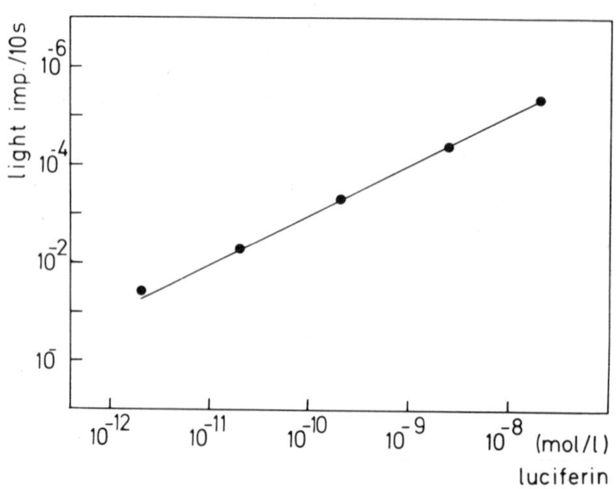

Fig. 3. Determination of D-luciferin in the luminometric assay.

Table 1. Kinetic constants of the luciferin derivatives

Enzyme	D-luciferin derivative	K_m (mol/l)	V_{max} ($mol.min^{-1}.mg^{-1}$)	k_{cat} (1/s)	k_{cat}/K_m ($l.mol^{-1}.s^{-1}$)
Carboxylic esterase	D-luciferin methyl ester	1.6×10^{-5}	10.6	11*	662
Carboxypeptidase A	D-luciferyl-L-phenyl alanine	3.5×10^{-3}	13.6	7.7*	2.2
Carboxypeptidase B	D-luciferyl-L-N^a-arginine	2.0×10^{-4}	12.0	7.0*	35
Arylsulfatase	D-luciferin-O-sulfate	7.1×10^{-5}	6.4	24*	338
Alkaline phosphatase	D-luciferin-O-phosphate	4.3×10^{-5}	610	1010*	23,500

* protein concentration was given by the manufacturers.

The most important point was whether the luciferin derivatives can be cleaved by enzymes and whether luminometrically active luciferin is indeed released by these reactions. As can be seen from the data in Table 1, which have been obtained by following the hydrolytic reactions by the luminometric assay, luciferin derivatives can be cleaved by appropriated enzymes. Esterification of luciferin with methanol lead to a compound which could be cleaved by carboxylic esterase. Luciferyl-L-phenylalanine was a substrate for carboxypeptidase A. Formation of luciferin-L-N -arginine lead to a substrate for carboxypeptidase B. Sulfatation of the phenolic hydroxyl group of luciferin resulted in a substrate which was cleaved by arylsulfatase. Luciferin-O-phosphate was a substrate for alkaline phosphatase.

The Michaelis-Menten constants and V_{max} values determined for the luciferin substrates and the respective enzymes were in a range typical for other substrates of these enzymes (Table 1). Especially alkaline phosphatase showed a very high catalytic rate with the substrate luciferin-O-phospate. The detection limits of the tests are between 10 fg per test (alkaline phosphatase) and 500 fg per test for carboxylicesterase (Figure 4).

The results demonstrate that the sensitivity of the bioluminescence substrates are decades lower than found with the usual substrates (10) and can be conveniently used for ultra sensitive enzymatic activity test. Thus active enzymes can be measured in very low concentrations.

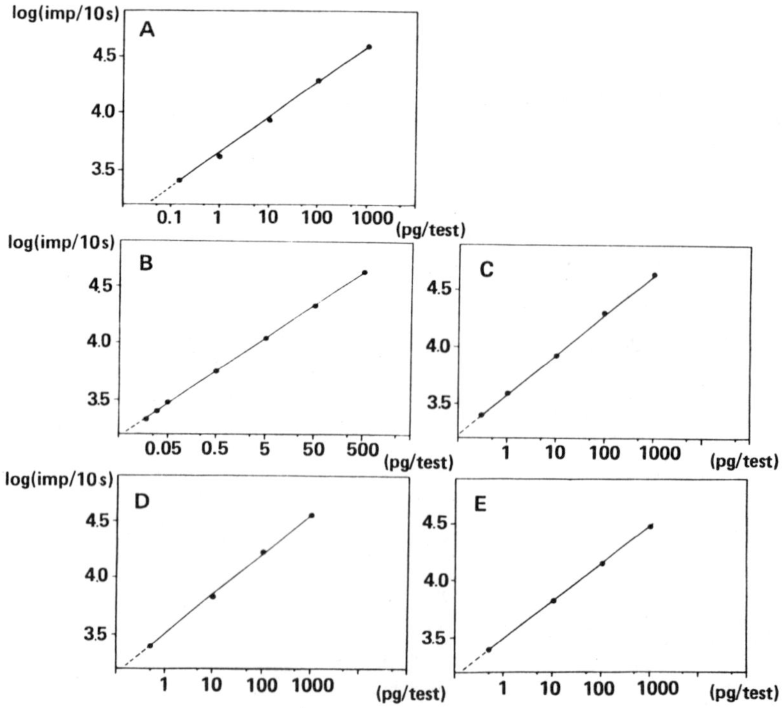

Fig. 4. Detection limits of D-luciferin derivatives in the enzymatic test systems. A, arylsulfatase; B, alkaline phosphatase; C, carboxylic esterase; D, carboxypeptidase A and E, carboxypeptidase B.

Alkaline phosphatase showed the best catalytic constants of all enzymes and can be detected down to 10 fg per test. These properties make this enzyme suitable for the application in bioluminescence-enhanced enzyme immunoassays. Bioluminescence-enhanced enzyme immunoassays for peptides and proteins have been elucidated.

Bradykinin, human urinary kallikrein and immunoglobulin G were conjugated to alkaline phosphatase according to established methods. The enzymatic activities of the conjugates after synthesis were sufficiently high for the intended aim.

Using these enzyme conjugates, a number of different variants of immunoassays with different detection systems have been developed for the assay of bradykinin, human urinary kallikrein and human urinary kallikrein antibody-titer determinations as examples. The principles of the assays are shown schematically in Figure 5. The bradykinin assay was performed as a competitive test (Figure 5d), whereas for human urinary kallikrein a competitive (Figure 5a), an immuno enzymometric assay (Figure 5b), and a sandwich antigen assay (Figure 5c) were established. Anti human urinary kallikrein titers were measured by sandwich antibody assay (Figure 5e). The assays were performed as described in (7).

The assay system was optimized with respect to enzyme immunoassay conditions (binding of antigen/antibody to microtiter plates, antibody-antigen binding, etc.) as previously described (7), and to optimal detection system conditions (concentration of substrates, temperatures, etc.). Luciferin liberation at 25°C and 37°C as well as the kinetics of the releasing reaction was also investigated. At 37°C luciferin was liberated at a higher rate. Optimal substrate concentrations for the detection systems were determined by adding increasing amounts of D-luciferin derivatives to the respective assay. As can be seen from the data, optimal substrate concentrations are in the range of about 10 μmol/l.

In Figure 6 to Figure 9 binding curves of different types of bioluminescence enhanced enzyme immunoassays for human urinary kallikrein, bradykinin and antibody titers (anti-human urinary kallikrein serum) are shown using different detection systems. The standard curves have been constructed by plotting the fraction of light impulses of the blank and the

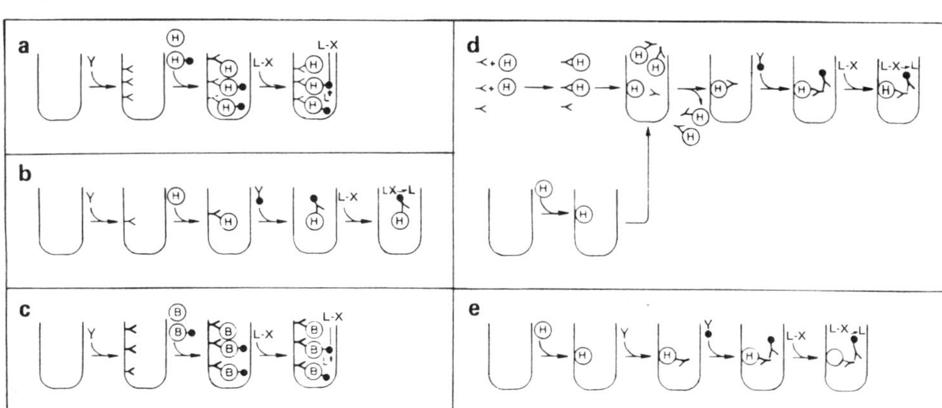

Fig. 5. Scheme of several bioluminescence enhanced enzyme immunoassays. Competitive assays a, c; sandwich assay b; immuno enzymometric assay d; sandwich antibody assay e.

samples - against the dose of antigens either in a logit-log mode (Figure 6 and 8) or in a linear-log mode (Figure 7 and 9). Sensitivity and the slope of the curves are varying in dependencey of the detection system and the type of enzyme immunoassay. The sensitivities of the bioluminescence enhanced enzyme immunoassays were defined according to the method described by Kaiser (11). The lowest concentration of human urinary kallikrein that produces a response greater than that caused in the absence of human urinary kallikrein was 1 fg per well in the immuno enzymometric assay (Figure 8), this corresponds to 5 pg/l. For bradykinin the lowest detectable concentration was 1 fg per well corresponding to 5 pg/l (Figure 6). Mass of antibodies which gave 50% binding in the assay were 200 fg per well corresponding to 1 ng/l specific anti-human urinary kallikrein immunoglobulin G.

The determination of precision (intra and inter assay coefficient of variation), specificity and recovery of the assays led to the same results as previously published for enzyme immunoassays with peroxidase as label (12, 13). The relative intraassay coefficient of variation (N between 15 and 25 for each assay; measured at different days) ranged from 2-6% for all assays and detection systems. The intraassay coefficient was calculated for all standard curve concentrations. The interassay coefficient of variation of identical samples (N between 8 and 22) was 4-12%.

The sensitive bioluminescence enhanced enzyme immunoassay developed for peptides (bradykinin), proteins (human urinary kallikrein) and antibody titer determinations should be especially suitable for studies in which highly sensitive methods are needed, e.g. for detection of monoclonal antibody synthesis in cell cultures, demonstration of in vitro translation products, etc. The high sensitivity of the bioluminescence enhanced enzyme immunoassay (1 pg/l; in general radio and enzyme immunoassays reach a sensitivity of 10 to 100 ng/l; (14)) could be demonstrated by employing the novel bioluminescence detection system for different types of enzyme immunoassay.

The bioluminescence-enhanced enzyme immunoassays described are useful for determination of antigens in a very low concentration range (down to 10^{-15} mol/l or 0.2 10^{-19} mol/test). The assay procedure is simple to perform, and the substrates are more than one year stable at -20°C in the dark. Substrate solutions can also be stored at -30°C for some weeks. The luminescence enhanced enzyme immunoassays and commonly available enzyme immunoassays.

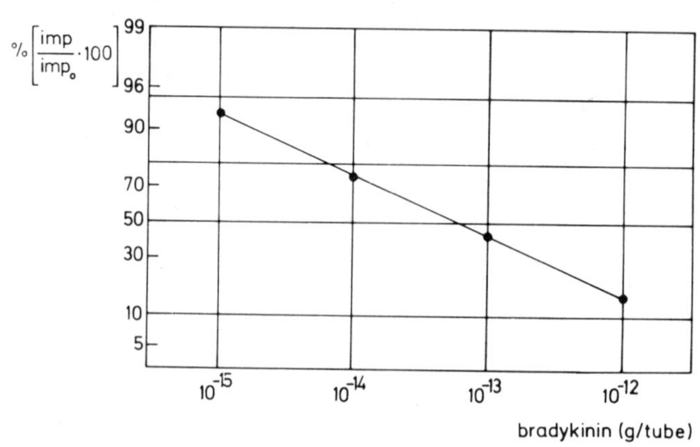

Fig. 6. Dose response curve of bradykinin in the competitive bradykinin bioluminescence enzyme immunoassay.

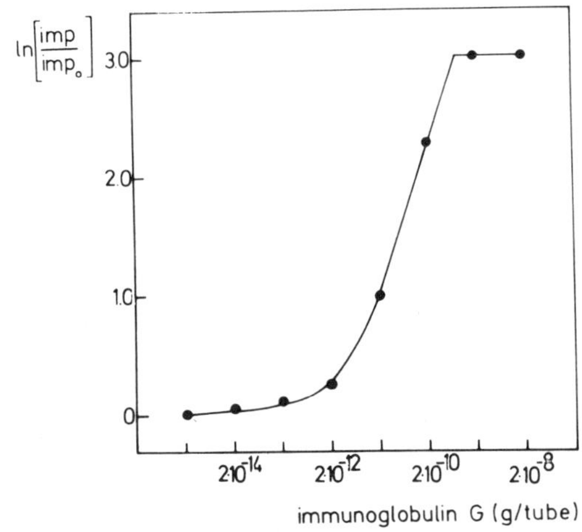

Fig. 7. Binding curve of rabbit anti-HUK IgG in the HUK sandwich antibody assay.

A further approach of the bioluminescence-enhanced detection system is its application in protein blotting (8). The principle of bioluminescent protein is shown in Figure 10.

After application of proteins to a nitrocellulose filter, either by direct application or by electric transfer, an antibody alkaline phosphatase conjugate is added. Protein-bound alkaline phosphatase conjugate liberates luciferin from the substrate luciferin-O-phosphate, and luciferin is oxidized by luciferase (Photinus pyralis) under light emission. Emitted light exposes a sensitive photographic film. Thus specifically bound alkaline phosphatase conjugate can be visualized after development.

In the detection step, alkaline phosphatase and luciferase (Photinus pyralis) have to be active both in the same buffer solution. The pH optima of these enzymes, pH 7.8 for luciferase (Photinus pyralis) and pH 9.6 for alkaline phosphatase are different. Based on the results depicted in Figure 11, a buffer of pH 8.0 was chosen, where alkaline phosphatase and luciferase are still sufficiently active.

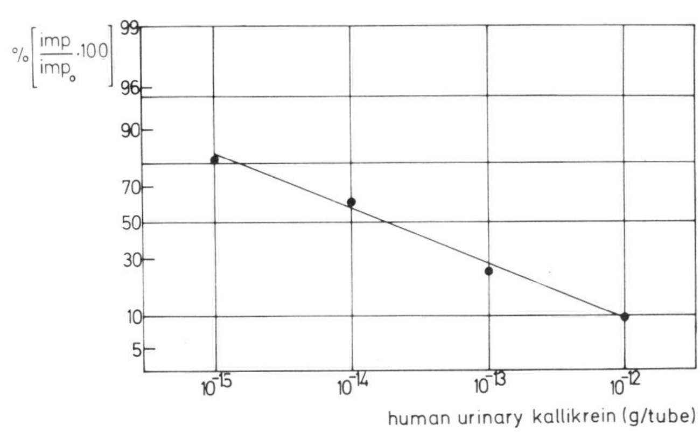

Fig. 8. Binding curve of HUK in the immuno enzymetric HUK assay.

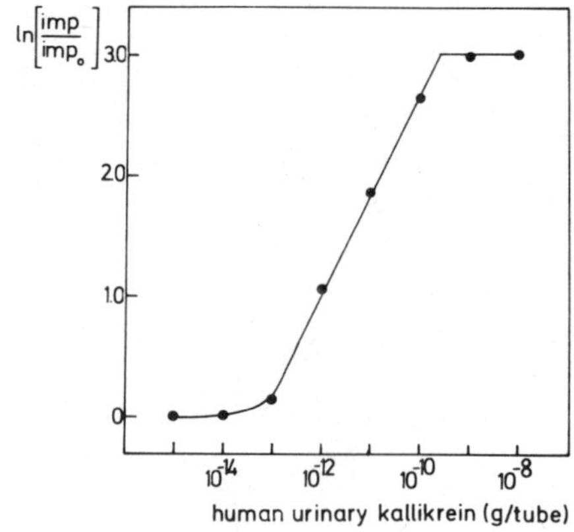

Fig. 9. Binding curve of HUK in the HUK sandwich antigen assay.

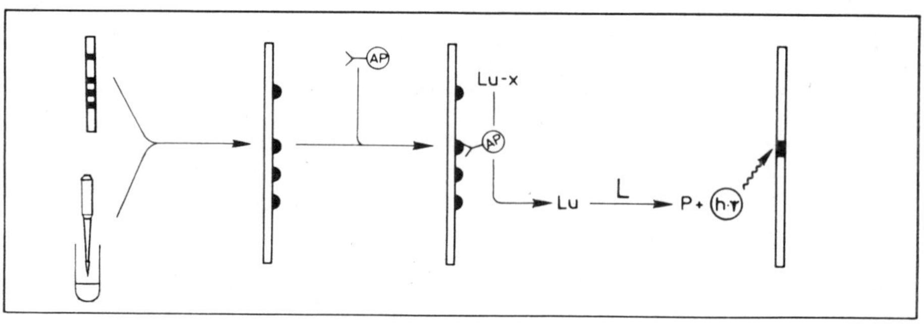

Fig. 10. Scheme of the bioluminescent-enhanced detection system of protein blotting.

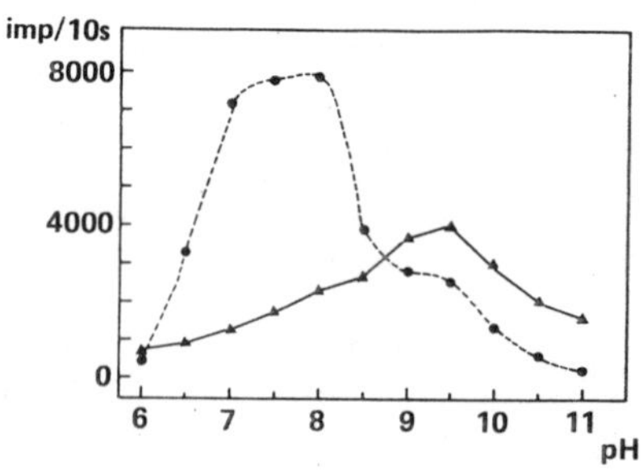

Fig. 11. pH optima of calf intestine alkaline phosphatase ( ) and firefly luciferase (o, Photinus pyralis).

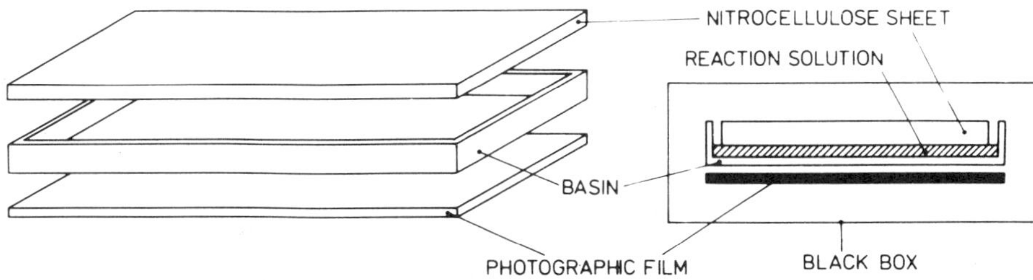

Fig. 12. Scheme of the device for the visualization of the bioluminescence-enhanced detection system.

Because of commercially non-available detection equipment, the detection has been performed using a self-constructed device as demonstrated in Figure 12.

Figure 13 an example of a photographic film, where even spots corresponding to 5 pg protein (rabbit immunoglobulin G) can be detected is shown.

The detection system has been successfully used for the following antigens: rabbit immunoglobulin G, human urinary kallikrein and porcine leukocyte elastase. The detection system is at present not optimized with respect to sensitivity of photographic film and exposure time, whereas enzyme activities have been stabilized for two hours using the detection buffer described above. Limits of detection determined for the present method are between 5 to 5000 pg (corresponding to 30 x 10^{-18} mol of rabbit immunoglobulin G, to 10^{-16} mol of human urinary kallikrein (15) and to 7 x 10^{-16} mol of porcine leukocyte elastase (16)) at an exposure time of 2 hours. Taking into account that the measurement of light is performed under non-optimal conditions (only exposure of photographic film, no use of electronical photomultipliers or scanners) the detection limits can be lowered by powers of ten even at a shorter exposure time of the film. By use of sensitive electronical detection systems, e.g. a photomultiplier or a luminometric scanner, protein samples could be measured by the same principle in the femtogram range (< 10^{-20} mol), as demonstrated recently for bioluminescence-enhanced immunoassays (6, 7).

Very sensitive detection systems are of special interest and need today. By their use many problems in science and diagnosis could be solved more rapidly and at an earlier stage, as the quick determination of low concentrations of proteins or peptides produced by recombinant techniques in microorganisms or culture broth (9) or the determination of low antibody titers in the first stages of antibody formation, e.g. in diseases like AIDS.

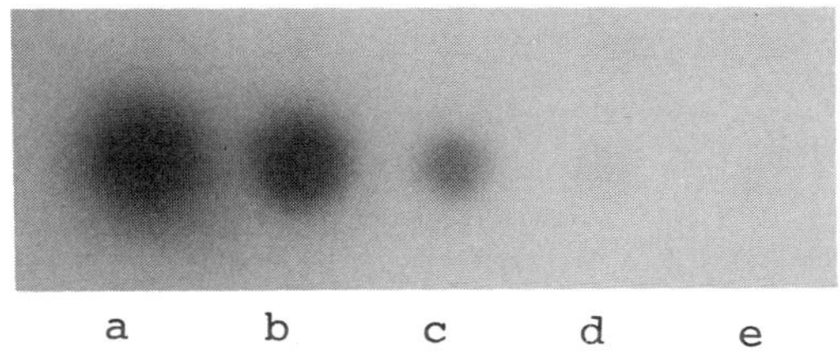

Fig. 13. Photographic detection of a protein blot on nitrocellulose filter. a, 5 ng; b, 500 pg; c, 50 pg; d, 5 pg; e, control, no transfer of protein.

Furthermore, sensitive nucleic acid hybridization techniques using a bioluminescence-enhanced detection system (substrate: D-luciferin-O-phosphate) are under investigation (9).

ACKNOWLEDGEMENT

We wish to thank Prof. Dr. H. Fritz for his support during this work.

REFERENCES

1. Burr JG. Chemi- and Bioluminescence. New York: Marcel Dekker, 1985.
2. DeLuca MA. Meth Enzymol 1978; 57.
3. Kricka LJ, Stanley PE, Thorpe GHG, Whitehead TP. Analytical Application of Bioluminescence and Chemiluminescence. New York: Academic Press, 1984.
4. Schölmerich J, Andreesen R, Kapp A, Ernst M, Woods WG. Bioluminescence and Chemiluminescence, New Perspectives. Chichester: John Wiley, 1987.
5. Miska W, Geiger R. Anal Biochem 1987; in press.
6. Miska W, Geiger R. J Clin Chem Clin Biochem 1987; 25: 23-30.
7. Geiger R, Miska W. J Clin Chem Clin Biochem 1987; 25: 31-38.
8. Hauber R, Geiger R. J Clin Chem Clin Biochem 1987; in press.
9. Hauber R, Geiger R. Personal communication.
10. Smith RE, Bissell ER, Mitchel AR, Pearson KW. Thromb Res 1980; 17: 393-402.
11. Kaiser H. Fresenius Z. Anal Chem 1965; 209: 1-18.
12. Geiger R, Miska W. Adv Exp Med Biol 1986; in press.
13. Franke M, Rohrschneider S, Geiger R. J Clin Chem Clin Biochem 1982; 20: 621-626.
14. Yalow RS. Trend Anal Chem 1982; 1: 128-132.
15. Geiger R, Stuckstedte U, Fritz H. Hoppe Seyler's Z Physiol Chem 1980; 361: 1003-1016.
16. Geiger R, Junk A, Jochum J. J Clin Chem Clin Biochem 1985; 23: 821.

PHAGOCYTE CHEMILUMINESCENCE AND IMMUNE CELL INTERACTION

M. Ernst and J. Ennen

Forschungsinstitut Borstel, Department of Immunology and Cell Biology, D-2061 Borstel, West Germany

In the course of some chemical reactions excited electronic states are generated which upon relaxation to the ground state emit light which according to its generation is called chemiluminescence (CL). Phagocytosis associated CL was first described by R. Allen (1) who reported on a low level CL derived from granulocytes during phagocytosis of bacteria. Today we know that the basis of phagocyte CL is the property of phagocytes to respond to appropriate stimuli with the so-called respiratory burst of oxygen consumption, this occurs independently of mitochondrial respiration but is closely associated with the univalent reduction of molecular oxygen to yield the superoxide anion and its many derivatives which play a major role in the bactericidal and tumoricidal action of phagocytes (for review see Klebanoff (2)) and which also can give rise to chemiluminescence (1, 3, 4). The primary oxygen reduction in phagocytes is provided by a transmembrane NADPH dependent oxidase complex the activation steps of which have been partially identified and reviewed in papers by Babior (5) and more recently by Hamilton and Adams (6). Some of the chemical reactions involved in the production of reactive oxygen intermediates (ROI) from phagocytes are given in the following chemical equations:

$$\text{NADPH} + 2\,O_2 \xrightarrow{\text{NADPH oxidase}} 2\,{}^{\cdot}O_2^- + \text{NADPH}^+ + H^+$$

Hydrogen peroxide is performed from two superoxide radicals by a dismutation reaction in which one radical is oxidized and the other is reduced:

$$2\,{}^{\cdot}O_2^- + 2\,H^+ \longrightarrow {}^1O_2 + H_2O_2$$

This dismutation can occur spontaneously or catalyzed by the enzyme superoxide dismutase, in the latter case yielding ground state molecular oxygen ${}^3O_2$ instead of singlet molecular oxigen ${}^1O_2$. Superoxide and hydrogen peroxide are not used directly by the phagocyte for microbial killing because H_2O_2 is only weakly microbicidal and O_2^- is completely innocuous. These compounds are used instead as starting materials for the production of the true microbicidal oxidants of phagocytes which fall in two classes: oxidizing radicals and oxidized halogens. The production of the former involves both starting compounds: In the presence of some heavy metal cations as katalyst (e.g. Fe^{+++}, Cu^{+}) superoxide can interact with hydrogen peroxide (Haber-Weiss-reaction) to form the toxic substance $^{\cdot}OH$ (hydroxyl radical) and singlet molecular oxygen:

$$^{\cdot}O_2^- + H_2O_2 \xrightarrow{\text{metal cation}} {}^1O_2 + {}^{\cdot}OH + OH^-$$

Granulocytes and to a lesser degree monocytes contain an abundance of myeloperoxidase (MPO) - positive granules which together with halide anions and hydrogen peroxide form a very efficient antimicrobial system toxic to a wide variety of organisms including bacteria, fungi, viruses, mycoplasma, or protozoa; it is also toxic to certain mammalian cells, e.g. erythrocytes, leukocytes, thrombocytes, and tumor cells. With chloride as the halide, the primary product appears to be hypochlorous acid or its ionized form, the hypochlorite anion which in turn can react with hydrogen peroxide to form singlet molecular oxygen.

$$H_2O_2 + Cl^- + H^+ \xrightarrow{\text{MPO}} H_2O + HOCL$$

$$H^+ + OCL^-$$

$$H_2O_2 + OCL^- \longrightarrow {}^1O_2 + H_2O + CL^-$$

Due to their oxidizing capacity all these reactive oxygen intermediates can induce a low level chemiluminescence in the presence of for example polysaccharide compounds from bacterial cell walls. However this "native" phagocyte chemiluminescence is very weak and very often below the detection limit of most of the CL measuring devices. Thus instead of measuring native CL directly today one takes advantage from some chemiluminescent indicator molecules which when added to phagocyte cultures allow to monitor the time course of ROI release from stimulated phagocytes due to their ROI dependent chemiluminescence. One of the most frequently used CL indicator molecules is luminol (5-amino-2,3-dihydro-1,4-phthalazine-dione) which yields considerable light emission when oxidized preferably in the presence of peroxidases.

Another CL indicator molecule is lucigenin (10,10'-dimethyl-9,9'-biacridinium dinitrate) which emits light best with ROI when it is reduced for example by the superoxide anion (3).

In the following we want to present examples for the application of luminol aided phagocyte chemiluminescence in the investigation of monocyte/lymphocyte interactions.

1. LYMPHOCYTE/MONOCYTE INTERACTION IN TUMOR CELL INDUCED MONOCYTE CHEMILUMINESCENCE

When mouse macrophages were incubated for 24 hours in the presence of macrophage activating factor(s) (MAF) they are primed to kill a variety of subsequently added tumor cell lines. We could demonstrate that this killing process is associated with the release of activated oxygen species such as O_2^- and H_2O_2 giving rise to a low level but significant luminol dependent chemiluminescence. In some but not all tumor target cell lines H_2O_2 but not the superoxide anion is the macrophage derived killing agent (7).

Since it had been claimed that human killer (NK) cells exert a luminol dependent chemiluminescence during coincubation with their tumor target cells (8) we investigated this phenomenon more thoroughly in order to clarify if NK cells could indeed release activated oxygen species which until that time was a unique property of phagocytes. By careful characterization and separation of the effector cell populations during enrichment of human NK cells from PBMC we found that human NK cells do not release any activated oxygen species which were detectable in the presence of luminol. However, we saw a significant tumor cell induced CL in NL cell enriched populations which contained only small amounts of monocytes (1%), in fact we observed a positive co-operation of target cell recognizing NK cells and CL mediating monocytes (9, 10).

In this paper we present data from experiments in which we used different lymphocyte subsets in order to characterize the surface phenotypes of those lymphocytes which act synergistically with monocytes in tumor cell induced monocyte CL. As effector cells we used cell sorter (cytofluorograf, Ortho, system 50) derived monocytes (due to light scatter-gated sorting of cells with high 90° light scatter signals), and Leulla-FITC, Leu7-FITC, OKT4-FITC, and OKT8-FITC labeled lymphocytes (due to the fluorescence signals of cells within the lymphocyte window defined by the appropriate forward light scatter and 90° light scatter signals. The purity of the sorted cell populations was more than 90% according to the reanalysis performed immediately after sorting. In the CL assays three different "effector cell" conditions were run in parallel in the 6 channel

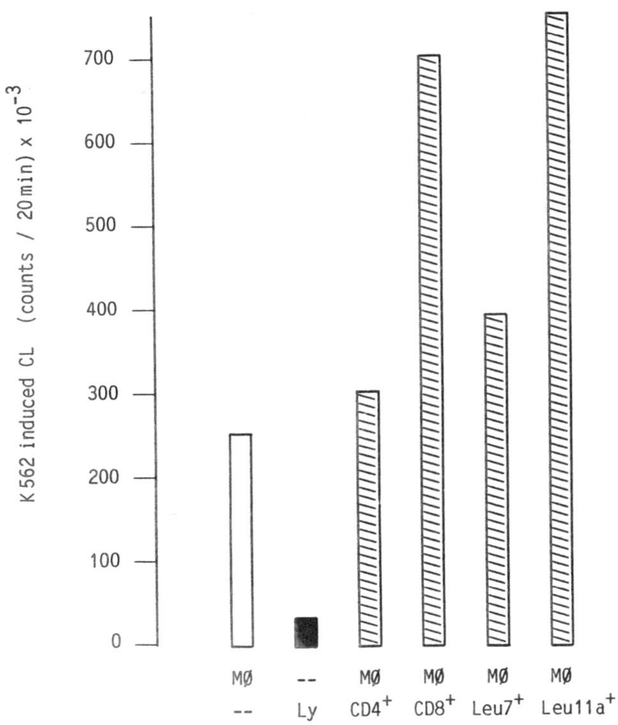

Fig. 1. K562-cells induced chemiluminescence in monocytes (MØ) alone (left most bar) and lymphocyte-subpopulations (Ly) alone, and in monocytes reconstituted with the distinct cell sorter derived lymphocyte subpopulations (MØ plus $CD4^+$, $CD8^+$, $Leu7^+$, or $Leulla^+$ lymphocytes).

chemiluminescence measuring device (Biolumat LB 9505, Berthold, Wildbad, FRG) for each lymphocyte subpopulation to be tested: one duplicate contained 10,000 monocytes alone, one duplicate contained 50,000 lymphocytes alone and one duplicate contained 10,000 monocytes + 50,000 of the respective lymphocyte subpopulation. All samples were triggered by the addition of 100,000 K562 tumor target cells. The results are shown in Figure 1. As expected from previous experiments Leu11a positive lymphocytes (NK cells) exert a clear synergistic activity with monocytes. In contrast $CD4^+$-lymphocytes do not seem to interact at all with K562 tumor cells triggered monocytes. $Leu7^+$ lymphocytes contribute to a minor extent to an increased K562 cells induced monocyte CL, however, $CD8^+$ lymphocytes like $Leu11a^+$ lymphocytes show as striking positive co-operation with the tumor cell induced monocyte CL.

Thus we conclude that by surface marker definition $CD8^+$ (suppressor/ cytotoxic) T cells and $Leu11a^+$ cells interact synergistically with monocytes in tumor cell induced monocyte CL activation whereas T helper cells ($CD4^+$ lymphocytes) do not interfere with tumor cell induced monocyte CL activation, and $Leu7^+$ lymphocytes which partially can belong to the $CD8^+$ and to the $Leu11a^+$ lymphocyte subset exert a moderate CL enhancing capacity for K562 cells triggered monocyte CL.

2. LYMPHOCYTE/MONOCYTE INTERACTION IN LYMPHOKINE INDUCED ENHANCEMENT OF FC-RECEPTOR MEDIATED MONOCYTE CHEMILUMINESCENCE

The regulation of immune functions is mediated by both by cell-cell-interactions and by soluble factors very often derived from immune cells themselves such as the lymphokines interleukin-2 (IL-2), interferon-gamma (IFN-gamma), macrophage activating factor(s) (MAF), tumor necrosis factor beta and the monokines interleukin-1 (IL-1), interferon-alpha (IFN-alpha), tumor necrosis factor alpha. Since phagocytes, especially monocytes and macrophages, play an essential accessory role in nearly all cellular immune responses where the above mentioned cytokines are released it seems very likely that monocytes and macrophages respond in a certain fashion to all these cytokines. To further study this question the effect of recombinant human IFN-gamma (rIFN-gamma) and of recombinant IL-2 (rIL-2) on the Fc-receptor mediated monocyte chemiluminescence was analyzed. Either highly purified monocytes alone or reconstituted mononuclear cell populations (monocytes + lymphocytes) were treated with either lymphokine. As CL stimulus we used particulate model immune complexes, namely sheep erythrocytes coated with specific rabbit IgG. These IgG coated sheep erythrocytes (E-IgG) are phagocytosed by normal monocytes, and they give rise to luminol aided CL, whereas non-IgG coated sheep erythrocytes are not phagocytosed and do not induce CL. Thus we used an Fc-receptor mediated phagocytosis associated CL assay for the examination of the monocytic functional capacity.

In a first set of experiments we incubated highly purified monocytes or monocytes + lymphocytes (the lymphocyte to monocyte ratio being 2 : 1) with rIFN-gamma in the concentration range of 0.05 U/ml to 5 U/ml for 48 hours, then the cells were washed and used for the E-IgG (1 Mio per assay) induced CL measurement in a monocyte concentration of 100,000 per 300 μl medium. A typical result of such an experiment is shown in Table 1. As expected from literature data dealing with the effect of IFN-gamma on the expression of Fc-receptors on monocytes (11), we found a dose dependent increase in Fc-receptor mediated monocyte CL when purified monocytes were preincubated with IFN-gamma. However, when IFN-gamma was added to monocytes reconstituted with T lymphocytes the subsequent Fc-receptor mediated monocyte CL was found at strikingly enhanced levels as compared to the CL from IFN-gamma treated purified monocyte alone. Thus Fc-receptor mediated

Table 1. Enhancement of recombinant interferon-gamma driven monocyte activation by co-cultivation with T lymphocytes

Cell population	rIFN-gamma (U/ml)	Fc-receptor mediated monocyte CL activity
Monocytes	none	3,516 ± 156
	0.05	4,431 ± 438
	0.5	5,380 ± 250
	5.0	7,935 ± 190
Monocytes + T lymphocytes	none	5,032 ± 490
	0.05	7,780 ± 1114
	0.5	7,729 ± 257
	5.0	17,865 ± 135

monocyte CL reflects an IFN-gamma induced T lymphocyte/monocyte interaction at IFN-gamma concentration levels which are below the detection level of IFN-gamma (7 U/ml) in the bioassay for antiviral activity.

The formation of IFN-gamma in peripheral blood mononuclear cells (MNC) can be induced by another lymphokine, namely interleukin-2 which, originally described as T-cell growth factor, has proven to exert more features than the single role as growth factor for T-lymphocytes. IL-2 acts as growth factor on T-cell blasts, on mitogen or antigen activated T-cells, and on activated B-cells as well. In addition, IL-2 has an immunoregulatory effect: IL-2 induces the expression of IL-2 receptors on T-cells; it induces the release of IFN-gamma from T-cells; it enhances the cytolytic activity of natural killer cells; it may induce indirectly the release of TNF-alpha from monocytes. It has been shown by the use of monoclonal antibodies that all cell types in which IL-2 has direct effects bear the IL-2 receptor on their membrane surface. This was demonstrated for activated T-cells, for activated NK cells and for activated B-cells. More recently it has been shown (12) that the IL-2 receptor is also expressed on the surface of activated cells of the monocytic cell line U937. We therefore asked the question whether IL-2 has any biological effect - directly or indirectly - on human peripheral blood monocytes. Using the monoclonal anti-IL-2-receptor antibody Tü69 (Clonab-IL-2R, Biotest) we could demonstrate by means of cytofluorografic analysis that 5% to 25% of blood monocytes of normal healthy individuals bear the IL-2 receptor.

Having this in mind we cultured monocyte enriched MNC for 48 hours in the presence of rIL-2 (6 - 600 U/ml) and then measured spontaneous monocyte derived CL, latex (polystyrene beads, 1.0 μ in diameter) and zymosan (particulate yeast cell wall extract) induced CL, and the E-IgG induced monocytic CL. At 60 and 600 U/ml of rIL-2 the E-IgG induced CL was increased significantly to an extent comparable to that obtained with low dose of rIFN-gamma. However, highly purified monocytes (purity more than 95%) showed only a moderate activation pattern following culture with these doses of rIL-2 indicating that the considerable increase in E-IgG triggered CL seen after preculture of MNC in rIL-2 containing medium was probably lymphocyte mediated. These data suggested that the rIL-2 effect on subsequent E-IgG triggered monocyte CL was dependent on the presence of T lymphocytes since the amount of the rIL-2 effect on highly purified monocytes could be explained by 2-3% residual T lymphocytes within the purified population. At this point we thought that the IL-2 effect reported so far might be due to an indirect effect of IFN-gamma. IL-2 could induce the release of IFN-gamma from the residual T-cells in our monocyte enriched MNC population. We checked this possibility by the use of an anti-IFN-gamma monoclonal antibody which blocked the IFN-gamma activity in our system. To our surprise we found that the simultaneous presence of the anti-IFN-gamma antibody and IL-2 during the 48 hours culture led to a further increase in the subsequently performed Fc-receptor mediated CL test. We, therefore, concluded that IL-2 induces a monocyte activating activity in monocyte enriched MNC, which is partially inhibited by IFN-gamma.

The next step in clarifying the target cell for IL-2 in our system was to use better defined cell populations. For this purpose we prepared highly purified monocytes (more than 95% esterase positive cells) and highly purified T cells (sheep erythrocyte rosetting MNC). We then added different amounts of T cells (addition of 5%, 10%, 30%, and 50% T-cells) to the highly purified monocytes and cultured these defined MNC populations for 48 hours with IL-2 (600 U/ml). The subsequently measured E-IgG stimulated monocyte CL revealed a clear cut T-cell dose-effect relationship of the IL-2 driven increase in the E-IgG triggered monocyte CL. With increasing proportions of T cells during the culture period with IL-2 we found an up to 7 fold increase of the subsequent CL response. From this result we conclude that this mode of IL-2 action on monocytes require a monocyte/T cell interaction.

Keeping these results in mind we then asked the question whether the monocyte activating activity of IL-2 was mediated only by the direct action of IL-2 on monocytes and T-cells and the cellular interaction of these or whether the release of a monocyte activating factor (MAF) might be induced by IL-2. Therefore we cultured for 48 hours highly purified monocytes and highly purified T-cells, and defined reconstituted monocyte - T-lymphocyte mixtures with and without IL-2. Then we harvested the corresponding supernatants and tested them for MAF-activity in our CL system using as

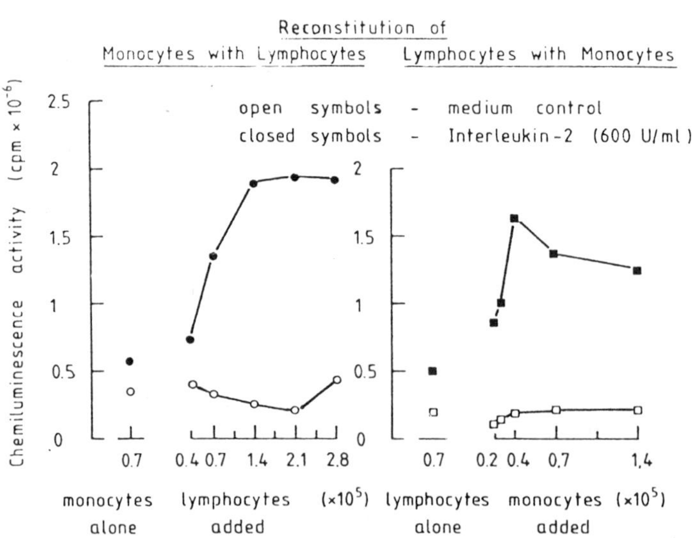

Fig. 2. Detection of soluble macrophage activating factor(s) (MAF) in supernatants of monocytes co-cultured with increasing amounts of T-cells (left) and vice versa (right), which were treated with 600 U/ml rIL-2 (48 h.) as detected by activation of fresh monocytes (indicator monocytes). Indicator monocytes were cultured with supernatants for 48 hours, and the E-IgG CL activity (counts/25 min.) was monitored.

indicator cells highly purified monocytes cultured for 48 hours with these supernatants. Subsequently these indicator monocytes were washed and assayed for Fc-receptor mediated CL. The results of such experiments are shown in Figure 2. In this graph is depicted the IL-2 induced MAF-activity in supernatants of purified monocytes reconstituted with increasing amounts of T-cells (left) and the IL-2 induced MAF-activity in supernatants of purified T-cells reconstituted with increasing amounts of monocytes (right). Both graphs show that IL-2 induces the release of high MAF-activity only in those mononuclear cell populations which contain both monocytes and T-cells, and maximal MAF release is achieved when the ratio of T-cells to monocytes is 2 : 1. The main conclusion from these results is nearly the same as from the preceding results: Neither T-cells alone nor monocytes alone do release this MAF-activity when triggered with IL-2. The IL-2 induced release of this MAF again requires a monocyte/T-cell interaction.

We then addressed the question whether there is still a direct effect of IL-2 on highly purified monocytes alone. Therefore we developed the experimental scheme which is shown in Figure 3. In a first culture step highly purified monocytes or highly purified T-cells were cultured for 18 hours with or without IL-2. Then the cells were washed and used in a second culture step as "culture primed cells": IL-2 primed monocytes, control monocytes, IL-2 primed T-cells, and control T-cells. In the second culture the monocytes were cultured for another 48 hours in normal culture medium either alone or they were co-cultured with control T-cells or IL-2 primed T-cells. The supernatants of these secondary cultures were collected and tested for IL-1 activity, for IL-2 activity, for MAF-activity, and for IFN-gamma activity (in a very sensitive microtiter ELISA system provided by Hoffmann La Roche, Basel, Switzerland). In addition the cellular CL-activity following E-IgG stimulation was measured. A representative example for the E-IgG triggered CL of monocytes following this secondary culture step is shown in Figure 4. In the upper part we see the CL traces of control monocytes alone (left, low CL), of control monocytes co-cultured with control T-cells (middle, low CL), and of control monocytes co-cultured with IL-2 primed T-cells (right, high CL). In the lower part of the E-IgG induced CL patterns of IL-2 primed monocytes alone (left, low CL), of IL-2 primed monocytes co-cultured with control T-cells (middle, high CL), and of IL-2 primed monocytes co-cultured with IL-2 primed T-cells (right, high CL) are shown. We recognize an IL-2 dependent increase of Fc-receptor mediated monocyte CL activity only under conditions when monocytes were co-cultured with T-cells in the secondary culture. The most astonishing feature of this set of experiments is that IL-2 primed monocytes alone do not show an increased Fc-receptor mediated CL, but when co-cultured in the secondary culture with control T-cells a strikingly enhanced Fc-receptor mediated monocyte CL response is observed. Thus far we conclude that IL-2 primed monocytes and IL-2 primed T-cells are induced by IL-2 to interact with T-cells or monocytes, respectively, and as a result of such an interaction an increased monocyte Fc-receptor function is detectable.

In order to test our hypothesis that IL-2 priming of purified monocytes is related to specific binding of IL-2 to the monocyte plasma membrane, we tried to inhibit the IL-2 priming of monocytes by means of the monoclonal anti-IL-2-receptor antibody Tü69. The result of these experiments is given in Table 2.

This table shows clearly that IL-2 priming of purified monocytes leads to a strikingly enhanced Fc-receptor mediated CL activity following the secondary co-culture with control T-cells (line 2 as compared to line 1) whereas by the simultaneous treatment of purified monocytes with IL-2 and the Tü69 monoclonal antibody the IL-2 priming of purified monocytes is

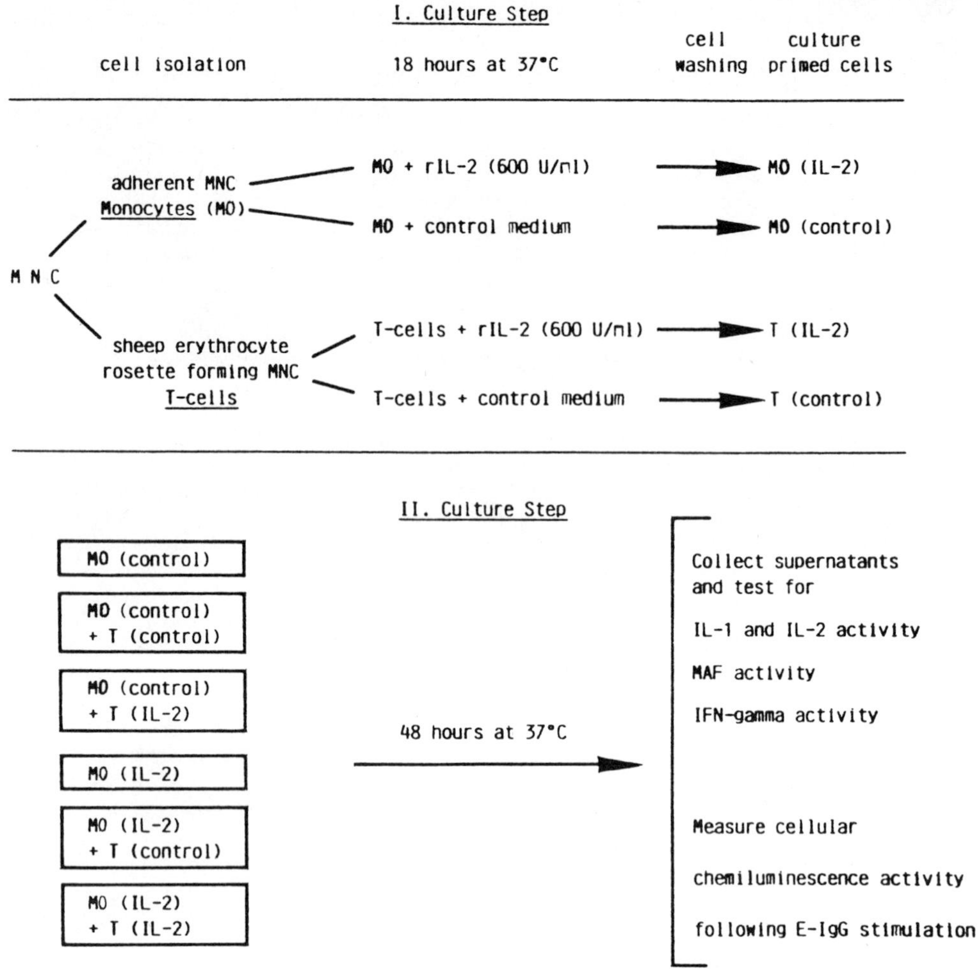

Fig. 3. Experimental scheme for the assessment of a direct IL-2 effect on highly purified monocytes.

Table 2. Influence of the monoclonal anti-IL-2-receptor antibody Tü69 on IL-2 priming of purified monocytes

Cell composition during the secondary culture	Fc-receptor mediated CL response following secondary secondary culture (counts/20 min.) x 10^{-3}
Contr. T-cells + contr. monocytes	1,794 ± 744
Contr. T-cells + IL-2 treated monocytes	8,200 ± 602
Contr. T-cells + Tü69 treated monocytes	2,239 ± 152
Contr. T-cells + IL-2 and Tü69 treated monocytes	2,263 ± 262

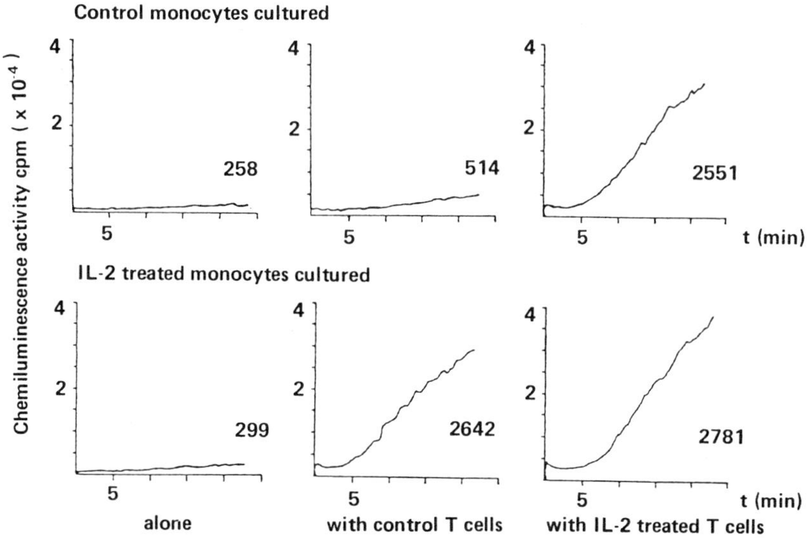

Fig. 4. E-IgG induced CL of control monocytes (upper panel) and of IL-2 primed monocytes (lower panel) following secondary culture alone (left), with control T-cells (middle), and with IL-2 primed T-cells.

totally abrogated (line 4 as compared to line 3 or 1). Thus we conclude that IL-2 priming of monocytes is IL-2 receptor dependent.

As to the cytokine activities in the supernatants of the secondary culture step (cf. experimental scheme in Figure 3) it is interesting to note that the IL-2 activity in all the supernatants tested (all conditions) was less than 0.5 U/ml, the MAF activities in these supernatants corresponded to the direct cellular chemiluminescence activity, the interferon-gamma activities were also low in all conditions (less than 0.3 U/ml). However we found some IL-1 activity in the supernatants of the secondary cultures of IL-2 primed monocytes, but no IL-1 activity in the supernatants of the secondary cultures of control monocytes. Thus, the spontaneous release of interleukin-1 may be one T-cell independent feature of IL-2 primed monocytes.

SUMMARY

1. Lymphocyte/monocyte interaction in tumor cell induced monocyte chemiluminescence (CL)

   The tumor cell induced monocyte CL is synergistically enhanced by the monocyte interaction with NK-cells, $CD8^+$ T-cells, and partially with $Leu7^+$ lymphocytes but not by the interaction with $CD4^+$ T helper cells.

2. Lymphocyte/monocyte interaction in interleukin-2 (IL-2) induced enhancement of Fc-receptor mediated monocyte chemiluminescence

- A proportion of 5 to 25% of fresh monocytes from healthy donors bears the IL-2 receptor on their membrane.

- IL-2 induces a monocyte/T cell interaction which leads to - enhanced Fc-receptor mediated monocyte chemiluminescence and the release of a factor into the supernatant with MAF activity which is not interferon-gamma.

The IL-2 induced MAF release is dependent on the T-cell/monocyte ratio (a ratio of 2 : 1 giving highest MAF activity).

IL-2 acts directly on monocytes
- IL-2 induces IL-1 production in monocytes
- IL-2 primes monocytes for T-cell/monocyte interaction
- IL-2 priming of monocytes is IL-2 receptor mediated, it can be inhibited by an anti-IL-2-receptor monoclonal antibody (Tü69, Biotest).

Our finding that IL-2 directly acts on monocytes is in line with a very recent report (12) in which the direct augmentation of the cytotoxicity of human monocytes by IL-2 is described. The biological significance of the IL-2-effect on monocytes has yet to be elaborated. However, our results imply a regulatory role for IL-2 in the function of monocytes. The IL-2 driven monocyte/T-cell interaction which enhances the Fc-receptor mediated oxidative burst activation in monocytes may account for a new phagocyte (and lymphocyte?) activation process.

ACKNOWLEDGEMENTS

The skillful technical assistance of Mss. R. Bergmann, J. Schenck, A. van Lunteren, and of Mr. R. Reichenbach is gratefully acknowledged.

REFERENCES

1. Allen RC, Stjernholm RL, Steele RH. Biochem Biophys Res Commun 1972; 47: 679-684.
2. Klebanoff SJ. In: Gallin JI, Fauci AS, eds. Advances in Host Defence Mechanisms. New York: Raven Press, 1982; Vol 1: 111.
3. Allen RC. In: Adam W, Cilento G, eds. Chemical and Biological Generation of Excited States. New York: Academic Press, 1982: 309.
4. Ernst M. Behring Institute Mitt 1980; 65: 55.
5. Babior BM. Blood 1984; 64: 959.
6. Hamilton TA, Adams DO. Immunology Today 1987; 8: 151.
7. Ernst M, Lang H, Fischer H, Lohmann-Matthes M-L, Staudinger H. In: De Luca M, McElroy P, eds. Bioluminescence and Chemiluminescence. New York: Academic Press, 1981: 609.
8. Helfand SL, Werkmeister J, Roder JC. J Exp Med 1982; 156: 492.
9. Ernst M, Lange A, Flad H-D, Havel A, Ulmer AJ. Eur J Immunol 1984; 14: 634.
10. Ernst M, Lange A, Havel A, Ennen J, Ulmer AJ, Flad H-D. In: Kricka LJ, Stanley PE, Thorpe GHG, Whitehead TP, eds. Analytical Applications of Bioluminescence and Chemiluminescence. London: Academic Press, 1984: 315.
11. Vogel SN, Friedmann RM. In: Vilcek J, De Maeyer E, eds. Interferons and the Immune System. Amsterdam: Elsevier, 1984: 35.
12. Malkovsky M, Loveland B, North M, et al. Nature 1987; 325: 262.

PHAGOCYTOSIS-INDEPENDENT CHEMILUMINESCENCE

Friedrich-Ernst Maly

Institute of Clinical Immunology, Inselspital Bern
Switzerland

INTRODUCTION

In 1972, cellular CL emission was first observed by Allen in phagocytic cells ingesting particles (1). From then on, the phenomena of phagocytosis and chemiluminescence were often linked conceptually. However, many chemiluminescence responses occur in the absence of particles. Rather, the principle of CL emission is the generation of electronically excited states by chemiexcitation, whose relaxation to ground state leads to emission of photons. In biological systems, only oxidation reactions liberate sufficient energy to generate excited states. Thus, chemiluminescence is an expression of an oxygenation reaction. The oxidants may be relatively simple derivatives of oxygen, like superoxide radical, hydroxyl radical, singlet oxygen, hydrogen peroxide and hypochlorite, also called "reactive oxygen species (ROS)". However, organic peroxides or hydroperoxides can also act as oxidising species in CL-yielding reactions. ROS can modify lipids, proteins and nucleic acids, and are therefore important mediators, e.g. of phagocytic anti-infectious activity, but also of inflammation and carcinogenesis.

Because of their sensitivity, CL assays employing "chemilumigenic probes" like luminol and lucigenin which are oxidisable substrates with a high CL quantum yield, are an attractive alternative approach to detection of ROS especially at low cell numbers, potentially even at the single cell level with the use of Pholasin (2).

In phagocytes, the most obvious way of ROS and CL generation is particle uptake, e.g. via surface receptors for Ig and complement. However, many other important mediators will activate the NAD(P)H-oxidase. Of the group of phagocytic cells including PMN, eosinophils, basophils, monocytes and different tissue macrophages, I shall concentrate on activation of PMN CL by a soluble complement activation product, C5a, and by several members of the family of lympho- and cytokines.

Further, several types of cells which do not phagocytose, have been claimed to produce ROS and/or show CL. This includes thymocytes, natural killer cells, epidermal cells, endothelial cells, glial cells and finally EBV-transformed human B lymphoblastoid cell lines. A general problem with freshly isolated cells is contamination with phagocytes, which - sometimes cooperating with non-phagocytic cells - are in fact the source of ROS, and of CL. Both in the case of thymocytes (3) and NK-cells (4), it has become clear that ROS generation and light emission do in fact derive from the phagocytes in the preparation.

However, in EBV-transformed B cell lines able to produce ROS (5) and CL (6) upon addition of phorbol ester, a contribution of phagocytes can be ruled out. We have studied such lines with the aim to characterize their ROS-generating system and have searched for other stimuli capable of inducing ROS and CL generation.

MATERIALS AND METHODS

The preparation of neutrophilic granulocytes (PMN), mononuclear cells (MNC), their culture-supernatants (MNC-SN) and the longterm culture of Epstein-Barr transformed human B lymphocytes (EBV-BLCL) have been described elsewhere (7, 8). CL kinetics were measured in a LB 9505 6-channel luminometer (Berthold, Wildbad, FRG). For simultaneous detection of CL from samples contained in microtiterplates, a single-photon-imaging device was employed (Figure 1) (9) (Hamamatsu C 1966 VIM, Hamamatsu, Tokyo, Japan). Descriptions of further materials and techniques used will be included when appropriate.

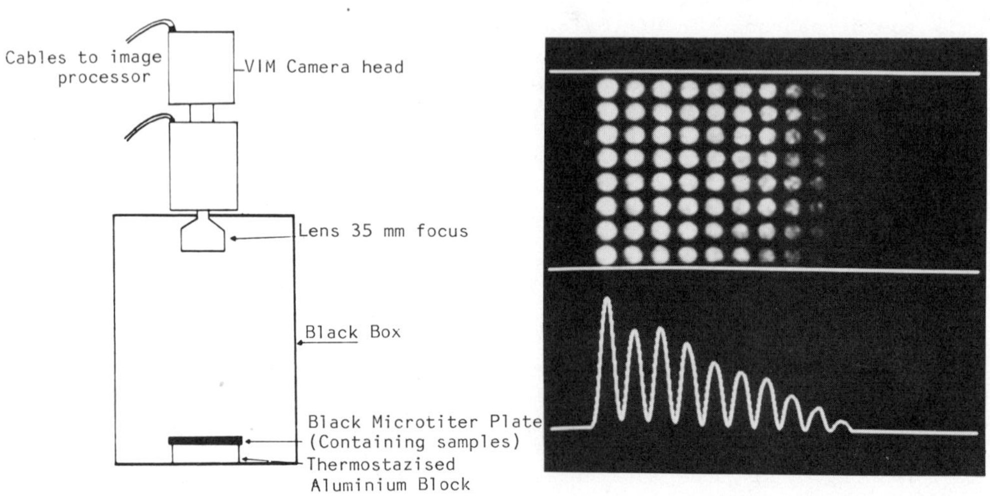

Fig. 1. Multiparallel quantitation of luminescence by single photon imaging. (Left): Schematic diagram of CL detection from microtiterplates (MTP). The "luminescent" image of the MTP is focused on the double-microchannelplate port of a Hamamatsu C 1966 VIM system. Photon counts accumulated over a counting period are stored in a frame memory of the image processor. (Right): Luminescent image (10 minutes) of ATP-dependent bioluminescence of ATP-CLS reagent (Boehringer Mannheim; FRG) in a MTP. ATP concentrations used (row 1 ... row 10): 10, 9, 8, 7, 6, 5, 4, 3, 2, 1 x 10^{-9} M/l; row 11: reagent blank (all total volume 100 μl); row 12: empty well. Histogram: average photon count/row (arbitrary units).

RESULTS AND DISCUSSION

Chemiluminescence response of human PMN to complement C5a

Investigations into induction of PMN CL by serum-derived factors were motivated by studies showing activation of PMN CL by sera from patients with rheumatoid arthritis or systemic lupus erythematosus (10). Apart from immune complexes, we reasoned that complement activation products should also be present in such sera. Since it had been shown that the purified complement split product C5a, a potent chemoattractant for PMN, was able to induce superoxide release from PMN (11), we investigated whether activation of serum complement in vitro would lead to a material capable of eliciting a PMN CL response.

In fact, after activation of the complement system in normal human pool serum by incubation with Zymosan (50 mg/ml for 1 hr at 37°C), marked stimulation of a very fast PMN CL was observed, while with the control serum (only heated, but without Zymosan addition), no CL response was observed. The size of the CL-response correlated well with levels of C5a peptide determined by radio-immunoassay (12), and purified C5a elicited a similar, but smaller CL response (Figure 2). Though non-specific serum factors were required to restore the C5a-induced CL-response to the level observed with Zymosan-activated serum, C5a (and also C5a des-arg) were the molecule(s) primarily responsible for the CL response induced by Zymosan-activated serum. PMN CL can thus be used to assess complement activation in vitro with a sensitivity similar to radio-immunoassay. Interestingly, presence of C5a in vivo apparently partially de-sensitizes granulocytes to in vitro-stimulation with C5a (13). Thus, it appears possible to monitor in-vivo complement activation by the ensuing down-regulation of the PMN receptors for C5a which shows as a reduced CL-response to C5a ex vivo.

Chemiluminescence response of human PMN to lympho- and cytokines

Lymphokines and cytokines, i.e. substances produced in low amounts by activated immune cells (e.g. T cells, B cells, monocytes, granulocytes), but also non-immune cells (e.g. endothelial cells, epidermoid cells) can influence and regulate the functions of an ever increasing spectrum of cell types.

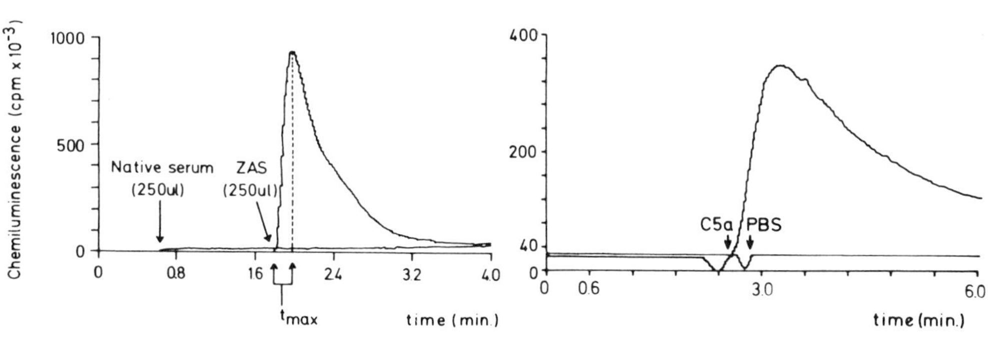

Fig. 2. Human (PMN) CL triggered by Zymosan-activated serum and purified C5a. 1 x 10^6 PMN were incubated in 250 μl minimal essential medium without phenol red (pH 7.2, 20 mM Hepes) with 100 um lucigenin (luci-mem). PMN were stimulated at 37°C with (left) Zymosan-activated serum (ZAS) or normal serum or (right) 10^{-7} M/l purified human C5a or phosphate buffered saline (PBS = solvent for C5a). CL was measured with a Berthold LB 9505.

We were interested in lympho- or cytokines which could activate the production of reactive oxygen species by PMN. In fact, mononuclear cells, if exposed to stimulants like bacterial lipopolysaccharide (LPS) or the mitogen phytohaemagglutinin (PHA) produced soluble mediators capable of stimulating PMN chemiluminescence. The mediator responsible for induction of PMN CL by culture supernatants of LPS-treated mononuclear cells, termed granulocyte chemiluminescence inducer (GCI) (7), was produced by monocytes, as could be shown by studies with MNC separated by elutriation. GCI-activity eluted from gel filtration at 50-60 kDa apparent molecular weight and exhibited protein nature, since cycloheximide inhibited its production and trypsin treatment diminished its activity. Further, GCI could be delineated from interleukin 1, which had been implicated as a stimulant for PMN degranulation and is also produced by LPS-treated monocytes.

Further, GCI-activities have been described in SN's from MNC and alveolar MO treated with silica particles and in supernatants of certain epidermoid cell lines. The former may be important in the pathogenesis of silica-induced pneumokoniosis: there is an influx of PMN into the lung in early phases of silicosis, and superoxide seems to be an important cofactor in the activation of the collagen-forming enzyme prolyl-hydroxylase. The latter may play a role in inflammatory skin diseases which present with a PMN infiltrate, like psoriasis.

While the further characterisation of the molecules responsible for GCI-activities is going on, a number of cytokines, known to be produced under such culture conditions, have been purified and cloned. Using recombinant human materials, it could be confirmed that interleukin 1 does not activate an oxidative burst in PMN. However, tumor necrosis factor (TNF) and lymphotoxin-alpha (LT) were found to be activators of PMN superoxide production and chemiluminescence, while granulocyte-macrophage-colony-stimulating factor (GM-CSF) and interferon-gamma (γ-IF) were inactive by themselves, but could enhance the response to a second stimulus applied shortly after (GM-CSF, 1-2h) or with a long time interval (γ-IF, 18-24h) (14). Therefore, it has to be analysed whether GCI-activity of diverse supernatants can be explained by their contents of TNF, LT and GM-CSF. Interestingly, GCI-activity of PHA-MNC-SN could be almost annihilated by treatment with anti-TNF and anti-GM-CSF, indicating that the GCI-activity of PHA-MNC-SN was in fact due to TNF and GM-CSF. On the contrary, GCI-activity of LPS-MNC-SN was only slightly inhibitable (by 10-15%) by such treatment, though PHA-MNC-SN was more active than LPS-MNC-SN. This indicates that, under appropriate stimulation, e.g. with LPS, mononuclear cells produce a mediator(s) different from TNF, LT and GM-CSF which possesses GCI-activity.

Thus, several cytokines are now recognised to possess GCI-activity, substantiating the concept of cytokine control of PMN free oxygen radical production. From the biological potency of the already cloned cytokines, i.e. TNF and LT, it can be expected that the mediator(s) responsible for as yet "unexplained" GCI-activity will also be biologically, and potentially clinically, interesting.

Chemiluminescence of non-phagocytic cells: EBV-transformed B lymphocytes as a model

Epstein-Barr virus activates and transforms resting human B lymphocytes into continuously growing cell lines, which are no more contaminated with phagocytic cells. Such EBV-B lymphocyte cell lines (EBV-BLCL) share immunological properties with monocytes: they can present antigen to T cells, produce an interleukin-1-like factor(s) and can also produce superoxide (5) and lucigenin-dependent chemiluminescence (6) in response to stimulation with phorbol ester. Since EBV-BLCL derived from individuals with chronic granulomatous disease, who do not possess a functioning superoxide generating system in the form of a NAD-(P)H-oxidase in their phagocytes, also do not show superoxide generation, it was concluded that superoxide production by EBV-BLCL from normal individuals had its basis in the expression of the phagocytic enzyme (5).

The phagocytic superoxide generating system, termed "NAD(P)H-oxidase", though not yet fully understood, is recognised to contain a low potential cytochrome B (Cyt B-245) and a flavoprotein which binds to and is functionally inhibited by diphenyl-iodonium (DPI) (15). Therefore, we analysed 2 "superoxide-producer" EBV-BLCL and 2 B cell tumor lines which do not produce superoxide, for the presence of these components.

Both "superoxide-producer" EBV-BLCL contained cytochrome B-245, as determined by redox-controlled spectrometry, as well as 44 kDA DPI-binding peptide (DPI-BP) visualized by binding of I^{125}-DPI to cellular proteins separated on SDS-gel. Further, superoxide production of EBV-BLCL was

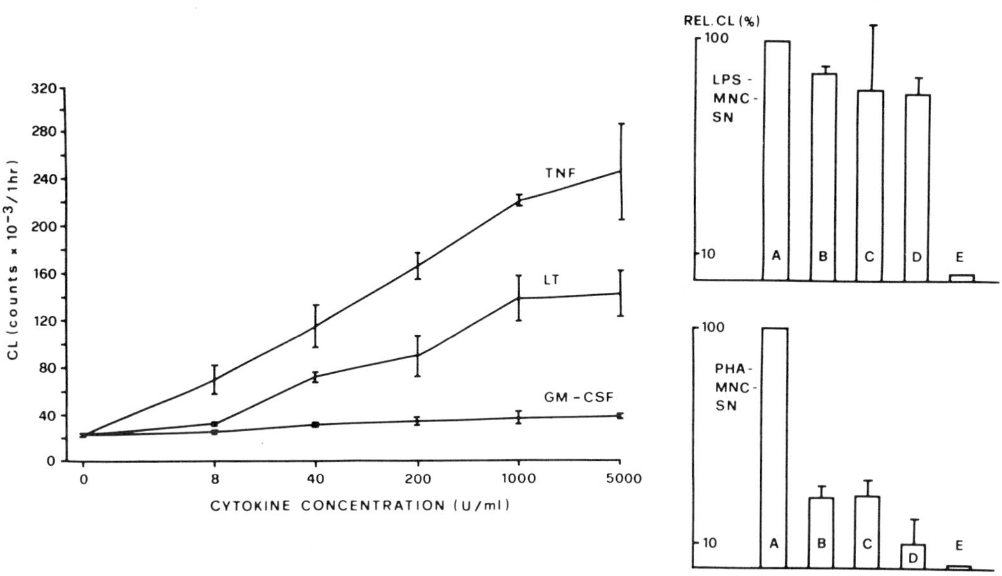

Fig. 3. Human granulocyte CL triggered by cytokines and mononuclear cell culture supernatants. (Left): Effects of recombinant human cytokines, 20,000 isolated PMN in 100 μl luci-mem dispensed into wells of a 96-well black MTP were stimulated with tumor necrosis factor (TNF, Genentech), lymphotoxin-α (LT, Genentech, San Francisco, USA) or granulocyte-macrophage colony stimulating factor (GM-CSF, Biogen, Geneva, Switzerland). 1 hr integrals of CL were determined simultaneously in triplicates from all samples by photonic imaging. Data from a typical experiment are shown as mean $\pm$ SD. (Right): Effect of anti-TNF and anti-GM-CSF on GCI-activity of LPS- or PHA-MNC-SN. Anti-TNF (1/100 F.D.) or anti-GM-CSF (1/200 F.D.) were added to samples 1 hr before assaying. A: untreated SN's; B: + anti-TNF; C: + anti-GM-CSF; D: + anti-TNF + anti-GM-CSF; E: medium control. Data shown are mean $\pm$ SD (n = 3). Absolute 1 hr CL integrals induced by: LPS-MNC-SN 58.9 $\pm$ 6.3 x 10^6; PHA-MNC-SN 872 $\pm$ 120 x 10^6; medium 2.9 $\pm$ 1.5 x 10^6 counts. Measurements were performed with 1 x 10^5 PMN in a Berthold LB 9505.

inhibited by DPI at doses comparable to those needed for inhibition of PMN superoxide production.

In the 2 "superoxide-non-producer"-lines, neither cytochrome B-245 nor the 44 kDA DPI-binding peptide could be detected, further strengthening the correlation of "superoxide-producer" status with the presence of Cyt B-245 and DPI-BP. These results make it likely that EBV-BLCL do in fact express the phagocytic NAD(P)H-oxidase or at least an electron transport chain very similar to it. Thus, our results are in line with the conclusions based on the findings with EBV-BLCL derived from CGD-patients (5). However, importantly, presence of EBV-genome in a B cell line is not necessarily accompanied by "superoxide-producer" status, as evidenced by the WIL-2 line. This leads to the question, whether "superoxide-producer"-EBV-BLCL are an EBV-specific phenomenon or whether normal B lymphocytes also express a phagocytic oxidase. Since "resting" B lymphocytes, isolated from peripheral blood, do neither contain Cyt B-245 nor produce superoxide, expression of a superoxide generating system would have to be confined to "activated" B cells, e.g. B lymphoblasts, which thus might be the physiological correlate of the "superoxide-producer"-EBV-BLCL.

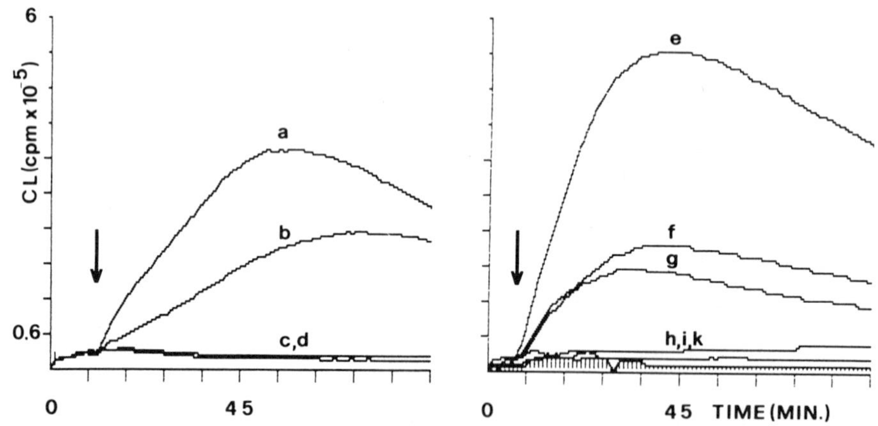

Fig. 4. Triggering of "Fl" EBV-BLCL via surface immunoglobulins. 200,000 "Fl" cells in 250 μl luci-mem in polystyrene plastic vials (Lumacuvette, 3M, Basel, Switzerland) were placed in a Berthold LB 9505 thermostated to 37°C. Data are original registrations from a typical experiment. (Left): Cells were stimulated with a) 50 μl of "pansorbin" 1/10 (suspension of staph. aureus Cowan I bearing protein A, Calbiochem, La Jolla, USA), b) 250 μl of "pansorbin" 1/10, with c) 50 μl of "sansorbin" 1/10 (staph. aureus devoid of protein A, Calbiochem, La Jolla, USA) or with 250 μl of "sansorbin" 1/10. (Right): In e), f) and h) plastic vials were coated with goat-anti-mouse-immunoglobulin antibody (Cappel, West Chester, PA, USA), while g), i) and k) were control-treated. After addition of "Fl" cells in lucimen, the following stimuli were applied: e) anti-human IgG at 1/20 final dilution, f) anti-human IgG at 1/100 F.D., g) phorbol myristate acetate 20 ng/ml, h) and i) medium (mem), k) anti-human IgG 1/20 F.D. (anti-human IgG: mouse monoclonal anti-human immunoglobulin antibody BS 16, Biotest, Dreieich, FRG).

Alternatively, the incorporated EBV-genome might lead to "aberrant" activation and expression of the genes coding for the components of the phagocytic superoxide generating system.

Experiments to identify a "physiological correlate" of the EBV-transformed B lymphocyte are currently in progress. However, since EBV infects and transforms B cells also in vivo and thus causes a spectrum of diseases (mononucleosis, chronic EBV-infection, burkitt-lymphoma), superoxide production by EBV-transformed B cells may conceivably also take place in vivo. However, so far phorbol ester has been the only substance able to stimulate O_2^- production in EBV-BLCL. Searching for a more physiological stimulus, we analysed whether EBV-BLCL show CL if triggered via their antigen receptor, i.e. surface immunoglobulin (sIg), which is one of the B cell's most important triggering structures. For example, in the "F1" line, addition of particles bearing protein A, which binds immunoglobulins and thus provides sIg-crosslinking, induced a CL response even higher than PMA. Further, a mouse monoclonal anti-IgG antibody was also effective, but only when additionally crosslinked by a solid-phase-bound anti-immunoglobulin antibody. Therefore, mere "binding" of sIgG seems not enough, rather "multi-point-crosslinking" appears needed to activate O_2^- production.

Taken together, EBV-BLCL constitute an example of expression of a phagocyte-like superoxide- and CL-generating system in lymphoid cells. They may be an in vitro analog of B cells at differentiation stages, in which they may be capable of antigen-triggered superoxide production, representing a "hybrid" between an antigen-specific lymphoid cell and a non-specific phagocyte.

CONCLUSIONS AND OUTLOOK

The data shown here represent only a few examples, but may show the potential of CL techniques in immunological research.

In the clinical field, studies of phagocyte CL have been mostly done using particulate or unphysiologic stimuli like Phorbol ester. On the basis of new insights into the fine tuning of phagocyte CL by endogenous mediators, a complementary and potentially fruitful approach seems possible:
The detection of altered phagocyte reactivity towards endogenous stimuli, e.g. lymphokines, complement products or neuropeptides, produced during a disease process, promises interesting applications. Changes in phagocyte function due to the presence of such mediators in the patient may thus be detected, which do not affect gross phagocyte function, as assessed with particulate stimuli and cannot be evaluated by measuring blood levels of mediators. Measuring phagocytic CL, induced by a panel of different soluble mediators, may be useful to monitor disease activity and possibly be of prognostic value in immunosuppressed patients, patients suffering from infections or sepsis, malignant diseases, autoimmune disorders or allergic diseases.

Finally, EBV-BLCL constitute a "prototype" of non-phagocytic cells carrying a phagocyte-like superoxide generating system whose activity can be easily monitored by CL. Though at present this field is only of basic research interest, identification of an in-vivo correlate, either under physiologic or pathologic conditions (e.g. auto-immune or immune deficiency states) would open new perspectives also in clinical research.

ACKNOWLEDGEMENTS

The studies presented were in part supported by grants from the Stiftung Volkswagenwerk (Hannover, FRG), Krebsforschung International (Frankfurt a.M., FRG) and the Suisse National Science Fund. Further I would like to thank Mrs. M. Bader for her patience and expert secretarial work.

REFERENCES

1. Allen RC, Stjernholm RL, Steele RH. Biochem Biophys Res Comm 1972; 47: 679-684.
2. Roberts PA, Campbell AK. In: Schoelmerich et al., ed. Bioluminescence and chemiluminescence. Wiley and Sons 1986: 133-136.
3. Weidemann MJ, Yin Foo D, Kolbuch-Braddon ME, Mitchell SV, Domzig W, Peterhans E. IBID; 1986: 109-112.
4. Domzig W, Maly FE. Immunobiol 1985; 169: 162-174.
5. Volkman DJ, Buescher ES, Gallin JI, Fauci AS. J Immunol 1984; 133: 3006-3009.
6. Maly FE, Brantschen St, De Weck AL. IBID; 1986: 105-108.
7. Maly FE, Kapp A, Walker C, De Weck AL. Lymphokine Research 1986; 5: 21-33.
8. Maly FE, Cross AR, Jones OTG, Dahinden CA, De Weck AL. Manuscript in preparation; 1987.
9. Maly FE, Urwyler A, Rolli HP, Dahinden CA, De Weck AL. Submitted to J Anal Biochem; 1987.
10. D'Onofrio CD, Maly FE, Fischer H, Maas D. Klin Wochenschr 1984; 62: 710-716.
11. Goldstein IM, Roos D, Kaplan HB, Weissmann G. J Clin Invest 1975; 56: 1155 ff.
12. Kapp A, Maly FE, Schoepf E. Arch Dermatol Res 1985; 278: 41-43.
13. Kapp A, Schoepf E. Acta Derm Venereol (Stockh.) 1986; 66: 285-289.
14. Perussia B, Kobayashi M, Rossi ME, Anegon I, Trinchieri G. J Immunol 1987; 138: 744-765.
15. Cross A, Jones OTG. Biochem J 1986; 237: 111-116.

CHAPTER 13
CLINICAL CHEMICAL ASPECTS OF AGEING

THE AGEING PROCESS: EPIDEMIOLOGY, THEORIES AND MODEL SYSTEMS

D.L. Knook

TNO Institute for Experimental Gerontology, Rijswijk, The Netherlands

1. AGEING AS A WORLDWIDE PHENOMENON

The ageing of the population is becoming a crucial problem. The developed countries have already seen a considerable increase in the proportion of their elderly during the last decades. Developing countries will face this increase in the very near future. The United Nations estimate that in the year 2000 two out of three of the world's 600 millions elders will be living in the developing regions of the world.

Two factors, which are independent of each other, play a decisive role in the "greying of the world". These factors are the continuously increasing life expectancy of individuals and a sharp decline in the absolute and relative number of children, or more general, the age group of 19 years of age and younger.

The greatest increase in life expectancy in the developed countries occurred during the first half of this century, and was mainly due to a sharp decline in infant mortality and an improvement of social and medical conditions. Presently these two factors largely contribute to the increase of the life expectancy in developing countries.
The tendency for an increase in life expectancy can be nicely illustrated with the total number of Dutch persons of 100 years of age and older. This number was 105 persons in 1970, whereas only 16 years later, in 1986, already more than 600 inhabitants of the Netherlands were 100 years of age and older.

The decrease in the percentage of youth in most western societies is perhaps as spectacular as the increase in average life span. 45% of the Dutch population in 1900 belonged to the age group of 0-19 years. By 1960, this percentage had declined to 36%, and the present day contribution of the youth to the total population is only 28%.

After the post-war baby boom, nearly all western countries had a sharp decline in birth rate, and this phenomenon is also observed in many developing countries.

The combination of the two factors mentioned above, viz. the increase in life expectancy and the decrease in number of youth, has resulted in a

Table 1. Persons of 65 years and older as percentage of total population in some western countries

Country	1960	1980	2000
Sweden	12.0	16.2	15.8
Austria	12.0	15.5	14.7
German Fed. Republic	10.8	15.1	15.5
United Kingdom	11.7	14.9	14.9
Norway	11.7	14.6	14.3
Denmark	10.6	14.3	14.6
France	11.6	13.7	14.2
Italy	9.3	13.5	16.4
The Netherlands	9.0	11.5	13.3
USA	9.2	11.2	12.2
Ireland	11.2	11.1	9.6

Source: C.B.S., 1982

strong growth in the proportion of elderly individuals in the population of many countries (see Table 1).

The group of individuals aged 75 and older occupies a special and demanding place in society, because these individuals often require substantial medical attendance and nursing. This group shows a worldwide increase. The projected annual growth rate until 2005 for the population aged 75 and over ranges from 1.56% for Italy to 2.94% for Japan, 3.36% for China and 3.97% for Brazil (1). The increase in the relative and absolute number of old and very old inhabitants confronts us with the paradox of medical knowledge and treatment. Prevention and treatment of diseases and disorders at earlier stages of life result in an increasingly number of elderly who, although partially, suffer from chronic degenerative diseases. Thus, as a consequence of the greying of the population not only age-dependent diseases, but also chronic disease processes pose a steadingly increasing health problem. This will have an important personal and socio-economic impact, as well as great consequences for the costs of health care. It is obvious that progressively larger amounts of money will be necessary for adequate medical treatment and for health care provisions.

2. THE BIOLOGICAL AGEING PROCESS AND AGEING THEORIES

Ageing is not a simple biological process. It consists rather of a whole series of complex changes occurring with time. It is not surprising, therefore, that the biological bases for the ageing process are still largely unknown. Due to the recent advances in biochemistry and molecular biology, it is only now possible that an attempt can be made to elucidate the mechanisms of the biological ageing process (2).
The numerous changes that take place during ageing occur at various levels of biological organization and a change at one level may have impact on a function at another. The key question is at what level of biological organizaton does the ageing process start. Is it at the level of molecules, cells, organs or the organism? Changes at the cellular level may have a significant impact on the physiological functions of an organ, but environmental influences, such as infectious diseases, and life style factors, such as nutrition, will probably also have their impact on the ageing process. The most important cellular, intracellular and

environmental factors influencing the ageing of an organism and their interrelations are summarized in Figure 1.

It is clear from Figure 1 that several physiological changes will have profound effects on the ageing process. The blood supply of organs can be mentioned as an example. The blood circulation decreases with age due to changes in the heart and blood vessels on which the process of atherosclerosis is superimposed. The organs will receive less blood as a result, and consequently the cells in the organs receive less oxygen and nutrients, which will have a depressive effect on their functional status.

Another important concomitant process of ageing is the decline in the efficiency of coordination by the control mechanisms of our body: the endocrine and central nervous systems. The involvement of hormones and neuronal signals in the complex interactions of different cells, tissues, and organs makes it evident that a failure of regulation will influence the ageing process.
Although age-related changes in a regulatory mechanism such as the hormone balance are well-known, the question remains whether these changes are the causes or the consequences of ageing.

These few examples of the many factors mentioned in Figure 1 emphasize that components change with ageing in living organisms and that most ageing changes manifest at the organismic level will result from a variety of factors. This does not exclude the possibility that the fundamental cause of ageing might result from changes in one distinct group of macromolecules and that a primary change in those molecules will cause numerous secondary and tertiary ageing changes at the various levels of biological organization. However, in judging the various ageing theories, one should keep in mind that perhaps one single ageing process does not exist, but

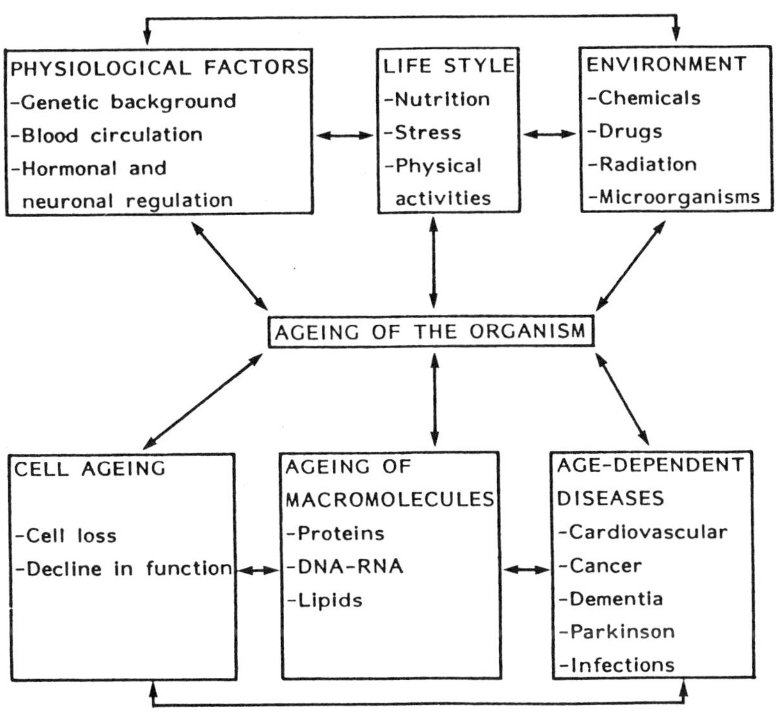

Fig. 1. Various components influencing the ageing of an organism.

that ageing of an organism may be the consequence of more than one primary process (2).

The majority of the various ageing theories can be grouped on the basis of the level of biological organization within an organism.

Organ theories are based on documented changes in organ functions with age. The basic idea in this group of theories is that a single organ system is responsible for the ageing of the total organism. It has been proposed that the immune system or the central nervous system may play a central role. Indeed, a wide variety of changes in immunological parameters has been observed with ageing. Some basic questions, however, remain unanswered, such as: are these age-related changes really due to the ageing process or do they result from the increase in diseases with old age? If they are not disease-related, do these changes really influence other bodily functions and what are the underlying primary causes for the ageing of the immune system itself.

A second group of theories ascribes the central role in the ageing process to a number of ageing changes occurring at the cellular level, mainly in biochemical processes. The most promising theories include:

- an age-related increase in damage of molecules caused by the action of free radicals;
- increase with aging in the cross-linking between macromolecules, especially collagen, but also DNA molecules;
- the accumulation of lipofuscin; lipofuscin or "age pigment" is considered to be an end product of lysosomal digestion and there is a good correlation between the age of a person and the amount of "age pigment" in long-lived cells in organs like the brain, the heart and the liver.

Experimental data demonstrate that these three processes, free radical damage, cross-linking and lipofuscin accumulation, occur and that their effects increase in severity with ageing. However, this does not prove that these processes can be considered as the primary causes of the ageing process. The presence of lipofuscin in old long-lived cells may be indicative of impaired cellular functioning or, it may also simply be the undigestible residue left after lysosomal activity. Thus the accumulation of lipofuscin does not prove that several other cellular ageing phenomena are caused by lipofuscin accumulation or by decreased capacity of the lysosomal digestion system. A comparable reasoning can be followed with regard to the increase in cross-linking or the accumulation of damage by free radicals.

In a third group of theories, attention is directed towards the genetic material in the cells because DNA forms the basis of the functioning of each cell and therefore of the whole organism. There are indeed several strong indications for an important role of DNA in cellular ageing. Each species of animals has its own inherited lifespan, which indicates a genetically controlled ageing process. Studies on the lifespan of identical twins suggest that also within the human species longevity is at least partly inherited. However, if the ageing process was completely genetically programmed, animals with the same genetic background would have the same lifespan. Strains of mice and rats inbred for over 20 generations and longer can be considered as genetically highly homogenous. Nevertheless, animals from such a strain born on the same day and kept under identical conditions show a comparable variation in lifespan as is found for non-inbred rats or mice.

According to one of the DNA theories on ageing the lifespan of cells is the result of a genetic programme and the lifespan of each species is

also genetically-programmed. In this concept, the cellular ageing process consists of various steps which occur according to a programme coded in the cellular DNA. Ageing is thus seen as the last series of steps of the processes of development and differentiation, as a selective switching off or on of a set of genes according to a programmed sequence.
Other DNA-based theories propose that the errors and damages occurring in the DNA may accumulate during the lifetime of the cells. This may lead to a decline in cellular functioning, which will finally negatively influence the functioning of the organism.
Experiments have shown in recent years that DNA molecules in old cells are remarkably intact. Today, much attention is devoted to the extensive cellular enzyme system capable to detect and repair damages in DNA molecules. It is possible that these DNA repair systems play an important role in determining the lifespan of cells, and thereby, of organisms.

3. DEVELOPMENT OF BIOMEDICAL GERONTOLOGY

The development of biomedical ageing research can be schematically divided into three stages, viz. an inventory stage, a theory-forming stage and a theory-testing stage. Although there is a chronological order in their origin, these stages are perpetual.

Inventory research has sometimes been compared to a "collecting of stamps". However, this stage in biomedical gerontology has been of great importance for obtaining a thematic view of the various aspects of the biological ageing process.
Nevertheless, one has to recognize that a great deal of experimental gerontological data collected in the past is of limited value as numerous investigators did not use healthy animals or the animals under study were not really aged.

Similar problems exist in human studies which demonstrated a strong decline in bodily functions in aged individuals. These data involving cross-sectional studies were often obtained from a mixture of healthy and diseased elderly individuals.

The establishment of base-time data in healthy elderly individuals is a specific problem that needs to be addressed. Screening of immunological parameters in healthy, elderly individuals is currently being performed under the auspices of the immunology programme of EURAGE, the concerted action programme on ageing and diseases of the European Communities.

The data obtained during the inventory stage has yielded a second, theory-forming stage. As already discussed above, numerous theories to explain the ageing process have been formulated which deal with age changes occurring at various levels of biological organisation.

The current explosive development in molecular biology and biochemistry, especially with regard to experimental studies performed at the genome level, now opens a third stage of biomedical gerontological research, the theory-testing stage. Today, there is increasing evidence that the biological ageing process is, at least partially, governed by the genetic information present in each cell. However, no documented data exist that prove that DNA contains the information to direct the cellular ageing process. It is also possible that the primary ageing process is related with the genetics information, but not identical with it. An example is that species-differences in lifespan might be caused by differences in cellular defense systems against the primary cause of the ageing process and these defense systems might in turn be regulated on basis of the genetic information.

4. MODEL SYSTEMS FOR AGEING RESEARCH

A large volume of data in human studies was collected during the inventory stage of the development of biomedical ageing research, which now appears to be of limited value for previously discussed reasons. These data were obtained in cross-sectional studies in which cohorts of volunteers or patients of different age classes were compared. Today, most investigators agree that the establishment of age-related clinical data in healthy individuals should be performed in longitudinal studies.

Longitudinal studies with healthy individuals have many drawbacks for experimental biomedical ageing research. Not only will the results of a specific experiment be known after long time periods, and will in the meantime the investigator age with his test persons, but also for obvious reasons many experiments cannot be performed with human volunteers. Therefore, non-human model systems are necessary for experimental biological and biomedical ageing research and for testing the various ageing theories. Non-human models can be employed since several basic ageing phenomena occurring at the molecular and cellular level are common to many, if not all, living organisms although they may manifest themselves differently. These animal models can be exploited in both longitudinal and cross-sectional studies. In the latter case, the influence of the so-called cohort effect will be minimal, since all test animals can be kept under identical conditions.

Experimental animals and other model systems for ageing studies should possess several of the following favourable characteristics (3):

a) a reasonably short lifespan;
b) convenient and not too expensive care;
c) genetic homogeneity;
d) absence of infectious diseases and known pathology;
e) controlled environmental situation, including diet;
f) adequate background of basic biological information.

Laboratory rats and mice meet nearly all of the above criteria of suitability for experimental models. However, one should realize that, in spite of equal environmental conditions, highly inbred rats of the same strain and age are not necessarily the same physiologically. Littermates will develop different pathological lesions with ageing and will show differences in lifespan. For these and other reasons, other model systems for ageing studies have been developed, such as insects, mainly Drosophila or the housefly Musca, free-living nematodes, general Turbatrix aceti and Caenorhabditis elegans, protozoa, mainly Paramecium aurelia, and cells in culture or isolated cells. The system of cells in culture has been most studied of these non-rodent model systems, especially in the past two decades. The main weakness of the system is the unproved relationship between "ageing" of cells in culture and the ageing process in vivo (3). Other cellular model systems for the study of ageing are human erythrocytes and isolated liver cells. Both long-lived postmitotic hepatocytes and relatively short-lived sinusoidal liver cells can be isolated from mammalian livers and offer a useful model systems to study ageing (4).

5. FUTURE PERSPECTIVES

Currently, the ageing process is still a largely unexplored area on the map of biological knowledge. New technological developments in molecular biology now offer the possibility to unravel the molecular and other biological causes of the ageing process.

However, spectacular developments should not be expected within the next few years. The advances in biomedical gerontology will be enormous in the long term, and may lead to an elucidation of the mechanisms of the biological ageing process. Next steps will consist of the design and application of procedures for the prevention and possible therapy of ageing changes. It is anticipated that parallel with the developments in fundamental biomedical gerontology, clues will be found to understand the relationship between the ageing process and age-related diseases.

It will be possible within a couple of years to perform an early diagnosis on basis of molecular changes at the genome level for age-related diseases, like Alzheimer dementia. This will greatly facilitate the application of possible therapies and the designing of new therapies or strategies for prevention of the disease.

It is of crucial importance to close the existing gap between basic medical and applied clinical sciences in order to apply the basic findings in biomedical gerontology to patients suffering from the diseases of old age.

Interest for gerontology, including biomedical and medical ageing research, is rapidly growing, but the current efforts in all areas of gerontological research are extremely small in comparison with the social and medical problems which confront our society as a result of the greying of the population. These problems will only increase in the immediate future.

A better understanding of the biological causes of the ageing process and of the relations between this basic process and age-dependent diseases, including Alzheimer disease, Parkinson disease, many forms of cancer, arteriosclerosis and senile cataract will facilitate a rational approach to a better medical care for the elderly. Therefore, the conditions should be created for further stimulation of this field of science in order to fulfill the specific needs of both elderly individuals and society. Only in this way can optimal health and health care for the elderly be guaranteed in the future.

REFERENCES

1. A Changing World. US Census Bureau, 1986.
2. Knook DL. Aging of cells: accident or programme? Endeavour New Series 1982; 6: 162-167.
3. Rothstein M. Biochemical approaches to aging. New York: Academic Press, 1982.
4. Knook DL. The Isolated hepatocyte: a cellular model for aging studies. Proc Soc Exp Biol Med 1980; 165: 170-177.

CHANGES DURING AGEING IN IMMUNOLOGY, PROTEINS, ENZYMES AND HORMONES

H.M. Hodkinson

Department of Geriatric Medicine, University College and Middlesex Medical School, St Pancreas Hospital
London NW1 OPE, England

INTRODUCTION

There is a large and growing body of information on changes in reference ranges with age. Individual laboratory analytes may show large or small changes though many show no changes at all. This paper will review some of the age changes affecting immunological tests, plasma proteins, serum enzymes and circulating levels of hormones and will address the possible mechanisms underlying these changes.

HORMONAL CHANGES

Some of the most striking changes are those affecting hormones. In the case of the sex hormones this is clearly linked to the fundamental developmental events of the menarche and menopause. In adult men there is no abrupt change to correspond with the menopause but there is a progressive decline of testosterone production with age and an increased conversion rate of testosterone to oestrogens.

Effectively the postmenopausal ovary no longer produces oestrogens. Circulating oestrogen levels fall markedly, the main one, oestradiol, falling to approximately a fifteenth of its premenopausal value. Correspondingly there is increased secretion of gonadotrophins by the pituitary in response to lower sex hormone levels and follicular stimulating hormone levels rise on average fifteen-fold and leuteinising hormone values five-fold (1). Gonadotrophin levels are maximally elevated some two to three years after the menopause following which there is a continued though modest age related decline though values are still far higher than premenopausal ones even in advanced old age.

These changes which are a direct result of gonadal ageing also have indirect effects on a number of other analytes through two principal mechanisms, firstly by effects on bone metabolism and secondly by alteration of anabolic effects, particularly on protein synthesis.

SEX HORMONE EFFECTS ON BONE

Oestrogens antagonise the trophic effects of parathormone on osteoclasts so that osteoclastic resorption of bone accelerates after the menopause and contributes to the development of postmenopausal osteoporosis. This has an important enzyme effect. Alkaline phosphatase has two main circulating isoenzymes derived from bone and liver respectively. The bone isoenzyme derives from osteoblasts and, as osteoblastic and osteoclastic activity are closely coupled, is produced in greater amounts whenever osteoclastic activity is increased. Thus alkaline phosphatase can be seen to change appreciably with age. There are high levels in children and particularly in adolescence corresponding to the rapid bone remodelling associated with the adolescent growth spurt, but then levels are stable in early adult life. After the menopause in women however there is an abrupt rise in alkaline phosphatase levels which remain higher than male values thereafter. This can clearly be ascribed to increased bone mobilisation due to the loss of oestrogenic inhibition of parathormone as there are similar rises in serum calcium and phoshphate, again with higher levels in females than males thereafter, and an increased urinary excretion of hydroxyproline, the main breakdown product of bone matrix collagen (2, 3).

SEX HORMONE EFFECTS ON PROTEIN SYNTHESIS

Levels of a number of serum proteins are clearly influenced by sex hormones, oestrogens usually having positive anabolic effects whilst androgens more often lead to lower levels (4). So, for example, serum levels of the vitamin D-binding protein Gc globulin, of thyroxine-binding globulin (TBG), sex hormone-binding globulin (SHBG), transortin and caeruloplasmin all show increased levels when oestrogens are administered whereas androgens decrease the levels of TBG and SHBG. Though there is clear evidence of anabolic effects of sex steroids on these carrier proteins, such effects may well be more general and may, for example, also apply to albumin, prealbumin and transferrin, all of which show declines with age (4). The sex hormone effects can well explain the rise in SHBG with age which is seen in men (1), given the fall in androgen levels and rise in oestrogens, both of which would favour increased synthesis. The small increase in TBG with age in men can be similarly explained but it is surprising that TBG shows no clear age trend in women though a fall might have been expected. Caeruloplasmin levels are higher in premenopausal women than in men but fall to male levels after the menopause as would be expected given the anabolic effect of oestrogens (4). However, no age changes have been described in the case of Gc-globulin and of transcortin in either sex. There is no current explanation for these inconsistencies.

Quite a different mechanism underlies the progressive rise with age of serum ferritin and its lower levels in women than men before the age of the menopause. Serum ferritin appears to reflect body iron stores (tissue ferritin and haemosiderin) and these rise progressively with age though this accumulation is slowed in premenopausal women because of iron loss due to menstrual bleeding (5).

OTHER HORMONAL CHANGES

The reference range for gastrin is substantially higher in old age (6). The higher values correlate with the presence of positive gastric parietal cell antibodies and these presumably are due to the occurrence of autoimmune atrophic gastritis which has a rising prevalence with age.

This is an example of what would appear to be another major mechanism underlying changes in reference ranges in old age, a rising prevalence of relevant disease which is undiagnosed or underdiagnosed in the reference population. Indeed to return to alkaline phosphatase, though the elevation with age in women is attributable to oestrogen lack, and falling androgens in men might explain the more modest rise in men, in both sexes occult disease is likely to make a contribution also, particularly Paget's disease which has a prevalence of the order of 5% in old age and which is often unrecognised unless a specific search is made by radiology or isotope bone scan. Similarly, though the falls in albumin and prealbumin with age might perhaps be due to falling levels of sex steroids, they are perhaps more likely to be due to the rising prevalence of occult disease, falls due to the negative acute phase response occurring in a very wide variety of conditions and such illness effects being very powerful.

Two other gut hormones show elevated ranges in old age, pancreatic polypeptide and motilin (6). These too may be disease effects given the high prevalence of gall bladder disease and gut pathology respectively.

Elevation of TSH values in old age, particularly in women, can also be attributed to disease, given the high prevalence of primary thyroid disorders including those of autoimmune aetiology - a situation comparable to that for gastrin.

IMMUNOLOGY

The reference range for white blood cell count is somewhat reduced in old age (7). There are rather confused data as to what age effect there may be on individual cell subtypes and on the function of white cells in old age and these difficulties relate to the inadequacies of present laboratory techniques but also the difficulties of ensuring that we are not mistaking disease effects for age effects.

Certainly clinical experience indicates that elderly patients are far less likely to mount a leucocytosis in respect to infection. The white blood cell count is thus of far less value in geriatric clinical practice as a marker of infection than it is in younger age groups particularly as body temperature responses are often also impaired. Fortunately, however, the acute phase response to infection, in contrast to fever and leucocytosis, appears to be fully preserved in old age (8). C-reactive protein estimations, particularly when available on a real-time basis, are thus of very great practical value in the clinical management as well as diagnosis of elderly patients (9). The ranges for the acute phase proteins are mostly unchanged in old age (10, 11).

The immunoglobulins show a general rise in old age which accounts for the steady rise in total globulins. There are specific rises in the IgG, IgA, IgD and IgE classes though IgM appears to fall somewhat (4). Here again, ranges are probably elevated because of the inclusion of individuals with occult disease rather than there being a true age effect - there is certainly an almost exponential rise in the prevalence of paraproteins with age.

ENZYMES

Apart from the changes in alkaline phosphatase already considered, most enzymes show little or no change in reference range with age. However, though there is no change of total activity, some enzymes do show a change in isoenzyme predominance, for example lactic acid dehydrogenase (LDH) and amylase, though the cause of these changes is not known (11).

CONCLUSION

Many analytes have unchanged reference ranges with age. Others to show changes though these are not usually large except for the sex hormones and their trophic hormone. Many of the other changes may be related to the anabolic consequences of sex hormone declines or to their effects on the skeleton. Some changes, for example those of ferritin, can be related to changes in body composition with age. Many changes however are most likely to be due to the accumulation of unrecognised diseases with age rather than a true ageing phenomenon.

REFERENCES

1. Kenny RA, Fotherby K. The sex steroids and trophic hormones. In: Hodkinson HM, ed. Clinical biochemistry of the elderly. Edinburgh: Churchill Livingstone, 1978: 246-258.
2. Crilly RG, Jones MM, Horsman A, Nordin BEC. Rise in plasma alkaline phosphatase at the menopause. Clinical Science 1980; 58: 341-342.
3. McPherson K, Healy MJR, Flynn FV, Piper KAJ, Garcia-Webb P. The effect of age, sex and other factors on blood chemistry in health. Clin Chim Acta 1978; 84: 373-397.
4. Freeman H, Cox ML. Plasma Proteins. In: Hodkinson HM, ed. Clinical biochemistry of the elderly. Edinburgh: Churchill Livingstone, 1984: 46-74.
5. Lynch SR, Lipschitz DA. Iron deficiency anaemia. In: Denham MJ, Chanarin I, eds. Blood disorders in the elderly. Edinburgh: Churchill Livingstone, 1985: 21-42.
6. Adrian TE, Bloom SR. Gut and pancreatic hormones in the elderly. In: Hodkinson HM, ed. Clinical biochemistry of the elderly. Edinburgh: Churchill Livingstone, 1984: 237-245.
7. Caird FI, Andrews GR, Gallie TB. The leucocyte count in old age. Age and Ageing 1972; 1: 239-244.
8. Kenny RA, Hodkinson HM, Cox ML, Caspi D, Pepys MB. Acute phase protein response to infection in elderly patients. Age and Ageing 1984; 13: 89-94.
9. Cox ML, Rudd AG, Gallimore R, Hodkinson HM, Pepys MB. Real-time measurement of C-reactive protein in the management of infection in the elderly. Age and Ageing 1986; 15: 257-266.
10. Cox ML, Freeman H, Hodkinson HM, Harrington MG. Serum protein in the elderly: Reference ranges I. J Clin Exp Gerontol 1983; 5: 201-206.
11. Cox ML, Freeman H, Hodkinson HM, Pepys MB, Ogle SJ. Serum protein in the elderly: Reference ranges II. J Clin Exp Gerontol 1983; 5: 295-302.

CLINICAL BIOCHEMISTRY OF CONNECTIVE TISSUE AND ITS ROLE IN AGEING

H. Greiling, R. Reinards* and K. Kleesiek**

*Department of Clinical Chemistry and Pathobiochemistry
Aachen University of Technology, D-5100 Aachen (FRG)
**Institute of Laboratory and Transfusion Medicine, Heart
Center North-rhine-Westfalia, D-4970 Bad Oeynhausen (FRG)

The connective tissue is a target organ in the body, in which the age-dependent changes can be seen with our eyes. The wrinkling skin is one of the evidences of the physiological ageing process, especially of the connective tissue.

The turgor and the skin elasticity are correlated with the chronological age and are caused by definite biochemical and morphological changes in the connective tissue. The father of gerontology, the Swiss Verzar, has firstly observed the age dependency of heat-induced shrinkage in rat tail tendons. On the basis of these observations, the cross-linkage theory was postulated, that ageing is caused by increased cross-linking of proteins, especially collagens.

Besides the various collagen types I to XII and the elastins, the so-called ground substance, which is composed of glycosaminoglycans and proteoglycans, is produced by specific cells in connective tissue. They are developed from a primitive mesenchymal cell. Specific cells, fibroblasts, chondrocytes etc. produce specific tissue types which have a characteristic composition of macromolecules, and also glycosaminoglycans and proteoglycans, which determine the biomechanical properties of the tissue.

There are age-dependent changes in proteoglycan structure, which are different in various connective tissue types. The proteoglycan subunit in cartilage consists of chondroitin sulfate and keratan sulfate chains connected to a protein core. Numerous proteoglycan subunits are aggregated by means of a hyaluronate chain. Since the early report of Kaplan and Meyer (1), which described for the first time age-dependent changes in the composition of the glycosaminoglycans in human costal cartilage, further studies have confirmed both in animal and human cartilage an age-dependent decrease of total chondroitin sulfate, combined with an increase in the ratio of chondroitin 6-sulfate to chondroitin 4-sulfate (2). In addition to the decrease in the total glycosaminoglycan concentration with age, the number of unsulfated disaccharide units present in chondroitin sulfate is reduced. Mathews and Glagov found in human costal cartilage a relative increase in keratan sulfate with age (3).

Roughley and White found that the proteoglycan content of the human articular cartilage decreases with age; the same is true for the size of the proteoglycan subunit (4). The results of these authors confirmed our findings concerning the ratio of chondroitin sulfate to keratan sulfate in human knee joint cartilage: an increase in keratan sulfate relative to chondroitin sulfate, and an increase in chondroitin 6-sulfate relative to chondroitin 4-sulfate. In addition, Roughley and White observed an increase of protein relative to glycosaminoglycans, and a decrease in serine and glycine in relation to an increase in arginine of the proteoglycan core protein with age. Similar results were also obtained by Inerot et al. in canine hip articular cartilage. They found, that the extracted proteoglycans were smaller in the older proteoglycan monomers (5). Ageing of human cartilage is also accompanied by an increase of the hyaluronate binding region and partially fragmented link proteins (6).

Also in the human aorta a large aggregating chondroitin sulfate proteoglycan, besides aortal proteoglycans with lower molecular weight and a high dermatan sulfate content with enriched iduronic acid is found (7). We could detect an increase of keratan sulfate and a decrease of chondroitin sulfate with increasing age (8).

Since the increasing deposition of calcium phosphate in each aortic segment correlates with an enhancement of the concentration and relative proportion of keratan sulfate, both topographically increasing from the aortic arch to the bifurcatio aortae, and with age, a kind of nucleating function may also be considered for keratan sulfate. As ion exchanging macromolecules the glycosaminoglycans exhibit different affinities to Ca^{2+}, which is high in chondroitin sulfates, especially chondroitin 6-sulfate, and lowest in keratan sulfate. A decrease in chondroitin sulfate content of the aortic proteoglycans in favour of keratan sulfate, which binds Ca^{2+} to a much lesser extent, would increase the portion of deimmobilized Ca^{2+}, which may give rise to insoluble calcium phosphate. Such a micro-precipitate could then serve as a crystallization nucleus.

Aorta proteoglycans are capable of interacting with low density lipoproteins. The higher the anionic part of the glycosaminoglycans the better the precipitation, that means the higher the insolubility of the LDL-glycosaminoglycan complex. Another important aspect is the age-related advanced glycosylation of connective tissue components f.i. collagen also in atherosclerosis. LDL is covalently bound by glycosylated collagen more than threefold as much as by normal collagen (9). Collagen from normal subjects shows an age-related increase in the accumulation of advanced glycosylation endproducts. In diabetes mellitus the thickening of basement membranes and intima are related to this phenomenon.

How can we explain the age-dependent changes of decreased chondroitin sulfate and increased keratan sulfate on a molecular level? Some studies indicate an age-associated depression in the biosynthesis of proteochondroitin sulfate in both animal and human cartilage. The enzyme activity, catalyzing the covalent attachment of xylose to the seryl residues of the core protein, has been suggested to be rate-limiting in the biosynthesis of the glycosaminoglycan chain of proteochondroitin sulfate. Therefore we studied the activity of UDP-xylose: proteoglycan core protein xylosyltransferase in the costal cartilage of young and senescent rats.

The activity of xylosyltransferase, measured in samples of old and young costal cartilage is decreased about 70% in the tissue of old rats, related to cartilage wet weight and protein. In old cartilage, there was a reduction of cell number, as checked by the decline of extractable DNA to about 30%. However, the decrease of the activity of xylosyltransferase in relation to the concentration of DNA in old cartilage was similar to that reported above (about 70%). The results suggest that the reduced xylosyltransferase activity might lead to a diminished concentration of proteochondroitin sulfate in the cartilage of old animals (10).

When the activity of xylosyltransferase is related to galactosamine as a measure of the concentration of chondroitin sulfate, no significant differences between young and old rats were detectable. Thus, there is a coordinate decrease of both xylosyltransferase activity and chondroitin sulfate content in the cartilage of old rats in comparison to that of young animals.

Our findings point to xylosyltransferase as a possible regulator of proteochondroitin sulfate synthesis at the posttranslational level. This is in agreement with the findings of Schwartz, which show the independence of proteochondroitin sulfate synthesis from the biosynthetic rate of the respective core protein (11). The rather short half-life of xylosyltransferase, as compared to other enzymes involved in the glycosaminoglycan chain synthesis, indicates the possible regulatory role of this enzyme in chondroitin sulfate proteoglycan production.

Since the reduction of xylosyltransferase activity in our experiments was also found, if related to the concentration of DNA, a loss of chondrocytes as a cause of this change is excluded. The activity of another cytosolic enzyme, LDH, was found to be nearly unaffected by the age of the animals, if related to DNA. Therefore it is conceivable that, due to the reduction of xylosyltransferase activity in old cartilage, the enzymatic attachment of xylose to the seryl residues of core protein becomes rate-limiting under such conditions. In accordance with the assumption, Inerot (5) has shown similar chain length (molecular weight) of chondroitin sulfate in old and young cartilage, and Honda (12) has presented evidence that the glycosaminoglycan chain elongation reaction is not impaired in the cartilage of senescent rats. Thus, the age-dependent diminution of total chondroitin sulfate is due to a reduced number of initiated carbohydrate side chains along the core protein, which is likely to be caused by a decreased xylosyltransferase in old tissue.

Our results do not answer the question of whether there is a decrease in the specific activity of this enzyme or a diminution of enzyme protein. The nature of the primary event that causes diminished xylosyltransferase activity is not yet known and needs further investigation. Of particular importance might be the changed composition of hormones with age, since Schiller has shown, that in diabetic thyroidectomized, and hypophysectomized rats, the reduced synthesis of proteoglycans is due to a

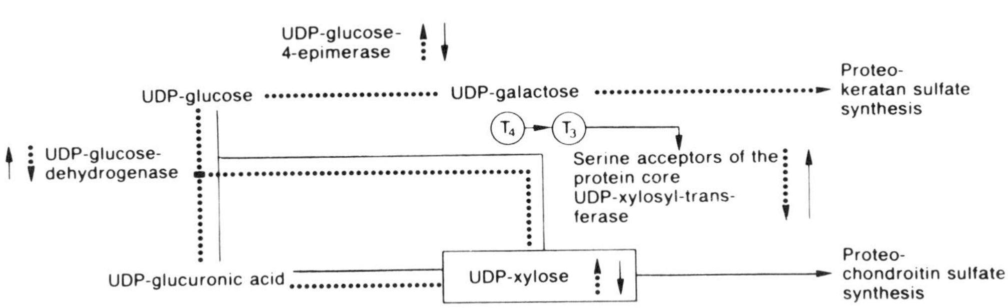

Fig. 1. Postulated action of thyroid hormones and their analogues on the distribution and biosynthesis of chondroitin sulfates and keratan sulfates in joint cartilage.

decreased activity of xylosyltransferase (13). She observed that in hormone-deficient animals, the enzymatic attachment of xylose to seryl residues of the core protein becomes the ratlimiting step in the synthesis of proteochondroitin sulfate. Other authors have shown that hormones which are involved in the sulfation of proteoglycans, e.g. somatomedin and thyroxine, especially affect the activity of xylosyltransferase.

With increasing age free thyroxine concentration in blood serum decreases (14). We found a stimulation of chondroitin sulfate synthesis by triiodothyronine in chondrocyte cultures. In the presence of rT_3, the antagonist of T_3, the galactosamine/glucosamine ratio diminishes, that means an increase of keratan sulfate synthesis. Also the incorporation of ^{14}C-serine into the core protein increases with T_3 concentration (15). The same occurs also with thyroxine (T_4) and 3',5'-diiodothyronine, that means an activation of chondroitin sulfate synthesis with T_4 and an inhibition with 3',5'-diiodothyronine. Also in human blood serum there is an increase of rT_3/T_3-ratio after 60 years.

We postulate the following regulatory mechanism (Figure 1). T_4 and T_3 increase the number of serine acceptors for xylose and therefore the xylosyltransferase activity. This results in a diminished UDP-xylose pool. UDP-xylose is a feedback inhibitor of UDP-glucose dehydrogenase. A decreased UDP-xylose concentration stimulates the UDP-glucose dehydrogenase and therefore the proteochondroitin sulfate synthesis. Decreased concentrations of T_4 and T_3 in blood serum found in older animals or an increase of 3',5'-diiodothyronine or rT_3 have the inverse effect and stimulate the proteokeratan sulfate synthesis, as it is seen in the elderly. Thus, the characteristic changes in connective tissue with increasing age can be explained from the regulatory enzyme level.

Is there also an age-dependent degradation of proteoglycans by proteinases? Elastase seems to cleave the link protein of proteoglycan aggregates and it is known to degrade the nonhelical terminal peptides of cartilage collagen molecules, which leads to the disruption of the main intermolecular cross links in collagen fibers and to the deterioration in the tensile strength of articular cartilage.

Various mechanisms are described that may cause an enzymatic destruction of the articular cartilage. An extracellular proteolytic activity of proteinases in the synovial system is inhibited by an excess of very potent proteinase inhibitors, α_1-proteinase inhibitor and α_2-macroglobulin, derived from the plasma. Thus, the occurrence of a relevant free proteinase activity in the synovial fluid is presumed to be unlikely.

Quantitation of elastase complexes to α_1-proteinase inhibitor was carried out by an enzyme-linked immunoassay. The mean concentration of elastase in the inflammatory synovial fluid was approximately 10,000 μg/l, compared with 41 μg/l in non-inflammatory synovial fluids (16). The concentration of elastase-α_1-proteinase inhibitor complex is elevated also in synovial fluids of "activated" degenerative joint diseases, which occur particularly in the elderly. But systematic studies are necessary to investigate a possible age dependency of the concentration of elastase in plasma or tissue.

Oberley has postulated that ageing occurs by damage of differentiated postmitotic cells (17). That means ageing in dividing cells is much slower than that of nondividing cells in human. The best evidence for this is the fact that the major manifestations of human ageing are heart diseases, brain dysfunction, vessel diseases, all of which are mainly postmitotic cell diseases, involving organs with very little growth potential in the adult. Oberley believes that the initial site of ageing is mitochondrial

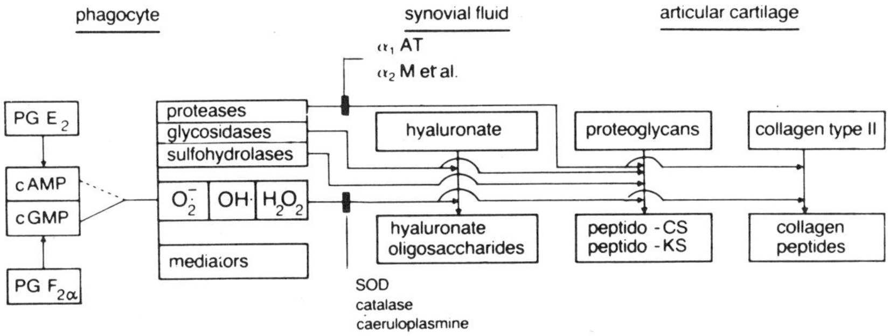

Fig. 2. Destruction of extracellular matrix components in inflammatory joint diseases and in age-dependent pathobiochemical processes. (α_1AT = α_1-proteinase inhibitor; α_2M = α_2 macroglobulin; CS = chondroitin sulfate; KS = keratan sulfate; SOD = superoxide dismutase).

DNA, not the nuclear DNA. He postulates that the damage of ageing cells is caused by oxygen radicals, which are produced about 5 to 10 times the amount in mitochondrial membranes compared to nuclear membranes (18). Cutler postulates so-called longevity genes (19). He has hypothesized that a number of systems may be involved in longevity, including detoxification processes, antioxidants, cellular renewal processes, redundancy mechanisms, specific metabolic rate control, gene control systems, relative enzyme

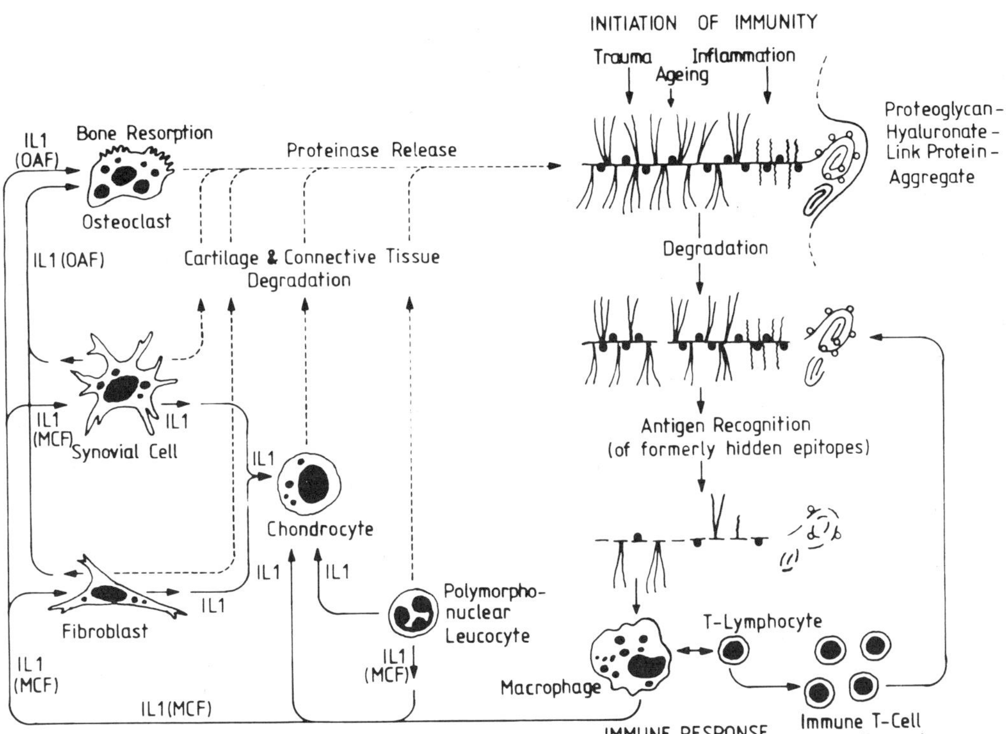

Fig. 3. Immunochemical age-dependent processes, which initiate immune T-cell involved interleukin 1 production. Interleukin 1 stimulates the release of proteinases from osteoclasts, chondrocytes, synovial cells and fibroblasts.

levels, and structural components of cells. In his opinion antioxidants determine human longevity. These are: superoxide dismutase, catalase, urate, α-tocopherol and carotenoids. Further important antioxidants are ascorbate, ceruloplasmin, and glucose.

We have shown that cartilage destruction by granulocytes can be inhibited "in vitro" by addition of superoxide dismutase. We developed a model for frustrated phagocytosis in which FMLP-activated granulocytes react with cartilage slices. We have shown an inhibition of the cartilage destruction by granulocytes in the presence of glycosaminoglycan polysulfate, a potent proteinase inhibitor, and superoxide dismutase.

Therefore we postulate that various proteinases and other lysosomal enzymes combined with superoxide radicals are responsible for the destruction of the proteoglycans and collagens especially I, II and III from cartilage. This process can be inhibited by α_1-proteinase inhibitor and α_2-macroglobulin, but it is also important to diminish the concentration of superoxide and hydroxyl radicals by superoxide dismutase and H_2O_2 by catalase (Figure 2).

Cutler has evaluated the mentioned antioxidants which determine human longevity and also the cartilage stabilisation in ageing. We know the role of immunochemical reactions in proteoglycan and collagen degradation.

The degradation of proteoglycans and collagens by proteinases and the regulatory role of interleukin 1 is shown in Figure 3. In the ageing process we have an accumulation of proteoglycan and collagen degradation products which can also react directly with macrophages leading to an age-dependent interleukin production which increases the degradation of proteoglycans and collagen in the cartilage. Also the decrease of T-helper cells with increasing age has an additional effect on the degradation of cartilage.

There are many unsolved problems in the connective tissue metabolism and more basical research in clinical biochemistry, which is necessary for a causal therapy in gerontology, is wanted for the future.

REFERENCES

1. Kaplan D, Meyer K. Nature 1959; 183: 1267.
2. Greiling H, Baumann G. In: Vogel HG, ed. Connective tissue and ageing. Amsterdam: Excerpta Medica, 1973: 160.
3. Mathews MB, Glagov S. J Clin Invest 1966; 45: 1103.
4. Roughley PJ, White RJ. J Biol Chem 1980; 255: 217.
5. Inerot SI, Heinegard D, Andell L, Olsson SE. Biochem J 1978; 169: 143.
6. Roughley PJ, Mort JS. Clin Sci 1986; 71: 337.
7. Heinegard D, Paulsson M. In: Piez KA, Reddi AH, eds. Extracellular matrix biochemistry New York: Elsevier, 1984: 299.
8. Stuhlsatz HW, Löffler H, Mohanaradhakrishnan V, Cosma S, Greiling H. J Clin Chem Clin Biochem 1982; 20: 713.
9. Brownlee M, Vlassara H, Cerami A. Diabetes 1985; 34: 938.
10. Wolf B, Gressner AM, Nevo Z, Greiling H. Mech Ageing Dev 1982; 19: 181.
11. Schwartz NB. J Biol Chem 1977; 252: 6316.
12. Honda A, Abe M, Nurata S, Mori Y. J Biochem 1979; 85: 519.
13. Schiller S. Biochim Biophys Acta 1976; 451: 457.
14. Wilke TJ. Clin Chem 1983; 29: 1428.
15. Greiling H, Kleesiek K, Reinards R. In: Kuettner KE, Schleyerbach R, Hascall VC, eds. Articular cartilage biochemistry. New York: Raven Press, 1986: 195.

16. Kleesiek K, Reinards R, Brackertz D, Neumann S, Lang H, Greiling H. Rheumatol Int 1986; 6: 161.
17. Oberley LW, Oberley TD. In: Johnson JE Jr, Walford R, Harman D, Miquel J, eds. Free radicals, ageing, and degenerative diseases. New York: Alan R. Liss, 1986: 325.
18. Patton SE, Rosen GM, Rackman E.; 1980.
19. Cutler RG. In: Johnson JE Jr, ed. Ageing and cell function. New York: Plenum Press, 1984: 1.

BIOCHEMICAL AND HISTOPATHOLOGICAL EXAMINATION OF THE SKIN: TOOL FOR PREDICTION OF CORONARY ATHEROSCLEROSIS

H. Bouissou, A. Aouidet and P.M. Valdiguié

Faculté de Médecine de Toulouse-Rangueil, Laboratoires d'Anatomie Pathologique et de Biochimie, Toulouse, France

Skin biopsy gives an excellent indication of the degree of atheroma of the coronary artery. Over the past ten years, various investigations on concomitant ageing of the dermis, the aorta and the coronary arteries have been carried out mainly by Bouissou et al. (1, 2).

They assessed that the status of the skin reflects that of the arteries with sharp relationship between atherosclerosis and skin alterations. The increase of cholesterol and apoprotein B in the skin is emphasized with ageing and also with myocardial infarction (MI) in young subjects under 40 years.

Moreover the prognostic value of skin biopsy is studied in assessing vascular risk in young healthy subjects with no clinical history, in obese subjects and in subjects whose father or mother while still young, have presented MI, arterial hypertension or insulin-dependent diabetes.

I. THE PRESENT KNOWLEDGE

I.1. Morphology of normal and of ageing skin

- Normal skin, called type 0, has a thick dermis with close-packed collagen bundles. It has numerous slender elastic fibres (called oxytalan fibres), a network of normal deep and elaunic elastic fibres, it has active fibroblasts with a marked ergastoplasmic reticulum, and an extended cytoskeleton. All these are in close relationship with the interstitium which they have secreted.
- Type I skin shows only a few elastic fibres in the superficial dermis.
- Type II skin is ageing skin. The collagen fibres are dispersed in abundant ground substance, and there are few bundles of collagen. There are no elastic fibres. The fibroblasts are not very active, with reduced ergastoplasm in a retracted cell with a dense cytoskeleton. The cell has little contact with the interstitium.

I.2. Morphology of the normal and ageing aorta and coronary arteries

- The young aorta has a thin intima, and a media of slender, closely-packed parallel elastic fibres. Between them are smooth muscle cells, little ground substance and some collagen fibres. With age, the intima becomes thicker and fibrous. The elastic fibres nearest the interior disappear. Glycosaminoglycans accumulate in the remaining elastic network. Collagen fibres become more and more abundant in the intima. The aorta becomes arteriosclerotic. There is a progressive overload in the intima-media areas of lipids originating from the blood, and atheroma develops.

- The coronary artery is a muscular artery consisting of an intima, a continuous internal elastic layer and a muscular media. With age, the intima becomes fibrous and considerably thicker; the internal elastic layer disappears in places. The media becomes fibrous, and its muscle cells disappear progressively.

In these two arteries, ageing is called normal (physiological arteriosclerosis) when it sets in slowly and late in life. It is called arteriosclerotic disease when it appears early and intensely. Morphologically, these two types of ageing do not differ.

I.3. Parallel ageing of the aorta, the coronary artery and the skin, based on autopsy findings. It shows the types of skin observed, and the aortic and coronary structures with normal and abnormal ageing.

Type 0 skin corresponds to the normally ageing aorta, whatever the age of the subject, more or less than 45 years. Some coronary arteries show abnormal ageing. This is because the coronary artery begins to get old very early, from the age of five years.

With type I skin, whatever the age of the subject, appears a number of abnormally aged aortas. Aged coronaries are of course even more numerous.

Table 1.

	Age	Number	C.C		
1	30 - 49 41.5 ± 4.3*	44	0.384 ± 0.068	0.001	0.01
				<P<	
	Y.C 43.8 ± 2*	19	0.440 ± 0.08	0.01	<P<
2	50 - 65 57.8 ± 4.7*	51	0.416 ± 0.059	NS p <0.001	0.02
	O.C 61 ± 8.4*	43	0.457 ± 0.051		
				NS	
3	+ 65 57.8 ± 5.3*	56	0.450 ± 0.070		

* Mean ± S.D.

Cutaneous cholesterol level (C.C) (mmol/100 gr fresh tissue) in young population with coronary heart disease (Y.C), in old population with coronary heart disease (O.C) and in three unaffected groups at different ages (1 - 2 - 3).

Type II skin, whatever the age of the subject, nearly always corresponds to abnormally aged aortas and coronary arteries.

These autopsy findings are corroborated in biopsies carried out in young and normal subjects, in elderly subjects, and in young subjects (aged under forty) with myocardial infarction or insulin-dependent diabetes. The percentage of aged skins is 12 in young subjects. It is 84, 78 and 88 percent respectively in coronary heart disease, diabetics and in the elderly. In practice, according to our statistics, type 0 skin may be seen at any age, but it is however very rare after the age of 40. Type I skin indicated normal ageing after the age of 30 but abnormal earlier. Type II skin is normal after the age of 40. Before this age, it indicates precocious aortic and coronary ageing.

I.4. Increase in cutaneous cholesterol

The Table 1 shows the increase of cutaneous cholesterol with age, the degree of atheroma of the aorta, and with myocardial infarction before the age of 40. On this point, we recall that cutaneous apoprotein B is a better marker of coronary condition than serum apoprotein B. We have shown this in coronary heart disease patients with normal coronarography and in vascular patients with abnormal coronarography. In fact, in a population with abnormal coronarography and a normal level of serum apoprotein B, cutaneous apoprotein B corresponds to the state of the arteries as shown by coronarography "Douste-Blazy P." (3).

II. EVALUATION BY SKIN BIOPSY OF THE STATUS OF THE CORONARY ARTERIES IN YOUNG POPULATIONS

II.1. Evaluation of ageing by histological study

Methods. Cutaneous biopsies were performed in 414 young subjects aged 20 ± 1.5 years. They were carried out under deep local anaesthesia on the inside of the upper arm. There was also detailed questioning on personal and family history, clinical examination, and serum and urine biological tests. The clinical and biological tests were negative in all subjects. Three hundred and ten had no hereditary antecedents. Ninety-six had direct

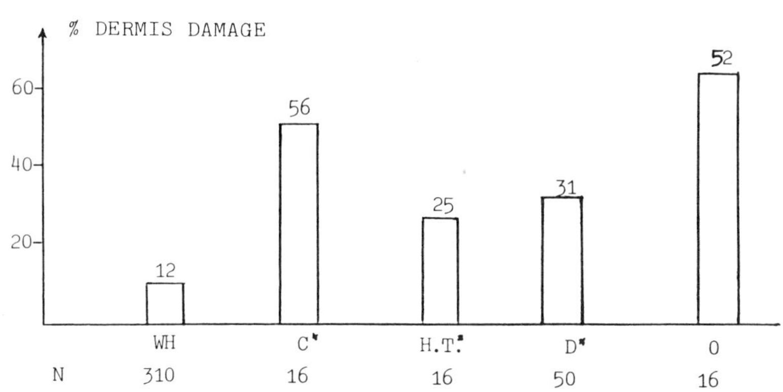

Fig. 1. Evaluation of cutaneous ageing in young populations (20 ± 1.5 years) without (WH) or with hereditary clinical history (C = coronary heart disease, H.T. = hypertension, D = diabetes) or with obesity.

antecedents (their father or mother was still young). Of these, 16 presented myocardial infarction, 16 severe arterial hypertension, and 59 insulin-dependent diabetes. We also studied 16 obese subjects aged 20 ± 1.5 years. In all four groups, the skin biopsy was systematically studied by light and electron microscopy.

Results. The Figure 1 summarizes the results obtained by light microscopy: 12 percent of subjects with no clinical history had abnormal skin. This percentage of aged skin was 56 percent in subjects with a history of myocardial infarction, 25% in those with arterial hypertension and 31% in those with diabetes. It was 52 percent in obese subjects. These abnormal skins were type I. For the age of the subjects, they should have been type O.

Ultramicroscopic study of the dermis of subjects with cardiovascular problems showed certain changes. These changes resembled those noted in aged subjects. The fibroblasts were globular and quiescent. The reticulum was not abundant. The cytoskeleton was dense. There was a loss of cohesion with the collagen bundles of the interstitium. These fibroblasts resembled those of the aged subject.
These ultramicroscopic lesions were observed in all subjects with a history and with type I skin. However, these fibroblasts of aged appearance were also seen in 15 percent of type O skin in these subjects. The total number of cases of abnormal skin is thus greater than the one detected by the usual optical histological methods.

Conclusion. Some deductions may be made from these observations in subjects with a hereditary cardiovascular history:

In these perfectly healthy subjects there is a high percentage of aged skin, in the absence of any clinical or histological sign. This precocious cutaneous ageing may be interpreted as the first sign of vascular ageing, which is still silent. It may correspond to the "morphological substrate of arteriosclerotic terrain".

The ultramicroscopic dermal lesions observed resemble those seen in normal or accelerated ageing.

In subjects who are young and healthy but who have direct cardiovascular antecedents, skin biopsy is a good screening test for the detection of atherosclerosis.

II.2. Evaluation of the degree of atheroma by estimation of cutaneous cholesterol

a) Cutaneous cholesterol and plasma lipoproteins cholesterol in young subjects with normal type O skin

In 16 volunteers, whose average age was 22 ± 1.5 years, the cutaneous cholesterol content was measured using Lieberman's method or enzymatic method (4). Triglyceride levels, total LDL and HDL cholesterol were assayed. The ratios of total to HDL cholesterol and LDL to HDL cholesterol were calculated. Triglyceride levels were found to be normal (0.7533 ± 0.0043). The mean cholesterol levels were: cutaneous, 0.355 ± 0.010 μmol per 100 mg skin; plasma, 5.563 ± 0.273 mmol/l; LDL, 4.034 ± 0.260 mmol/l; HDL, 1.394 ± 0.106 mmol/l. No correlation existed between the level of cutaneous cholesterol and the other cholesterol levels (total, HDL, and LDL cholesterol). A correlation was noted between the level of cutaneous cholesterol and the ratio of total to HDL cholesterol (2 = 0.063, $p < 0.01$) and of LDL to HDL cholesterol (2 = 0.64, $p < 0.01$), "Bouissou H." (5).

Table 2. Cutaneous cholesterol content in young subjects (22 ± 1.5 y) with type O skin and with or without hereditary coronary heart disease

	N	Mean + E.S.M.	
without	74	0.265 ± 0.162	p <0.001
with	14	0.314 ± 0.105	

In spite of the fact that all the subjects seemed identical in view of an anatomo-pathological study, variations were in fact found in the levels of cutaneous cholesterol. However the cutaneous cholesterol, which is stable for one subject, is independent of the LDL cholesterol and the HDL cholesterol considered separately.

The significant correlation found between cutaneous cholesterol and the ratio of total HDL cholesterol and of LDL to HDL cholesterol confirms the very important role of these two fractions in depositing cutaneous cholesterol. Indeed, the two correlations are positive, supporting the theory of antagonistic actions of the LDL, which participate in cholesterol deposition, and of the HDL, which favour its return to the liver.

This study is all the more interesting in that it was carried out on young subjects who had a normal skin and no personal or hereditary antecedents. For 14 of the 16 subjects, there was perfect correlation. The four subjects with a high level of cutaneous cholesterol had an increased risk factor total cholesterol/HDL cholesterol.

These results confirm the usefulness of calculating the total cholesterol/HDL cholesterol ratio, which gives a good indication of the cholesterol levels in the skin.

Thus the interest of these two parameters, in conjunction with histological examination, is confirmed for the detection of early atherosclerosis and/or arteriosclerosis.

b) Cutaneous cholesterol in young subjects with or without hereditary coronary heart disease (CHD)

The Table 2 presents the levels of cutaneous cholesterol in 74 young and healthy subjects with no antecedents, and in 14 young subjects with cardiovascular antecedents. The difference between them is highly significant at $p < 0.001$.

II.3 <u>The cutaneous apoprotein B</u>

Its quantification is somewhat delicate because the extraction from the fresh skin samples requires pulverization under liquid nitrogen and then homogenization with Triton X100 in Tris HCl buffer to solubilize the protein.

The estimation can be done either by electroimmunoassay (6) or by enzymoimmunoassay (7).

The miniaturization of this technique is under development. We hope in a not to long future to be able to estimate the skin cholesterol and apoprotein B in population with vascular risk and so to promote a new test for health screening.

ACKNOWLEDGEMENTS

The valuable assistance and advices from Mrs Pieraggi, Julian, Solera and Ms De Graeve and Thiers are gratefully acknowledged.

REFERENCES

1. Bouissou H. Atherosclerosis 1974; 18: 449.
2. Bouissou H. Gerontology 1981; 27: 99.
3. Douste-Blazy P. Brit Heart J 1985; 54: 452.
4. Aouidet A. Clin Chem 1983; 29: 2001.
5. Bouissou H. Biomedicine 1979; 31: 236.
6. De Graeve J. Atherosclerosis 1984; 52: 301.
7. Fievet G. Clin Chem 1984; 30: 98.

CHAPTER 14
LABORATORY MANAGEMENT AND COST EFFECTIVENESS

CAN THE QUALITY OF LABORATORY TESTING BE MAINTAINED AT REDUCED COST?

PMG Broughton

Wolfson Research Laboratories, Queen Elizabeth Medical Centre, Birmingham B15 2TH, UK.

In Britain and many other countries health care administrators are looking for ways of saving money without at the same time reducing the standards of patient care. This has led to what has been called a crusade for cutting laboratory costs. Clinical chemistry has attracted particular attention, largely because of the widespread belief that it is expensive and that a significant proportion of tests are unnecessary. To meet this situation, several different strategies have been proposed (1) for reducing the number of unnecessary tests. Undoubtedly money can be saved by eliminating unnecessary tests, assuming that these can be identified, but the financial saving is much less than that predicted by a simple proportion (2).

Laboratories are now facing growing demands from their users at a time of increasingly severe financial restriction, and in these circumstances there is a danger that quality will deteriorate. The problem can be summarised as

More work, better quality, same cost
or
Same work, same quality, less cost
or
More work, less quality, less cost.

PERCEPTIONS OF QUALITY AND COST

The 4 groups of people most closely concerned with this question are the laboratory, the clinician, the health care administrator, and the patient. All have different perceptions of both quality and cost.

To the laboratory, quality usually means analytical reliability, mainly accuracy and precision, and cost refers to the annual laboratory expenditure. Part of this expenditure goes on quality control programmes which are designed to maintain accuracy and precision at desired levels. The annual cost of these programmes probably varies widely, and one estimate (3) put this at about 25% of the total cost of laboratory consumables. Is it possible to reduce this expenditure without at the same time reducing quality? Although there seems to be no published data, the relationship between the two probably resembles that shown in Figure 1.

It is very unlikely that laboratory A would try and titrate its quality control - that is reduce expenditure and see when quality started to deteriorate. However, laboratories A and B spend different amounts but achieve similar levels of quality. Do they agree on the level of quality which they require? There are widely differing views on desirable standards of performance (4), but in practice most laboratory workers probably regard the best that they can achieve as being the acceptable or desirable level. However, laboratory C probably regards its level of quality as satisfactory, and the clinicians it serves may be quite satisfied because the quality of a laboratory involves many factors other than analytical accuracy and precision. But it is doubtful whether laboratories A or B would deliberately reduce their level of quality to that of C in order to save money.

However, if they did this, or it was forced upon them for financial reasons, clinical care would eventually suffer, although it is impossible to predict when this might happen. The clinician would then be led to repeating tests unnecessarily, and ordering other investigations in order to explain poor quality laboratory results. Thus, the laboratory would save money on its quality control but there would be additional expenditure in other areas which the laboratory would probably not know about, particularly if the cost fell on someone else. It is therefore necessary to look at needs and costs in a wider context. Laboratory scientists are in some danger of developing ever more accurate and precise methods for tests which have little or no clinical usefulness, and need to consider not only the efficiency of the analytical process but the effectiveness of the results - that is, how they are used by the clinician.

The clinician sees both quality and cost in a rather different way. He expects the results to be reliable so that he can act upon it with confidence. Although he may use different terminology, he requires accuracy so that he can interpret the result against reference ranges and his previous experience, and precision so that he can reliably monitor trends in individual patients (5). In some circumstances he can tolerate what may seem to be large errors because he has other information about the patient which is not known to the laboratory. He expects the laboratory to prevent biological variation or pre-analytical error from confusing his interpretation. Most importantly, he expects the result to be relevant -

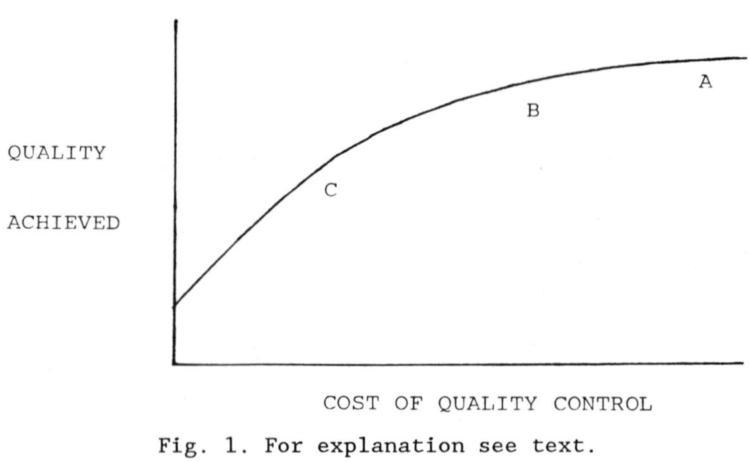

Fig. 1. For explanation see text.

that is, it must be the right test done at the right time, and it must contribute in some way to the making of a clinical decision (6). For the clinician quality must ultimately refer to the clinical care of patients, and he will judge the quality of laboratory tests by their relevance and usefulness in making a clinical decision.

Laboratory workers tend to believe that speed is only essential in emergency situations, but to the clinician speed is part of the quality of service if it enables him to make decisions more rapidly. He will see cost in relation to the use of his own time. If results can be produced more rapidly he is unlikely to be influenced by complaints that this is costing the laboratory more money, particularly when the cost of a laboratory test is trivial compared with that of an X-ray or half an hour in the operating theatre. It is easy to forget that an inexpensive laboratory test may lead to clinical actions, such as drug treatment or an operation, which are very much more expensive.

Clinicians are now becoming more involved with the cost of laboratory testing, for example through the introduction of Diagnostic Related Groups in the United States and elsewhere. In Britain hospital laboratories have until now been given an annual budget and expected to provide a comprehensive service with this limited amount of money. Recently a number of experiments have been started in which some of this money is distributed amongst the clinicians who use the laboratory. So the clinician is given a budget which he can, at least in theory, spend on what he likes - more laboratory tests, more radiology or more nurses. The objective of this is to make the clinician more accountable and directly responsible for expenditure on the care of his patients. It is of course hoped that this will lead to a reduction in the number of inappropriate laboratory tests and thereby save money, but it is too early to predict whether it will be effective.

In some countries the health care administrator regards the laboratory as an important source of income for the hospital, but in Britain and elsewhere the laboratory is a cost centre, although a small one in comparison with the total cost of running a hospital. Administrators may want to know whether every hospital needs its own laboratory, why expensive automated equipment is only used for a few hours each day, and why laboratory services cannot be radically reorganised, with one factory-like laboratory serving a whole town. Like the clinician, the administrator will judge the quality of medical care, and the contribution of laboratory tests to this, in terms of morbidity and mortality. He will ask whether increased laboratory testing will lead to a reduction in bed occupancy, which means cost. These concepts are usually unfamiliar to the laboratory, which does not have this information with which to judge the effectiveness of its work. The administrator is not primarily interested in the cost per test, but in the total expenditure of the laboratory and how much is devoted to specialities, such as obstetrics or internal medicine. He will judge this against the competing demands of other services in the hospital. What will be the financial effect on the laboratory of performing say 100 liver transplants per year or opening an additional 50 hospital beds for psychiatric patients? At present, reliable information for this sort of financial planning is usually not available, and it is in the laboratory's interests to try and provide it.

The patient is not concerned with the wider issues facing the administrator, but only with his own personal health. He will judge the quality of a laboratory test or any other aspect of medical care by whether it leads to an improvement in his health or provides him with reassurance - the faster the better. Delay not only means prolonged uncertainty and worry, but loss of earnings. Ultimately he will have to pay, either directly or through taxes or insurance, but for most patients the cost of laboratory tests is not a problem where their personal health is concerned.

TESTING NEAR THE PATIENT

A good example of these differing attitudes to quality and cost is provided by tests done at the bedside, in the doctor's office, or by the patient at home (5). Many laboratory staff are concerned at the poor quality of some of these tests, sometimes with some justification. However, this is likely to be a temporary phase, and improved technology will very soon make these tests more foolproof, as it has with blood and breath alcohol measurements, which are probably the commonest tests performed close to the patient. The laboratory is also worried at the diversion of financial resources away from the laboratory, particularly when equipment on wards or in clinics is only used occasionally. The clinician seems to be unconvinced by either of these arguments. He probably sees bedside testing as a way of improving his own efficiency, and more rapid results save his time. They may also increase his income, which is another issue entirely. The administrator is probably not sure that tests done outside the laboratory will improve patient care. They may cost a little more but are unlikely to reduce the cost of the laboratory. However, they will save money if they keep the patient out of hospital. And finally, as far as the patient is concerned, he gets personal attention, a rapid answer and a quicker clinical decision. To him that is quality.

When tests are done nearer the patient it should be easier for the clinician to relate them to the clinical state of the patient. Consequently the tests done should be more relevant, and unnecessary tests less likely to be requested. It remains to be seen whether this will lead to a general change in requesting patterns for work done in the laboratory.

REQUESTING PATTERNS

Both the relevance and the utilisation of laboratory tests varies widely, with large differences in test-request patterns between different physicians, hospitals, and countries.

AVERAGE NUMBER OF SOME TESTS IN 3 CANADIAN AND 3 BRITISH TEACHING HOSPITALS (expressed as tests per 1000 inpatient days)

	Britain			Canada		
	A	B	C	D	E	F
Potassium	164	161	126	512	338	295
Glucose	75	42	53	606	491	453
Cholesterol	8	12	19	60	24	273
Creatine kinase	1	20	2	48	28	39

(ref. 7)

In this example there are differences between laboratories within each country in the numbers of potassium, glucose, cholesterol and creatine kinase measurements, but very much larger differences between countries. These and other published data do not seem to reflect any differences in morbidity, mortality or patient care. Although there are probably similar differences in other branches of laboratory medicine and in radiology, it is clinical chemistry which seems to attract most attention.

A further example is provided by the use of enzyme assays in patients with suspected myocardial infarction.

AVERAGE NUMBER OF ENZYME ASSAYS PER PATIENT IN A BRITISH AND A CANADIAN CARDIAC CARE UNIT

Myocardial infarction (mean bed stay)	Britain				
	AST	ALT	CK	LD	Total
No (2.43 days)	2.03	2.04	2.01	2.01	8.09
Yes (3.00 days)	2.17	2.17	2.17	2.17	8.68
	Canada				
	CK	CKIso	LD	LDIso	Total
No (2.57 days)	3.55	1.64	3.48	3.48	12.15
Yes (4.74 days)	10.17	8.98	9.89	9.86	38.90

(ref. 7)

In these 2 hospitals there were large differences in the number of enzyme assays performed per patient in cardiac care units. For patients without an infarct, an average of 8 enzyme tests per patient were performed in Britain but 12 in Canada; for patients with an infarct, the averages were nearly 9 in Britain against 39 in Canada. In addition, different enzyme tests were used in the 2 laboratories. In a small local survey of 15 clinical chemistry laboratories in Britain, it was found that in cases of suspected myocardial infarction each laboratory measured between 1 and 3 enzymes, but altogether a total of 6 different enzymes was measured in 10 different combinations. Clearly there is no consensus on the most effective enzyme assays to use in these circumstances. The clinician cannot be entirely blamed for these differences, because the laboratory probably decides which enzyme assays to offer and in what combination. Instead of asking why clinicians order unnecessary or inappropriate tests, it might be better to ask why laboratories do them. Although there is plenty of information in the literature of the effectiveness of tests such as enzyme assays in suspected myocardial infarction, it often seems to be ignored.

CONCLUSIONS

The problems of cost and quality affect laboratories, clinicians and administrators in many countries, but at present they all seem to be using different approaches to basically similar problems, with varying degrees of success. With a few notable exceptions, professional organisations, both nationally and internationally and in medicine and in the laboratory field, have not done much about this. There are 4 practical steps which clinical chemists could take.

1. Useless test should be completely deleted from the laboratory repertoire. Candidates would include the following:
Cl^- and HCO_3^- in an electrolyte profile
Cardiac enzyme profiles
Lactate dehydrogenase (total)
ALT or AST, not both
Fe, transferrin (except in haemochromatosis)

Bromsulphthalein clearance
Plasma urea (replace by creatinine)
Urea clearance
Creatinine clearance
Oxo and oxogenic steroids
Faecal fat
Urine pregnancy oestrogens

Although there is unlikely to be complete agreement about all tests on this list, most clinical chemists would probably agree that some of these are useless. The best approach would be through local discussions: a laboratory is more likely to delete a test from its repertoire if it can be shown that another laboratory has done this successfully without affecting patient care.

2. Since financial resources will continue to be limited, a new test should wherever possible only be introduced when an old one can be deleted. The new test should be shown to have a better diagnostic sensitivity and specificity than the old one.

3. Test usage should be monitored - per ward, per clinician and if possible per patient - as part of internal quality control. If this differs or changes between doctors, laboratories or hospitals, investigate the reasons, and make sure that the clinician has the information.

4. Use a consensus approach to the development of test protocols for use in common clinical conditions, such as suspected myocardial infarction or preoperative laboratory tests (8). At present most clinicians and laboratories use protocols (e.g. profiles), but these may be ill-defined and there is no agreement within either group on which is most cost-effective. Protocols have obvious disadvantages, in that they might be applied rigidly and lead to an increase in the number of unnecessary tests. Nevertheless, where protocols have been published, or could be formulated without much difficulty, it is important that they should be studied experimentally at a local level to see whether they are cost-effective; obviously this must be done in close collaboration with clinicians. Any laboratory which rejects this appproach needs to find some other way of improving (or proving) its cost-effectiveness.

These problems are universal, and clinical chemists, with the active support of IFCC, must take the initiative in tackling a neglected area of laboratory management - that is, the cost-effectiveness of laboratory testing. Clinical chemists are being challenged on both the cost of the service they provide and its quality as judged by its influence on patient care. We must respond to this challenge if clinical chemistry is to be allowed to continue to develop and prosper.

REFERENCES

1. Fraser CG, Woodford FP. Strategies to modify the test-requesting patterns of clinicians. Ann Clin Biochem 1987; 24: 223-31.
2. Winkelman JW, Hill RB. Clinical laboratory response to reduced funding. JAMA 1984; 252: 2435-40.
3. Tydeman J, Morrison JI, Hardwick DF, Cassidy PA. The cost of quality control procedures in the clinical laboratory. Am J Clin Pathol 1982; 77: 528-33.
4. Fraser CG. Desirable performance standards for clinical chemisty tests. Adv Clin Chem 1983; 23: 299-339.
5. Broughton PMG, Buckley BM. Performance requirements of tests performed nearer the patient. Scan J Clin Lab Invest 1987; 47: 99-104.

6. Hardwick DF, Ian Morrison J, Tydeman J, Cassidy PA, Chase WH. Laboratory costs and utilization: a framework for analysis and policy design. J Med Educ 1981; 56: 307-15.
7. Henderson AR, Gardner MD. Patients with suspected myocardial infarction: their test request patterns for clinical biochemistry in a British and a Canadian cardiac care unit. J Clin Pathol 1986; 39: 803-7.
8. Blery C, Szatan M, Fourgeaux B, Charpak Y, Darne B, Chastand CL. Evaluation of a protocol for selective ordering of preoperative tests. Lancet 1986; 1: 139-41.

MANAGEMENT AND ORGANIZATION OF CLINICAL CHEMISTRY AND LABORATORY DIAGNOSTICS IN GDR

H.-J. Thiele* and W. Dummler**

*Researchinstitute of Medical Diagnostics, Dresden;
**Institute of Pathobiochemistry, Wilhelm Pieck University Rostock, GDR

Biochemistry as well as Pathological and Clinical Biochemistry having important traditions in our country:
Leonor Michaelis postulated, in cooperation with Maud Menten in the late twentieth basic rules of enzyme kinetics, the so-called Michaelis-Menten-equation.
Karl Lohmann described the energy-rich-phosphates, especially ATP, for the first time.
Otto Warburg found the basic principles for kinetic measurement of NAD- and NADP-dependent enzymes.

The clinical biochemists of the GDR have close connections to these scientific traditions. For this reason the Society of Clinical Chemistry and Laboratory Diagnostics established the Leonor-Michaelis-Award for outstanding results in the field.

The German Democratic Republic as a result of the II. World War is one of the smaller countries in the very heart of Europe with 16,64 millions of inhabitants. It is one of the socialistic countries in Europe with an area of 108,333 km^2. The capital is Berlin, celebrating its 750th anniversary this year. There are 15 counties with 219 districts and 7553 towns and villages. The supreme organ of government is the People's Chamber of the GDR with 500 deputies. There are 5 political parties.

All political power comes from the people, lead by the working class.

The health service of the GDR is embedded in the socioeconomic structure of this socialistic country. The constitution clearly defines the principles of our health policy. Specifically it is based on Article 35.

The Constitution of the GDR stipulates in Article 35:

(1) All citizens of the German Democratic Republic shall be entitled to the protection of their health and working capacity.
(2) This right shall be guaranteed by the systematic improvement of working and living conditions, public health meausures, a comprehensive social policy, and the promotion of physical culture, school and mass sport, and outdoor recreation.

(3) Material security, medical aid and medicines, and other medical services shall be provided free of charge in the event of illness or accident on the basis of a social insurance system.

These principles of our health policy are set out in more detail in various pieces of legislation.

The central tenets are as follows:

- The entire health service is run and developed by the State on planned lines. Hospitals, sanatoria, health centres and clinics are for the most part publicly owned and financed from the public treasury. The vast majority of doctors, nurses, medical laboratory technicians and other paramedical staff receive their salaries from the State.
- Everyone is financially secure in illness, disablement and old age. This underlines the socialists State's humanitarian concern to enable citizens to retain their health, fitness and optimism to a ripe old age.
- All medical services - preventive, therapeutic or follow-up - are provided free of charge, from vaccinations, drugs, artificial limbs and ambulance services to residential treatment in sanatoria and in case of complicated surgery. They are available to everyone regardless of social status, place of residence or other circumstances.
- Every patient is free to consult the doctors of his or her own choice.
- A widely ramified system of health care, which ranges from the family doctor to the skilled medical specialist, from outpatient treatment to specialist consultations, from small hospitals to highly specialized facilities, guarantees that every patient is treated with the necessary expertise and maximum success.

Advances in medical science at home and abroad are rapidly translated into practical improvements benefiting the individual patient.

- Preventive medicine is becoming more and more important. Immunization schemes and screening programmes are now routine procedure. Within the existing comprehensive system of health care the citizen and the State have to play a joint role. As educational standards and public awareness of the individual's responsibilities increase, people are displaying a greater concern for their own health, e.g. by adopting a more suitable life-style.

Two large religious communities of the GDR are making a valuable contribution to the care of patients and disabled. All together, there are 486 charitable institutions (including hospitals) maintained by the Protestantic and 152 administered by the Roman Catholic Church. The churches are working in close partnership with the state healthe services.

On 11 December of each year, the date of birth of Robert Koch, the most famous German bacteriologist, the government pays tribute to all working in health care.

The pattern of responsibility for health care organization is as follows:

- National level
  People's Chamber (Parliament)
  Council of Ministers (Government)
  Ministry of Health

Minister
- County level
County Assembly (elected body)
County Council (executive arm)
Health and Social Services Department
County Officer of Health
- District level
District Assembly (elected body)
District Council (executive arm)
Health and Social Services Department
District Officer of Health.

Under the Local Government Act of 1973 the county and district councils supervise the whole economic, social and cultural development in their local area. In close touch with the general public and public organizations they administer health and social policies at the local level.

The annual budgetary expenditure on health and social services of the country (not including social insurance benefits) climbed from 5 milliards to 12.4 milliards mark between 1965 and 1985, representing 5.3 per cent of national income in 1985. In addition to these funds from the national budget considerable financial resources are made available by firms and other institutions for purposes of health care. In the same year social insurance, administered by the Confederation of Free German Trade Unions, spent 24.1 milliards mark for pension, sickness benefit, drugs, appliance and physiotherapy. This provided conditions for a high standard for medical and social services which are to be further improved.

In the GDR there are many differently structured institutions:
9 teaching hospitals with 116 units and institutes in the framework of 6 universities (Berlin, Greifswald, Halle, Jena, Leipzig, Rostock) and in addition of 3 medical academies (Dresden, Erfurt, Magdeburg)
7 central hospitals or institutes
49 county hospitals and specialist hospitals
238 district hospitals
112 municipal hospitals
581 health centres (Polikliniken) incl. 132 industrial health centres
988 outpatient clinics (Ambulatorien) incl. 319 industrial clinics
3606 medical practices incl. 1978 industrial first-aid posts staffed with doctors
927 dental practices
594 doctors in private practice
787 dentists in private practice
6789 district nurses' posts and first-aid posts staffed with nurses incl. 1352 first-aid posts in industry.

The responsibility of the teaching hospitals rests upon the Minister of Higher Education but he is obliged by law to follow the general and special rules and directions, given by the Minister of Health in order of the People's Chamber. Within the context of the overall development of science and technology the Minister of Health is also responsible for the development of medical science and the integrated administration and planning of medical research. For this purpose he closely cooperates with the Minister of Higher Education and the President of the GDR Academy of Sciences. The Minister is assisted by a body of scientific advisers and a Medical Science Council made up of distinguished scientists drawn from the various branches of medicine and related fields.

In 1985 there was the total number of health and social service workers 531,000 therefrom

37,943 physicians (22.8 per 10,000 population)
11,757 dentists
3,783 pharmacists
11,600 other high educated staff

169,112 hospital beds (101.6 per 10,000 population).

At present every physician requisits about 5000 tests per year from the clinical laboratories in the average.

In 1986 in the clinical laboratories we had to perform about 200 millions of tests. They are supplied by microbiological, toxicological, histo- and cytological and other examinations performed in special institutions.

From year to year the total number of required tests rises by 3 to 5 per cent. Much effort is necessary to fulfill this.

Of fundamental value for the systematic development of clinical laboratory services in our country were the recommendations of a WHO-expert commission on Health Laboratory Services in 1962. It is a matter of course that we adapted and updated these recommendations sometimes in the last 25 years, but generally, some of these ideas are valid up to now.

So there is a national system of three levels of clinical laboratories recommended by a board of specialists and directed by the Minister of Health in 1968 supplemented by microbiological, pathological-anatomical, blood transfusion and other laboratory services (Table 1).

On county level a special courier system by car guarantees fast transportation of material to the central clinical, microbiological and other laboratories. The retransmission of testresults follows generally by telex.

Table 1. Clinical laboratories (1986) = 195.6 Mio tests =

Level	Characteristics	Number	Responsible for about
Type I 48.9 mio tests	outpatient clinics health centres (small) municipal hospitals (<100 beds)	1108	20 tests
Type II 94.9 mio tests	district hospitals municipal hospitals (>100 beds) health centres (big)	401	50 tests
Type III 51.8 mio tests	teaching hospitals central hospitals county hospitals	69	120 tests
	Clinical laboratories	1578 total	

1986 (tests in thousands)

Clinical Biochemistry	118,951.5
qualitative	46,865.9
quantitative	71,905.9
- proteins	21,379.4
- electrolytes	19,809.1
- enzymes	15,958.2
- lipids	2,401.9
- hormones (excl. Nucl. med.)	1,053.3
- intermediates (glucose, creatinine, etc.)	58,087.8
Haematology	47,032.5
Haemostaseology	4,046.5
Transfusion serology	7,865.7

19 of the biggest laboratories are supported by laboratory information (and management) systems (LIS), partly in direct connection to telex, since ten or more years. At present a new LIS is under preparation for much more laboratories in the county.

About 15 per cent of all quantitative tests are carried out by mechanized systems, approximately 12 per cent of all tests are performed as emergency, so called Cito-tests. The total number of these orders is still raising from year to year. It is guaranteed, that every districts at least one laboratory is still working around the clock.

Range of tests (in thousands)

Test	GDR (total)	GDR (%)
1. glucose, qual. in urine	17,015.5	9.4
2. glucose, quant. in urine	14,803.8	8.2
3. protein, qual. in urine	14,226.7	7.9
4. haemoglobin in blood	11,865.2	6.6
5. leukocyte count	10,563.2	5.9
6. urinary sediment	9,480.2	5.3
7. haematocrit	9,244.4	5.1
8. acidity, qual. in urine	5,317.9	3.0
9. ALAT in serum	4,608.3	2.6
10. creatinine in serum	4,380.0	2.4
11. glucose, quant. in serum	4,105.2	2.3
12. cell differentiation in blood	3,688.4	2.1
13. urobilinogen, qual. in blood	3,248.9	1.8
14. sodium in serum	3,178.6	1.8
15. acetone, qual. in urine	3,162.7	1.8
16. ASAT in serum	3,147.7	1.8
17. MCHC	2,933.5	1.6
18. potassium in serum	2,878.8	1.6
19. cell count in urine	2,715.6	1.5
20. protein, quant. in serum	2,681.4	1.5
21. AP in serum	2,067.5	1.2
22. bilirubin, quant. in serum	2,055.2	1.1
23. platelet count	1,885.3	1.0
24. calcium in serum	1,885.3	1.0
25. thromboplastin time (Quick)	1,830.3	1.0

An annually standardized report of each laboratory provides the actual information in detail about situation and trends of workload, which is the basis for planning and further development of all resources of the laboratories for the Ministry of Health.

About 20 per cent of all tests requested are performed for diabetes diagnostics and monitoring, that means 40 millions of tests. In comparison there are more than 650,000 patients with diabetes mellitus under control, this is 4 per cent of the total population.

In the last years the service for diabetics was extended by introducing a dry chemistry test system for blood glucose, developed and manufactured by scientific institutions and national firms of the GDR. It is internationally the first transparent testfilm with properties comparable to those of Boehringer/Mannheim, Ames-Miles and others.

Maternal and child services are provided by about 900 antenatal clinics where expectant mothers are examined periodically. More than 99 per cent of all deliveries take place in obstetrical hospital wards. While they are still in the maternity hospital, 3.6 millions of new-born babies have been examined for phenylketonuria (PKU by Guthrie-test) since 1969. The incidence of PKU is in the GDR 1 : 9,900. The incidence of so-called permanent mild hyperphenylalaninaemia is 1 : 98,000.

In addition we are introducing a mass-screening for hypothyreosis by new-born-testing of elevated TSH at the 5th day of life with RIA and in the future by an EIA with monoclonal antibodies also developed in our country. As a result of the first 100,000 tests, in this moment the incidence seems to be 1 : 3,800. This corresponds to experiences in other European countries. Generalized testing of mucoviscidosis accomplished for several years was interrupted in 1983 because of minimal therapeutic consequences.

In preparation is a screening for neural tube defects and anencephalies by AFP-testing of expectant mothers in the 16 to the end of 18 week of gestation. It is calculated, that the total testing of all pregnant women will start in 1989. Up to now there were 20,000 women tested. The incidence seems to be 1 : 685.
In addition to these activities there is some effort for stepwise introduction of genomic diagnostics in research and high specialized care.

RESPONSIBILITIES

From the point of view of the Ministry of Health there are three institutions which are responsible for main aspects of clinical laboratory diagnostics.

- National Institute of Drugs (IFAR), Berlin
  - standardization of methods
  - quality control
  - evaluation of instrumentation, testkits, etc.
- Researchinstitute of Medical Diagnostics (FMD), Dresden
  - organization, management (incl. laboratory information systems)
  - research, development (cooperation with the industries)
- Academy of Postgraduate Medical Education (AfÄF), Berlin (Chair of Clinical Chemistry and Laboratory Diagnostics).

From the scientific point of view we have in the GDR two additional organizations which are responsible for scientific life at a high level in its complexity. These organizations are:

firstly the Society of Clinical Chemistry and Laboratory Diagnostics of the GDR and

secondly (in a free translation of the official name) the National Committee on Standardization of Diagnostic Laboratory Methods.

This Committee on Standardization does not belong to the Society but works under the direct supervision of the Ministry of Health of the GDR. But, of course, the cooperation between the Committee and the different Medical-Scientific Societies is very close (Microbiology, Toxicology, Haematology and Blood Transfusion, and, of course, Clinical Chemistry).

The Committee on Standardization, which has been founded in 1965, is responsible for the complete spectrum of problems concerning the general strategy of standardization and quality control in medical diagnostic laboratories (of course, the methods to be used in the field of research are not standardized!) and for the experimental work in collaboration with the IFAR. Under the supervision of the Committee and the IFAR there are working about 15 groups, which are charged with the experimental work necessary for the elaboration of standardized protocols (method comparison, optimization etc.). Each working group consists of 15 to 20 members from all over the country, and anybody interested in this kind of scientific work can become member of a group, according to his personal intentions.

When the protocol is finished, it is published and open for discussion at the national level. All comments and critical remarks are seriously checked rather often with the result of a modification of the protocol. Finally the protocol is validated by the Minister of Health and introduced into the Pharmacopoeia of the GDR, Part "Diagnostic Laboratory Methods", that means, those protocols are obligatory for all medical diagnostic laboratories. The responsibility for the final part of the procedure that is: publication of the drafts, redaction of the final texts, conformity with other standards, introduction into the Pharmacopoeia, lies in the hands of the IFAR.
At present we have about 300 standardized methods.

From time to time, that is in general every five years, the protocols are checked whether they are still useful or not. Obsolete methods are then replaced by new ones. But it has to be emphasized that the main criteria whether a method is obsolete or not is if any new method is more efficient according to diagnostic information and also/or more efficient from the viewpoint of laboratory technology and organization (automation, miniaturization) (Table 2).

In addition I should like to mention that the responsibility for the organization and supervision of all activities in the field of quality

Table 2. Standardization (1985)

- Chemical and physical methods	72
- Toxicologic-chemical methods	10
- Immunohaematologic methods	12
- Haemostaseologic methods	17
- Cytomorphological methods	15
- Microbiological and serologic methods	20
- Mycological methods	5
- Parasitological methods	10

control lies also in the hands of the IFAR which cooperates with 15 County Laboratories of Quality Control.

Quality control (by legislation)
- external quality control
  - national scheme 2-3 x per year (resp. National Institute of Drugs (IFAR), Berlin)
  - county scheme 5-10 x per year (resp. County Laboratories of Quality Control (BLQ))
- internal quality control (daily).

The national requests to analytical quality are very pretentious but they are a challenge for reliable performance of laboratory work.

The Researchinstitute of Medical Diagnostics is directed to the Ministry of Health. It is responsible for the elaboration of recommendations for territorial and internal laboratory organization, for a scientific comparison of the performances and for further development of Clinical Laboratory Diagnostics in the GDR. FMD is supported by a board of advisers and supports and advises on his part the relevant industry in direction of instrumentation and test kit development.

FMD is ordered for a national project of research and development in the field of Clinical Biochemistry in cooperation with high skilled institutes and laboratories in the county. FMD is the software house for laboratory information (management) systems.

As topics development are seen
- Immunoassay
- Ultramicroanalytics
- Instrumentation
- Selftesting
- Therapeutic drug monitoring
- Automation (robotics)
- Laboratory information systems
- Diagnostic strategies (stepwise diagnostics)
- Expertsystems.

Permanent postgraduate qualification is of special importance for development of medical care for high educated as well as for technical staff. The responsibility for this, including Pathological and Clinical Biochemistry and Laboratory Diagnostics, has the Academy of Postgraduate Medical Education.

Postgraduate education
- for graduates (medicine, chemistry)
  4 years
  - Facharzt für Pathologische und Klinische Biochemie
  - Fachwissenschaftler für Klinische Chemie und Laboratoriums-diagnostik
- for technicians
  1-2 years
  Fachassistentin für Klinische Biochemie.

At present about 1,000 physicians and chemists are still working in clinical laboratories and scientific institutions. They acquire their special qualification in a four years course of postgraduate education, including hospitations in high skilled institutions etc. Also for the technical staff the possibility of specialization is given in addition to the three years course of basic qualification.

As a result of close cooperation between all institutions mentioned above the development of Laboratory Diagnostics in the GDR can be assessed as well and the system is in the majority of questions equal to clinical demands. But considering the permanent increase of demands and costs, also in the future big effort is necessary to adapt the Clinical Laboratory Diagnostics to over all limited possibilities and funds.

In order to attain an increase of diagnostic effectiveness (including the introduction of new tests) faster than the increase of costs, we believe that the development of diagnostic strategies, as well as the determination of the diagnostic validity of tests and the continuous discussion and cooperation between clinical chemists and clinicians on the proper choice of appropriate tests according to the diagnostic problem are the main roads for our work in organization and management of clinical laboratory services in the near future.

THE ORGANIZATION OF THE NATIONAL HEALTH SERVICE AND ITS CONSEQUENCE FOR THE CLINICAL CHEMISTRY LABORATORY IN SWEDEN

Carl-Henric de Verdier

Department of Clinical Chemistry, University Hospital
S-751 85 Uppsala, Sweden

THE SYSTEM OF MEDICAL CARE IN SWEDEN

Some basic facts about the organization and responsibility for health and medical care in Sweden are presented in Table 1.

Average life span may be used as an index of the health situation in a country. In Iceland it is 77.1 y closely followed by Japan, Sweden and Norway all above 76 y. Australia, USA, Denmark, France and Finland lay in the interval 74.1 - 74.9 y. For comparison China has the numerical value of 65 y. The development during the last century for different age groups is demonstrated in Figure 1 which shows a prosperous decline in mortality rate for babies.

The age pyramid of the Swedish population with a large proportion of the population of high ages combined with a rather negative attitude among young people against nursing - an attitude which is enforced by the unwillingness of local health authorities to pay competitive salaries.

This recruitment difficulties are primarily valid for personnel working in wards but have some bearings also with regard to personnel in clinical laboratories.

Table 1. Public health & medical care in Sweden

Under Swedish law the responsibility for providing the public with both in-patient and ambulatory care in the event of illness, injury or maternity rests with the county or municipal authorities. 23 selfgoverning counties and 3 autonomous municipalities.
Each county is governed by a County Council.
The health service is mainly financed by local income tax but is subject to national legislation and Government control.
Planning and supervision are the responsibility of the National Board of Health and Welfare.

Private health care exists on a very limited scale. Only 5% of the physicians work full-time in private practice. The cost of different types of public medical care is divided into the following parts: Short-term care 51%, long-term care 23%, psychiatric care 13% and outpatient primary care 12%.

In Sweden, as in most other West European countries, there has been a clear tendency towards increased allocation of resources to primary care. The hospitals with their traditionally strong position have opposed, arguing that primary care will not reduce the need for hospital care.

In Sweden the distribution of the 8.3 million inhabitants is uneven, with a dense population in and around the main cities Stockholm, Gothenburg and Malmö and a sparse population in the northwestern parts of the country. Most of the counties have a too small population to allow investments in staff, localities and equipment for very specialized clinical activities such as neurosurgery, thoracic surgery, burn centres etc. For this reason counties are collaborating within a larger area called a hospital region.

In general it is university hospitals that act as regional hospitals. Figure 2 shows the counties with the corresponding regions. The principal three levels of medical care are indicated in Figure 3 with the most specialized medical diagnosis and treatment in the regional hospitals in the top. A hospital region corresponds to a population of about one million and a regional hospital has a size between one and two thousand beds.

How much does medical care in Sweden cost compared with other countries? A comparison for the last 15 years between the United States and Sweden, where the cost is expressed as per cent of GNP shows that the

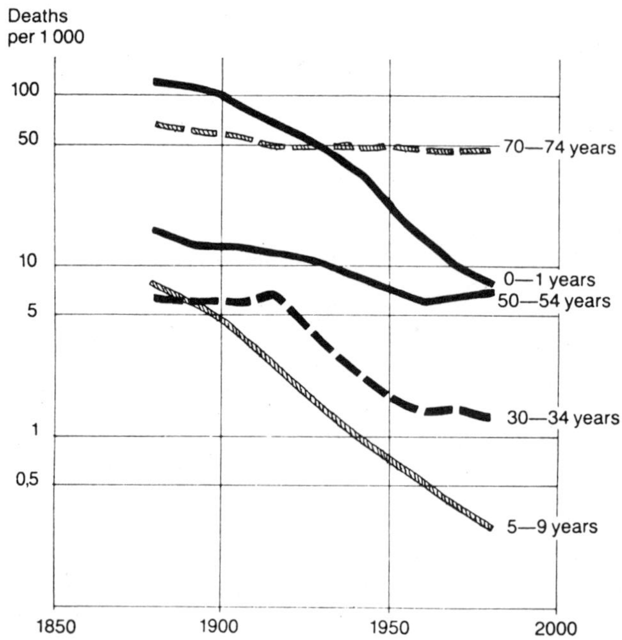

Fig. 1. Changes in mortality rate during the last one hundred years in Sweden.

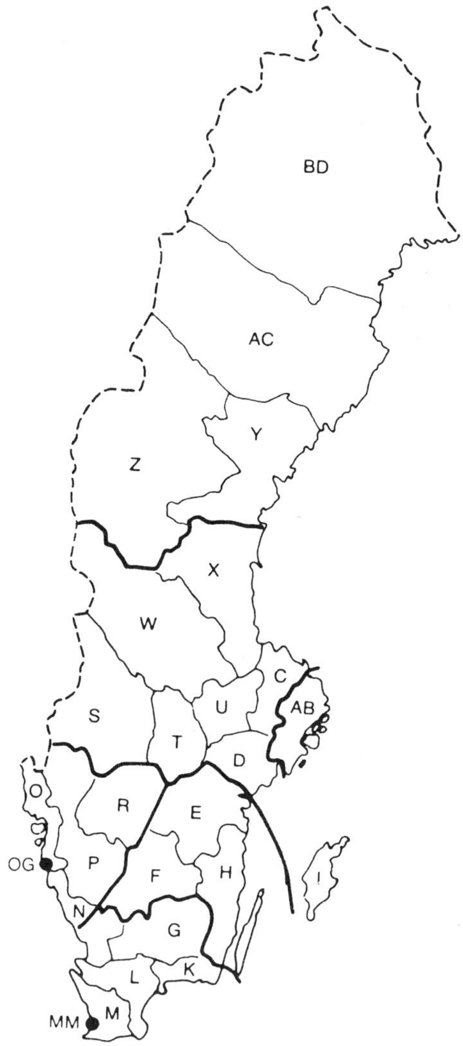

Fig. 2. Sweden with counties and seven hospital regions.

values and the earlier increasing tendencies are similar for both countries but that Sweden has started to level off a year earlier or in 1981 instead of in 1982. The maximal value for Sweden is 9.7% or 1% lower than for the United States. The most likely explanation for the stationary cost during the last years is an increased cost consciousness and attempts to start to reduce costs. Many larger hospitals in Sweden as in the United States now start to analyze costs and production by means computer systems in about the same way as in industry.

IMPACT ON THE MEDICAL CARE SYSTEM ON ORGANIZATION OF CLINICAL CHEMISTRY

Seven factors of importance can be listed as follows:

1. The principle of equity in health care.

Fig. 3. For explanation see text.

2. Uneven distribution of the population.
3. The county (population 54 - 1,500 thousand) is the basic unit for the public health care system.
4. The counties co-operative within hospital region (population about 1 million).
5. The private sector is small.
6. A high degree of hospital-based medical care.
7. Continued extension of primary care.

These seven factors can be commented upon in the following ways:

Ad 1. The effect of the principle of equity is difficult to interpret. It may lead to a large number of specimens to investigate but there is also a risk that a publicly owned organization will be less service-minded than a private one. There has not recently been any reimbursement system that has favoured the laboratory or its hospital. Different investigations a few years ago showed that about 4% of the expenditure for medical care was used for clinical chemistry, which in Sweden also includes haematology. The comparatively low labour costs in the clinical chemical laboratories compared with nursing wards, and the increase in salaries in relation to equipment and materials, will mean that the given figure of 4% will probably have decreased.

Ad 2. The uneven distribution of the population will course problem in counties with a sparse population but can be delt with by organizing an effective transportation system.

Ad 3. Having the county as the main organizer for clinical laboratories implies that there is little competition, but on the other hand it means that there are rather large laboratory units which can be made more efficient by automation and computerization.

Ad 4. The regional hospital organization is considered to be effective and of great help for patients who need more advanced medical diagnosis and treatment. In the more harsh economic climate there is a tendency, however, to restrict the referral of patients to regional hospitals.

Ad 5. Discussed under 3.

Ad 6. This is favourable for clinical chemistry which generally has its main unit at the central hospital as the base for the organization within the county.

Ad 7. The politicians in Sweden responsible for allocation of money to the health care system have decided like those in most other European countries to extend the primary care. The clinical chemists reacted accordingly and organized decentralized laboratories or helped the clinicians at the primary care centres to organize laboratories. The National Board of Health and Welfare and the Federation of County Councils have not issued any regulations about the organization and the responsibility for such types of activities, although the clinical chemists have requested it.

NORDIC COLLABORATION WITHIN THE AREA OF CLINICAL CHEMISTRY

The structure of medical care is fairly similar in the Nordic countries and there are also great parallelisms between the educational programs for clinical chemistry for physicians or for technicians in the four countries. The authorities for medical care in the countries have been convinced to allocate the equivalent of 140 thousand US $ annually to Nordic clinical chemical laboratories for development of new techniques or procedures or the background for these. I will comment on some of the main publications that have been produced during the decade in which NORDKEM has been in action.

The first one describes how the quality requirement for analytical procedures should be estimated - not in general terms but based on real examples. The second one describes quality control in Scandinavia but also quality control procedures which assure, with a given probability, a predecided quality level. The other two publications are in Scandinavian languages and describe the role of clinical chemistry in primary care and a proposed standard for annual laboratory reports. Some further activities have also been devoted to laboratory management and as a consequence a course for laboratory directors has recently been organized.

SOME THOUGHTS FOR THE FUTURE

A main concern for the future for the medical care in Sweden is the economical situation if it is also then will be paid mainly by taxes. The county council tax shows an almost linear rise during the last 25 years and it is considered as an almost impossible political situation to rise it further. As a continuation I may perhaps be allowed to describe the prospects of clinical chemistry as they can be perceived in a Nordic twilight. The prospects are not good unless actions are taken. Such possible actions are summarized in Table 2.

Table 2. Actions to improve the future of clinical chemistry

1. Stop peeling off new subspecialities.
2. Assume responsibility for clinical chemistry in primary care and other decentralized areas.
3. Convert clinical chemistry gradually from a "body fluid science" to a "science of metabolism and function of organs and cells".
4. Promote an efficient management.
5. Collaborate with other clinical laboratory disciplines as the methodological boundaries become indistinct.

It is very important that new and often scientifically interesting areas are not given away. These responsibilities include a lot of work but it is necessary to assume them in order to retain resources.

Point 3 is self-explanatory. Many new techniques are now ready to be incorporate into clinical chemistry e.g. microanalytical techniques and biodynamic models for cell suspensions, new methods for the study of the biochemistry and function of cell membranes, NMR and DNA-technologies.

Point 4 does not mean that it should be as cheap as possible. Analytical results and when necessary interpretations of sufficiently high quality should arrive at the requesting clinicians without too much delay.

Seen from the point of view of hospital management the development of similar technologies within clinical chemistry, clinical immunology, clinical cytology etc. may require a more integrated utilization of the resources if this is economically justified.

THE ORGANISATION OF THE NATIONAL HEALTH CARE AND ITS CONSEQUENCES ON THE CLINICAL CHEMISTRY LABORATORY IN NIGERIA

Popo A. Akinyanju

Clinalabs Nig LTD., Lagos, Nigeria

Nigeria, with its population of about 100 million people is the largest country in Black Africa. Even though Nigeria operates an oil-based economy, the vast majority of her people are peasants who live in rural areas far-away from modern medical facilities. The health services of Nigeria have evolved through a series of historical developments including a succession of policies and plans which had been introduced by previous administrations. The health services, in the current state, are judged to be unsatisfactory and inadequate in meeting the needs and demands of the public as reflected by the low state of health of the population.

HISTORICAL BACKGROUND

The public health services in Nigeria originated from the British Army Medical Services. During the colonial era, government offered to treat the local civil servants and their relatives and eventually the local population living close by government stations. The colonial medical service then developed and provided free medical service to the Army and the Colonial Service Officers. Any treatment offered to the local population was only incidental. Thereafter various religious bodies and private agencies established hospitals, dispensaries and maternity centres in different parts of the country. The first attempt at planning ahead for the development of health services in Nigeria took place in 1946 as part of the exercise which resulted in the ten-year plan for Development and Welfare (1946-56). The proponents of the plan were mainly expatriate officials since Nigeria was still a colonial territory. Even though this was not an integrated development plan and was not related to any overall economic target, it however was modest and realistic and it eventually served as the basis for subsequent health plans. Since independence in 1960, health policies have been enunciated in various forms within the overall embrace of National Development Plans. The 2nd National Development Plan (1970-74) aimed at correcting some of the deficiencies of the previous plan. Both plans were similar in the sense that the priority was 95% curative and 5% preventive and supportive. The first deliberate attempt to draw up a comprehensive National Health policy embracing issues as Manpower Development, Medical Research, health management, comprehensive health care services based on basic health care was in the 1975-80 period. Even the 1980-86 policy was a slight departure from the previous one. This is evidenced in the large number of defects and inadequacies now recognised such as:

a. Only about 35% of the population had access to modern health care services.
b. Remarkably high investment on curative services to the detriment of preventive services.
c. Poor and inefficient management resulting in waste and inability to meet goals.
d. Lack of basic health statistics, hindering planning monitoring and evaluation.
e. Inadequate financial resource allocation.
f. Basic infrastructure and logistic support either defective or totally absent.

Table 1 shows estimates obtained from sample surveys in 1982, a reflection of poor state of health of the population. Infant mortality is 10 times as high as in developed countries. Most of the deaths and serious illnesses which occur among Nigerians are due to conditions which are easily preventable, or which can be treated with simple remedies. In recognition of all the defects and failures of previous plans, the Federal Government of Nigeria therefore enumerated a National Health Policy to achieve health for all based on the national philosophy of social justice and equity. The government adopted Primary Health Care as the means of achieving this goal.

Primary Health Care as defined in the Alma Ata Declaration of 1978 is "Health care, based on practical, scientifically sound and socially acceptable aims to individuals and families in the community and through their full participation and at a cost that the community and country can afford to maintain at every stage of their development in the spirit of self-reliance and self-determination. It forms an integral part both of the country's health system, of which it is the central junction, and main focus of the overall social and economic development of the community. It is the first level of contact of individuals, the family and community with the National Health System, bringing health care as close as possible to where people live and work, and constitutes the first element of a continuing health care process".

STRATEGY - PRIMARY HEALTH CARE

The goal of the national health policy, is a level of health that will enable all Nigerians achieve socially and economically productive lives. The health services based on primary health care include at least the following:

a. Immunisation against major infectious diseases.
b. Adequate supply of water and basic sanitation.
c. Maternal and Child Care.
d. Health maintenance (Food and nutrition).
e. Provision of essential drugs and supplies.
f. Injury prevention and management.
g. Control of endemic diseases.

In order to meet the health needs of the rest of the population, the Federal and State Governments decided, in a co-ordinated manner, to support a system of National Health Care developed at 3 levels.

Primary health care providing services of preventive, curative, promotive and rehabilitative nature. This is the entry point of the total health care system. At this level, there is contact between conventional

Table 1. State of health of Nigerians 1982 survey

Crude death rate	-	16 per 1000 population
Crude birth rate	-	50 per 1000 population
Childhood mortality	-	144 per 1000 children (1-4 yrs)
Infant mortality	-	85 per 1000 live births
Life expectancy at birth	-	50 years

medicine and traditional health care which is strongly bound to local culture and beliefs. Consequently, there is need for a lot of mutual understanding and collaboration in order to enlist the confidence of the people whom they seek to cure, or prevent from contacting disease. At this level, there are no standard laboratories. Each Primary Health Clinic has a small room designated as laboratory, and is expected to offer such tests as Routine Urinalysis, Haemoglobin, stool examination for parasites and maybe blood glucose. The laboratory is manned by a laboratory assistant trained in a school of health technology. There are presently 24 such schools.

Secondary Health Care will provide specialised services to patients referred from the PHC level, through out-patient or in-patient services. These secondary level services are available at District or General Hospitals which also serve as administrative headquarters supervising the peripheral units. Supportive services such as laboratory, radiology, blood bank and physiotherapy are available at this level. The present policy of the Federal Government is to aid the State Governments to improve the existing hospitals and laboratories. Tertiary Health Care consists of highly specialised services provided mainly by University Teaching Hospitals and other special hospitals. They are at the apex of the referral system and hence in some cases selected centres have been encouraged to develop special clinical expertise and have been supported by advanced technology. The Clinical laboratories in Teaching Hospitals are expected to be the peer laboratories. Unfortunately, in several cases they have been unable to play leading roles due to inadequate or sometimes inappropriate equipping. In 1986 four Teaching Hospitals were designated as centres of excellence in Immunology and Communicable diseases, Neuroscience, Cancer and Oncology and Cardiothoracic disorders. It is sad to remark that the laboratories in some of these Teaching Hospitals are still operating at

Table 2. Hospitals in Nigeria

Univ. Teaching	General	Private
14 (7180)	764 (48058)	1436 (12751)
Orthopaedic	Armed forces	Paediatric
3 (731)	12 (2356)	1 (92)

(-) no. of beds

pedestrian levels inspite of huge sums of money invested by Government in attempting to uplift the quality of expertise in these institutions. Table 2 shows the number of hospitals in Nigeria. Preliminary data from an ongoing survey reveal that majority of these hospitals (both private and government owned) have only token laboratories where only the bearest investigations are performed.

STRATEGY AND MANAGEMENT

The main thrust of the strategy rests with the Federal Ministry of Health which is responsible for policy making, formulation of legislation, co-ordination, continuous assessment, definition of standards for materials in respect of delivery of health care, monitoring and evaluating implementation, giving support to State and Local Government for sectoral activities. The Federal Ministry of Health has set up mechanisms to:

1. Seek cooperation on health with other nations.
2. Collaborate with United Nations agencies, Organisation of Africa Unity and West Africa Health Community on regional and global health care improvement strategies.
3. Work closely with other developing countries especially neighbouring states within the region.
4. Co-ordinate actively with other countries in intervention programmes for control of communicable diseases.

In a majority of instances the government has recorded notable successes in its broad strategy. It is remarkable however that even though Nigeria has received a generous amount of aid from International Health Organisations (drugs, vaccines, etc.) over the years there has been no major beneficial impact to the laboratory either in terms of technology or expertise. There is pressing need for International organisations to stimulate awareness in this supposedly "easily dispensable" specialty. There are 20 states in Nigeria. Certainly it would not be too much to ask International Health Organisations to donate, basic photometers, small centrifuges and microscopes to the Nigerian government specifically for use in Primary Health Clinics. Another area of difficulty has been that of communication and collaboration with our immediate neighbours who are mostly Francophone. The language barrier and the unfavorable long distance communication facilities make it extremely cumbersome to exchange ideas, information and personnel.

The State Ministries of health are buttressed by the Federal Ministry so that they become the directing and co-ordinating authority on health work within the State. Some of their responsibilities include:

1. Review of the functions and mechanisms of Health Institutions especially at the Primary Health Care and 1st referral level.
2. Provide transport and communication to facilitate the efficient running of the referral system.
3. Ensure adequate logistic support for distribution of supplies and equipment.

FINANCING

The Federal Government policy on health care financing places high priority on PHC with particular reference to under-served areas and groups. Effort is being made to redistribute financial allocation with a view to place more emphasis on promotive and preventive services.

The government is determined that capital development will be limited to (a) completion and rehabilitation of existing facilities and (b)

Table 3. Government subventions to teaching hospitals

Years	Naira Million	US.D Million
1984	118	169
1985	128	178
1986	139	112
1987	99*	25**

* Effect of depressed economy
** Effect of Naira devaluation

construction of small health centres in unserved areas. However, because of the poor state of the economy and the draconian austerity measures introduced by government, sufficient funds have been difficult to find even for small projects.

Table 3 illustrates the levels of government subventions to Teaching Hospitals over the past four years. Government subvention reached a peak in 1986. In 1987 there was a 29% drop in government subvention. The accompanying devaluation of the Nigerian currency, Naira, presents an even worse picture in hard currency terms, the 1987 subvention representing 22% of the previous year's figure. The hard crunch of currency insufficiency is hard felt even at the priority level of Primary Health Care. Unfortunately, in any situation where funds are insufficient, the laboratory is the first discipline to be starved.

Cost recovery schemes have been initiated. Users pay for curative services, but preventive services are heavily subsidised generally. The socially and economically disadvantaged segments of the population receive public assistance. In government hospitals, free care is given to children, the aged and patients with cancer. In all other cases minimal fees are collected for each service rendered in the hospital. Fees are charged for drugs, x-ray, laboratory investigations etc. So far, the cost recovery is only a minute portion of actual government expenditure. It is extremely difficult to successfully operate cost recovery schemes in an environment where a majority of the people are low income workers, who exist on survival wages.

Furthermore, some of the services which they are required to pay, for example, laboratory tests, are highly cost intensive operations. The abyssmal disparity between wages and costs of some laboratory materials is illustrated in Tables 4 and 5.

Government has encouraged the private sector to participate actively in the Health Care sector especially at the Secondary level. There are presently several private and missionary hospitals and health clinics in Nigeria. Even though the participation of the private sector has increased availability of health facilities, the cost of services is astronomical.

MANPOWER

The broad policy of government is to ensure that medical, nursing, public health and other schools of health sciences include in their education programmes the philosophy of "Health for All", the principles of PHC and the essentials of the managerial process for national health development. The next table illustrates the rate of training of doctors and

Table 4. Wage structure in Nigeria

Years	Minimum/month		Univ.Prof./month	
	Naira	US.D	Naira	US.D
1982	120	180	1000	1500
1985	125	125	1000	1000
1987	150	37.5	1200	300

laboratory technologists in Nigeria (Table 6). Between 1960 and the present time, there has been a dramatic and commendable increase in the number of medical doctors and laboratory personnel trained in Nigeria. Commendable though the effort is, we are still far short of the total number that would be required to take health expertise into the rural areas. Presently a high percentage of these personnel work in the urban centres where they have facilities like running water, electricity, cars and good roads. There are very few incentives offered to personnel working in rural areas - the effect of this is obvious. The laboratories are more unfortunate than the Clinical departments. There is a dearth of intermediate and junior laboratory manpower. Up till a few years ago, most laboratories were manned by trained technologists headed by a Clinician or a Ph.D. Lab. Director. The absence of middle level manpower created a vacum which no one was interested to fill especially as technologists and laboratory heads were perennially jockeying for leadership of the laboratory. Presently, there are 24 established schools with 2 years programmes for training certified Laboratory Assistants and Technicians, and it is hoped that in a few years from now, the laboratories in rural area health centres would have adequate manpower to perform at least the basic investigations.

Laboratory medicine in general and Clinical Chemistry in particular, seem to attract fewer candidates than the Clinical courses. In the Medical laboratory schools there are more students registering for Microbiology than for Chemistry. Table 7 reveals the relative unpopularity of laboratory disciplines even at post-graduate levels. Obviously, it is much easier and eminently cheaper for a physician to start up a private clinic than for his counterpart Clinical Chemist to set up a laboratory. The consequence of shortage of high level professionals in the laboratory is poor quality leadership especially by unqualified medical personnel, who have no knowledge of the principles, processes and skills necessary to manage complex laboratories. Therefore there is now a steady stream of "Impostors" like surgeons, physicians not to talk of dentists and ophthalmologists in some cases usurping the roles of laboratory director when vacancies do not exist in their own specialties. The Nigerian Association of Clinical Chemists is presently working on recommendations to government, of broad-based incentives that they hope would attract more candidates into the specialty. In this regard, the International Health Organisations would be of immense help in offering advice and support to our government.

Another area of manpower deficiency is that of equipment maintenance engineers or technicians. All too often basic equipment are committed to waste because of extremely minor faults. The Federal Government in 1983 established and equipped at great cost, a Hospital Equipment Maintenance School. So far, administrative difficulties have prevented the smooth functioning of this very important school. It is unreasonable to expect foreign instrument manufacturers to fly engineers thousands of miles, simply to spend 5 minutes effecting a simple repair. A consistent resolve

Table 5. Costs of test controls

Test	Cost/ml Naira	US.D
Normal electrolyte/enz.	12.8	3.2
Chemistry serum control	15.6	3.9
Abnormal enz. control	27.6	6.9

to implement the existing policy on training of maintenance engineers and technicians is needed.

TECHNOLOGY

The Nigerian government has assumed a positive posture with regards to technology for the Health Care system. Firstly, it has declared to select the most cost-effective and appropriate technology for use at all levels. Secondly, there is the determination to reduce importation and to encourage local manufacture, thus stemming foreign currency expenditure and creating employment. The definition of appropriate technology in our context must be guided by a lot of caution and bold realism. We must purchase not only the technology that we can conveniently afford, but also that which can be operated, maintained and repaired by locally available expertise.

Table 6. Registered doctors and medical laboratory technologists in Nigeria

Year	Doctors	Population/ doctor	Lab. techs.	Population/ lab. tech.
1960	1079	47,300	30	1.7 million
1970	2683	24,500	113	582,000
1980	8037	10,500	1330	64,710
1986	16003	6,200	2882	34,400
1990	(22000)	(5,000)	(4900)	(22,400)

( ) Projected Figure.

Table 7. Registrants for residency programme at Lagos University Teaching Hospital 1971-86

Surgery	-	114	Medicine	-	86
Paediatrics	-	44	Clinical chemistry	-	0
Public health	-	13	Microbiology	-	0
Morbid anatomy	-	13	Radiology	-	4
Haematology	-	4	Obs. & Gynae	-	72

Furthermore, the sophistication of the technology must be in consonance with supportive infrastructure existing at the site of use. Part of Government intention is to develop local capability to produce essential drugs, vaccines and dressings, and to maintain surveillance on the quality of locally produced items.

The single greatest hindrance to attaining government objectives is finance. The acute shortage of funds in the health sector converts government aspiration to a mirage.

DISCUSSION

In spite of all the apparent good intentions of government with respect to improvement of the technological facilities supporting health care delivery, there has been a blatant and sometimes unbelievable disregard for establishment of new laboratory facilities or improvement of existing ones, even at the tertiary level where laboratories are supposed to function at their best. As far back as 1978, the Federal Government conceived the idea of establishing 5 independent Comprehensive Diagnostic Centres in Nigeria. In 1983, two million Naira was allocated to each centre for purchase of equipment, and in fact equipment for one centre in Lagos was purchased and has since 1984 remained in the crates in which they were transported to Nigeria. The objectives of the Diagnostic Centre which would apart from other things, incorporate an ultramodern laboratory were:

1. Health screening for the general public.
2. Pre-employment medical examination for civil servants.
3. Diagnostic investigations of referred patients.
4. Savings on government spending on Senior Civil Servants who routinely travelled abroad annually for medical check-up and treatment.

There are several reasons for the collapse of this admirable project. They are:

a. Lack of interest in the project by the supervising administrators who annually enjoyed not only the services of diagnostic centres in Britain and Germany, but also the luxury of foreign travel funded by government.
b. Lack of co-ordination between the relevant ministries involved.
c. Insufficient involvement of professionals in the planning and implementation.

FUNCTIONAL HEALTH UNITS AT VARIOUS LEVELS

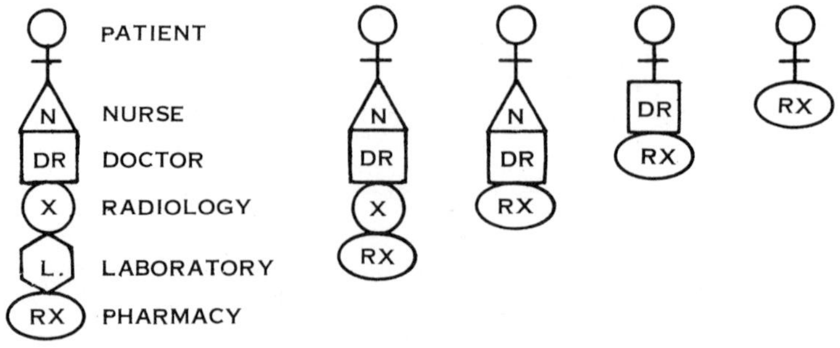

Fig. 1. For explanation see text.

d. Lack of enthusiasm by Health Ministry Officials in projects not involving large scale drug purchase or extensive inpatient facilities.

The laboratory with the services it offers seem to be bludgeoned, disregarded, ignored, relegated and misunderstood by every sector of the population. For example, most private firms offer free health care to all their staff, except that their doctors are often instructed not to request laboratory tests for Junior Staff. Unfortunately, the Trade Unions that protect the welfare of these workers, in demanding better health care, only clamour for more and more drugs. They have never been concerned by the fact that, accurate diagnosis and proper monitoring of disease might not be possible, without the aid of the laboratory. The public in general misunderstand the role of the laboratory in the Health Care system.

Figure 1 illustrates the priority rating of the five major health disciplines to a typical patient. The first dispensable discipline is the laboratory, followed by radiology. Thereafter comes the nurse and, rather surprisingly, the supposed leader of the team, the doctor. The average patient would rather dash to a pharmacist to complain of an ailment and purchase drugs immediately, rather than go through his doctor who might send him on the arduous journey to a laboratory or an x-ray centre. It is remarkable that in Nigeria there are no regulations forbidding the purchase of drugs without prescription.

The market for the services of the laboratory is highly restricted since doctors are the only consumers of the services. The relationship between the laboratory and doctors practising at the 3 levels of health care make an amusing study. At the tertiary level, laboratories have sufficient recognition and work, since most of the patients attending institutions at this level are either very ill, literate or affluent. At the secondary level, the out-patient clinics are usually bursting at the seams. The few doctors available are grossly overburdened with work and the patients, after standing in seemingly unending queues, in hot crowded waiting rooms, easily develop the "Fast Food Store" syndrome. They wish to spend as short a time as possible with the doctor, get a prescription, collect some drugs or take an injection, and off they go. Of course, the doctors often oblige. Here, the laboratories do not get as busy as one would expect. In any case, they are not even equipped to handle large numbers of specimens. At the primary level, the degree of illiteracy and the presence of the traditional doctor who examines, diagnoses and treats without the help of x-ray, laboratory or pharmacy, place the clinical laboratory in the category of a luxury facility.

It has been argued that to achieve Health for All, what the developing countries need to do is eradicate endemic and parasitic disease and implement an effective Primary Health Care in all rural areas. If this is so, then the question arises - Should Clinical Chemists in the 3rd World strive to attain comparable training, knowledge and expertise as their counterparts in developed countries? Certainly the opportunities differ on both sides, but the impact and role of Clinical Chemists in the Health Care delivery of their particular communities should bear comparable relevance.

Table 8. Some diseases affecting development of Clin-Chem. laboratory services in Nigeria

1.	Financial	-	Kwashiorkor
2.	Manpower	-	Anemia
3.	Diarrhea of inappropriate technology		

Since the duty of a Clinical Chemist is essentially to provide relevant service, i.e. service that can be understood, appreciated and beneficially utilised by its consumers, he must be sufficiently innovative to design and implement test programmes and methods that will blend with, and complement community needs and aspirations. Additionally, he must be alert to the tremendous upsurge in technological advance in the developed world and also concern himself with the world wide changing patterns of disease. It is illogical to think that certain diseases are confined to the developed world and would not affect developing countries.

Since Primary Health Care is a priority in Nigeria, and it is well recognised that there will be little need for Clinical Chemistry investigations at this level due to the patterns of prevailing disease, Clinical Chemists are therefore provided with a challenge. We need to define more closely how the Nigerian Clinical Chemist could become more useful and relevant to the needs of his society at Primary Health Care level. We cannot afford to propagate the notion that Clinical Chemistry is only of significance to the "Affluent" in urban centres. We are therefore faced with some hard choices. It would be unwise to abandon specialisation in favour of rural health work, yet, we would be doomed if we failed to fortify ourselves with the understanding of modern day advances like Immunology, Dry Chemistry Systems, Microchip technology, and Computerisation. We must be careful not to confine ourselves to the role of "solution mixers".

In analysing the causes of poor development of the Clinical Chemistry laboratory in Nigeria, there are several factors to be considered. Table 9 only illustrates 3 of them. The Clinical Chemist is virtually powerless in curing the first disease - lack of finance. He can be instrumental in curing the manpower anemia. I believe that he should assert himself boldly and dispassionately to put a stop to the third disease - diarrhea of inappropriate technology.

CONCLUSION

I admit that aid to developing countries is laudable, but I contend that unless aid is a means to an end of suffering and not merely a temporary or political relief, it could become counter-productive because it might lure the beneficiaries into a state of false security. The time has now come for International Health Organisations and developed nations to shift from their present position of regular aid-providers to that of collaborating partners in the permanent development and establishment of health sector technology in developing countries.

ACKNOWLEDGEMENT

I am grateful to Dr. B.A. Dada, Chief Consultant Statistician and Dr. Okunsanya, Senior Consultant of the Federal Ministry of Health, Lagos for their kind assistance.

CHAPTER 15
LIPIDS

Clinical significance of apolipoprotein measurements
J. Fruchart, J.M. Bard, C. Fievet, P. Duriez, and P. Puchois

Metabolic and endocrine aspects of cardiovascular risk factors. Observations on lipid metabolism in two "experiments of nature" and one experiment of human nature
J.A. Gevers Leuven, L.W. Hessel, M. Smit, and L. Havekes

CLINICAL SIGNIFICANCE OF LIPOPROTEIN PARTICLES

J.C. Fruchart, J.M. Bard, C. Fievet, P. Duriez and P. Puchois

SERLIA, Institut Pasteur, 1 rue du Pr. Calmette, 59019 Lille Cédex, France
Département Lipides Lipoprotéines et U. INSERM 279, Faculté de Pharmacie, 3 rue du Pr. Laguesse, 59045 Lille Cédex France

INTRODUCTION

Apolipoproteins, the protein components of the lipoproteins have a variety of structural and metabolic roles related to the metabolism of lipids. They are involved in determining the transfer of lipids from one lipoprotein particle to another and act as cofactors, activators and inhibitors of enzymes involved in the synthesis and degradation of lipoprotein particles (1). Additionally, the apolipoproteins are necessary for the recognition by cell-surface receptor sites which specifically promote the catabolism of the various lipoprotein fractions (2).

This led to their use as specific markers for classifying lipoprotein species. According to the classification system proposed by P. Alaupovic (3, 4) plasma lipoproteins consist of a mixture of particles containing lipids with one, two ore more apolipoproteins.

The quantification of these discrete lipoprotein particles causes important analytical problems. Due to their specificity and elegance the immunological procedures represent a logical choice for technical innovation in this field. Here we describe new methodologies for the molecular analysis of lipoprotein particles and discuss the clinical significance of this approach.

METHODS

Mapping of lipoprotein particles was obtained with an enzyme-linked differential antibody immunosorbent assay (5, 6). Stable mouse or rat hybridoma cell lines which produces monoclonal antibodies to human apolipoproteins B, A-I, A-II, C-III, E, (a) have been developed. These antibodies have been purified and characterized. We have partially localized the epitopes they recognize. We have also produced oligoclonal antibodies which are specific of certain regions of human apolipoproteins (7).

(a)

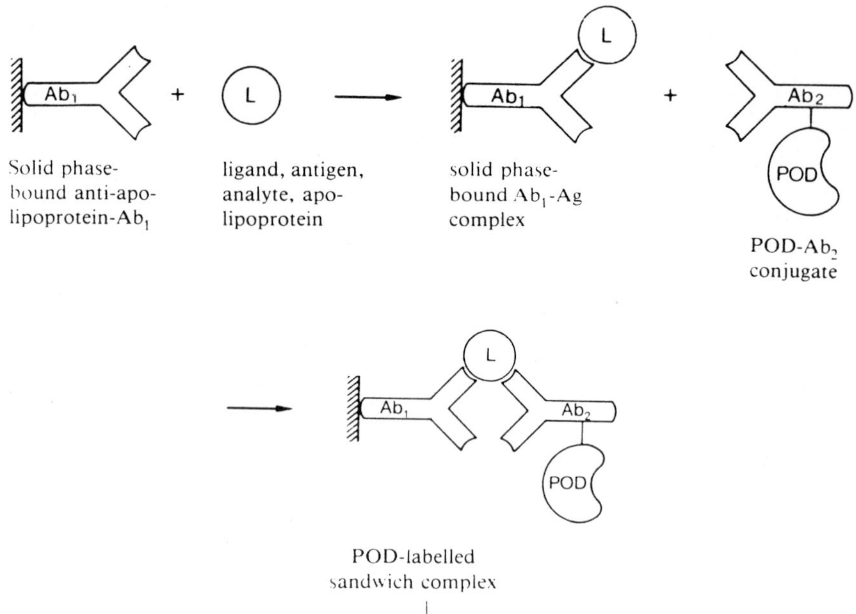

(b) $$2\ 1{,}2\text{-Phenylenediamine} + 2\,H_2O_2 \longrightarrow 2{,}2'\text{-diamino-azobenzene} + 4\,H_2O\,.$$

Figure 1: Principle.

Principle

Immunoenzymometric assay was performed with coated microtiter plates as the solid phase antibody and enzyme-labelled antibodies as the conjugate (Figure 1).

Lipoprotein particles react with immobilized antibody Ab1 which recognizes an epitope present on each lipoprotein particle (pan-antibody) or a specific epitope present on one species and absent from another. The solid phase is washed to remove unbound particles. Excess enzyme-labelled antibodies Ab2, which can be directed against the apolipoprotein recognized by Ab1 or against another apolipoprotein present on the same particle, are added and react with bound antigens. The solid phase is then washed to remove unbound labelled antibodies. In the enzymatic indicator reaction solid phase bound enzyme is measured; it is directly related to the epitope expression of one or two apolipoproteins on the surface of lipoprotein particles.

Separation of lipoprotein particles according to their apolipoprotein composition

Using a two site differential immunoenzymometric assay, we were able to measure in whole sera different apolipoprotein associations. For example, we have used this method for measuring apo A-I associated and unassociated with apo A-II in plasma (8). To directly determine associated apo A-I, we coated microtiter plates with antibody to apo A-II, blocked the nonspecific binding sites, and incubated the plate with plasma, immobilizing the lipoprotein particles containing both apo A-II and apo A-I. The unbound constituents of plasma were washed away, peroxidase-labeled antibody to apo A-I was added, the plate rewashed, peroxidase substrate added, and the resulting color measured. Apo A-I unassociated with apo A-II was evaluated by subtracting the concentration of associated apo A-I from the total apo A-I concentration.

We have also recently applied this procedure to measure apo B associated and nonassociated with apo C-III, apo E and apo (a).

Separation of lipoprotein particles according to the epitope expression of one apolipoprotein to their surface

Monoclonal or oligoclonal antibodies can be used as Ab1 in our immunoenzymometric assay and allow the separation of lipoproteins that might be similar by physiochemical criteria but different according to the epitope expression of one apolipoprotein on their surface (6, 9).

RESULTS AND DISCUSSION

Measurement of lipoprotein particles according to their apolipoprotein composition. Clinical significance

Separation of HDL particles. Apo A-I exists in at least two different types of lipoprotein particles in human plasma (3). More than half of the total apo A-I is associated with apo A-II; the rest occurs in lipoproteins that contain no apo A-II. These two populations of particles apparently represent two metabolically distinct pools of apo A-I containing lipoproteins and we have shown recently that cholesterol efflux in culture adipose cells is mediated by LpA-I particles but not by LpA-I:A-II particles, a difference that could be clinically important (10).

Both lipoprotein forms of apo A-I were detected mainly in HDL but also in all major density classes. The relative proportion of LpA-I was higher in HDL_2 than in HDL_3 (8).

In order to elucidate to what degree the HDL decrease observed in coronary artery disease affects these two types of lipoprotein particles, LpA-I and LpA-I:A-II were measured in plasma from one hundred and fifty male subjects matched for age and clinical data (11). Fifty patients referred as the CAD group had significant coronary disease with 50% or higher grade stenosis in one or more major coronary arteries. Fifty patients referred as the no CAD group had normal coronary angiograms. The other 50 subjects were healthy volunteers. Only subjects with normal plasma lipid levels were included in this study.

Results showed that the lower apo A-I levels for CAD subjects were reflecting in fact a decrease of LpA-I particles (Table 1). This observation provides strong support for the proposal that the decrease of HDL in CAD affects LpA-I and not LpA-I:A-II. These results could be related to other studies where the decrease of HDL level in CAD has been ascribed to HDL_2 subfraction richer in LpA-I than HDL_3.

The precise relationship between plasma LpA-I particle concentrations and the development of atherosclerosis remains to be elucidated. However, two potential mechanisms are worthy of consideration. First, it may be that the appearance of LpA-I particles is a reflection of triglyceride rich particle lipolysis in the plasma. This proposal would link atherosclerotic risk to defective clearance of plasma triglycerides and certainly is consistent with the negative correlation which we found between LpA-I particle concentration and plasma triglycerides. Alternatively, it is possible that the anti-atherogenic potential of LpA-I particles may reside in their putative role in reverse cholesterol transport. Indeed we have shown recently that the efflux of sterol from cultured cells was mediated

Table 1: Lipoprotein and apolipoprotein levels in angiographically proven CAD subjects (Group I), angiographically proven CAD free subjects (Group II) and controls (Group III)

Levels (mg/dl)	Group I	Group II	Group III
Total cholesterol	178.3±49.5	189.7±41.8	190.9±43.1
Triglycerides	103.6±36.8	108.2±39.3	101.8±34.2
HDL cholesterol	43.8±13.5	55.1±12.6**	59.2±23.6**
Total apo A-I	104.5±19.5	112.3±20.4	130.9±28.0**
Apo A-I bound to apo A-II	74.8±17.2	71.5±13.0	79.4±14.3
Apo A-I not bound to apo A-II	29.6±17.1	40.7±16.7*	51.5±22.2**
Apo A-I not bound to apo A-II/Apo A-I[a]	27.6±14.6	35.5±10.6**	38.0±10.7**

* Significant difference at the 0.005 level of significance
** Significant difference at the 0.001 level of significance
[a] Values are expressed in percent

by LpA-I (10). Recently Albers et al. reported that LCAT and CETP are mainly transported by LpA-I (12). Together with our present data, these studies suggest that LpA-I might be involved in reverse cholesterol transport.

An other example illustrates the clinical interest of this approach. Due to its effect on HDL cholesterol, it has been suggested that moderate alcohol consumption may be protective against coronary artery disease (13). In order to evaluate alcohol effect on apo A-I containing particles, we have measured LpA-I and LpA-I:A-II in plasma from three hundred and fifty male subjects, matched for age and clinical data (14).

These subjects have been divided in five groups according to their alcohol consumption. This classification made by a dietician has been checked using GGT values.

Results confirmed that alcohol consumption increases HDL cholesterol, total apo A-I and apo A-II but more remarkable is the fact that alcohol increases LpA-I:A-II and decreases LpA-I. These opposite variations are dose dependent and the differences have been found highly significant. So, our findings indicate that if alcohol consumption has an anti-atherogenic effect, it could not be through the increase of HDL. This study emphasizes also that the study of lipoprotein particles defined by their apolipoprotein composition are promising for future clinical, pharmacological and epidemiological studies.

Separation of apo B containing particles. Apo B exists in different types of particles in human plasma. In normal fasting plasma, more than 50% of the total apo B occurs in particles which contain a single apolipoprotein (LpB). The rest is associated with other apolipoproteins. The physicochemically defined lipoproteins as VLDL, IDL or LDL were found to be heterogeneous with respect to this concept and a particle as LpB may occur in any segment of the density spectrum depending on the composition and content of its lipid complement. Reliable quantification of different apo B containing lipoproteins is essential for further clarification of measuring apo B.

Lipoprotein particles containing B and C-III, B and E or B and (a) can be quantified using an enzyme linked differential antibody immunosorbent assay. Monoclonal antibodies against C-III, E and (a) were used as solid phase antibody and apo B present in the retained particles was evaluated using peroxidase labelled anti apo B monoclonal antibodies.

Quantification of lipoproteins containing apolipoproteins C-III and B (LpC-III:B) or E and B (LpE:B). In a recent study (15), we have demonstrated that type III dyslipoproteinemia is characterized by a drastically increased concentration of LpE:B and slightly higher levels of LpC-III:B. To a less extent, these lipoproteins are also increased in type IIa dyslipoproteinemia (16).

Since accelerated atherosclerosis occurs in long term dialysis patients, we have quantified apo B containing lipoprotein particles in 49 patients and 49 matched controls (17). Total apo B was significantly reduced in patients with chronic renal failure when compared to controls but, in contrast, lipoprotein particles containing B and C-III or B and (a) were highly increased (17).

Measurement of lipoprotein particles according to the epitope expression of one apolipoprotein to their surface

Use of monoclonal antibodies for the detection of genetic polymorphism in human apolipoproteins. We recently developed a monoclonal antibody against LDL called BIP 45 which interacted with apo B with different binding affinities (9). A non competitive enzyme-linked immunoassay was used to measure the amount of BIP 45 epitopes and total apo B in a group of 244 male healthy volunteers and 11 families. Binding ratio R was defined as the amount of apparent apo B bound to monoclonal BIP 45 divided by the amount bound to polyclonal antibody. In the reference population, the profile of apparent apo B concentration determined by BIP 45 is clearly not monophasic. 114 plasmas (46.7%) with ratio $R < 0.6$ bound BIP 45 very poorly and were designated weak reactors. 109 plasmas (44.7%) has $0.6 < R < 1.8$ and were designated intermediate reactors. 21 plasmas (8.6%) with $R > 1.8$ were designated strong reactors. Competitive ELISA method confirmed that apo B containing particles from different subjects exhibited low, intermediate and high affinity to the antibodies.

The existence of three binding patterns to BIP 45 could be explained by the existence of two common apo B alleles designated BIP(+) and BIP(-). Allele BIP(+) would code for the epitope recognized by the antibody. Thus subjects with the high affinity for BIP 45 (strong reactors) would be homozygote (genotype BIP(+), BIP(+)); subjects with low affinity for BIP 45 would be homozygote (genotype BIP(-), BIP(-)) and subjects with intermediate affinity would be heterozygote (genotype BIP(-), BIP(+)). Statistical analysis indicated that the experimental reference population is in agreement with the Hardy-Weinberg equilibrium. So the distribution of the apo B BIP 45 is compatible with the genetic transmission of 2 co-dominant alleles.

One interesting finding of this study is that the gene frequency for BIP(-) BIP(+) allele detected by BIP 45 corresponds to the gene frequency at the Ag(g)/Ag(c) locus that was previously found by Butler using human antisera in immunodiffusion or passive haemagglutination methods (18). This led us to examine with family studies if the binding patterns of apo B containing lipoprotein particles against BIP 45 are really genetically transitted and led us to study the relation between these binding patterns and the Ag system.

The inheritance of apo B reactivity against BIP 45 was investigated with the sandwich or the competitive ELISA methods in eleven families. In 5 families, all parents were found to be weak reactors and the 13 children were found to be weak reactors. In 3 families, one parent was found to be an intermediate reactor and the other one a weak reactor; then 3 out of the 7 children were found to be intermediate reactors; 1 out of the 8 children was found to be a weak reactor, 4 intermediate reactors and 3 strong reactors. These data support the hypothesis that BIP 45 identifies an epitope on apo B determined by 2 apo B alleles which are inherited in a co-dominant fashion.

The immunoreactivity against BIP 45 of 30 plasmas classified in the Ag system was studied with the competitive ELISA method. Results suggest that BIP 45 is an anti Ag(c) reacting strongly with the apo B carrying the Ag(c) factor but weakly with the apo B containing the Ag(g) factor. The human Ag(c) factor seems to be highly immunogenic in mice since Tikkanen et al. have reported that an other anti apo B monoclonal antibody called MB 19 specifically binds to the factor. The competitive binding study of BIP 45 and MB 19 was performed and results indicated that BIP 45 is specifically and highly displaced by MB 19. So, the two antibodies recognize closely related expression on apo B.

The apo B genetic polymorphism detected by BIP 45 is not associated with a particular lipoprotein level in the reference population. However, by testing a population undergoing coronary angiography, we have found that number of patients with intermediate binding was increased in patients with coronary artery disease (9).

Use of monoclonal antibodies for the evaluation of apo B epitope's expression on lipoproteins. As any single epitope of an apolipoprotein may be more or less exposed, depending on which particle one is examining, Mabs separate lipoproteins that might be similar by physicochemical criteria. Three selected Mabs to apolipoprotein B, called BL3, BL5 and BL7 and previously characterized (19, 20) have been tested. If these Mabs are coated, the use of labelled polyclonal antibodies to apolipoprotein B allows to probe the expression of apolipoprotein B epitopes on lipoproteins in different situations. To evaluate the potential correlation between specific epitopes on B lipoproteins subpopulations and predisposition to coronary artery disease, we have quantitated lipoprotein particles recognized by BL3, BL5 and BL7 in normal and angiographed patients.

Male subjects tested, matched for age and clinical data were:

- 42 patients with atypical chest pain and normal coronaries;
- 101 patients with coronary artery disease as indicated by fifty per cent or higher grade stenosis in one or more major coronary arteries;
- 104 volunteers.

Patients with coronary artery disease compared to controls had a significant increase in all B particles but probit discriminant analysis, performed according to Albert and Chapelle (21) showed that particles screened by antibody BL3 contribute to a better discrimination of the three groups (20, 22) (Table 2). LpBL3 represent the parameter with the best predictive power and allows alone a good classification of 77% of the patients. So, it is possible that some subpopulations of apo B containing lipoproteins are more abundant in atherosclerotic patients and can be distinguished immunochemically from others.

Use of an oligoclonal antibody for the direct evaluation of pro-apolipoproteins in serum. The fact that synthetic peptides copying a part

Table 2: Univariate predictive powers of apolipoprotein band of lipoprotein particles recognized by different monoclonal antibodies

	Predictive power X2	Error rate (%)
LpBL3	25.02	23.46
LpBL5	20.56	29.63
LpBL7	13.76	37.04
Apo B	10.10	39.02

LpBL3: Particles screened by antibody BL3
LpBL5: Particles screened by antibody BL5
LpBL7: Particles screened by antibody BL7

of a protein sequence can elicit antibodies capable of reacting with the whole protein has received considerable attention in the recent years. Although this approach is suitable for any protein, even those sequence has only been deduced from recombinant DNA studies, it may be put to particular effect as a means of differentiating closely related proteins which vary from each other in only part of their primary sequence. In such cases, immunization with the complete protein will elicit cross-reactive antibodies, with the use of synthetic peptides which mimics the most divergent domains will allow the production of specific antibodies (7). The determination of proproteins as pro-apo A-I is particularly relevant to this method (23).

It is known that pro-apo A-I represents the intracellular form of the protein which is converted in plasma to apo A-I, the main protein of HDL, commonly considered as a negative marker of risk of coronary heart disease. Determination of pro-apo A-I might therefore improve our understanding of processes which lead to low plasma levels of HDL. Pro-apo A-I is secreted by human intestine and liver and is two charges units more basic than the major plasma apo A-I. It contains a six amino-acid-long-N-terminal extension with the sequence Arg-His-Phe-Trp-Gln-Gln. In human plasma this pro-apo A-I represents only 2 to 5% of the total apo A-I since it is rapidly converted into the "mature" form by proteolytic removal of the hexapeptide prosequence. Until now, levels of pro-apo A-I have been determined by measurement of radioactivity after two dimensional electrophoresis of labeled delipidated lipoproteins. This technique presents disadvantages. It's a complex technique, time consuming which requires the use of radioisotopes.

We have used an oligoclonal antibody for the direct evaluation of human pro-apo A-I in serum. Although antiprotein-antibodies have in some cases been successfully raised using short peptide of only 5 or 6 residues in length, our experience has shown that longer peptides are better candidates.

So, we decided to elongate the [1-6] peptide with 3 amino acids [7-9] found in mature apo A-I. Using this approach, the immunizing peptide then consisted of 9 amino acids. Since only 3 of these were in common with the leader tripeptide of apo A-I, it was considered unlikely that the nonapeptide would elicit antibodies which would cross react with the mature A-I protein to any significant extend.

The peptide was coupled to tetanus toxoid using carbodimide activation of the alpha-carboxyl-group of Pro_9 or the side chains of Asp_7 and Glu_8.

In each case, the linkage occurred at the C-terminus of the peptide and the N-terminus of the peptide was thus exposed in a similar manner as in the native protein. Rabbit immunization with this conjugated peptide [1-9] resulted in the production of specific anti pro-apo A-I antibodies as demonstrated by immunoblotting after isoelectrofusing of apo HDL (23).

The anti pro-apo A-I antibody prepared enabled us to develop an immunoassay of pro-apo A-I. In this non competitive enzymeimmunoassay, oligoclonal anti pro-apo A-I antibodies were coated to the plate and polyclonal anti apo A-I were used as labelled antibodies. This assay requires a delipidation step, indicating that, when localized inside lipoproteins, the NH_2-terminus of pro-apo A-I is not accessible to antipeptide antibodies. The epitope however can be exposed by delipidation. When varying amounts of apo HDL standard were added to delipidated plasma of known pro-apo A-I, recoveries ranged from 94.6 to 106.5%. The intra- and inter-assay coefficients of variation were respectively 3.8% and 7.9%. The average level of pro-apo A-I in the plasma of normolipidemic subjects determined by ELISA was 88 ± 15 μg/ml. Non significant difference was observed between female and male subjects.

The mean ratio of pro-apo A-I was 7.1%. This is higher than the ratio found by others. A possible explanation for this discrepancy is the difference in the assay method. Values usually found by others were determined by technique which may allow in vitro pro-apo A-I conversion and degradation to occur.

Our enzymeimmunoassay allows rapid, simple and sensitive quantitation of pro-apo A-I on a large scale. We have applied this assay in different physiopathological states where pro-apo A-I synthesis and conversion could be impaired or accelerated. In the sample of a Tangier patient the ratio of pro-apo A-I to apo A-I was 29%. In an other hand, this ratio was lower in patients with cirrhosis (mean value: 3.5%; n = 30).

CONCLUSION

Lipoproteins represent a class of extremely heterogeneous immunogens. According to Alaupovic's concept, they consist of a mixture of particles differing by their apolipoprotein composition and reflecting the totality of metabolic reactions of all components of the system.

However, a difficulty with this classification system is the reversible nature of the interaction of all apolipoproteins except apo B which particles composition may occur during the isolation of lipoproteins. Moreover, the dissociation of apolipoproteins from lipoprotein particles as well as the mixed or self-association of different apolipoproteins cause difficulties for accurate interpretation of the results obtained on current techniques used in the separation of lipoproteins.

Antibody mapping of lipoprotein particles with an immunoenzymometric assay offers the unique opportunity to separate particles in a mild way and limits the possible risks for alteration of lipoproteins.

Another feature of apolipoproteins, dealing with their molecular properties, pose important analytical problems for the quantitation of lipoprotein particles. Apolipoproteins have major changes in molecular structure and epitope expression as a function of the environment. Lipids have a very important influence for maintaining epitopes in a given conformation. The presence of other apolipoproteins may also affect epitope expression. The development of techniques to produce highly specific antibodies and their use in immunoenzymometric assay allow to separate particles with an epitope expressed on the surface from those in which the epitope is masked or altered in conformation.

Although numerous studies have confirmed the immunochemical heterogeneity of human plasma lipoproteins, the origins, fate, possible inter-conversions and clinical significance of lipoprotein species separated according to their apolipoprotein composition remain obscure.

The potential correlation that we have shown between lipoprotein subpopulations and predisposition to coronary artery disease seems to confirm that apolipoproteins should be used as specific markers for the definition and classification of plasma lipoproteins.

REFERENCES

1. Mahley RW, Innerarity TL, Rall SC, Weisgraber KH. Plasma lipoproteins: apolipoprotein structure and function. J Lipid Res 1984; 25: 1277-1291.
2. Gianturco SH, Via DP, Gotto AM, Bradley WA. Lipoproteins receptors. Role of receptors in biology and medicine, 1986; 219/16: 167-192.
3. Alaupovic P. The physicochemical and immunological heterogeneity of human plasma high-density lipoproteins. In: Miller NE, Miller GJ, eds. Clinical and Metabolic Aspects of High-Density Lipoproteins. Amsterdam: Elseviers Science Publishers, 1984.
4. Alaupovic P. The concepts, classification systems and nomenclature of human plasma lipoproteins. In: Lewis LA, Opplt JJ, eds. Lipoproteins: Basic Principles and Concepts. Boca Raton, Florida: CRC Press, Inc, 1980.
5. Fruchart JC, Fievet C, Puchois P. Apo-lipoproteins. In: Bergmeyer HU, ed. Methods of Enzymatic Analysis. Weinheim: VCH, 1985.
6. Fruchart JC. Polyclonal, oligoclonal and monoclonal antibody mapping of lipoprotein particles. In: Fidge NH, Nestel PJ, eds. Atherosclerosis VII. Elseviers Science Publishers B.V., 1986.
7. Gesquière JC, Delpierre C, Tartar A, Cachera C, Equagoo K, Fruchart JC. Synthetic peptides as alternative antigens in the production of antibodies against human apolipoproteins. Clin Chem 1985; 31/5: 784.
8. Koren E, Puchois P, Alaupovic P, Fesmire J, Kandoussi A, Fruchart JC. Quantification of two different types of apolipoprotein A-I containing lipoprotein particles in plasma by enzyme-linked differential-antibody immunosorbent assay. Clin Chem 1987; 33: 38-43.
9. Duriez P, Vu Dac N, Koffigan M, et al. Detection of human apolipoprotein B polymorphic species with one monoclonal antibody (BIP 45) against low density lipoprotein. Atherosclerosis 1987; in press.
10. Barbaras R, Puchois P, Fruchart JC, Ailhaud G. Cholesterol efflux from cultured adipose cells is mediated by LpA-I particles but not by LpA-I:A-II particles. Biochem Biophys Res Commun 1987; 142/1: 63-69.
11. Puchois P, Kandoussi A, Fievet P, et al. Apolipoprotein A-I containing lipoproteins in coronary artery disease. Atherosclerosis 1987; in press.
12. Cheung MC, Wolf AC, Blum KD, Tollepson JH, Albers JJ. Distribution and localization of lecithin: cholesterol acyltransferase and cholesteryl ester transfer activity in A-I containing lipoproteins. J Lipid Res 1986; 27: 1135.
13. Ferrence RG, Truscott S, Whitehead PC. Drinking and the prevention of coronary heart disease: findings, issues and public health policy. J Studies Alcohol 1986: 47/5.

14. Puchois P, Ghalim N, Demarquilly C, Zylberberg G, Fruchart JC. Effect of alcohol consumption on lipoprotein particles which contain apo A-I and apo A-II (LpA-I:A-II) and apo A-I but not A-II (LpA-I). The American Heart Association's, 59th Scientific Sessions, Dallas, 17-20 novembre 1986. Circulation 1986; 74/III: 383.
15. Fruchart JC, Davignon J, Bard JM, Grothe AM, Richard A, Fievet C. Effect of fenofibrate treatment on type III hyperlipoproteinemia. Am J Med 1987; in press.
16. Fruchart JC, Fievet C, Bard JM, et al. Effect of fenofibrate on lipoprotein particles predictive of coronary atherosclerosis. IX International Symposium on Drugs Affecting Lipid Metabolism, Florence, 22-25 octobre 1986; in press.
17. Parsy D, Dracon M, Cachera C, et al. Lipoprotein abnormalities in chronic hemodialysis patients. Nephrology Dialysis Transplantation 1987; in press.
18. Duriez P, Butler R, Tikkanen MJ, et al. A monoclonal antibody (BIP 45) detects AG (c, g) polymorphism of human apolipoprotein B. J Immunol Meth 1987; in press.
19. Salmon S, Goldstein S, Pastier D, et al. Monoclonal antibodies to low density lipoprotein used for the study of low- and very-low-density lipoproteins, in "ELISA" and immunoprecipitation technics. Biochem Biophys Res Commun 1984; 125/2: 704-711.
20. Fievet C, Demarquilly C, Luyeye I, Fievet P, Moschetto Y, Fruchart JC. Utilisation d'anticorps polyclonaux et monoclonaux pour le dépistage de l'athérosclérose. Intérêt de nouveaux marqueurs. Ann Biol Clin 1985; 43/4: 500-504.
21. Albert A, Chappele JP. Stepwise probit discrimination with specific application to short-term prognosis in acute myocardial infarction. Comput Biomed Res 1981; 14: 391-398.
22. Fruchart JC, Parra H, Kandoussi A, Fievet C, Lablanche JM, Bertrand M. Monoclonal antibody mapping of lipoprotein particles in the prediction of coronary atherosclerosis. The American Heart Association's, 58th Scientific Sessions, Washington, 11-14 novembre 1985. Circulation 1985; 72/4: 92.
23. Barkia A, Martin C, Puchois P, et al. Enzyme linked immunosorbant assay for human proapolipoprotein A-I using specific antibodies against synthetic peptide. J Lipid Res 1987; in press.

METABOLIC AND ENDOCRINE ASPECTS OF CARDIOVASCULAR RISK FACTORS: OBSERVATIONS ON LIPID METABOLISM IN TWO "EXPERIMENTS OF NATURE" AND ONE EXPERIMENT OF HUMAN CULTURE

J.A. Gevers Leuven *, L.W. Hessel **, M. Smit **, and L. Havekes *

TNO Gaubius Institute, Leiden, The Netherlands**
Sylvius Laboratories, Dept. of human Genetics, Leiden, The Netherlands**

INTRODUCTION

Amongst 250 possible risk factors (1), only a few can be assigned as causing atherosclerotic disease. The others are mere indicators, i.e. correlates or consequences of the disease.

In the first part of this chapter, attention will be paid to two correctable causative factors; an elevated level of cholesterol in circulating Low Density Lipoprotein (LDL) and an elevated level in circulating Intermediate Density Lipoprotein (IDL). Both LDL and IDL if increased, have a deleterious effect on the arterial wall. This effect is sadly demonstrated by Familial Hypercholesterolemia (FH) and Familial Dysbetalipoproteinemia (FD). These two inborn errors of metabolism show us how malfunctioning of one or two factors can cause atherosclerosis in man: in FH a receptor defect exists by which the LDL clearance is hampered (2) and in FD a ligand defect is present (mutated apo E) by which the clearance of lipoprotein core remnants, mainly IDL, is decreased (3). Both disorders are informative about normal lipoprotein metabolism and can therefore be called "Experiments of Nature".

In the second part we describe an "Experiment of Human Culture" where lipoprotein metabolism is altered by an initiative of healthy subjects, e.g. by taking oral contraceptives. It is well known that women have an advantage over men in terms of their prognosis for ischaemic heart disease. This has been attributed to estrogens, as the advantage disappears after menopause. On the other hand, the use of high dose oral contraceptives is associated with an increase in myocardial infarctions (4, 5). The effects of sex hormones on cardiovascular prognosis inspired our work on the metabolic basis of the association, in particular on the effect of oral contraceptives on lipid metabolism. In this study we investigated healthy women who started using these hormones.

In short, we present a case history of a patient with homozygous FH, a description of a group of patients with FD and a description of the effect of two low dose oral contraceptive preparations in healthy female subjects.

PART 1

1. About Familial Hypercholesterolemia (FH)

Up till now, more than ten mutations in the cellular LDL-receptor protein have been described that lead to an insufficiency of the LDL receptors. The term "Homozygous FH (HmFH)" may refer to two situations, namely:

1. the molecular picture i.e. coincidence of exactly the same mutation on each chromosome (e.g. with consanguinous parents);
2. the clinical and cellular picture which may be a result of the coincidence of two defects on the same locus, irrespective of whether these defects are the same or different ("compound heterozygosity").

The term "pseudo homozygous FH" is reserved for patients with either normal LDL receptor activity having the clinical picture (6), or for patients with the picture and with secondarily altered receptor function e.g. due to circulating anti-LDL-receptor antibodies (7). Higgins and co-workers have described a patient with normal receptor activity but with abnormal, not binding LDL (8).

The mutations can be classified into four classes (2):

1. receptor synthesis defect;
2. receptor transport defect (the translation products remain in the endoplasmatic reticulum);
3. receptor binding defect;
4. receptor clustering defect.

It is clear that the majority of HmFH patients will be (compound) heterozygous in molecular terms and this may explain in part the heterogeneity in the clinical picture.

Treatment of FH is directed at the lowering of cholesterol levels but HmFH patients are generally resistant to the usual dietary and pharmacological regimens. Effective treatments are lifelong plasmapheresis (9, 10) or specific LDL removal from plasma by LDL apheresis, using a column with a specific LDL-binding matrix (11) or employing Heparin-induced Extracorporeal LDL-Precipitation ("HELP") (12). Another possibility is portacaval shunting or even liver transplantation (13). The use of HMGCoA Reductase inhibitors, such as MK 733 (14) is not promising for HmFH patients, as the main effect of these drugs is exerted via induction of functioning LDL receptors.

CASE REPORT

Our patient was 10 years old when she was first seen in the clinic. Since infancy she had growing and disfiguring xanthomas in the skin of her elbows, buttocks, knees, and in many tendons and ligaments (see Figure 6). These xanthomas contained foam cells, i.e. macrophages loaded with lipid; especially with cholesterol esters. She had a heart murmur indicating an aorta stenosis as a consequence of severe atherosclerosis. When she was 13, she had no angina pectoris but serious signs of myocardial ischaemia were present in the exercise electrocardiogram. Meanwhile, her serum cholesterol levels ranged between 17.9-29.3 mmol/l (mean 22.1) with normal triglyceride levels. Both parents had hypercholesterolemia.

Figures 1-4 give more documentation of this patient. Figure 1 shows, that the ultracentrifugation fraction with a density between 1.019 and

1.063 g/ml, known as LDL, was able to suppress the cholesterol synthesis in normal fibroblasts. No difference was found between normal and HmFH LDL in this respect. However, in HmFH-fibroblasts normal LDL suppressed cholesterol synthesis only by 20% whereas HmFH-LDL did so by more than 70%. The first conclusion to be drawn from this experiment is, that the patient was not totally receptor negative. The second conclusion: her FH-LDL is different from normal LDL. We wondered what could cause the difference between normal and HmFH LDL in the ability of cholesterol suppression.

Figure 2 shows the results of an ELISA using anti-apolipoprotein antibodies. It is clear that HmFH-LDL contained apo E. It was also present in the LDL of a heterozygous FH patient but it was virtually absent in normal LDL (16).

Figure 3 shows, that all subfractions of LDL of this patient contained this apo E. In Figure 4 the apo E containing lipoproteins are shown. They were obtained from the LDL-density range behaved differently from LDL in immunoelectrophoresis. It thus seemed improbable that the apo E was present on the LDL particles themselves (as defined in metabolic terms). Rather, it was present on other lipoproteins having the same density range as LDL. This could be "Lipoprotein-E" or partly HDL-1 in the higher density fractions and partly IDL in the lower density fractions. As the lipoprotein carried no apo A-1, as revealed by the ELISA study (Figure 2c), the possibility of HDL-1 was excluded, leaving IDL as one probable candidate. The relatively large suppression of cholesterol synthesis in HmFH fibroblasts by HmFH-LDL was explained by the presence of apolipoprotein E, which is known to bind several times better to the LDL-receptor than does apo B (17).

Since the usual and some experimental therapeutic measures failed to ameliorate the picture (19), we performed a portocaval shunting operation according to Starzl (20).

Table 1 shows the effect of this operation on some parameters of lipid metabolism. Cholesterol levels were lowered by 41% and LDL-apo B by 26%. HDL and Apo A-1 increased and apo E decreased considerably.

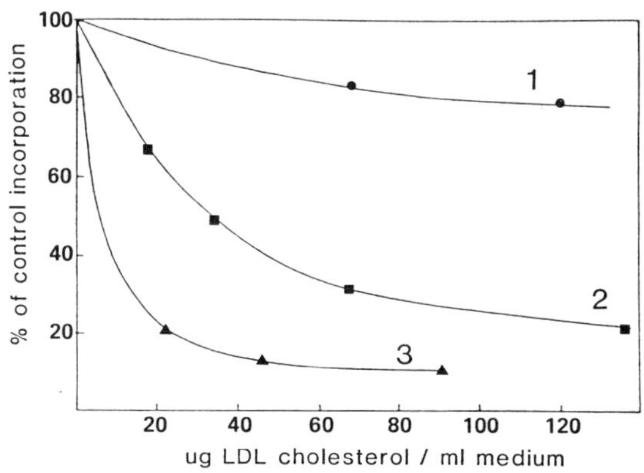

Fig. 1. The effect of increasing LDL cholesterol concentration on cholesterol synthesis in fibroblasts of a normal subject and from a patient with HmFH.
1. Patient fibroblasts with normal LDL.
2. Patient fibroblasts with patient LDL.
3. Normal fibroblasts with normal LDL or patient LDL (overlapping completely) (15).

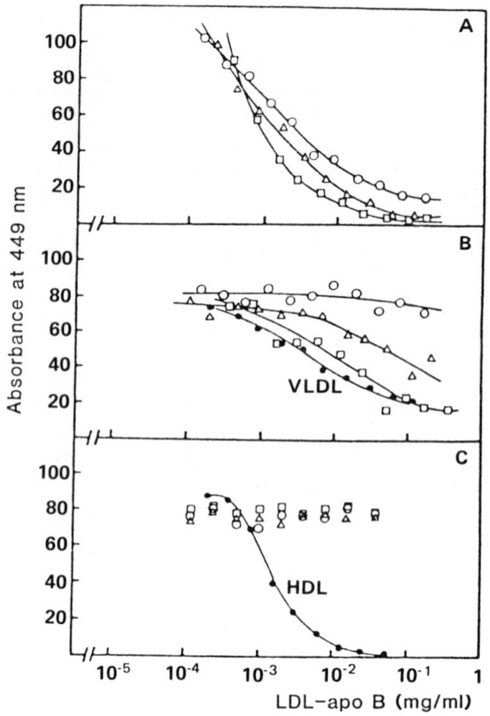

Fig. 2. Determination of apolipoproteins B, E and A-1 in normal and patient LDL density fraction by the ELISA technique. The LDL preparations (density 1.023 to 1.045 g/ml) were isolated by sequential ultracentrifugation.
(A) LDL coated; anti-apolipoprotein B serum. (B) Delipidated VLDL coated; anti-apolipoprotein E antiserum. (C) Delipidated HDL coated; anti-apolipoprotein A-1 antiserum. o---o, LDL isolated from a normal subject; □---□, LDL isolated from the homozygous familial hypercholesterolemic patient; Δ---Δ, LDL isolated from the heterozygous familial hypercholesterolemic patient; •-•, VLDL (in B) or HDL (in C) added instead of the respective LDL preparations (15).

In Figure 5 it is shown that the decrease in apo E is caused by a drop in apo E containing lipoproteins in the LDL density region. It should also be noted that the cholesterol concentration in the fractions at the lesser density side of the LDL peak had been diminished. The LDL peak itself was lower.

She had no apparent side effects from the operation. Her menses returned 3 weeks after operation and showed no difference compared with before. Her serum amino acid pattern remained normal and her performance at school did not reveal any symptom of hepatic encephalopathy. Her xanthomata decreased dramatically, as shown in Figure 6.

We conclude that in this girl with receptor defective HmFH the LDL density fraction contained apo E. This fact may explain why this fraction could suppress cholesterol synthesis in the patient's own fibroblasts in vitro more powerfully than normal did LDL. Portocaval shunt was successful and was accompanied by disappearance of apo E from the LDL density class.

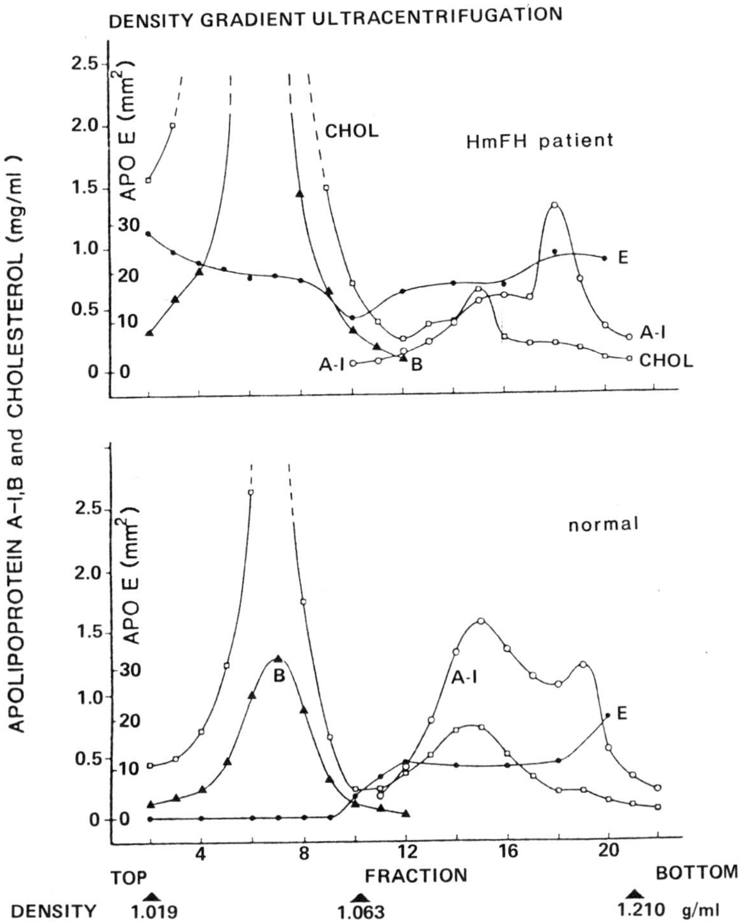

Fig. 3. The concentration profile of cholesterol apo B and Apo A_I and the relative concentration profile of apo E along a density gradient after ultracentrifugation, according to Redgrave et al. (18). Cholesterol -□-; apo B - Δ -; apo A_I - o - (mg/dl); apo E - ● - (mm^2 of RID).

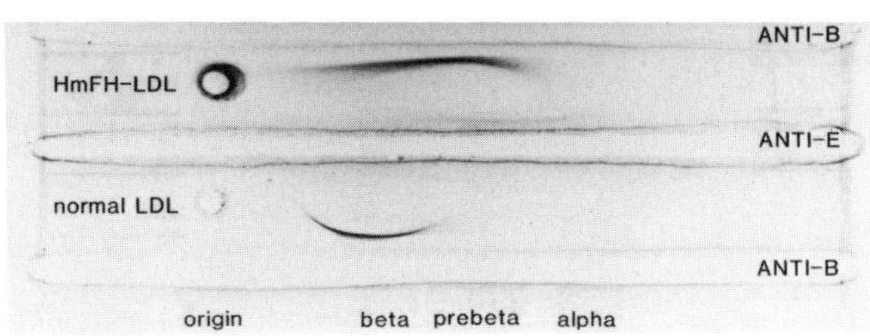

Fig. 4. Immunoelectrophoretogram of LDL (density 1.019 - 1.063 fraction) from the patient (HmFH-LDL) and of a normal control subject. Monospecific antisera against apo B and apo E were used for precipitation.

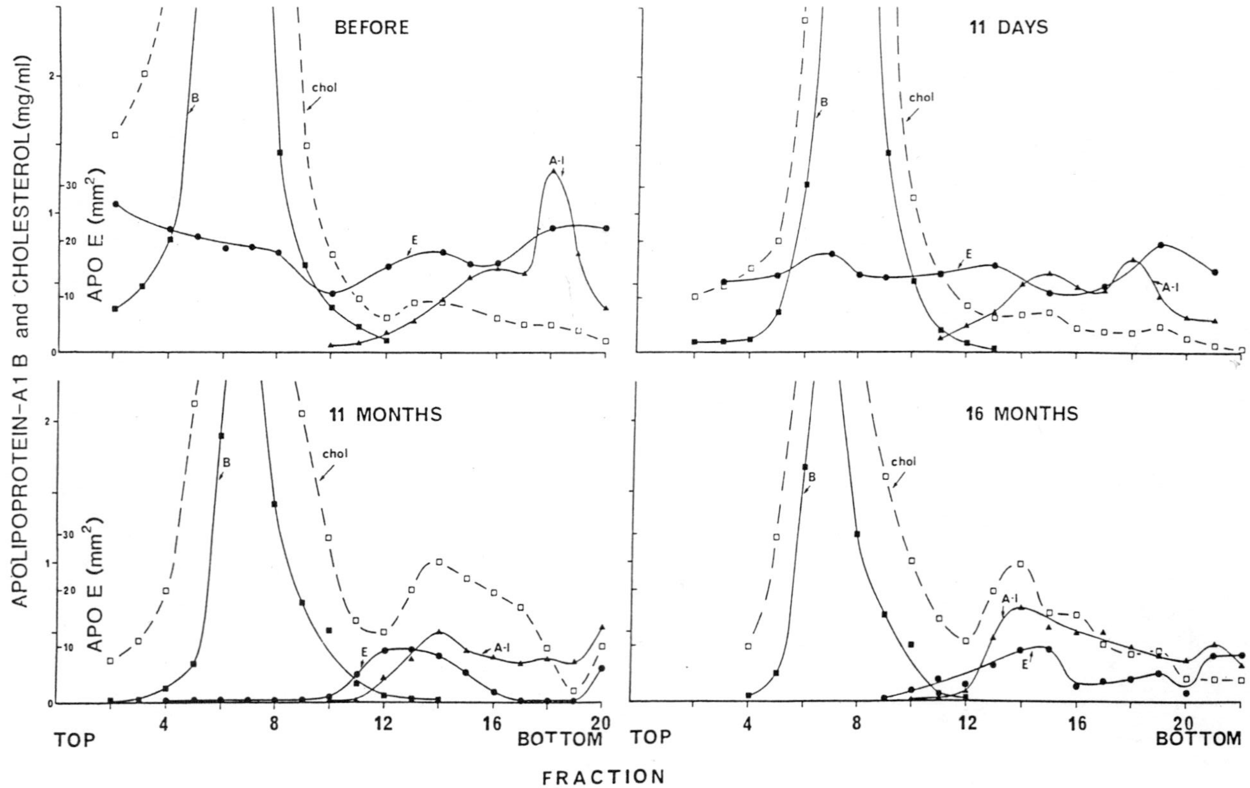

Fig. 5. Density gradient ultracentrifugation of plasma of a patient homozygous for Familial Hypercholecterolemia before and after protocaval shunt operation.
- □ - cholesterol; - ■ - apo B; - ● - apo E; - ▲ - apo A-1.

Table 1. Lipid and lipoprotein levels of HmFH patient MA before and after portocaval shunt operation

time	chol mmol/l	HDL-chol mmol/l	LDL-apo B mg/dl	apo A-I mg/dl	apo E % of pool*
before	21.9	0.85	259	79	178
11 days after	15.9		210	52	148
11 months after	13.4	1.03	198	109	115
16 months after	12.8	1.19	185	108	80

* Mixed plasmas from 5 apparently healthy donors.

2. About Familial Dysbetalipoproteinemia (FD)

FD was formerly called Hyperlipoproteinemia type III according to Fredrickson (3). Other names were "Broad Beta Disease" and "Floating Beta Disease". All these names were abandoned because they refer to sera and not primarily to patients.

The classical diagnostic criteria are based on signs and symptoms which result from accumulation of chylomicron- and VLDL-remnants in the plasma. Thus a good name for the disease would have been: "Remnant Removal Disease" but as it is not feasible to measure the remnant clearance rate in each patient, we still use the classical criteria for definition of the disease, i.e.:

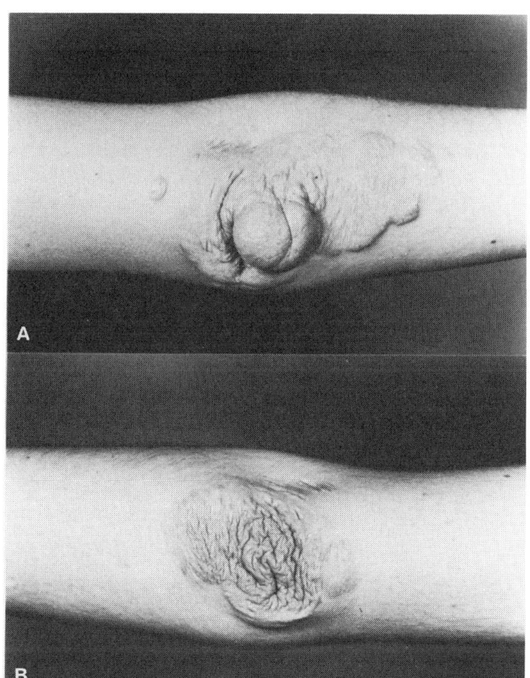

Fig. 6. Tuberous xanthomas on the right elbow of a 13 year old patient with HmFH before (A) and 2 years after portocaval shunt operation (B).

1. Xanthochromia striata palmaris. This is a yellowish colouring of the palm creases (21, 22). Sometimes these patients also have larger palmar xanthomas outside the creases. Moreover, tuberous xanthomas are often found on the elbows.
2. In the lipid stained free boundary electrophoretogram of whole serum a broad beta band is found instead of well separated beta- and prebetabands.
3. Lipoproteins with a density below 1.006 g/ml in postabsorptive serum have a beta mobility (so-called "floating beta's"). These lipoproteins have an elevated ratio of cholesterol over triglyceride.
4. The serum cholesterol- and triglyceride levels are increased.

None of these criteria has been used without questions: Morganroth et al. (23) found xanthochromia striata palmaris in only half of the patients. Stuyt et al. (24) found that 14 out of 19 patients of different families with FD had the symptom. Schneider (25) and Kameda et al. (26) described patients having a broad beta band in the absence of hyperlipoproteinemia and xanthomas, and Masket's patients had a broad beta band in only half of the cases (27).

In this paragraph, patients who had the classical picture of FD as defined by the four criteria mentioned above are compared with hyperlipidemic patients without the necessary characteristics (HLP-non FD). In this comparison the classical picture was taken as a "golden standard" to evaluate a newer clinical discriminator, namely a certain apolipoprotein E-phenotype.

Apo E is required for normal clearance of chylomicron- and VLDL-remnants by the liver (28). Figure 7 gives a schematic representation of the metabolic routes involved here.

Apo E shows a polymorphism in terms of iso-electric point (pI), of which the isoform with the lowest pI, apo E2, shows a defective binding to the LDL-receptor. Homozygosity of this abnormality causes a decreased clearance of remnants. This may cause disease only if the production of remnants is increased at the same time. Thus FD is the result of the coincidence of two factors: one is a ligand defect and the other is overproduction of VLDL. The latter abnormality causes the expression of the disease and is called "Hyperlipidemia Factor". Other precipitating factors for the disease are hypothyroidism, overweight, alcoholism, systemic lupus erythematodes etc.

In the normal population, three alleles on the same locus are known, corresponding with genes each coding for one of the isoforms of apo E,

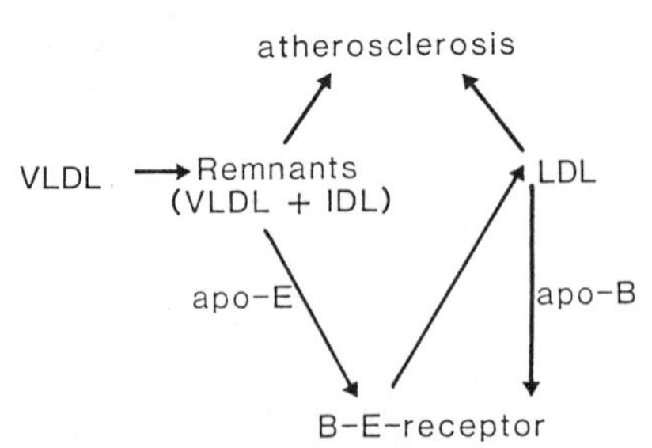

Fig. 7. Schematic view of the formation and clearance of two classes of atherogenic lipoproteins, VLDL remnants and LDL. Apo E is involved in the clearance of remnants, apo B is involved in the clearance of LDL.

called apo E4, E3 and E2 in order of decreasing pI. Thus, three homozygote forms are found, designated as E4/4, E3/3 and E2/2 and three heterozygous forms: E4/2, E4/3 and E3/2. The prevalence of E2/2 in the normal population is about 1%. In patients with manifest FD, however, the prevalence of apo E2/2 is near 100%. Since the prevalence of FD is only 1 : 2500, apparently, only a few percent of the E2/2 population has the "hyperlipidemia factor". In other words, E2/2 may cause an overproduction-hyperlipidemia to develop into FD.

The question arises: Is E2/2 an absolute prerequisite for FD? Are there false positives, i.e. patients with HLP-non-FD with apo E2/2, and are there false negatives, i.e. patients with FD without apo E2/2, i.e. with apo E3/2 or E3/3, 4/3, etc.? It is conceivable that other mutations can also cause insufficient binding with the receptor. They are not necessarily to be found by IEF. Conversely, it is not certain whether all apo E2 isoforms are defective in binding to the receptor. The false positives and false negatives in this respect are informative for the structure-function relationship of apo E.

To answer the questions raised in the previous paragraph, Table 2 shows the mutations in apo E described elsewhere, whether or not they are coexistent with FD. They are all exceedingly rare. If an isoform has a higher pI than apo E4 it receives a higher number, e.g. E5. If lower than E2, it receives a lower number, e.g. E1.

Screening for these mutants is tedious. It requires ultracentrifugation and other laborious steps to separate apo VLDL from other plasma consistuents?. Therefore, we developed a slabgel IEF followed by immunoblotting. This method is suitable for large scale use (40). Figure 8 shows the results in 19 subjects.

It can be seen that there are more than the one or two bands that may be expected on the basis of genetics. This is due to post-translational modification of the apo E molecules by sialic acid residues. These sialylized apo E-forms (so-called "minor forms") may superimpose over the focused unsialylized gene products and this may blur the phenotype. However, there are rules to guide: the first band, reading from basic to acid regions, consists always of a non-sialylized apo E isoform. Homozygous

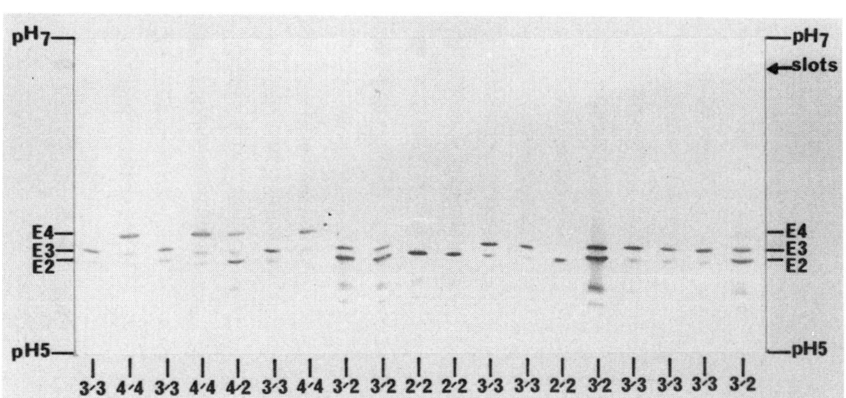

Fig. 8. Apo E immunoblot of isoelectric focusing slab gels applies in parallel with 19 different serum samples that had been delipidated with chloroform-methanol. Polyclonal anti-apo E antibodies were used as first antibody. The serum sample had been stored at -20°C for less than 2 weeks (40).

Table 2. Mutations in apo E

	Accompanied by HLP III	Primary structure of apo E3 wild type and its mutants 99	112	127	142	145	146	152	158..299
					docking site				
E3 (wild type)[a]	no	ala	cys	gly	arg	arg	lys	ala	arg
E4 (Cys_{112} arg)[b]	no		arg						
E2 (arg_{158} Cys)[c,a]	yes								cys
E2* =E2 (arg_{145} Cys)[a]	yes					cys			
E2**=E2 (lys_{146} gln)[d]	yes						gln		
E3 (ala_{99} thr)[e] (ala_{152} pro)	no	thr						pro	
E1 (gly_{127} asp)[f] (arg_{158} cys)	yes			asp					cys
E3 (Cys_{112} arg)[g] (arg_{142} cys)	yes		arg		cys				
E3 Leiden	yes		arg						
E1 Bethesda [h]	yes	contained 2 cystein residues							
E3 [i]	yes	contained 1 cystein residue							
E5 [j]	no	early atherosclerosis							
E7 [k]	no	early atherosclerosis							

a) 29, b) 30, c) 31, d) 32, e) 33, f) 34, g) 35, h) 36, i) 37, j) 38, k) 39.

pictures are easy to identify as their gene products always focus as a first band stronger than the others. If the first band is weaker than the following one, the picture is suggestive for heterozygosity, apo E3/2 and 4/3. In apo E4/2 the band between E4 and E2 is weaker than the other two. Apo E1 consists of sialylation products only (with rare exceptions, see Table 2).

An elegant technique to find point-mutants in cysteine residues is cysteamine modification (41), which causes all apo E isoforms containing cysteine to obtain a higher pI. Apo E2, containing two cysteine residues, acquires two extra basic groups and moves to the apo E4 place. Apo E3 does the same, taking only one step and apo E4 remains in its place.

One of our patients who fulfilled all criteria for FD had phenotype apo E3/3. This means he was false negative for this test (42). On cysteamine modification only small part of this apo E3 moved to the apo E4 place suggesting that part of the apo E3 either had one cysteine inaccessible for modification or that the patient was heterozygous for apo E3-wild type and a mutated kind of apo E3 in which cysteine was absent. We isolated apo E from VLDL, which contained only trace amounts of the cysteamine modifyable isoform. Amino acid analysis showed that the mutated protein did not, indeed, contain any cysteine, so this patient was probably heterozygous for a new mutant. The isolated apo E3 from VLDL did not bind to the receptor, as did apo E2 of another patient with FD (Figure 9, 42).

We concluded that this patient had a mutant apo E called apo E3 Leiden, which could be held responsible for the disease (15). The precise

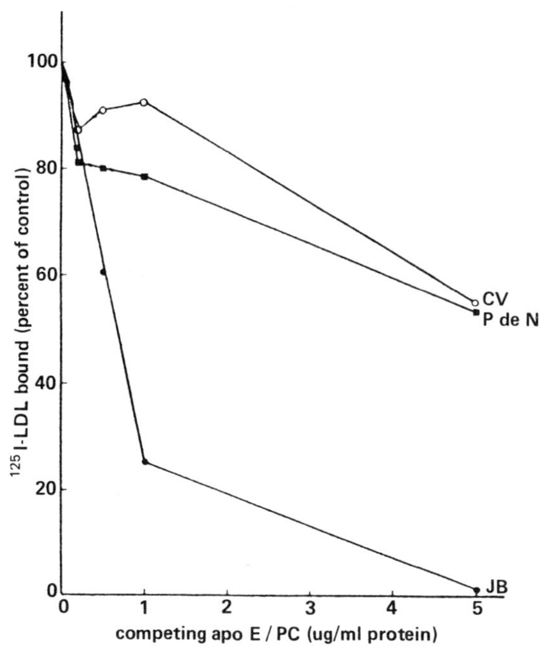

Fig. 9. Ability of apo E/PC complexes to compete with ^{125}I-labelled LDL (120 cpm/ng LDL protein) for binding to the specific LDL receptor on cultured human fibroblasts. The 100% value (high affinity binding of ^{125}I-LDL in the absence of apo E/PC complexes) was 30 ng of ^{125}I-LDL per mg cell protein. All measurements were performed in triplicate. ●-●: patient J.B., type IV; ■-■: patient P. de N., type III; o-o: patient C.V., propositus (42).

nature of this mutation is under investigation. The results found in this patient show that some patients may have genuine FD having an apo E phenotype indistinguishable from phenotype apo E3/3.

Three other probands with FD demonstrated apo E phenotype E3/2. In their families two of them had one sib with FD so we were able to study 5 patients with this combination. They were, like the proband with apo E3 Leiden, false negative for FD. We conclude that among hyperlipidemic patients the phenotype apo E2/2 does not have a 100% sensitivity for the diagnosis FD.

False positives (patients having apo E2/2 but showing HLP-non-FD) have also been found, so the criterium is also not 100% specific.

Chait and collaborators (43) showed that in FD the VLDL triglyceride production is increased. This has been confirmed by Stuyt (44), who found a doubled plasma triglyceride turnover rate in FD patients compared with normals. Family studies revealed that a close linkage exists between manifest FD and apo E2/2, so it can be postulated that if the hyperlipidemic-factor is at least partly genetic, its gene may be close to the apo E gene. This hypothesis was tested using recombinant-DNA techniques. The DNA of FD patients and of HLP-non-FD patients were digested with restriction enzymes. The Southern blots were hybridized with an apo E c-DNA probe. A Restriction Fragment Length Polymorphism (RFLP) was found using the enzyme HpaI (Figure 10).

The fragments reacting with the apo E probe were called H1 (60 kb) and H2 (20 kb). It appeared that the allele frequency of the fragment H2 was 0.95 in FD-patients and only 0.39 in controls (45): 37 of 39 patients with FD and with apo E2/2 phenotype have genotype H2H2 and 2 had H1H2, whereas of the normolipidemic subjects only 4 of 27 had H2H2; 13 had H1H2 and 10 had the H1H1 genotype. We postulated that the HpaI polymorphic site is in linkage disequilibrium with a gene involved in the generation of hyperlipidemia. We have localized the polymorphic site between the apo E and apo C-1 genes (Smit et al., in press). We are now sequencing the region around the polymorphic HpaI site.

It is by now probable that a specific genetic alteration is responsible for the hyperlipoproteinemia found in patients with the phenotype apo E2/2 (FD). Perhaps this polymorphic gene is somehow operative in many other patients with enhanced atherogenesis.

PART 2

About an "Experiment of Human Culture"

The use of oral contraceptives is a reliable method for family planning. However, in the sixties and seventies an association was found between the use of these hormones and the incidence of myocardial infarction. It has been hypothesized that the progestagenic influence might have been causal here (46). On the other hand, the use of estrogens has been given credit for prevention of heart attacks in postmenopausal age. However, the subject is controversial (47, 48). Several sex hormone sensitive factors may conceivably be involved here. Table 3 summarizes the direction of their change during treatment with androgens and estrogens.

It is known that LDL enhances atherogenesis and that elevated plasma HDL levels are associated with a better prognosis. As sex hormone-analogues have a substantial influence on these lipoproteins (49), we studied their change during use of two low dose oral contraceptive preparations, in relation to the changes that were induced in classical parameters of sex

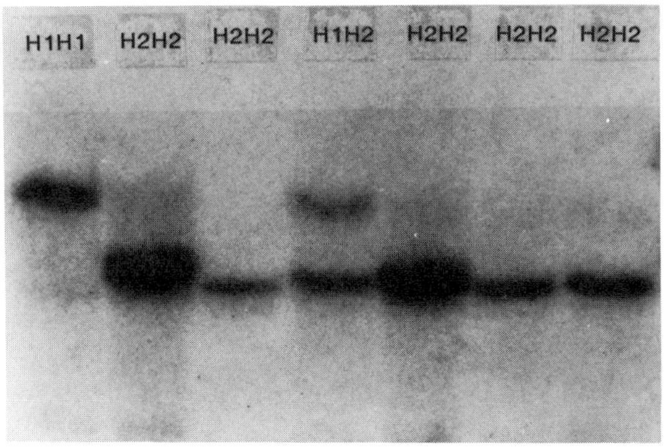

Fig. 10. Southern blot analysis of the Hpal polymorphism detected with apo E cDNA. The hybridisation pattern obtained from a Hpal digest of 10 μg DNA from six unrelated individuals is shown.

hormone-effects on the liver, Sex Hormone Binding Globulin concentration and ceruloplasmin activity. Furthermore, we also investigated the metabolic background of these changes by measuring the post heparin lipolytic activities (PHLA) which are said to influence the HDL level when estrogens or androgens are given separately (50). Table 3 gives the direction of the effects.

50 Apparently healthy women aged 21 ± 2 years (mean ± S.D.) completed this study. After randomization 26 received a preparation containing 37.5 μg ethinyl-estradiol (EE) and 750 μg lynestrenol (LYN). The other group received a preparation containing 30 μg EE and 150 μg levonorgestrel (LNG). A control group constituted 25 women not intending to take oral contraceptives. Blood samples were collected before and after three cycles. More detailed information is described elsewhere (51, 52).

Figure 11 shows the effect of these oral contraceptives on the concentration of lipoprotein constituents in plasma. The change in the hormone sensitive parameters shows that no androgenic effect could be

Table 3. Effect of agents with androgenic or estrogenic properties in the concentration or activity of various parameters in human plasma, according to the literature

	androgenic	estrogenic
SHBG	↓	↑
Cerulo	=	↑
HDL-chol	↓	↑
Apo AI	↓	↑
Apo AII	↓	↑
PH-LPL	=	=
PH-LLA	↑	↓

Abbreviations: SHBG = Sex Hormone Binding Protein; Cerulo = Ceruloplasmin activity; HDL chol = High Density Lipoprotein cholesterol; Apo AI = Apolipoprotein AI; PH-LPL = postheparine lipoprotein lipase activity; PH-LLA = postheparine liver lipolytic activity.

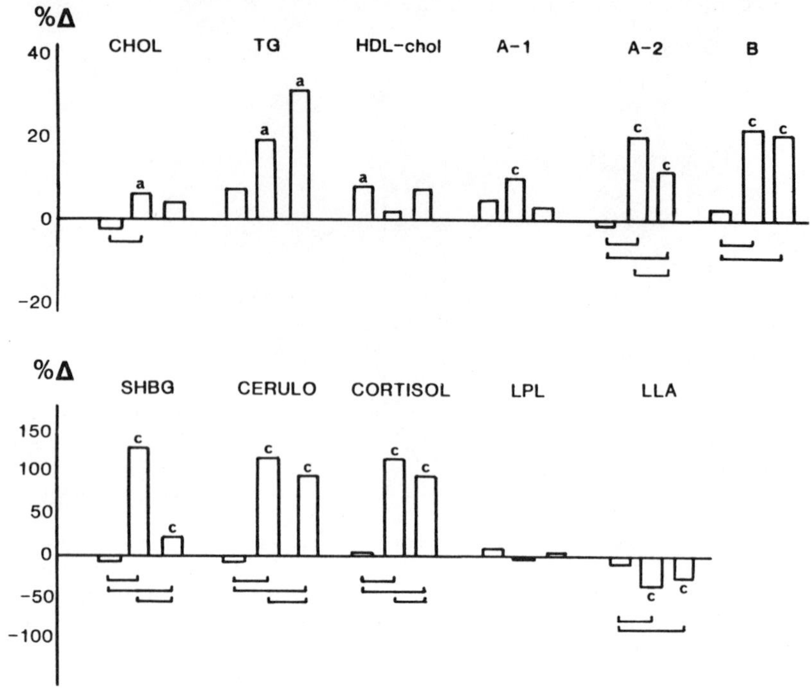

Fig. 11. Effect of two low-dose oral contraceptives on various hormone sensitive parameters. Each triplet of bars represents the mean percentual change of a parameter i.e. of controls (left bar); of women on LYN-EE (middle bar) and of women on LNG-EE (right bar). Abbreviations in Table 3.
a = change significance $p < 0.05$; c = change significance $p < 0.001$.
⌴ = change significantly different from corresponding change in the other group as indicated.

discerned, or that if an androgenic effect were present, it was at least dominated by the estrogenic effect in both preparations. The difference between the pills is evident in SHBG and in apo A2 but it is absent when looking at LDL-apo B, cortisol, the post heparin lipolytic activity of liver-lipase (LLA) and ceruloplasmin activity (although the small difference in increase during the use of both pills was statistically significant). This is surprising, as Kuusi and co-workers found a strong influence of androgens (increase) and of estrogens (decrease) on LLA. The liver is involved in the catabolism of HDL, probably via LLA (53). None of the estrogen/androgen-sensitive parameters correlated with each other.

In conclusion, both preparations had an estrogenic dominance over the possible androgenic effects. The changes in the sex hormone sensitive parameters were in the same direction, but the extent of some of these changes did not reflect the expectations based on the difference in the apparent androgen/estrogen balance.

REFERENCES

1. Hopkins PN, Williams RR. Atherosclerosis 1981; 40: 1-52.
2. Brown MS, Goldstein JL. Science 1986; 232: 34.

3. Brown MS, Goldstein JL, Fredrickson DS. In: Stanbury JB, Wyngaarden JB, Fredrickson DS, Goldstein JL, Brown MS, eds. The metabolic basis of inherited disease. New York: McGraw-Hill Book Company, 1983; Vol. 5: 655.
4. Stadel BV. New Engl J Med 1981; 305: 612.
5. Stadel BV. New Engl J Med 1981; 305: 672.
6. Morganroth J, Levy RI, McMahon AE, Gotto AM. J Pediatrics 1974; 85: 639.
7. Corsini A, Loma P, Sommariva D, Fumagalli R, Catapamo AL. Eur J Clin Invest 1986; 78: 940.
8. Higgins MJP, Lecamwasam DS, Galton DJ. The Lancet 2, 1975: 737.
9. De Gennes J-L, Touraine R, Maunand B, Truffert J, Laudat Ph. Société Médicale des Hôspitaux de Paris 1967; 118: 1377.
10. Thompson GR, Lowenthal R, Myant NB. The Lancet 1, 1975: 1208.
11. Saal SD, Parker TS, Gordon BR, et al. Am J Med 1986; 80: 583.
12. Eisenhauer T, Armstrong VW, Wieland H, Fuchs C, Scheler F, Seidel D. Klin Wochenschr 1987; 65: 161.
13. Starzl TE, Bahnson HT, Hardesty RL, et al. The Lancet 1, 1984: 1382.
14. Mol MJTM, Erkelens DW, Gevers Leuven JA, Schouten JA. The Lancet 2, 1986: 936.
15. Havekes L, De Wit E, Gevers Leuven JA, et al. Hum Genet 1986; 73: 157.
16. Havekes L, Vermeer BJ, De Wit E, et al. Biochim Biophys Acta 1980; 617: 529.
17. Innerarity TL, Pitas RE, Mahley RW. J Biol Chem 1979; 254: 4186.
18. Redgrave TG, Roberts DCK, West CE. Anal Biochem 1975; 65: 42.
19. Hessel LW, Gevers Leuven JA, Havekes L. In: Carlson LA, Olsson AG, eds. New York: Raven Press, 1984: 163.
20. Starzl TE, Putnam CW, Chase HP, Porter KA. The Lancet 2, 1973: 940.
21. Polano MK. Dermatologica 1974; 149: 1.
22. Vermeer BJ, Van Gent CM, Goslings B, Polano MK. Brit J Dermatol 1979; 100: 657.
23. Morganroth J, Levy RI, Fredrickson DS. Ann Int Med 1975; 82: 158.
24. Stuyt PMJ, Demacker PNM, Van 't Laar A. Eur J Clin Invest 1984; 14: 219.
25. Schneider J, Maurer M, Kaffarnik H. Klin Wochenschr 1974; 52: 941.
26. Kameda K, Matsuzawa Y, Kubo M, et al. Atherosclerosis 1984; 51: 241.
27. Masket BH, Levy RI, Fredrickson DS. J Lab Clin Med 1973; 81: 794.
28. Sherrill BC, Innerarity TL, Mahley RW. J Biol Chem 1980; 255: 1804.
29. Rall SC, Weisgraber KH, Innerarity TL, Mahley RW. Proc Natl Acad Sci USA 1982; 79: 4696.
30. Weisgraber KH, Innerarity TL, Mahley RW. J Biol Chem 1982; 257: 2518.
31. Schneider WJ, Kovanen PT, Brown MS, et al. J Clin Invest 1981; 68: 1075.
32. Rall SC, Weisgraber KH, Innerarity TL, Mahley RW. J Clin Invest 1983; 71: 1023.
33. McLean JW, Elshourbagy NA, Chang DJ, Mahley RW, Taylor JM. J Biol Chem 1984; 259: 6498.
34. Weisgraber KH, Rall SC, Innerarity TL, Mahley RW. J Clin Invest 1984; 73: 1024.
35. Innerarity TL, Weisgraber KH, Arnold KS, Rall SC, Mahley RW. J Biol Chem 1984; 259: 7261.
36. Ghiselli G, Gregg RE, Brewer HB. Biochim Biophys Acta 1984; 794: 333.
37. Havel RJ, Kotite L, Kane JP, Tun P, Bersot T. J Clin Invest 1983; 72: 379.
38. Yamamura T, Yamamoto A, Hiramori K, Nambu S. Atherosclerosis 1984; 50: 159.
39. Yamamura T, Yamamoto A, Sumiyoshi T, Hiramori K, Nishioeda Y, Nambu S. J Clin Invest 1984; 74: 1229.
40. Havekes LM, De Knijff P, Beisiegel U, Havinga J, Smit M, Klasen E. J Lipid Res 1987; 28: 455.

41. Weisgraber KH, Rall SC, Mahley RW. J Biol Chem 1981; 256: 9077.
42. Havekes LM, Gevers Leuven JA, Van Corven E, De Wit E, Emeis JJ. Eur J Clin Invest 1984; 14: 7.
43. Chait A, Albers JJ, Brunzell JD. Eur J Clin Invest 1980; 10: 17.
44. Stuyt PJM. Clinical and metabolic studies in type III hyperlipoproteinemia. Thesis. Krips Repro Meppel, 1982.
45. Klasen EC, Talmud PJ, Havekes L, et al. Hum Genet 1987; 75: 244.
46. Kay CR. Am J Obstet Gynecol 1982; 142: 762.
47. Stampfer MJ, Willett WC, Colditz GA, Rosner B, Speizer FE, Hennekens CH. New Engl J Med 1985; 313: 1044.
48. Wilson PWF, Garrison RJ, Castelli WP. New Engl J Med 1985; 313: 1038.
49. Schaefer EJ, Foster DM, Zech LA, Lindgren FT, Brewer HB, Levy RI. J Clin Endocrinol Metab 1983; 57: 262.
50. Hafner SM, Kurhwaha, Forter DM, Appelbaum D, Hazzard WR. Metabolism 1983; 32: 413.
51. Gevers Leuven JA, Kluft C, Bertina RM, Hessel LW. J Lab Clin Med 1987; 109: 631.
52. Gevers Leuven JA, Havekes L, Van der Kooij-Pontier HA, et al. Metabolism 1984; 33: 1039.
53. Jansen H, Van Tol A, Hülsman WC. Biochem Biophys Res Commun 1980; 92: 53.

CHAPTER 16
SYMPOSIUM FLOWCYTOMETRY

Clinical flowcytometry: history and horizons
T.V. Shankey, R.C. Flanigan, and C.J. Herman

Current and future prospects of flowcytometry in clinical hematology
H.J. Tanke and R.A. Langelaar

CLINICAL FLOW CYTOMETRY: HISTORY AND HORIZONS

T.V. Shankey, R.C. Flanigan and C.J. Herman

Departments of Pathology and Urology, Loyola University
Stritch School of Medicine, Maywood, Illinois 60153, USA

A. BACKGROUND

Flow cytometry represents an important technology for the clinical laboratory. Its earliest routine clinical application involved impedance based cytometers, or Coulter counters, which remain the single most widely used technology for enumeration of peripheral blood cell populations. With the advent of flow cytometers capable of automated analysis of blood cells, using histochemical staining and absorbtion measurements on individual cells (e.g. Hemalog instruments), or laser based flow cytometers using the specificity provided by monoclonal antibodies, flow cytometry has become a central technology for clinical hematology. In this review, we will provide an overview of the currently used clinical applications of flow cytometry, and will try to provide some insight to developments in the technology in the research setting, which are likely to have a significant impact on clinical medicine in both the near term, and in the more distant future.

At the beginning, it is useful to briefly describe a flow cytometer, in its most basic form. A typical flow cytometer produces a stream of cells, which pass through a detection or analysis zone one cell at a time. At some point in the flowing stream, cells are usually illuminated with coherent (i.e. laser) or noncoherent (i.e. mercury arc source) light, and detectors measure a number of different types of signals produced by individual cells. Some considerations of these different signals will be discussed later. It should be noted here, that the analysis of individual cells is not restricted to optical or light induced signals. Other applications have included microwave irradiation of individual cells (1), which provides a means to enumerate subpopulations of erythrocytes without additional reagents. Both the type of signal measured, and the degree of accuracy necessary for different kinds of measurements, place different demands on the type of cell stream used. In addition, the type of irradiation (or illumination), and the types of reagents used (fluorescence, histochemistries) to analyze different cell populations, all place constraints on the design of flow cytometers. As such, the early impedance based blood cell counters simply pulled cells through the detection orifice by vacuum, while laser based flow cytometers used to quantitate cellular DNA must place cells in a uniform position in order to ensure identical illumination of each cell. Many flow cytometer systems used to quantitate cellular DNA, or to provide highly sensitive systems to measure low level fluorescence enclose the flow stream in a quartz flow cell. In contrast, flow cytometers designed to sort cells have generally used analysis with the cell stream in air, a configuration that generally allows faster analysis and sort rates with current instrumentation. It should be noted that the demands generated by these other applications have resulted in recent improvements in fluidics and stream shaping flow cells used in newer clinical hematology instruments.

The origins and development of flow cytometry are closely related to microscope based cytophotometric analysis technology. The pioneering studies in the 1930's by Torbjörn Caspersson and his colleagues, utilized quantitative microscope cytophotometry to demonstrate the importance of quantitative measurement of nucleic acids and proteins within individual cells. Caspersson's studies resulted in a better understanding of the biology of normal and abnormal cells, including the landmark demonstration that the nucleic acid content of cells doubled during cell division (2), and that cells from human cancers frequently contain increased levels of nucleic acids compared to their normal cellular counterparts. A major impetus for development of microscope based quantitative cytology instrumentation came from the work of Papanicolaou and Traut (3), who established the importance of cellular morphology in exfoliative cytology, as a practical tool for cancer detection and prognosis.

The development of flow based quantitative cytology was given a major impetus, as noted above, with the development by W. Coulter of the impedance based blood cell counter (4). This flow cytometer is composed of a tube with a small hole drilled in one side. The tube is placed in a weakly conducting solution, such as isotonic saline, and a DC potential is established between the inside and outside of the tube by two electrodes. Any particle, such as a cell, passing through the small hole, will alter the field in direct proportion to the volume of the particle. However, not all particles take the same path through the orifice, and since the field within the orifice is not uniform, identical particles taking different paths will generate different signals. Furthermore, cells which re-enter the DC field after passing through the orifice produce broadening of the impedance peak, seen as a shoulder on the right side of the peak. As noted earlier, improved impedance counters focus the cell stream, using a laminar flow sheath system which sends cells through the orifice in a nearly uniform path, resulting in improved resolution. Thus, newer clinical hematology instruments (Toa, Sysmex) have taken advantage of improved focusing of the cell stream, with improved resolution to differentiate macrocytic platelets from microcytic erythrocytes.

The next landmark in the subsequent development of flow cytometry came with the important work of Louis Kamentsky who developed instrumentation capable of simultaneous measurement of multiple cellular parameters. While his initial work in cellular analysis involved microscope based image analysis, his fluidic cell sorter (5) established that cell size and nucleic acid content are key parameters for the classification of cancers. In addition, this approach provided the means to isolate, or sort out specific populations of cells for further study. At about the same time, Mack Fulwyler at Los Alamos National Labs, developed an impedance based flow cytometer that could also sort cells, using a modification of the technology developed by Richard Sweet for ink jet printing (6). This same technology forms the basis for all commercial cell sorting flow cytometers currently in use.

The subsequent development of flow cytometry instrumentation can simplistically be divided into DNA and antibody directed fluorescence schools. The DNA school, established by Caspersson's initial work, led to Kamentsky's cytometers, as noted above, first with IBM, and later at

Table 1. Parameters Available for Flow Cytometric Analysis

Intrinsic Signals	Extrinsic Signals (Need Reagent)
Cell Volume (Impedance)	DNA/RNA Content (Quantitative Dye)
Cell Diameter (Light Scatter)	Cell Surface Glycopeptides
Granularity (90° Scatter)	Intracellular Polypeptides
Absorbance (Heme, Lipofuscins)	Cellular pH/Activation
Fluorescence (Tryptophan)	Enzyme Function
Raman Scattering	cDNA/cRNA Site Specific Probes

Bio/Physics Systems, which he founded. There, he developed the Cytograf and Cytofluorograf which became the commercial instruments sold by Ortho. A microscope based flow cytometer for DNA measurements developed by Dittrich and Gohde (7), became the basis for the Phywe instrument. This low cost flow cytometer played an important role in establishing the clinical importance of precise measurements of DNA content in human cancers.

Any discussion of the antibody directed school of flow cytometer instrumentation must begin with the work of Coons (8) who developed the fluorescent antibody technique and established its use in fluorescence microscopy. A major impetus for the use of fluorescent antibodies in flow cytometry came from the work of the Herzenbergs at Stanford University, first with a modification of Kamentsky's instrument (9), and later with a much improved instrument equipped with an argon-ion laser (10). This later instrument became the FACS I, the first flow cytometer sold by Becton Dickinson. The key addition of the laser allowed for the first time the detection and quantitation of relatively weak signals from fluorescent antibody stained cells. A number of laboratories involved in important research in cellular immunology soon saw the power of fluorescent antibody staining and cell sorting, and much of the subsequent demand (and commercial response) was for these types of instruments. With the subsequent development and commercialization of monoclonal antibodies, particularly those directed at enumerating subpopulations of human lymphocytes (11), much of the clinical emphasis, at least in the USA, has been on immunophenotyping for leukemias and lymphomas. Ironically, it is only recently that many clinical labs in the USA have rediscovered the importance of DNA measurements for the prognosis and monitoring of patients with solid tumours.

As indicated in Table 1, a light based flow cytometer can provide a number of important measurements on individual cells, without the need to use any reagents (intrinsic signals). These include important parameters such as cell size, which is directly proportional to the amount of light scattered at near-forward (2-15°) angles, and cell granularity, which is proportional to the amount of light scattered at roughly 90°. The important discriminating power of flow cytometry is greatly enhanced, however, with the use of a variety of reagents, which bind to specific types of cells, or to specific parts or compartments of individual cells. A complete list of such reagents is not provided here. Rather, we include the types of stains or reagents which we believe will provide important measurements of clinical significance. These include DNA and RNA binding dyes, monoclonal and polyclonal antibodies specific for cell surface and intracellular glycopeptides, and cDNA (or cRNA) probes which can bind in situ to specific sequences on single stranded DNA or RNA, under appropriately controlled conditions. It should be noted that an additional parameter, not frequently used in flow cytometry, is detection of change of any measured parameter with time. As an example, rates of enzyme reaction have been measured using an adaptation of commercial hematology instrumentation (12).

B. BASIC HEMATOLOGY AND BEYOND

We present here, a brief sampling of current and near term applications of laser based flow cytometry instrumentation in the clinical laboratory. As indicated in Table 1, passing unstained peripheral blood cells through a cytometer provides forward and 90° light scatter signals from each cell. This information alone allows enumeration of erythrocytes, lymphocytes, monocytes, and granulocytes from unstained whole blood. Further enumeration of subpopulations of these cells is provided by the use of specific probes, either as enzyme histochemistries, or using labelled antibodies. The histochemical approach was pioneered commercially by Technicon with the Hemalog instruments, which most recently can use enzyme linked monoclonal antibodies to enumerate lymphocyte, granulocyte and monocyte subpopulations. In contrast, laser based flow cytometers generally use fluorescent antibodies to enumerate these same cell populations. The most common clinical application here is for immunophenotyping, or enumerating cell populations from peripheral blood, or other tissues, from leukemia and lymphoma patients, or patients with a variety of other conditions, such as AIDS, autoimmune diseases, trauma, inherited immunodeficiencies, and monitoring allograft rejection. While these represent the most frequent applications, other useful clinical applications include the clonal excess technique (13, 14) for early detection of recurrence of lymphomas, detection of platelet autoantibodies, and reticulocyte enumeration. While several flow cytometric techniques have previously been used for reticulocyte counting, the recent development of the thiazole orange dye (15) has allowed simple and reproducible enumeration of reticulocytes. The technique used requires 5 - 10 μl of whole blood, directly stained with thiazole dye. The sample can be analyzed after one hour, and remains stable for over one week at 4°C. The results obtained by this flow cytometric technique are generally in agreement with manual (new methylene blue) reticulocyte counts, and the flow cytometric method rapidly counts statistically significant numbers of cells.

Much of the emphasis of our own work in clinical flow cytometry is in DNA content and proliferation, or cell cycle analysis of human tumours, with the focus on urologic tumours. The relevance of DNA content of cell cycle analysis for human tumours is well summarized by Barlogie et al. (16), demonstrating that both the extent of DNA aneuploidy, and the percentage of tumour cells in cycle (S-phase index) correlate well with tumour grade, and frequently with survival. DNA content analysis of low grade bladder tumours (Grade I) usually show a single peak, identical to diploid (2c) human cells. In addition, few cells from these tumours are proliferating, as shown by the lack of cells either in the S-phase area, or in G_2 or M. However, high grade bladder tumours (Grade III) frequently show a major cell population with greater than 2c DNA, and sometimes show significant numbers of cells in S-phase.

We would point out that this same type of DNA analysis is now commonly performed with archival material, using nuclei isolated from fixed, paraffin embedded tissue, following a technique originally described by Hedley et al. (17). The advantage of this technique is the ability to perform retrospective studies on relatively large numbers of patients. This allows establishment of clinically important markers in a shorter time than would be needed for prospective studies.

Again, both diploid (or pseudodiploid) and aneuploid tumours can be identified by this technique. Our own studies of DNA content analysis of nuclei taken from archival material from low grade tumours has indicated that a small percentage (<25%) demonstrate an aneuploid cell population by flow cytometric DNA analysis. This approach allows selection of aneuploid tumours for further study, to determine if there are other cellular markers which may be useful either to identify tumour cells, or to predict the course of the disease in individual patients.

It is frequently difficult to differentiate tumour from normal cells by DNA analysis alone. This is best seen in histograms which show a single peak with 2c DNA content. In this case, it is impossible to rule out the possibility that an aneuploid peak with near 2c DNA content may be hidden within the normal cell, 2c DNA peak. Another possibility is the presence of a minor cell population, which represents the tumour cells, hidden within the cycling normal cells, in S or G_2/M phases. The earliest flow cytometric solution to this classic "signal to noise" problem was provided by Braylan (18) who analyzed lymph nodes from patients with B-cell lymphomas using simultaneous DNA analysis and cell surface staining with fluorescein labeled anti-immunoglobulin antibodies. This technique allows DNA analysis to be gated on B-cells only (surface Ig positive cells). In patients with advanced stage lymphoma, this represents the leukemic clone.

A similar situation exists in many other non-lymphoid tumours, where gating the DNA analysis on tumour cells would greatly simplify both DNA content and cell cycle analysis. Antibodies to specific intermediate filaments were first utilized by the group at Nijmegen to limit DNA analysis in flow cytometric studies to cells of the appropriate tissue type (19). This approach takes advantage of the observation that most, if not all tumours, maintain the expression of the same type of intermediate filament as found in their normal cellular counterpart. Thus, Ramaekers et al. used antibodies to appropriate cytokeratins (intermediate filaments found only in epithelial cells) to limit the DNA analysis to epithelium, in the analysis of endometrial adenocarcinoma and bladder carcinoma. By gating only on these cytokeratin positive cells, it is possible to determine the number of tumour cells in S-phase. In the single parameter DNA analysis these same cycling tumour cells would be lost in the 4c DNA peak of the normal G_2/M cells.

An alternative procedure to provide two parameter DNA analysis involves the staining of nuclear proteins, either specifically, using antibodies to nuclear proteins which increase during specific portions of the cell cycle, or nonspecifically, using fluorocein-isothiocyanate (FITC) to stain total nuclear proteins. The nonspecific staining technique, developed by Roti Roti (20) involves reaction of isolated nuclei with FITC and a DNA binding dye. Dual parameter analysis of DNA plus nuclear proteins allows the cell cycle to be broken down into major subcompartments. Although not shown here, the analysis of some tumour cell nuclei by this technique demonstrates aneuploid cell cycle compartments which are readily separated from diploid cells (Shankey, unpublished).

Specific nuclear protein staining techniques have been developed (21), which allow two parameter DNA analysis with quantitative staining of proliferation specific nuclear polypeptides, using labelled monoclonal antibodies. Demonstration of the importance of this technique by Bauer et al. (22), shows cell cycle and DNA content analysis of aneuploid tumour cells taken from a colon adenocarcinoma. It should be noted that these data are obtained from archival, fixed and embedded tissue, demonstrating that a variety of cell specific probes can be quantitated as part of a retrospective study using paraffin blocks.

Table 2. Projected Clinical Applications of Flow Cytometry

Cellular Analysis	Particle and Molecular Analysis
Steroid Receptors (Estrogen, Androgen) Oncogene Products Flow Karyotyping DNA/RNA In Situ Hybridization	Particle Counting Immunoassays Molecular Detection/Identification

C. FUTURE CLINICAL APPLICATIONS OF FLOW CYTOMETRY

We end this review with a brief overview of some near term clinical applications, and some conjectures about the clinical cytometry lab of the future. A brief summary of our thoughts on applications, for both periods, is shown in Table 2. For the near term applications, we start in looking at the current "hot" topics in research labs investigating clinically important areas of cancer biology. As indicated in Table 2, much emphasis has been placed recently on understanding the roles of oncogenes and their polypeptide products in cancer cell biology. We would focus here on the Ha-ras oncogene, due to its current level of interest and potential role in bladder cancer. Numerous reports have documented an increase in the level of Ha-ras gene product, both in freshly isolated human bladder tumours (23) and in bladder tumour cell lines (24). A report that the degree of invasiveness of human colon cancer correlates with the level of Ha-ras product (25), has led to speculation that the level of Ha-ras product may also correlate with invasiveness of bladder cancers. While no report of flow cytometric analysis of primary bladder cancers is published as of the writing of this review, flow cytometric DNA and Ha-ras product quantitation of bladder tumour cell lines has been reported (26). We speculate that soon many other reports will analyze DNA content with simultaneous quantitation of numerous oncogene products using flow cytometry. Any comment or prediction of the eventual clinical relevance of these activities would seem highly conjectural at this time.

Another potentially useful clinical marker for flow cytometric analysis of some cancers is the simultaneous analysis of DNA content and steroid receptors. At present, biochemical determinations of tumour steroid receptor content, using radioactive ligand binding assays or more recent enzyme immunoassays, give no direct measurement of the amount of receptor on tumour cells. Rather, these techniques measure the receptor content of an aggregate amount of tissue, which hopefully contains mostly tumour cells. With the recent development of monoclonal antibodies to cellular estrogen receptors, it is now possible to measure DNA plus estrogen receptor content for breast cancer cell lines using flow cytometry (27). While this technique could have significant clinical importance in the near future, it should be noted that image analysis systems have also demonstrated their utility in performing this same analysis. In addition, image analysis techniques allow a knowledgeable cytopathologist to identify tumour cells for receptor content analysis. In contrast, flow cytometry techniques will only differentiate normal from tumour cells if the tumour is sufficiently aneuploid by DNA analysis, or if other markers can be used for flow cytometric assays of diploid, or near diploid tumours.

As also indicated in Table 2, flow cytometric karyotyping is an additional technique with potential clinical applications. Unfortunately, the technology for quantitative and reproducible isolation of all human chromosomes is not developed sufficiently for routine clinical use. However, dual parameter DNA analysis of human chromosomes stained with two DNA binding dyes, resolves most of the individual chromosomes, and allows identification of a clinically important disease, Downs Syndrome, for which there is now a significant demand for classic karyotyping. Routine clinical use of flow karyotyping will require techniques which resolve all normal chromosomes, in addition to techniques which allow identification of specific translocations, deletions, and inversions. The technologies which will probably provide the tools for detecting these chromosomal changes are in situ cDNA probes, possibly combined with techniques to identify centromeres (as an index point to identify different chromosomes). Flow cytometric techniques have already demonstrated considerable promise in this direction, including a time of flight (or slit-scan) technique to analyze chromosome size and centromere position (28). In addition, the work of Trask et al. (29) has demonstrated the potential of flow cytometric analysis of in situ cDNA probes as a means to detect specific genotypic markers, and possibly phenotypic expression as well. However, image analysis techniques may well surpass flow cytometry in the area of chromosome analysis, for routine clinical use.

Finally, as noted in Table 2, flow technologies could have clinically important applications in the analysis of cellular products, in addition to their central role in quantitative cytology. Applications such as non-isotopic immunoassays using fluorescent particles (30) could provide the sensitivity needed for many clinical immunoassay applications in a nonhomogeneous format.

REFERENCES

1. Leif RC, Schwartz CM, Rodriguez M, et al. Two-dimensional impedance studies of BSA buoyant density separated human erythrocytes. Cytometry 1985; 6: 13-21.
2. Caspersson T, Schultz J. Nucleic acid metabolism of the chromosomes in relation to gene reproduction. Nature 1938; 142: 294-5.
3. Papanicolaou GN, Traut HF. The diagnostic value of vaginal smears in carcinoma of the uterus. Am J Obst Gynecol 1941; 42: 193-206.
4. Coulter WH. High speed automatic blood cell counter and cell size analyzer. Proc Natl Electronics Conf 1956; 12: 1034-42.
5. Kamentsky LA, Melamed MR. Spectrophotometric cell sorter. Science 1967; 156: 1364-5.
6. Sweet RG. High frequency recording with electrostatically deflected ink jets. Rev Sci Instrum 1965; 36: 131-9.
7. Dittrich W, Gohde W. Impulsfluorometrie bei Einzellen in suspensionen. Z Naturforsch (B) 1969; 24: 360-1.
8. Coons AH, Creech HJ, Jones RN. Immunological properties of an antibody containing a fluorescent group. Proc Soc Exptl Biol Med 1941; 47: 200-2.
9. Hulett HR, Bonner WA, Barrett J, Herzenberg LA. Cell sorting: automated separation of mammalian cells as a function of intracellular fluorescence. Science 1969; 166: 747-9.
10. Bonner WA, Hulett HR, Sweet RG, Herzenberg LA. Fluorescence activated cell sorting. Rev Sci Instrum 1972; 43: 404-9.
11. Kung PC, Goldstein G, Reinherz EL, Schlossman SF. Monoclonal antibodies defining distinctive human T cell surface antigens. Science 1979; 206: 347-9.
12. Beumer T, Pennings A, Beck H, Wessels J, Haanen C, Herman C. .Continuous measurement and analysis of staining kinetics by flow cytometry. Cytometry 1983; 4: 244-9.

13. Ault KA. Detection of small numbers of monoclonal B lymphocytes in the blood of patients with lymphoma. N Eng J Med 1979; 300: 1401-5.
14. Ligler FS, Vitetta ES, Smith RG, et al. An immunologic approach for the detection of tumour cells in the peripheral blood of patients with malignant lymphoma: implications for the diagnosis of minimal disease. J Immunol 1979; 123: 1123-6.
15. Lee LG, Chen C-H, Chiu LA. Thiazole orange: a new dye for reticulocyte analysis. Cytometry 1986; 7: 508-17.
16. Barlogie B, Raber MN, Schumann J, et al. Flow Cytometry in clinical cancer research. Cancer Res 1983; 43: 3982-97.
17. Hedley DW, Friedlander ML, Taylor IW, Rugg CA, Musgrove EA. Method for analysis of cellular DNA content of paraffin-embedded pathological material using flow cytometry. J Histochem Cytochem 1983; 31: 1333-5.
18. Kruth HS, Braylan RC, Benson NA, Nourse VA. Simultaneous analysis of DNA and cell surface immunoglobulin in human B-cell lymphomas by flow cytometry. Cancer Res 1981; 41: 4895-99.
19. Ramaekers FCS, Beck HLM, Feitz WFJ, et al. Application of antibodies to intermediate filament proteins as tissue-specific probes in the flow cytometric analysis of complex tumours. Anal Quant Cytol Histol 1986; 8: 271-80.
20. Roti Roti JL, Higashikubo E, Blair OC, Uygur N. Cell cycle position and nuclear protein content. Cytometry 1982; 3: 91-6.
21. Clevenger CV, Bauer KD, Epstein AL. A method for simultaneous nuclear immunofluorescence and DNA content quantitation using monoclonal antibodies and flow cytometry. Cytometry 1985; 6: 208-14.
22. Bauer KD, Clevenger CV, Endow RK, Murad T, Epstein AL, Scarpelli DG. Simultaneous nuclear antigen and DNA content quantitation using paraffin-embedded colonic tissue and multiparameter flow cytometry. Cancer Res 1986; 46: 2428-34.
23. Fujita J, Yoshida O, Yuasa Y, Rhim JS, Hatanaka M, Aaronson SA. Ha-ras oncogenes are activated by somatic alterations in human urinary tract tumours. Nature 1984; 310: 464-6.
24. Der CJ, Krontiris TG, Cooper GM. Transforming genes of human bladder and lung carcinoma cell lines are homologous to the ras genes of Harvey and Kirsten sarcoma viruses. Proc Natl Acad Sci (USA) 1982; 79: 3637-40.
25. Thor A, Horan Hand P, Wunderlich D, Caruso A, Muraro R, Schlom J. Monoclonal antibodies define differential ras gene expression in malignant and benign colonic diseases. Nature 1984; 311: 562-5.
26. Czerniak B, Herz F, Wersto R, Koss LG. Expression of Ha-ras oncogene p-21 protein in regulation of the cell cycle of cultured human tumour cells. Am J Pathol 1987; 126: 411-16.
27. Shenkenberg T, Dressler LG, Greene GL, McGuire WL. Estrogen receptor measurement by flow cytometry: multiparameter analysis should identify high risk breast cancer patients. Proc ASCO 1986; 5: 10 (Abstr).
28. Lucas JN, Gray JW. Centromeric index versus DNA content flow karyotypes of human chromosomes measured by means of slit-scan flow cytometry. Cytometry 1987; 8: 273-9.
29. Trask B, Van den Engh G, Landegent J, In de Wal NJ, Van der Ploeg M. Detection of DNA sequences in nuclei in suspension by in situ hybridization and dual beam flow cytometry. Science 1985; 230: 1401-3.
30. Saunders GC, Jett JH, Martin JC. Ultrasensitive flow cytometric fluoroimmunoassays. ISRD Ann Report Feb 1985: 31-33.

CURRENT AND FUTURE PROSPECTS OF FLOW CYTOMETRY IN CLINICAL HAEMATOLOGY

Hans J. Tanke and Roland A. Langelaar

Department of Cytochemistry and Cytometry, Medical Faculty
University of Leiden, The Netherlands

INTRODUCTION

Flow cytometry represents a relatively new technique that has been developed for the quantitative analysis of individual cells and cellular constituents. The principle of flow cytometry is that cells or cellular components in aqueous suspension are passed through a sensing region where optical or electrical signals are generated and measured. The typical analysis rate of commercial instruments is in the order of several thousand objects per second. Cells are generally fluorescently stained, although non-fluorescing dyes in sufficient concentration can be measured as well on the basis of axial light absorption. Staining is not required for measurements of cell size by light scatter or electrical resistance. The reproducibility of fluorescence measurements is 2% or better, and the detection limit of most commercial instruments is 2000-3000 molecules of fluorescein per cell. A schematic representation of a typical flow cytometer is seen in Figure 1. The cells are hydrodynamically forced with a constant speed of 5-10 m/sec to a region onto which a high intensity light source, usually a laser or high pressure arc lamp, is focussed to generate light scatter signals and fluorescence emission. Properly oriented photodetectors collect a fraction of the signals and generate electrical signals proportional to the optical signals. These signals are accumulated, analysed and usually presented as single or multi-parameter frequency distributions (histograms). Manual interpretation or mathematical analysis of these histograms provides biological information of the measured cell population.

Some flow cytometers allow the physical separation of objects from the main fluid stream, on the basis of the signals measured. This process is called cell sorting and accomplished by means of microdroplets generated by a transducer connected to the nozzle. Droplets that contain the objects of interest may be electrostatistically charged and deflected in an electric field downstream.

Flow cytometers are able to measure simultaneously several parameters from one object. A typical number of parameters for clinically used instruments is 3-5; however, larger experimental systems allow measurements of up to 8 parameters. After detection and amplification analog signals are first digitized using proper analog-to-digital (AD) converters. Digitized data may be displayed directly using univariate or bivariate frequency distributions (histograms). The display of more than 2 parameters (multivariate data) becomes very complex. A third parameter may be displayed in the vertical direction. The result is the display of a cube containing dots, each point representing the location of a parameter triplet of a correponding cell in this cube. Multiparameter data are preferably recorded in list mode. This means that the correlated data are digitized and directly stored on disc or tape, after the sample is being analysed. Selection of the most discriminative parameters as well as gated analysis of the data may then be performed later.

General reviews and textbooks on flow cytometry and cell sorting are given by (1-3).

The properties discussed above make flow cytometers ideal instruments for investigating a relatively large number of objects for up to 8 variables. Clinical applications of flow cytometry have been reported over the last two decades. However, a decisive step forward with respect to routine clinical application has been the introduction of microcomputers for signal handling and manipulation, and for operation control. Flow cytometers have developed from complicated research instruments towards user-friendly laboratory systems. At present, flow cytometry facilities are found in most university hospitals and are increasingly used for clinical applications.

CLINICAL APPLICATIONS OF FLOW CYTOMETRY

Clinical applications of flow cytometry are numerous and rapidly increasing. The earliest clinical applications of flow cytometry were in the field of haematology, where Coulter counters were introduced for the counting of blood cells such as erythrocytes, leucocytes and platelets (4).

Later on, measurements of light scatter and axial light loss were performed for differential counting of leucocytes based on differences in peroxidase and esterase staining (5). At present flow cytometric analysis

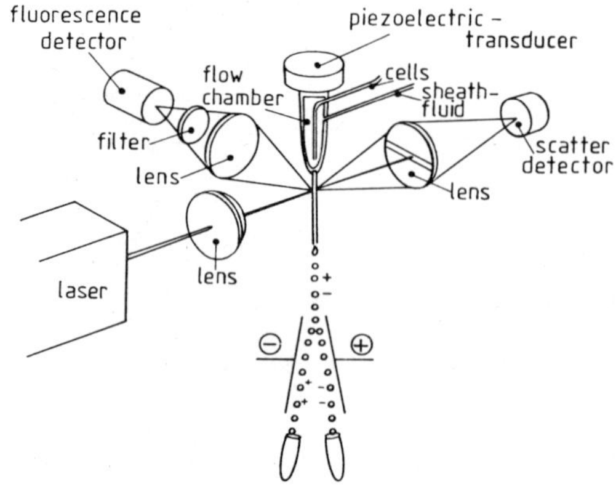

Fig. 1. Schematic configuration of a flow cytometer with orthogonal axis of excitation, fluid stream and photodetection system, allowing electrostatic deflection of selectively charged droplets that contain the cells of interest.

is increasingly being used in (immuno)haematology for the counting of reticulocytes in peripheral blood, for immunological typing of lymphocytes and for the characterization of leukemias with panels of monoclonal antibodies.

The number of reticulocytes (normal values around 2%) in the peripheral blood reflects the erythropoietic activity of the bone marrow, which is of considerable interest in case of anemia and in oncological patients treated with cytostatic drugs that affect bone marrow proliferation. However, accurate counting of especially low numbers of reticulocytes (1% or lower) form a major problem in routine haematology, due to the large statistical errors involved in determining such percentages. Flow cytometric analysis of reticulocytes, fluorescently stained for RNA with Pyronin Y (6), has proven to be of clinical value. Within one minute 100,000 cells are readily analysed and the results are plotted as RNA histograms (Figure 2).

The percentage of reticulocytes as well as the age distribution (younger reticulocytes contain more RNA than older ones) can be calculated from these profiles (7). Besides Pyronin Y other dyes such as Acridine Orange (8), Thioflavine T (9) and Thiazole Orange (10) have been proposed. Flow cytometric analysis of reticulocytes is more reproducible than manual evaluation (coefficient of variation approximately 20% versus errors up to 100% as reported for hand counts), and is sensitive enough to detect changes as a consequence of chemotherapy (11). Wider clinical application of this method requires the use of appropriate standards, such as microspheres with defined amounts of RNA, that can be mixed with blood and used as internal standards to define the reticulocyte population. The use of such an arbitrary definition of a reticulocyte has the advantage that the outcome of the test becomes more or less independent of the type of flow cytometer and fluorescent RNA staining dye that is used. At present, normal values have to be determined for each staining procedure, type of instrument and mathematical analysis procedure.

The classification of leukemias has been based on morphology of the cells and their cytochemical staining properties. Morphological classification may be difficult, requires experienced hematologists and

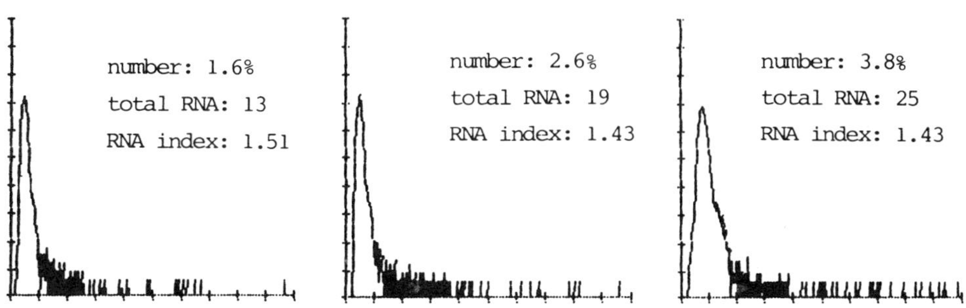

Fig. 2. Examples of fluorescence histograms obtained by flow cytometry of Pyronin Y stained red blood cells. Abscissa: relative Pyronin Y fluorescence (RNA content of the reticulocyte); ordinate: number of objects (logarithmically). Note the presence of the major population in the lower channels, representing mature red blood cells, and reticulocytes of different maturation stages (RNA content) in the higher channels. Parameters related to the degree of maturation such as RNA content and RNA index can be calculated from these histograms (see also reference 7).

remains subjective. Recently various monoclonal antibodies have become available, by which the type and origin of the leukemia can be determined accurately. Especially when combinations of different reagents are used, leukemias can be immunologically characterized (12, 13). At present panels of monoclonal antibodies, appropriate controls and software procedure for decision making are available for rapid and reproducible classification. Although such typing can also be performed manually using a microscope, flow cytometry is preferred since it is fast (and therefore relatively cheap), unbiased and allows the storage of list mode data for each patient. This latter aspect may be important to study changes in the characteristics in time during therapy.

Another major application of flow cytometry has been the analysis of DNA ploidy. Rapid and reliable methods have been developed for the preparation of cell suspensions from biopsy material. These cells are then fluorescently stained for DNA and analysed by flow cytometry (14). Cells with abnormal DNA content are often observed in solid tumours and leukemias. Besides diagnostic values, DNA ploidy of the tumour has been evaluated, especially in solid tumours, as a prognostic parameter. Generally, tumours with diploid or peri-diploid stemlines have a better prognosis in comparison with tumours with stemlines of aneuploid high DNA content, although this may depend on the type of tumour and the applied therapy (for example see Figure 3).

A major disadvantage of prognostic studies in oncology is the long time of follow-up (sometimes up to 10 years) that is required before the real clinical value of such prognostic parameters can be statistically verified. The use of paraffin-embedded archival material for flow cytometric analysis of DNA content, originally described by Hedley and colleagues (15) allows retrospective analysis of a well documented material with known follow-up. Moreover, it has also been possible to study the expression of onco-proteins in paraffin-embedded material (16), although their diagnostic and prognostic importance for the various tumours is not clear yet.

As in solid tumours aneuploidy in chromosome number in leukemia generally is associated with poor prognosis. The exception is the ALL, the most frequent form of leukemia in children. About one third of these leukemias are hyperdiploid and surprisingly have better prognosis than the diploid cases (17).

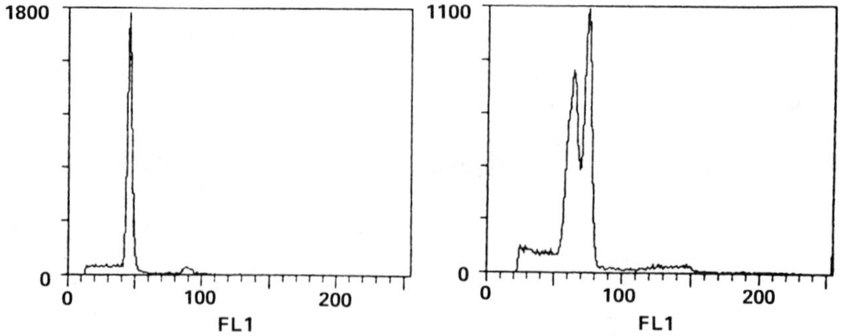

Fig. 3. Examples of DNA ploidy distributions of solid tumours as measured by flow cytometry. Left: Example of a normal diploid distribution; right: Example of an aneuploid DNA distribution. Abscissa: DNA content (ethidium fluorescence); ordinate: number of events.

A last clinical application of flow cytometry is the measurements of subpopulations of mononuclear cells, in particular T cells, in transplantation immunology. The steadily increasing number of monoclonal antibodies by which mononuclear blood cells, with somewhat different functions, can be recognized, allows characterization of the immune system in more detail and more easily than before. Although studies of lymphocyte subsets can also be done using microscopic techniques, flow cytometry is at present the method of choice. Manual counting of a few hundred cells is not only time consuming and therefore expensive, but also leads to large statistical errors. Both can be overcome by rapid objective flow cytometric analysis of 10,000 cells in a few seconds. Flow cytometric evaluation becomes almost a prerequisite if subsets have to be identified on the basis of three- or four-colour immunofluorescence.

The so-called T4/T8 ratio (roughly related to T helper cells/T suppressor cells) has been extensively studied in renal and bone marrow transplantation (BMT). Pre-transplant T4/T8 ratios were found to correlate with renal graft survival in adult patients on low-dose corticosteroid therapy: low ratios correlated significantly with poor graft survival (18). This was however not observed in patients treated with cyclosporin A. The study of T4 and T8 cells in patients treated with methotrexate for BMT showed a lower repopulation rate for T8 cells in comparison with T4 cells. However, patients that developed at least grade 2 graft versus host disease (GVHD) showed a much faster repopulation rate of the T8 subpopulation, in comparison with those patients that developed no or only grade 1 GVHD. Measurement of T4/T8 ratio at day 12 therefore allowed the identification of patients at risk for severe GVHD (19). As in renal transplantation, this phenomenon was not observed in patients treated with cyclosporin A.

Recently developed monoclonal antibodies allow the specific recognition of the various types of cytotoxic lymphocytes: natural killer cells (non-T cells), HLA class 1 restricted cytotoxic T cells, and HLA non-restricted T cells (20). It is to be expected that with this type of reagent, more refined information about the immune status of transplant patients can be obtained.

FUTURE PROSPECTS OF FLOW CYTOMETRY

In the past decennium, two developments have significantly influenced the applicability of flow cytometry for both experimental studies and clinical applications. First, the hybridoma technique by which mono-specific antibodies can be prepared in almost unlimited amounts, has facilitated the quantitative studies of a variety of macromolecules and cell biological processes that were hardly possible before that time. Secondly, the use of microcomputers for operation of flow cytometers and for adequate data handling and presentation, has been clearly established. Their application have changed flow cytometers into user friendly instruments, that can be operated easily even by relatively unexperienced personnel. This aspect has considerably stimulated the use of flow cytometers in clinical laboratories.

Further use of monoclonal antibodies and specific fluorescent dyes, as well as further perfection of the instrumentation including lowering the costs of operation, will enlarge the contribution of flow cytometry in experimental and clinical studies. A number of clinical tests such as counting of reticulocytes and platelets, phenotyping of lymphocytes and immunological characterization of leukemia, which are presently still performed manually in many clinical centers, are expected to be carried out using flow cytometry.

It should be noted that also in the field of image analysis significant progress along these lines has been made recently. Image analysis has and will profit in a similar way from the availability of a variety of immunological reagents and powerful microcomputers, and is now fast enough to analyse a few hundred cells/second. For a number of clinical applications image analysis may therefore compete with flow cytometry in the near future. However, a major advantage of flow cytometry is its flexibility in sorting live cells and cellular constituents, which will ensure its unique position in biomedicine.

REFERENCES

1. Bonner WA, Hulett HR, Sweet RG, Herzenberg LA. Fluorescence activated cell sorting. Reviews of Scientific Instrumentation 1972; 43: 404-09.
2. Melamed MR, Mullaney PF, Mendelsohn ML. Flow cytometry and sorting. New York: John Wiley, 1979.
3. Van Dilla MA, Dean PN, Laerum OD, Melamed MR. Flow cytometry: instrumentation and data analysis. Ploem JS, ed. Series on Analytical Cytology. London: Academic Press, 1985.
4. Coulter WH. Means for counting particles suspended in a fluid. US Patent no. 2,656,508. Filed August 27, 1949. Issued October 20, 1953.
5. Ansley HR, Ornstein L. Enzyme histochemistry and differential white cell counts on the Technicon Hemalog D. Advances in Automated Analysis 1971; 1: 437-443.
6. Tanke HJ, Nieuwenhuis IAB, Koper GJM, Slats JCM, Ploem JS. Flow cytometry of human reticulocytes based on RNA fluorescence. Cytometry 1980; 1: 313-20.
7. Tanke HJ, Rothbarth PH, Vossen JMJJ, Koper GJM, Ploem JS. Flow cytometry of reticulocytes applied to clinical haematology. Blood 1983; 61 6: 1091-7.
8. Natale PJ. Method and reagent for quantitative determination of reticulocytes and platelets in whole blood. US Patent 4,336,029; 1982.
9. Sage BH, O'Connel JP, Mercolino TJ. A rapid vital staining procedure for flow cytometric analysis of human reticulocytes. Cytometry 1983; 4: 222-7.
10. Lee L, Chen CH, Chiu LA. Thiazole orange: A new dye for reticulocyte analysis. Cytometry 1986; 7: 508-517.
11. Tanke HJ, Van Vianen PH, Emiliani FMF, et al. Changes in erythropoiesis due to radiation or chemotherapy as studied by flow cytometric determination of peripheral blood reticulocytes. Histochemistry 1986; 84: 544-548.
12. Foon KA, Schroff RW, Gale RP. Surface markers on leukemia and lymphoma cells; recent advances. Blood 1982; 60: 1-19.
13. Majidic O, Bettelheim P, Stockinger H, et al. M2, a novel myelomonocytic cell surface antigen and its distribution on leukemic cells. International Journal of Cancer 1984; 33: 617-22.
14. Vindelov LL, Christensen IJ, Nissen NI. A detergent-trypsin method for the preparation of nuclei for flow cytometric DNA analysis. Cytometry 1983; 3: 323-7.
15. Hedley DW, Friedlander ML, Taylor IW, Rugg CA, Musgrove EA. Method for analysis of cellular DNA content of paraffin embedded pathologic material using flow cytometry. Journal of Histochemistry and Cytochemistry 1983; 31: 1333-5.
16. Watson JV, Sikora KE, Evan GI. A simultaneous flow cytometric assay for c-myc oncoprotein and cellular DNA in nuclei from paraffin-embedded material. Journal of Immunological Methods 1985; 83: 179-192.
17. Smets LA, Slater RM, Behrendt H, Van 't Veer MB, Homan-Blok J. Phenotypic and karyotypic properties of hyperdiploid acute lymphoblastic leukemia of childhood. British Journal of Haematology 1985; 61: 113-123.

18. Van Es A, Tanke HJ, Baldwin III WM, Oljans PJ, Ploem JS, Van Es LA. Ratios of T lymphocyte subpopulations predict survival of cadaveric renal allografts in adult patients on low-dose corticosteroid therapy. Clinical Experimental Immunology 1983; 52: 13-20.
19. Gratama JW, Naipal AMIH, Oljans PJ, et al. Early shifts in the ratio between T4+ and T8+ T lymphocytes correlates with the occurrence of acute graft-versus-host disease. Blood 1984; 63: 1416-23.
20. Lanier LL, Philips JH. Evidence for three types of human cytotoxic lymphocytes. Immunology Today 1986; 7: 132-134.

CHAPTER 17
CLINICAL CHEMISTRY AND NUTRITION

ASSESSMENT OF THE B-VITAMIN STATUS: SOME BIOCHEMICAL AND CLINICAL ASPECTS

H. van den Berg

Dept. of Clinical Biochemistry, TNO-CIVO Toxicology and Nutrition Institute, Utrechtseweg 48, Zeist, the Netherlands

INTRODUCTION

The vitamins represent a mixed group of organic substances designated as such not by their chemical characteristics but by their function (1). They participate in specific metabolic reactions and are essential for optimal growth and development of the organism. Vitamins are required only in trace amounts and do not provide energy. Traditionally they are divided into the fat-soluble and water-soluble vitamins. The latter group comprises the so called B-vitamins (Table 1) and vitamin C. In this paper some biochemical and clinical aspects of the B-vitamins are considered.

B-vitamins have their most important biochemical function as co-enzymes or as prosthetic groups of a large variety of enzymes in the metabolism of carbohydrates, fats and proteins, and in the erythropoiesis. Some characteristic co-enzymatic functions are summarized in Table 2.

For some B-vitamins other, non-cofactor functions have emerged. PLP plays a role in enzyme stabilization and modulates the steroid hormone action by an (in-vitro) reduction of the activation of cytosolic steroid receptors, i.e. it inhibits the binding of the cytosolic steroid receptor complex with nuclear DNA (2). Thiamin (as thiamin triphosphate) plays a specific role in peripheral nerve conduction, probably independent of its co-enzyme function (3).

APPROACHES FOR ASSESSMENT OF THE B-VITAMIN STATUS

The vitamin status can be defined as the total amount of a particular vitamin present in the body and available to catalyse vitamin dependent biochemical (biological) processes. A chronic inadequate dietary vitamin intake will ultimately result in vitamin deficiency. Vitamin deficiency, but also toxicity (e.g. vitamin B6) is accompanied by more or less specific clinical symptoms. The manifestation of characteristic clinical deficiency symptoms is preceded by a period during which body stores are depleted.

Pure nutritional vitamin deficiencies have virtually disappeared, at least in industrialized (western) societies, although certain population groups may be at higher risk to develop vitamin deficiency. To these risk

Table 1. Nomenclature of the B-vitamins

Trivial name	Generic descriptor[1]	Old designation(s)
Vitamin B1	Thiamin	Aneurin
Vitamin B2	Riboflavin	Vitamin G, lactoflavin
Vitamin B6	Vitamin B6 (3-hydroxy-2-methylpyridine derivatives)	Adermin
Vitamin B12	Vitamin B12 (cyanocobalamins)	
Folates	Folacin	Vitamin Bc, or M
Niacin	Niacin (pyridine 3-carboxylic acid derivatives)	Vitamin PP Vitamin B5
Pantothenic acid	Pantothenic acid	Vitamin B3
Biotin	Biotin	Vitamin H, co-enzyme R

[1]IUNS Committee on Nomenclature (J Nutr 1980; 110: 8-16)

groups belong individuals with aberrant dietary habits, such as macrobiotics and vegetarians (vitamin B12), long-term users of imbalanced health food diets, and individuals with increased needs, such as pregnant and lactating women.

Disease may be the cause and the consequence of a vitamin deficiency. In patients a 'conditional' (secondary) vitamin deficiency may develop, even in spite of an apparently adequate dietary intake, as a result of specific pathology, and/or drug therapy (4, 5). Malabsorption syndromes, organ failure or chronic disease may affect vitamin absorption and metabolism, resulting in an increase in vitamin need (4, 6). Assessment of the vitamin status is therefore relevant in clinical decision making. The relation between B-vitamin status and disease will be discussed in more detail in the last paragraph.

Assessment of the vitamin status can in principle be approached at three different levels (7):

- assessment of dietary intake
- clinical examination
- biochemical assessment.

Dietary assessment

Dietary vitamin intake can be calculated from dietary interviews and be compared with Recommended Dietary Allowances (RDA). However, there are some problems with this dietary approach at the level of the individual (7, 8). First of all RDA's are intended to be used for evaluation of the vitamin intake of population groups, and not for individual assessment. RDA's are developed to allow for individual variation. Therefore a margin of safety of two standard deviations above the estimated average requirement is included in the RDA. Hence RDA's represent an average consumption across time and not a daily intake rate. An individual may on average consume an intake less than the RDA, which is fully compatible with normal metabolic needs. On the other hand individuals consuming the RDA are not assured of an adequate status. Another complicating factor is that bioavailability and nutrient-nutrient interactions are not considered. Dietary assessment can therefore be useful to identify individuals with an inadequate dietary pattern, but not for classification of individuals.

Table 2. Some important biochemical reactions involving B-vitamin coenzymes

Vitamin	Co-enzyme form(s)	Biochemical reactions
Thiamin	Thiamin diphosphate (ThDP)	Oxidative decarboxylation
Riboflavin	Flavin mononucleotide (FMN) Flavin adenine dinucleotide (FAD)	Redox reactions
Vitamin B6	Pyridoxal-5'-phosphate (PLP)	Amino acid metabolism
Vitamin B12	Methylcobalamin Deoxyadenosyl B12	Homocysteine transmethylase Methylmalonyl-CoA mutase
Folacin	Substituted tetrahydrofolate (THF)-polyglutamates	C_1-transfer
Niacin	Nicotinamide di-nucleotide phosphate (NAD(P))	Redox reactions
Pantothenic acid	4-Phospho-Pantotheine	Acyl group exchange (part of CoA and ACP)
Biotin	Biotin	(Trans-)Carboxylase reactions

Based upon estimates of the minimal (daily) requirement, i.e. the amount necessary to prevent clinical disorders, recommended daily allowances (RDA) have been set by National Nutrition Councils and other authorities. These RDA's allow for interindividual variation in vitamin need and are considered to be sufficient for the 'maintenance of health in nearly all people' (WHO/FAO, 1974). RDA's are therefore intended to be used for evaluation of dietary (vitamin) intake of population groups. Although dietary assessment can be helpful to detect people 'at risk' for developing a vitamin deficiency (e.g. aberrant dietary pattern) classification of individuals with this technique is not permitted.

Clinical assessment

The clinical manifestations of B-vitamin deficiencies are well-documented. However, these specific symptoms are apparent only at a relatively late stage (9). The first clinical signs of hypovitaminemia are rather vague and non-specific like fatigue, depression, loss of weight, loss of appetite, increased irritability, etc. (1, 10). Careful clinical examination and anamnesis including drug history and dietary habits can provide valuable information and may initiate laboratory investigation to confirm depletion of vitamin stores and to show biochemical abnormalities.

Biochemical assessment

Biochemical assessment will in principle give specific information on the adequacy of the vitamin stores in the body, and thereby confirm subclinical deficiency, i.e. detect biochemical 'abnormalities' at an early stage before clinical symptoms appear (9).

LABORATORY METHODS

At present a wide range of laboratory methods is available for biochemical assessment of the B-vitamin status (11, 12). Some currently used methods for routine assessment of the B-vitamin status are summarized in Table 3.

These methods can roughly be divided into two categories:

1. Direct ('static') methods measuring the specific vitamin or its metabolites in biological matrices (e.g. blood or urine)
2. Indirect (functional or 'dynamic') methods measuring the in-vitro or in-vivo response of a vitamin related biochemical parameter (e.g. enzyme activity) or physiological function.

Direct methods

Direct tests are especially useful in the assessment of available vitamin stores. However, a prerequisite for an accurate direct biochemical test is that a relationship has been established between the content in blood or urine and its bodystores or related physiological function. Unfortunately, most of the data available on this relationship is based upon animal studies.

Information on the stores in human body is scarce. In general, it is assumed that these stores are relatively small and sufficient to maintain health for only about 6 to 10 weeks in case of complete dietary deprivation. In this respect vitamin B12 is an exception as significant amounts can be stored (1-10 mg), sufficient for at least one year. Some data are available from experimental depletion studies with human volunteers.

In our Institute we performed a vitamin depletion study in which volunteers were fed a diet containing about 35% of the RDA for respectively thiamin, riboflavin and vitamin B6 for a period of 10 weeks. Within 4 weeks vitamin blood levels fell below the lower limit of the reference range, established with apparently healthy blood donors, concurrently with a decrease in the aerobic capacity of the volunteers (13).

In his classical study on folacin depletion in man Herbert (14) observed a decrease of serum folacin already within the first twenty days. Megaloblastic staining (bone marrow) and hypersegmentation in the peripheral blood smear were found after 35 and 50 days, respectively. Red cell folacin concentrations below 20 ng/ml were observed after ca. 100 days, just before megaloblastic anemia developed (day 135).

Small variations in the daily vitamin intake hardly affect the vitamin levels in blood. However, the urinary excretion of B-vitamins more or less reflects recent vitamin intake and is therefore of limited value in the assessment of the vitamin status in individuals. Some B-vitamins are about equally distributed between plasma and the erythrocytes, such as vitamin B6, biotin, pantothenic acid and niacin. Thiamin, riboflavin and folacin are mainly present in the erythrocytes, and vitamin B12 nearly only in plasma. In some cases determination in plasma is considered more sensitive or specific, like for folacin and PLP, in spite of the higher content in erythrocytes.

Homeostatic regulation mechanisms may affect the vitamin blood level, or the urinary excretion. Under certain conditions, like in disease, during drug treatment (including hormonal contraceptives) or in certain physiological conditions, such as in pregnancy or in old age, the distribution of the vitamin between tissues and the circulation might be affected and should be considered with the selection as well as the interpretation of tests (15, 16).

Table 3. Methods for routine assessment of the B-vitamin status in humans

Vitamin	Parameter	Method[1]	Reference range[2]
Thiamin	Thiamin (B)	HPLC-Fl	98-117 nmol/l
	Thiamin excretion (U)	HPLC-Fl	> 65 μg/g creat.[3]
	ETK-stimulation test (E)	Enz.	α-ETK: 0.96-1.19
Riboflavin	FAD (B)	HPLC-Fl	0.24-0.44 μmol/l
	Riboflavin excretion (U)	HPLC-Fl	> 65 μg/g creat.[3]
	EGR-stimulation test (E)	Enz.	α-EGR: 0.96-1.30
Vitamin B6	PLP (B,P)	REA, HPLC	20-89 nmol/l (P)
	4-pyridoxic acid excretion (U)	HPLC-Fl	> 300 μg/g creat.[3]
	EGOT-stimulation test (E)	Enz.	α-EGOT: 1.32-2.23
Vitamin B12	Vit. B12 ('true' cobalamin)(S)	CPB	135-550 pmol/l
Folacin	Folacin (S,E)	CPB	4.7-17.9 nmol/l (S)
Niacin	Nicotinic acid (+ amide)(B)	MB	37-63 μmol/l
	N-Me-nicotinamide (NMN)/ N-Me-2-pyridone (NMP) excretion (U)		NMN/NMP-ratio > 1.0[3]
Biotin	D(+)Biotin (B,P)	MB, CPB	0.5-2.3 nmol/l (B)
Pantothenic acid (3)	Pantothenate (B)	MB, RIA	0.7-2.1 μmol/l
	Pantothenate excretion (U)		> 2.0 mg/g creat.[3]

(B): Blood; (S): Serum; (P): Plasma; (E): Erythrocytes
[1] HPLC-Fl: High performance liquid chromatography with fluorometric detection; Enz.: Enzymatic assay; CPB: Competitive Protein Binding (radio) assay; RIA: Radioimmunoassay; MB: Microbiological assay.
[2] Adult blood donors (P2.5 - P97.5 range), established at CIVO-TNO.
[3] Adapted from Sauberlich, Skala & Dowdy (11)

Functional tests

Solomons and Allen defined functional indicators of nutritional status as "diagnostic tests to determine the sufficiency of host nutriture to permit cells, tissues, organs, anatomical systems or the host him/herself to perform optimally the intended, nutrient-dependent biological function" (17). In an excellent review article (17) they summarized the (potentially) available functional tests according to experimental approach. The various categories, elaborated for the B-vitamins, are presented in Table 4.

Enzyme stimulation tests are nowadays frequently used as in-vitro functional tests. In these tests the activity of a vitamin dependent enzyme in a red cell hemolysate is measured before and after the addition of an excess of its co-enzyme. These tests seem not to be affected by (small) variations in the recent intake of the vitamins.

For the assessment of the thiamin status the erythrocyte transketolase (ETK) stimulation test was first described by Brin (18). For riboflavin and vitamin B6 the glutathion reductase (EGR) stimulation test and the glutamate-oxaloacetate transaminase (EGOT) stimulation test are used, respectively. Also the glutamate-pyruvate transaminase (EGPT) is used for vitamin B6 status assessment. However, the activity of this enzyme in (human) red cells is much lower, and far less stable than EGOT (19). Several modifications of these assays have been described (20). Although

Table 4. Classification of functional indices by experimental approach (according to Solomons and Allen, 1983)

In vitro tests of in vivo function	
- Enzyme stimulation tests	B1, B2, B6
- d-Uridine suppression test	B12, Folacin
Induced responses and load tests in vivo	
- Amino acid load test	B6, B12, Folacin
- Glucose load test	B1
Spontaneous in vivo response	
- neutrophylic hypersegmentation	Folacin
- physical performance	B1, B2, B6
- host defense	B6
- nerve function	B1, B12
- mental performance	?
Response of individual or population	
- birthweight	Folacin, B6

the holo-enzyme levels change as a result of vitamin excess and deficiency, also non-nutritional factors, like drug-use and disease may influence the enzymes (12, 19). The ETK apo-enzyme seems more labile, in-vivo as well as in-vitro, as compared with the EGOT- and EGR-apo-enzymes (19).

Although these in-vitro stimulation tests are considered as functional tests at the cellular level, they hardly reflect a biological or physiological function. Therefore, there is an increasing interest in the development of functional (biological) parameters for evaluation of the nutritional status. In-vivo load tests, such as measurement of the excretion of xanthurenic acid after an oral load of tryptophan for assessment of the vitamin B6 status, the histidine load test (or FIGLU test) for assessment of the folacin status, and the measurement of methylmalonic acid after an oral valine load have been proven very useful. However, most functional tests, except the enzyme stimulation tests, are not very suitable for routine assessment.

Functional (in-vivo) tests can be of great value however, in establishing and optimalization of interpretation criteria, or cut-off points, for the direct biochemical parameters, and for evaluation of the various parameters (see below).

The choice of the test depends on many factors, especially on the aim of the study. Tests reliable in population studies can be inappropriate for individuals, i.e. in clinical decision-making. Monitoring of patients may necessitate other tests than those suitable for classification of patients as adequate or deficient. Basic knowledge of vitamin metabolism (kinetics), and on the specificity and sensitivity (see below) of the various tests is required in the selection and with the interpretation of an appropriate test. The Group of European Nutritionists (GEN), as well as the IFCC Expert Panel on the Laboratory Assessment of Nutritional Status (EPLANS) consider the standardization of biochemical indicators of the nutritional status. Because nutritional status assessment necessitates special expertise and facilities, and for reasons of standardization, efficiency and quality control, concentration of these analyses in specialized centres appears recommendable.

INTERPRETATION OF TESTS

Classification of patients in risk categories requires a careful selection of interpretation criteria or cut-off points. These cut-off points can be derived in various ways. Commonly cut-off points are derived from the distribution of a given parameter in a population of apparently healthy volunteers or blood donors. To improve the validity several investigators have tried to establish cut-off points from reference values for selected individuals in good health and with well described living and eating habits (21), or for vitamin-supplemented groups (22).

The problem with these cut-off points is that they are purely statistical, not related to a physiological or biological function. Especially in the interpretation of a single measurement for an individual the use of statistical cut-off points may result in improper classification (7).

For proper classification of individuals as adequately nourished or vitamin deficient cut-off points should be established using functional criteria. The sensitivity (i.e. the chance that the test is 'positive' if a deficiency is present) and specificity (i.e. the chance that the test is 'negative' if the vitamin status is adequate) should be evaluated by determining the distribution of the test results for persons with and without established deficiency.

For instance, Kuriyama (23) examined whole blood thiamin and ETK stimulation in beri-beri patients. As a result of a significant overlap in the distribution of values these patients could not be clearly separated from the controls. The highest predictive (prognostic) value was obtained when both parameters were combined. Such an overlap can result from the fact that clinical symptoms used for diagnosis are irreversible, while biochemical values could recover again. Using the earlier, but less specific symptoms for classification of individuals may give problems as well. Buzina (24) evaluated the riboflavin status using the EGR-stimulation test in a population of children (N = 280), with a 20% incidence of subclinical riboflavin deficiency, evidenced by the presence of angular stomatitis. Only 6 out of 18 children with an EGR-stimulation ratio (α-EGR) > 1.30 showed clinical symptoms. Using these data, the sensitivity and specificity of the EGR test can be calculated as 17 and 95%, respectively. Choosing a cut-off level of 1.20, sensitivity increases up to 37%, but specificity decreases to 82%. For an 'ideal' test the sum of specificity and sensitivity should equal 200% (25). From these data it can be concluded that the EGR test is not a very accurate test to classify individuals with and without angular stomatitis. On the other hand, symptoms of angular stomatitis might not be very specific indicators for riboflavin deficiency.

The estimation of the true prevalence of B-vitamin deficiency remains a problem. Functional indices, such as physical and mental performance or work capacity, may show large variation in response. There are some examples where in-vivo functional tests rather than clinical criteria were used to establish and validate cut-off points. The tryptophan load test was used by Buzina and Brubacher (26) to establish cut-off points for the EGOT stimulation test.

Hypersegmentation of neutrophilic granulocytes is used as a functional index for the folacin status and to establish a cut-off level for serum folacin (27). There is an urgent need to develop more sensitive and practical functional tests of the vitamin status. As yet the reliability of vitamin status assessment of an individual can be improved by combining different tests and performing longitudinal investigations, rather than a single measurement. The various tests in the assessment of the same vitamin may reflect different levels of vitamin metabolism and function and will allow a better 'scaling' (or classification) of the individual's nutriture.

THE RELATIONSHIP BETWEEN B-VITAMINS AND DISEASE

The relationship between B-vitamin nutritional status and metabolic disease/pathology is rather complex. Vitamin deficiency, but also vitamin toxicity (e.g. vitamin B6), is accompanied by more and less specific clinical symptoms and physiological changes. On the other hand, as already indicated, disease may result in a conditional vitamin deficiency. Moreover, adaptive mechanisms may induce (secondary) changes in the metabolism of the vitamins, e.g. a change in distribution between compartments suggesting a vitamin depletion (or repletion) rather than a different control. This phenomenon is especially found for drugs affecting the vitamin binding proteins, e.g. hormones (5).

In several studies a relatively high incidence of 'abnormal' values of the parameters of the vitamin status were observed in groups of randomly selected hospital in-patients (28, 29). In our department at CIVO-TNO in about 20% of the patient samples (± 50,000 requests/year) we observe results outside our reference range. The important question is: does this represent 'true' vitamin deficiencies, or just abnormal biochemical tests, and are these deficiencies caused by or a consequence of the disease?

It is beyond the scope of this lecture to deal with all aspects of the relation between morbidity/mortality and the B-vitamins, and of the vitamin-drug interactions. Excellent reviews have appeared (4-6). Table 5 summarizes some of most important causative and conditional vitamin deficiencies according to the 'type' of the disease.
Some of the most relevant relationships are shortly discussed below.

B-vitamins and diseases of the gastrointestinal tract

Although most B-vitamins are absorbed mainly by passive diffusion, gastrointestinal disorders may lead to secondary vitamin deficiencies due to impaired digestion and absorption. This is often observed for vitamin B12 and folacin as specific binding proteins (e.g. Intrinsic Factor) or enzymatic processes (deconjugation of folacin polyglutamates) are involved. Therefore gastrectomy and ileal resection will result in vitamin B12 and folacin deficiency. Vitamin deficiency (esp. of vit B12) may result from coeliac disease, pancreatic insufficiency and cystic fibrosis. Folacin deficiency (and probably also of pantothenic acid and niacin) leads to changes in the structure and function of the GI-tract and thereby in impaired (vit B12) absorption. Achlorhydria and atrophic gastritis may impair the absorption of biotin, vitamin B12 and folacin, especially in advancing age (16).

B-vitamins and neurological disease

Peripheral neuropathy (e.g. in alcoholism), and other neurological disorders may be caused by a deficiency of several B-vitamins (thiamin, vitamin B6, B12, folacin and panthothenic acid). However, the specific mechanisms involved are largely unknown. Wernicke's encephalopathy is the most common manifestation of thiamin deficiency. However, genetic factors may be involved. Blass and Gibson (30) reported an abnormality in the binding of ThDP to transketolase in fibroblasts from Wernicke patients.

More recently sensory neuropathy and ataxia were found in persons taking megadoses (2-6 g/day) of vitamin B6 for long periods (2 - 40 months) (31).

Table 5. Relationship between B-vitamins and disease

Clinical disorder	Causative	Conditional
Gastrointestinal tract	FA[1], Panth, Niacin	Malabsorption syndromes (B12, FA, Biot)
Neurological disease	B1, B6, B12, FA Panth, Niacin	Alcohol abuse (B1)
Hemapoietic disease	B12, FA, B2, B6	Inborn errors of B12 and folacin metabolism; Drug therapy; alcohol abuse (FA)
Hepatic disease	B1	Liver cirrhosis (B1, B6) Hepatic failure (B6)
Renal disease	?	Uremia(B6) Hemodialysis; CAPD (B6, FA)
Cardiac failure	B1, FA, B6 (?)	--
Chronic disease	?	Cancer (Chemo-/radio-therapy) (B12, FA); Vitamin dep. inborn errors (B1, B2, B6, B12, FA, Biot); Asthma (B6)

1 Folacin

B-vitamins and hemapoietic disease

Vitamin B12, folacin and riboflavin as well as vitamin B6 are involved in the heme synthesis and erythropoiesis. Deficiencies of these vitamins leads to anemia and characteristic morphological changes in the peripheral blood smear. Vitamin B12 deficiency results in a 'trapping' of methylfolate and a defective synthesis of polyglutamates, and thereby in a (secondary) folacin deficiency (32).

B-vitamins and hepatic disease

The liver is one of the major organs for vitamin storage and metabolism. Chronic liver disease may therefore result in secondary vitamin deficiencies. Patients with liver cirrhosis have reduced PLP plasma levels, due to increased hepatic dephosphorylation. On the other hand, increased plasma PLP levels were found in fulminant hepatic failure, probably as a result of a release of the vitamin by necrotic hepatic tissue. The stage of the disease seems therefore important. In the first stage of the disease increased blood levels are observed, followed by lowered levels when stores are depleted. Thiamin deficiency has been reported in chronic liver disease in alcoholism, and as a result of hepato-renal syndrome and severe lactacidosis.

B-vitamins and renal disease

In spite of an apparently adequate intake vitamin B6 deficiency is often observed in uremic patients undergoing hemodialysis. This deficiency is probably due to an increased clearance of PLP (33). Hemodialysis and peritoneal dialysis (CAPD) may lead to higher vitamin B6 and folacin needs due to an increased loss through the dialysate (34).

B-vitamins and chronic diseases and metabolic disorders

Although malignant disease may be associated with anorexia and cachexia, the B-vitamin status seems fairly adequate in cancer patients. Radiotherapy and chemotherapy may affect the vitamin status. We observed a decrease in the levels of vitamin B12 and folacin during radiotherapy in patients with cancer of the uterus, bladder and prostate, while chemotherapy with cytostatic drugs resulted in decreased folacin levels (35). Vitamin responsive inborn errors of metabolism have been described for thiamin (e.g. Leigh's disease), vitamin B6 (e.g. homocystinuria), biotin (biotinidase deficiency), vitamin B12 and folacin metabolism (36).

Effect of drugs and alcohol

Drugs may affect vitamin metabolism in many ways, for example by impairing vitamin intake and absorption, by affecting the vitamin distribution and metabolism, or by interference of the coenzyme binding (5). Chronic alcoholics are usually malnourished and have multiple vitamin deficiencies, especially of thiamin, vitamin B6 and folacin (4). These deficiencies result from decreased dietary intake, decreased intestinal absorption (thiamin, folacin) and altered vitamin metabolism as a result of impaired liver function.

CONCLUSIONS

- Disease may be the cause as well the consequence of vitamin deficiency.
- Biochemical assessment will in principle give specific information on the adequacy of vitamin stores in the body, and confirm subclinical deficiency at an early stage, before clinical symptoms appear.
- Nowadays, a wide variety of laboratory methods for the assessment of the B-vitamin status is available. Which parameter is the best depends on many factors, and on the type of study. However, (international) standardization of analytical techniques is needed.
- For proper classification of individuals practical and sensitive functional tests have to be developed. These can be used to evaluate 'direct' methods for determination of the vitamin status, and to optimize interpretation criteria.

REFERENCES

1. Marks J. Vitamins, Their role in medical practice. Lancaster England: MTP Press Ltd., 1985.
2. DiSorbo DM, Litwack G. Biochem Biophys Res Commun 1981; 99: 1203-08.
3. Itokawa Y. J Appl Nutr 1979; 29: 5-16.
4. Nutrition, Metabolic and clinical applications. In: Hodges RE, ed. Human Nutrition, A comprehensive treatise. New York: Plenum Press, 1979: Volume 4.
5. Roe DA. Nutr Rev 1984; 42: 141-154.
6. Dickerson JWT. Biblthca Nutr Dieta 1985; 35: 44-52.
7. Underwood BA. Nutr Rev Suppl May 1986: 213-24.
8. Beaton GH. Nutr Rev 1986; 44: 349-58.
9. Brubacher G. Biblthca Nutr Dieta 1979; 28: 176-83.
10. Selhub J, Rosenberg IH. In: Wright RA, Heymsfield S, McManus CB, eds. Nutritional Assessment. Boston: Blackwell Scient. Publ. Inc., 1984.

11. Sauberlich HE, Skala JH, Dowdy RP. Laboratory tests for the assessment of nutritional status, 2nd ed. Ohio: CRC Press Inc., 1976.
12. Bamji MS. In Briggs M, ed. Vitamins in Human Biology and Medicine. Ohio: CRC Press Inc., 1981.
13. Herbert V. Trans Ass Amer Phys 1962; 75: 307-20.
14. Van der Beek E, Van Dokkum W, Schrijver J, et al. Am J Clin Nutr; (accepted for publ.).
15. Van den Berg H. Proc Int Symp. Vitamins and minerals in pregnancy and lactation, Innsbruck, 1986; in press.
16. Suter PM, Russell RM. Am J Clin Nutr 1987; 45: 501-12.
17. Solomons NW, Allen LH. Nutr Rev 1983; 41: 33-50.
18. Brin M, Tai M, Ostahever AS, et al. Nutr 1960; 71: 273-81.
19. Thurnham DI. Proc Nutr Soc 1981; 40: 155-163.
20. Bayoumi RA, Rosalki SB. Clin Chem 1976; 22: 327-335.
21. Herbeth B, et al. Clin Chem 1986; 32: 1756-1759.
22. Kübler W. Biblthca Nutr Dieta 1980; 30: 17-29.
23. Kuriyama M, Mizuma A, Yokomine R, Igata A, Otuji Y. Clin Chim Acta 1980; 108: 169-177.
24. Buzina R, Subotianec K. Proc GEN Workshop on Nutritional status assessment of individuals and population groups, Gargnano (It), 1984.
25. Gambino SR. In: Albertini A, Ekins RP, Galen RS, eds. Cost/Benefit and predictive value of Radioimmunoassay. Amsterdam: Elsevier Science Publ., 1984: 3-8.
26. Buzina R, Brubacher G. In: Koishi H, ed. Proc. Xth Internat. Congress of Nutr. Kyoto: Victory Sha Press, 1976: 469-73.
27. Pietrzik K. In: Himberg JJ, et al, eds. B-vitamins in Medicine. Braunschweig/Wiesbaden, FRG: Vieweg & Sohn, 1986: 31-43.
28. Lemoine A, Le Devehat C, Codaccioni JL, et al. Am J Clin Nutr 1980; 33: 2595-2600.
29. Leevy CM. Frank O, Baker H. Am J Clin Nutr 1965; 17: 259.
30. Blass JP, Gibson GE. N Engl J Med 1977; 25: 1367-1370.
31. Schaumberg H, Kaplan J, Windebank A, et al. N Engl J Med 1983; 309: 445-48.
32. Shane B, Stokstad ELR. Ann Rev Nutr 1985; 5: 115-141.
33. Lacour B, Parry C, Drüeke T, et al. Clin Chim Acta 1983; 127: 205-215.
34. Boeschoten EW, Schrijver J, Krediet RW, et al. Neurology, Dialysis and Transplantation; in press.
35. Schreurs WHP, Odink J, Egger RJ, Wedel M, Bruning PF. Internat J Vit Res 1985; 55: 425-432.
36. Bartlett K. Adv Clin Chem 1983; 23: 142-199.

GROWTH FACTORS IN MALNUTRITION

M.V.L. Du Caju

Dept. of Pediatrics, University of Antwerp, B 2610 Antwerp - Belgium

Normal growth in the animal and in the human is the result of multiple interactions between hormonal and nutritional factors and the target cells. Amongst the hormones involved in the regulation of postnatal growth pituitary growth hormone (GH) is one of the major stimulatory components. However, GH is not directly effective on growth cartilage but induces the generation of somatomedins (SM) or insulin-like growth factors (IGF) (1, 2).

SM/IGF's are well defined polypeptide growth factors which are mainly under GH control and which exert varying degrees of growth-promoting and insulin-like effects on cartilage and on a variety of other cells both "in vitro" and "in vivo" (3, 4).

In the human two distinct SM/IGF's have been isolated. SM/IGF I is a one-chain polypeptide with a Mr of 7646 and composed of 70 amino acid residues, while IGF II has as Mr of 7471 with 67 amino acid residues. Both growth factors are mainly produced by the liver but minor quantities are secreted by various other organs and cell systems on which SM/IGF's themselves are acting. SM/IGF I appears to be the major growth factor in postnatal life, whereas IGF II seems to be more related to fetal growth.

In normal nutritional conditions SM/IGF I concentrations, measured by radioimmunoassay, and total SM/IGF bioactivity, estimated by the incorporation of ^{35}S-sulphate in cartilage segments, mostly mimic the level of GH secretion, whereas immunoassayable IGF II levels are scarcely influenced by GH.

Malnutrition in children and in animals is mostly associated with growth deceleration in spite of paradoxical normal or elevated serum GH concentrations (5). This review summarizes the data on SM/IGF's in serum during acute fasting and chronic malnutrition and during subsequent nutritional rehabilitation. The data on serum SM/IGF I concentration indicate that this parameter may be helpful for monitoring the response of malnourished patients to nutritional therapy.

STUDIES ON SERUM SM/IGF BIOACTIVITY

In children with chronic malnutrition of different origin SM/IGF bioactivity is significantly depressed (6-8). Similar findings are observed in patients with Kwashiorkor, in which serum SM/IGF bioactivity rises during recovery and is correlated with changes in serum albumin and transferrin (9, 10). In a larger number of severely malnourished children Hintz et al. (11) demonstrated that the progressive normalization of the SM/IGF activity during refeeding was inversely correlated with the GH values.

Acute fasting in animals also decreases bioassayable SM/IGF's but nutritionally deprived rats are not able to increase their serum SM/IGF's in response to GH injections (12-14). The protein content of the diet plays an important role in the restoration of SM/IGF's during refeeding (15).

Mixing normal serum with that from malnourished children results in a dose-dependent lowering of the apparent SM/IGF activity, suggesting the presence of an inhibitory substance of SM/IGF in severe malnutrition (7). This inhibitory activity disappears progressively during refeeding (11).

STUDIES ON IMMUNOASSAYABLE SM/IGF I AND IGF II CONCENTRATIONS

SM/IGF I levels are undoubtedly decreased in serum of acute fasted animals (16, 17) and return to normal during nutritional repletion. These changes in SM/IGF I concentration are related to alterations in hepatic GH binding sites (18, 19) indicating changes in the generation of SM/IGF I.

Similarly, SM/IGF I levels are diminished both in obese and normal weight adults after a short time fasting and normalize again with improving daily nitrogen balance (20, 21).

The reduction of IGF II levels during acute fasting is less pronounced than that of SM/IGF I concentrations. Fasting also markedly inhibits the response of both SM/IGF I and IGF II to the administration of GH in GH deficient patients (22).

In Kwashiorkor and marasmus SM/IGF I and IGF II concentrations are low and return to normal with refeeding (23).

These observations indicate that the decrease in SM/IGF bioactivity observed in malnutrition and acute fasting are at least partially due to a reduction in immunoreactive SM/IGF I and IGF II but do not diminish the possible importance of bioassay inhibitors.

The inverse relationship between GH levels and SM/IGF I concentrations are in favour for the hypothesis that SM/IGF's are involved in the regulation of the secretion of GH through a negative feedback mechanism. More direct evidence for the role of the SM/IGF's on the GH secretion is provided by "in vivo" experiments in which the administration of SM/IGF I in rats suppressed the release of GH (24, 25).

In obesity SM/IGF bioactivity and SM/IGF I concentrations are perfectly normal in spite of low GH levels (7, 26).

STUDIES ON SM/IGF INHIBITORS

The nature of the inhibitor of SM/IGF bioactivity in severe malnutrition has not yet been well defined. The inhibitory activity in serum of malnourished children and acutely fasted rats is heat-labile (27, 28). In starved rat serum the inhibitor has a molecular weight between 27,000 and 40,000, is acid-stable and destroyed by trypsin (29). In diabetic rat serum several circulating inhibitors of SM/IGF's with a Mr of 250,000, 24,000 and 940 respectively, have been observed (30).

In normal human serum we have isolated a low molecular weight fraction with a Mr of 1200 which strongly inhibits the growth promoting activity and insulin-like activity in various tissues and cells, including cartilage, fibroblasts and adipose tissue. This inhibition is dose related, reversible and non-toxic. This serum fraction contains protein material but is still heterogeneous as shown by thinlayer chromatography and ion-exchange chromatography (31). Inhibitors of SM/IGF's with a similar molecular weight have been observed in serum of uremic patients but are not further characterized (32-34).

From these observations it remains possible that some of these inhibitors may be involved in conditions as in malnutrition, offering a mechanism for limiting the utilization of the available nutrients in order to preserve them for other vital processes.

CLINICAL USE OF SM/IGF I MEASUREMENTS IN MALNUTRITION

The measurement of serum SM/IGF I has been proposed by several authors as a useful means for monitoring the response of malnourished patients to nutritional therapy.

Isley et al. (21) observed a good correlation between changes in SM/IGF I and daily nitrogen balance during fasting and refeeding. They demonstrated that both protein and energy intake is required to achieve normalization of SM/IGF I levels during refeeding. Clemmons et al. (35) have shown that SM/IGF I concentrations are rising significantly by day 2 of nutritional support therapy and are normalized by day 10, without significant changes in serum prealbumin, retinal binding protein or transferrin. The measurement of SM/IGF I is a more convenient method than that of nitrogen balance during short term nutritional repletion.

REFERENCES

1. Salmon WD, Daughaday WH. J Lab Clin Med 1957; 49: 825-836.
2. Daughaday WH, Hall K, Raben MS, Salmon WD, Van den Brande JL, Van Wyk JJ. Nature 1972; 235: 107.
3. Froesch ER, Burgi H, Ramseier EB, Bally P, Labhart A. J Clin Invest 1963; 42: 1816-1834.
4. Van Buul-Offers S, Van den Brande JL. Acta Endocrinol (Copenhagen), 1979; 92: 242-257.
5. Pimstone BL, Wittmann W, Hansen JDL. Lancet 1966; 2: 779-780.
6. Du Caju MVL, Van den Brande JL. Acta Paed Scand 1973; 62: 96.
7. Van den Brande JL, Du Caju MVL. In: Raiti S. Advances in human growth hormone research. DHEW Publication, 1974b; n° (NIH) 74-612: 98-126.
8. Rappaport R, Prevot C, Czernichow P. Acta Paed Scand 1980; 69: 37-41.
9. Grant DB, Hambley J, Becker D, Pimstone BL. Arch Dis Child 1973; 48: 596-600.
10. Mohan PS, Jaya Rao KS. Arch Dis Childh 1979; 54: 62-64.
11. Hintz RL, Suskind R, Amatayakul K, Thanangkul O, Olson R. J Pediatr 1978; 92: 153-156.
12. Salmon WD. In: Pecile, Müller, eds. Growth and Growth hormone. Amsterdam: Excerpta Medica, 1972: 180-191.
13. Phillips LS, Young HS. Diabetes 1976; 25: 516-527.

14. Phillips LS, Orawski AT, Belosky DC. Endocrinology 1978; 103: 121-127.
15. Yeh JK, Aloia JF. Metabolism 1978; 27: 507-509.
16. Takano K, Hizuka N, Shizume K, Hasumi Y, Kogawa M, Tsushima T. Endocrinology 1980; 107: 1614-1619.
17. Eigenmann JE, De Bruijne JJ, Froesch ER. Acta Endocrinol (Copenhagen), 1985; 108: 161-166.
18. Baxter RC, Bryson JM, Turtle JR. Metabolism 1981; 30: 1086-1090.
19. Maes M, Underwood LE, Ketelslegers JM. Mol Cell Endocrinol 1984; 37: 301-309.
20. Clemmons DR, Klibanski A, Underwood LE, et al. Clin Endocrinol Metab 1981; 53: 1247-1250.
21. Isley W, Underwood LE, Clemmons DR. J Clin Invest 1983; 71: 175-182.
22. Merimee TJ, Zapf J, Froesch ER. J Clin Endocrinol Metab 1982; 55: 999-1002.
23. Soliman AT, Hassan AEHI, Aref MK, Hintz RL, Rosenfeld RG, Rogol AD. Ped Res 1986; 20: 1122-1130.
24. Abe H, Molitch ME, Van Wijk JJ, Underwood LE. Endocrinology 1983; 113: 1319-1324.
25. Tannenbaum GS, Guyda HJ, Posner BI. Science 1983; 220: 77-79.
26. Caufriez A, Golstein J, Lebrun P, Herchuelz A, Furlanetto R, Copinschi G. Clin Endocrinol 1984; 20: 65-70.
27. Van den Brande JL, Van Buul S, Heinrich U, Van Roon F, Zurcher T, Van Steirtegem AC. In: Luft R, Hall K, eds. Advances in metabolic disorders. Academic Press, 1975: 171-181.
28. Price DA, Wit JM, Van Buul-Offers S, et al. Endocrinology 1979; 105: 851-861.
29. Salmon WD, Holladay LA, Burkhalter VJ. Endocrinology 1983; 112: 360-370.
30. Phillips LS, Bajaj VR, Fusco AC, Matheson CK. Diabetes 1983; 32: 1117-1125.
31. Du Caju MVL. Thesis. University of Antwerp, 1986.
32. Dzurik R, Hupkova V, Cernacek P, Valovicova E, Niederland TR. Clin Chim Acta 1973; 46: 77-83.
33. Bozovic LJ, Bostrom H, Bozovic M, Lundin L. Acta Endocrinol (Copenhagen), 1975; Suppl. 199: 117.
34. Phillips LS, Fusco AC, Unterman TG, Del Greco F. J Clin Endocrinol Metab 1984; 59: 764-772.
35. Clemmons RD, Underwood LE, Dickenson RN, et al. Am J Clin Nutrition 1985; 41: 191-198.

THE USE OF BIOCHEMICAL PARAMETERS TO EVALUATE AND TO ESTIMATE THE NUTRITIONAL STATUS OF KIDNEY PATIENTS

J. Bergström

Department of Renal Medicine, Karolinska Institute, Huddinge University Hospital, Stockholm, Sweden

THE UREMIC STATE

Uremia is a toxic syndrome caused by severe glomerular insufficiency, associated with disturbances in tubular and endocrine functions of the kidneys. It is characterized by retention of toxic metabolites, derived mainly from protein metabolism, associated with changes in volume and electrolyte composition of the body fluids and excess or deficiency of various hormones.

The primary event in urine formation is the glomerular ultrafiltration process. The ultrafiltrate is reduced in volume and modified as to composition by the metabolic activity of the tubular cells. The final urine that is the end result of these processes contains urea and other nitrogenous metabolites in high concentration, and has an appropriate volume and electrolyte composition to maintain homeostasis. Clearly, the development of the uremic state is primarily the consequence of severe glomerular insufficiency, which is a prerequisite for retention of toxic, nitrogenous metabolites. However, in acute as well as in chronic renal failure, severe glomerular insufficiency is associated always with secondary disturbances in tubular functions, which may explain such aspects of uremia as hyperkalemia and metabolic acidosis.

The functional renoprival state results in the loss of endocrine functions of the kidney, insufficient production of erythropoietin contributing to renal anemia, and lack of active vitamin D (1,25-OH-cholecalciferol) leading to deficient intestinal calcium absorption and osteomalacia. Moreover, the normal kidney has an important role in catabolizing various peptide hormones, and loss of these abilities may lead to endocrine disturbances (1).

Patients with acute or chronic end-stage renal failure may exhibit a number of toxic uremic symptoms and signs even when water, electrolyte and acid base balance are apparently fully controlled.

Symptoms from the central nervous system (fatigue, somnolence, coma, cramps) and gastrointestinal tract (nausea, vomiting) predominate. Other typical symptoms and signs of uremia are anemia, bleeding tendency, itching, peripheral neuropathy, susceptibility to infection (impaired cellular immune response), and various metabolic disturbances such as glucose intolerance, lipid abnormalities and abnormal metabolism of protein and amino acids.

The most obvious explanation provided for this diversity in symptomatology is that uremia is a generalized form of autointoxication by metabolites, which are normally excreted by the kidneys (2). There is strong evidence that at least some of these metabolites are products of protein metabolism, since increased protein intake may promote toxic symptoms, especially nausea and vomiting, whereas protein restriction alleviates these symptoms, generally associated with a decrease in blood urea concentration.

CARBOHYDRATE INTOLERANCE

The existence of carbohydrate intolerance in patients with advanced chronic renal failure has been recognized since many years. Clinically this abnormality is characterized by normal or marginally increased fasting blood glucose levels and normal or moderately increased fasting insulin concentrations. Although this metabolic abnormality very seldom causes any trouble in the clinical management of uremic patients, it may have important consequences with respect to protein and lipid metabolism and may be a major risk factor for the development of atherosclerosis. Tissue insensitivity, primarily in muscle, to the action of insulin, however, is the predominant factor contributing to the glucose intolerance in most patients with chronic failure (3).

HYPERLIPIDEMIA AND ATHEROSCLEROSIS

Hypertriglyceridemia, mainly due to increase in very low density (VLDL) triglycerides and low density (LDL) triglycerides, is a common finding in uremia. Total cholesterol is generally normal but the distribution is abnormal with a raised cholesterol: triglyceride ratio less in VLDL and a lowered ratio in LDL. High density protein (HDL) cholesterol and triglycerides are low (4). All these lipoprotein disturbances are considered to be potential risk factors for atherosclerosis. Several abnormalities in apolipoprotein composition of plasma lipoproteins have been identified in uremic patients (5).

DISTURBANCES IN AMINO ACID AND PROTEIN METABOLISM

Protein malnutrition

It is well recognized that patients with advanced renal failure have a tendency towards negative nitrogen balance and muscle wasting. This may be a consequence of poor food intake because of anorexia caused by uremic toxicity. Therapeutic use of low protein diets for amelioration of symptoms of uremic toxicity may be hazardous if the minimum requirements for protein and essential amino acids of the uremic patient are not fulfilled (6).

In chronic renal failure protein metabolism may also be affected by several factors such as the uremic intoxication (per se), hormonal derangements, alterations in the metabolism of amino acids and other nutrients, secondary to uremia or renal failure.

Underlying or superimposed infections are common and may lead to profound hypercatabolism (7). Extreme catabolism is most probably a major contributing factor to the persistently high mortality in patients with acute renal failure after trauma or surgery. The protein catabolic rate may amount to 150-200 g of protein/24 h (8), and the energy requirements are increased (9).

Several reports have documented that malnutrition is frequently present in patients with chronic renal failure treated with intermittent hemodialysis (IHD) (for references, see 12). An important contributing factor may be that hemodialysis procedure per se is a strong catabolic stimulus due to loss of amino acids in the dialysate (10-13 g per hemodialysis) (10) and to catabolic processes induced by interaction between blood and the membranes in the artificial kidney (11).

Protein malnutrition is also common in patients treated with continuous ambulatory peritoneal dialysis (CAPD) (for references, see 12). Loss of protein and amino acids in the dialysate and superimposed infection (peritonitis) may be contributing factors. In both IHD and CAPD patients the protein requirements are considerably higher than in healthy individuals and non-dialyzed chronic uremic patients (12).

Plasma and muscle free amino acids

There are numerous reports of abnormal plasma amino acid concentrations in patients with chronic renal failure (13, 14). Among the consistent findings are high concentrations of several non-essential amino acids and low concentrations of essential amino acids, including the branched chain amino acids valine, isoleucine and leucine. The plasma concentration of tyrosine is low and the phenylalanine/tyrosine ratio is high. Many of the plasma amino acid abnormalities found in uremia are similar to those observed in protein malnutrition and it has been suggested that they are in part attributable to dietary inadequacy.

The largest pool of free amino acids, however, is not in the extracellular pool but in skeletal muscle. In untreated uremic patients a typical muscle intracellular amino acid pattern is observed with low concentration of valine but normal concentrations of leucine and isoleucine and low concentrations of threonine, lysine and histidine. Plasma and intracellular levels of the posttranscriptionally modified 1-methylhistidine, 3-methylhistidine, hydroxylysine and hydroxyproline are high (15) due to decreased or absent urinary excretion. Abnormal intracellular amino acid concentrations in muscle have also been observed in IDH patients (16) and to a lesser degree in CAPD patients (17). Patients on low protein diet, IHD and CAPD have also low muscle taurine (15, 17).

The low tyrosine concentration in uremia and the resulting high phenylalanine/tyrosine ratio, however, can at least partly be explained by impaired conversion of phenylalanine to tyrosine, which has been attributed to inhibition, by some uremic toxins, of the phenylalanine hydrolase activity (18).

Certain disturbances in amino acid metabolism, such as insufficient production and low plasma levels of histidine, tyrosine, and serine may be due to insufficient synthesis in the diseased kidneys.

The pathogenesis of the abnormal branched chain amino acid patterns in muscle and plasma is not known. One can only speculate whether factors related to uremic intoxication per se or hormonal alterations contribute to the valine depletion and the abnormal distribution of isoleucine and leucine across the cell membrane.

There is evidence that cysteinsulphinic acid decarboxylase, a key enzyme in taurine synthesis, is depleted in uremia (19). Low intake and increased rate of degradation of taurine may also be factors of importance for taurine depletion.

EVALUATION OF THE NUTRITIONAL STATUS IN PATIENTS WITH RENAL FAILURE

Malnutrition caused by uremia and its sequeles is a serious condition which is strongly associated with morbidity and mortality (20). Early detection of malnutrition is therefore of utmost importance. Anthropometric measurements (body weight, skinfold thickness, arm muscle circumference, isometric muscle strength etc.), evaluation of body composition by radioisotope methods (total water, intracellular water, total K, total N etc.) and immunological methods (lymphocyte counts, skin tests for delayed hypersensitivity) have been used for evaluation of nutritional status in renal failure patients (21, 22). In the following the most common biochemical methods will be briefly summarized and commented on.

Urea and creatinine

Both urea and creatinine are excreted mainly by glomerular filtration and their concentrations increase in serum as renal failure progresses. The ratio urea/creatinine in serum and the urea appearance rate in the urine (24 h urea excretion) reflect dietary protein intake in stable patients with chronic renal failure and can be used to assess the compliance to low protein diet. Urea appearance rate in IHD patients can be estimated by applying pharmacokinetic modelling and be used for nutritional assessment (23).

Serum proteins

The most commonly used serum protein for measuring nutritional status in uremic patients are serum albumin and transferrin (21, 22). Transferrin seems to be more sensitive as an indicator of malnutrition but may be affected by other factors such as iron deficiency or iron loading. Complement factor C3, C1Q (but not C4), cholinesterase, pseudocholinesterase have also been used for nutritional assessment in uremic patients (21). Pre-albumin, retinol-binding protein and ribonuclease are less suitable since they are increased, mainly due to impaired degradation by the kidneys.

Plasma and muscle amino acids

Low plasma essential amino acids and a low ratio of essential to non-essential amino acids may be signs of protein malnutrition. By following the plasma amino acid pattern repeatedly it has been possible to show beneficial effects of nutritional interventions. Among the individual amino acids valine shows the highest correlation with anthropometric parameters and seems to best reflect the nutritional status of the patient (24).

Intracellular free amino acids in muscle reflects more accurately than plasma aminograms specific deficiencies of individual amino acids in the uremic state. By supplementing low protein diets with amino acid mixtures designed to compensate for these deficiencies (high valine, presence of histidine and tyrosine) the abnormalities in muscle free amino acids in uremia have been normalized or minimized (15).

Muscle protein RNA and DNA

The skeletal musculature contains the largest pool of protein which in situations of catabolic stress and increased gluconeogenesis releases amino acids which are deaminated in the liver or are used for synthesis of liver-derived proteins. Muscle protein soluble in weak alkali (alkali soluble protein, ASP) in relation to DNA is an estimation of the amount of protein per cell unit and should give the best estimation of the protein status of the individual. Guarnieri et al. (22) have shown that ASP/DNA is low in all categories of patients with chronic uremia; the same seems to be true in hypercatabolic patients with acute renal failure (own unpublished observations). The RNA/DNA ratio is also low in chronic uremia and is associate with the low ASP/DNA ratio, probably reflecting reduced protein synthesis in the uremic state.

REFERENCES

1. Diamond JR, Yoburn DC. Nonoliguric acute renal failure. Arch Intern Med 1982; 142: 1882-1884.
2. Bergström J, Fürst P. Uraemic toxins. In: Drukker W, Parsons FM, Maher JF, eds. Replacement of renal function by dialysis. The Hague: Martinus Nijhoff's Publishers, 1983: 354-390.
3. DeFronzo RA, Alvestrand A, Smith D, Hendler R, Hendler E, Wahren J. Insulin resistance in uremia. J Clin Invest 1981; 67: 563-568.
4. Norbeck HE, Carlson L. The uremic dyslipoproteinemia: Its characteristics and relations to clinical factors. Acta Med Scand 1981; 289: 489-504.
5. Attman PO. Dietary treatment in uremia. Renal function, protein and lipid metabolism. Thesis. Näringsforskning 1979; 23: 40-59.
6. Bergström J. Discovery and rediscovery of low protein diet. Clin Nephrol 1984; 21: 29-35.
7. Grodstein GP, Blumenkrantz MJ, Kopple JD. Nutritional and metabolic response to catabolic stress in uremia. Am J Clin Nutr 1980; 33: 1411-1416.
8. Feinstein EI, Blumenkrantz MJ, Healy M, et al. Clinical and metabolic responses to parenteral nutrition in acute renal failure. Medicine 1981; 60: 124-137.
9. Mault JR, Bartlett RH, Dechert RE, Clark SF, Swartz RD. Starvation. A major contribution to mortality in acute renal failure? Trans Am Soc Artif Internal Organs 1983; 29: 390-395.
10. Wolfson M, Jones MR, Kopple JD. Amino acid losses during hemodialysis with infusion of amino acids and glucose. Kidney Int 1982; 21: 500-506.
11. Bergström J, Alvestrand A, Gutierrez A. Acute and chronic metabolic effects of hemodialysis. In: Smeby LC, Jörstad S, Wideröe TE, eds. Immune and metabolic aspects of therapeutic blood purification systems. Basel: S Karger, 1986: 254-273.
12. Bergström J. Protein catabolic factors in patients on renal replacement therapy. Blood Purif 1985; 3: 215-236.
13. Gulyassy PF, Aviram A, Peters JH. Evaluation of amino acid and protein requirements in chronic uremia. Arch Intern Med 1970; 126: 855-859.
14. Young GA, Parsons FM. Plasma amino acid imbalance in patients with chronic failure on intermittent dialysis. Clin Chim Acta 1970; 27: 491-496.
15. Alvestrand A, Bergström J, Fürst P. Plasma and muscle free amino acids in uremia: influence of nutrition with amino acids. Clin Nephrol 1982; 18: 297-305.
16. Alvestrand A, Fürst P, Bergström J. Intracellular amino acids in uremias. Kidney Int 24 1983; Suppl 16: S9-16.
17. Lindholm B, Alvestrand A, Bergström J. Muscle free amino acids in patients treated with CAPD. In: Maher JF, Winchester JF, eds. Frontiers in peritoneal dialysis. New York: Field, Rich and Assoc Inc, 1986: 400-404.
18. Young GA, Parsons FM. Impairment of phenylalanine hydroxylation in chronic renal insufficiency. Clin Sci Mol Med 1973; 48: 89-97.

19. Bergström J, Qureshi GA, Rashed Qureshi A. Inhibition of cysteine sulphinic acid decarboxylase (CSD) as a cause of taurine (TAU) depletion in uremia. Abstract. Berlin: XXIVth Congr European Dialysis and Transplant Assoc, October 1987.
20. Acchiardo SR, Moore LW, Latour PA. Malnutrition as the main factor in morbidity and mortality of hemodialysis patients. Kidney Int 24, 1983; Suppl 16: S199-203.
21. Blumenkrantz MJ, Kopple JD, Gutman RA, et al. Methods for assessing nutritional status of patients with renal failure. Am J Clin Nutr 1980; 33: 1567-1585.
22. Guarnieri G, Toigo G, Situlin R, et al. Muscle biopsy studies in chronically uremic patients: Evidence for malnutrition. Kidney Int 24, 1983; Suppl 16: S187-193.
23. Borah MF, Schoenfeld PY, Gotch FA, Sargent JA, Wolfson M, Humphreys MH. Nitrogen balance during intermittent dialysis therapy of uremia. Kidney Int 1978; 14: 491-500.
24. Young GA, Swanepoel CR, Croft MR, Hobson SM, Parsons FM. Anthropometry and plasma valine, amino acids, and proteins in the nutritional assessment of hemodialysis patients. Kidney Int 1982; 21: 492-499.

CLINICAL BIOCHEMICAL PARAMETERS TO BE USED IN GENERAL AS INDICATORS OF MALNUTRITION OF CHILDREN IN DEVELOPING COUNTRIES

Hoàng Van Son

Department of Clinical Chemistry, Vietnam - Sweden children's Hospital, Hanoi, Vietnam

CHARACTERISTICS OF MALNUTRITION OF CHILDREN IN DEVELOPING COUNTRIES

In the 1980's, malnutrition is still increasing in most parts of the world. Malnutrition is the most important cause of morbidity and mortality in young children, every hour 10 children die due to malnutrition in developing countries (1). Protein-energy malnutrition (PEM) is the most important form of malnutrition in developing countries, there after are nutritional anaemia, xerophthalmia due to vitamin A deficiency, rickets and iodine deficiency.

Increasing low birthweight infants contribute a special feature to the picture of malnutrition in many parts of the world. 90% of 19 millions of low birthweight children per year were born in developing countries, among them 10 millions were born in India, Bangladesh and Pakistan (2).

The vicious cycle between malnutrition and infection is another important feature in developing countries where both these diseases are widespread and often co-existent. Various bacterial, viral infections especially diarrhoea, tuberculosis, measles and parasitic diseases during childhood precede and aggravate malnutrition. Recurrent episodes of infections result in a serious disturbance of the nutritional status. This predisposes the child to develop further attacks of infections by lowering the immune status. Thus a vicious cycle is running (Figure 1).

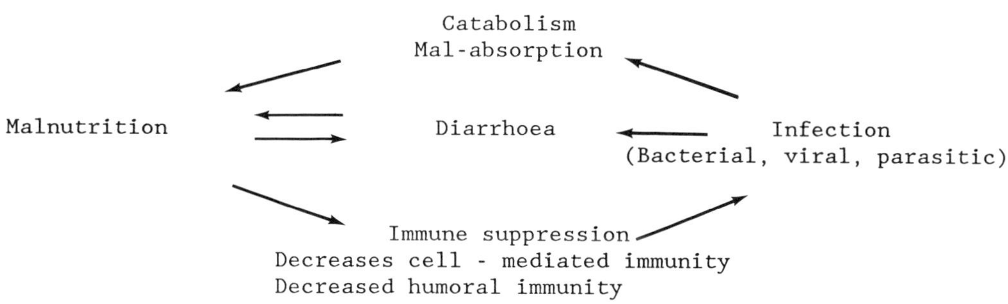

Fig. 1. The vicious cycle malnutrition - infection

A close relationship between malnutrition and diarrhoea has also been clearly shown. The data collected from hospitals demonstrated the important role of diarrhoea in about 70% of children suffering from severe PEM. Thao and Khanh (3) using scanning electronmicroscopy have shown striking changes in the jejunal mucosa of children suffering from Kwashiorkor. Severe villous atrophy with only convolutions or ridges have been seen, the crypts of Lieberkuhn were elongated and tortuous. In marasmic children the villi were partly atrophied. Garcia in Mexico (4) has also reported that the jejunal mucosa is abnormal and may be completely flat and devoid of villi. Therefore, the absorption of nutrients is greatly decreased, which in turn aggravates malnutrition. The alterations in gut function can lead to bacterial overgrowth of the small intestine (5). This may set a vicious cycle of gut dysfunction, followed by bacterial overgrowth, further damaging the gut and altering nutrient availability. Recently, in 1985, the study in Teknang (Bangladesh) (6) has also demonstrated the close relationship between diarrhoea and malnutrition.

In developing countries malaria, hookworm and ascaris infestations are the main causes of anaemia, especially of iron deficiency anaemia. Some tabus, food habits and bad quality of foods contribute to nutritional anaemia.

According to WHO's data there are 34-36 countries where vitamin A deficiency is a social problem of health. It is one of the four most important causes of blindness in the world. Every year, in Asia alone, more than half a million of children go blind due to vitamin A deficiency. More than 2/3 of them die after some weeks, 7 millions of children have symptoms of moderate deficiency of vitamin A. A close relationship between PEM and xerophthalmia has been clearly demonstrated.

The shortage and bad quality of foods is another problem of malnutrition in developing countries. Foods being stocked for long time, contaminated by moisture, fungi, worm, lack of hygiene and sanitation, of clean drinking water, lack of knowledge of child feeding, early weaning, tabus, ... all these factors contribute to malnutrition in children.

There is a big contrast between the extensivenss of malnutrition especially in rural areas and the sometimes very serious lack of equipments, machines, reagents and laboratory techniques. Routine techniques couldn't be used in large rural, mountain areas with no electricity, and even sometimes no water. Simple, cheap and easy to use techniques suitable in particular at grass root levels should be required. Thus, biochemical analysis for primary health care should be intensively and quickly studied. The criteria of evaluation of techniques such as: exactitude, sensitivity, reproducibility should be seen on the basis of real conditions of developing countries.

CLINICAL BIOCHEMICAL PARAMETERS USED AS INDICATORS OF MALNUTRITION OF CHILDREN IN DEVELOPING COUNTRIES

Protein - energy malnutrition (PEM)

PEM is the most important form of malnutrition in developing countries with highest morbidity and mortality. Therefore, biochemical parameters for investigation of protein in malnutrition are very important.

Albumine in serum might be considered as a good criterium, though there are still many contradictions. Whitehead (7) supposed that at the present time the most effective laboratory method of screening for PEM is the serum albumin. If it decreases to 3 g% and lower, all signs of

Kwashiorkor appear. McLaren et al. (8) have proposed a scoring system for severe forms of PEM (Table 1). In Vietnam, Uyên (9) reported that the concentrations of serum total protein of children suffering from Kwashiorkor were very low, half of the normal values. Hoa (10) has confirmed this phenomenon.

However some authors (12, 13) considered albumin as insensitive. As indicators of severity, prognosis and treatment of malnutrition proteins with shorter half-life would seem to be more sensitive. While the half-life of albumin is 20 days, it is only 8 days for transferrin, which could help in evaluation of subclinical malnutrition. The rapid turnover of plasma thyroxine - binding prealbumin (TBPA) (2 days) and of retinol - binding protein (RBP) (12 hours) should be more sensitive to short term changes in diet. From their study in Senegal, Ingenbleek et al. (14) concluded: measurement of TBPA (or RBP) is proposed as a method for the detection of pre - Kwashiorkor and early marasmus. Comparing a variety of methods in a study of 15 malnourished children and 73 healthy children in Nigeria. Ogunshina and Hussain (15) found that plasma TBPA was the only laboratory test that could grade malnourished patients, it is a good indice of mild PEM. In Vietnam, Nga (16) also found that PEM patients had TBPA much lower than those of normal children. More recently, in 1987, Son (17) has reported low concentrations of serum transferrin of children suffering from PEM. Similar results were reported by McFarlane et al. (12) in Nigeria.

Urea index is one indicator of nitrogen metabolism, and the percentage of urea nitrogen in total nitrogen is related to the rate of nitrogen metabolism, it depends upon the quantity of protein in foods. The urinary ratio N urea/creatinine decreases when the supply of protein is diminishing.

Table 1. Scoring system for severe forms of PEM (after McLaren, Pellet and Read, 1967)

Signs present	Points
Oedema	3
Dermatosis	2
Oedema plus dermatosis	6
Hair change	1
Hepatomegaly	1
Serum albumin (or serum total protein) (g/100 ml)	
< 1,00 (< 3,25)	7
1,00 - 1,49 (3,25 - 3,99)	6
1,50 - 1,99 (4,00 - 4,74)	5
2,00 - 2,49 (4,75 - 5,49)	4
2,50 - 2,99 (5,50 - 6,24)	3
3,00 - 3,49 (6,25 - 6,99)	2
3,50 - 3,99 (7,00 - 7,74)	1
> 4,00 (> 7,75)	0

score = sum of points. 0 - 3 = marasmus;
4 - 8 = marasmic Kwashiorkor; 9 - 15 = Kwashiorkor

both serum albumin and total protein were low in children with PEM. Khôi et al. (11) studying 303 children under 5 years of age in a town have reported that 6.6% of children had serum albumin levels < 3g%, while the group of 12-36 months of age (1.8%) being in the weaning period had pathological levels.

As a major nutrient group amino acids are important indicators of malnutrition. An extended proteolysis in all tissues occurs during starvation. The breakdown of muscle protein is the main source of free amino acids during starvation. The essential amino acids, particularly branched chain ones, being used locally for energy in muscles (leucine, isoleucine and valine), fall in untreated forms of PEM, while the non-essential amino acid glycine is rising, and the ratio valine/glycine consequently decreases. But there is no difference in amino acid patterns between marasmus and kwashiorkor patients. The concentration of amino acids and its pattern do not correlate with severity. Xuyên (18) has shown that the essential amino acids leucine, isoleucine, phenylalanine, valine, methionine were very low in kwashiorkor patients, but the level of alinine (non-essential) was normal. So the ratio non-essential/essential was very high. Among amino acids 3-methylhistidine (3-MeHis), a component of actin and myosin, which is mainly from skeletal muscle, has been considered by Young and Munro (19) as a good indicator for evaluation of the muscle protein turnover during starvation. But Heymsfield (20) supposed that 3-MeHis may be not applicable in most cases of malnutrition due to its variable turnover rate. According to Heymsfield urinary creatinine should be good for the evaluation of muscle mass of the body and good for the prognosis of PEM in children. Viteri and Alvarado (21) proposed the creatinine/height index in the estimation of protein status in PEM children. Solomons et al. in Guatemala (22) suggested the creatinine/height index as indicator for evaluation of the diet regime of children. The studies of Picou et al. in Jamaica (23), of Anasuya and Rao in India (24), of Whitehead in Uganda (25) demonstrated that at the hydroxyproline excretion, the hydroxyproline index may measure the turnover of collagen in the body and may be a good indicator of PEM. Iyengar and Rao in India (26) have shown that urinary total nitrogen, urea nitrogen and urea index increased with the low supply of protein. Urea index and 3-MeHis may differentiate malnourished children having negative nitrogen balance from those having positive nitrogen balance.

According to Scott and Berger (27, 28, 29) albumin, transferrin and amino acids are difficult to use in the evaluation of nutritional status in infancy. These authors have suggested plasma alkaline ribonuclease to be a good indicator in following of children's malnutrition and at comparison of different feeding diets.

Besides protein, glycaemia and cholesterolemia have also been studied in malnutrition. Blood glucose is low in malnutrition, and it seems to correlate with reduced liver glycogen. It is associated with an increased glucose-6-phosphatase activity in the liver. Muscle glycogen is also low. Dinh et al. (30) have shown low concentrations of cholesterol, phospholipides, triglycerides and essential fatty acids in malnourished children, but no differences of values of total fatty acids and non-essential fatty acids between malnourished and normal children. After one month of treatment the levels of linoleic acid became normal, but homogammalinoleic acid levels were normal only after 6 months of treatment.

Nutritional anaemia

Besides PEM, nutritional anaemia is very common in developing countries.

Serum ferritin, iron, total iron-binding capacity (TIBC) should be used if possible, but very simple tests such as haemoglobin and hematocrit can be performed easily in developing countries.

Simmons et al. (31) reported that in Jamaica, 69.1% of children of preschool age had haemoglobin levels below 11 g%, 46.3% below 10 g% and

10.9% below 8 g%. Lowest values have been found in children of 6-23 months being in weaning period. Main cause of anaemia was malnutrition, same patterns in rural and urban areas were found. In English-speaking Caribbean countries and Suriname (32), the main cause of nutritional anaemia was iron deficiency probably due to an inadequate intake with a low absorption of iron. The percentages of preschool children having Hb levels below 11 g% were: In Jamaica: 76%, in Cayman Islands: 42%, in Guyana: 41%, in Saint Lucia: 14.3%.

The percentages of schoolage children having Hb levels below 12 g% were: in Middle Caicos Islands: 100%, in Grenada: 65%, in Guyana: 57%.

In a filed study of 149 preschool children of 5 communes in Mexico, Deway (33) found 14% of them having low Hb levels, 16% having low hematocrit and 30% having low transferrin saturation.

The frequency of anaemia in 2 villages in Bangladesh (6) was 42%, and 40% of population had iron deficiency. One supposed that the bangladeshi food may inhibit the absorption of iron.

In one mountain area of South Vietnam, Kiêt (34) showed the highest frequency of nutritional anaemia (Hb levels below 10 g% (22.7%) in children being in weaning period.

MacDougall et al. (35) have used Hb, iron, TIBC and ferritin studying 48 children suffering from PEM in South Africa. Most of them had low values of iron-Hb. Hb, TIBC and ferritin correlated with the success of treatment.

Xerophthalmia

It is very striking to note that five children go blind every hour, five children go partially blind every hour in developing countries (1). So, the biochemical analysis for detecting xerophthalmia are very important, among them plasma vitamin A and RBP are the most sensitive indicators. Plasma vitamin A levels fall below normal limits before functional and structural changes occur.

On the basis of extensive epidemiological studies of 14,238 Vietnamese preschool children, Giây and Nhân (36) have estimated that the frequency of xerophthalmia was 0.78%, in which 0.08% was with active corneal lesions (X2, X3).

Santos et al. (37) studying 10,922 children aged 0-12 years old in Brazil, have demonstrated a close relationship between PEM and xerophthalmia in preschool children.

Tarwojo et al. (38) supposed that diet may be an important factor in the genesis of xerophthalmia in Indonesia. Serum vitamin A levels of the cases with active corneal lesions were 11.6 microgram/100 ml, of the cases with Bitot's spots were only 8.5 microgram/100 ml. 23% of 100 bangladeshi children suffering from severe PEM had xerophthalmia and were malnourished (40). Among them 22% had corneal lesions (X3A, X3B). Anh et al. (41) have shown that the serum RBP levels of children having clinical signs of vitamin A deficiency were very low, and 47% of the "normal" children (without clinical signs) has also low levels of RBP. Bhaskaram in India (42) have got following results: serum vitamin A of children suffering from moderate PEM were 8.3 ± 0.72 microgram/100 ml, from severe PEM were: 7.7 ± 0.76 microgram/100 ml.

Some other diseases and analysis

The techniques of estimation of vitamin D and its metabolites 25(OH)D3 are complicate and not easy to do, but some routine analysis as alkaline phosphatase, phosphorus and calcium could be performed in developing countries. The activity of alkaline phosphatase may increase even before the onset of rickets. Hai et al. (43) have shown that rickets often accompanied malnutrition in very young infants, 32% of them had high activity of alkaline phosphatase while their calcium and phosphorus levels were low.

Sedrani in Saudi Arabia (44) found low levels of 25(OH)D3 and phosphorus, high activity of alkaline phosphatase in rachitic children. After treatment these indicators changed significantly.

In the rice eating countries of South East Asia, there is increasing risk of thiamin deficiency because of the increasing use of small milling machines and bad quality of stocked rice. Erythrocyte transketolase is very sensitive but complicate to measure, but blood pyruvate should be used as indicator of beri beri in developing countries. Son et al. (45) found high levels of blood pyruvate in 20 cases of beri beri which became normal after treatment.

In some developing countries such as in Central Africa and New Guinea, iodine deficiency occurs, and endemic cretinism was found. Serum protein - bound iodine (PBI) is a good indicator of this disease, while serum thyroid stimulating hormone, T3 and T4 may be not suitable in developing countries.

Some other analysis would be mentioned in the list of biochemical techniques to be used in malnutrition. A close relationship between malnutrition and diarrhoea has been demonstrated, therefore the measurement of serum potassium, sodium and chloride should be suitable. Children suffering from PEM are often potassium deficient. Dien et al. (46) reported decreased potassium concentrations in Kwashiorkor children, lower than those in marasmic patients.

In recent years, it has been shown that the hair content of trace elements in particular zinc, may provide a reliable means of detecting deficiency at an early stage. But the techniques of analysis are still not widely used in developing countries.

The impairment of the immune status in malnutrition has been showed. In children suffering from PEM, cell-mediated and humoral immunity decreases. Hoa (47) found that in PEM, T and B lymphocytes, lymphocyte transformation and macrophage diminished, total complement, C3 and IgG decreased, but IgA and IgM increased. Smale et al. (48) also found that C3 and IgG decreased significantly with starvation and returned to baseline with repletion.

SUGGESTED CLINICAL BIOCHEMICAL ANALYSIS FOR MALNUTRITION IN DEVELOPING COUNTRIES

The clinical biochemical methods applying in developing countries should be suitable for the requirement of detecting malnutrition in a large population, simple and easy to do with inexpensive apparatus, but accurate and may be sensitive at some extent. Three broad groups should be suggested (Table 2):

1. Group of screening tests for routine field work.
2. Group of basic tests.
3. Group of important tests for appropriate conditions.

Table 2. Suggested clinical biochemical analysis for malnutrition

Screening tests	Serum albumin (or serum total protein), haemoglobin
Basic tests	Serum albumin, TBPA, RBP, haemoglobin, iron
Important tests	Serum TBPA, transferrin, RBP, amino acids, glycaemia, vitamin A, alkaline phosphatase, calcium, phosphorus, iron, TIBC, urinary N urea/creatinine

The biochemical analysis should be performed and interpreted together with anthropometric assessments which are often accurate and easy to do in field. Some anthropometric indicators may be useful: weight/age, length/age and weight/length.

Some haematological techniques such as hematocrit, haemoglobin concentration, should be done together also to complete the picture of nutritional status.

REFERENCES

1. Indirabai K. Prevention of blinding malnutrition in India. J Trop Pediatr 1986; 32: 317-318.
2. Hofvander I. International comparisons of postnatal growth of low birthweight infants with special references to differences between developing and affluent countries. Acta Paediatr Scand 1982; suppl 296: 14-18.
3. Thao PB, Khánh NG, Nguyên PN, Mán NV, Lôc NK. Study of jejunal mucosa of malnourished children using scanning electronmicroscopy. Proceedings of the scientific meeting of the faculty of Medicine of Hanoi 1985. Hanoi, faculty of Medicine, 1985: 63-65.
4. Garcia S. Malabsorption and malnutrition in Mexico. Am J Clin Nutr 1968; 21: 1066.
5. Neale G, Gompartz D, Schonsby H, Tabaqchall S, Booth CC. The metabolic and nutritional consequences of bacterial overgrowth in the small intestine. Am J Clin Nutr 1972; 25: 1409.
6. Hall A. Report of nutrition working group. Annual Report 1985, Dhaka, International centre for diarrhaeal disease research, 1986: 22-29.
7. Whitehead RF, Coward WA, Lunn PG. Serum albumin concentration and the onset of Kwashiorkor. Lancet 1973; i: 63.
8. McLaren DS, Burman D. Text book of pediatric nutrition, 2nd ed. Edinburgh: Churchill Livingstone, 1982: 106.
9. Uyên LT. Alteration of some biological indicators in malnourished children. Pract Med (in Vietnamese) 1982; 28: 4-7.
10. Hoa VD, Hiên VT, Diên DN, Quy T. Serum protein of children suffering from mild, moderate and severe protein - energy malnutrition. Proceedings of the scientific meeting of the faculty of Medicine of Hanoi 1985. Hanoi, faculty of Medicine, 1985: 67.
11. Khôi HH, Binh HX, Thành LT. Evaluation of protein status in children by measuring of serum albumin. Pract Med (in Vietnamese) 1986; 32: 37-45.
12. McFarlane H, Ogbeide MI, Reddy S, et al. Biochemical assessment of protein-calorie nutrition. Lancet 1969, i: 392-395.
13. Starker PM, Gump FE, Askanazi J, Elwyn DH, Kinney JM. Serum albumin levels as an index of nutritional support. Surgery 1982; 91: 194-199.

14. Ingenbleek Y, Van den Schrieck HG, De Nayer P, De Visscher M. Albumin, transferrin and the thyroxine-binding prealbumin, retinol-binding protein (TBPA-RBP) complex in assessment of malnutrition. Clin Chim Acta 1975; 63: 61-67.
15. Ogunshina SO, Hussain MA. Plasma thyroxine binding prealbumin as an index of mild protein-energy malnutrition in Nigerian children. Am J Clin Nutr 1980; 33: 794-800.
16. Nga QL, Thuy LN, Tuyêt PY, Nga TT, Hà NM. Biochemical characteristics of some serum proteins related to nutritional status of children. Med Biochem (in Vietnamese) 1985: 28-32.
17. Son HV, Nga QL, Yén DN, Hà NM. Serum transferrin of children suffering from protein-energy malnutrition. Pediatrics (in Vietnamese) 1987: 1-6.
18. Xuyên NK. Changes of plasma amino acids concentrations in children suffering from protein-energy malnutrition. Pract Med (in Vietnamese) 1982; 28: 7-10.
19. Young VR, Murto HN. N-methylhistidine (3-methylhistidine) and muscle protein turnover: an overview. Fed Proc Fed Am Soc Exp Biol 1978; 37: 2291.
20. Heymsfield SB, Artega C, McManus C, Smith J, Moffitt S. Measurement of muscle mass in humans: validity of the 24-hour urinary creatinine method. Am J Clin Nutr 1983; 37: 478-494.
21. Viteri FE, Alvarado J. The creatinine-height index: its use in the estimation of the degree of protein depletion and repletion in protein calorie malnourished children. Pediatrics 1970; 46: 696-706.
22. Solomons NW, Torun B, Caballero B. Intact whole cow's milk is equivalent to lactose-hydrolysed milk for the early recovery from protein-energy malnutrition. Am J Clin Nutr 1982; 35: 830.
23. Picou D, Alleyne GAO, Waterlow JC, Seakins A. Hydroxyproline and creatinine excretion in protein-depleted infants. Biochem J 1965; 95: 18.
24. Anasuya A, Rao BSN. Urinary excretion of hydroxyproline in Kwashiorkor. Lancet 1966; 1: 94-95.
25. Whitehead RG. Biochemical tests in differential diagnosis of protein and calorie deficiencies. Arch Dis Child 1967; 42: 479-484.
26. Iyengar AK, Rao BSN. Effect of varying energy and protein intakes on some biochemical parameters of protein metabolism. Am J Clin Nutr 1982; 35: 733-740.
27. Scott PH, Berger HM, Wharton BA. A critical assessment of plasma alkaline ribonuclease as an indicator of protein nutritional status in infancy. Ann Clin Biochem 1984; 21: 357-362.
28. Scott PH, Berger HM, Kenward C, et al. Effect of gestational age and intrauterine nutrition on plasma transferrin and iron in the newborn baby. Arch Dis Child 1975; 50: 796-798.
29. Scott PH, Berger HM, Wharton BA. The effect of protein quality on plasma amino acid levels in low birthweight babies. In: Amino acid analysis. Chicester: Ellis Horwood, 1981: 188-194.
30. Dinh NL, Phuöc DT, Béréziat G. The role of essential fatty acids in the treatment of malnutrition. Proceedings of the International conference on applied nutrition, Hanoi, 25-29/4/1986. Hanoi: Institute of Nutrition, 1986: 61-63.
31. Simmons WK, Jutsum PJ, Fox K, et al. A survey of the anaemia status of preschool age children and pregnant and lactating women in Jamaica. Am J Clin Nutr 1982; 35: 319-326.
32. Simmons WK, Gurney JM. Nutritional anaemia in the English-speaking Caribbean and Suriname. Am J Clin Nutr 1982; 35: 327-337.
33. Dewey KG. Nutrition survey in Tabasco, Mexico: nutritional status of preschool children. Am J Clin Nutr 1983; 37: 1010-1019.
34. Kiêt DP, Truc DB. Anaemia of Édé children. Investigations and studies of the Institute for protection of children's health 1980-1985. Hanoi, 1986: 67.

35. MacDougall LG, Moodley G, Eyberg C, Quirk M. Mechanisms of anaemia in protein-energy malnutrition in Johannesburg. Am J Clin Nutr 1982; 35: 229-235.
36. Giay T, Nhân NT, Khôi HH. Control of xerophthalmia and vitamin A deficiency in Vietnam. Proceedings of the seminar on Vitamin A deficiency and xerophthalmia, Hanoi, 28-29/10/1986. Hanoi: Institute of Nutrition, 1986: 2-10.
37. Santos LMP, Dricot JM, Asciutti LS, d'Ans CD. Xerophthalmia in the state of Paraiba, northeast of Brazil: clinical findings. Am J Clin Nutr, 1983; 38: 139-144.
38. Tarwotjo I, Sommer A, Soegiharto T, Susanto D, Muhilal. Dietary practices and xerophthalmia among Indonesian children. Am J Clin Nutr 1982; 35: 574-581.
39. Brown KH, Gilman RH, Gaffar A, et al. Infection associated with severe protein-calorie malnutrition in hospitalized infants and children. Nutr Res 1981; 1: 33.
40. Nga TT, Há DN, Ninh QT. Some remarks about clinical manifestations of vitamin A deficiency induced xerophthalmia in children suffering from severe malnutrition at the Institute for protection of children's health in 1982. Proceedings of the seminar on Vitamin A deficiency and xerophthalmia. Hanoi, 28-29/10/1986. Hanoi: Institute of Nutrition, 1986: 40-43.
41. Ahn HM, Châu LB, Khôi HH. Assessment of vitamin A deficiency in preschool children by determination of retinol-binding protein (RBP) in serum. Proceedings of the seminar on Vitamin A deficiency and xerophthalmia. Hanoi, 28-29/10/1986. Hanoi, Institute of Nutrition, 1986: 35-39.
42. Bhaskaram P, Reddy V, Raj S, Bhatnagar RC. Effect of measles on the nutritional status of preschool children. J of trop Med and Hyg 1984; 87: 21-25.
43. Hâi LT, Trá LN, Diên DN. Clinical and laboratory characteristics of rachitic children suffering from malnutrition. Proceedings of the scientific meeting of the Faculty of Medicine of Hanoi, 1985. Hanoi: Faculty of Medicine, 1985: 109-110.
44. Sedrani SH, Elidrissy AWTH, EL Arabi KM. Sunlight and vitamin D status in normal Saudi subjects. Am J Clin Nutr 1984; 38: 129-132.
45. Son HV, Phúc HH, Yên DN, Haot LT, Khôi HH, Anh HM. Blood pyruvic acid and thiamine deficiency. Med Biochem (in Vietnamese) 1986: 23-32.
46. Diên DN, Tuân DM. Some clinical and biological characteristics of severe malnourished children. Investigations and studies of the Institute for protection of children's health 1980-1985. Hanoi, 1986: 6.
47. Hoa VD. Some indices of the immune status of Vietnamese normal and severe malnourished children. Thesis. Hanoi: Faculty of Medicine of Hanoi, 1983.
48. Smale BT, Mullen JL, Hobbs CL, Buzby GP, Rosato EF. Serum protein response to acute dietary manipulation. J Surg Res 1980; 28: 379-388.

CHAPTER 18
MASS-SPECTROMETRY

PRINCIPLES AND APPLICATIONS OF MASS SPECTROMETRY IN CLINICAL CHEMISTRY

S.P. Markey

Laboratory of Clinical Science, NIMH, Bethesda, MD 20892, USA

INTRODUCTION

Why should clinical chemists now accept mass spectrometry as another routine analytical laboratory tool? The task of clinical chemistry is to measure, as precisely and accurately as is meaningful, in a timely and cost-effective manner, substances which are known to reflect human health and disease. Many of these substances are defined by their primary molecular architecture and are thus candidates for any measurement technique which is chemical structure specific; other substances are defined by their function (enzymes, infective agents) and are indirectly measurable by determining the change in the quantity of a substrate or product with time. Mass spectrometers are the most accurate and precise physical measurement tools in analytical chemistry. However, in the past, mass spectrometers have been neither cost effective nor capable of timely results, and have been peripheral to the routine needs of most clinical laboratories. It is the purpose here to describe the physical principles of mass spectrometric analyses, and highlight changes which have made new instrumentation appropriate for clinical laboratories.

INSTRUMENTATION

There now exist dozens of varieties and models of commercial mass spectrometers, all of which share several common features. First, in order for a molecule to be directed in space, it must be ionized so that electric or magnetic fields will influence its motion. Second, in order for that ionized molecule to be measured, it must be in a vacuum sufficient that it will complete the mass separation and detection processes unaffected by collisions with background gas molecules. Thus, all mass spectrometers must have means of forming ions either in a high vacuum assembly or a means of transmitting ions from atmospheric pressure into a vacuum chamber. In either case, a prominent and major physical characteristic of all mass spectrometers, especially their initial price and maintenance, is the required high vacuum system and its associated control electronics.

Beside the vacuum system and the requirement to ionize, there are functional elements of all mass spectrometers which are logical categories for discussion.

- Sample introduction system - discrete or continuous
  - Gas, liquid, solid
  - Chromatographic - gas, liquid
- Ionizers
  - High pressure - atmospheric pressure ionization
  - Medium pressure - thermospray, chemical ionization (CI), fast atom bombardment (FAB)
  - Low pressure - electron impact (EI), secondary ion, plasma or laser desorption
- Mass Separators - high or low resolution
  - Magnetic, quadrupole, time-of-flight, ion trap, ion cyclotron resonance, tandem mass analyzers
- Detection/Control
  - Computer recording; data processing, instrument control.

The selection of any of the specific elements in the above categories effects dramatically the other instrument characteristics and requirements as described below. While the list above is extensive, the specific commercial instruments which meet the demands of the clinical chemist are limited, and consequently the scope of this review has been limited to the subjects most relevant to the clinical laboratory. Texts on mass spectrometry will detail on each of these topics are recommended for further information (1, 2, 3).

SAMPLE INTRODUCTION SYSTEMS

Because mass spectrometers operate at high vacuum (10^{-5} to 10^{-8} torr) obtained by using oil diffusion or turbomolecular pumps backed by mechanical oil pumps, much of the success and cost of a spectrometer is determined by the vacuum system design. This is especially true for sample introduction systems - the mean by which the analyst delivers successive samples for mass analysis.

Discrete, manually controlled or automated systems for delivery of gases, liquids, or solids into high vacuum have been highly developed, but are of limited applicability in clinical chemistry. Respiratory and blood gas mass analyzers are one special variety which have been commercialized for on-line monitoring of anaesthetics, CO_2, N_2, and O_2 (4). Multiple patients can be sampled in sequence at rates allowing breath-by-breath measurement using a single, simple, desk-top spectrometer pre-calibrated to measure fixed masses with sensitivity standards interposed. The instrumentation runs under computer control continuously, 24 hr/day for months at a time because only gases are admitted; the spectrometer operates in a static mode, not requiring scanning; and the detectors are simple, fixed gain amplifiers. The task of such a spectrometer, being highly repetitive and thoroughly defined, has enabled manufacturers to refine these products to high standards of reliability.

A second highly automated gas analyzer is the isotope ratio mass spectrometer, designed for accurate and precise determination of $^{2}H/^{1}H$, $^{13}C/^{12}C$; $^{15}N/^{14}N$, and $^{18}O/^{16}O$. Like the respiratory gas analyzer, these are highly developed spectrometers with automated repetitive gas inlet systems. In clinical research centers, numerous metabolic tests have been devised using a labeled substrate (e.g. ^{13}C-enriched glucose, ^{13}C-labeled glycerides, ^{13}C-amino acids, ^{2}H and ^{18}O enriched water) and then trapping expired $^{13}CO_2$ or sampling body fluids to determine the patient's rate of substrate utilization (5, 6, 7). Clinical tests formerly performed with radiolabeled substrates have been adapted for stable isotopes over the past 20 years, and the instrumentation exists to measure isotope ratios reproducibly, although it has not found as large a market as the respiratory gas analyzer.

Inlet systems for liquids and solids, while appropriate for many industrial and research tasks with mass spectrometers used for organic compound analysis, are of marginal interest in clinical chemistry because mass analysis of unseparated liquids or solids is not discriminatory. The resulting spectra are composites of all of the components present, dominated by those of highest concentration which are, invariably, of least diagnostic interest for clinical analyses. This status may change if tandem mass analyzers, described below, are developed for routine usage.

The use of on-line chromatographic inlet systems, either gas or liquid, comprise the major sample introduction systems of interest to the clinical chemist. Gas chromatography-mass spectrometry (GC-MS) has evolved recently to the standards demanded of clinical laboratory instrumentation, in part, because fused silica capillary columns have simplified interfacing the atmospheric pressure chromatographic separation to the high vacuum of the ionization chamber (8). Also, fused silica columns have improved separation and mixture resolution further extending the utility of gas phase analyses. Columns are extended through a simple heated tube directly into the iron source. The gas load used with analytical columns can be accommodated entirely by the mass spectrometer's pumping system without requiring a helium separation device or other plumbing which formerly made GC-MS an artful practice. In addition to column technology, accompanying improvements in GC auto-injectors, and development of bench-top mass spectrometers have made these units common in clinical laboratories.

Liquid chromatography-mass spectrometry (LC-MS) is a relatively recent development when compared to the almost 25 years of GC-MS evolution. Over the past five to seven years, LC-MS systems have been changing from mechanical to direct vaporization methods, and there are expectations that interface designs will continue to improve (9). Difficulties in interfacing LC and MS are presented by the LC solvent and the nature of the analytes. In GC, 1 mL $min.^{-1}$ or less of helium is readily accommodated by high vacuum systems and the analytes are inherently vaporizable or have been derivatized chemically for that purpose. In contrast, LC solvents are commonly aqueous buffers with organic modifiers, and the analytes of interest are frequently polar, non-volatile, thermally-labile substances. Further, the gas load from vaporized liquids is about 1000-fold that from GC. Mechanical moving belt systems, onto which the LC effluent is continuously sprayed or vaporized, were the first successful commercial systems and remain a most useful solution for many problems. The sample, deposited in a discrete volume on an inert belt surface, is transported through differentially pumped chambers into high vacuum where it can be vaporized and ionized by a variety of thermal techniques. Of particular value is the fact that the moving belt offers a means to automate simply the delivery of samples of solids dissolved in liquids, whether or not separated by LC. Most laboratories have been reluctant to accept the moving belt solution to LC-MS because it appears mechanically complex, fragile, and expensive to operate and maintain. However, in those laboratories which have persisted, the belt has proven itself highly reliable, and there may well be circumstances in which it is the most cost-effective interface.

Most current LC-MS interest is directed toward the thermospray interface rather than the moving belt (10). This interface combines both sample introduction and ionization, accommodating aqueous solvents with volatile buffer salts at flow rates of 1 mL $min.^{-1}$, and contains no moving parts. The LC effluent is directed into a heated capillary tube from which it vaporizes into a vacuum chamber. Ions, formed by gas phase processes, are continuously sampled by extraction lenses into a mass spectrometer. The thermospray interface represents a simplification of the first effects to sample vaporized LC effluents in a mass spectrometer, most of which used very elaborate schemes to handle aqueous solutions. When it was realized that analytically useful ions were present in the vaporized buffered aqueous solutions, particularly NH_4OAc/methanol mixtures, Vestal and colleagues devised and commercialized thermospray interfaces which have been adapted to all types of organic mass spectrometers (10). Critical to routine applicability of thermospray interfaces are the "tuning" of vaporizer temperature with respect to the analyte, the eluting solvent composition, and the solvent flow and instrument pumping system. As a consequence, conditions published for the analysis of a particular compound on one instrument may not be translatable to another laboratory. Measurement and optimization of all these variables are areas of current development which will determine how readily acceptable thermospray LC-MS will be in clinical laboratories. However, this technique is most appealing for repetitive analyses of polar substances in solution, and like GC-MS, is being incorporated on fully automated bench-top spectrometers.

IONIZERS

Ionizers are discussed with regard to the pressure domain in which they operate because the vacuum design defines which introduction systems and mass separators are compatible. For example, the atmospheric pressure ionization source requires a very expensive differential pumping system in order to transfer ions from atmospheric pressure to 10^{-7} torr for mass analysis. Medium pressure sources still require differential pumping, but because the ionization region may be at 10^{-3} torr, a lens with a 1-2 mm hole may be adequate to separate the ionization region from the mass analyzer. Typically both regions have diffusion or turbomolecular pumps, and the pressure is raised only during sample analysis. Low pressure ionization sources can be built on a single pumped chamber, and consequently are components of the least expensive mass spectrometers.

Electron impact ionization, the most common low pressure technique, uses electrons drawn from a heated tungsten or rhenium filament across a chamber to a collector. By maintaining a voltage difference of >12-15 volts (typically 70 V) between the filament and chamber, neutral molecules in the chamber will be exposed to an electron beam with sufficient energy to cause ionization by removing one of an analyte molecule's electrons, usually a non-bonded outer shell electron from a heteroatom such as N, O, halogen, or S. This event may leave the resulting ion, the molecular ion ($M^{+\cdot}$), with sufficient excess energy that bond breakage results, with the formation of thermodynamically more stable fragment ions. The mass of the fragment ion differs from that of the molecular ion by the mass of the resulting uncharged neutral or radical.

$$M + e \longrightarrow M^{+\cdot} + 2e$$

$$M^{+\cdot} \longrightarrow A^{+} + B.$$

The components required to produce electron impact ionization are well defined and have been thoroughly engineered. Filament assemblies are rugged, lasting months or years, and the temperature controlled stainless steel ionization chambers do not require frequent cleaning or maintenance because they are operated in an inherently clean environment. Most importantly, the process of electron ionization is highly reproducible - the same ionization and fragmentation characteristics will be displayed by a given molecular structure on every mass spectrometer. It is this characteristic which first made EI-MS valuable for organic structure determination, and subsequently has been exploited for compiling libraries of fragmentation patterns, as "molecular fingerprints" (11). While electron impact ionization is the accepted basis of qualitative and quantitative organic mass spectrometry, it may produce such extensive fragmentation that only low mass fragment ions, lacking in complete structure specificity are present. An example is the compound ephedrine (MW 165), which does not display fragment ions above m/z 100 in its EI spectrum because the ionized molecule can best stabilize by forming a positive ion at m/z 58 and a benzylic radical (12).

Chemical ionization (CI) is an alternative "soft" ionization technique. CI uses electron impact to ionize an added reagent gas like methane or ammonia at 10^{-3} torr, and then secondary ion-molecule collisions generate clouds of ions which will chemically ionize analytes. In the case of ephedrine in a CI source, proton transfer from methane reagent gas results in a spectrum with a prominent protonated molecular ion (MH^+) at m/z 166 and a major ion for the loss of water m/z 148 (12). CI provides useful complimentary information for many of those compounds which fragment extensively on EI, but at a cost of some instrumental complexity, due to the requirement to differentially pump the ion source and analyzer segments of the spectrometer.

Both EI and CI ionization require that the analyte molecule be in the vapor phase in the ion source, a condition which cannot be met by many polar and thermally labile biomolecules. Surface desorption/ionization techniques have greatly extended mass spectrometry capability in recent years (13). By directing high energy particles either from an ion/atom gun, nuclear disintegration of ^{252}Cf, or a burst of laser energy at a sample on a target, all types of organic molecules can be desorbed and ionized. If an ion or atom beam is directed at a thin film of analyte in a viscous liquid such as glycerol, desorption/ionization is enhanced and a stable beam of cationated or protonated ions can be measured. This latter technique, termed fast atom bombardment (FAB), has gained wide acceptance because it is a reproducible ionization method and readily retro-fitted to spectrometers with differential pumping systems (14). All of the desorption/ionization techniques produce ions with relatively little accompanying fragmentation. As a result, they are more likely to be applied to clinical chemistry problems, when coupled with tandem mass spectrometers, to give the required specificity of analysis (3).

MASS SEPARATION

Magnetic deflection of ions in vacuum, the method employed in the earliest mass spectrometers in the 1920's, is still one of the most frequently encountered mass separation techniques. Lighter ions will be deflected more than heavier ions, when accelerated (by formation in a field several kV above ground) into a magnetic field. If a fixed detector is placed at one position, an ion of a given m/z can be focused on the detector by increasing or decreasing the magnetic field strength, or by varying the acceleration voltage which propelled the ions into the field. Magnetic deflection mass spectrometers frequently use electric field energy focusing of ions, in series with magnetic deflection, in order to correct for differences in initial kinetic energy which the ions have at the moment of acceleration. Such double focusing spectrometers have greater ability to separate ions of varying m/z, termed "resolving power" or "resolution". Ions of different elemental composition, but the same nominal mass, may be distinguished by accurate measurement of their fractional mass, since each element has a characteristic fractional mass relative to carbon, 12.000 (e.g., H-1.00782, O-15.9949). The enhanced resolution of a double focusing magnetic deflection analyzer increases the specificity of measurement in biomedical problems, because it is possible to discriminate between elemental compositions which have the same nominal mass but differ in fractional mass. For example, the acetyl ion C_2H_3O (43.0184) can be distinguished from the hydrocarbon (43.0547), amine C_2H_5N (43.0421), diazo $CH_3N_2^+$ (43.0296), nitroso CHNO (43.0058), and fluorocarbon C_2F (42.9984) ions by a double focusing spectrometer with a resolution of >3500.

The quadrupole mass analyzer is the type most often encountered in clinical chemistry laboratories, especially in bench-top units. The physical principle of mass separation is one of mass filtration rather than trajectory discrimination, as in the magnetic deflection spectrometer. An ion in an electric field will move toward a surface of opposite charge; by rapidly alternating the polarity of four surrounding surfaces using radiofrequences, an ion may be balanced and held in the center of that field. If the frequency is too slow or fast, the ion will collide with one of the surfaces. Ions of a given m/z injected at low energy into a field defined by four rods (quadrupoles) will either traverse the length of the rod assembly with a spiralling trajectory, or will describe an unstable trajectory and never reach the detector assembly. Scanning a spectrum is possible by systematically changing the magnitude of the direct current/ radiofrequency voltages in a fixed ratio - a task readily accomplished in milliseconds.

The comparison of magnetic deflection vs. quadrupole mass analyzers does not produce immutable truths, since the commercial technology being applied to the manufacture of both varieties is rapidly changing. Most instrument manufacturers offer both types, distinguishing between them by their options and prices. Both magnetic and quadrupole instruments are fully computer controlled, fast scanning, and easily used. Both use the same types of inlets and ionization sources, but quadrupole sources can operate at either ground potential or a few volts (<20) off ground, whereas magnetic sector instruments require ionization sources to be several thousand volts from ground. The consequence is that construction of quadrupole inlet systems tends to be simpler. In general, the cost of building and maintaining a small quadrupole analyzer is less than that of a comparable magnetic analyzer for low resolution analysis (i.e., unit m/z resolution), and thus the bench top units being marketed for clinical chemistry use quadrupole mass filters. For high stability and reliability, permanent magnetic mass analyzers with fixed collectors are most successful in isotope ratio and respiratory gas analysis. For laboratories requiring high performance and high resolution, double focusing magnetic deflection instruments dominate the market.

The use of tandem mass analyzers is becoming increasingly popular, and as engineering refinements are made these instruments may gain acceptance in clinical chemistry (3). The advantage of tandem mass analysis is that complex mixtures can be analyzed without prior separation. The use of soft ionization techniques, such as FAB, chiefly produces molecular ions. The molecular ion of a particular analyte can be separated from the mixture in a first mass analyzer, fragmented by collision with an inert gas in an intermediate chamber, and the characteristic mass fragments analyzed and detected in the second mass analyzer. Tandem MS, or MS/MS, is like GC/MS or LC/MS in that the first stage performs compound separation, and the second is used for characterization or quantification. The advantage of MS/MS is speed - since only milliseconds are required for the separation step. As a consequence, overall multi-component analyses can be completed in seconds. The disadvantage of MS/MS is cost and instrument sophistication. However, the second generation of tandem quadrupole analyzers are now being marketed and demonstrate considerable instrument versatility and simplification in comparison to the early more costly versions.

DETECTION/CONTROL

Mass separated ions are detected by electron multipliers which are sensitive to small numbers of ions either by ion counting or analog amplification techniques. Overall sensitivity in mass spectrometry is limited by the ionization and mass separation stages which may deliver to the detector, at best, one ion per 10^5-10^8 molecules. Consequently, 10^{-15} to 10^{-18} moles represent the minimum detectable quantity for the most favorable cases.

Computer recording and the complete automation of spectrometers has made the resulting instrumentation more suitable for clinical laboratories. On any of several commercial GC/MS and LC/MS systems, an operator can initiate an auto-tuning procedure, select a programmed analytical sequence, and then leave a rack of samples to cycle through the injection, chromatography, ionization, mass analysis, recording, and data processing reporting. For high performance and high sensitivity analyses, much of the work still requires skilled chemists or chemical technicians, especially when determining conditions for new assays and suitable derivatization procedures. Small computer data systems have become integral to the mass spectrometer, and as with other instrumentation, the complete digitization of instrument control has meant generally increased reliability and ease of operation.

CLINICAL CHEMICAL APPLICATION OF MASS SPECTROMETRY

The major areas of clinical chemical interest in mass spectrometry are in metabolic disease diagnosis (organic acidemias (15, 16), steroid hormone abnormalities (17)), drug monitoring (18) (emergency medicine, therapeutic drug analysis, forensic toxicology, chemotherapeutic monitoring), infectious disease diagnosis (19), and the establishment of standard reference methods and materials (20, 21).

Two examples are illustrative of the capabilities of GC/MS and LC/MS systems. The screening of physiological fluids for traces of abused drugs has become a routine task for clinical laboratories in the U.S. The recommendation of the National Institute of Drug Abuse is that GC/MS be required to confirm positive screening assays determined by immunochemical or chromatographic means. The U.S. Department of Defense has defined specific GC/MS identification protocols requiring the observation of at least three ions within certain intensity ranges in order to confirm radioimmunoassay results. The advantage of mass spectrometry over all other chemical analytical techniques is that a stable isotope labeled analog or isotopomer of the substance to be quantified can be added to the biofluid to serve as an accurate internal standard. From screening procedures, a tentative identification or drug class is targeted, and the relevant deuterated drug internal standard isotopomer added to the sample prior to extraction and workup for GC/MS. The individual protocols for commonly encountered abused drugs (cocaine, tetrahydrocanabinol, phencyclidine, etc.) have been well described in the literature (18), and instrument manufacturers are incorporating chemical and chromatographic methods into software. The results of one laboratory performing 300 RIA assays per week and confirming positive findings by GC/MS has been reported by Spiehler and Sedgwick (22). The rate of confirmed positives by GC/MS for cocaine (62%), opiates (77%), phencyclidine (54%), and barbiturates (57%) when double antibody precipitation was employed is not impressive. The false positives by RIA were due to interfering substances, and not lower detection limits of GC/MS. In one case, an RIA false negative for PCP was reinvestigated due to clinical impressions of the patient's behavior, and GC/MS confirmed high levels of PCP. Thus, GC/MS confirmation of immunoassay results is not only technically feasible, but mandatory for accurate drug testing programs.

No similar drug confirmation studies or clinical assays have been reported using LC/MS thus far, although laboratories are performing drug identifications by LC/MS (23). In general, this is because LC/MS is relatively recent, there are well established and sensitive procedures for specific compounds by GC/MS, and LC/MS is better suited to those polar compounds which cannot be analyzed conveniently by GC/MS. One research example is the separation and quantification of choline and acetylcholine from mouse brain homogenates (24). Using deuterated internal standards, linear standard curves were constructed over the 30 pmol-30 nmol range, and sensitive determinations were made in mouse brain homogenates, comparable to those reported by GC/MS but far simpler in methodology. It is anticipated that similar simplified assays will be developed for LC/MS which will be of direct applicability to clinical chemistry.

CONCLUSIONS

Clinical chemists should reconsider mass spectrometric techniques for routine assays, especially those suited to GC/MS bench top instrumentation. Years of developmental research have resulted in methods and instrumentation which can be employed in a routine and cost effective manner by skilled chemical technicians. Future developments in LC/MS and MS/MS promise additional mass spectrometric instrumentation of considerable promise for clinical chemistry.

REFERENCES

1. Gaskell S. Mass Spectrometry in Biomedical Research. Wiley, 1986.
2. Watson JT. Introduction to Mass Spectrometry, 2nd edition. Raven Press, 1985,
3. In: McLafferty FW, ed. Tandem Mass Spectrometry. New York: John Wiley, 1983.
4. Smidt U. Respiratory Mass Spectrometry. In: Waller GR, Dermer OC, eds. Biochemical Applications of Mass Spectrometry, Chapter 22. New York: J. Wiley, 1980: 703-712.
5. Caprioli RM, Bier DM. Use of Stable Isotopes. In: Waller GR, Dermer OC, eds. Biochemical Applications of Mass Spectrometry, Chapter 27. New York: J. Wiley, 1980: 895-925.
6. Irving CS, Thomas MR, Malphus EW, et al. Lysine and protein metabolism in young women. J Clin Invest 1986; 77: 1321-1331.
7. Schoeller DA. Energy expenditure from doubly labeled water: some fundamental considerations in human. Am J Clin Nutr 1983; 38: 999-1005.
8. Message GM. Practical Aspects of Gas Chromatography/Mass Spectrometry. New York: J. Wiley, 1984.
9. Vestal MC. High performance liquid chromatography-mass spectrometry. Science 1984; 226: 275-281.
10. Blakley CR, Vestal ML. Thermospray interface for liquid chromatography/mass spectrometry. Anal Chem 1983; 55: 750-754.
11. McLafferty FW. Interpretation of Mass Spectrum, 3rd edition. Mill Valley, CA: Univ Science Books, 1980.
12. Fales HM. Newer Ionization Techniques. In: Milne GWA, ed. Mass Spectrometry: Techniques and Applications. New York: J. Wiley, 1971: 179-215.
13. Delgass WN, Cooks RG. Focal points in mass spectrometry. Science 1987; 235: 545-553.

14. Rinehart KL JR. Fast atom bombardment mass spectrometry. Science 1982; 218: 254-260.
15. Goodman SI, Markey SP. Diagnosis of Organic Academias in Gas Chromatography-Mass Spectrometry. New York: AR Liss, 1981.
16. Chalmers RA, Lawson AM. Organic Acids in Man. London: Chapman and Hall, 1982.
17. Shackleton CHL, Chai W. Mass spectrometry-application to steroid and peptide research. Endocrine Rev 1985; 6: 441-486.
18. Foltz RL, Fentiman AF, Foltz RB. GC/MS Assays for Abused Drugs in Body Fluids. NIDA Research Monograph, 1980; 32. Washington: U.S. Government Printing Office.
19. Brondz I, Olson I. Microbial chemotaxonomy. Chromatography, electrophoresis and relevant profiling techniques. J Chrom Biomed Appl 1986; 379: 367-411.
20. Lawson AM, Gaskell SJ, Hjelm M. Methodological aspects on quantitative mass spectrometry used for accuracy control in clinical chemistry. J Clin Chem Biochem 1985; 23: 433-441.
21. DeLeenheer AP, Lefevere MF, Lambert WE, Colinet ES. Isotope dilution mass spectrometry in clinical chemistry. Adv Clin Chem 1985; 24: 111-161.
22. Spiehler VR, Sedgwick P. Radioimmunoassay screening and GC/MS confirmation of whole blood samples for drugs of abuse. J Analyt Toxicol 1985; 9: 63-66.
23. Covey TR, Henion JD. High performance liquid chromatography in veterinary toxicology. J Liquid Chromatogr 1984; Suppl 2: 205-315.
24. Liberato DJ, Yergey AL, Weintraub ST. Separation and quantification of choline and acetylcholine by thermospray liquid chromatography/mass spectrometry. Biomed Environ Mass Spectrom 1986; 13: 171-174.

MODERN MASS SPECTROMETRIC TECHNIQUES APPLIED TO THE ANALYSIS OF STEROIDS

Simon J. Gaskell

Department of Medicine and Center for Experimental Therapeutics, Baylor College of Medicine, Houston, Texas, USA

In the late 1950's, Reed reported the use of direct-insertion probe mass spectrometry (MS) for the study of steroid hydrocarbon structures, as an example of the application of MS in organic chemistry (1, 2). The importance of MS in biomedical analyses, however, emerged only with the development of the mixture analysis capabilities of combined gas chromatography (GC)-MS. Sjovall and Vihko (3) reported the first GC-MS of steroid derivatives shortly after the development of the jet separator by Ryhage (4).

Combined GC-MS remains the mainstay of clinical/biomedical analysis of steroids by MS but a number of important refinements have been made. The use of open-tubular columns (5), now routine, has greatly improved mixture separations. Profiling of, for example, urinary steroids may now in many cases be achieved by capillary GC with flame ionization detection, though elegant procedures of computerized GC-MS data processing have been developed (6, 7).

Selected ion monitoring (SIM) achieves limits of detection in the sub-picogram range (in favorable instances). The precision and potential accuracy of SIM determinations of steroids in physiological fluids, particularly when isotopically-labeled internal standards are used for quantification, has prompted their use as reference procedures for the assessment of routine assays (8, 9). Where necessary, the selectivity of detection of SIM may be improved by the use of chemical ionization or by monitoring specific fragmentations occurring in the first field-free region of a double-focusing mass spectrometer (10) or in the collision region of a tandem instrument (11).

These and related analytical techniques have been the subject of review (12, 13, 14).

Despite the elegance of existing techniques for the clinical analysis of steroids, the motivation for the development of new methods is two-fold. First, the speed and convenience of analysis may be improved if the MS analysis permits the simplification or elimination of pre-instrumental procedures of fractionation and derivatization. Second, direct MS analysis of steroid conjugates, without the need for prior hydrolysis, is desirable. In this paper, two of the newer MS techniques are discussed in relation to steroid analysis, namely liquid chromatography (LC)-MS and fast atom bombardment (FAB)-MS.

LIQUID CHROMATOGRAPHY-MASS SPECTROMETRY OF STEROIDS

The use of liquid chromatography-mass spectrometry (LC-MS) has several potential benefits for steroid analysis. For unconjugated steroids, the removal of the need for derivatization will simplify experimental procedures. Furthermore, the performance of the chromatographic separation at low temperatures will reduce problems of thermally-induced breakdown or adsorbtive losses. Conjugated steroids are generally difficult or impossible to analyze by gas phase methods; thus, LC-MS of steroid sulfates and glucuronides is particularly attractive.

LC-MS has not yet been widely applied in steroid analysis but its use is increasing; in most instances, the thermospray (TSP) interface (15) has been used. Watson et al. (16) reported TSP LC-MS data for a number of steroid mono- and disulfates. The solvent was acetonitrile/water; no buffer salts were required to effect TSP ionization. Singly and doubly charged molecular anions, for the mono- and disulfates respectively, were prominent and detection limits during SIM of these ions were claimed to be of the order of 100 pg.

Molecular anions are also prominent in the spectra recorded during TSP LC-MS of steroid glucuronides (17, 18). The particular value of the LC-MS approach is illustrated by the separate determination of the 3- and 17-glucuronides of 5α-androstane-3α, 17α-diol (18).

Watson et al. (19) have also reported TSP spectra for several unconjugated steroids; $[M+H]^+$ or $[M+H-H_2O]^+$ ions were prominent. For analytes which are not ionized in the LC liquid phase, the principal mode of ionization is considered to be gas phase chemical ionization where the reagent ions are generated from the solvent buffer during the thermospray process (20, 21).

Quantitative analyses for cortisol using TSP LC-MS have been developed by Yergey (personal communication) and by Gaskell et al. (22). In the latter example, serum (1 ml) was supplemented with [$^2H_3$]cortisol (500 ng) and extracted by immunoadsorption. A portion of the total extract was analyzed by TSP LC-MS with selected ion monitoring of m/z 363 and 366, corresponding to $[M+H]^+$ ions of the analyte and internal standard, respectively. In a comparison of the procedure with a GC-MS assay of the bis-methyl oxime, tris-trimethylsilyl ether derivative (of serum extracts after HPLC fractionation), the following conclusions were drawn:

a) The sensitivity of TSP LC-MS detection was inferior, by about two orders of magnitude, to the GC-MS sensitivity.

b) The precision of the LC-MS method was poorer. Thus, replicate (n = 6) analyses of a single serum sample gave coefficients of variation of 7% and 1.2% for the LC-MS and GC-MS procedures, respectively. The difference was in part attributable to the poorer sensitivity and in part to the noisy signal which is characteristic of ion current traces obtained by TSP MS.

c) Nevertheless, the agreement between the two sets of data for serum extracts was satisfactory. Thus, linear regression (n = 14) of LC-MS data (y) on GC-MS data (x) gave the equation:
$y = 0.934x + 12.4$ ng/ml, with a correlation coefficient of 0.968.

The advantage of the LC-MS method lay in the direct analysis of a serum extract, without the need for prior fractionation or derivatization. Recent developments in TSP MS suggest that the approach should be pursued further. The incorporation of a discharge ionization source in the TSP interface has markedly improved both the intensity and the stability of the signal. McDowall et al. (23), for example, have reported the SIM analysis of 100 pg of corticosterone with an excellent signal/noise ratio.

FAST ATOM BOMBARDMENT MASS SPECTROMETRY OF STEROIDS

Fast atom bombardment (FAB) MS and related techniques of secondary ion MS (SIMS) have enabled the analyses of a variety of polar and labile compounds of biomedical importance. The potential of these techniques in the analysis of steroid conjugates was rapidly recognized in several laboratories. Positive and negative ion spectra of steroid sulfates and glucuronides have been reported (24, 25, 26). The negative ion mode is generally preferred since (for mono-conjugates) the spectra are independent of the cationic species present.

The sensitivity of FAB MS analysis of the cortisol derivative was at least an order of magnitude greater than that achieved for the parent steroid (though modest by comparison with electron impact MS of a suitable derivative). The principal high mass ion, m/z 496, corresponds to the intact cationic derivative.

Shackleton and Straub (25) investigated the application of liquid SIMS to the profiling of urinary steroid glucuronides. Steroids were extracted using octadecylsilyl silica cartridges; total extracts were desalted and analyzed directly or first separated into conjugate fractions. Liquid SIMS profiles of urinary glucuronides yielded information equivalent, in most respects, to that derived from GC analysis of the glucuronide fraction after hydrolysis and derivatization. Discrimination between isomeric steroid glucuronides is not possible using a SIMS analysis of this type. Studies of steroid glucuronides by FAB and tandem MS (27) have suggested that daughter ion spectra of $[M+H]^+$ ions subjected to collisionally-activated decomposition (CAD) may shown significant differences between isomers. A detailed study of androsterone and etiocholanolone glucuronides, however, revealed the risk of non-reproducibility arising from variations in the energetics of ion formation in the FAB ion source (28). Further work in this area is required.

There are presently few examples of quantitative analyses of steroids by FAB MS (26, 29, 30). Two difficulties must be addressed for the highest sensitivity work:

a) The analyte signal intensity may be affected by other solutes in the FAB liquid matrix with greater surface activity. Thus, adequate sample purification procedures are required and internal standardization using isotopically-labeled analogues is generally essential for rigorous quantification.

b) The high background usually observed during FAB MS, yielding a signal at every m/z value, reduces effective sensitivity and both precision and accuracy of quantification.

The selectivity of MS detection is therefore important. One approach to the improvement of selectivity is the use of tandem MS. Thus, determination of an analyte based on detection of a characteristic parent and a daughter, obtained following collisionally-activated decomposition (CAD), provides an extra element of selectivity compared with conventional

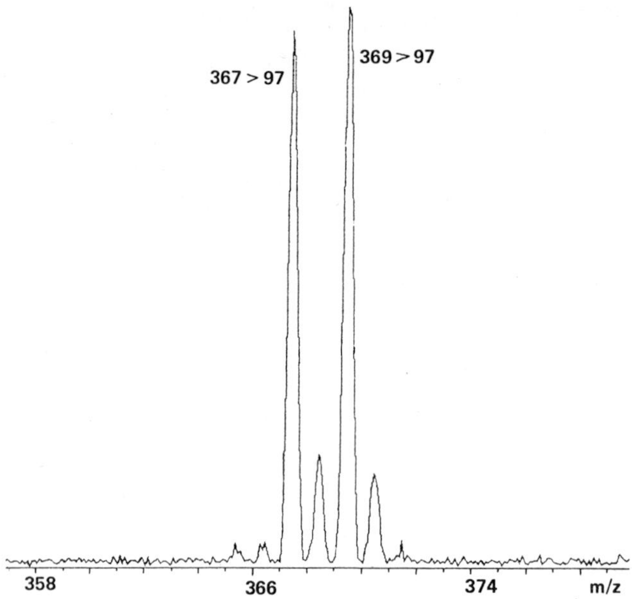

Fig. 1. Analysis of an extract of blood serum by FAB/tandem MS: narrow mass range scan of parent ions yielding daughters of m/z 97 following collisional activation. The concentration of DHAS was determined to be 834 ng/ml.

selected ion monitoring. The sections that follow provide two examples of this approach.

Analyses for serum dehydroepiandrosterone sulfate using fab/tandem ms

The high sensitivity of detection of steroid sulfates by FAB MS suggests a procedure for their quantification in biological fluids. A procedure has been developed for dehydroepiandrosterone sulfate (DHEA-S); details are provided elsewhere (31) and only a brief summary is therefore given here. Serum was supplemented with $[^2H_3]$DHEA-S and extracted by immunoadsorption. Direct analysis of the total extract by negative ion FAB MS indicated that the expected signals at m/z 367 and 369, corresponding to the DHAS and $[^2H_2]$DHEA-S anions, respectively, were subject to interference from other components of the extract and from the glycerol matrix itself.

Tandem mass spectrometry experiments were performed using a tandem hybrid (EBQQ) instrument (VG 7070HSQ), in which collisionally-activated decomposition (CAD) occurred in the first quadrupole (Q1). Steroid sulfate anions gave, on CAD, high yields of a common daughter ion of m/z 97 (HSO_4^-). A quantitative procedure was established, therefore, in which the quadrupole mass analyzer (Q2) was set to transmit m/z 97 while the magnet (B) was scanned across a mass range of m/z 355-380, yielding a spectrum of parent ions (including m/z 367 and 369) which gave m/z 97 on CAD. A typical result for a serum extract is shown in Figure 1.

After analysis by FAB/tandem MS, serum extracts were converted to the heptafluorobutyrate derivative (effecting displacement of the sulfate group (32)). GC-MS analyses with SIM of $[M-C_3F_7\ COOH]^{+\cdot}$ ions provided a second set of quantitative data for the serum extracts. The comparison of FAB/tandem MS and GC-MS data showed generally good agreement. Some negative bias is expected in the FAB/tandem MS data since the detection procedure will not discriminate between androsterone sulfate (or its isomers) and

[$^2H_2$]DHAS, the isotopically labeled analogue of the analyte available for this work. Nevertheless, the quality of the data suggests a continuing value of the FAB/tandem MS approach to the quantification of steroid conjugates.

Analyses for serum cortisol using fab/tandem ms

The steroid sulfates are excellent analytes for FAB MS by virtue of their ionic nature; the FAB MS process favors the analysis of pre-charged species. For the analysis of neutral analytes, derivatization reactions which introduce pre-charged groups will improve sensitivity. This has been demonstrated for the analysis of cationic hydrazone derivatives of corticosteroids by SIMS (33) and FAB MS (34). This approach has been evaluated for the determination of serum cortisol, following conversion to a pre-charged derivative by reaction with Girard's P reagent (1-[carboxymethyl]pyridinium chloride hydrazide).

CAD of m/z 496 in the first quadrupole of a hybrid (EBQQ) tandem mass spectrometer yielded a major daughter ion of m/z 417, corresponding to loss of a pyridine molecule from the derivative group. An ion of m/z 80, attributable to protonated pyridine, was also observed.

Quantitative analyses were performed by dual reaction monitoring of m/z 496→417 and m/z 499→420, the latter reaction being associated with [$^2H_3$]cortisol, which was used as internal standard. Mixtures of cortisol (50-450 ng) and [$^2H_3$]cortisol (300 ng) were converted to the hydrazone derivative and analyzed by selected reaction monitoring to give the standard curve. Figure 2 shows the analysis of a serum fraction obtained by extraction and HPLC separation prior to derivatization. The signal intensity is clearly satisfactory but the form of data acquisition does not permit an assessment of the level of the background signal. A preliminary evaluation of the quantitative validity of the procedure indicated that precision was poorer than that achieved in an established GC-MS method (reflecting, at least in part, a lesser sensitivity). The agreement, however, between the FAB/tandem MS and GC-MS methods was generally satisfactory; in a group of six serum samples, the discrepancy between the

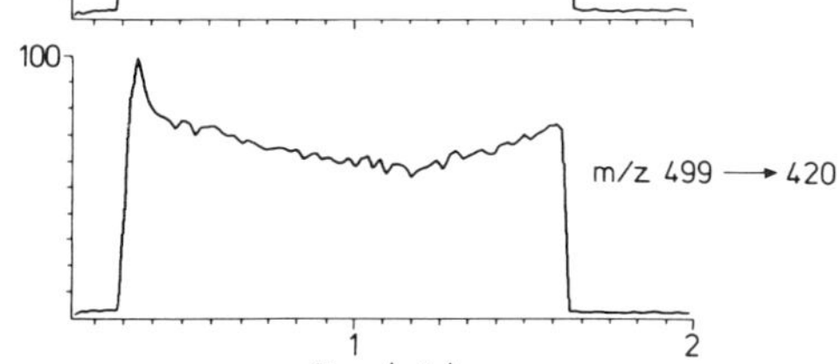

Fig. 2. FAB/tandem MS of a serum fraction with dual reaction monitoring of m/z 496→417 and m/z 499→420. The concentration of cortisol was determined to be 265 ng/ml.

two sets of data was 12% or less, with the exception of one sample where poor reproducibility of the FAB result was observed.

This approach requires further validation as a quantitative method but may be of value as an alternative, and complementary, method to the established GC-MS procedure. In contrast to the LC-MS assay described above, however, no advantage accrues with respect to simplification of the analytical procedure or elimination of the derivatization step. Pilot experiments in which total serum extracts were analyzed, after derivatization but without HPLC fractionation, indicated severe suppression of the steroid signal, associated with other extract constituents.

CONCLUSIONS

It is not intended that the emphasis in this paper on newer techniques of MS analysis of steroids should detract from the continuing value of established methods. The importance of GC-electron impact MS analysis of well-chosen derivatives of steroids continues by virtue of the elegance of the mixture separations, the sensitivity of analysis (particularly when SIM is used) and the diagnostic utility of the MS fragmentation which may be referred to a very extensive body of reference data.

The development of new MS techniques in the field of steroid analysis has complemented or refined established methods rather than opened new areas of application. Thus, direct MS analysis of steroid sulfates, for example, is now possible, avoiding the previous need for hydrolysis of the sulfate group and subsequent derivatization. Further advances in the direct analysis of conjugates may be expected, for example by application of the continuous-flow FAB technique, with its potential for LC-MS coupling (35, 36).

ACKNOWLEDGMENT

Much of the author's work described here was performed at the Tenovus Institute for Cancer Research, Cardiff, U.K. The generous support of the Tenovus Organization is gratefully acknowledged.

REFERENCES

1. DeMayo P, Reed RI. Chem Ind (London) 1956: 1481-1483.
2. Reed RI. J Chem Soc 1958: 2432-2436.
3. Sjovall J, Vihko R. Steroids 1966; 7: 447-458.
4. Ryhage R. Anal Chem 1964; 36: 759-764.
5. Vollmin JA. Chromatographia 1970; 3: 238-241.
6. Setchell KDR, Alme B, Axelson M, Sjovall J. J Steroid Biochem 1976; 7: 615-629.
7. Vrbanac J, Braselton WE, Holland JF, Sweeley CC. J Chromatog 1982; 239: 265-276.
8. Siekmann L. In: De Leenheer AP, Roncucci RR, Van Peteghem C, eds. Quantitative Mass Spectrometry in Life Sciences II. Amsterdam: Elsevier, 1978: 3-16.
9. Bjorkhem I, Lantto O, Svensson L. In: Wilson DW, Gaskell SJ, Kemp K, eds. Quality Control in Clinical Endocrinology. Cardiff UK: Alpha Omega Publishing, 1980: 61-66.
10. Gaskell SJ, Millington DS. Biomed Mass Spectrom 1978; 5: 557-558.
11. Gaskell SJ, Porter CJ, Green BN. Biomed Mass Spectrom 1985; 12: 139-141.
12. Sjovall J, Axelson M. Vitam Horm 1982; 39: 31-144.

13. Gaskell SJ. In: Glick D, ed. Methods of Biochemical Analysis. New York: John Wiley, 1983; Vol 29: 385-434.
14. Shackleton CHL. Endocrine Rev 1985; 6: 441-486.
15. Blakley CR, Vestal ML. Anal Chem 1983; 55: 750-754.
16. Watson D, Murray S, Taylor GW. Biomed Environ Mass Spectrom 1985; 12: 610-615.
17. Watson D, Murray S, Taylor GW. Biomed Environ Mass Spectrom 1986; 13: 65-69.
18. Shackleton CHL, Gaskell SJ, Liberato DJ. In: Jaeger H, ed. Recent Advances in Capillary Chromatography. Heuthig Press; in press.
19. Watson D, Taylor GW, Laird S, Vinson GP. Biochem J 1987; 242: 109-114.
20. Bursey MM, Parker CE, Smith RW, Gaskell SJ. Anal Chem 1985; 56: 2597-2599.
21. Alexander AJ, Kebarle P. Anal Chem 1986; 58: 471-478.
22. Gaskell SJ, Rollins K, Smith RW, Parker CE. Biomed Environ Mass Spectrom 1987; in press.
23. McDowall MA, Merren TO, Smith DC, Swain D. American Lab 1987: 74-88.
24. Liehr JG, Beckner CF, Ballatore AM, Caprioli RM. Steroids 1982; 39: 399-405.
25. Shackleton CHL, Straub KM. Steroids 1982; 40: 35-51.
26. Gaskell SJ, Brownsey BG, Brooks PW, Green BN. Biomed Mass Spectrom 1983; 10: 215-219.
27. Guenat C, Gaskell SJ. In: Benninghoven A, Colton RJ, Simons DS, Werner HW, eds. Secondary Ion Mass Spectrometry: SIMS V. Berlin: Springer Verlag, 1986: 524-526.
28. Cole RB, Guenat CR, Gaskell SJ. Anal Chem 1987; 59: 1139-1144.
29. Shackleton CHL, Chai W. In: Burlingame AL, Castagnoli N, eds. Mass Spectrometry in the Health and Life Sciences. Amsterdam: Elsevier, 1985: 491-504.
30. Moneti G, Agati G, Giovannini MG, et al. Clin Chem 1985; 31: 46-49.
31. Gaskell SJ. Biomed Environ Mass Spectrom 1987; in press.
32. Murray S, Baillie TA. Biomed Mass Spectrom 1979; 6: 82-88.
33. Ross MM, Kidwell DA, Colton RJ. Int J Mass Spectrom Ion Proc 1985; 63: 141-148.
34. DiDonato GC, Busch K. Biomed Mass Spectrom 1985; 12: 364-366.
35. Ito Y, Takeuchi T, Ishii D, Goto M. J Chromatogr 1985; 346: 161-166.
36. Caprioli RM, Fan T, Cottrell JS. Anal Chem 1986; 58: 2949-2956.

MASS SPECTROMETRY AND THE DIAGNOSIS AND STUDY OF METABOLIC DISEASE

R.A. Chalmers

Perinatal and Child Health, MRC Clinical Research Centre
Harrow, U.K.

The application of mass spectrometry to the diagnosis and study of metabolic disease can be particularly exemplified by studies on disorders of organic acid metabolism (the organic acidurias). The diagnosis and the investigation of such disorders depends, particularly in young infants and neonates, on the study of metabolites in body fluids, especially urine. The complexity of organic acids and their metabolites in body fluids has made their identification and detailed study very dependent upon the use of mass spectrometry and the increase in the identification of disorders of this kind has paralleled the application of mass spectrometry to this field (1): prior to 1966 relatively few organic acidurias were known and considerable overlap with amino acidopathies occurred. Since that time, there has been a steady increase in the number of known disorders to more than 50 at the present date.

These organic acidurias are characterised by the accumulation in body fluids of non-amino carboxylic acids that may contain other functional groups, for example hydroxyl and keto, and may include non-amino nitrogens as in the pyrrolidone and pyrrolidine carboxylic acids. Many of the acids are excreted as conjugates, for example with glycine or with glucuronic acid and these compounds are included in this classification. Organic acids or their acyl CoA esters occur as intermediates in the metabolism of amino acids, lipids and carbohydrates as well as in the tricarboxylic acid (Kreb's) cycle itself: enzyme defects can occur at any stage in these pathways leading to accumulation and alternative metabolism of the intermediates, often with devastating biochemical and clinical effects. This can be illustrated by reference to the metabolism of amino acids, where in a general metabolic pathway, after removal of the α-amino nitrogen, all the following intermediates in the pathway taking the carbon skeleton to the tricarboxylic acid cycle are organic acids or acyl CoA esters. The latter compounds contain high energy bonds and the CoA moiety is readily exchangeable with L-carnitine; the acylcarnitines have assumed greater importance in the study of organic acid metabolism and mass spectrometry has proved invaluable in their identification and these compounds are thus also considered here.

A typical metabolic pathway is that of L-isoleucine in which propionyl CoA occurs as an intermediate prior to further metabolism to succinyl CoA and entry into the tricarboxylic acid cycle. A defect in the metabolism of propionyl CoA leads to its accumulation with consequent serious biochemical and clinical effects, resulting in secondary disorders in pyruvate oxidation and metabolism, causing lactic aciduria and hypoglycaemia, in glycine cleavage causing hyperglycinaemia, in the urea cycle causing hyperammonaemia and in the tricarboxylic acid cycle itself leading to general loss of energy production with increased glycolysis in an attempt to overcome this, thereby exacerbating glycogen depletion and lactic aciduria. The propionyl CoA is metabolised alternatively to a wide range of acidic compounds, the complexity of which necessitates chromatography for their separation and mass spectrometry for their identification.

These acidic metabolites are highly polar, hydrophilic and thermally labile compounds that are also relatively involatile and must be extracted from the body fluids, usually with use of the solvents or anion exchange methods, the extracts dried and the extracted acids converted into volatile, non-polar and thermally stable derivatives for chromatography and GC-MS. The most regularly employed derivatising agents are the trimethylsilylating reagents that replace any active hydrogen in the molecule concerned with a trimethylsilyl (TMS) group, forming TMS esters with carboxylic acids, TMS ethers with hydroxyl groups and N-TMS derivatives with nitrogen-containing compounds. Keto acids are usually stabilised prior to trimethylsilylation by formation of an O-alkyl oxime derivative. These derivatives are usually studied by use of electron impact mass spectrometry to produce characteristic spectra for identification in metabolic profiling and diagnostic studies, with specific fragment ions being used for accurate and unambiguous quantification using selected ion monitoring.

The concept of metabolic profiling using a GC-MS (and computer) system was introduced originally by Dalgliesh and his co-workers in Horning's Laboratory (2) and has been extensively used in both diagnostic and detailed biochemical studies. Continuous repetitive scanning with the mass spectrometer of the components eluting from the gas chromatograph produces a large volume of spectral data which can be stored on a computer for subsequent data processing. Several hundred spectra are from a single analysis when relatively slow scanning speeds of, for example, 1 second per decade are used and several thousand spectra result when the faster scanning speeds that are necessary for repetitive scanning of components when fused silica capillary columns are employed. Retrospective processing of the spectral data with automatic background subtraction and library search procedures allows individual component identification with searching speed increased by the use of specialised sub libraries of spectra, by use of 'pre-search' algorithms that utilise only the 8 most prominent ions and by searching through relatively small numbers of spectra pre-selected by use of retention index 'windows'. Diagnosis and search for specific metabolites or groups of metabolites may be simplified still further by use of extracted ion profiles: for example use of a small selected group of ions at m/z 57, 83, 200, 219, 247 and 287 will facilitate the identification of the most characteristic metabolites that are diagnostic of organic acid disorders of amino acid metabolism while providing indicators of other disorders, for example a lactic aciduria, that then requires use of a second group of extracted ions for further clarification. Thus in propionic acidaemia, the presence of methylcitrate (m/z 287), 3-hydroxypropionate (m/z 219), and propionylglycine (m/z 57) among other metabolites provides an easily recognised 'fingerprint' that characterises this particular disorder and clearly distinguishes it from other related disorders for example multicarboxylase deficiency. In the latter disorder, in which the activities of all three mitochondrial carboxylase enzymes are deficient, the presence of lactate (m/z 219), 3-hydroxyisovalerate (m/z 247), and tiglylglycine and 3-methylcrotonylglycine (m/z 83) in addition to the above metabolites provides a clear diagnosis of multicarboxylase deficiency (3) (Figure 1).

When previously unidentified components occur in metabolic analyses, the use of high resolution mass spectrometry with accurate mass measurements may permit the structure of the unknown compound to be derived by consideration of the fragmentation pattern and precise elemental composition of the fragment ions obtained from their accurate masses. Final synthesis of the required standard and comparison of the low resolution mass spectra of the synthesised compound with that in the biological fluid extract provides the absolute identification and this is exemplified by the characterisation of 5-hydroxyhexanoic acid in the urine of patients with dicarboxylic aciduria (4).

It is useful at this stage to consider the accurate quantification of metabolites using GC-MS. Conventional gas chromatography with appropriate internal standards generally provides much of the necessary quantitative data, particularly for urine and blood samples. Even here, the presence of overlapping peaks, even on capillary gas chromatography, necessitates the use of a more specific method of quantification and the use of mass spectrometry with selected ion monitoring (SIM) is indispensible. In this process the mass spectrometer is focussed only on pre-selected ions allowing much greater times for measurement and hence increased sensitivity as well as specificity. An area where such analyses have proved essential is in the prenatal diagnosis of metabolic diseases by analysis of amniotic fluid for metabolites excreted into the fluid by the fetus, this occurring from around 9 weeks' gestation onwards with some 10-15 mls of urine being produced per day by 15 weeks.

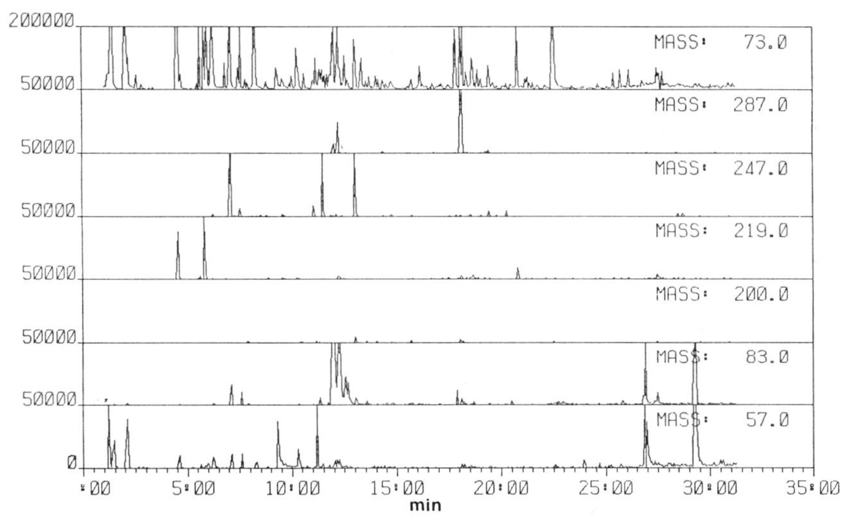

Fig. 1. Extracted ion profile of TMS organic acids extracted from the urine of a patient with multicarboxylase deficiency caused by holocarboxylase synthetase deficiency showing the characteristic 'fingerprint' of this disorder with increases in intensities of ions at correct retention indices for methylcitrate (m/z 287), 3-hydroxyisovalerate (m/z 247), 3-hydroxypropionate and lactate (m/z 219), tiglylglycine and 3-methylcrotonylglycine (m/z 283) and propionylglycine (m/z 57). The trace at m/z 273 produces an equivalent of the total ion chromatogram for the TMS organic acids.

Both EI and CI may be used for quantitative SIM work but for TMS organic acid analysis, EI is preferred and characteristic fragment ions are employed, for example m/z 287 and m/z 479 (M-15) for methylcitrate in the prenatal diagnosis of propionic acidaemia: in practice more than one ion is monitored simultaneously including that of an appropriate internal standard and rapid sequential peak switching is employed. The use of an internal standard is essential for quantitative SIM work and ideally this should be a compound that is chemically similar, with similar extraction and derivatisation properties to the metabolite of interest. Thus in the prenatal diagnosis of glutaric aciduria in which glutarate is the metabolite of interest, 3-methylglutarate is a useful internal standard, selected in part for the above reasons and because it elutes from the gas chromatograph reasonably close to glutarate and has a characteristic ion at m/z 275, again close to that of TMS-glutarate (m/z 261). Although use of deuterated internal standards has been advocated and possibly provides the greatest accuracy, isotope effects may be observed, particularly when capillary GC columns are employed and experience has shown that in practice, excellent results are obtained with use of chemically-similar standards or analogues. In a prenatal diagnosis of this kind, when an affected fetus is present and excreting abnormal metabolites into the amniotic fluid, simultaneous increase in the intensity of the characteristic ions at the correct retention index occurs providing a highly specific diagnosis and assay and, with comparison to a standard curve prepared simultaneously, accurate and precise quantification can be achieved. Such diagnoses have become very reliable with experience and the potential exists for the prenatal diagnosis of many of the known organic acidurias and other metabolic diseases in which the fetus is biochemically abnormal in utero.

Another area of metabolic disease research in which the use of mass spectrometry has proved invaluable is in the study of acylcarnitines: L-carnitine (3-hydroxytrimethylamino butyrate or butyrobetaine) is a highly polar, basic molecule as are the acylcarnitines in which the hydroxy group on the carnitine is esterfied by the acyl moiety.

$$(CH3)_3 - \overset{+}{N} - \underset{OH}{CH_2} - CH - CH_2 - COOH \qquad \text{carnitine}$$

$$(CH3)_3 - \overset{+}{N} - \underset{OCOR}{CH_2} - CH - CH_2 - COOH \qquad \text{acylcarnitine}$$

These compounds are not amenable to gas chromatography, unless the positively charged trimethylamino group is destroyed, and other methods are required for their analysis. The method of choice for identification is fast atom bombardment (FAB) mass spectrometry: in this process the sample or suitable extract is applied to a steel target probe coated with a relatively volatile matrix such as glycerol and is then bombarded with a fast stream of neutral heavy atoms of inert gas, for example xenon, that have been previously energised to 6-8 kV translational energy. The bombardment produces a sputtering effect on the surface of the glycerol with some of the displaced molecules of the matrix and sample becoming ionised in the process and hence analysable using a mass spectrometer: both negative and positive ions may be produced but with the already positively charged acylcarnitines, positive ion FAB-MS is used in their analysis. The major ion produced is the molecular ion but specific fragment or daughter ions are also produced of which that at M-59 is of most interest since this retains the acyl moiety and is thus specific of the parent acylcarnitine. Use of linked scanning techniques at constant $^B/E$ (magnetic/electrostatic) ratio facilitates study of these characteristic ions and the use of accurate mass measurements is also valuable in identification of new compounds.

Patients with organic acidurias excrete increased quantities of acylcarnitines into their urine, this excretion being enhanced still further if additional exogenous L-carnitine is administered. Use of FAB-MS has permitted the identification of a number of acylcarnitines characteristic of several organic acidurias, for example propionylcarnitine in propionic acidaemia and methylmalonic aciduria, isovalerylcarnitine in isovaleric acidaemia and octanoylcarnitine and hexanoylcarnitine in medium-chain acyl CoA dehydrogenase deficiency. In the latter disorder, more detailed study of the excreted acylcarnitines has shown the apparent specificity of the medium-chain acyl CoA dehydrogenase enzyme in man, with only saturated C_6 to unsaturated C_{10} acylcarnitines being observed. Similarly, study of patients with 3-hydroxy-3-methylglutaric aciduria has shown the acylcarnitines excreted include 3-hydroxyisovalerylcarnitine and 3-methylglutarylcarnitine providing for the first time evidence that both hydroxyacyl and carboxyacyl (dicarboxylic) moieties will form carnitine esters (5).

FAB-MS can be used to quantify these acylcarnitines, with use of stable isotope dilution analysis using deuterated internal standards and HPLC-MS provides a valuable method for characterisation and quantification of isomeric acylcarnitines, with use of a thermospray interface between the liquid chromatograph and the mass spectrometer source (6).

In conclusion, the use of mass spectrometry in the study of metabolic diseases has proved invaluable, in diagnosis including prenatal diagnosis, in the characterisation of previously unrecorded metabolites and the elucidation and study of metabolic pathways, including the use in vivo of substrates labelled with stable isotopes, which time has not permitted detailed consideration, and in the study of new polar metabolites such as the acylcarnitines with more specialised techniques. Most of these studies, particularly diagnostic and clinical investigational, have been made with use of low resolution mass spectrometry and with the ready availability of small, inexpensive and relatively simple GC-MS instruments (mass selective detector and ion trap), these methods come within the scope and abilities of most larger or regional clinical biochemistry laboratories. Mass spectrometry will certainly have an increasingly important and regular place in modern clinical biochemistry, opening up this area to much wider clinical application in the future.

REFERENCES

1. Chalmers RA, Lawson AM. Organic acids in Man. The Analytical Chemistry, Biochemistry and Diagnosis of the Organic Acidurias. London: Chapman & Hall, 1982.
2. Dalgliesh CE, Horning EC, Horning MG, Knox KL, Yarger K. A gas-liquid chromatographic procedure for separating a wide range of metabolites occurring in urine and tissue extracts. Biochem J 1966; 101: 792.
3. Chalmers RA. In: Holton JB, ed. The inherited Metabolic Diseases. Edinburgh and London: Churchill-Livingstone; in the press.
4. Chalmers RA, Lawson AM. Identification of 5-hydroxyhexanoic acid in the urine of twin siblings with a Reye's-like syndrome associated with dicarboxylic aciduria and hypoglycaemia and with similarities to Jamaican Vomiting Sickness. Biomed Mass Spectrom 1979; 6: 444.

5. Chalmers RA, Tracey BM, Stacey TE, Cheng KN, Madigan MJ, Lawson AM. The use of fast atom bombardment mass spectrometry to identify and study urinary acylcarnitines in disorders of organic acid metabolism. Biochem Soc Trans 1986; 14: 967.
6. Yergey AL, Liberato DJ, Millington DS. Thermospray liquid chromatography/mass spectrometry for the analysis of L-carnitine and its short-chain acyl derivatives. Anal Biochem 1984; 139: 278.

ISOTOPE DILUTION - MASS SPECTROMETRY AS A REFERENCE METHOD

L. Siekmann, R. Kruse and G. Röhle

Institut für Klinische Biochemie, Universität Bonn
F.R. Germany

The accuracy of clinical chemical testing is today evaluated by means of both internal and external quality control using appropriate control sera. For each analyte in a control material more than one analytical method exists and, unfortunately, the results of various methods and of various test kits applied to one and the same control material differ to a greater or lesser extent. This situation is particularly untenable considering the fact that only one value e.g. creatinine concentration in serum can be the "true" one. It is obvious that any progress towards improving the comparability of analytical results from different laboratories is hampered as long as methods with a known or unknown bias are accepted.

This becomes also apparent when external quality control is carried out. Since different method-dependent target values exist, several different evaluations must be performed for each analyte; this means that each participant's result is compared to his method's target value and judged according to the corresponding method-dependent tolerance range. It seems quite strange to an analytical chemist that for an analyte which is represented by a well-defined molecular species, different concentration values in the sample have, more or less officially, been accepted. In future, therefore, it will be necessary to examine as far as possible the accuracy of our clinical chemical methods and to replace inaccurate analytical procedures with more reliable ones. The only way to ascertain the specificity and accuracy of a clinical chemical method is to compare it with a procedure that has been shown to yield accurate results.

In view of this, clinical chemistry has established its own hierarchy of methods. In this hierarchical scheme, daily routine methods are the lowest order; either they have a known bias, which is accepted by the user, or their inaccuracy is not even known. Reference methods are the second rank in the system and are defined as methods which, after exhaustive investigation, have been shown to have neglibible inaccuracy in comparison with their imprecision. The definitive methods then take the highest rank; they are procedures which after exhaustive investigation have been found to have no known source of inaccuracy or ambiguity.
Whenever possible, the accuracy of routine methods should be traceable to the accuracy of reference methods, and these in turn be compared with definitive methods.

An analytical principle which appeared to be most suitable for the development of definitive or reference methods is isotope-dilution mass spectrometry (ID-MS). This technique, which was introduced in clinical chemical analyses as early as 1970 for the measurement of steroid hormones using deuterated oestrogens as internal standards (1), has, over the last 15 years, proven to be the most powerful tool for a high accuracy in the analysis of many compounds. On the basis of this analytical principle much developmental work has been done by Dr. Björkhem's group in Sweden, by Dr. Gaskell in Cardiff, by Prof. DeLeenheer's laboratory in Gent and by the National Bureau of Standards in Washington. Some of the recent developments in, and results of, ID-MS as practiced by our laboratory will be presented in the following.

The analytical principle of ID-MS may be demonstrated using the measurement of creatinine (2) as an example: A fixed amount of labelled creatinine is added to the serum sample, and the labelled and the non-labelled creatinine are equilibrated at room temperature. The substances are then absorbed on an ion exchange material. After washing with water, the creatinine and the labelled internal standard are eluted by ammonia solution, and the isolated creatinine is then reacted to form the trimethylsilyl derivative. The reaction products are injected into a fused silica capillary column for combined GC-MS, and the m/z values 332 and 329 - characteristic for molecular masses of the labelled and the non-labelled analyte - are monitored continuously during gas chromatography. The analytical results are calculated from the isotope ratios determined in a sample and a series of standards containing defined mixtures of the labelled and the non-labelled analyte.

A non-linear calibration function (3) is used which takes into consideration the fact that small amounts of the non-labelled substance are present in the labelled compound and vice versa, which means that the non-labelled substance, due to the natural isotope composition of the chemical elements, contributes to the recording of the labelled compound.

Selected ion recording carried out after a serum sample has been processed demonstrates that, due to a high specificity of the mass spectrometric detection, no interfering substances can be observed during gas chromatography as shown in Figure 1. Although the sample, extracted and chromatographically cleaned, still contains hundreds of accompanying components from the biological matrix in addition to creatinine and the labelled compound, it is only these two that show up during gas chromatography when the mass spectrometric detector is adjusted to these specific masses.

The accuracy of this analytical process is achieved by means of the high specificity that the mass selective process entails as well as by means of the exact control of recovery that the analytical procedure affords using the principle of isotope dilution.

It seemed reasonable to set up reference or definitive methods first for those analytes which show a particularly wide scatter of method-dependent target values. On the basis of our own experience with an external quality control scheme, this is the case for the parameters creatinine, uric acid, cholesterol, total glycerol as well as for all hormone determinations.

The analytical procedures for the measurement of uric acid (4), cholesterol (5) and total glycerol (6) are similar to the above-mentioned methods for creatinine and will therefore not be described here in detail. A comparison of the results obtained by the ID-MS technique with those of the method-dependent target values still in use for the evaluating quality

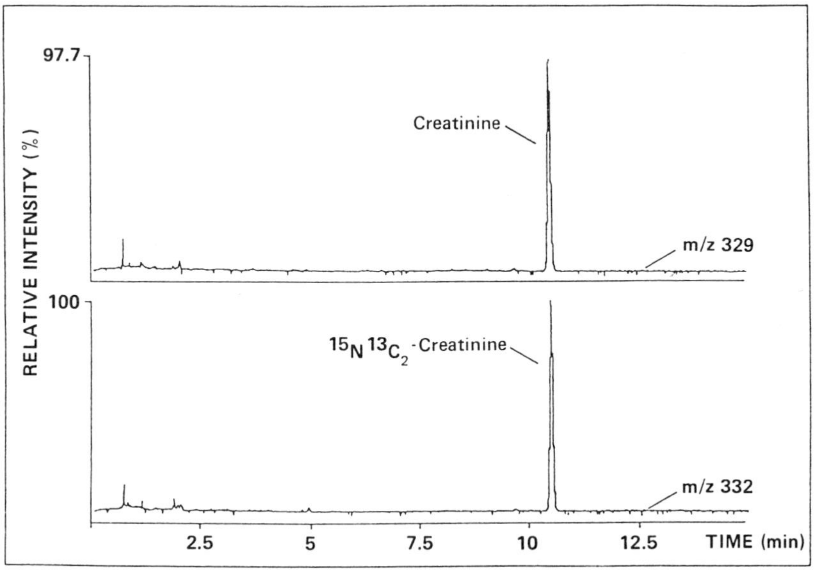

Fig. 1. Selected ion monitoring of creatinine (upper panel) and labelled creatinine (lower panel) after processing a serum sample.

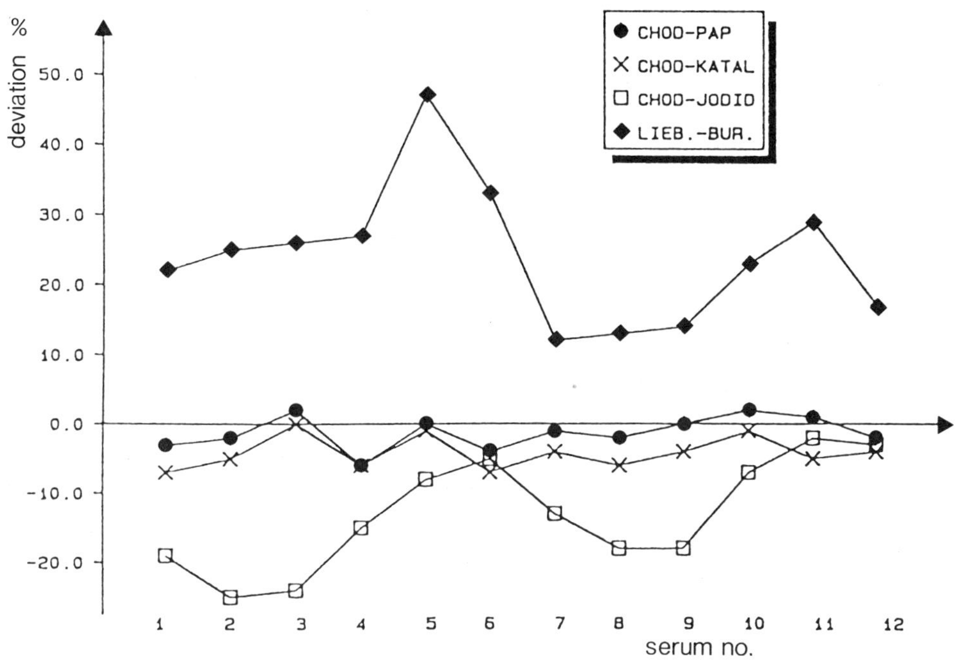

Fig. 2. Deviation of the method-dependent target values from ID-MS reference method values in external quality control of cholesterol. Methods: CHOD-PAP = Cholesteroloxidase-p-Aminophenazon, CHOD-KATAL = CHOD-Katalase, CHOD-JODID = CHOD-Jodide, LIEB.-BUR. = Liebermann-Burchard.

control surveys, demonstrates that the Liebermann-Burchard reaction gave up to 50% too high and the CHOD-Iodine method up to 20% too low values as shown in Figure 2. Only the CHOD-PAP and the CHOD-Katalase method agree with the ID-MS results.

These examples clearly demonstrate that at present there is an unacceptable wide scatter of method-dependent target values for many clinical chemical parameters. In order to improve accuracy in clinical chemistry it is absolutely essential to replace the method-dependent target with reference method values.

At this point it is essential to note that the use of a highly specific and accurate method like ID-MS is only one of the essential factors when setting up reference method values for external quality control. A further important prerequisite is the use of a primary standard of extremely high purity. For a variety of clinical chemical parameters, e.g. creatinine, uric acid, cholesterol and cortisol, well-characterized certified reference materials are available from the National Bureau of Standards (Washington).

Furthermore it is necessary to use carefully calibrated equipment for sampling the serum and for the preparation of primary standard solutions. To establish a reference method value, it is necessary to repeat the analytical procedure several times. The analyses should be carried out independently on at least three different occasions; each time a new standard solution has to be prepared. Control materials for internal accuracy control must be included in each analytical series. (Useful reference materials based on human serum matrix are available for several analytes from the BCR in Brussels and the National Bureau of Standards in Washington). Finally, the precision of these results obtained on different occasions should meet the requirement of a reference method value, which means that imprecision should be lower than that of the most precise routine method.

Whilst for routine clinical chemical parameters the determination of reference method values by ID-MS has been introduced only in recent years, the procedure was used as early as 1977 for the external quality control of hormone determinations.

A manufacturer's list of thyroxine target concentrations in a serum pool indicates that the 24 target values obtained for the same sample using RIA vary considerably from one test kit to the other. This is probably due to the different qualities of the antibodies and reagents used in the various commercial test kits. To carry out external quality control of thyroxine determinations, for instance, this would mean having 24 different target values and 24 different YOUDEN-diagrams to evaluate the participants, an impractical procedure. Until 1977 no reliable criteria were available to decide whether a hormone concentration obtained by RIA or EIA was accurate or not. It was imperative to establish a methodology to provide the basis for developing reference or definitive methods. Consequently, since 1977, reference method values for the evaluation of collaborative surveys of steroid hormones and, more recently, of thyroxine have been determined by ID-MS.

The analytical principle involved here can be demonstrated using the measurement of testosterone in human serum as an example (3): To a serum sample containing about 5 ng testosterone, 5 ng testosterone labelled with two 13C-atoms is added. The two steroids are extracted and cleaned by column chromatography on Sephadex LH-20. Then the heptafluorobutyric ester derivative is formed and the derivative then injected into a capillary gas chromatography column, the end of which is coupled to the mass

Table 1. Precision of ID-MS for the measurement of various analytes

Analyte	Derivative	Coefficient of variation
Creatinine	TMS	0.5%
Total Cholesterol	TMS	0.8%
Uric Acid	TMS	0.4%
Total Glycerol	HFB	0.6%
Aldosterone	Acetal-HFB	2.0%
Cortisol	MO-HFB	1.0%
Progesterone	HFB	1.3%
17-OH-Progesterone	HFB	1.6%
Oestradiol-17β	HFB	1.3%
Oestriol	PFP	1.2%
Testosterone	HFB	0.8%
Thyroxine	Me-TFA	1.9%

TMS = trimethylsilyl, HFB = heptafluorobutyric ester, TFA = trifluoroacetic ester, PFP = pentafluoropropionic ester, Me = methyl ester, MO = methoxime

spectrometer. The instrument is set to the two masses 680 and 682, each of which is characteristic, one for the testosterone derivative and the other for the corresponding, isotopically-labelled compound. The analytic results are calculated from the isotope ratios determined in a sample and in a series of standards containing defined mixtures of the labelled and the non-labelled steroid. In the meantime, reference methods using the technique of ID-MS have been developed for eight low-molecular hormones in addition to four routine parameters. As a measure for the precision of the procedures, the coefficients of variation are listed in Table 1.

The imprecision, in addition to the methodological variation, reflects a small, but inevitable, vial-to-vial variability in the lyophilized serum pools. This means that the methodological imprecision itself may be even lower.

The methods for the determination of other hormones are similar to the one described here for the measurement of testosterone (7, 8).

Reference method values for external quality control obtained by ID-MS have become the touchstone for analyses in clinical endocrinology in the F.R. Germany, and the following examples show how reliable, or rather how unreliable, the determinations of low-molecular hormones can be.

As a measure for the accuracy of an analysis, we can take the percentage deviation between the participants' medians in collaborative surveys and the values established by mass spectrometry as shown in Figure 3. On the upper lefthand side the figure shows the percentage deviations of the medians for 6 control samples that were sent out in the year 1978. One can see that the readings showed medians that deviated as much as 130% from the reference-method value. What was determined here does not in fact deserve the name of aldosterone. Up to 1983 there was little change in this situation. Not until the last years has a tendency toward more accurate results come about. Since then, medians have deviated from the mass spectrometric reference method value by 30% at the most.

Looking at the way accuracy in cortisol analysis has developed over the years, it is striking that in 1978, when external quality control was

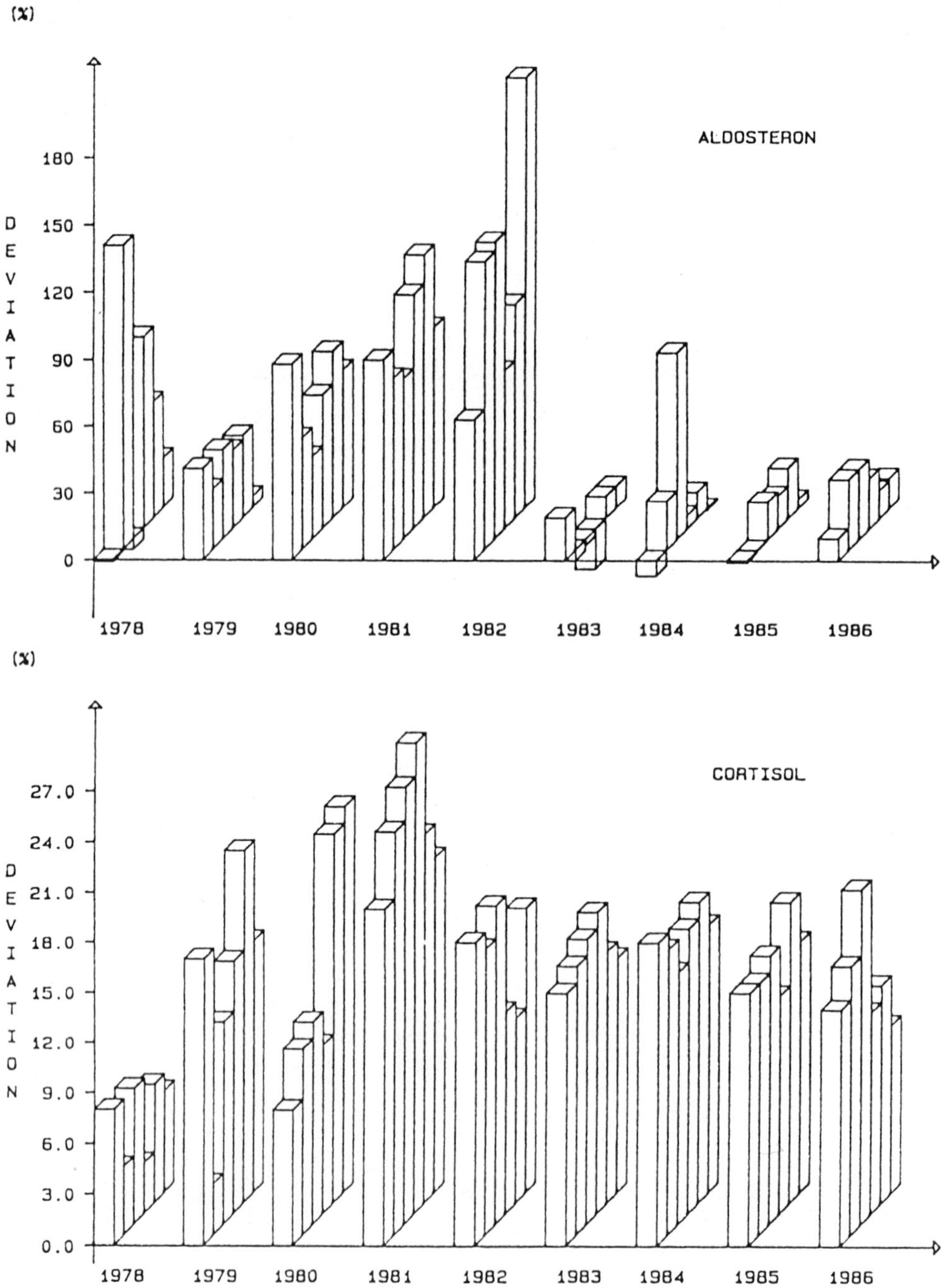

Fig. 3. Deviation of the median values of participants' results from ID-MS reference method values in external quality control of aldosterone (upper panel) and cortisol (lower panel).

just beginning, by far the best results were achieved, as demonstrated in the lower part of the Figure 3. Up to 1978, plasma pools were used in the collaborative surveys which had first been freed from all steroids by means of activated charcoal; then, defined amounts of the hormones were added to the pools. In treating the serum matrix with activated charcoal, a great many substances had obviously been eliminated that might have interfered with the immunological determinations. Since 1979, serum pools not previously treated with activated charcoal have been in increasing use. This has resulted in a noticeable rise in the participants' medians, since unknown components from the biological matrix appear to lead to cross-reactions when cortisol is immunologically determined. In time, results of cortisol routine methods have improved somewhat - in accuracy as well as in comparability. But they remain, still today, 15% above the mass spectrometric reference method values.

During the early years of external quality control, the accuracy for unconjugated oestriol in serum proved to be astonishingly good as can be seen from Figure 4. This changed drastically toward the end of 1981. Especially when the control samples contained conjugated oestriol, the medians of the participants rose significantly higher than the mass spectrometric reference method values. As it turned out, it was just at that time that a kit manufacturer who dominated the oestriol-determination market started using a new antibody which obviously gave rise to cross-reactions with the conjugated steroid. Meanwhile a small group of participants who used their own laboratory methods to determine oestriol, continued to produce results which accorded with mass spectrometric values. It proved possible to convince the manufacturer that this state of affairs needed correcting and, with this as main factor, results have greatly improved these last 2 years.

For thyroxine, success in establishing a mass spectrometric reference method did not come about until 1985. Figure 5 shows the median deviations between the mass spectrometric reference method values and groups of

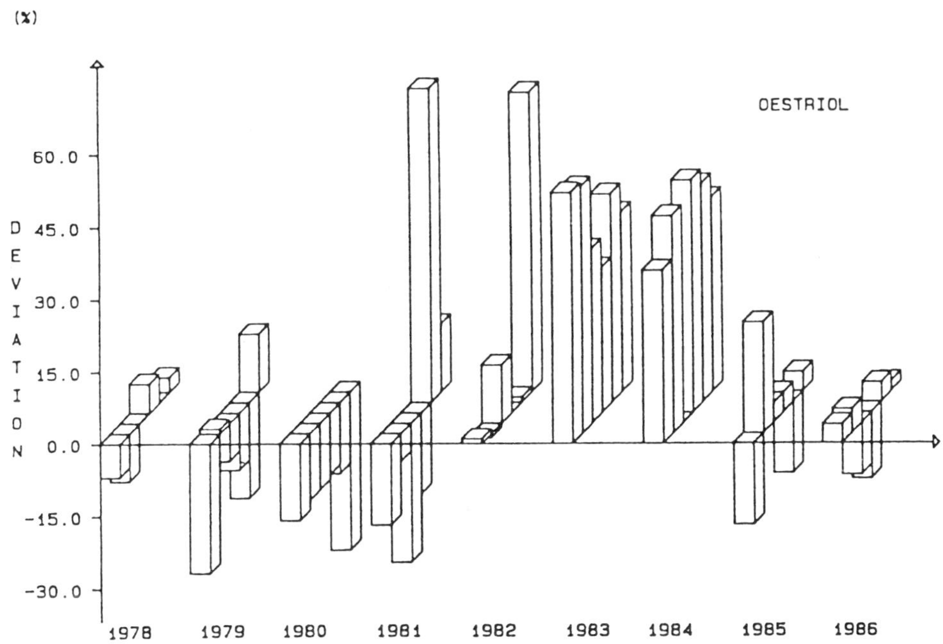

Fig. 4. Deviation of the medians of participants' results from ID-MS reference method values in external quality control of oestriol.

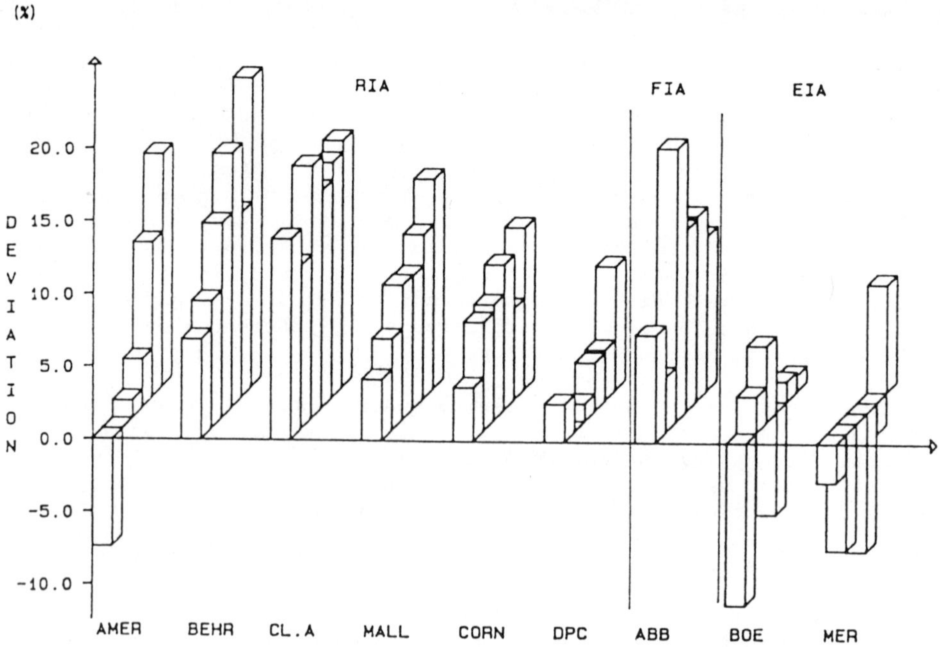

Fig. 5. Deviation of the medians of participants' results using various test kits from ID-MS reference method values in external quality control of thyroxine.

participants using different test kits. One can see for example that the results of enzyme and fluorescence immunoassay methods from Merck (MER), Boehringer (BOE) and Abbott (ABB) are no worse than the radioimmunological tests from Amersham (AMER), Behring (BEHR), Clinical Assays (CL.A.), Mallinckrodt (MALL), Corning and Diagnostic Products (DPC).

The situation as it stands now can be summed up by saying that over the last ten years there has been a clear trend toward improved accuracy and better comparability for measuring hormones. However, it should be noted that collaborative surveys, which for hormone determinations are carried out on a voluntary basis, have only included interested laboratories. In Western Germany, where for routine chemical parameters a participation in collaborative surveys is already legally mandatory, the law governing calibration is currently under review by the legislature. In 1988 not only will routine clinical chemistry but also the determination of several hormones have to undergo the test of external quality control. Moreover, method-dependent target values will be generally replaced by reference method values. By introducing reference methods and in particular ID-MS into external quality control it can be assumed that clinical chemical testing will improve in accuracy.

REFERENCES

1. Hoppen HO, Siekmann L, Breuer H. Identifizierung von Aminosäure- und Peptidconjugaten von Oestrogenen mit Hilfe der Massenspektrometrie. Z Anal Chem 1970; 252: 299-304.
2. Siekmann L. Determination of creatinine in human serum by isotope dilution-mass spectrometry. J Clin Chem Clin Biochem 1985; 23: 137-44.

3. Siekmann L. Determination of steroid hormones by the use of isotope dilution-mass spectrometry: a definitive method in clinical chemistry. J Steroid Biochem 1979; 11: 117-23.
4. Siekmann L. Determination of uric acid in human serum by isotope dilution-mass spectrometry. J Clin Chem Clin Biochem 1985; 23: 129-35.
5. Siekman L. Hüskes KP, Breuer H. Determination of cholesterol in serum using mass fragmentography - a reference method in clinical chemistry. Z Anal Chemie 1976; 279: 145-6.
6. Siekmann L, Schönfelder A, Siekmann A. Isotope dilution-mass spectrometry of total glycerol in human serum - a reference method in clinical chemistry. Z Anal Chemie 1986; 324: 280-1.
7. Siekmann L, Breuer H. Determination of cortisol in human plasma by isotope dilution-mass spectrometry. J Chem Clin Biochem 1982; 20: 883-92.
8. Siekmann L. Determination of oestradiol-17β in human serum by isotope dilution-mass spectrometry. J Clin Chem Biochem 1984; 22: 551-7.

CHAPTER 19
QUALITY ASSURANCE

EXPERT SYSTEMS FOR THE CLINICAL INTERPRETATION OF LABORATORY REPORTS

P.J. Compton

Garvan Institute of Medical Research, St. Vincent's Hospital
Sydney 2010, Australia

INTRODUCTION

In 1986 a review by Buchanan (1) suggested that there were only 60 expert systems in routine use in any application and of these only four were medical systems (2-5). Buchanan may well have missed some systems, for medical expert systems is a very active area. In particular there is widespread interest and activity in the area of laboratory expert systems as suggested by Spackman and Connelly's informal survey (6) and the programme for the 6th International Meeting on Clinical Laboratory Organisation and Management, immediately preceding this Congress. Rather than attempting to review such a rapidly developing area I propose to provide an introduction to the topic based on experience with one of the four systems mentioned by Buchanan (1), GARVAN-ES1 (5) which provides for the automatic clinical interpretation of reports from a thyroid laboratory.

WHAT ARE EXPERT SYSTEMS?

Artificial intelligence is the area of computer research which attempts to produce computer programs capable of mimicking intelligent human behaviour. The most difficult area in this research is emulating common sense while most progress has been made with Expert Systems which attempt to mimic the decision making of recognised human experts working within their specific domain of expertise.

Various definitions of expert systems include extra features such as the ability to explain reasoning, but the emulation of human expert judgement is the central function of an expert system. This key concept can be highlighted by contrasting expert systems with methods of automatic clinical interpretation of laboratory reports based on multivariate statistics. Such methodologies generally attempt to find a 'gold standard', some method of independently identifying the patient diagnosis; they then go back and see if there is any way the laboratory data can be used to enable the various diagnoses to be identified and discriminated. The goal of such systems is to find features in the data not obvious to the human eye. The expert system on the other hand attempts only to look at the data in the same way as the expert. Expert systems may be built automatically from patients archives in a way superficially similar to statistical

LABORATORY REPORT

John Citizen
1 The Avenue
Smalltown

Date 10/8/86
Ref: Dr. A. Doctor
The Surgery
Small Town

Medicare No 007
Sex Male
D.O.B 1/1/60
Request ID 0001

ASSAY	RESULT		REF RANGE
FTI	150		65 - 155
T3	3.8	nmol/L	1.2 - 2.8 nmol/L
hTSH	<0.07	mU/L	0.2 - 6.0 mU/L

COMMENT: Elevated T3 and suppressed TSH consistent with T3 toxicosis

LABORATORY REPORT

Jane Citizen
1 The Avenue
Smalltown

Date 10/8/86
Ref: Dr. A. Doctor
St. Margaret's Hospital

Medicare No 008
Sex Female
D.O.B 1/1/60
Request ID 0002

ASSAY	RESULT		REF RANGE
FTI	150		65 - 155
T3	3.8	nmol/L	1.2 - 2.8 nmol/L
hTSH	0.3	mU/L	0.2 - 6.0 mU/L

COMMENT: Elevated T3 consistent with increased binding protein

Fig. 1. Example laboratory reports. The data in the second report which is circled distinguishes it from the first report. St. Margaret's is a maternity hospital which suggests that the patient is almost certainly pregnant with elevated binding protein. The patient age and sex alone (lightly circled) suggest the patient may have elevated binding protein.

systems (see below), but the aim is quite different, the expert system builder attempts to find in the data the factors the experts must have used to provide the expert interpretation that is stored in the archives. The statistical system builder on the other hand uses an independent source of the diagnosis (e.g. general medical records) and then attempts to determine whether the laboratory data may be able to be used to identify the same diagnosis. Generally statistical techniques consider only broad classes of diagnosis, such as hypothyroidism, euthyroidism and hyperthyroidism (7) for thyroid diagnosis while an expert system, such as GARVAN-ES1, emulating an expert provides 56 thyroid interpretations. As a result statistical methods do not compare to experts (8) but at the same time suggest powerful new ways of looking at data and possibly reducing diagnostic costs (9).

The explicit aim of emulating an expert is fundamental to expert systems and is the reason for the limitations of some 'algorithmic' systems rather than their computing methodology. McConnel et al. (10) describe a system for thyroid report interpretation which could well be an expert system if it took into account patient age, the origin of the referral and the referring doctor's comments, as well as laboratory data and sex. The human expert uses all such information and if a computer program does not, there are occasions on which it will be non expert in its interpretations. Figure 1 (and Figure 3) provides an example of the necessity of using all the information used by an expert. As long as an expert system fulfills the fundamental aim of emulating an expert there are few restrictions on formalism and methodology, for example PUFF (2) the oldest medical system in routine use is now written in BASIC, but such an approach makes knowledge upgrading very difficult.

EXPERT SYSTEMS IN MEDICINE

Medicine, with its central emphasis on the diagnostic process, has been one of the main areas of expert system research. MYCIN (11), INTERNIST

(12), PIP (13) and CASNET (14) are just some of the better known experiments. These programs have involved prodigious effort; the largest, INTERNIST (now CADUCEUS), at present has knowledge of of 572 diseases and 4,500 signs and symptoms (15). They are widely known with the MYCIN project alone having over 80 publications and are milestones for research in the area. These systems have performed well when tested against physicians (12) but none of them are in routine use (1). Schwartz et al. (16) in a recent review of the area, consider that medical expert systems will not find real clinical use in medicine till beyond the year 2000 and that considerable technical advances need to be made. It is important to note that all the above systems, as well as the review of Schwartz et al. (16) are concerned with clinical diagnosis, using the computer as a consultant. The fundamental difficulty with clinical diagnosis, is that the diagnostic process is interwoven with the infinitely rich and essentially open ended domain of common sense (17), an area of much artificial intelligence research but as yet little real success. The second problem is that with a consultation system, a human must choose to use it. As Engle and Flehinger (18) suggest, a system is unlikely to be reliable until it is widely used and the gaps in it's knowledge filled, but clinicians are unwilling to use a system which has gaps in its knowledge.

Medawar (19) has reminded researchers that, "....research is surely the art of the soluble" and therefore it would seem more appropriate to try and seek domains for medical expert systems where they can run automatically during the diagnostic cycle, and where the knowledge domain is closed; that is where all the information that could be used by the expert can be identified and made available to the computer and where there is no opportunity for the expert to seek new data by any human interaction with the patient. Automatic clinical interpretation of laboratory reports fulfills these requirements and this is no doubt the reason that three (2, 3, 5) of the four medical expert systems identified as being in routine use by Buchanan (1) are concerned with interpretation of laboratory data. The fourth, ONCOCIN (4), also is not concerned with clinical diagnosis, but with advising physicians regarding oncology protocols. A report interpretation expert system runs automatically when the laboratory computer generates a report. The data that an expert might use in assigning a clinical comment to a report, actual assay results, patient age and sex, origin of the referral and referring doctor's brief notes on the request can be made available to the computer. A further advantage of such a system is that, depending on legal requirements, laboratory reports still need to be signed by the expert, providing a perfect opportunity to identify and correct gaps in the systems knowledge. Checking that comments are correct is a much simpler operation than identifying and drafting appropriate interpretations and the experts at the Garvan Institute who sign reports are delighted with this change in work load. Interpretations are of course more consistent when using a computer.

A variant on purely providing a clinical interpretation is the application of organising sequential testing. Van Lente et al. (20) describe the use of EXPERT to build an expert system to advise on secondary testing and clinically important interpretations for a routine screening program. Such an application is logically equivalent to providing a clinical interpretation which advises whether to add to, or subtract from, the tests ordered by the clinician, with both cost and diagnostic benefit, depending of course on local legal constraints on who can request tests.

It is important to note that diagnosis from laboratory results is a different application to providing a clinical interpretation of laboratory reports. INTERNIST (21) has been used experimentally to diagnose some complex cases purely from laboratory data, omitting clinical data. In this application the aim is the correct diagnosis of the patient rather than the emulation of the expert adding a clinical interpretation to the report. What the study illustrates is that in a very complex knowledge base derived from physician interview, there is sufficient redundancy in the knowledge for diagnoses to be made with some data missing.

EXPERT SYSTEM SHELLS

A classic difference between expert systems and ordinary programming languages is that expert systems technology separates the knowledge base from the 'inference engine', the way in which the system reasons about the data it is given using the knowledge it has. The term 'shell' is used to describe the computer language, or more correctly the software environment, in which an expert system executes. In normal computer language the programmer is completely free to code both the rules and conditions (or knowledge) that the program may require and the way in which the program runs, while with an expert system shell, the programmer or knowledge engineer, mainly codes in the knowledge; how the system will use this knowledge to arrive at conclusions from data is largely beyond the knowledge engineer's control and has been built into the shell. There are also languages such as LISP and Prolog which are designed specifically to deal with knowledge and can simplify the process of building sophisticated shells and expert systems. One purpose of the separation of knowledge from the programming details is to hopefully make the knowledge more accessible. Shells use different ways of organising and accessing knowledge, such as 'frames', 'blackboards', 'production rules' etc. (22, 23) but the differences between these approaches are not relevant to the discussion here. Feigenbaum (15) has emphasised the 'knowledge principle', that expert systems work primarily because they know a lot about a specific domain rather than because of particular features of a shell. Discussion here will focus on production rules since they are highly suitable for capturing the type of knowledge which experts provide for report interpretation and because a shell to handle production rules can be fairly easily set up in a conventional language such as C. A production rule is essentially an IF ... THEN rule: If certain conditions are true then certain conclusions are true or carry out certain actions. Figure 2 provides an example of the production rule used to identify the interpretation for the case in Figure 1.

As well as varieties in the way knowledge can be stored there are many ways in which the inference engine can work. The two most basic designs for production rule systems are 'backward chaining' and 'forward chaining'.

```
RULE (300)

IF (T3 is high
   and (FTI is normal or FTI is low)
   and TT4 isnt high
   and TSH is normal
   and (pregnant or SOURCE("STMW"))
   and not...)
   or .....
THEN DIAGNOSIS ("Elevated T3 consistent
                with increased binding protein")
```

Fig. 2. A production rule from GARVAN-ES1 which identifies one type of high binding profile. An earlier rule which concludes a patient from "STMW" (St. Margaret's Hospital (Figure 1)) is very likely to be pregnant has been incorporated here. Other conditions in the rule have been omitted for simplicity.

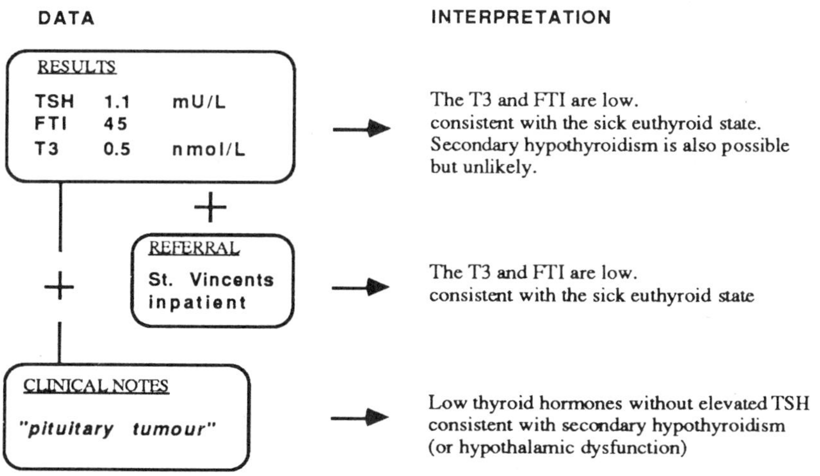

Fig. 3. Probabilistic wording in an interpretation.
The first interpretation which indicates two possibilities is from the laboratory data alone. The extra data that the patient is sick enough to be an inpatient, or has a pituitary abnormality provides information to be able to discriminate between the two interpretations.

With backward chaining the system has a list of hypotheses and asks for data to identify the correct hypothesis. The system's rules determine the relative likelihood of the various hypotheses after each new datum is entered, which in turn is the basis of which datum is asked for next. Backward chaining systems are highly suitable for consultation systems such as MYCIN (23) as they guide the interview and data collection. Forward chaining is the appropriate design for a laboratory interpretation system. All the relevant data can be supplied to the expert system which then searches through its rule base to find rules which the data satisfy. Multiple rules may fire, with the early rules affirming conclusions which are themselves conditions for later rules to fire.

A much discussed element of many inference engines is probabilistic reasoning. The purpose of probabilistic reasoning is to weight data and hypotheses. Some data are more or less certain than others and hypotheses can be more or less likely. There have been many strategies to deal with probability, for example, the unsuitability of Baye's theorem for complex domains led the MYCIN group to develop 'certainty factors' (23). However experts find it extremely difficult to provide information in terms of numerical probabilities, so strategies have been developed to use probability ranges (23, 24, 25). The underlying purpose of probabilistic reasoning is to deal with insufficient knowledge, with cases for which the expert has not provided explicit rules. Shortlife (personal communication) considers the choice of certainty factors for MYCIN was unimportant, the key factor in a successful system again being the amount of knowledge the system has. Feigenbaum's 'knowledge principle' again; the amount of knowledge a system has is the prime determinant of whether it will work.

For the report interpretation domain, not only it is feasible to build in sufficient knowledge not to require probability at all, but experts seem to naturally provide rules which do not require probabilistic reasoning. That is, they suggest probability in the wording of the interpretation rather than indicating the strength of a connection between data and an interpretation (see example Figure 3). They also consider data in very

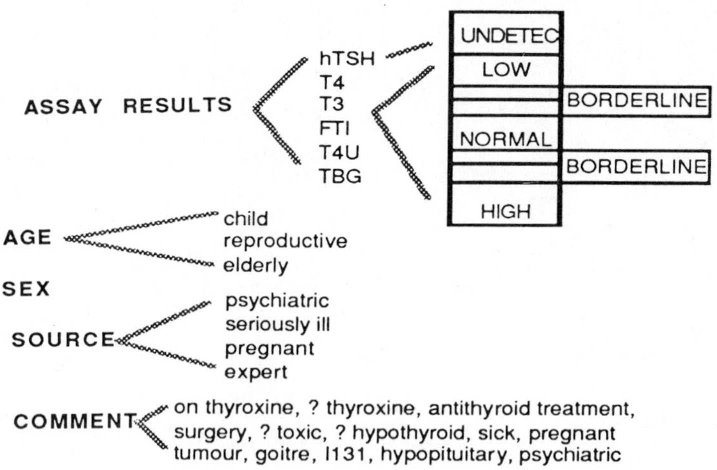

Fig. 4. The data categories used in GARVAN-ES1 rules. Hormone levels can belong in up to two categories; e.g. a T3 result migt be normal or normal but borderline low, low but borderline low or low. Age reduces to three categories and relevant referral sources to four. Referring doctors's free text clinical comments reduce to the classes shown.

coarse categories rather than probabilistically. It has been sufficient to capture thyroid interpretation expertise by classifying data in broad categories such as low, normal, borderline etc. (see below and Figure 4). This is consistent with our earlier observations of errors in reference limits which do not appear to bother clinicians (26) because they use such information in a hypothetico-deductive fashion (27).

When expert systems are actually used, it becomes apparent that the most important issues are how knowledge is entered into the system and how it is maintained and further developed. Consequently as expert systems become more widely used the issue of knowledge acquisition has become an increasing focus for research. The American Association for Artificial Intelligence now sponsors an annual workshop entitled the "Knowledge acquisition for knowledge-based systems workshop". The first workshop was held in November 1986 and one theme that emerged was the development of knowledge engineering tools tailored to specific domains (W. Buntine personal communication). TEREISIAS (23) is an example of such a system designed fo use with MYCIN.

ESSENTIAL REQUIREMENTS FOR A REPORT INTERPRETATION EXPERT SYSYEM (THE GARVAN-ES1 EXPERIENCE)

Data availability

The most basic requirement for a report interpretation expert system is having all data available on computer. Nearly all laboratory systems will have available the actual assay results, the patient age and sex and the source, or where the patient has come from. If the human expert who provides an interpretative comment for reports does not look at the original referral, then it is not necessary to store the referring doctor's brief notes. However if this information is used, it is essential that it be stored on computer for use by the expert system. As this information is frequently omitted by the referring clinician the system must have rules to cope with data in the presence or absence of such comments. One simple approach to dealing with referring doctor's notes is to allow the data entry personnel to enter the notes as free text. The expert system then searches this text for keywords, and synonyms and misspellings of the keywords. Figure 5 shows a sample rule for GARVAN-ES1. This system seems fairly robust, but could be enhanced by prompting the clerk with the expansions of the recognised keywords etc.

```
RULE (500.0001)
IF  not COMMENT ("not on t4")
    and (COMMENT ("t4,t3,thyroxin,throxin,thryoxin,replacement")
    of COMMENT ("hypo rx,orox,treated hypo,dxrt,rx"))
THEN on_t4 NOW TRUE
```

Fig. 5. A sample rule from GARVAN-ES1 used to identify if the referring doctors's notes imply the patient is on thyroxine.

Inference engine

The requirements for an inference for a laboratory report interpretation system are forward chaining and categorical reasoning. Rather than using an artificial intelligence language which may not be available on a laboratory computer a simple shell to provide the type of inferencing required can be developed in C. C is a widely used computer language which allows for fairly transportable programs. It also allows the use of a preprocessor which enables English like rules to be entered into the program (e.g. Figure 3) with the preprocessor translating the rules to actual C code. Figure 6 provides a simple example from the preprocessor used in GARVAN-ES1. In this sort of system the rules are compiled each time they are changed but there is an advantage of speed in the final system.

Rules

GARVAN-ES1 is set up so that the first rules that fire categorise the analyte levels into high low borderline etc. (see Figure 4) and the system then knows that 'TSH is high' for example. The system then searches the stored referring doctor's notes for keywords (see Figure 5). If a match is found a variable such as 'on_t4' may be set true. The program then proceeds through the main body of rules to see if the conditions required by any of the rules are met resulting in either an intermediate variable being set true or an interpretation being arrived at. The program proceeds through all the rules so it is possible (but undesirable) for more than one interpretation to be made.

Checking

As noted above all reports generated by GARVAN-ES1 are signed by an endocrinologist or senior biochemist before issue. As well as avoiding medico legal problems, this check provides an opportunity to pick up cases which the system has not interpreted or has not interpreted correctly. When GARVAN-ES1 was introduced into routine use, 6% of cases required additions or modifications to the rules. This has now dropped to 0.3% of cases rejected, a twenty fold improvement. It is possible that the accuracy rate is not quite as high as this since the experts signing reports may correct an interpretation without referring the case for knowledge engineering, however since the experts take some pleasure in finding lacunae in the

```
# define  IF              if (
# define  THEN            ){trace(rule);
# define  SOURCE (A)      cmpbuf(patnt_info->source, A, lenstr (A))
# define  DIAGNOSIS (A)   diagnosis(patnt_info, 0, A);}
```

Fig. 6. A sample of simple preprocessor code used to translate rules to program code.

system's knowledge, 0.3% is probably fairly accurate. The few cases it missed are more likely to be fairly routine interpretations for unusual but often obvious hormonal profiles rather than for very rare interpretations. As Feigenbaum has pointed out (5), this brittleness is a consequence of the knowledge principle. An expert system may know a great deal, but its performance falls off abruptly at the limits of its knowledge, whereas a human can fall back on reasoning from underlying models and so his performance degrades more gracefully. Therefore although a trainee endocrinologist may not need close supervision when signing reports (he generally knows when he doesn't know), an expert system's interpretations must be checked. As more experience is gained with expert systems this checking may not be necessary, but it seems prudent for the present. In case this seems unduly negative, I should note that an apparent 99.7% agreement with experts is extremely high and experts much prefer to rapidly check interpretations for correctness, than identify and compose interpretations. The system also outperforms the experts on occasion. When the rules are examined after an interpretation has been challenged sometimes the expert has to concede that the expert system has looked at the data more carefully and that its interpretation is more appropriate. Such disagreements are fairly minor however; it is apparent that the system now contains fairly complete and esoteric thyroid knowledge.

The data base of 'cornerstone cases'

Each time a case is rejected by the expert signing the report, rules are modified or added so that the correct interpretation will be produced in the future. The case is also added to a data base. The cases in this data base are 'cornerstone cases', each of them has required changes to the system's knowledge so that essentially they define the system's knowledge. This data base is of fundamental importance when the system's knowledge is upgraded. Frequently when a rule is added, the new rule will subsume or contradict some other rule in the system, the new rule is too broad. The same problems can arise when an existing rule in the system is modified, or alternatively the modified rule may now miss cases it used to correctly interpret. When rules are modified all the cases in the data base are run through the expert system and the program automatically flags any case for which the interpretation changes. The knowledge engineer continues with the rule adjustments till all the cases in the data base are given their original interpretation, and the new case is correctly interpreted as well. This way the knowledge in the system grows incrementally, building on what is already known rather than chaotically displacing it.

Knowledge engineering tools

The essential knowledge engineering tool is the data base of cornerstone cases. The system can also be run in a manual entry rather than automatic batch mode. It can then be queried for a list of the intermediate variables set and the rules used. The data base checking program also provides a rule trace, so that the rules firing inappropriately can be identified. It also alerts the knowledge engineer to redundant interpretations, and provides simple suggestions as to whether a rule should be narrowed or widened etc. A trace of which rules use which variables and expressions can be obtained. These tools are the minimum required to deal with a laboratory domain.

GARVAN-ES1, PRESENT STATUS

GARVAN-ES1 interprets about 6,000 cases a year but since cumulative reports are issued as results become available the system probably does about 10,000 interpretations a year, however the time it adds to a report print run is negligible. The program is small, being able to run in a 64K job space with minimal overlaying and is written in Whitesmith's C; it presently runs on a DEC PDP11/73 under the TSX Plus multiuser operating system. The interface between the expert system in C and the normal assay programs in BASIC is provided by each of the programs reading and writing files in the appropriate formats. The programs run automatically under the

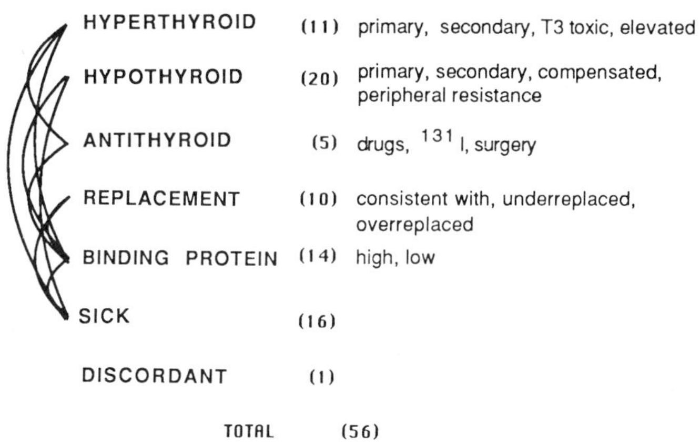

Fig. 7. The categories of interpretation produced by GARVAN-ES1. The number of times each category of interpretation appears in the 56 made is indicated. The connecting lines indicate which categories can occur together in the same interpretation.

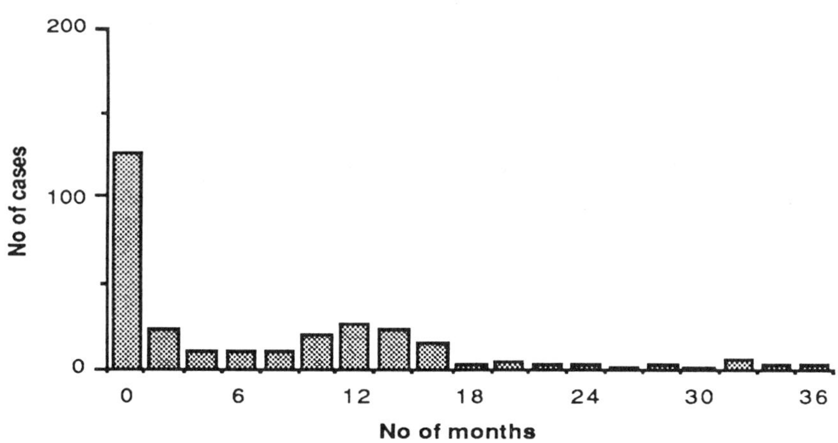

Fig. 8. Number of cases added each two months to the data base of 'cornerstone' cases. These are cases whose interpretation (or lack of interpretation) by GARVAN-ES1 has been picked up as inappropriate by the expert signing reports. The current rejection rate is 0.3%.

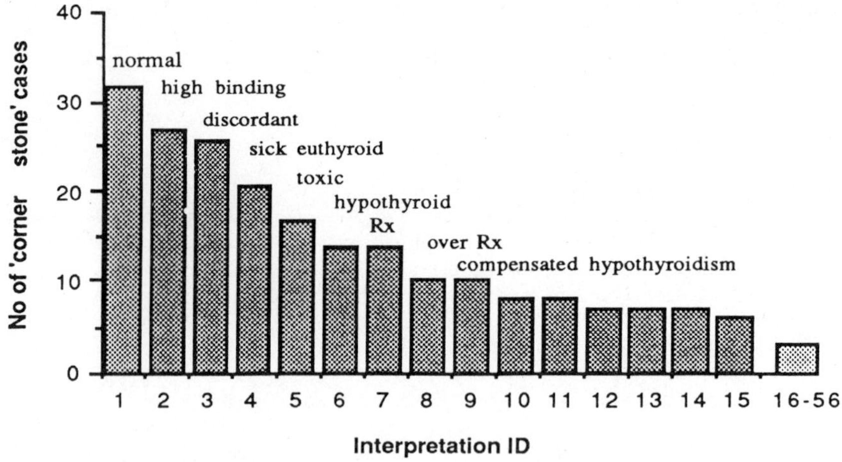

Fig. 9. The number of cases in the data base of 'cornerstone cases' for each interpretation. These are not interpretation categories as in Figure 7 but individual interpretations. Only the first nine are named and the number of cases for interpretations 16 to 56 has been averaged.

control of a monitor command file at report generation time. The system analyses only single samples, not multiple time samples as produced from a TRH test.

The data that the system takes into account in its interpretations is illustrated in summary form in Figure 4, while the range of interpretations it produces is summarised in Figure 7. The present interpretations are restricted to two lines of 64 characters each due to constraints of the laboratory reporting programs, however there is no reason why longer interpretations could not be used, to be sent specifically to more junior hospital staff. The system already has the converse feature in providing very brief comments to some experts who we assume would not wish to have their reports interpreted.

The system was introduced into routine use in mid 1984 after having shown to be 94% accurate on 950 test cases. It is now 99.7% accurate in terms of the number of reports which are rejected and require modifications to the rules to be made. The increase in accuracy is illustrated in Figure 8, where it can be seen that the system's performance is now stable. During this time the size of the rule base has increased by about 80%. However more than half this increase has been due to refining existing rules rather than creating new rules for novel interpretations.

This requirement for continual refining of the rules also shows up in the range of cases in the data base which highlights a major problem but one of our most interesting observations using the system. Figure 9 illustrates the number of cases in the data base for each interpretation. Most of the 56 interpretations are represented by only two or three cases in the data base, but over 30 cases have had to be accumulated for the system to learn the hormonal profiles due to increased binding. Even for simple thyrotoxicosis or primary hypothyroidism the system has made 15 odd mistakes each before reaching its present level of competence. The 30 odd normal cases are in the data base because at some stage they have been misdiagnosed; the system does not at present provide an interpretation if the hormonal profile is normal. It seems that multiple examples are required since when experts are asked for the rules for an interpretation, rather than give the reasons why they came to a given interpretation, they justify the interpretation instead. That is, implicitly or explicitly, they provide the reasons why the interpretation cannot be some likely alternative. This is an interesting confirmation of Karl Popper's theory of knowledge that you can never prove a theory or belief with data, rather you disprove the alternative hypotheses and your belief is the hypothesis that remains (28). This philosophy seems to underlie not only the problems with knowledge engineering but a further reason why laboratory report interpretation has been a successful domain for expert systems. Compared to other domains, report interpretation is particularly suitable for obtaining rules on a case by case basis which Popper's analysis would suggest is a more natural way for the expert to provide justifications, i.e. rules, for his interpretations. The converse is that complete expert systems evolve slowly, with knowledge engineering dependant on the rate at which the various diagnostic profiles show up in data.

THE FUTURE

One way of overcoming the knowledge engineering problem is by inductive learning from examples and we have been conducting experiments to see if an expert system identical to GARVAN-ES1 can be built automatically from archival data (29). The algorithm we are using is C4, a descendant of the ID3 algorithm (30) which is widely used in commercial inductive learning systems. C4 copes with missing and noisy data, continuous variables and outputs expert knowledge as decision trees. The data used were from Garvan archives and had been run through GARVAN-ES1 to provide categorised interpretations. To date, using data from some 4000 cases divided 3 to 1 between training and test cases we have been able to a build a system which has 93% agreement with GARVAN-ES1. Interestingly although C4 was given no information as to reference limits its cutoff values were almost identical to reference limits. The experiment also picked up some errors in GARVAN-ES1's knowledge which had been missed.

It would seem that at least a very useful first draft expert system can be built by inductive techniques if data bases with categorised interpretations are available. However unless the data bases are very large it would seem likely that hand encoded knowledge as above will still be required. With the present research interest in the area it is possible that new and more robust ways of obtaining knowledge from experts may be developed and it is likely that these will be domain specific. It is likely also, that fourth generation software engineering techniques may be adaptable to knowledge engineering (31). We are exploring this possibility for laboratory expert systems.

CONCLUSION

There is no doubt that expert systems can provide automated clinical interpretation of reports which mimic that provided by the relevant experts and so provide a high level of quality assurance for diagnostic information being passed onto clinicians. The subtlety of GARVAN-ES1's interpretation of thyroid laboratory reports would suggest that more complex domains should be amenable to expert system interpretation, but there is no doubt that smaller and more self-contained domains are easier. The only new data that perhaps some laboratories will have to ensure are entered into their computer systems are the referring doctor's notes. The major new type of knowledge that will have to be faced in some applications is that required to deal with sequential samples (32).

Knowledge engineering is a difficult but soluble problem with routine checking of laboratory reports providing an ideal opportunity for identifying problems, and in the next few years new approaches to knowledge engineering are to be expected. Inductive learning already provides a basis for at least first draft systems.

ACKNOWLEDGEMENTS

The development of GARVAN-ES1 has been a collaborative effort. Mr. Kim Horn, now of Ausonics Pty Ltd., worked on GARVAN-ES1 for his MSc. He developed the shell and carried out much of the initial knowledge engineering. Dr. Les Lazarus, director of the Garvan Institute provided the impetus and finance for the project as well as being the ultimate arbiter of interpretations and rules. Professor Ross Quinlan, Head of the School of Computing Sciences at New South Wales Institute of Technology, Sydney 2007, Australia, suggested important strategies for the project and has been responsible for the inductive learning experiments. The time and expert knowledge provided by Dr. Ken Ho and other Garvan endocrinologists and the support provided by Dr. Ross Vining and the Garvan Assay Service has been invaluable. Mr. Bob Jansen of the CSIRO Division of Information Technology, North Ryde 2113, Australia is working on the application of software engineering techniques to laboratory expert systems.

REFERENCES

1. Buchanan BG. Expert systems: working systems and the research literature. Expert Systems 1986; 3 (1): 32-51.
2. Kunz JC, Fallat RJ, McClung DH, et al. A physiological rule-based system for interpreting pulmonary function test results. In: Proc Comp Crit Care Pulmonary Med, 1979: 375-9.
3. Weiss SM, Kulokowski CA, Galen RS. Developing microprocessor based expert models for instrument interpretation. In: Proceedings of the 7th International Joint Conference on Artificial Intelligence Aug. 24-28. Los Altos, Calif. Morgan Kaufmann, 1981: 853-5.
4. Kent DL, Shortliffe EH, Carlson RW, et al. Improvements in data collection through physician use of a computer-based chemotherapy treatment consultant. J Clin Oncol 1985; 3: 1409-17.
5. Horn KA, Compton P, Lazarus L, Quinlan JR. An expert computer system for the interpretation of thyroid assays in a clinical laboratory. Aust Comp J 1985; 17 (1): 7-11.
6. Spackman KA, Connelly DP. Knowledge-based systems in laboratory medicine and pathology A review and survey of the field. Arch Pathol Lab Med 1987; 111 (2): 116-9.
7. Coomans D, Broeckaert M, Jonckheer M, Massart DL. Comparison of multivariate discrimination techniques for clinical data - application to the thyroid functional state. Meth Inform Med 1983; 22: 93-101.
8. McNeely MDD. Computerized interpretation of laboratory tests: an overview of systems, basic principles and logic techniques. Clin Biochem (Canada) 1983; 16 (2): 141-6.
9. Boyd J. Perspectives on the use of chemometrics in laboratory medicine. Clin Chem 1986; 32 (9): 1726-33.
10. McConnel TH, Ashworth CT, Ashworth RD, Nielsen CR. Algorithm-derived, computer-generated interpretative comments in the reporting of laboratory tests. Am J Clin Path 1979; 72 (1): 32-41.
11. Shortliffe EH. Computer based medical consultations: MYCIN. New York: American Elsevier, 1976.
12. Miller RA, Pople HE Jr, Meyers JD. Internist-1, an experimental computer-based diagnostic consultant for general internal medicine. N Eng J Med 1982; 307 (8): 468-76.
13. Pauker SG, Gorry GA, Kassirer JP, Schwartz WB. Towards the simulation of clinical cognition: taking a present illness by computer. Am J Med 1976; 60: 981-95.

14. Weiss SM, Kulikowski CA, Amarel S, Safir A. A model-based method for computer-aided medical decision making. Art Int 1978; 11: 148-72.
15. Feigenbaum EA. Knowledge processing: from file servers to knowledge servers. In: Proc 3rd Aust Conference on the Applications of Expert Systems, May 13-15. Sydney: New South Wales Institute of Technology, 1987: 1-9.
16. Schwartz WB, Patil RS, Szolovitis P. Artificial intelligence in medicine: where do we stand. N Engl J Med 1987; 316: 685-8.
17. Blois MS. Clinical judgement and computers. N Engl J Med 1980; 303: 192-197.
18. Engle RL, Flehinger BJ. Why expert systems for medical diagnosis are not being generally used: a valedictory opinion. Bull NY Acad Med 1987; 63 (2): 193-198.
19. Medawar P. The art of the soluble: creativity and originality in science. Harmondsworth, UK: Penguin, 1969.
20. Van Lente F, Castellani W, Chou D, Matzen RN, Galen RS. Application of the EXPERT consultation system to accelerated laboratory testing and interpretation. Clin Chem 1986; 32 (9): 1719-25.
21. Myers JD. The computer as a diagnostic consultant with emphasis on use of laboratory data. Clin Chem 1986; 32 (9): 1714-8.
22. Hayes-Roth F, Waterman DA, Lenat DB, eds. Building expert systems. Reading, Mass: Addison Wesley, 1983.
23. Buchanan BG, Shortliffe EH, eds. Rule-based expert systems: The MYCIN experiments of the Stanford heuristic programming project. Reading, Mass: Addison Wesley, 1984.
24. Shafer G. A mathematical theory of evidence. Princeton, NJ: Princeton University Press, 1976.
25. Quinlan JR. INFERNO: a cautious approach to uncertain inference. Computer Journal 1983; 26 (3): 255-69.
26. Compton PJ, Stuart MC, Lazarus L. Error in laboratory reference limits as shown in a collaborative quality-assurance program. Clin Chem 1986; 32 (5): 845-9.
27. Griner PF, Mayewski RJ, Mushlin AI, Greenland P. Selection and interpretation of diagnostic tests and procedures. Ann Intern Med 1981; 944: 553-92.
28. Popper K. Conjectures and refutations. London: Routledge and Keagen Paul, 1963.
29. Quinlan JR, Compton PJ, Horn KA, Lazarus L. Inductive knowledge acquisition: a case study. In: Quinlan JR ed. Applications of expert systems. Reading, UK: Addison Wesley, 1987: 159-73.
30. Quinlan JR. Discovering rules by induction from large collections of examples. In: Michie D, ed. Expert Systems in the micro electronic age. Edinburgh University Press, 1979: 168-201.
31. Jansen B. Software engineering of expert systems. Technical report TR-FD-87-02. Sydney: SCIRO Division of Information Technology, 1987.
32. Coiera E. Modelling clinical expectation over time: the BREW expert system for fluid and electrolyte therapy in intensive care. In: Proc 3rd Aust Conference on the Applications of Expert Systems, May 13-15. Sydney: New South Wales Institute of Technology, 1987: 103-123.

QUALITY CONTROL IN CLINICAL LABORATORIES

D. Stamm

Department of Clinical Chemistry, Max-Planck-Institut für Psychiatrie, D-8000 Munich, Federal Republic of Germany

0. INTRODUCTION

Numerous systems for quality control in clinical laboratories have been developed and employed over the past 30 years (1-4).

A major component of such systems is control of the reliability of the analysis, and statistical quality control is the tool used to monitor both the precision and the accuracy of the results. The purpose is to ensure that imprecision and inaccuracy, which are the numerical values of precision and accuracy, do not exceed certain limits.

Two important problems related to this aspect of quality control have not yet been solved satisfactorily:

1. How to ensure the accuracy of the analytical results.
2. How to select the limits for quality control so that they fulfil the medical requirements.

A rational approach to solving these problems is the subject of my talk today.

The state of the art for ensuring accuracy in clinical chemistry is the reference method concept. In the U.S.A. this concept was developed as the National Reference System in Clinical Chemistry (NRSCC), and the system has been partially realized (5, 6).

In the Federal Republic of Germany this reference method concept is currently being introduced in the new "Guidelines for Quality Control of Quantitative Analyses in Medical Laboratories" (7). These Guidelines were developed in connection with the 1985 revision of the Calibration Act (8) and approved by the Federal Medical Association in January 1987. They have the power of a law.

When these Guidelines were being prepared the question arose of a well-founded procedure for specifying the maximum acceptable imprecision and inaccuracy for the various quantities that are determined quantitatively in clinical laboratories.

1. ENSURING THE ACCURACY OF ANALYTICAL RESULTS

Accuracy is defined unambiguously in the Recommendations of the EP-NPQC of the IFCC (1978) (3).

Accuracy: Agreement between the mean estimate of a quantity and its true value.

Inaccuracy: Numerical difference between the mean of a set of replicate measurements and the true value. This difference (positive or negative) may be expressed in the units in which the quantity is measured, or as a percentage of the true value.

In order to guarantee accuracy, the basic principles of measurement must be adhered to.

1.1 Basic principles of measurement

Quantitative analysis is a measurement process. To measure means to quantify a certain property of a material. The property must be well defined.

For a measurement we need:

1. A unit of measure
   which gives us a scale for quantification (e.g. appropriate SI units).

2. A material realization of the unit
   This is the calibration material of "reference material".

3. An analytical system
   that produces a signal by means of a transducer; by comparing the signal from the material to be measured with that from the calibration material the analytical result can be calculated.

In such cases measurement thus consists of comparing the signal from a calibration material (also called a reference material), which is a well-defined realization of the property to be quantified, with the signal from the specimen to be measured.

In this comparison, a mathematical function (analytical function) is used to calculate the property of the specimen that is to be quantified from the relationship of the signals.

The accuracy of the measurement depends to a very large extent on the definition of the property and the realization of this property in the reference material (9, 10).

The signal from the specimen to be measured can contain components that do not come from the property to be measured but rather from other major and minor components of the specimen, i.e. the matrix. This depends on the analytical specificity of the analytical method used.

1.2 Reference materials

A reference material is a well-defined realization of a unit of measure (measuring unit).

Example: The ultrapure D-glucose prepared as Standard Reference Material (SRM) No. 917 by the National Bureau of Standards (NBS), which in water solution - is a realization of the property "mass concentration" of glucose.

The ISO definition of a reference material is not well enough differentiated with regard to the properties of the materials and the resulting possible uses of the materials. This definition is not helpful in clinical chemistry; rather, it causes confusion.

The NBS in the U.S.A. has proposed a hierarchy of reference materials that can be used in clinical chemistry. In the NBS Proposal reference materials are classified on the basis of the analytical methods that are used in determining the property value.

For theoretical reasons and also for reasons stemming from practical experience in the use of such materials, it is better to differentiate between calibration materials and control specimens (9). The optimal characteristics of the two types of materials if they are to serve their stated purpose well are quite different and in fact complementary. In the recommendations of the NCCLS in the U.S.A. (10, 11) and of DIN in the Federal Republic of Germany this important distinction has been made, a distinction which for didactic reasons is to be strongly recommended.

When measurements are actually made, the reference materials serve as calibrators. They are used as the basis of the analytical function (or calibration function), which in turn is used to transform the measured signal into an analytical result.

A distinction is made between primary and secondary calibrators. The calibrator should be as well defined as possible.

1.3 Reference methods in the hierarchy of analytical methods

The metrologists have proposed a hierarchy of methods with increasing accuracy (3, 5, 6, 9) for relating the properties of a reference material to a system of units. For theoretical and practical reasons it is helpful to differentiate between definitive methods and reference methods and between these two and routine methods.

Routine methods too must be thoroughly evaluated not only after development and prior to publication but also whenever they are introduced in well-staffed laboratories (3, 12). Comparisons of methods using statistical methods or results of interlaboratory surveys are not a substitute for such an evaluation.

The reference method values obtained with reference methods are a good estimate of the "true value"; therefore they are needed in the control of accuracy (13).

At a request of the German Society for Clinical Chemistry, reference methods are being taken from the literature or developed and then tested for various quantities. So far reference methods are available and in use for 17 quantities (Table 1).

A report on our experience with these reference methods was given at a symposium held in June of last year at the International Conference on Biochemical Analysis in Munich (31).

In a comparative study, reference method values were determined for many control materials to be used in quality control according to the Guidelines of the Federal Medical Association. In the same control specimens, method-dependent assigned values were also determined according

Table 1. Quantities for which the German Society for Clinical Chemistry already has reference methods

Substrates	
S-calcium (14, 15)	S-creatinine (21-23)
S-chloride (16)	S-lithium (drug monitoring) (15)
S-cholesterol (17)	S-sodium (16)
S-glucose (18)	S-triglycerides (24)
S-uric acid (19, 20)	S-total protein (25)
S-potassium (16)	
Hormones	
S-aldosterone (26, 27)	S-estriol, unconjugated (27, 30)
S-cortisol (28)	S-progesterone (27)
S-estradiol-17β (27, 29, 30)	S-testosterone (27)

to the standard protocol (32). Good agreement was found between the reference method values and the assigned values obtained with reliable routine methods.

1.4 Basic principles of quality control of analyses

Control specimen systems have proven effective for quality control (1-4). In such systems control specimens are introduced into the runs of analyses, these runs consisting of one or more calibrators and the patient specimens. The analytical results for these control specimens are regarded as a representative sample of that particular run of analyses and are subjected to a statistical test of precision and accuracy.

This test is effective, however, only if the following basic rules (9) are followed:

1. Calibration materials and control specimens must be used completely independently of one another.

2. Calibration materials should be as highly purified and well defined as the state of the art allows. Ideally, solutions would be prepared using pure, well-defined solvents only.

3. The matrix of the control specimens should be as much like the matrix of the specimens to be examined as possible so that procedural deficiencies and interference factors due to the matrix of the specimens can be monitored.

1.5 Prerequisites for quantitative analyses

The essential prerequisites for quantitative analyses are:

1. The calibration materials must be well defined and characterized (9). The other components in the calibration material must not contribute (positively or negatively) to the measurement signal, i.e. the signal must reflect the true value of the analyte and nothing else.

The calibration function or the basis for calculation must be clearly specified and the way it was established must be reproducible and controllable.

2. The analytical system, consisting of reagents and equipment and the manner in which the analysis is to be performed, must be clearly described, and it must be possible to monitor all three components of the system.

3. Effective quality control must be possible and must be performed (3, 4), and the control limits to be met must be appropriate to the medical requirements.

Quantitative analyses in the true sense are performed only when these prerequisites have been met. Otherwise the investigations are suitable for the purpose of orientation only, a fact that must be taken into consideration when clinical chemical findings (13) obtained in this way are interpreted.

The most common errors made are: (a) The necessary prerequisites for calibration are not fulfilled, and (b) accuracy control as part of quality control is not performed in accordance with the IFCC Recommendations (3).

2. RULES FOR ESTABLISHING CONTROL LIMITS

The medical requirements are that the maximum allowable imprecision and maximum allowable inaccuracy should be selected in such a way that the resulting proportion of incorrect classifications in the medical assessment (13) of the analytical results is as small as possible.

2.1 The goal of the clinical laboratory investigation

The goal of a clinical chemical investigation is not achieved simply by obtaining a reliable analytical result. Rather, the goal of such an investigation is a clinical chemical finding. This finding is arrived at in a series of four steps (13).

1. Preparations for analysis.
2. Analysis, including reliability check.
3. Analytical assessment.
4. Medical assessment.

The clinical chemist is involved in all four steps. He or she must give instructions on how each step is to be performed and plan control procedures. In this process there are many and varied interrelationships; in the present paper only one aspect is discussed, the effect of the reliability of the analytical procedure on the medical assessment and consequently on the reliability of the finding. The important questions of biological influence factors and analytical interference factors are not included because they have been treated extensively elsewhere (33).

2.2 Medical assessment

A measurement, for example S-glucose (substance concentration) = 9 mmol/l, is not by itself an adequate basis for a medical decision or action. Rather, this analytical result must first be subjected to a medical assessment. This involves consideration of information about the patient and about previous steps in the investigation, for instance the fact that the patient had not had anything to eat or that he had breakfast 1½ hours before specimen collection.

There are several different ways of making the medical assessment (13), the choice depending on the reason the analysis was performed.

1. Longitudinal assessment
   This is a comparison with earlier results from the same patient (comparison of two or more analytical results).

2. Transverse assessment
   This is a comparison with reference values (36) from a reference population of individuals in a specified state of health or illness:

   - Comparison with a reference interval
   - Comparison with a decision limit
   - Comparison with a therapeutic range.

In this medical assessment the number of incorrect classifications because of the imprecision and inaccuracy of the analytical procedure should be kept as small as possible. The maximum allowable imprecision and inaccuracy for quality control of the analytical procedure must be specified with this in mind.

The effects of these limits on the number of incorrect assignments and the resulting rules for specifying the limits so that the medical requirements are met are illustrated with a simple, easy-to-follow example. The example used is transverse assessment by comparison with a reference interval; analogous rules have been established for the other types of medical assessment.

In this illustration of the effects of imprecision and inaccuracy, a reference interval is used that is based on reference values with a symmetrical distribution that does not differ markedly from a normal distribution. Analogous rules can be established for asymmetrically distributed reference values.

2.3 Selecting the limits: "Comparison with a reference interval" - an example

If the reference interval includes 95% of the reference values from subjects in a defined state of health, then as a rule 2.5% of the reference values are above this interval and 2.5% below. Consequently, 2.5% of the results assessed by comparison with this reference interval will be false positive. The width of the reference interval (RI) is equivalent to about 4 standard deviations of the reference values (4 s_{RV}). The components (13) of the standard deviation of the reference values (s_{RV}) are

$$s_{RV} = \sqrt{s_B^2 + s_{AD}^2 + s_{other}^2} ,$$

where s_B is the biological variation, s_{AD} the between-day analytical imprecision and s_{other} the variation from other sources (e.g. from specimen collection and preparation).

The biological variation and the coefficient of variation (CV) can be quite different for different quantities (e.g. the CV ranging from 2% for S-sodium to 25% for the S-triglycerides).

The relative imprecision of the analytical procedure is constant over the range of measurement considered in the example used here (34). For this

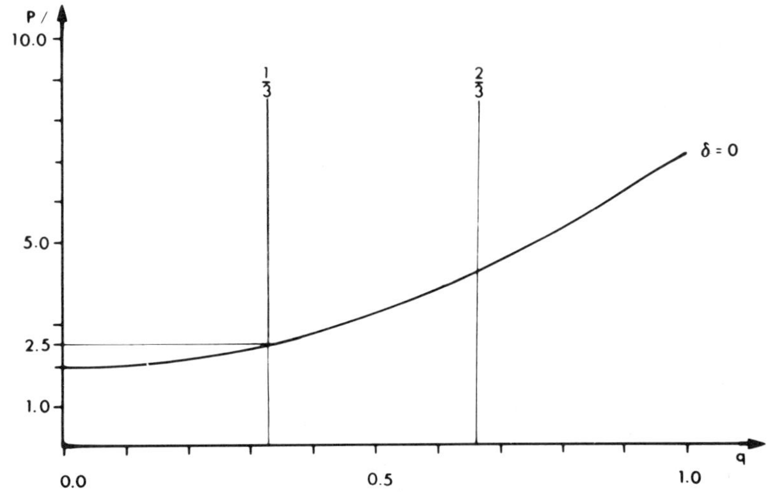

Fig. 1. Effect of analytical imprecision on the percentage of incorrect assignments in the medical assessment. P%, probability of an incorrect assignment (false positive or false negative) on one side of the reference interval; q, relationship between standard deviation of the measurement (s_{AD}) and biological variation (s_B), $q = s_{AD}/s_B$; δ, inaccuracy

reason it is legitimate to use the relative measures of variation and deviation in the following illustrations.

2.3.1 Effect of imprecision

The acceptable relationship between the analytical variation and the biological variation should be specified in such a way that the analytical variation does not produce any major increase in the variation of the reference values. This means (Stange, 1970; 35) that

$$s_{AD} \leq 1/3\ s_B\ ,$$

where S_{AD} is the between-day imprecision of the analytical results and s_B is the biological variation. Under these conditions, the latter is about 1/4 the width of the reference interval ($s_B \approx 1/4$ RI). From this we have

$$s_{AD} \leq 1/12\ \text{RI}\ .$$

The Figure 1 shows the procedure for estimating the increase in false positive assignments when there is an increase in imprecision for a normally distributed biological quantity.

Assuming the above, the percentage of false positive assignments on one side of the reference interval increases from 2.5% if $q = 1/3$ to 7.5% if $q = 1$; i.e. it increases by a factor of 3.

2.3.2 Effect of inaccuracy

The effect of inaccuracy on the percentage of false assignments can be illustrated in an analogous manner. The Figure 2 shows that even if inaccuracy ($\delta < 0$) is no more than 1/12 RI, the percentage of false assignments - assuming optimal imprecision - increases by almost a factor

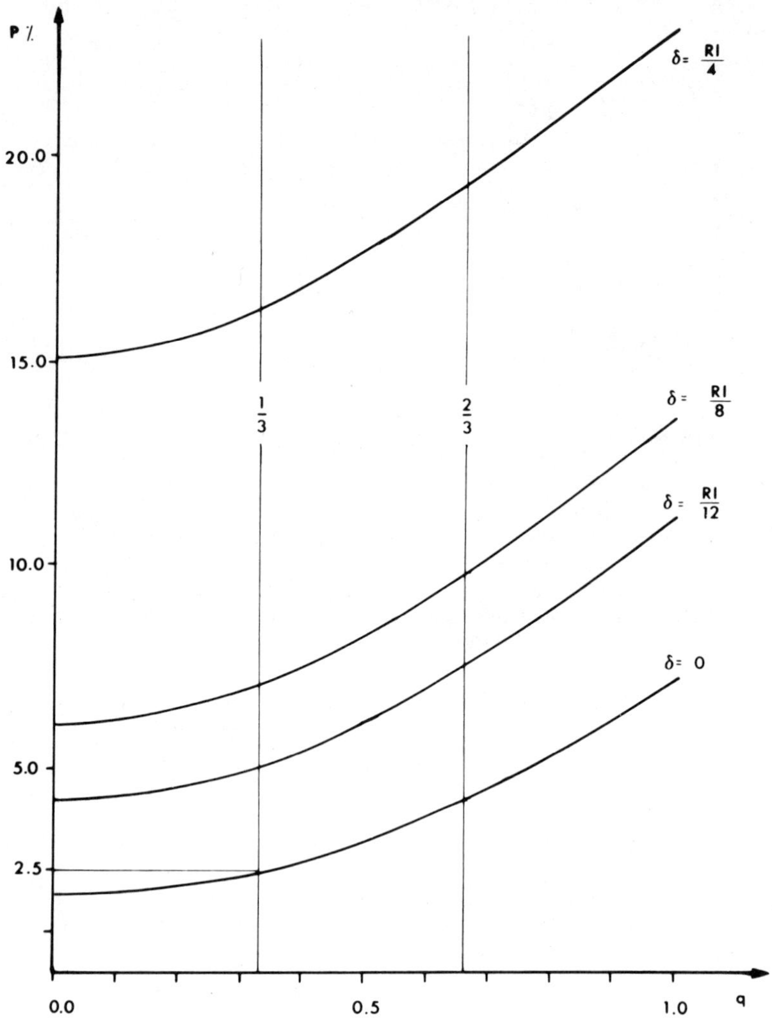

Fig. 2. Effect of analytical inaccuracy in relation to the level of imprecision on the percentage of incorrect assignments in the medical assessment. RI, width of the reference interval; see also key to Figure 1.

of 2. If inaccuracy is about equal to the standard deviation of the reference values ($s_{RV} \approx 1/4$ RI), the percentage of false assignments even assuming optimal imprecision increases to more than 16%. The opposite, i.e. an increase in false negative assignments, occurs for $\delta < 0$. The effect of inaccuracy on false assignment is so great that everything conceivable must be done to keep inaccuracy as small as possible.

2.3.3 Medical requirements

Many attempts have been made to set performance standards for clinical laboratory investigations (2). In the following, an attempt is made to derive the level of reliability required in such investigations from the medical requirements.

To repeat, by fulfilling the medical requirements is meant keeping the percentage of false assignments in the medical assessment of analytical results within predictable and medically relevant limits by specifying appropriate quality control criteria:

1. Maximum allowable imprecision

$$s_{AD} \leq 1/3\ s_B \approx 1/3\ .\ 1/4\ RI = 1/12\ RI$$

2. Maximum allowable inaccuracy

$$\delta \leq 1/12\ RI.$$

In everyday laboratory routine, control of accuracy is often based on a single determination on an accuracy control specimen. For this purpose the maximum allowable deviation of an individual result (a) from the location parameter (reference method value or method-dependent assigned value) has been fixed at 2 s_{AD} for the random error and 1 s_{AD} for the systematic error, or a total of 3 s_{AD}.

This maximum allowable deviation from the location parameter is also used as the decision limit in the short-term interlaboratory surveys.

As already shown in Figures 1 and 2, percentages of false assignments can increase for quantities where these specifications cannot be met. For most quantities imprecision can be kept below the maximum allowable level, but for a good number of quantities there are problems in keeping within the maximum allowable inaccuracy. As evaluations of assigned values for control specimens (32) and interlaboratory survey results have shown, for some quantities the systematic differences between results obtained within different analytical systems are so large that they lead to substantial percentages of diagnostically and therapeutic relevant false classifications. For the reasons just given, a solution is urgently needed to the problem of how to ensure that the accuracy of clinical chemical analyses is within the specified limits. On the one hand clinical chemists must recognize this problem and find a solution, and on the other they must grasp the eminent importance of accuracy for the effectiveness of their medical actions.

2.4 Limits for other types of medical assessment

For the other types of medical assessment, the critical difference is used in establishing rules for determining the medical requirements.

2.4.1 Longitudinal assessment

Critical difference (13): Two analytical results x_1 and x_2 are significantly different with a confidence level of 95% if the absolute value of the difference D between them is larger than the critical difference D_k, where

$$D_k \geq 2\sqrt{2}\ s_{AD} = 2.8\ s_{AD}$$

and

s_{AD} = the between-day analytical variation.

The prerequisites are:

1. Not too much deviation from a normal distribution.
2. The same s_{AD}.
3. s_{AD} is calculated from enough measurements.

If these prerequisites are not met, then theoretically the confidence level can drop to 75%. In such cases it is advisable to multiply s_{AD} by 4 rather than by 2.8; then the probability increases to 88% in the worst case.

The biological influence factors and interference factors (33) must still be taken into account, however.

Assuming the worst possible situation, it is advisable to specify the maximum acceptable imprecision as follows:

$$D = | x_1 - x_2 |; \qquad D = 4 s_{AD}$$

$$s_{AD} \leq 1/4 D$$

2.4.2 Limits for a comparison with a decision limit

A medical assessment is made by comparison with a decision limit in, for example, an oral glucose tolerance test or when therapeutic ranges are used. Therapeutic ranges are specified by two decision limits, these limits frequently having been determined from different populations. In comparing an analytical result x_1 with a decision limit x_{DL}, the question may arise of whether x_1 is significantly above s_{DL} (considering only analytical factors).

This assessment can be made by using the critical difference and assuming the worst possible situation. Then x_1 is significantly above the decision limit if

$$D = | x_1 - x_{DL} | \geq 4 s_{AD} .$$

It follows that the maximum allowable imprecision is then

$$s_{AD} \leq 1/4 D_{min} ,$$

where D_{min} is the smallest deviation from the limit of the therapeutic range that, based on clinical observations and measurements during determination of the limit of the therapeutic range, can still be clearly differentiated.

3. CONCLUSIONS

In closing I want to summarize three things, the current situation, what needs to be done, and who should do it.

Currently, method-dependent analytical results usually serve as the basis for control of accuracy and precision. For many quantities, however, such results can be quite inaccurate and thus will not provide an adequate check on the results they are intended to control. The best solution to this problem is the introduction and regular use of a control system that is based on the reference method concept. In such a system the limits for the maximum allowable imprecision and inaccuracy for each quantity are specified only after considering the medical requirements. In my talk today I have outlined the work that has already been done to develop and implement such a system. Now the IFCC as the international organization of scientists in clinical chemistry should go to work immediately on the most urgent unsolved problems and after clarifying the theory of calibration develop international recommendations. The realization of these recommendations should then be left to the national societies with their differing scientific, technical and economic situations. An assessment should be made of the effects on patient care, research and teaching of the measures and limits recommended by the individual societies.

4. REFERENCES

1. Levey S, Jennings ER. Amer J Clin Path 1950; 20: 1059-1066.
2. Fraser CG. Advances in Clinical Chemistry 1983; 23: 299-339.
3. Büttner J, Borth R, Boutwell JH, Broughton PMG, Bowyer RC. Approved Recommendations on Quality Control in Clinical Chemistry. Part 1 - Part 6. Expert Panel on Nomenclature and Principles of Quality Control in Clinical Chemistry. In: Saris N-E, ed. International Federation of Clinical Chemistry, Recommendations and Related Documents, Volume 1, 1978-1983. Berlin, New York: Walter de Gruyter, 1984: 48-89.
4. Stamm D. Experience with quality control in clinical chemistry. Z Lebensm Unters Forsch 1986; 182: 372-384.
5. Boutwell JH. A National understanding for the development of reference materials and methods for Clinical Chemistry. Proceedings of a Conference. American Association for Clinical Chemistry, Washington DC; 1978: 465 pp.
6. Tietz NW. Clin Chem 1979; 25: 833-839.
7. Richtlinienen der Bundesärztekammer zur Qualitätssicherung in Medizinischen Laboratorien vom 16. Januar 1987. Deutsches Ärzteblatt, 1987; Vol 84: in press.
8. Neufassung des Eichgesetzes vom 22 Februar 1985. Bundesgesetzblatt 1985; Teil I: 410-422.
9. Stamm D. J Clin Chem Clin Biochem 1979; 17: 283-297.
10. NCCLS: Tentative guidelines for calibration materials in clinical chemistry, 1982; Vol 2, No. 17: 499-526.
11. NCCLS: Tentative guidelines for calibration materials in clinical chemistry, 1982; Vol 2, No. 18: 527-554.
12. Stamm D. J Clin Chem Clin Biochem 1979; 17: 280-282.
13. Stamm D. J Clin Chem Clin Biochem 1982; 20: 817-824.
14. Külpmann WR. Dtsch Ges f Klin Chem Mitteilungen 1983; 4: 25-26.
15. Külpmann WR, Buchholz R, Leupold C, Maibaum P. J Clin Chem Clin Biochem 1985; 23: 542.
16. Külpmann WR, Lagemann J, Sander R, Maibaum P. J Clin Chem Clin Biochem 1985; 23: 865-874.
17. Siekmann L, Hüskes KP, Breuer H. Z Anal Chem 1976; 279: 145-146.
18. Neese JW, Duncan P, Bayse D, Robinson M, Cooper T, Stewart CH. HEW Publication No. (CDC) 1976; 77-8330.
19. Siekmann L. J Clin Chem Clin Biochem 1985; 23: 129-135.
20. Siekmann J, Siekmann A, Mackrodt D, Johnen C. Z Anal Chem 1986; 324: 279-280.
21. Schumann G, Büttner J. Z Anal Chem 1986; 324: 278-279.
22. Schumann G. Z Anal Chem 1986; 324: 209-211.
23. Siekmann L. J Clin Chem Clin Biochem 1985; 23: 137-144.
24. Siekmann L, Schönfelder A, Siekmann A. Z Anal Chem 1986; 324: 280-281.
25. Külpmann WR, Biesz B, Maibaum P. J Clin Chem Clin Biochem 1985; 23: 606.
26. Siekmann L, Spiegelhalder B, Breuer H. Z Anal Chem 1972; 261: 377-381.
27. Siekmann L. J Steroid Biochem 1979; 11: 117-123.
28. Siekmann L, Breuer H. J Clin Chem Clin Biochem 1982; 20: 883-892.
29. Siekmann L. J Clin Chem Clin Biochem 1984; 22: 551-557.
30. Siekmann L. Z Anal Chem 1978; 280: 122-123.
31. Experiences with Reference Methods (D. Stamm, Chairman), with contributions of Haeckel R, Külpmann WR, Siekmann L, Röhle G, Schumann G, Stamm D. Z Anal Chem 1986; 324: 204-211.
32. Hansert E, Stamm D. J Clin Chem Clin Biochem 1980; 18: 461-490.
33. Keller H, Guder WG, Hansert E, Stamm D. J Clin Chem Clin Biochem 1985; 23: 3-6.

34. Röhle G, Voigt U, Siekmann L. Commun Lab Med 1986; 2: 131-138.
35. Stange K. Angewandte Statistik, Erster Teil. Heidelberg, Berlin, New York: Springer-Verlag, 1970: 182-185.

CHAPTER 20
HUMAN INFERTILITY

CLINICAL ASPECTS OF FEMALE INFERTILITY

P. Devroey and A.C. van Steirteghem

Centre for Reproductive Medicine - Vrije Universiteit Brussel
Laarbeeklaan 101 - 1090 Brussels - Belgium

INTRODUCTION

About 15% of the couples are infertile and need medical assistance for reproduction. In about half of those 15% the infertility is caused by a female pathology. In the female, several causes, alone or in association can be present: blocked Fallopian tubes, anovulation, endometriosis, immunological problems and primary ovarian failure. In some patients no apparent cause for the infertility can be detected, they have so-called unexplained or idiopathic infertility.

Ovulation induction in anovulatory patients is a widely accepted procedure and will not be discussed in the present contribution.

Since the first successful in vitro fertilization (IVF) pregnancy in 1978 the problem of blocked Fallopian tubes is theoretically solved (1). The IVF laboratory takes over the role of the Fallopian tube for 2 or 3 days.

IVF can also be used for unexplained infertility, endometriosis as well as for infertility due to the presence of sperm antibodies.

Other techniques such as gamete intra-Fallopian transfer (GIFT) (2, 3) and zygote intra-Fallopian transfer (ZIFT) (4) can be done when at least one Fallopian tube is healthy.

The birth of the first baby after replacement of a frozen-thawed embryo has opened a new field of treatment (5).

Oocyte (embryo) donation in primary ovarian failure has solved the problem of absent ovaries (6). The use of oocyte donation solves also the problem of those patients, who can not use their oocytes for genetical reasons.

We present our 1986 results of infertility treatment by in vitro fertilization, gamete intra-Fallopian transfer, cryopreservation of human embryos and oocyte (embryo) donation.

MATERIALS AND METHODS

IVF techniques are closely related to the successful use of superovulation. Several drugs are available alone or in association: clomiphene citrate (Clomid^{R}, Merell), human menopausal gonadotrophins (Humegon^{R}, Organon; Pergonal^{R}, Serono), LH-RH analogue (Suprefact^{R}, Hoechst) and purified FSH (Metrodin^{R}, Serono).

Adequate follicle stimulation is assessed by the measurement of 17 β-estradiol (E2), luteinizing hormone (LH), follicle stimulating hormone (FSH), progesterone (P) and ultrasound evaluation. In the presence of follicles of at least 18 mm of diameter on ultrasound human chorionic gonadotrophins are injected. (Pregnyl^{R}, Organon; Profasi^{R}, Serono).

Oocyte retrieval is done laparoscopically or by direct puncture under ultrasound guidance. Mature oocytes are inseminated with 10,000 to 50,000 capacitated sperm cells (7). If normal fertilization and cleavage occurs embryo transfer will be done. To minimize the risk of multiple pregnancy no more than 3 embryos are replaced into the uterine cavity.

The remaining embryos are cryopreserved for later use (8, 9, 10, 11). If one healthy Fallopian tube is present, three mature oocytes and about 5,000 good motile sperm cells are replaced under laparoscopic vision in the distal part of a healthy tube.

Oocyte and embryo donation is done in primary ovarian failure and in patients with genetic risks.

Donated oocytes are inseminated with husband's sperm and placed as embryos into the uterine cavity of the recipient. If the husband is azoospermic fresh or frozen donor sperm is used i.e. embryo donation or prenatal adoption.

Donated oocytes are obtained from volunteers, from patients undergoing tubal sterilisation or as excess oocytes from patients undergoing in vitro fertilization.

In primary ovarian failure, estradiol valerate (Progynova^{R}, Schering) and natural progesterone (Progesterone Federa S.V.; Sterop S.V., Brussels) is administered to mimic a natural cycle and to obtain secretory endometrium (12, 13, 14, 15, 16).

RESULTS

In vitro fertilization and embryo transfer are necessarily used in patients with blocked Fallopian tubes but can also be used in patients with unexplained infertility (17). Superovulation induces the development of several ovarian follicles. In 1986 the mean number of retrieved oocytes per laparoscopy in our centre was 6.2 (1,818 oocytes retrieved in 278 retrievals). Five hundred thirteen embryos were replaced in 220 patients and 290 embryos were cryopreserved for later use. In the presence of blocked Fallopian tubes or ovarian adhesions the mean number of retrieved oocytes was 5.7/laparoscopy (738 oocytes in 129 retrievals). When the oocyte retrieval is done under ultrasound guidance the mean number of oocyte retrieved was 4.1 (552 oocytes in 136 retrievals).

The retrieval procedure did not alter the pregnancy rate 24.3% (via laparoscopy) and 26.6% (via ultrasound).

In patients with at least one healthy Fallopian tube the GIFT procedure was performed. Indications were unexplained infertility, endometriosis and male infertility. There were 1,691 oocytes retrieved in 192 GIFT trials (average 8.2), 28.1% of pregnancies were obtained. Upto 3 oocytes were placed into the Fallopian tube, the remaining were inseminated, cultured in vitro and cryopreserved, if they developed normally; 420 embryos could be frozen.

If patients failed to conceive cryopreserved embryos were replaced in subsequent natural cycles. Twenty patients became pregnant i.e. 15% per embryo replacement.

In patients with primary ovarian failure 9 pregnancies were obtained out of 27 oocyte or embryo donations. In normal cycling women, requiring oocyte or embryo donation 11 pregnancies were obtained out of 45 attempts.

DISCUSSION

In patients with tubal pathology the obtained pregnancy rates were approximately 25% per replacement. Those figures were similar with the spontaneous pregnancy rate in nature. An important difference is that in nature generally the pregnancy is established with one embryo, following the sequence of one oocyte, one embryo and a singleton pregnancy. In IVF this sequence changes to x-oocytes retrieved, 3 embryos replaced and a singleton pregnancy. We must further notice that in IVF the number of retrieved oocytes and available embryos for replacement is unpredictable. Cryopreservation of human embryos solves the problem of the remaining embryos. It is generally accepted that the replacement of three embryos is ethically advisable. Pregnancy rate increases with the number of embryos replaced: 13% with 1 embryo, 21.8% with 2 embryos and 32.8% with 3 embryos.

If we consider the obstetric risk of multiple pregnancies and the neonatal complications of prematurity, it is unacceptable to replace more than three embryos. For the same reasons it is advisable not to replace more than three oocytes in the GIFT procedure.

Although we replaced only three embryos in IVF and 3 oocytes in GIFT, twin pregnancies were obtained in 20% of the cases. In a series of 320 pregnancies, three tripleton pregnancies were observed.

The routine stimulation regime with clomid and hMG was satisfactory in only 85% of the started cycles. We had to cancel the remaining 15% for several reasons: premature LH-rises, premature and tonic elevated progesterone, an inadequate follicle growth. The combination of LH-RH analogues and hMG-hCG reduced the cancellation rate to approximately 2% (18).

Since the first reported pregnancy in a patient with primary ovarian failure in 1984 it has been proven that exogenous steroid replacement therapy makes atrophic endometrium receptive for implantation. Our series indicate that the artificial secretory phase of a substituted cycle is at least as receptive as the natural cycle. Nine pregnancies out of 27 replacements were obtained. In the future the implantation rate could be improved by giving other drugs in the substituted cycles.

CONCLUSION

The development of in vitro fertilization and related procedures improves the pregnancy rate in tubal, idiopathic infertility and in endometriosis. Cryopreservation of the remaining human embryos and their replacement in a subsequent cycle increased the chance of pregnancy and facilitates the ovum donation program by circumventing the asynchrony between donor and recipient. In primary ovarian failure exogenous sex steroid therapy make endometrium receptive for implantation.

REFERENCES

1. Edwards RG, Steptoe PC, Purdy JM. Establishing full-term human pregnancies using cleaving embryos grown in vitro. Br J Obstet Gynaec 1980; 87: 737-756.
2. Asch RH, Ellsworth LR, Balmaceda JP, Wong PC. Pregnancy after translaparoscopic gamete intra-Fallopian tranfer. Lancet 1984; ii: 1034-1035.
3. Braeckmans P, Devroey P, Camus M, et al. Gamete intra-Fallopian transfer: evaluation of 100 consecutive trials. Human Reproduction 1987; 2: 201-207.
4. Devroey P, Braeckmans P, Smitz J, Van Waesberghe L, Wisanto A, Van Steirteghem AC. Pregnancy after translaparoscopic zygote intra-Fallopian transfer in a patient with sperm antibodies. Lancet 1986; i: 1329.
5. Trounson A, Mohr L. Human pregnancy following cryopreservation, thawing and transfer of an eight-cell embryo. Nature 1983; 305: 707-709.
6. Lutjen P, Trounson A, Leeton J, Findlay J, Wood C, Renou P. The establishment and maintenance of pregnancies using in vitro fertilization and embryo donation in a patient with primary ovarian failure. Nature 1984; 307: 174-175.
7. Van Steirteghem AC, Devroey P, Heip J, et al. Monitoring of patients in an in vitro fertilization and embryo transfer program. In: Bettoci S, Carenza L, Loverro G, Sadurny G, eds. Human In Vitro Fertilization and Embyro Transfer and Early Embryo Development. Rome, Italy: CIC-Edizioni Internationali, 187-200.
8. Van Steirteghem AC, Van den Abbeel E, Braeckmans P, et al. Cryopreservation of human embryos and its application in an oocyte and embryo donation program. In: Proceedings of the Workshop on Embryos and Oocyte Freezing. Collection Fondation Marcel Mérieux, 1986: 107-122.
9. Van Steirteghem AC, Van den Abbeel E, Braeckmans P, et al. Pregnancy with a frozen-thawed embryo in a woman with primary ovarian failure. The New England J Med 1987; 317: 113.
10. Van Steirteghem AC, Van den Abbeel E, Camus M, et al. Cryopreservation of human embryos obtained after gamete intra-Fallopian transfer and/or in vitro fertilization. In: Human Reproduction; in press.
11. Lasalle B, Testart J, Renard J. Human Embryo features influence the success of cryopreservation with use of 1.2 propanediol. Fertil Steril 1985; 44: 645-651.
12. Devroey P, Braeckmans P, Camus M, et al. Pregnancies after replacement of fresh and frozen-thawed embryos in a donation program. In: Feichtinger W, Kemeter P, eds. Future aspects in Human In Vitro Fertilization. Berlin, Heidelberg: Springer-Verlag, 1987: 133-137.
13. Devroey P, Smitz J, Wisanto A, et al. Primary ovarian failure: embryo donation after substitution therapy. Annals of the New York Academy of Sciences; in press.
14. Leeton J, Harman J. Attitudes toward Egg Donation of thirty-four women who donated during their in vitro fertilization treatment. J Vitro Fert Embryo Transfer 1986; 3: 374-378.
15. Navot D, Laufer N, Kopolovic J, et al. Artificially induced endometrial cycles and establishment of pregnancies in the absence of ovaries. N Engl J Med 1986; 314: 806-811.

16. Noyes RW, Hertig AT, Rock J. Dating the endometrial biopsy. Fertil Steril 1950; 1: 3-25.
17. Naaktgeboren N, Devroey P, Traey E, Wisanto A, Van Steirteghem AC. Success of In Vitro Fertilization and Embryo Transfer in Relation to the Cause of infertility. Eur J Fertil Steril 1985; 16: 281-287.
18. Smitz J, Devroey P, Braeckmans P, et al. Management of failed cycles in an IVF/GIFT program with the combination of a GnRH analogue and hMG. Human Reproduction 1987; 2, 4: 309-314.

GIFT: BASIC AND CLINICAL CONCEPTS OF A NEW TREATMENT FOR INFERTILITY OF VARIOUS CASES

R.H. Asch

Department of OB/GYN, University of California, Irvine
101 The City Drive South, Orange, CA. 92668, USA

In 1978, the first baby was born from in vitro fertilization and embryo transfer (IVF-ET), in Oldham, England (1). For Steptoe and Edwards, this fact represented the successful culmination of 10 years of hard work and research. For biologists, embryologists and researchers in Reproductive Medicine, this represented the beginning of a new era involving the handling of human gametes and embryos. It also resulted in improvement of such previously used therapies as intrauterine insemination (IUI) (2), and the development of new ones such as ovum transfer (OT) (3), gamete intrauterine transfer (4), and Gamete Intrafallopian Transfer (GIFT) (5).

IVF and ET were originally designed as a therapy for patients with severely damaged or absent fallopian tubes; however, more recently, the use of IVF has been extended to the treatment of couples with infertility of different etiologies in whom traditional forms of therapy have failed. An important number of these patients have normal tubes, theoretically capable of normal gamete transport (Figure 1) and presenting a favorable milieu for early embryonic development. The GIFT technique involves the placement of both gametes, sperm and oocytes, into the ampullary portion of the fallopian tube and was developed as an alternative to in vitro fertilization in patients with normal tubes (Figure 2) (6, 7).

This study summarizes our experience with animal models and the first five series of 115 patients that underwent GIFT.

I. GIFT IN EXPERIMENTAL ANIMALS

A. Materials and Methods

1. Induction of Follicular Development: We utilized 25 regularly cycling rhesus monkeys (Macaca Mulatta). Twenty received human menopausal gonadotropin (hMG-Pergonal[R], Serono Laboratories, Braintree, MA), 1/2 ampule (37.5 I.U. LH and 37.5 I.U. FSH) intramusculary (I.M.) from day 2 of the menstrual cycle on (Group I). Serial daily serum estradiol (E2) measurements by radioimmunoassay were started on day 5 of the cycle. Laparoscopy was performed on day 9. Human chorionic gonadotropin (hCG, Profasi[R], Serono Laboratories, Braintree, MA) 1000 I.U. was given in I.M. Thirty-six hours later, laparotomy, follicular aspiration and gamete

transfer were performed. Group II considered of 5 regularly cycling animals that did not receive hormonal treatment. GIFT was performed with oocytes obtained from stimulated animals (Group I).

2. Semen Preparation: Semen was collected 3 hours prior to surgery by electro-ejaculation in 1 ml of TALP-HEPES containing bovine serum albumin (BSA, 3 mg/ml). The semen was washed 2 times by suspension in 10 ml TALP-HEPES and centrifugation for 10 minutes at 300G. After the second wash, the sperm were counted and resuspended in TALP-HEPES to a concentration of 10 million/ml. The final specimen was then incubated for 1 hour at 37°C.

3. Surgery: Laparotomy was carried out under general anesthesia with Ketamine HCL (5 mg/kg). Follicular aspiration was performed using a 20G needle. Oocytes recovered were placed in individual culture dishes and assessed for maturity based on the criteria described by Bavister et al. (8). Immediately after assessment, mature oocytes were transferred into a 10-15 μl drop containing 100,000 motile sperm. Gametes were gently aspirated from the culture dishes using a Tom catheter (Monoject, St. Louis, MO) and transferred into the mid-ampullary portion of the fallopian tube through the fimbriated end.

4. Post-transfer Follow-up: In order to support the luteal phase and to avoid the risk of luteal insufficiency due to follicular aspiration, all animals received progesterone in oil 5 mg I.M., daily from day 4 after transfer until the occurrence of menses or 4 weeks after the pregnancy was documented.

Pregnancy was confirmed by serial detection of serum mCG (macaque chorionic gonadotropin), measured every third day starting on day 7 post-GIFT. Pregnancy was also confirmed using the urinary Macaque chorionic gonadotropin kit (National Institute of Health) and the presence of an intrauterine gestational sac determined by ultrasound and uterine/rectal palpation.

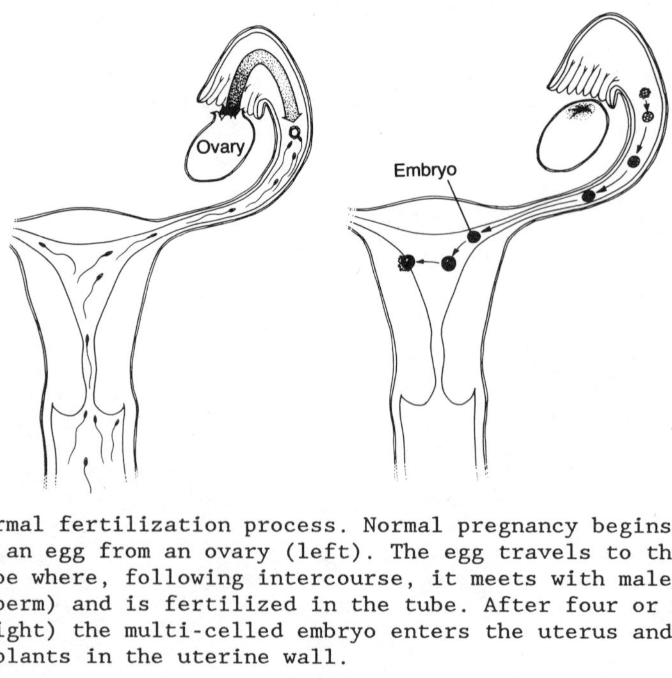

Fig. 1. Normal fertilization process. Normal pregnancy begins with release of an egg from an ovary (left). The egg travels to the fallopian tube where, following intercourse, it meets with male cells (sperm) and is fertilized in the tube. After four or five days (right) the multi-celled embryo enters the uterus and eventually implants in the uterine wall.

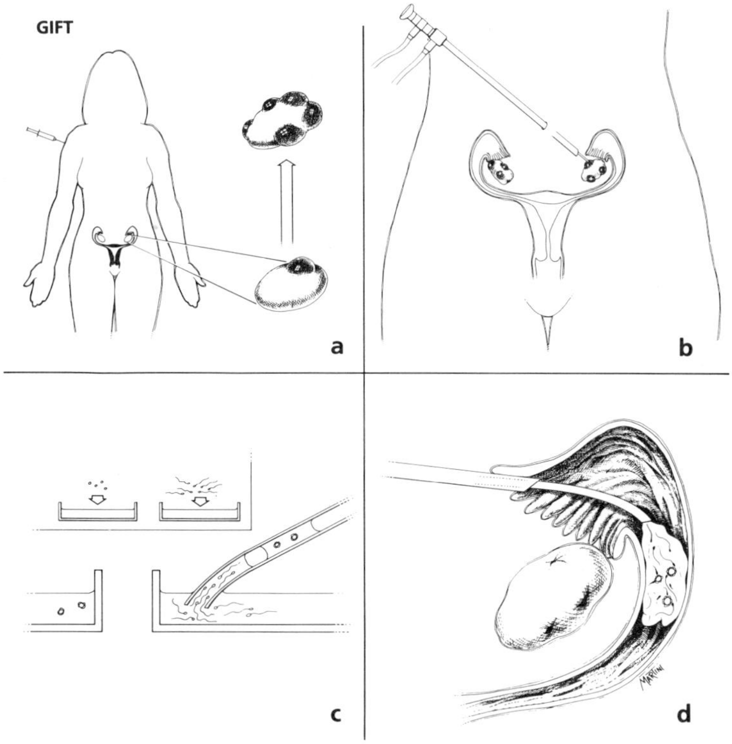

Fig. 2. GIFT (Gamete Intra-Fallopian Transfer). A: Induction of follicular development. B: Follicular aspiration. C: Catheter loading with both gametes. D: Gamete transfer into the tubes.

B. Results

Six animals were considered not to have ovarian stimulation and were dropped from the study. Nineteen animals underwent laparotomy for follicular aspiration and GIFT. Pregnancy was confirmed in 6 animals, and in one animal that received oocytes from another animal, a synchronized recipient (see Table 1).

C. Discussion

The pregnancy rate obtained in rhesus monkeys is encouraging and shows that these animals represent a suitable species for further study of the

Table 1. Pregnancy outcome in rhesus monkeys

Animal	Group	Pregnancy Outcome
4	2	miscarriage
5	1	miscarriage
6	1	miscarriage: twins 73 days post GIFT
8	1	normal delivery
11	1	miscarriage
12	1	miscarriage

GIFT technique. We do not have a good explanation for the high rate of early pregnancy loss in this short series (5 of the 6 pregnant monkeys miscarried). The animals utilized for the experiments had been used in previous experiments that included endocrinologic and surgical manipulations of their reproductive system. This could have negatively influenced their capability to carry a pregnancy to term. One of the animals had spontaneous abortion of day 71 post-GIFT (comparable to the second trimester in humans). This animal carried a twin pregnancy which is extremely rare in this species (1 : 1000 livebirths), so this could have also influenced the outcome of the pregnancy.

The effects of ovarian hyperstimulation on implantation and embryo development remains to be elucidated; however, this animal model could be ideal to study this problem.

II. GIFT IN HUMANS

A. Materials and Methods

One hundred and fifteen women had the GIFT procedure. Patients were treated by series according to the following protocol:

1. Induction of follicular development: In order to rule out the presence of ovarian cysts, a pelvic ultrasound was performed on day 3 of the menstrual cycle (ADR 4000 S/L, Temple AZ) and clomiphene citrate (Serophene, Serono Laboratories, Randolph, MA) was given in a dose of 50 mg every 12 hours from day 3 to 7 of the cycle. From day 6 of the menstrual cycle patients received daily intramuscular (I.M.) injections (150 I.U.) of human menopausal gonadotropins (hMG) (Pergonal^R, Serono Laboratories, Randolph, MA). Follicular maturation was assessed by daily ultrasound (U/S) and serum estradiol levels by specific radioimmunoassay (RIA) (antibody against E2, Radioassay Systems, Inc. Cat No. 1586, Carson CA., and 3H-E2 NET 517, New England Nuclear, Boston, MA). Human chorionic gonadotropin, (hCG) (Profasi^R, Serono Laboratories, Randolph, MA) was given when two or more follicles reached 16 mm on ultrasound and serum E2 levels reached 700 pg/ml. Thirty-six hours after hCG injection, patients underwent the GIFT procedure.

2. Surgical procedure: Follicular aspiration and gamete transfer were done through laparoscopy or minilaparotomy (suprapubic Pfannestiel incision of 3 cm). The total procedure lasted approximately 45 minutes, and was performed under general anesthesia.

3. Semen analysis: Semen was obtained by masturbation 2 hours prior to surgery and allowed to liquefy for 30 minutes. The samples were washed with culture medium: TALP-HEPES with 0.5% human albumin (ratio 1 : 3) and centrifuged for 10 minutes at 300G. The supernatant was discarded and 0.5 to 1 ml of medium layered onto the pellet. Sperm were allowed to swim out of the pellet for 45 to 60 minutes in a water bath at 37°C. The medium was then transferred to a tube where the sperm concentration was adjusted in order to obtain a final count of 10 x 10^6 motile sperm per ml. Before and after the preparation, the volume, count, motility, progression and morphology were recorded according to the World Health Organization criteria (9). The volume was adjusted to yield a sperm concentration of 100,000 motile sperm in a volume less than or equal to 25 μl.

4. Follicular aspiration: Follicular aspiration, the first step of the surgical procedure, was performed using a 18 cm, 14G steel needle (mini-laparotomy) or 52 cm (laporoscopy) (Figure 3). Needles were connected to a

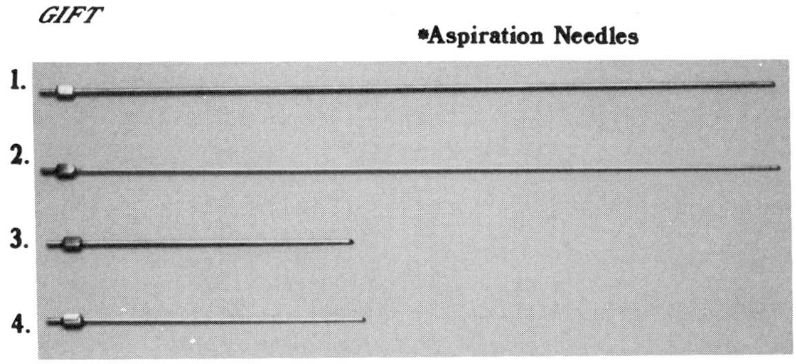

	Length (cm)	O.D. (mm)	I.D. (mm)
1.	45	28	21
2.	45	21	16
3.	18.5	28	21
4.	18.5	21	15

* Stainless steel
40° Bevel

Fig. 3. GIFT Aspiration Needles*
Specification of needles used for follicular aspiration; 1-2: laparoscopy; 3-4: laparotomy

wall suction system that never exceeded 120 mm Hg (De Lee suction catheter with mucus trap (ARGYLE), Division of Sherwood Medical, St. Louis, MO). Each follicle was aspirated into an individual tube and taken to the lab adjacent to the OR, then oocytes were identified under a dissecting microscope. Recovered oocytes were rinsed and placed in individual culture dishes (35 mm, Petri dish, Corning 25,000) containing CMRL-1060 culture medium (Gibco 3201535, Grand Island, NY) overlayed with paraffin oil (Fischer 0-121). Oocytes were held in an incubator at an atmosphere of 5% CO_2 and air at 37°C until classification.

Oocytes were classified from 1 to 6 (Figure 4) with an inverted microscope (differential interference contrast- Nomarski Optics) as follows:

(1) Immature oocyte with germinal vesicle.
(2) Immature, germinal vesicle not visible. Thick cumulus oophorus and corona.
(3) Partially dispersed cumulus oophorus and corona.
(4) Dispersed cumulus oophorus and corona radiata. No visible polar body.
(5) Well dispersed cumulus oophorus and corona radiata. Presence of first polar body.
(6) Atretric oocytes, degenerated or broken zona pellucida.

A maximum of 4 oocytes were transferred (2 to each tube). In terms of the transfer, we classified the cases as Type I: when one or more of the oocytes transferred were grade 5 and the remainder were grade 4; Type II: when all oocytes transferred were grade 4; Type III: when 1 to 2 oocytes classified as grade 4 and the remainder classified as grade 3 or less were transferred; and Type IV: when only oocytes classified as grade 3 or less were transferred.

5. Catheter preparation and transfer: A Tomcat catheter (Sovereign Tomcat 8890-704003; Monoject Division, Sherwood Medical, St. Louis, MO) (minilaparotomy) or a Deseret Intracath catheter (No. 3132, The Deseret Co., Sandy, UT) (laparoscopy) with a 1 cc Hamilton syringe was used for the transfer. The catheter was loaded with oocytes and 100,000 motile sperm in

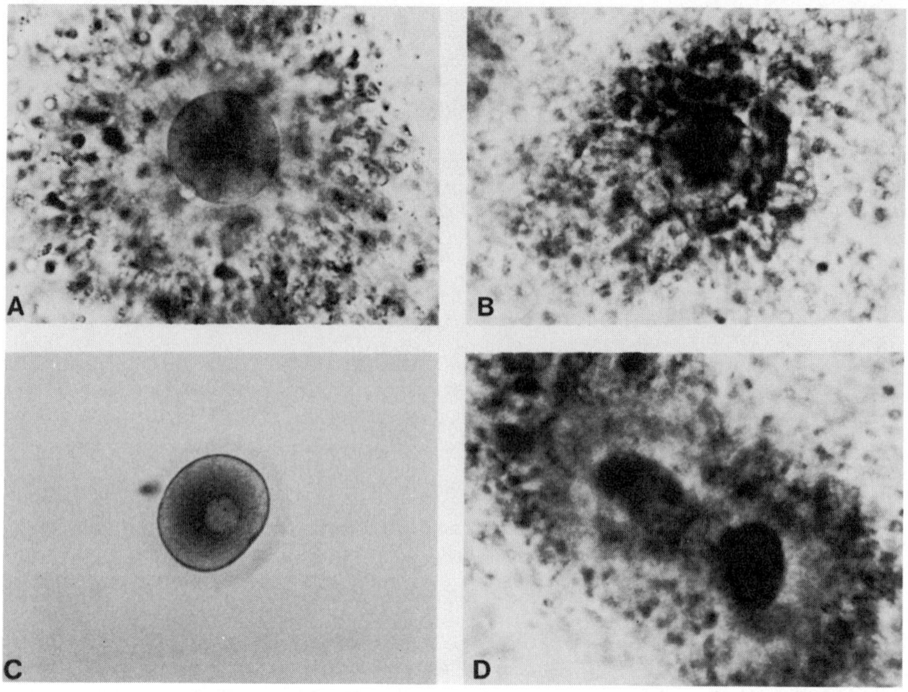

Fig. 4. Different types of oocytes obtained from follicular aspiration in patients heated with CC or hMG. A: Oocyte type 5: cumulus and corona dispersed. Presence of polar body (8 o'clock). B: Oocyte type 3: cumulus and corona tight (6 o'clock); cumulus and corona dispersed (10-6 o'clock). C: Oocyte type 1: germinal vesicle visible (no cumulus or corona). D: Oocyte degenerated.

total volume of 50 μl or less. Both gametes (oocytes and sperm) remained separated by an air space of 4 μl. The transfer was performed by introducing the catheter 10 to 25 mm into the fimbriated end of the fallopian tube.

The contents were then gently expelled into the tube. The same procedure was repeated for the second fallopian tube. After the transfer, the catheters were flushed with culture medium and the contents examined under the dissecting microscope to insure that no oocytes remained inside of the catheters or attached to their wall (Figure 5-6 show different types of transfers catheters).

6. Follow-up post transfer: Patients were sent home either the same day or the day following surgery, depending on their recovery. All patients received progesterone in oil 12.5 mg/I.M., daily from day 4 after transfer until up to 8 weeks of gestation or negative pregnancy tests. Serum βhCG levels were measured by radioimmunoassay serially from day 7 after transfer and pelvic ultrasound scans were performed three to five weeks post-GIFT.

B. Results

One hundred and fifteen patients between 25 and 42 years old (x ± S.D.: 33.8 ± 6.3) underwent the GIFT procedure. Table 2 shows the distribution by age and correlations with pregnancies. Duration of infertility was found to be between 3 and 19 years (x ± S.D.: 5.84 ± 3 >75); primary: 83% and secondary: 17%. Table 3 correlates the link between

Table 2. Effect of age on pregnancy rate

Age	Number of Cases	Pregnancies Number	Pregnancies Percentage
25-27	4	2	50.0
28-30	28	10	30.0
31-33	35	6	17.1
34-36	24	10	37.5
37-39	22	7	31.8
40-42	2	1	50.0
Total	115	36	31.3

diagnosis and term pregnancies. Patients with multiple causes of infertility represent a group with male factor associated with anovulation (n = 5). Of the patients with endometriosis, curiously, most of the pregnancies occurred in the cases where the disease was most severe. Table 4 correlates the type of transfer and pregnancies. Almost all the pregnancies occurred when oocytes 4 to 5 were transferred. (Types I, II, or II of gamete transfer). The best results were obtained with oocytes grade 5 (44.4%) (Type I). Five pregnancies were reported with transfer Type III oocytes. In one of the patients that only received grade 3 oocytes, a gestational sac was observed by ultrasound, but she spontaneously miscarried during the first trimester. The other patient received 2 grade 4 oocytes and one grade 3 oocyte. Ultrasound revealed 3 gestational sacs and she recently delivered 3 healthy babies.

While compiling the criteria of semen analysis, the motility showed the most substantial changes after the sperm preparation.

The distribution of pregnancies after the GIFT procedure was as follows: 23 deliveries, 10 spontaneous miscarriages during the first trimester and one during the second trimester (16th week). Two ectopic pregnancies were diagnosed. Of the 36 pregnancies, 10 patients carried twins (8 deliveries and 2 miscarriages) and one patient recently delivered triplets.

C. Discussion

The first patient that underwent GIFT had a diagnosis of unexplained infertility. The highest number of cases and pregnancies are in this group. However, indications for this procedure are now widespread and include: male factor, endometriosis and premature ovarian failure. Taking into consideration the type of oocytes recovered, we believe that the protocol for ovulation induction is very effective since the number of type 4 and 5 oocytes recovered was greater than 60%. As seen in IVF, the day of hCG administration depends upon different criteria. In this series, hCG administration depended on ultrasound and hormonal criteria. This stimulation protocol was considered successful because mature oocytes were retrieved (Type 4 and 5).

Semen samples were prepared according to the technique described by Ericsson (10). This preparation improved sperm motility and is very effective in significantly reducing the microbes present in human semen, as we demonstrated in a previous study (11). The number of sperm

Table 3. Results by etiology

Etiology	Number of Cases	Pregnancies Number	Percentage
Unexplained infertility	52	22	42
Male factor	30	4	13
Endometriosis*	13	5	38
Periadnexial adhesions	7	2	28
Multiple factors*	7	1	14
Premature ovarian failure+	2	2	100
Immunologic	2	0	0
Cervical factor	2	0	0
Total	115	36	31.3

* see text - Results
+ Oocytes donated

transferred per tube, 100,000 was decided according to most IVF protocols. Pregnancy rates obtained in this series (31.3%) surpassed that currently obtained with IVF (10-20%) (12) and demonstrated that GIFT is highly successful in the treatment of infertility. The incidence of miscarriages and ectopic pregnancies is comparable to that found in the infertile population.

The success obtained with the GIFT procedure in patients with unexplained infertility may be related to the fact that this technique is surpassing problems not well clarified as are: gamete transport, luteinization of unruptured follicles, and entrapped oocytes (13).

The GIFT technique, an alternative to IVF, mimics the physiologic process that normally leads to gestation (14). However, it is not envisioned that GIFT will completely replace IVF. In some cases where surgical findings (tubal damage), make transfer to the gamete to the tubes impossible, IVF may be a complement to GIFT.

In addition, extra oocytes not transferred during the GIFT technique, could be inseminated in vitro. If fertilized, they could be theoretically be transferred within the same cycle (GIFT or IVF-ET combined) or cryopreserved and transferred in future cycles (GIFT-IVF-Cryopreservation).

We reported two pregnancies after GIFT procedure (2 attempts in one of the patients), in two women with diagnosis of premature ovarian failure, transferring oocytes domated and sperm preparation of the recipients' husbands. These patients received steroid replacement to mimic normal events of an endometrial cycle, and subsequently to replace the function of the corpus luteum until the stage of the luteo-placental shift. Although there is a report in the literature of a pregnancy obtained after an IVF in a patient with premature ovarian failure (15), these 2 pregnancies represent the first cases after the GIFT procedure.

Table 4. Correlation between quality of oocytes transferred and pregnancies (clinical phase)

Transfer*	Number of cases	Pregnancies	Percentage
A. Type I	27	12	44.4
B. Type II	57	18	31.5
C. Type III	26	5	19.2
D. Type IV+	5	1	20.0

* See text - Materials and methods
\+ See text - Results

IN SUMMARY

GIFT is a new technique that may aid in the treatment of infertility when conventional methods of treatment have failed and one that could become an alternative to IVF when at least one fallopian tube is patent. The GIFT technique involves the placement of motile sperm and mature oocytes into the fallopian tube. This technique is more physiologic and could be more acceptable to groups opposed to any artificial methods of conception.

The experimental phase of the GIFT technique was performed using 25 rhesus monkeys, divided into two groups. Twenty (Group I) received hMG to induce follicular development and five (Group II) served as controls, receiving no hormonal therapy. All twenty-five animals underwent laparotomy for GIFT. As a result, we obtained 6 pregnancies (1 set of twins), one of them with a term delivery and 5 miscarriages (1 in the second portion of the pregnancy). The clinical phase of GIFT was carried out in 115 patients divided into a total of five series. Patients received clomiphene citrate and hMG for follicular development (monitored daily with ultrasound) and underwent daily sperm estradiol determinations. Human chorionic gonadotropin was given when two follicles measured > 16 mm by ultrasound and/or serum estradiol measured $\geq$ 700 pg/ml. Surgery was carried out 36 hours later. The four best oocytes obtained were loaded into a

GIFT

Transfer

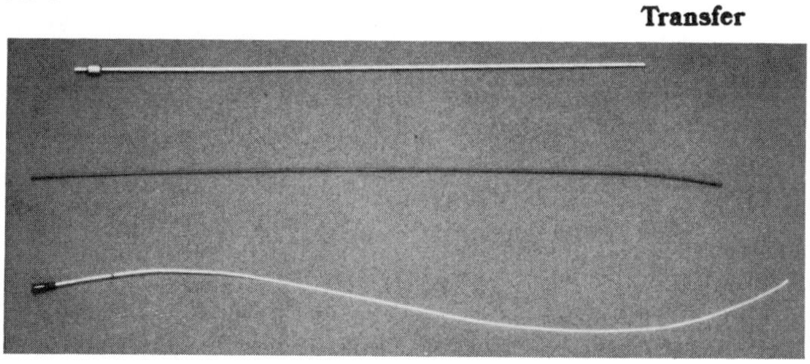

	Material	O.D.(mm)	I.D.(mm)	Length (cm)	Use
1.	Steel	2.8	2.1	45	Needle
2.	Teflon	2.0	1.8	55	Sleeve
3.	Teflon	1.7	0.8	60	Catheter

Fig. 5. Specifications of different components of kit used for gamete transfer

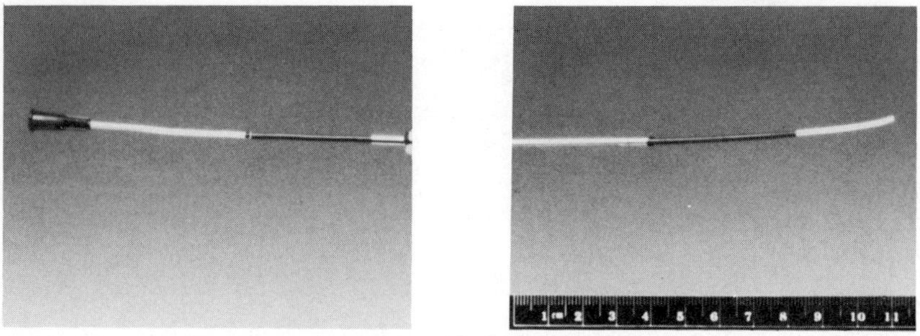

Fig. 6. Shows proximal and distal end of all kit components assembled.

catheter with the sperm preparation (obtained 2 hours prior to surgery). Of the 115 cases, there were 36 pregnancies, which resulted as follows: 23 deliveries, 10 miscarriages (9 in the first trimester and 1 in the second trimester), and 2 tubal pregnancies.

REFERENCES

1. Steptoe PC, Edwards RG. Birth after implantation of a human embryo. Lancet 1978; 2: 366.
2. Barwin BN. Intrauterine insemination of husband's semen. J Reprod Fertil 1974; 36: 101.
3. Buster JE, Bustillo M, Thornecroft JH, et al. Non-surgical transfer of an in vivo fertilized donated ovum to an infertile patient. Lancet 1983; 1: 816.
4. Craft J, Djahanbakch O, McLeod F, Bernard A, Green S, Twigg H. Human pregnancy following oocyte and sperm transfer to the uterus. Lancet 1982; 1:2: 1031.
5. Asch RH, Ellsworth LR, Balmaceda JP, Wong PC. Pregnancy after gamete intrafallopian transfer. Lancet 1984; 2: 1034.
6. Asch RH, Ellsworth LR, Balmaceda JP, Wong PC. Gamete intrafallopian transfer (GIFT): A new treatment for infertility. Int J Fertil 1985; 30: 41.
7. Asch RH, Balmaceda JP, Ellsworth LR, Wong PC. Preliminary experiences with gamete intrafallopian transfer. Fertil Steril 1985; 45: 366.
8. Bavister DB, Boatman DE, Leibfried L, Loose M, Vernon MW. Fertilization and cleavage of rhesus monkey oocyte in vitro. Biol Reprod 1983; 28: 983.
9. World Health Organization: Laboratory manual for the examination of human semen and semen-cervical mucus interaction. Edited by Belsey MA, Eliason R, Gallegos AJ. Singapore Press Concern, 1980.
10. Ericsson RJ. Isolation and storage of progressively motile human sperm. Andrologia 1972; 9: 111.
11. Wong PC, Balmaceda JP, Blance JD, Gibbs RS, Asch RH. Sperm washing technique removes microorganisms from human semen. Fertil Steril 1986; 45: 97.
12. Wood C, Trounson A. Current state and future of IVF. Clin Obstet Gynaecol 1985; 12: 753.
13. Dhont M, Serreyn R, Duvivier P, Vanluchene E, DeBoever J, Van der Kerckhove D. Ovulation stigma and concentration of progesterone and estradiol in peritoneal fluid: Relation with fertility and endometriosis. Fertil Steril 1984; 41: 872.

14. Kreitman O, Hodgen ED. Low tubal ovum transfer: An alternative to in vitro fertilization. Fertil Steril 1980; 34: 375.
15. Lutjen P, Trounson A, Leeton J, Findlay J, Wood C, Renou P. The establishment and maintenance of pregnancy using in vitro fertilization and embryo donation in a patient with premature ovarian failure. Nature 1984; 307: 174.

FUTURE POSSIBILITIES IN REPRODUCTIVE MEDICINE

Iain T. Cameron and Alan O. Trounson

Department of Obstetrics and Gynaecology and Centre for Early Human Development, Monash University, Melbourne Australia

In both sexes, efficient reproductive function is primarily dependent upon complex inter-relationships between the gonads and central nervous system. Gonadotropin releasing hormone (GnRH) from the arcuate nucleus of the hypothalamus is transported to the median eminence where it is released in pulsatile fashion into the pituitary portal vascular bed (1, 2). The resulting stimulation of the pituitary gonadotropins, follicle stimulating hormone (FSH) and luteinizing hormone (LH), has specific gonadal effects in both the male and the female, and these central events are in turn regulated by certain feedback mechanisms from the gonads themselves (Figure 1).

Recent developments in molecular biology, particularly in relation to recombinant DNA techniques, have provided an invaluable investigative approach to reproductive physiology. In addition, other advances in the field of peptide synthesis for example, have allowed the application of hormone agonists as therapeutic agents.

This article will review some of the more recent advances in reproductive medicine, and consider their possible future applications.

ENDOCRINOLOGY

GnRH analogues

The isolation and identification of the structure of GnRH led to the development of synthetic analogues for therapeutic application (3, 4, 5). A variety of long-acting compounds were synthesised with the aim of achieving prolonged gonadotropin release, however it was found that repeated administration of these agents resulted in a paradoxical inhibition of FSH and LH activity (6, 7).

Table 1 lists some of the currently used GnRH agonists. D-amino acid substitution in position 6 is common to most of these compounds, and many are nonapeptides, with replacement of the position 10 glycine with an ethylamide group. Such modifications result in both an increased binding affinity to pituitary GnRH receptors (8), and a degree of resistance to enzymatic degradation (9).

Table 1. GnRH agonists, listed in ascending order of potency. (Adapted from 7).

Structure	Proprietary name	Method of Administration
(D-Leu^6, Pro^9 NEt) LHRH	Leuprolide, Leuprorelin	Injection
(D-Trp^6) LHRH	Tryptorelin	Injection, injectable polymer microspheres
(D-Ser(tBU)^6, Pro^9 NEt) LHRH	Buserelin	Injection, nasal spray, polymer implant
(D-Nal (2)^6, Aza-Gly^{10}) LHRH	Nafarelin	Nasal spray, injectable polymer microspheres

The mechanism of action of the GnRH agonists relies on the mandatory pulsatile nature of GnRH release. If a continuous infusion of GnRH is administered, the pituitary becomes relatively insensitive to further stimulation, resulting in "down-regulation" and a fall in FSH and LH release (10). Indeed, this refractoriness can also be seen after the administration of single pulses of GnRH in women with hypothalamic amenorrhoea (11). The consequent inhibition of gonadotropin release prevents normal folliculogenesis and blocks ovulation. In men, both sperm production and testosterone secretion are suppressed.

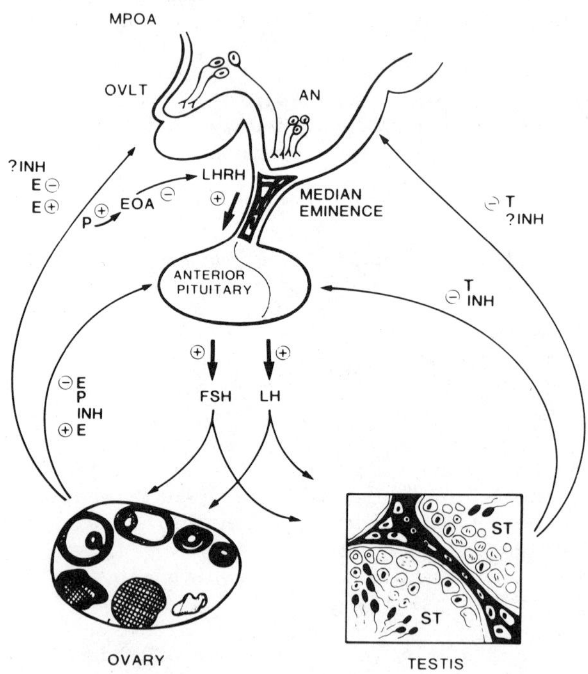

Fig. 1. The hypothalamus-pituitary-ovarian/testicular axis. (MPOA = medial preoptic area, OVLT = organ vasculosum lamina terminalis, AN = arcuate nucleus, EOA = endogenous opioid activity, E = estradiol, P = progesterone, T = testosterone, INH = inhibin, ST = semeniferous tubules, + = stimulation, - = inhibition).

Table 2. GnRH analogues: clinical indications

1. ANTIFERTILITY :	Ovulation inhibition Luteolysis Spermatogenesis inhibition
2. GONADAL SUPPRESSION:	Endometriosis Uterine fibroids Dysfunctional uterine bleeding Adjunct to ovulation induction/IVF Premenstrual syndrome Hirsutism Precocious puberty Malignant disease - breast and prostate carcinoma

The degree of this suppression of reproductive function depends upon the dose of drug given and the route of administration. Inactivation by the gastrointestinal tract has necessitated the development of alternative methods of therapy. Intranasal spray has proved to be effective (12), but if a more consistent inhibition of gonadal steroids is required, parenteral administration is preferable. The use of depot preparations has also been investigated to simplify continuous drug delivery (13, 14).

Thus the GnRH agonists offer a technique for reversible medical castration which has the potential for therapeutic application in a variety of situations (Table 2). The required degree of gonadal suppression will depend upon the clinical situation. For example, partial impairment of the hypothalamo-pituitary-ovarian axis will be adequate to disrupt ovulation, providing a contraceptive method with a reduced incidence of hypoestrogenic side effects. On the other hand, more complete gonadal steroid inhibition is required for the treatment of precocious puberty or breast and prostate carcinoma.

The development of more potent compounds and improved methods of administration will promote the therapeutic application of the GnRH analogues. There is likely to be considerable future potential for the use of these agents in conjuction with exogenous gonadotropins for ovulation induction (15), and for superovulation in in-vitro fertilization (IVF) programmes (16, 17). Advantages of this approach in IVF treatment cycles include the abolition of LH-induced thecal androgen production, which may impair folliculogenesis (18). In addition, besides preventing endogenous LH surge-associated "out of hours" oocyte retrieval, such treatment may improve IVF success rates. The administration of buserelin has been investigated in 44 women in whom previous attempts at IVF had failed to obtain more than one oocyte (MacLachlan VB, Healy DL, unpublished data). The median number of oocytes retrieved from 29 patients receiving the GnRH analogue and human menopausal gonadotropin (hMG) was 4 (range 0.19), with 0 (0.5) being obtained from 15 women treated with clomiphene citrate and hMG. The fertilisation rate was 76% in each group. Eighty-four percent of women receiving buserelin and hMG underwent the transfer of 3 embryos, whereas this was only true of 13% in the clomiphene/hMG group. A total of 3 pregnancies ensued, all from triple embryo transfers in the buserelin-treated women.

Inhibin

The existence of a seminal pituitary-inhibiting agent was first postulated following the observation of castration-induced pituitary-cell hyperplasia (19), and the name "inhibin" was suggested nine years later to describe an active principle from bovine testes that could inhibit the appearance of the pituitary "castration cells" (20). However, the structure of this putative inhibitor of pituitary FSH release has only recently been described.

Inhibin has been purified from bovine (21) and porcine follicular fluid (22, 23, 24). Cloning and sequencing of its peptide chains has subsequently been performed (25, 26) and the circulating compounds has been measured (27, 28).

The inhibin molecule comprises 2 peptide chains-α or A (~20 kDa) and β or B (~13 kDa) - linked by 2 disulphide bridges. It appears to have a molecular weight of 58 kDa in follicular fluid, but circulates as a 31 kDa entity. Furthermore, in the pig and human, inhibin exists in 2 forms (A and B) which result from differences in the amino acid sequence of the smaller β subunit (29, 30).

Although inhibin has been demonstrated as a circulating peptide hormone with specific inhibitory effects on pituitary FSH release, the discovery of homology between inhibin and various growth regulating factors, along with the existence of its peptide chains in forms other than the original AB entity, has suggested additional regulatory roles for the compound (Figure 2).

Structural homology was first noted between the inhibin β subunit and Transforming Growth Factor TGF-β (31), a 25 kDa homodimer which exists in both normal and neoplastic tissues. The transforming growth factors are mitogenic proteins which induce anchorage independent growth in mammalian cells and TGF-β itself has been shown to have either stimulatory or inhibitory effects on growth in different experimental systems.

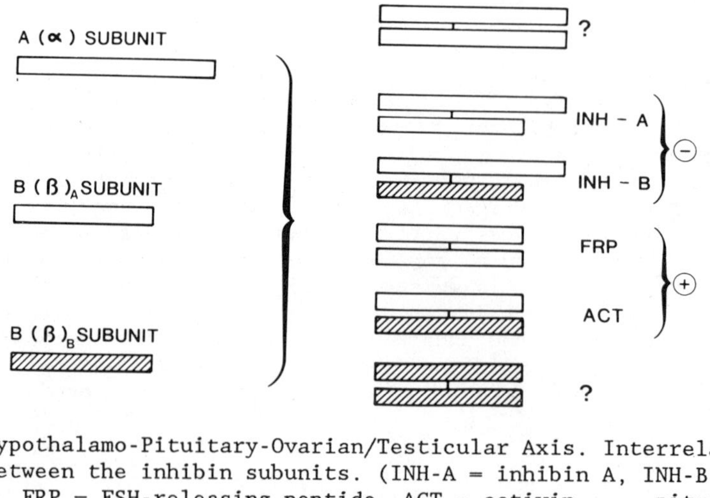

Fig. 2. Hypothalamo-Pituitary-Ovarian/Testicular Axis. Interrelationships between the inhibin subunits. (INH-A = inhibin A, INH-B = inhibin B, FRP = FSH-releasing peptide, ACT = activin, - = pituitary FSH inhibition, + = pituitary FSH stimulation). Adapted and redrawn from 30.

The structures of bovine and human Mullerian Inhibitory Substance (MIS) - a Sertoli cell glycoprotein which induces Mullerian duct regression in the male embryo - have recently been derived (32). Again, a marked structural similarity has been demonstrated between the C-terminal region of MIS and both the A and B subunits of inhibin, and TGF-β.

A further homology has been reported between TGF-β and a pattern-controlling gene in Drosophilia (33). Coupled with the discovery that circulating inhibin can be detected in early pregnancy in women without gonads who have conceived following the transfer of embryos derived from donated oocytes (34), such findings would suggest a role for inhibin, or a structurally closely related compound, in the regulation of early embryonic developement.

In addition to its physiological role, potential clinical applications of inhibin should be considered. The specific inhibition of FSH could provide an approach to fertility regulation in both sexes (35). Although a reduction of FSH release should impair spermatogenesis in the male, consistent azospermia, or oligospermia incompatible with fertilisation, would need to be achieved. Assuming that this proves practicable, the concomitant administration of testosterone may also be necessary, due to the potential inhibitory effect of inhibin on LH. As a female contraceptive agent, inhibin could be administered repeatedly on alternate weeks to prevent the recruitment and selection of a dominant follicle and suppress ovulation. Alternatively, continuous administration in the early follicular phase should result in an inadequate luteal phase. In the latter case, a consistent reduction in conception rates would have to be demonstrated, and variation in individual responses could lead to menstrual irregularities. In the former situation, any rebound increase in FSH release during the inhibin-free period of drug administration could result in sufficient stimulus for folliculogenesis to achieve ovulation before the next dose of drug is given.

The measurement of inhibin could have practical application as a diagnostic marker in superovulation treatment for IVF, for although estradiol is widely used to assess the response to treatment, this provides an index of thecal rather than granulosa cell function (36). Inhibin offers the opportunity to monitor granulosa cell activity more specifically, and a direct correlation has been observed in plasma between estradiol and inhibin concentrations in women receiving clomiphene citrate and hMG (28).

Finally the use of an agent such as an inhibin vaccine to augment gonadotropin stimulation of folliculogenesis could have commercial application for increasing ovulation rates in various animal species.

Progesterone Receptor Antagonists

In the first trimester of pregnancy, therapeutic abortion is most often performed by vacuum aspiration of the uterus under general anaesthesia (Office of Population Censuses and Surveys, MHSO, London, 1985). Although the procedure is safe and effective (37), complications are related to both gestational age, and surgical and anaesthetic techniques (38). Consequently, and with the additional desire of many women to avoid operative treatment (39), there has been much recent interest in the development of medical agents to interrupt pregnancy at an early stage ("menstruation induction").

The prostaglandins were the first agents to be used on a wide scale for medical abortion, however the instability of the naturally-occurring compounds, and a high incidence of gastrointestinal side effects has limited their acceptability (40, 41).

The development of the anti-progesterone agents has offered an alternative approach to menstrual induction (42). As a means of disrupting pregnancy, progesterone withdrawal is highly effective. At early gestations the major progesterone source is the corpus luteum, and the importance of this structure in the maintenance of pregnancy was elegantly demonstrated in luteectomised rabbits using porcine corpus luteum extracts (43). Subsequently Csapo and co-workers confirmed these findings in women, and established that the effects of luteectomy in early pregnancy could be reversed by exogenous progesterone (44, 45).

Clinical studies have confirmed the safety of the progesterone receptor antagonist RU486 (17-hydroxy-11-(4-dimethyl aminophenyl)-17-(1-prop-1-ynyl) estra-4, 9-dien-3-one, mifepristone) for menstrual induction, but the 60% incidence of complete abortion has been disappointing (46). A combined therapy using both RU486 and prostaglandin may be more effective, and complete abortion rates of 95% have been reported using RU486 with sulprostone (47) or gemeprost (48). Although an improved success rate of 85% has recently been reported using RU486 alone to induce menses within 10 days of the missed period (49), the augmentation of uterine contractility with exogenous prostanoids appears beneficial for expedient complete uterine evacuation.

Besides providing a safe alternative to surgery for early abortion, agents such as RU486 have potential application as "once-a-month" fertility regulators. In principle, the agents would be administered on a regular basis to induce menses, thereby avoiding the necessity for the prolonged administration of synthetic steroids. However, stringent follow up of all patients would be required to ensure treatment success. In addition, asynchrony between the ovarian and menstrual cycles, resulting from the delay to ovulation in the first post-abortion cycle, represents a major constraint to the use of these agents on a routine basis (50, 51).

EMBRYOS AND GAMETES

Oocyte donation

Advances in IVF technology have led to the development of oocyte donation as an infertility treatment option for women with non-functioning or absent gonads, and for women not wishing to use their own gametes because of inheritable disease (52).

Patients with hypergonadotropic hypogonadism are commenced on a cyclical steroid replacement regimen using estradiol valerate and progesterone in order to mimic the normal ovarian cycle endocrine profile (53). When an egg becomes available for donation, most commonly from another woman going through the standard IVF programme, it is fertilised with the recipient's husband's sperm, and transferred into the uterus between day 16 and 21 of the recipient's cycle. Should the donor and recipient cycles be asynchronous, the embryos are frozen and subsequently replaced at the appropriate time.

In the Monash University donor oocyte programme, five pregnancies in women with non-functioning gonads have now resulted in the birth of six healthy infants (one twin pregnancy). However one pregnancy following the transfer of a frozen-thawed embryo ended in spontaneous abortion at 12 weeks. Additionally, six pregnancies have been obtained with donor oocytes in the group of women not using their own gametes. Of these, three pregnancies have aborted spontaneously in the first trimester, two have successfully reached term and one is ongoing without complication.

Ovum donation for IVF therefore offers an effective treatment for these women. Furthermore, such programmes present a unique opportunity to study various aspects of early pregnancy. Firstly, that successful pregnancy can be maintained in the human by the administration of estradiol and progesterone alone in the first trimester is confirmed. Secondly, there is the chance to investigate the endocrinology of early pregnancy in the absence of confounding ovarian factors, and preliminary studies have demonstrated the production of inhibin by the conceptus in these women (34). Finally, the developmental requirements for a "receptive" uterus for implantation can be assessed using varying dose regimens of steroid priming. It is likely that uterine receptivity is an important variable for successful implantation (54), and undoubtedly standard cyclical steroid replacement stimulates good uterine receptivity, in that successful pregnancy can occur. Further studies are in progress to define the ultrastructural appearance of the receptive endometrium (55).

The most recent developments with future potential in oocyte donation programmes are in relation to the modification of steroid replacement regimens to overcome the need for synchrony between donor and recipient. The daily administration of 6-8 mg estradiol valerate for up to 4 weeks, with the commencement of gestogen on the day prior to oocyte retrieval has produced seven pregnancies in 17 women (56). This would suggest that the incremental increase in estradiol secretion in the first half of the spontaneous menstrual cycle is not a prerequisite for the production of secretory change in the uterus compatible with implantation.

Embryonic Biopsy

Second trimester amniocentesis has become a recognised method of antenatal diagnosis. Along with developments in ultrasound technology, it is possible to detect accurately conditions such as neural tube defects or chromosomal abnormalities, for which therapeutic abortion can be offered as a treatment option. Earlier diagnosis would be preferable though, and the subsequent evolution of chorion villous biopsy now permits many investigations in the first trimester (57). Furthermore, the development of DNA probes for markers of inheritable diseases such as cystic fibrosis or Huntington's Chorea has expanded the therapeutic potential of antenatal diagnosis.

However, the need for therapeutic abortion could be reduced if pre-implantation embryos generated by IFV could be screened for genetic abnormality by embryonic biopsy (58). The removal of a single cell from a 4-cell mouse embryo can now be achieved with successful in-vitro growth of 90-95% of the remaining 3-cell embryos to the blastocyst stage, and further in vivo studies are in progress (Wilton LJ, personal communication).

CONCLUSION: ETHICAL CONSIDERATIONS

Recent advances in reproductive medicine have opened up many areas for potential therapeutic application. However, the stage has been reached in the new reproductive technology where society has demanded that constraints be placed upon scientific research. Such constraints have applied particularly in the field of IVF and the associated embryo research.

Various bodies have been formed to assess the acceptability of reproductive research. One of the suggestions of the Warnock Committee was that the conceptus produced by IVF should not, unless transferred to a woman's uterus, be allowed to develop beyond 14 days, when the primitive streak first appears (59). On the other hand, the Infertility (Medical Procedures) Act 1984 (Victoria) has suggested that no embryo should be created unless it is intended to replace it in-utero after IVF, although the debate continues as to whether it will be acceptable to experiment on the very early human embryo before pronuclear fusion.

The past decade has seen major advances in the field of human reproduction, both in the understanding of basic pathophysiology, and in the clinical application of new technology. Nevertheless these advances have also led to the emergence of important ethical considerations. The stance of society in balancing the constraints imposed upon scientific research against the many potential benefits, will have a significant impact on future possibilities in reproductive medicine.

ACKNOWLEDGEMENTS

We thank Patricia Martin and Chris Hales for assistance with typing and illustrations respectively. ITC is in receipt of a National Health and Medical Research Council Medical Postgraduate Scholarship.

REFERENCES

1. Knobil E. The neuroendocrine control of the menstrual cycle. Rec Prog Horm Res 1980; 36: 53-88.
2. Sealey JE, Cholst I, Glorioso N, et al. Sequential changes in plasma luteinizing hormone and plasma protein during the menstrual cycle. J Clin Endocrinol and Metab 1987; 65: 1-5.
3. Matsuo H, Baba Y, Nair RM, Arimura A, Schally AV. Structure of the porcine LH- and FSH-releasing hormone. I. The proposed amino acid sequence. Biochem Biophys Res Comm 1971; 43: 1334-9.
4. Matsuo H, Arimura A, Nair RM, Schally AV. Synthesis of the porcine LH- and FSH-releasing hormone by the solid-phase method. Biochem Biophys Res Comm 1971; 45: 822-7.
5. Burgus R, Butcher M, Amoss M, et al. Primary structure of the ovine hypothalamic luteinizing hormone-releasing factor (LRF) (LH-hypothalamus-LRF-gas chromatography-mass spectrometrydecapeptide-Edman degradation). Proc Nat Acad Sci USA 1972; 69: 278-82.
6. Fraser HM, Dewart PJ. Control of the menstrual cycle by LHRH and its analogues. Bibliogr Reprod 1985; 46: C1-C12.
7. Posillico JT, Srikanta S, Eisenbarth G, Quaranta V, Kajiji S, Brown EM. Binding of monoclonal antibody (4F2) to its cell surface antigen on dispersed adenomatous parathyroid cells raises cytosolic calcium and inhibits parathyroid hormone secretion. J Clin Endocrinol and Metab 1987; 64: 43-50.
8. Loumaye E, Naor Z, Catt KJ. Binding affinity and biological activity of gonadotropin releasing hormone agonists in isolated pituitary cells. Endocrinol 1982; 111: 730-6.
9. Swift AD, Crighton DB. Relative activity, plasma elimination and tissue degradation of synthetic luteinizing hormone releasing hormone and certain of its analogues. J Endocrinol 1979; 80: 141-52.
10. Belchetz PE, Plant TM, Nakai Y, et al. Hypophysial responses to continuous and intermittent delivery of hypothalamic gonadotropin-releasing hormone. Science 1978; 202: 631-33.
11. Lambalk CB, Schoemaker J, Van Rees GP, et al. Short-term pituitary desensitization to luteinizing hormone-releasing hormone (LH-RH) after pulsatile LH-RH administration in women with amenorrhea of supra-pituitary origin. Fertil Steril 1987; 47: 385-90.
12. Petri W, Seidel R, Sandow J. A pharmaceutical approach to long-term therapy with peptides. In: Labrie F, Belanger A, Dupont A, eds. LHRH and its analogues, basic and clinical aspects. Elseviers Science Publishers B.V., 1984: 63-76.

13. Thomas EJ, Jenkins J, Lenton EA, Cooke ID. Endocrine effects of goserelin, a new depot luteinising hormone releasing hormone agonist. Brit Med J 1986; 292: 1407-8.
14. West CP, Baird DT. Suppression of ovarian activity by Zoladex depot (ICI 118630), a long-acting luteinizing hormone releasing hormone agonist analogue. Clin Endocrinol 1987; 26: 213-20.
15. Fleming R, Haxton MJ, Hamilton MP, et al. Successful treatment of infertile women with oligomenorrhoea using a combination of an LHRH agonist and exogenous gonadotrophins. Brit J Obstet Gynaecol 1985; 92: 369-73.
16. Porter RN, Smith W, Craft IL, Abdulwahid NA, Jacobs HS. Induction of ovulation for in-vitro fertilisation using buserelin and gonadotropins [letter]. Lancet 1984; ii: 1284-5.
17. Fleming R, Coutts JR. Induction of multiple follicular growth in normally menstruating women with endogenous gonadotropin suppression. Fertil Steril 1986; 45: 226-30.
18. Stanger JD, Yovich JL. Reduced in-vitro fertilization of human oocytes from patients with raised basal luteinizing hormone levels during the follicular phase. Brit J Obstet Gynaecol 1985; 92: 385-93.
19. Maclachlan VB, Healy DL. Unpublished data.
20. Quagliariello G. On muscle contraction. QJ Exp Physiol 1923; 13: 209-10.
21. McGullagh DR. Dual endocrine activity of the testis. Science 1932; 76: 19.
22. Robertson DM, Foulds LM, Leversha L, et al. Isolation of inhibin from bovine follicular fluid. Biochem Biophys Res Comm 1985; 126: 220-6.
23. Robertson DM, De Vos FL, Foulds LM, et al. Isolation of a 31 kDa form of inhibin from bovine follicular fluid. Mol Cell Endocrinol 1986; 44: 271-7.
24. Ling N, Ying SY, Ueno N, Esch F, Denoroy L, Guillemin R. Isolation and partial characterization of a Mr 32,000 protein with inhibin activity from porcine follicular fluid. Proc Nat Acad Sci USA 1985; 82: 7217-21.
25. Miyamoto K, Hasegawa Y, Fukuda M, et al. Isolation of porcine follicular fluid inhibin of 32K daltons. Biochem Biophys Res Comm 1985; 129: 396-403.
26. Mason AJ, Hayflick JS, Ling N, et al. Complementary DNA sequences of ovarian follicular fluid inhibin show precursor structure and homology with transforming growth factor-beta. Nature 1985; 318: 659-63.
27. Forage RG, Ring JM, Brown RW, et al. Cloning and sequence analysis of cDNA species coding for the two subunits of inhibin from bovine follicular fluid. Proc Nat Acad Sci USA 1986; 83: 3091-5.
28. McLachlan RI, Robertson DM, Burger HG, De Kretser DM. The radioimmunoassay of bovine and human follicular fluid and serum inhibin. Mol Cell Endocrinol 1986; 46: 175-85.
29. McLachlan RI, Robertson DM, Healy DL, De Kretser DM, Burger HG. Plasma inhibin levels during gonadotropin-induced ovarian hyperstimulation for IVF: a new index of follicular function? Lancet 1986; i: 1233-4.
30. Tsonis CG, Sharpe RM. Dual gonadal control of follicle-stimulating hormone [news]. Nature 1986; 321: 724-5.
31. Papapoulos SE, Frolich M, Mudde AH, Harinck JI, V.d. Berg H, Bijvoet DL. Serum osteocalcin in Paget's disease of bone: basal concentrations and response to bisphosphonate treatment. J Clin Endocrinol and Metab 1987; 65: 89-94.
32. Derynck R, Jarrett JA, Chen EY, et al. Human transforming growth factor-beta complementary DNA sequence and expression in normal and transformed cells. Nature 1985; 316: 701-5.
33. Cate RL, Mattaliano RJ, Hession C, et al. Isolation of the bovine and human genes for Muellerian inhibiting substance and expression of the human gene in animal cells. Cell 1986; 45: 685-98.

34. Padgett RW, St. Johnston RD, Gelbart WM. A transcript from a Drosophila pattern gene predicts a protein homologous to the transforming growth factor-beta family. Nature 1987; 325: 81-4.
35. McLachlan RI, Robertson DM, Healy DL, Burger HG, De Kretser DM. Circulating immunoreactive inhibin levels during the normal human menstrual cycle. J Clin Endocrinol and Metab 1987; 65: 954-61.
36. Findlay J. The nature of inhibin and its use in the regulation of fertility and diagnosis of infertility. Fertil Steril 1986; 46: 770-83.
37. McNatty KP, Markris A, DeGrazia C, Osathanondh R, Ryan KJ. The production of progesterone, androgens, and estrogens by granulosa cells, thecal tissue, and stromal tissue from human ovaries in vitro. J Clin Endocrinol Metab 1979; 49: 687-99.
38. Office of Population Censuses and Surveys, HMSO, London, 1985.
39. Paintin DB. Legal abortion in England and Wales. CIBA Foundation Symposium 1985; 115: 4-20.
40. Grimes DA, Cates W Jr. Complications from legally-induced abortion; a review. Obstet Gynecol Survey 1979; 34: 177-91.
41. Rosen AS, Nystedt L, Bygdeman M, Lundstrøm V. Acceptability of a nonsurgical method to terminate very early pregnancy in comparison to vacuum aspiration. Contraception 1979; 19: 107-17.
42. Karim SM, Filshie GM. Therapeutic abortion using prostaglandin F2alpha. Lancet 1970; i: 157-9.
43. Mocsary P, Csapo AI. Menstrual induction with PGF2 alpha and PGE2. Prostaglandin 1975; 10: 545-7.
44. Healy DL, Fraser HM. The antiprogesterones are coming: menses induction, abortion, and labour? [Editorial]. Brit Med J 1985; 290: 580-1.
45. Proc Soc Exp Biol Med 1930; 27: 403.
46. Csapo AI, Pulkkinen MD, Wiest WG. Effects of luteectomy and progesterone replacement therapy in early pregnant patients. Am J Obstet Gynecol 1973; 115: 759-65.
47. Csapo AI, Pulkkinen MD. Indispensability of the human corpus luteum in the maintenance of early pregnancy. Luteectomy evidence. Obstet Gynecol Survey 1978; 33: 69-81.
48. Kovacs L, Sas M, Resch BA, et al. Termination of very early pregnancy by RU 486, an antiprogestational compound. Contraception 1984; 29: 339-410.
49. Swahn ML, Cekan S, Wang G, Lundstrøm V, Bygdeman M. Pharmacokinetic and clinical studies of RU 486 for fertility regulation. In: Bailieu EE, Segal SJ, eds. The antiprogestin steroid RU 486 and human fertility control. 1985: 249-258.
50. Cameron IT, Michie AF, Baird DT. Therapeutic abortion in early pregnancy with antiprogesterogen RU 486 alone or in combination with prostaglandin analogue (gemeprost). Contraception 1986; 34: 459-68.
51. Couzinet B, Le Strat N, Ulmann A, Baulieu EE, Schaison G. Termination of early pregnancy by the progesterone antagonist RU 486 (Mifepristone). New Engl J Med 1986; 315: 1565-70.
52. Hodgen GD. Pregnancy prevention by intravaginal delivery of a progesterone antagonist: RU 486 tampon for menstrual induction and absorption. Fertil Steril 1985; 44: 263-7.
53. Baird DT, Cameron IT. Menstrual induction: surgery versus prostaglandins. CIBA Foundation Symposium 1985; 115: 178-91.
54. Chan CL, Cameron IT, Findlay JK, et al. Oocyte donation and in vitro fertilization for hypergonadotropic hypogonadism: clinical state of the art. Obstet Gynecol Survey 1987; 42: 350-62.
55. Lutjen PJ, Findlay JK, Trounson AO, Leeton JF, Chan LK. Effect on plasma gonadotropins of cyclic steroid replacement in women with premature ovaria failure. J Clin Endocrinol Metab 1986; 62: 419-23.
56. Rogers PA, Milne BJ, Trounson AO. A model to show human uterine

receptivity and embryo viability following ovarian stimulation for in vitro fertilization. J IVF* and ET 1986; 3: 93-8.
57. Murphy CR, Dwarte DM. Increase in cholesterol in the apical plasma membrane of uterine epithelial cells during early pregnancy in the rat. Acta Anatomica 1987; 128: 76-9.
58. Serhal P, Craft I. Simplified treatment for ovum donation [letter]. Lancet 1987; i: 687-8.
59. Asch R, Balmaceda J, Ord T, et al. Oocyte donation and gamete intrafallopian transfer as treatment for premature ovaria failure [letter]. Lancet 1987; i: 687.
60. Durand P. Chorion villus sampling in Europe. Human Reprod 1986; 1: 341-4.
61. Proc Aust Soc Rep Biol 1986; 18: 31.
62. Wilton LJ, personal communication.
63. Warnock Report HMSO. London, 1984.

CHAPTER 21
CLINICAL IMMUNOLOGY

THE MULTICHAIN INTERLEUKIN-2 RECEPTOR: FROM THE GENE TO THE BEDSIDE

Thomas A. Waldmann, Carolyn Goldman and Mitsuro Tsudo

The Metabolism Branch, National Cancer Institute, National Institutes of Health, Bethesda, Maryland 20892

T cells encompass an array of cell subsets that includes those that mediate important regulatory functions, such as help or suppression, as well as cells that are directly involved in effector functions, such as the cytotoxic destruction of antigen-bearing cells and the production of soluble products termed lymphokines. The success of these responses requires that human T cells change from a resting to an activated state. The activation of T cells requires two sets of signals from cell surface receptors to the nucleus. The first signal is initiated when appropriately processed and presented foreign antigen interacts with the 90-kD polymorphic heterodimeric T-cell surface receptor for the specific antigen. Following the interaction of the antigen presented in the context of products of the major histocompatibility locus and the macrophage-derived interleukin-1 with the antigen receptor, T cells express the gene encoding the lymphokine interleukin-2 (1, 2). To exert its biological effect, IL-2 must interact with specific high-affinity membrane receptors. Resting T cells do not express high-affinity IL-2 receptors, but receptors are rapidly expressed on T cells after activation with an antigen or mitogen (3, 4). Thus, the growth factor IL-2 and its receptor are absent in resting T cells, but after activation the genes for both proteins become expressed. Although the interaction of appropriately presented antigen with its specific receptor confers specificity for a given immune response, the interaction of IL-2 high-affinity IL-2 receptors determines its magnitude and duration.

The presence of receptors for IL-2 was first suggested by the observation that lectin- or alloantigen-activated T cells, but not resting T or B cells, could remove IL-2 from conditioned medium. Subsequently, Robb et al. (3), utilizing purified, biosynthetically labeled IL-2, demonstrated specific, saturable, high-affinity binding sites on IL-2-dependent T-cell lines as well as mitogen- and alloantigen-activated T cells. Further progress in the analysis of the structure, function, and expression of the human IL-2 receptor was greatly facilitated by the production of the anti-Tac monoclonal antibody by Uchiyama, Broder, and Waldman (5). We have demonstrated that anti-Tac recognizes the human receptor for IL-2 (6, 7).

Initial quantitative receptor binding studies employing radiolabeled anti-Tac and radiolabeled IL-2 suggested that activated T cells and IL-2-dependent T-cell lines express 5- to 20-fold more binding sites for the Tac antibody than for IL-2 (8, 9). Employing high concentrations of IL-2, Robb et al. (10) resolved these differences by demonstrating two affinity classes of IL-2 receptors. On various cell populations, 5-15% of the IL-2 receptors had a binding affinity for IL-2 in the range of 10^{-11} to 10^{-12} M, whereas the remaining receptors bound IL-2 at a much lower affinity, approximately 10^{-8} or 10^{-9} M. The structural basis for the difference in IL-2 binding affinity is discussed below. As outlined below, we have utilized the anti-Tac monoclonal antibody and radiolabeled IL-2 in crosslinking studies to: (i) define multiple IL-2 binding peptides that participate in the human receptor for IL-2; (ii) molecularly clone cDNAs for the 55-kD peptide of the human IL-2 receptor; (iii) define the cellular distribution of IL-2 receptors; (iv) determine the immunoregulatory effects that require the interaction of IL-2 with its receptor; (v) analyze disorders of IL-2 receptor expression on leukemic cells; and (vi) develop protocols for the therapy of patients with IL-2 receptor-expressing adult T-cell leukemia and autoimmune disorders and for individuals receiving organ allografts.

CHEMICAL CHARACTERIZATION OF THE MULTICHAIN IL-2 RECEPTOR

The IL-2 binding receptor peptide identified by the anti-Tac monoclonal on PHA-activated normal lymphocytes was shown to be a 55-kD glycoprotein (6, 7). Leonard and co-workers (6, 7) defined the post-translational processing of this 55-kD glycoprotein by employing a combination of pulse-chase and tunicamycin experiments. The IL-2 receptor was shown to be composed of a 33-kD peptide precursor following cleavage of the hydrophobic leader sequence. This precursor was co-translationally N-glycosylated to 35- and 37-kD forms. After a 1-hour chase, the 55-kD mature form of the receptor appeared, suggesting that O-linked carbohydrate was added to the IL-2 receptor. Furthermore, the IL-2 receptor was shown to be sulfated (11) and phosphorylated on a serine residue (12).

There were a series of unresolved questions concerning the IL-2 receptor that were difficult to answer when only the 55-kD Tac peptide was considered. These questions included: (i) what is the structural explanation for the great difference in affinity between high- and low-affinity receptors; (ii) how, in light of the short cytoplasmic tail of 13 amino acids (see below), are the receptor signals transduced to the nucleus; and (iii) how do certain Tac-negative cells (e.g., natural killer and lymphokine-activated killer cells) make nonproliferative responses to IL-2. One observation that helped resolve these questions was that Tac cDNA could reconstitute high-affinity receptors only when transfected into lymphoid cells (13) but not when transfected into nonlymphoid cells (14). Furthermore, there was a conversion of low-affinity IL-2 receptors to a high-affinity form following the fusion of cell membranes of L cells transfected with cDNA, encoding the p55 murine counterpart of the Tac peptide, with membranes from human T cells (13a). These two observations supported the view that cofactors are present on T cells that combine with the p55 peptide to create high-affinity receptors. However, the nature of these putative cofactors and their role in the generation of high-affinity receptors was not defined in these studies.

Very recently, we (15) and others (16) reported a new non-Tac IL-2 binding peptide with a M_r of 75,000 (p75). Using crosslinking methodology, we demonstrated the p75 peptide on MLA 144, a gibbon ape T-cell line that does not express the Tac antigen but manifests 6,800 low-affinity (K_d = 14 nM) IL-2 binding sites per cell. The p75 peptide was also identified in addition to the Tac peptide (p55) on both HUT 102, a HTLV-I-induced T-cell leukemia line, and PHA-activated lymphoblasts. These cell populations express both high-affinity and low-affinity receptors. We proposed a multichain model for the high-affinity IL-2 receptor on which an independently existing Tac of p75 peptide would represent low-affinity receptors, whereas high-affinity receptors would be expressed when both peptides are expressed and associated in a receptor complex (15). To test this working hypothesis, a variety of T-cell lines were examined for IL-2 binding and were subjected to IL-2 crosslinking studies to determine if there was a correlation between the affinity of IL-2 binding and the IL-2 binding peptides expressed. In these studies, cell lines bearing either the p55 Tac or the p75 peptide alone manifested low-affinity IL-2 binding, whereas a cell line bearing both peptides manifested both high- and low-affinity receptors. Furthermore, after the internalization of labeled IL-2 through high-affinity receptors, the p75 peptide could not be detected by crosslinking studies. Finally, fusion of cell membranes from low-affinity IL-2 binding cell lines bearing the Tac peptide alone (MT-1) with membranes from a cell line bearing the p75 peptide alone (MLA 144) generated hybrid membranes bearing high-affinity receptors (17). These studies support a multichain model for the high-affinity IL-2 receptor in which an independently existing Tac or p75 peptide would represent low-affinity receptors, whereas high-affinity receptors would be expressed when both peptides are present and associated in a receptor complex (15).

In addition to the p55 and p75 IL-2 binding peptides, flow cytometric resonance energy transfer measurements support the association of a 95-kD peptide, termed T27, with the 55-kD Tac peptide (18). The role of this peptide has not been defined.

MOLECULAR CLONING OF cDNAs FOR THE HUMAN 55-kD TAC IL-2 RECEPTOR PEPTIDE

Three groups (19, 20, 21) have succeeded in cloning cDNAs for the IL-2 receptor protein. The deduced amino acid sequence of the IL-2 receptor indicates that this peptide is composed of 251 amino acids, as well as a 21 amino acid signal peptide. The receptor contains two potential N-linked glycosylation sites and multiple possible O-linked carbohydrate sites. Finally, there is a single hydrophobic membrane region of 19 amino acids and a very short (13 amino acid) cytoplasmic domain. Potential phosphate acceptor sites (serine and threonine, but not tyrosine) are present within the intracytoplasmic domain. However, the cytoplasmic domain of the IL-2 receptor identified by anti-Tac appears to be too small for enzymatic function. Thus, this receptor differs from other known growth factor receptors that have large intracytoplasmic domains with thyrosine kinase activity. However, the p75 peptide associated with the Tac peptide may play a critical role in the transduction of the IL-2 signal to the nucleus.

Leonard and co-workers (22) have demonstrated that the single gene encoding the IL-2 receptor consists of 8 exons on chromosome 10p14. However, mRNAs of two different sizes (approximately 1500 and 3500 bases long) have been identified. These classes of mRNA differ because of the utilization of two or more polyadenylation signals (20). Receptor gene transcription is initiated at two principal sites in normal activated T lymphocytes (22). Furthermore, sequence analyses of the cloned DNAs also indicate that alternative mRNA splicing may delete a 216-base pair segment in the center of the protein coding sequence encoded by the fourth exon (20, 22). Using expression studies of cDNAs in COS-1 cells, Leonard and co-workers (20) demonstrated that the unspliced but not the spliced form of the mRNA was translated into the cell surface receptor that binds IL-2 and the anti-Tac monoclonal antibody.

DISTRIBUTION OF IL-2 RECEPTORS

The majority of resting T cells, B cells, or monocytes in the circulation do not display the 55-kD peptide of the IL-2 receptor. Specifically, less than 5% of freshly isolated, unstimulated human peripheral blood T lymphocytes react with the anti-Tac monoclonal antibody. The majority of T lymphocytes, however, can be induced to express IL-2 receptors by interaction with lectins, monoclonal antibodies to the T-cell antigen receptor complex, or alloantigen stimulation.

In addition, normal peripheral blood B lymphocytes activated by Staphylococcus aureus Cowan I organisms, pokeweed mitogen, phorbol myristic acetate, or anti-μ immunoglobulins can be induced to express IL-2 receptors (23). In addition, cloned Epstein-Barr virus-transformed human B-cell lines derived from Tac-positive, activated, normal B cells continued to express the Tac antigen in long-term cultures. Thus, certain activated normal B cells display the Tac antigen and manifest high-affinity receptors for IL-2.

IL-2 receptors identified with the anti-Tac monoclonal antibody have been detected on cells of the monocyte-macrophage series, including cultured monocytes, Kupffer cells of the liver, cultured lung macrophages, Langerhans' cells of the skin, and Reed-Sternberg cells in Hodgkin's disease (24, 25).

Rubin et al. (26) demonstrated that activated normal peripheral blood mononuclear cells and certain lines of T- and B-cell origin release a soluble form of the IL-2 receptors into the culture medium. Using an enzyme-linked immunosorbent assay with two monoclonal antibodies that recognize distinct epitopes on the human IL-2 receptor, they showed that normal individuals have measurable amounts of IL-2 receptors in their plasma and that certain lymphoreticular malignancies are associated with elevated plasma levels of this receptor. The release of soluble IL-2 receptors appears to be a consequence of cellular activation of various cell types that may play a role in the regulation of the immune response. Furthermore, the analysis of plasma levels of IL-2 receptors may provide a new approach to the analysis of lymphocyte activation in vivo.

The p75 IL-2 binding peptide is expressed along with the 55-kD Tac peptide on activated T and B lymphocytes. Furthermore, it is expressed on certain circulating cells that do not express the Tac antigen. It has been known that Tac-nonexpressing large granular lymphocytes can be stimulated by IL-2 to enhance natural killer activity and to generate cytotoxic cells termed LAK or lymphokine-activated killer cells that can lyse natural killer-resistant tumor targets (27, 28). Using crosslinking methodology with radiolabeled IL-2, we demonstrated that normal large granular lymphocytes and leukemic large granular lymphocytes from all individuals tested expressed the p75 IL-2 binding peptide but did not express the Tac peptide.

LYMPHOCYTE FUNCTIONS THAT ARE REGULATED BY THE INTERACTION OF IL-2 WITH ITS RECEPTOR

The anti-Tac monoclonal antibody has been used to define those lymphocyte functions that require an interaction of IL-2 with the 55-kD inducible receptor on activated T and B lymphocytes. The addition of anti-Tac to cultures of human peripheral blood mononuclear cells inhibited a variety of immune reactions. Anti-Tac profoundly inhibited the proliferation of T lymphocytes stimulated by soluble antigens and by cell surface antigens (autologous and allogenic mixed lymphocyte reactions).

Anti-Tac was also shown to inhibit a series of T-cell functions, including the generation of both cytotoxic and suppressor T lymphocytes in allogeneic cell cultures, but did not inhibit their action once generated. In contrast to the action of T cells, anti-Tac did not inhibit the IL-2 induced activation of large granular lymphocytes into effective natural killer cells. As noted above, such cells express the p75 but not the 55-kD Tac peptide. Furthermore, upregulation of the expression of Tac mRNA and Tac peptide by IL-2 has been demonstrated for a number of cell types, including some that initially express few if any Tac molecules (23, 29). The addition of IL-2 to such Tac-negative cells, including large granular lymphocytic leukemia cells, augmented transcription of the Tac gene and induced the Tac peptide (31). Neither the IL-2-induced activation of large granular lymphocytes nor the upregulation of Tac gene expression was inhibited by the addition of anti-Tac. These results strongly suggest that the p75 peptide is responsible for IL-2-induced activation of large granular lymphocytes and that the p75 peptide alone can mediate an IL-2 signal. Thus, the p75 peptide may play an important role in the IL-2-mediated immune response not only by participating with the Tac peptide in the formation of the high-affinity receptor complex on T cells but also by contributing to the initial triggering of large granular lymphocyte activation so that these cells become efficient natural killer and lymphokine-activated killer cells.

DISORDERS OF IL-2 RECEPTOR EXPRESSION IN ADULT T-CELL LEUKEMIA

A distinct form of mature T-cell leukemia was defined by Takasuki and co-workers (32) and termed adult T-cell leukemia (ATL). ATL is a malignant proliferation of mature T cells that have a propensity to infiltrate the skin. Cases of ATL are associated with hypercalcemia and usually have a very aggressive course. The ATL cases are clustered within families and geographically, occurring in the southwest of Japan, the Caribbean basin, and in certain areas of Africa. HTLV-I has been shown to be a primary etiologic agent in ATL (33). All the populations of leukemic cells we have examined from patients with HTLV-I-associated ATL expressed the Tac antigen (34). The expression of IL-2 receptors on ATL cells differs from that of normal T cells. First, unlike normal T cells, ATL cells do not require prior activation to express IL-2 receptors. Furthermore, using a ^{3}H-labeled anti-Tac receptor assay, HTLV-I-infected leukemic T-cell lines characteristically expressed 5- to 10-fold more receptors per cell (270,000-1,000,000) than did maximally PHA-stimulated T lymphoblasts (30,000-60,000). In addition, whereas normal human T lymphocytes maintained in culture with IL-2 demonstrate a rapid decline in receptor number, adult ATL lines do not show a similar decline. It is conceivable that the constant presence of high numbers of IL-2 receptors on ATL cells may play a role in the pathogenesis of uncontrolled growth of these malignant T cells.

As noted above, T-cell leukemias caused by HTLV-I, as well as all T-cell and B-cell lines infected with HTLV-I, universally express large numbers of IL-2 receptors. An analysis of this virus and its protein products suggests a potential mechanism for this association between HTLV-I and IL-2 receptor expression. The complete sequence of HTLV-I has been determined by Seiki and colleagues (35). In addition to the presence of typical long terminal repeats (LTRs), gag, pol, and env genes, retroviral gene sequences common to other groups of retroviruses, HTLV-I and -II were shown to contain an additional genomic region between env and the 3′ LTR referred to as pX or more recently as tat. Sodroski and colleagues (36) demonstrated that this pX or tat region encodes a 42-kD protein, now termed the tat protein, that is essential for viral replication. The mRNA for this protein is produced by a double splicing event. These authors demonstrated that the tat protein acts on a receptor region within the LTR of HTLV-I, stimulating transcription. This tat protein may also play a central role in directly or indirectly increasing the transcription of host genes such as the IL-2 and the IL-2 receptor genes involved in T-cell activation and HTLV-I-mediated T-cell leukemogenesis.

THE IL-2 RECEPTOR AS A TARGET FOR THERAPY IN PATIENTS WITH ATL AND PATIENTS WITH AUTOIMMUNE DISORDERS AND INDIVIDUALS RECEIVING ORGAN ALLOGRAFTS

The observation that ATL cells constitutively express large numbers of IL-2 receptors identified by the anti-Tac monoclonal antibody, whereas normal resting cells and their precursors do not, provides the scientific basis for therapeutic trials using agents to eliminate the IL-2 receptor-expressing cells. Such agents could theoretically eliminate Tac-expressing leukemic cells or activated T cells involved in other disease states while retaining the mature normal T cells and their precursors that express the full repertoire for T-cell immune responses. The agents that have been used or are being prepared include: (i) unmodified anti-Tac monoclonal; (ii) toxin (e.g., A chain of ricin toxin, Pseudomonas toxin) conjugates of anti-Tac; (iii) conjugates of alpha- and beta-emitting isotopes (e.g., bismuth-212 and yttrium-90) with anti-Tac; (iv) antibodies, prepared by recombinant DNA techniques, that express the antigen-binding domains of anti-Tac associated with the constant domains of human immunoglobulin light and heavy chains; and (v) recombinant peptides, with the ligand IL-2 associated with a toxin.

We have initiated a clinical trial to evaluate the efficacy of intravenously administering anti-Tac monoclonal antibody in the treatment of patients with ATL (37). None of the nine patients treated suffered any untoward reactions and none produced antibodies to the mouse immunoglobulin or to the idiotype of the anti-Tac monoclonal. Three of the patients had a temporary marked reduction in the number of circulating leukemia cells or complete remission following anti-Tac therapy. In one of these patients, therapy was followed by a 5-month remission, as assessed by routine hematological tests, immunofluorescence analysis of circulating T cells, and molecular genetic analysis of arrangement of the genes encoding the β chain of the T-cell antigen receptor as well as the genes of the retrovirus HTLV-I. Following the 5-month remission the patient's disease relapsed, but a new course of anti-Tac infusions was followed by a virtual disappearance of skin lesions and an over 80% reduction in the number of circulating leukemic cells. Two months subsequently, leukemic cells were again demonstrable in the circulation. At this time, although the leukemic cells remained Tac positive and bound anti-Tac in vivo, the leukemia was no longer responsive to infusions of anti-Tac and the patient required chemotherapy. This patient may have had the smouldering form of ATL wherein the leukemic T cells may still require IL-2 for their proliferation. Alternatively, the clinical responses may have been mediated by host cytotoxic cells reacting with the tumor cells bearing the anti-Tac mouse immunoglobulin on their surface by such mechanisms as antibody-dependent cellular cytotoxicity.

These therapeutic studies have been extended in vitro by examining the efficacy of toxins coupled to anti-Tac to selectively inhibit protein synthesis and viability of Tac-positive ATL lines. The addition of anti-Tac antibody coupled to Pseudomonas exotoxin inhibited protein synthesis by Tac-expressing HUT 102-B2 cells, but not that by the Tac-negative acute T-cell line MOLT-4, which does not express the Tac antigen (38).

The action of toxin conjugates of monoclonal antibodies depends on their ability to be internalized by the cell and released into the cytoplasm. Anti-Tac bound to IL-2 receptors on leukemic cells is internalized slowly into coated pits and then endosomic vesicles. Furthermore, the toxin conjugate does not pass easily from the endosome to the cytosol, as required for its action on elongation factor 2. To circumvent these limitations, an alternative cytotoxic reagent was developed that could be conjugated to anti-Tac and that was effective when bound to the surface of leukemic cells. It was shown that bismuth-212 (^{212}Bi), an alpha-emitting radionuclide conjugated to anti-Tac by use of a bifunctional chelate, was well suited for this role (39). Activity levels of 0.5 μCi or the equivalent of 12 rad/ml of alpha radiation targeted by ^{212}Bi-anti-Tac eliminated greater than 98% of the proliferative capacity of the HUT 102-B2 cells, with only a modest effect on IL-2 receptor-negative lines. This specific cytotoxicity was blocked by excess unlabeled anti-Tac, but not by human IgG. Thus, ^{212}Bi-anti-Tac is a potentially effective and specific immunocytotoxic agent for the elimination of IL-2 receptor-positive cells.

In addition to its use in the therapy of patients with ATL, antibodies to the IL-2 receptors are being evaluated as potential therapeutic agents to eliminate activated IL-2 receptor-expressing T cells in other clinical states, including certain autoimmune disorders and in protocols involving organ allografts. The rationale for the use of anti-Tac in patients with the disease aplastic anemia is derived from the work of Zoumbos and co-workers (40) who have demonstrated that select patients with aplastic anemia have increased numbers of circulating Tac-positive cells. In this group of patients, the Tac-positive but not Tac-negative T cells were shown to inhibit hematopoiesis when cocultured with normal bone marrow cells. Furthermore, we have demonstrated that anti-Tac inhibits the generation of activated suppressor T cells (Oh-ishi and Waldmann, unpublished observations). Studies have been initiated to define the value of anti-Tac in the therapy of patients with aplastic anemia. The rationale for the use of an antibody to IL-2 receptors in recipients of renal and cardiac allografts is that anti-Tac inhibits the proliferation of T cells to foreign histocompatibility antigens expressed on the donor organs and prevents the generation of cytotoxic T cells in allogeneic cell cocultures. Furthermore, in studies by Kirkman and co-workers (41), the survival of renal and cardiac allografts was prolonged in rodent recipients treated with an anti-IL-2 receptor monoclonal antibody. Thus, the development of monoclonal antibodies directed toward the IL-2 receptor expressed on ATL cells, on autoreactive T cells of certain patients with autoimmune disorders, and on host T cells responding to foreign histocompatibility antigens on organ allografts may permit the development of rational new therapeutic approaches in these clinical conditions.

SUMMARY

Antigen-induced activation of resting T cells induces the synthesis of IL-2, as well as the expression of specific cell surface receptors for this lymphokine. There are at least two forms of the cellular receptors for IL-2, one with a very high affinity and the other with a lower affinity. We have identified two IL-2 binding peptides, a 55-kD peptide reactive with the anti-Tac monoclonal antibody and a 75-kD non-Tac IL-2 binding peptide. Cell lines bearing either the p55, Tac, or the p75 peptide alone manifested low-affinity IL-2 binding, whereas cell lines bearing both peptides manifested both high- and low-affinity receptors. Fusion of cell membranes from low-affinity IL-2 binding cells bearing the Tac peptide alone with membranes from a cell line bearing the p75 peptide alone generated hybrid membranes bearing high-affinity receptors. We propose a multichain model for the high-affinity IL-2 receptor in which both the p55 Tac and the p75 IL-2 binding peptides are associated in a receptor complex. The p75 peptide is the receptor for IL-2 on large granular lymphocytes and is sufficient for the IL-2 activation of these cells. In contrast to resting T cells, human T cell lymphotropic virus I-associated adult T-cell leukemia cells constitutively express large numbers of IL-2 receptors. Because IL-2 receptors are present on the malignant T cells but not on normal resting cells, clinical trials have been initiated in which patients with adult T-cell leukemia are being treated with either unmodified or toxin-conjugated forms of anti-Tac monoclonal antibody directed toward this growth factor receptor.

REFERENCES

1. Morgan DA, Ruscetti FW, Gallo RC. Science 1976; 193: 1007-1008.
2. Smith KA. Immunol Rev 1980; 51: 337-357.
3. Robb RJ, Munck A, Smith KA. J Exp Med 1981; 154: 1455-1474.
4. Waldmann TA. Science 1986; 232: 727-732.
5. Uchiyama T, Broder P, Waldmann TA. J Immunol 1981; 126: 1393-1397.
6. Leonard WJ, Depper JM, Uchiyama T, Smith KA, Waldmann TA, Greene WC. Nature (London) 1982; 300: 267-269.
7. Leonard WJ, Depper JM, Robb RJ, Waldmann TA, Greene WC. Proc Natl Acad Sci USA 1983; 80: 6957-6961.
8. Depper JM, Leonard WJ, Kronke M, Waldmann TA, Greene WC. J Immunol 1984a; 133: 1691-1695.
9. Depper JM, Leonard WJ, Kronke M, et al. J Immunol 1984b; 133: 3054-3061.
10. Robb RJ, Greene WC, Rusk CM. J Exp Med 1984; 160: 1126-1146.
11. Leonard WJ, Depper JM, Waldmann TA, Greene WC. In: Greaves M, ed. Receptors and Recognition. London: Chapman and Hall, 1984a; Vol. 17: 45-66.
12. Shackelford DA, Trowbridge IS. J Biol Chem 1984; 259: 11706-11712.
13. Hatakeyama M, Minamoto S, Uchiyama T, Hardy RR, Yamada G, Taniguchi T. Nature (London) 1985; 318: 467-470.
13a. Robb RJ, Proc Natl Acad Sci USA, 1986; 83: 3992-3996.
14. Greene WC, Robb RJ, Svetlik PB, Rusk CM, Depper JM, Leonard WJ. J Exp Med 1985; 162: 363-368.
15. Tsudo M, Kozak RW, Goldman CK, Waldmann TA. Proc Natl Acad Sci USA 1986; 83: 9694-9699.
16. Sharon M, Klausner RD, Cullen BR, Chizzonite R, Leonard WJ. Science 1986; 234: 859-863.
17. Tsudo M, Kozak RW, Goldman CK, Waldmann TA. Proc Natl Acad Sci USA 1987b; in press.
18. Szollosi J, Damjanovich S, Goldman CK, et al. Proc Natl Acad Sci USA 1987; in press.
19. Cosman D, Cerreti DP, Larsen A, et al. Nature (London) 1984; 321: 768-771.
20. Leonard WJ, Depper JM, Crabtree GR, et al. Nature (London) 1984b; 311: 626-631.
21. Nikaido T, Shimizu N, Ishida N, et al. Nature (London) 1984; 311: 631-635.
22. Leonard WJ, Depper JM, Kronke M, et al. Science 1985; 230: 633-639.
23. Waldman TA, Goldman CK, Robb RJ, et al. J Exp Med 1984a; 160: 1450-1466.
24. Hermann F, Cannistra SA, Levine H, Griffin JD. J Exp Med 1985; 162: 1111-1116.
25. Holter W, Goldman CK, Casabo L, Nelson DL, Greene WC, Waldmann TA. J Immunol 1987; 138: 2917-2922.
26. Rubin LA, Kurman CC, Biddison WE, Goldmann ND, Nelson DL. Hybridoma 1985; 4: 91-102.
27. Ortaldo JR, Mason AT, Gerard JP, et al. J Immunol 1984; 133: 779-783.
28. Grimm EA, Rosenberg SA. In: Lymphokine. New York: Academic Press, 1984; Vol. 9: 279-311.

29. Reem G, Yeh NH. Science 1984; 225: 429-430.
30. Depper JM, Leonard WJ, Drogula C, Kronke M, Waldmann TA, Greene WC. Proc Natl Acad Sci USA 1985; 82: 4230-4234.
31. Tsudo M, Goldman CK, Bongiovanni KF, et al. Proc Natl Acad Sci USA 1987a; in press.
32. Takasuki K, Uchiyama T, Sagawa K, Yodoi J. In: Seno S, Takaku F, Irino S, eds. Topics in Hematology. Amsterdam: Excerpta Medica, 1977: 73-77.
33. Poiesz BJ, Ruscetti FW, Gazdar AF, Bunn PA, Minna JD, Gallo RC. Proc Natl Acad Sci USA 1980; 77: 7415-7419.
34. Waldman TA, Greene WC, Sarin PS, et al. J Clin Invest 1984b; 73: 1711-1718.
35. Seiki M, Hattori S, Hirayama Y, Yoshida M. Proc Natl Acad Sci USA 1983; 80: 3618-3622.
36. Sodroski JG, Rosen CA, Haseltine WA. Science 1984; 225: 381-385.
37. Waldmann TA, Longo DL, Leonard WJ, et al. Cancer Res 1985; 45: 4559s-4562s.
38. Fitzgerald D, Waldmann WJ, Willingham MC, Pastan I. J Clin Invest 1984; 74: 966-971.
39. Kozak RW, Atcher RW, Gansow OA, Friedmann AM, Waldmann TA. Proc Natl Acad Sci USA 1986; 83: 474-478.
40. Zoumbos N, Gascon P, Trost S, Djeu J, Young N. N Engl J Med 1985; 312: 257-265.
41. Kirkman RL, Barrett LV, Gaulton GN, Kelly VE, Yhier A, Strom TB. J Exp Med 1985; 162: 358-362.

NEW CONCEPTS IN THE ETIOPATHOGENESIS OF SYSTEMIC LUPUS ERYTHEMATOSUS

Jean-Paul Viard, Laurent Jacob and Jean-François Bach

INSERM U 25, Hôpital Necker, 161, rue de Sèvres
75730 Paris Cedex 15, France

DEFINITION AND ANIMAL MODELS

Systemic lupus erythematosus (SLE) is an autoimmune disease characterized clinically by multiple organ involvement and biologically by the production of a wide spectrum of autoantibodies directed against different cell types and cellular antigens, notably nuclear antigens. Polyclonal B cell hyperactivation is considered as the immunological cornerstone of the disease, both in humans and animal models (1).

Several mouse strains, presenting with clinical symptoms and immunological abnormalities closely resembling those of human SLE have been extensively studied (2). The New-Zealand Black (NZB) mouse and its hybrids, the (NZBxNZW)F_1 mouse (B/W) or the (NZBxSWR)F_1 mouse (NSF1) (3, 4), develop with age, particularly in females, an autoimmune syndrome characterized by severe glomerulonephritis and autoimmune haemolytic anemia. The MRL/Mp-lpr/lpr model (MRL/l) suffers from diffuse T cell proliferation and multiple organ involvement, including glomerulonephritis, arthritis and Sjögren's syndrome, in both sexes. In BXSB mice, males are precociously ill with a typical lupus-like disease. More recently, the Palmerston-North and the viable Motheaten mouse have been reported as developing SLE-like features, the latter being a model of severe and multiple autoimmune lesions due to the expansion of Lyl + B cells (5). None of the murine models exactly parallels human SLE. These strains are genetically heterogeneous, and the course of their disease is influenced by distinct accelerating factors. In addition, they differ by their autoantibody profile: if all lupus-prone mice produce anti-DNA antibodies, anti-Sm antibodies, for instance, are found exclusively in MRL mice. Nevertheless, these animal models are still useful tools for investigation effector and etiologic factors and to test new therapeutic approaches in SLE.

EFFECTOR MECHANISMS: THE ROLE OF ANTI-DNA ANTIBODIES

The most specific autoantibodies in SLE are directed against double-stranded DNA (ds DNA). Antinative DNA antibodies are found in 60-90% of SLE patients and are usually not found at high titers in other inflammatory or autoimmune diseases (1).

Several lines of evidence have led to the conception that anti-DNA antibodies could play a major role in the pathogenesis of SLE. First, there is at least a rough positive correlation between the severity of the disease and immunological abnormalities, including the level of anti-DNA antibodies. Studies have shown that there is a relation between high titers of anti-ds DNA antibodies and severe glomerulonephritis (6). Conversely, patients in remission state generally have normal immunological tests, and a rise of the level of anti-DNA antibodies at a given time often has a good predictive value of relapse (7). However, these correlations are anything but absolute, since SLE patients without antinative DNA antibodies are rare but well known to clinicians (1), and in some SLE patients, an elevation of anti-DNA antibodies titers is not followed by, or associated with a clinical flare of the disease (7, 8).

Several studies have tried to demonstrate more directly the responsibility of anti-ds DNA antibodies in the glomerular lesions of SLE. The classical results of Lambert and Dixon in B/W mice (9), and of Koffler et al. (10) and Krishnan and Kaplan (11) in man showed the presence of DNA, anti-DNA antibodies and complement in kidney eluates, suggesting that DNA-anti-DNA immune complexes were responsible for tissular damage. Other investigators looked for circulating DNA-anti-DNA immune complexes in the sera of lupus patients, but these studies have had contradictory results (12-14). The fact that DNA can bind to the glomerular membrane with a high affinity has led to the alternative hypothesis of the in situ formation of immune complexes, occurring after the binding of the putative antigen to the glomerulus (15, 16).

The lack of an absolute correlation between anti-DNA antibodies as a whole and disease activity has opened the question whether a particular population of anti-DNA antibodies could be considered as pathogenic. Several experimental results show that complement-fixing IgG antibodies, and particularly IgG2b antibodies, cationic in charge and bearing certain idiotypes are especially nephritogenic (3, 4, 17-19). In addition, the administration of anti-idiotype antibodies directed against anti-DNA antibodies has been shown to temper the evolution of the disease in B/W mice (20).

However, in any case, it should be emphasized that the biochemical characterization of DNA in the immune complexes has always been elusive, and that the activity of the disease is not reflected by the level of circulating immune complexes, implicating that DNA-anti-DNA complexes may not be the only antigen-antibody system involved in SLE, if it is involved at all (21). Interestingly, immunologic manipulation of the Sm-anti-Sm system in MRL/l mice can influence the course of the disease and the level of antinuclear autoantibodies (22, 23).

In addition one must consider the paradox that exists between the high titers and frequency of anti-ds DNA antibodies in the sera of SLE mice and patients on the one hand, and the poor immunogenicity of ds DNA on the other hand (24). Indeed, the study of monoclonal anti-DNA antibodies, obtained from lupus-prone mice or SLE patients has led to the hypothesis that the relevant target of anti-ds DNA antibodies could be a different molecule. Monoclonal anti-DNA antibodies have been shown to cross-react with a wide range of unrelated biochemical structures such as phospholipids (25), proteoglycans (26), bacterial antigens (27, 28), IgG (29), or a cell-surface protein which has been termed Lupus Associated Membrane Protein (LAMP) (30).

Finally, if autoantibodies reacting with DNA are undoubtedly a part of the immune disorders characteristic of SLE, and if some of these antibodies are related to the activity of the disease, one can still wonder whether their target autoantigen is DNA itself or some unrelated molecule.

ETIOLOGIC FACTORS: ENVIRONMENTAL, HORMONAL AND IMMUNOLOGICAL CLUES

Environmental factors

Attention has been paid to the potential role of environmental factors in the triggering of immunological abnormalities in SLE, and the hypothesis of viral infection in the lupus-prone mice strains has been extensively studied. B/W, MRL/l, and BXSB mice all present with high expression of the retroviral glycoprotein gp70, and immune complexes constituted of gp70 and anti-gp70 antibodies have been demonstrated in these strains. Anti-gp70 antibodies which were thought to be specific of lupus-prone strains (2) seems to be a part of non-specific polycloncal activation (31). Nevertheless, anti-DNA and anti-gp70 responses are genetically closely interdependent (32). Similarly to anti-DNA antibodies, anti-gp70 antibodies are associated with disease activity and glomerulonephritis, but evidence is lacking for a direct pathogenic role of these antibodies.

Hormonal factors

The influence of hormonal factors in the development of SLE is clearly shown by the higher incidence of SLE in females than in males in B/W mice (2) and humans, and by the frequent exacerbation of the disease during pregnancy or following oral contraception (1). In addition, abnormalities of sex hormone metabolism, such as an increased production of estrogenic metabolites or low plasma androgen levels, have been shown in SLE female patients (33). In young castrated B/W males, partial testosterone replacement or thymosin can abolish the disease-accelerating effect of castration (34).

Immunological factors

The production of anti-DNA antibodies is a part of the polyclonal autoimmune activation in SLE. In murine models, arguments for a primitive B cell or T cell abnormality have been proposed, according to the strain studied (2). However, recent studies indicate that common cellular abnormalities can be found in genetically distinct lupus-prone strains. Studying three different models of SLE, namely NSF1, B/W, and MRL+/+ mice, Datta et al. (35) found that all three non-lpr strains manifested a common cellular defect specifying the production of pathogenic (cationic) anti-DNA IgG antibodies, at the age they begin to develop glomerulonephritis. Spleen cells from these mice contained two populations of T helper cells that were essential for inducing B cells to produce cationic IgG class autoantibodies to both ss and ds DNA: one population was of the classical L3T4+ Lyt2- phenotype, whereas, the second population was L3T4- Lyt2-, which is similar to the phenotype of proliferative T cells in MRL/l mice (2).

MONOCLONAL ANTI-DNA ANTIBODIES RECOGNIZE A CELL-SURFACE PROTEIN: THE LAMP HYPOTHESIS

Several monoclonal anti-ds DNA antibodies, derived from fusion of lupus B/W mice spleen cells with a non-secreting hybridoma line, have been shown to recognize a cell-surface antigen in immunofluorescence studies. Non-specific Fc binding has been ruled out, and this structure has been shown to be a protein (30) which was termed Lupus Associated Membrane Protein (LAMP). Thus, the antibody secreted by the hybridoma PME77 which is an IgG2b, Kappa immunoglobulin, strictly specific for ds DNA and the B helix conformation (36), recognizes the cell-surface protein LAMP on the membrane of different cell types involved in SLE pathogenesis, such as red cells, platelets, B and T lymphocytes, neuronal cells, and glomeruli. In contrast, LAMP is not detected on the surface of pancreas, intestine and liver cells (37). In immunoblot analysis, using a membrane preparation of Raji cells as antigen, and the monocloncal antibody PME77, LAMP is defined by five major polypeptides of 14, 16, 17, 33 and 34 kDA (36). On the contrary to ds DNA, LAMP is immunogenic, since a polyclonal anti-LAMP antiserum has been raised in a rabbit, interestingly without cross-reactivity with DNA (38), which shows that the major immunogenic epitope(s) on LAMP is (are) distinct from the epitope(s) cross-reacting with DNA. Most importantly, LAMP is modified at the surface of B/W and MRL/l lupus mice spleen cells (39). On these cells, LAMP displays an increased resistance to enzymatic treatment by trypsin, elastase and papain, as shown by the appearance in these conditions of a single polypeptide of 55 kDA. This could potentially explain why sera from these mice contain anti-LAMP antibodies: a modification of the LAMP primary structure, or alternatively of the membrane microenvironment, could trigger auto-immunization against the modified self antigen. Besides, these data indicate that the different LAMP polypeptides probably derive from a single protein. This hypothesis is supported by the observation that micropurification of the rabbit polyclonal anti LAMP antiserum on the 17 kDA polypeptide gave reactivity with the whole set of polypeptides (38).

Anti-LAMP antibodies have been regularly found in the sera of lupus-prone B/W and MRL/l mice and of twenty-five patients with active SLE, and not in sera from non autoimmune mice or patients with other inflammatory rheumatic diseases or inactive SLE (40). Importantly, two patients, who had active SLE according to the 1982 ARA classification, but who were negative for anti-DNA antibodies, as assessed by Farr assay, solid phase radioimmunoassay and fluorescence on Crithidia luciliae had anti-LAMP antibodies. Moreover, anti-LAMP antibodies have been looked for in the sera of the same five patients at different times of their disease. In these tests, the presence of anti-LAMP antibodies showed a better correlation with the clinical activity of SLE than antinative DNA antibodies. It has also been demonstrated that IgGs eluted from kidneys of lupus MRL/l mice contained a strong anti-LAMP activity in immunoblot analysis (39). In addition, spleen cells from MRL/l and not non-autoimmune mice are able to spontaneously produce anti-LAMP antibodies in vitro (41).

Taken together, these results allow to propose that LAMP may be an alternative to ds DNA as target antigen of "anti-DNA antibodies", and we believe that there is good evidence for a potential pathogenic role of anti-LAMP antibodies in SLE.

THERAPY OF SLE

Corticosteroid therapy does not prove satisfactory for all SLE patients and is not devoid of major side effects when administered chronically at high dosage. As an alternative, the immunosuppressive agent cyclophosphamide has been tested both in murine and human SLE. Cyclophosphamide can prevent the disease in MRL/l mice, if initiated before the onset (2). In SLE patients, Austin et al. showed that intravenous infusions of cyclophosphamide every three months plus low-dose prednisone can have good results on severe SLE glomerulonephritis, reducing the risk of end-stage renal failure (42). Other treatments like total lymphoid irradiation have been shown to reduce immunological abnormalities, both in mice (43) and man (44) but have also demonstrated high toxicity and life-threatening complications (44). Based on the demonstration of the sex hormone dependence of SLE in B/W mice and man, therapeutic trials with danazol or cyproterone acetate (CA) have had interesting results on the clinical activity of the disease. CA therapy has been tested in an open trial on seven female patients (45). In these cases, reduction in the number of clinical flares was associated with significant improvement of the estradiol:testosterone ratio, without noticeable side effects.

The undecapeptide of fungal origin cyclosporin has been shown to improve life span of B/W mice and to decrease anti-DNA antibody production in male mice of this strain (46). Cyclosporin has also been given to twelve patients presenting with a severe form of SLE, all of them being either steroid resistant or steroid dependent, and having severe side effects of steroid therapy (21). Given initially at a dosage of 5 to 10 mg/kg/day in five cases, and at 5 mg/kg/day in seven cases, this treatment could only be continued at full dosage for six months in eight patients, because of side effects of high dosage. Of these eight patients, all but one improved clinically. Noticeably, two patients with severe central neurological involvement recovered completely and returned to normal life. In the patients with glomerulonephritis, hematuria and proteinuria regressed. Iterative kidney biopsies could be performed in four patients, and three of them had proliferative glomerulonephritis. Under cyclosporin therapy, activity of the disease clearly declined in two and remained unchanged in one. The fibrosis index remained unchanged in three and worsened in one, while no vascular lesions were noted. Concurrently, the steroid dosage could be substantially tapered. In contrast, immunological parameters were not influenced by cyclosporin therapy: levels of antinuclear antibodies, anti-DNA antibodies, complement fractions, and false positive syphilis tests when present, remained unchanged. Paradoxically, plasma IgG levels were significantly increased, as a probable result of steroid dosage diminution. In MRL/l mice, cyclosporin therapy has shown the same discrepancy between clinical improvement and serological status quo (47).

In murine models of SLE, therapy by immunological manipulations with anti-idiotype antibodies have given good but transient results on disease course and levels of autoantibodies (20). Administration of monoclonal anti-L3T4 antibodies in B/W mice has also proven efficient on life span and glomerulonephritis (48).

REFERENCES

1. Bach JF. Pathology of Immune Complexes. Systemic Lupus Erythematosus. In: Bach JF, ed. New York: Wiley Medical Publishers, 1982.
2. Theofilopoulos AN, Dixon FJ. Murine models of systemic lupus erythematosus. Adv Immunol 1985; 37: 269-390.
3. Gavalchin J, Seder RA, Datta SK. The NZB x SWR model of lupus nephritis. I. Cross-reactive idiotypes of monoclonal anti-DNA antibodies in relation to antigenic specificity, charge, and allotype. Identification of interconnected idiotype families inherited from the normal SWR and the autoimmune NZB parents. J Immunol 1987; 128: 128-137.
4. Gavalchin J, Datta SK. The NZB x SWR model of lupus nephritis. II. Autoantibodies deposited in renal lesions show a distinctive and restricted idiotypic diversity. J Immunol 1987; 128: 138-48.
5. Sidman CL, Shultz LD, Hardy RR, Hayakawa K, Herzenberg LA. Production of immunoglobulin isotypes by Ly-1 + B cells in viable Motheaten and normal mice. Science 1986; 232: 1423-5.
6. Tron F, Bach FJ. Relationships between antibodies to native DNA and glomerulonephritis in systemic lupus erythematosus. Clin Exp Immunol 1977; 28: 426-32.

7. Tron F, Jungers P, Droz D, Bach JF. Valeur des tests immunologiques dans la surveillance du traitement du lupus érythémateux disséminé. Nouv Presse Méd 1980; 9: 2319-23.
8. Lightfoot RW, Hughes GRV. Significance of persisting serologic abnormalities in systemic lupus erythematosus. Arthritis Rheum 1976; 19: 837-43.
9. Lambert PH, Dixon FJ. Pathogenesis of glomerulonephritis of NZB/W mice. J Exp Med 1968; 127: 507-22.
10. Koffler D, Agnello V, Kunkel HG. Polynucleotide immune complexes in serum and glomeruli of patients with systemic lupus erythematosus. Am J Pathol 1974; 74: 109-24.
11. Krishnan CS, Kaplan MH. Immunopathologic studies of systemic lupus erythematosus. II. Antinuclear reaction of globulin eluted from homogenates and isolated glomeruli of kidneys from patients with lupus nephritis. J Clin Invest 1967; 46: 569-79.
12. Bruneau C, Benveniste J. Circulating DNA-anti-DNA complexes in systemic lupus erythematosus. Detection and characterization by ultracentrifugation. J Clin Invest 1979; 64: 191-8.
13. Tron F, Letarte J, Roque-Antunes MC, Lesavre P. Specific detection of DNA-anti-DNA immune complexes in human SLE sera using murine monoclonal antibody. Clin Exp Immunol 1982; 49: 481-7.
14. Izui S, Lambert PH, Miescher PA. Failure to detect circulating DNA-anti-DNA complexes in systemic lupus erythematosus. Clin Exp Immunol 1977; 30: 388-92.
15. Izui S, Lambert PH, Miescher PA. In vitro demonstration of a particular affinity of glomerular basement membrane and collagen for DNA: a possible basis for a local formation of DNA-anti-DNA complexes in systemic lupus erythematosus. J Exp Med 1977; 144: 428-43.
16. Izui S, Lambert PH, Türler H, Miescher PA. Features of systemic lupus erythematosus in mice injected with bacterial lipopolysaccharides. Identification of circulating DNA and renal localization of DNA-anti-DNA complexes. J Exp Med 1977; 145: 1115-30.
17. Hahn BH. Characteristics of pathogenic subpopulations of antibodies to DNA. Arthritis Rheum 1982; 25: 747-52.
18. Gavalchin J, Nicklas JA, Eastcott JW, et al. Lupus prone (SWR x NZB)F1 mice produce potentially nephritogenic autoantibodies inherited from the normal SWR parent. J Immunol 1985; 134: 885-94.
19. Livneh A, Halpern A, Lazo A, Halpern R, Diamond B. Monoclonal antibody to a cross-reactive idiotype on cationic human anti-DNA antibodies expressing lambda light chains: a new reagent to identify a potentially differential pathogenic subset. J Immunol 1987; 138: 123-7.
20. Hahn BH, Ebling FM. Suppression of murine lupus nephritis by administration of an anti-idiotypic antibody to anti-DNA. J Immunol 1984; 132: 187-90.
21. Bach JF, Jacob L, Feutren G, Tron F. The questionable role of anti-DNA antibodies in the pathogenesis of systemic lupus erythematosus. In: Schwartz RS, Rose NR, eds. Autoimmunity: Experimental and Clinical Aspects. Ann NY Acad Sci 1986; 475: 231-40.
22. Brennan FM, Williams DG, Bovill D, Stocks MR, Maini RN. Administration of monoclonal anti-Sm antibody prolongs the survival and renal function of MRL-lpr/lpr mice. Clin Exp Immunol 1986; 65: 42-50.
23. Stocks MR, Williams DG, Maini RN. Differential induction of lupus associated antinuclear autoantibodies in MRL mice by monoclonal anti-Sm antibodies. Clin Exp Immunol 1987; 67: 492-9.
24. Stollar BD. Nucleic Acid Antigens. In: Sela M, ed. The antigens. London: Academic Press, 1973; Vol. 1: 1-84.
25. Lafer EM, Rauch J, Andrzejewski C, et al. Polyspecific monoclonal lupus autoantibodies reactive with both polynucleotides and phospholipids. J Exp Med 1981; 153: 897-909.

26. Faaber P, Capel PJA, Rijke GPM, Vierwinden G, Van de Putte LBA, Koene RAP. Crossreactivity of anti-DNA antibodies with proteoglycans. Clin Exp Immunol 1984; 55: 502-8.
27. Naparstek Y, Duggan D, Schattner A, et al. Immunochemical similarities between monoclonal antibacterial Waldenström's macroglobulins and monoclonal anti-DNA lupus autoantibodies. J Exp. Med 1985; 161: 1525-38.
28. Schoenfeld Y, Vilner Y, Coates ARM, et al. Monoclonal anti-tuberculosis antibodies react with DNA, and monoclonal anti-DNA autoantibodies react with Mycobacterium tuberculosis. Clin Exp Immunol 1986; 66: 255-61.
29. Darwin BS, Grudier JP, Klatt CL, Pisetsky DS. IgG antinuclear antibodies with cross-reactive rheumatoid factor activity. J Immunol 1986; 137: 3796-801.
30. Jacob L, Tron F, Bach JF, Louvard D. A monoclonal anti-DNA antibody also binds to cell-surface protein(s). Proc Natl Acad Sci USA 1984; 81: 3843-5.
31. Dixon, FL. Murine lupus. A model for human autoimmunity. Arthritis Rheum 1985; 28: 1081-8.
32. Shirai T, Ohta K, Kohno A, et al. Naturally occurring antibody response to DNA is associated with the response to retroviral gp70 in autoimmune New Zealand mice. Arthritis Rheum 1986; 29: 242-50.
33. Jungers P, Nahoul K, Pelissier C, Dougados M, Tron F, Bach JF. Low plasma androgens in women with active or quiescent systemic lupus erythematosus. Arthritis Rheum 1982; 25: 454-7.
34. Melez KA, Deleargyros N, Bellanti JA, Goldstein AL, Smathers P, Steinberg AD. Effect of partial testosterone replacement or thymosin on anti-DNA in castrated (NZB x NZW)F1 males. Clin Immunol Immunopathol 1987; 42: 319-27.
35. Datta SK, Patel H, Berry D. Induction of a cationic shift in IgG anti-DNA autoantibodies. Role of T helper cells with classical and novel phenotype in three models of lupus nephritis. J Exp Med 1987; 165: 1252-68.
36. Jacob L, Tron F. Monoclonal anti-deoxyribonucleic antibodies. I. Isotope and specificity studies. J Immunol 1982; 128: 895-8.
37. Jacob L, Léty MA, Louvard D, Bach JF. Binding of a monoclonal anti-DNA autoantibody to identical protein(s) present at the surface of several human cell types involved in lupus pathogenesis. J Clin Invest 1985; 75: 315-7.
38. Jacob L, Léty MA, Bach JF, Louvard D. Human systemic lupus erythematosus sera contain antibodies against cell-surface protein(s) that shar(s) epitope(s) with DNA. Proc Natl Acad Sci USA 1986; 83: 6970-4.
39. Jacob L, Léty MA, Monteiro RC, Jacob F, Bach JF, Louvard D. Altered cell-surface protein(s), crossreactive with DNA, on spleen cells of autoimmune lupic mice. Proc Natl Acad Sci USA 1987; 84: 1361-3.
40. Jacob L, Léty MA, Choquette D, et al. Presence of antibodies against a cell-surface protein, cross-reactive with DNA, in systemic lupus erythematosus: a marker of the disease. Proc Natl Acad Sci USA 1987; 84: 2956-9.
41. Viard JP, Bach JF, Jacob L. Splenocytes from MRL/Mp/lpr/lpr mice spontaneously produce antibodies against a cell-surface protein cross-reacting with DNA. Clin Immunol Immunopathol; in press.
42. Austin HA, Klippel JH, Balow JE, et al. Therapy of lupus nephritis. Controlled trial of prednisone and cytotoxic drugs. N Engl J Med 1986; 314: 614-9.
43. Kotzin BL, Arndt R, Okada S, Ward R, Thach AB, Strober S. Treatment of NZB/NZW mice with total lymphoid irradiation: long-lasting suppression of disease without generalized immune suppression. J Immunol 1986; 136: 3259-65.

44. Ben-Chetrit E, Gross DJ, Braverman A, et al. Total lymphoid irradiation in refractory systemic lupus erythematosus. Ann Intern Med 1986; 105: 58-60.
45. Jungers P, Kutenn F, Liote F, et al. Hormonal modulation in systemic lupus erythematosus. Preliminary clinical and hormonal results with cyptroterone acetate. Arthritis Rheum 1985; 28: 1243-50.
46. Israel-Biet D, Noël LH, Bach MA, Dardenne M, Bach JF. Marked reduction of DNA antibody production and glomerulopathy in thymulin (FTS-Zn) or cyclosporin A treated (NZB x NZW)F1 mice. Clin Exp Immunol 1983; 54: 359-65.
47. Mountz JD, Smith HR, Wilder RL, Reeves JP, Steinberg AD. Cs-A therapy in MRL-lpr/lpr mice: amelioration of immunopathology despite autoantibody production. J Immunol 1987; 138: 157-63.
48. Wofsy D, Seaman WE. Reversal of advanced murine lupus in NZB/NZW F1 mice by treatment with monoclonal antibody to L3T4. J Immunol 1987; 138: 3247-53.

IGE-MEDIATED HYPERSENSITIVITY

R.C. Aalberse

Central Laboratory of the Netherlands Red Cross Blood Transfusion Service and Laboratory for Clinical and Experimental Immunology, University of Amsterdam, The Netherlands

IgE is a special immunoglobulin for at least two reasons:

1. Mast cells are triggered very efficiently via IgE antibodies.

2. A strikingly high fraction of the total IgE in some patients is directed against trivial, otherwise completely harmless glycoproteins in "dusty" material in the environment (pollen, mite fecal material in house dust, animal danders). These antigens are usually referred to as "allergens".

In this presentation I will briefly discuss these two aspects of IgE, with the emphasis on the second point, because of its relevance for clinical chemistry. Most of the material discussed here is covered to a greater or lesser extent in immunology- and allergy textbooks, for example in the beautifully illustrated book of Roitt, Brostof and Male (1). References given in this book are not repeated here.

MAST CELLS AND IgE RECEPTORS

The mast cell family has at least three members: connective tissue mast cells, mucosal mast cells and basophilic leukocytes. These cells differ in many respects, but have two important characteristics in common: a high affinity IgE receptor on the cell surface, and the presence of histamine-containing granules in the cytoplasm. The binding of IgE to the IgE receptor does not seem to have any effect on the cells. However, these cells can be triggered by allergen, if the IgE bound to the receptors has a sufficiently high affinity for this allergen. The allergen is assumed to cross-link IgE antibodies on the cell surface and thus bring the IgE-receptor molecules in close contact. This then is the signal for the cell to release granules and to start the synthesis and release of mediators from lipids in the cell membrane: leukotrienes, prostaglandines and, depending on the cell type, platelet-activating factor.

IMMUNOLOGICAL CONSEQUENCES OF THE CROSS-LINKING HYPOTHESIS

This cross-linking hypothesis implies, that per allergen molecule at least two IgE binding sites have to be available. Furthermore, per cell a sufficient number of IgE molecules with a high affinity for the same allergen molecule should be present.

Originally it was assumed that allergen molecules would most likely be polymeric in order to provide the same epitope (= antigenic determinant) repeatedly. This certainly would make it easier to understand the efficiency of the cross-linking by allergen. However, most of the well-characterised allergens do not appear to be polymeric under normal conditions and do not seem to have a clearly repeating structure.

For the immune system this would imply that it has to produce IgE antibodies to at least two different epitopes on each allergen.

Another point that has implications for the IgE immune response is the high affinity of the IgE receptor for monomeric IgE. Binding of IgE to this receptor does not require allergen. This is in marked contrast to the binding of IgG to the Fc-gamma receptor on phagocytic cells. In the latter situation, monomeric IgG binds very weakly; in the presence of antigen, however, complexes will be formed that contain multiple IgG molecules. The Fc-gamma receptor will efficiently bind these complexed IgG molecules. In the latter situation, the antigen acts to select antibodies with a suitable specificity.

In the case of the IgE receptor, no such selection by antigen is possible. The cell surface is occupied with IgE antibodies of all specificities present in the body fluids.

The immunological consequence of these two points is, that, in order to have an efficient IgE/mast cell system, the IgE-response should be highly selective and react to only a few antigens and disregard most other antigens. Whenever the IgE-system reacts, however, it is important that more than one B-cell clone reacts. One would therefore expect a tight regulation at the antigen level rather than at the epitope level.

THE MEASUREMENT OF TOTAL AND ALLERGEN-SPECIFIC IgE

As discussed above, the peculiar way the IgE-system is controlled results in low total IgE-levels but relatively high antigen-specific IgE-levels. It is not exceptional to find that in a serum more than 20% of the total IgE is directed against a single antigen, i.e. removal of IgE antibodies by means of a solid-phase coupled antigen often results in a significant reduction of the total IgE level. For IgG antibodies it is quite exceptional to find more than 1% of the total IgG to be directed to a single antigen.

The consequences are:

1. sera with IgE antibodies to several antigens tend to have elevated total IgE levels. This is convenient, because it helps to identify patients with allergy: patients with an elevated total IgE level are likely to have IgE antibodies to identifiable antigens. It should be stressed that this information is to be used in a statistical way and is of limited use for assessing individual patients: the chance of finding specific IgE antibodies increase with increasing serum IgE levels. The actual specificity of the antibodies depends obviously on the exposure of the patient. In Western Europe inhalant allergens,

notably house dust mites, pollen from grasses and trees, and pets are the most likely candidates as the allergen source. In tropical areas, parasites like Schistosoma mansoni are often implicated.

2. The assay of antigen-specific IgE is to some extent technically easier than the assay for antigen-specific IgG. As in other assays, "background" or "non-specific binding" is a major concern in antibody assays. Detecting 10 ng/ml IgE antibody in 100 ng/ml total IgE is obviously much simpler than detecting 10 ng/ml IgG antibody in 10,000,000 ng/ml total IgG. However, even if the problem of non-specific binding is in general less with IgE antibody assays than with IgG antibody assays, non-specific binding may become important when sera are tested with a markedly elevated total IgE level, such as sera from patients with severe atopic eczema. This non-specific binding is more pronounced with some allergens than with other.

TECHNICAL ASPECTS OF THE RADIO-ALLERGO-SORBENT TEST (RAST) (2)

Because of the high affinity of IgE for the mast cell receptor, low levels of allergen-specific IgE in the circulation may already be clinically relevant. Therefore, assays for allergen-specific IgE needs to be very sensitive; antibody levels of 1 ng/ml should be detectable. The most commonly used method is a solid-phase immunoassay, the RAST, which uses ^{125}I-labeled anti-IgE, or a similar test based on the use of an enzyme-label. For this test two reagents are needed: an insolubilised allergen and the labeled anti-IgE. The first reagent is used to extract the antibodies from the serum, the second reagent is then used to quantitate the amount of IgE extracted.

The allergosorbent is usually prepared by coupling the allergen to CNBr-activated polysaccharides (paper discs or agarose beads). For some allergens non-covalent adsorption onto polystyrene microtiter wells also works well, but for many allergens the latter procedures proves to be unsatisfactory. Even with the polysaccharide supports it is not uncommon to find that the allergosorbent is not potent enough to extract all antibodies from the serum that is tested. This can easily be verified by retesting the depleted serum. This is a major reason why it is less than trivial to express the results of these antibody assays in meaningful units.

With respect to the other reagent, the labeled anti-IgE antibodies, the use of monoclonal reagents is often suggested to be the one and final answer to all problems. From the producers point of view monoclonal antibodies definitely are a major step forward, because it is so much easier to obtain purified antibodies. With the polyclonal reagents it was very desirable if not essential to use affinity-purified antibodies. The IgE required for this affinity purification was scarce and often from the same source as the anti-IgE used for immunization. In the latter situation it is difficult to get rid of anti-idiotypic antibodies, i.e. antibodies that will react exclusively with the IgE used for immunization but not with other, "normal" IgE. The availability of monoclonal antibodies makes affinity purification unnecessary and this is clearly a major advantage. Otherwise there is no evidence that monoclonal antibodies are better than good polyclonal reagents. Several commercially available monoclonal antibodies are in fact less useful than the "old" polycloncal reagents. One problem is that of sensitivity or avidity: how easy is it to detect picogram quantities of IgE on a solid phase. In this respect monoclonal antibodies usually score lower than polyclonal antibodies. But even with respect to specificity monoclonal antibodies may have a disadvantage: their specificity may be too high. A monoclonal antibodies exists that discriminates between IgE-kappa and IgE-lambda (3). Another monoclonal anti-IgE was found to react with an epitope on the IgE molecule that is absent from the IgE of some people from Eastern Asian countries (4). In both these examples it was an all-or-none effect. With other reagents the effect may be more subtle: a slightly stronger reaction with one type of IgE than with another type. The conclusion as far as the specificity is concerned, is, that it is essential to analyze the response of a serum panel containing samples from every ethnic population that is likely to be present in the patient population and compare in a quantitative way the monoclonal reagent with that of a good polyclonal reagent. This is particularly important for the non-competitive two-site assays that use one monoclonal antibody as capturing reagent and another monoclonal antibody as the detecting reagent. It is a frustrating thought that it is impossible to really prove a particular reagent to be directed to a non-allotypic epitope. It is however possible to minimise this risk by the above-mentioned procedure and by including a mixture of several antibodies rather than a single monoclonal reagent in the assay. In the case of the two-site assay this would imply a mixture of monoclonal antibodies as capturing reagent and a mixture of other monoclonal reagents as detecting reagent.

QUANTITATION OF IgE ANTIBODY MEASUREMENTS

The serologic procedures with respect to IgE-mediated hypersensitivity are now well established. The RAST or closely related techniques are almost universally used. The reagents are available but not fully standardised. This is particularly true for the allergens. A major obstacle for standardization of the IgE antibody measurement is the huge number of potentially relevant allergenic molecules and the variability of the biologic source materials.

Absolute quantitation would be possible for IgE antibodies to purified allergens. Only a few allergens have been purified and these purified allergens represent only a small fraction of the total allergenic potential of a particular extract.

Alternatively, one could use monoclonal antibodies to allergens, which are steadily becoming available and prove to be useful tools for the measurement of IgE antibodies to allergenic molecules. For this application they are used as capturing reagents in an indirect RAST procedure (5). For routine purposes however this would not seem to be a practical approach for the reasons indicated above.

It can easily be demonstrated that absolute quantitation of antibodies to a mixture of antigens (which is what all allergen extracts are) is hard to achieve. The next best approach is to aim at a reproducible result. This requires a tight control of the allergen source materials, extraction procedures and preparation of the allergosorbent. The consequence is, however, that the test results between allergens cannot be readily compared; moreover, different test procedures may give different results even if the results are related to a reference serum for that particular allergen.

Since it is likely that many different assays for IgE antibodies to allergens will soon appear on the market, each with different allergen source materials and different test procedures, it seems prudent not to over-emphasise the quantitative value of the test result, but to stress the qualitative or semi-quantitative nature of result.

CONCLUSIONS

For the clinical laboratory the most important consequence of the progress achieved in the field of the IgE-mediated allergies are related to the IgE antibody assay. These antibody assays have a few peculiarities which make them different from the IgG/IgM antibody assays as used in microbiology and auto-immunology. These are partially a reflection of the biologic properties of the IgE-mast cell system, and partially a consequence of the complex nature of the antigens involved: the allergens.

REFERENCES

1. Roitt I, Brostoff J, Male D. Immunology. Churchill and Livingstone and Cower, 1985.
2. Wide L, Bennich H, Johansson SGO. Diagnosis of allergy by an in vitro test for allergen antibodies. Lancet 1967; ii: 1105-1107.
3. Magnusson CGM, Aalberse RC, Johansson SGO. Monoclonal antibodies against human IgE. Int Arch Allergy appl Immunol 1986; 80: 329-332.
4. Van Loghem E, Aalberse RC, Matsumoto H. A genetic marker of human IgE heavy chains, Em (1). Vox Sang 1984; 46: 195-206.
5. Aalberse RC. Monoclonal antibodies in allergen standardization. Arbeiten aus dem Paul Ehrlich Inst. Stuttgart: Gustav Fischer Verlag, 1983; 78: 137-140.

RECENT ADVANCES IN OUR KNOWLEDGE OF COMPLEMENT

P.J. Lachmann

MRC Mechanisms in Tumour Immunity Unit, Cambridge, England

INTRODUCTION

There is no sign of slackening in the pace of complement research. It is likely that there are at the present time more scientists working on this topic than there have ever been. An indication may be that 320 abstracts have been submitted for the 1987 complement workshop. It is therefore impossible and might be tedious to try to list all the advances that have been made in the last five or ten years. What follows is a personal selection of some selected topics.

FAMILIES OF COMPLEMENT COMPONENTS

It was first pointed out about eight years ago (1) that the complement system could be regarded as a series of gene duplicates where the corresponding components typically in the classical and alternative and membrane attack pathways were members of the same family of proteins and originated by gene duplication. At the time the evidence for this view was based on "gross" protein chemistry and resemblance in reaction patterns but, with the coming of molecular biological techniques which have allowed the great majority of complement components to be cloned and fully sequenced, this concept has been found to be entirely valid and indeed there are more such families than had been suspected and their relationships are more complex than had been believed.

THE FAMILY WITH REPEATING 60 AMINO ACID DOMAINS

An entirely new superfamily has recently been recognised that shares variable numbers of a consensus 60 amino acid repeat (2). This family was originally believed to be made up of those proteins that bound C3b and C4b: Factor H, C4 binding protein and CR1.

To these were added in due course Factor B and C2 which have analogous binding properties; and CR2 (which although it binds to C3dg and not to C3b appears to be remarkably similar to CR1); and, more recently, the decay accelerating factor (DAF) which also is C3b binding. However the family includes some complement components - Clr (and almost certainly also its gene duplicate Cls) which is not C3b binding - and several proteins - beta-2-glycoprotein-2, IL-2 receptor and coagulation factor XIII which are not complement components at all.

The evolutionary significance of all this is unclear. It is however particularly interesting that C2 and Factor B as well as Clr all have domains containing this consensus repeat as well as domains entirely typical for serine esterases and therefore are obviously compound proteins derived from two quite separate families.

THE SERPIN FAMILY

Another protein family that has relevance to the complement system is that described as the Serpins (3). This family of serum proteinase inhibitors has many members but its more important plasma protein constituents are alpha-1-antitrypsin, antithrombin 3 and the Cl inhibitor. The last is an important regulatory protein in the complement cascade. Serpins share a common structure containing a stressed loop which contains the binding site and they have the interesting property of binding their enzyme by one particular amino acid residue ('the bait' residue) which gives specificity; and the bond N-terminal to which becomes cleaved after interaction with the enzyme. Following such cleavage, the serpin totally alters its conformation and becomes antigenically strikingly heat stable (4) which forms the basis of an assay for cleaved serpins on the circulation. It has been demonstrated by Ziccardi (5) that the Cl inhibitor is required to keep Cl in inactivated form and even heterozygous deficiency of Cl inhibitor is associated with the disease of hereditary angio-oedema, a condition where attacks of enzyme activation at an extravascular site causes local exhaustion of Cl-inhibitor with consequent autocatalytic activation of Cl and of kallikrein. Generation of kinins and analogous fragments from complement components gives rise to the syndrome of angio-oedema. In recent years particular interest has been taken in those pedigrees that make abnormal proteins rather than no protein and it has been demonstrated that these abnormal proteins have irregular inhibitory profiles in regard to the enzymes with which they react. It would be valuable to know the structural basis of this variation and, indeed the first lesion in such a molecule has now been identified (6) and is a mutation in the bait residue itself - from arginine to histidine.

THE PIPLC FAMILY AND PAROXYSMAL NOCTURNAL HAEMOGLOBINURIA

A further family of proteins has recently come to light that bears a relation to the complement system. These are proteins that are anchored in membranes by an anchoring mechanism that is sensitive to phosphatidylinositol specific phospholipase C. There are a number of these proteins, including the trypanosomal variant protein on which much work has been done, but the interest in relation to complement applies to the decay accelerating factor (DAF) and the homologous restriction factor (HRF) both of which appear to be anchored in membranes by this mechanism. DAF is a membrane protein that controls the assembly of C3 convertases and HRF is a C9 and C8 binding protein that protects cells from lysis by homologous complement. These are two important examples of the membrane associated control proteins of the complement system whose importance has only recently been recognised. However, interest derives particularly from the existence of the disease, Paroxysmal Nocturnal Heamoglobinuria (PNH), a clonal proliferation of red cells that lack both DAF and HRF and also acetylcholinesterase and alkaline phosphatase, two further cell proteins that are membrane bound in the same way. Red blood cells in PNH are exquisitively sensitive to lysis by the patient's own alternative complement pathway giving rise to an haemolytic anaemia. Although the basic abnormality is not worked out, it is now fairly apparent that this curious disease, where a number of molecules are absent from the cell membrane, probably does have a single abnormality in some element of the so-called "PIPLC anchor". The molecular basis of the defect is still eagerly sought.

THE LFA1/CR3/FAMILY OF PROTEINS

Yet another new family of proteins has a bearing on complement function. This is a group of cell membrane proteins sharing a common light chain and having distinctive heavy chains. There are three (and possibly four) members of this family: LFA1, CR3, the molecule known as P150,95; and, possibly, the platelet membrane protein GP2b,3a. In relation to the complement, the interest is in CR3. This molecule is the iC3b receptor and is of importance in that it seems to be the receptor whose ligand binding fires the respiratory burst in polymorphs. Its importance is again emphasised by the existence of subjects that lack this whole family of proteins usually on the basis of making an unstable beta chain. These children have a characteristic immunity deficiency with indolent skin ulcers usually associated with staphylococcal infection and a tendency to get staphylococcal septicaemias. They also heal their skin ulcers slowly and they have characteristic gum problems. They are deficient in LFA1 on their lymphocytes and in, particularly, CR3 and P150,95 on monocytes and polymorphs. It is interesting that their dominant clinical abnormality is associated with impaired polymorph function. Whether this impairment is entirely due to CR3 binding of complement or whether the ability of this receptor also to bind carbohydrates (yeast cell walls, betaglucan) is undetermined.

COMPLEMENT GENETICS AND COMPLEMENT FUNCTION IN VIVO

The advent of molecular biological techniques in complement research has led to a great increase in our understanding of the genetic organisation of this complex system. It is now known that complement components are widely scattered throughout the genome and that, in addition to the cluster in the MHC, there is a cluster on chromosome 1q that includes CR1 and CR2 as well as Factor H and C4 binding protein. Otherwise the complement system is remarkable for the number of chromosomes on which components are found.

A most elegant study, using genetic studies to define complement function in vivo was done by traditional phenotypic techniques (7). This was a study comparing the genetic characters of the survivors of Dutch immigrants to Surinam who had survived major epidemics of typhoid and Yellow Fever with the modern Dutch population. Of 26 loci tested, only four showed marked allele frequency changes. These were GM, HLA-B, GLO and C3. Of these changes the increase in the allele frequency of C3F is the most impressive. This is because it is the less common allele which becomes appreciably more frequent in the survivor population and because C3F is an entirely Caucasian gene and admixture in Surinam with either Amerindian or Black genes would have diluted the effect rather than enhanced it. This study demonstrates therefore that C3 and a particular allele of C3 is of importance in resistance to infection.

INFECTION AND COMPLEMENT DEFICIENCY

The data on C3 is borne out by the study of isolated complement component deficiencies. It has been found that deficiencies of C3 are associated with immunity deficiencies similar to those seen in the antibody deficiency syndrome. However, by far the commonest infection found in complement deficiency is with Neisseria and in particular meningococcal meningitis. This is found in many complement deficiencies and is almost the only infection found in patients with deficiencies of the membrane attack complex. The reason is not far to seek. Neisseria require to be killed in the plasma since they can survive and are transported on or in polymorphs.

COMPLEMENT DEFICIENCY AND IMMUNE COMPLEX DISEASE

A more surprising finding is the strong association of complement deficiencies, particularly those of the early classical pathway components with immune complex disease and particularly with autoimmune immune complex diseases like systemic lupus erythematosus. This is clearly physiologically associated with complement deficiency rather than being due to a linked gene and an increased susceptibility to these diseases is seen even in heterozygotes for a complement deficiency. The mechanisms responsible for this association are not immediately apparent but it seems likely that the following play a part:

1. Complement is required to keep immune complexes in a soluble form rather than in an insoluble state.
2. Soluble complexes adequately coated with complement are carried in the circulation predominantly bound on red cell CR1. This prevents access to the vascular endothelium and prevents sequestration at peripheral sites. Most such red cell bound complexes are removed in the liver where the red cells pass through macrophage lined sinusoids.

Under conditions of complement deficiency, this mechanism fails and the complexes are carried in the plasma from which they are much more rapidly removed at multiple peripheral sites where they give rise to inflammation and generate feedback autoantibody formation and the formation of more immune complexes. It seems extremely likely that this effector failure is the principal genetic predisposition to systemic lupus erythematosus and that any other changes found in these diseases are secondary (8).

REFERENCES

1. Lachmann PJ. An evolutionary view of the complement system. Behring Inst Mitt 1979; 63: 25-37.
2. Reid KBM, Bentley DR, Campbell RD, et al. Complement system proteins which interact with C3b and C4b. A superfamily of structurally related protein. Immunology Today 1986; 7: 230-234.
3. Carrell RW, Boswell DR. Serpins: the superfamily of plasma serine proteinase inhibitors. In: Barrett, Salvesen, eds. Proteinase Inhibitors. Elsevier Sci Publishers BV, 1986.
4. Pemberton PA, Harrison RA, Lachmann PJ, Rabson A, Walport M, Carrell RW. Analysis of Serpin inactivation in disease. Abstract submitted to XIIth International Complement Workshop, Chamonix, September 1987.
5. Ziccardi RJ. A new role for C1-inhibitor in homeostatis: control of activation of the first component of human complement. J Immunol 1982; 128: 2505-2508.
6. Aulak KS, Pemberton PA, Rosen FS, Carrell RW, Lachmann PJ, Harrison RA. The variant protein C1-INH(At) contains an Arg -> His substitution at the reactive centre P1 residue. Abstract submitted to XIIth International Complement Workshop, Chamonix, September 1987.

7. De Vries RRP, Meerakhan P, Bernini LF, Van Loghem E, Van Rood JJ. Genetic control of survival in epidemics. J Immunogenetics 1979; 6: 271-287.
8. Lachmann PJ, Walport MJ. Deficiency of the effector mechanisms of the immune response and autoimmunity. Ciba Found Symp 1987; in press.

CHAPTER 22
REGULATORY PEPTIDES

γ-ENDORPHIN WITH DEVIANT BIOLOGICAL ACTIVITY: A MOLECULAR MARKER FOR SCHIZOPHRENIA?

V.M. Wiegant, J.P.H. Burbach, O. Gaffori, J. Verhoef, G.L. Kovács, W.M.A. Verhoeven, J.M. van Ree and D. De Wied

Rudolf Magnus Institute for Pharmacology, Medical Faculty
University of Utrecht, Utrecht, The Netherlands

INTRODUCTION

In the mid seventies, the search for an endogenous ligand for the opiate receptor lead to the discovery of pentapeptide molecules with opiate-like activities, the enkephalins. Numerous other endogenous opioid peptides - now termed 'endorphins' - have subsequently been identified. They belong to three large families, each identified by an own precursor molecule, e.g. pro-enkephalin, pro-dynorphin and pro-opiomelanocortin (POMC). The full sequence of these precursors has been elucidated using recombinant DNA techniques. Their unique structure, anatomy, post-translational processing, and possible functions and relations to opioid receptor subtypes have been subject of several recent review articles (1-4).

Once the presence of endorphins in the brain was established, their involvement in a variety of psychiatric disorders was hypothesized, mainly based on their pharmacological effects in animal models. A role for endorphins has been postulated in schizophrenia, depression, mania, addiction, autism, etc. In this respect, the attention has mainly been focussed on peptides related to β-endorphin (βE), that are derived from POMC.

Psychiatric disorders may arise from dysfunctions in the brain, in particular in cortical and limbic areas. These are therefore the tissues of choice for biochemical research on the postulated role of endorphins in psychopathology. Major drawbacks in this respect are the lack of suitable animal models for mental illnesses, and the inaccessability of the human brain for biochemical studies. Most in vivo studies in humans concerning endorphins and psychopathology are therefore restricted to neuroendocrine parameters, such as peptide levels in peripheral blood or cerebrospinal fluid. Although this approach has provided valuable information, for instance in depression, it should be kept in mind that a disturbance in neuroendrocrine functions in fact is not necessarily a reflection of the brain disorder that causes the psychopathology, but may well be a secondary phenomenon, and that brain dysfunctions leading to psychopathology may not be expressed in neuroendocrine disturbances.

Another avenue to investigate disorders of the human brain is the study of post mortem tissue. A major advantage of this approach is that the diseased tissue itself is subject of investigation. Disadvantages are, however, the limited availability of post mortem brain, the fact that many variables of possible importance are beyond control of the investigator, and that details on the history of the patients are usually not sufficiently available. In this survey evidence will be reviewed on the role of γ-endorphin (γE) and related POMC-derived peptides in schizophrenia, with a focus on studies with post mortem human brain tissue.

POMC, A MULTIFACTORIAL PRECURSOR MOLECULE

POMC is a 31k MW protein. It contains the full amino acid sequence of ACTH and β-LPH, and an 16k N-terminal part with a third (modified) copy of ACTH-(4-10) (5, 6). POMC is synthesized in the corticotrophs of the anterior lobe of the pituitary, the cells of the intermediate pituitary (melanotrophs), in neuronal cells in the brain, in certain cells of the lymphoid system, and in a number of other tissues (7-10). In the brain POMC containing neurons are found in two distinct cell groups. The major group is located in the hypothalamic nucleus arcuatus. These cells send out axons rostrally that synaps in the septal area and caudally in the midbrain central grey. A smaller cell group, with as yet unknown projection areas, is present in the nucleus tractus solitarii (8, 9).

Once synthesized at the rough endoplasmic reticulum, POMC is translocated to the Colgi apparatus, and subsequently cleaved at specific recognition sites by peptidases into biologically active peptides. Thus, ACTH, β-LPH and the 16k-fragment are formed, and stored in secretory vesicles, where they have been found co-localised (11). During maturation of the vesicles, the peptides are subject to further proteolytic cleavage, and to covalent modifications (e.g. acetylation). These processing steps dramatically affect the biological activity of the molecules. For example, ACTH-(1-39) (corticotropin) can be cleaved and amidated to form ACTH-(1-13)-NH_2, which then can be acetylated yielding α-MSH, a peptide with potent melanotropic activty. Similar processing of β-LPH (β-lipotropin) generates the opioid βE-(1-31) and an acetylated form of this peptide, which is completely devoid of opiate-like activity (12, 13). When maturation is completed, the vesicles contain a cocktail of peptides with different biological properties, that upon release can be expressed by interaction of the peptides with specific receptors located at effector cells.

Interestingly, POMC is processed differently in various tissues. Whereas in the anterior lobe of the pituitary β-LPH appears to be the predominant product, and βE is mainly present as unmodified βE-(1-31), the intermediate lobe contains little intact POMC and β-LPH sized material, but mainly acetylated and shortened forms of βE, such as βE-(1-27), βE-(1-26), βE-(1-17) (γE) and βE(1-16) (αE) (14-17). Like in the intermediate pituitary, little β-LPH sized material is found in the brain, and POMC is almost completely cleaved to βE in this tissue. In its turn, βE is further processed to a considerable extent, and fragments of βE and their acetylated forms, including α- and γ-endorphins are formed (15-17).

Thus, POMC is a multifactorial precursor. The characteristics of the peptides generated from POMC depend on the presence and activity of the enzymes involved in its processing. Consequently, the quality of the organism's response to POMC cell activity is not only specified at the level of the peptide receptor, but also determined at the level of the POMC cell, by the enzymatic machinery that governs the processing of the precursor.

γE AND SCHIZOPHRENIA: THE HYPOTHESIS

The processing of βE to γE and αE has important consequences for the biological activity of the peptide. Not only are the opioid properties of βE partly lost in αE and γE, but also specific, novel central nervous system (CNS) activities are unvealed. These are evidenced by the effects of the peptides in a variety of neuropharmacological test systems in rats, including active and passive avoidance behavior, various grip tests and dopamine induced motor activities (18-20). In these paradigms, the effect of αE resemble those of psychostimulants (e.g. amphetamine), drugs that can induce psychotic symptoms in humans. The effects of γE resemble in certain aspects those of neuroleptics (e.g. haloperidol). These drugs are clinically used as antipsychotics. Interestingly, these actions of αE and γE can be dissociated from the opioid activities of the peptides. For instance, the effects of αE on active avoidance behavior are not blocked by the opiate receptor antagonist naltrexone. In addition, the des-Tyr^1-(DT-), Des-enkephalin- (DE-) and N^{α}-acetyl- (Ac-) forms of γE, that are completely devoid of opiate-like properties, exert a similar neuroleptic-like action as γE (18, 20, 21).

Thus, α and γ-type endorphins have been neuropharmacologically characterized as functional antagonists with respect to their CNS activities. A balanced expression of αE and γE-type activities may be of importance for the homeostatic control of adaptive brain functions. In view of the neuroleptic-like properties of γ-type endorphins in particular, it was postulated, that a relative deficiency in γE-type activity in the brain (or conversely a relative abundance of αE-type activity) may be an aethiological factor in psychopathologies for which neuroleptics are beneficial e.g. schizophrenia (22).

CLINICAL STUDIES WITH γ-TYPE ENDORPHINS

In parallel with endocrine disorders involving impaired production of hormones that are successfully treated with substitution therapy (e.g. diabetes mellitus; diabetes insipidus; pituitary hypofunctions), a number of clinical trials was performed, in which schizophrenic patients were treated with γ-type endorphins. In these studies, so far a total of 43 patients were treated with DTγE and 36 patients were treated with DEγE (23, 20). The design of the studies (except one) was double blind, either placebo controlled or cross-over. The patients were diagnosed as suffering from schizophrenia or schizoaffective psychosis, based on the Research Diagnostic Criteria, DSM III criteria for schizophrenia, the Present State Examination Criteria, and the course of illness. Of the total of 79 patients, 16 did not show a response to treatment as assessed with the Brief Psychiatric Rating Scale (response < 20%), 25 showed a slight response (20 - 50%), 20 displayed a moderate response (50 - 80%) and 18 others a marked response (> 80%). Only three of the 20 patients treated with placebo showed a slight response, while the others did not respond. Thus, in about half of the patients a clinically obvious therapeutic effect of DTγE or DEγE was observed (response > 50%).

In studies by others, including a total of 87 patients, 24 showed a beneficial response to DTγE (24-33). In one of two trials with DEγE by others, the peptide induced the same antipsychotic effect as haloperidol (34). In the other trial, a significant improvement was observed with the highest of the three doses DEγE used (35). These data suggest, that γ-type endorphins have antipsychotic properties in a number of schizophrenic patients.

ENDORPHIN CONCENTRATIONS IN POST MORTEM BRAIN OF SCHIZOPHRENIC PATIENTS

A variety of dysfunctions of the POMC system can be envisaged, that may result in a relative deficiency in γE-type activity in the brain. These include malfunctions in the enzymatic machinery involved in the processing of βE, the generation and the degradation of γ-type endorphins. Such defects may be expressed in altered tissue concentrations of endogenous βE fragments. We have analysed post mortem hypothalamic tissue of 28 schizophrenic patients and of 22 control subjects without known psychopathology, and have determined the levels of αE, βE and γE by specific radioimmunoassays in HPLC fractionated extracts of the tissue (Wiegant et al., submitted). The hypothalamus was selected for this study, as it contains the POMC synthesizing neurons that project to limbic and cortical brain regions, and contains the highest concentration of POMC peptides in the brain. In tissue obtained from schizophrenic patients, the concentration of both αE and γE was significantly higher than in controls (+ 70 and + 50% respectively). Interestingly, no difference was found in the concentration of βE, the precursor of α- and γ- endorphins. The increase in αE and γE therefore seems to be rather selective, and may reflect a dysfunction in the metabolism of βE, possibly the generation of αE and γE in the brain of the patients. Indeed, others have shown using in vitro methods, that the activity of brain enzymes involved in the fragmentation of βE and the generation of γ-type endorphins, is altered in schizophrenia (36, 37).

An important question pertains to the relationship of these differences to the psychopathology of the patients. As in all post mortem studies in humans, ante mortem pharmacotherapy concerns a variable that is beyond the control of the investigator. Most patients included in our study did receive medication of some type, and in so far known, all patients but one received neuroleptics at the time of death. The possibility that neuroleptics in fact have distorted the results should therefore be seriously considered. Chronic treatment of rats with neuroleptics affects the POMC system in the brain (38, 39). However, such treatment not only increases the hypothalamic concentration of αE and γE, but elevates that of βE to a similar extent (Sweep et al., in preparation). Therefore it seems, that the selective increase of αE and γE in schizophrenics is not simply explained by ante morten treatment of the patients with neuroleptics.

DEVIANT BIOLOGICAL ACTIVITY EXPRESSED BY γE FROM SCHIZOPHRENICS

The activity of γE and related peptides as neuroleptic-like principles is associated with the C-terminal (6-17) amino acid sequence of the peptide, as evidenced by structure activity studies in animals (21). Modifications in that part of the molecule generally destroy its neuroleptic-like activity. An alteration in the structure of γE in schizophrenics, leading to impaired bioactivity of the peptide, thus could underly the postulated deficit in γE-type activity in the brain of such patients. To investigate this possibility, we have studied the biological activity of HPLC fractions containing γE obtained from post mortem tissue of schizophrenic patients and control subjects. One-trial learning passive avoidance behavior in rats was used as the bioassay. In this paradigm, neuroleptics and peptides possessing neuroleptic-like activity such as γE, DTγE and DEγE attenuate the retention of the avoidance response (18, 40). This effect is not only observed after peripheral (sc) administration, but also when the peptides are injected in pg doses topically in the brain, into the nucleus accumbens (41). When animals treated with synthetic γE, or with γE fractions obtained from hypothalamic tissue of controls (pooled from 22 subjects) were tested for passive avoidance retention, a similar dose dependent inhibition was found with both treatments. However, γE containing fractions isolated from hypothalami of schizophrenics (pooled from 28 patients) did not display such activity, but were either inactive, or induced the opposite effect in certain doses. In a subsequent experiment, γE containing fractions from single hypothalami of 12 other patients were tested individually, and again the deviant biological activity was found (Wiegant et al., in preparation).

In these studies, HPLC was used to fractionate tissue extracts, and the γE containing fractions were identified based on HPLC retention time and reactivity in a specific γE radioimmunoassay. In addition to immunoreactive γE, however, such fractions contain other tissue constituents, and may even be contaminated with drugs taken by the patients prior to death. For a number of reasons derived from additional experiments, it is, however, unlikely that substances other than γE (for instance contaminating drugs) are responsible for the deviant bioactivity found in schizophrenics.

1) No or deviant bioactivity was found in hypothalamic extracts of 12 individual patients, of which at least 2 did not receive neuroleptics at the time of death.

2) γE fractions prepared from individual pituitaries of 2 of these patients and of 4 other patients displayed no or deviant bioactivity similar to the fractions isolated from hypothalamus. At least one of these patients was not on neuroleptics.

3) γE generated in vitro by incubation of the βE containing HPLC fractions prepared from hypothalamic or pituitary tissue of patients in the presence of Cathepsin D (an enzyme that cleaves the Leu^{16}-Phe^{17} bond in βE), and subsequently isolated by HPLC of the incubate, also displayed no or deviant bioactivity.

4) γE purified to homogeneity from a single schizophrenic pituitary by consecutive Sephadex G-50 chromatography, reverse phase HPLC and paired ion HPLC, did not possess bioactivity when assayed in the passive avoidance paradigm. These data strongly suggest, that the deviant bioactivity found in γE fractions of schizophrenic hypothalamus, reflects intrinsic properties of the γE molecule, and is likely related to the structure of the peptide. Such a structural defect can arise either from a mutation in the γE coding region of the POMC gene and then is to be detected in the amino acid composition or sequence of the peptide, or from a posttranslational modification. Although amino acid substitutions, in particular those in the C-terminal region of γ-type endorphins generally render the peptide biologically inactive, they do not necessarily affect the properties of the peptide in HPLC and RIA systems used in this study (Wiegant et al., unpublished results). Thus, it remains possible that a structurally abnormal form of γE could be responsible for the observed deviant bioactivity in schizophrenics. Therefore, the amino acid composition of biologically inactive γE, purified from a single schizophrenic pituitary was determined. No differences were found with similarly purified γE from control pituitary. In addition, analysis of the γE coding region of pituitary POMC mRNA from two other pituitaries of schizophrenics did not show differences in nucleotide sequence as compared to controls (42). These results rule out the first possibility, namely that a mutation in the POMC gene is responsible for the biologically defective γE in schizophrenia. Posttranslational modifications of the peptide are neither detected after hydrolizing conditions, such as used for instance for the analysis of the amino acid composition, nor are they detectable at the level of the POMC mRNA. Thus, the possibility that such a modification is responsible

for the deviant bioactivity still remains open, and is presently under investigation.

CONCLUDING REMARKS

The data reviewed in this survey provide evidence that: a) γ-type endorphins exert neuroleptic-like actions in various animal models: b) γ-type endorphins have intrinsic antipsychotic properties in certain schizophrenic patients; c) the metabolism of βE, in brain tissue of schizophrenics differs from normal, leading to an increased accumulation of γE; d) γE isolated from tissues of schizophrenics expresses deviant biological activity. These data support the hypothesis that a deficit in γE-type activity in the brain underlies psychopathology in a certain category of schizophrenic patients. It is tempting to speculate that, in absence of γE with normal bioactivity, feedback mechanisms are triggered that are responsible for an increased generation of γE in the brain of schizophrenics.

The beneficial effect of γ-type endorphins in schizophrenic patients is associated with a higher incidence of certain antigens of the HLA system in these patients (43, 44), indicating that genetic factors encoded within the HLA genomic region are associated with the therapeutic effect of the peptides. The deviant biological activity of γE is expressed in two ontogenetically different tissues, the hypothalamus and the pituitary. This suggests, that the defect is genetically determined and may also be detectable in other POMC-expressing tissues that are better accessible in vivo than brain or pituitary. This would offer perspectives for the development of diagnostic methods, that allow a therapy directed classification of certain schizophrenics.

REFERENCES

1. O'Donohue TL, Dorsa DM. Peptides 1982; 3: 1039-1072.
2. Liotta AS, Krieger DT. In: Krieger DT, Brownstein MJ, Martin JB, eds. Brain Peptides. New York: J Wiley and Sons, 1984: 613.
3. Akil H, Watson SJ, Young E, Lewis ME, Khatchaturian H, Walker JM. Ann Rev Neurosci 1984; 7: 223-255.
4. Khatchaturian H, Lewis ME, Schafer MK-H, Watson SJ. TINS 1985; 8: 111-119.
5. Roberts JL, Seeburg PH, Shine J, Herbert E, Baxter JD, Goodman HM. Proc Natl Acad Sci USA 1979; 76: 2153-2157.
6. Nakanishi S, Inoue A, Kita T, et al. Nature 1979; 278: 423-427.
7. Bloom FE, Battenberg E, Rossier J, et al. Life Sci 1977; 20: 43-48.
8. Watson SJ, Akil H, Richard CW, Barchas JD. Nature 1978; 275: 226-228.
9. Finley JCW, Lindström P, Petrusz P. Neuroendocrinology 1981; 33: 28.
10. Blalock JE. Clin Endocrinol 1985; 22: 823-827.
11. Pelletier G, Leclerc R, LaBrie F, Cote J, Chretien M, Lis M. Endocrinology 1977; 100: 770.
12. Akil H, Young E, Watson SJ, Coy D. Peptides 1981; 2: 289-292.
13. Deakin JF, Dostrovsky JO, Smyth DG. Biochem J 1981; 189: 501-506.
14. Mains RE, Eipper BA. J Biol Chem 1979; 253: 651-656.
15. Zacharian S, Smyth DG. Nature 1982; 296: 250-252.
16. Wiegant VM, Verhoef J, Burbach JPH, Van Amerongen A. Life Sci 1983; 33: 125-128.
17. Wiegant VM, Verhoef J, Burbach JPH, et al. Life Sci 1985; 36: 2277-2285.
18. De Wied D, Bohus B, Van Ree JM, Urban I. J Pharmacol Exp Ther 1978a; 204: 570-580.

19. De Wied D, Kovács GL, Bohus B, Van Ree JM, Greven HM. Eur J Pharmacol 1978b; 49: 427-436.
20. Van Ree JM, Verhoeven WMA, De Wied D. Progr Brain Res 1987; 72: 249-268.
21. De Wied D, Van Ree JM, Greven HM. Life Sci 1980; 26: 1575-1579.
22. De Wied D. TINS 1979; 2: 79-82.
23. Verhoeven WMA, Van Praag HM, Van Ree JM, De Wied D. Arch Gen Psychiatry 1979; 36: 294-298.
24. Bourgeois M, Laforge E, Myard J, Blayac J, Lemoine J. Ann Med Psychol 1980; 138: 1112-1119.
25. Casey DE, Korsgaard S, Gerlach J, Jørgensen A, Simmelsgaard H. Arch Gen Psychiatry 1981; 38: 158-160.
26. Emrich HM, Zaudig M, Von Zerssen D, Kissling W, Dirlich G, Herz A. In: Emrich HM, ed. Modern Problems in Pharmacopsychiatry. Basel: Karger, 1981; Vol. 17: 279-286.
27. Fink M, Papakostas Y, Lee J, Meehan Th, Johnson L. In: Perris C, Struwe G, Jansson B, eds. Biological Psychiatry. Amsterdam: Elsevier, 1981: 398-401.
28. Korsgaard S, Casey DE, Gerlach J. Psychopharmacology 1982; 78: 285-286.
29. Manchanda HY, Hirsch SR. Psychol Med 1981; 11: 401-404.
30. Meltzer HY, Busch DA, Tricou BJ, Robertson A. Psychiatry Res 1982a; 6: 313-326.
31. Meltzer HY, Busch DA, Lee J, Papacostas Y. Psychopharmacol Bull 1982b; 18: 44-47.
32. Tamminga CA, Thighe PJ, Chase ThN, DeFraites EG, Schaffer MH. Arch Gen Psychiatry 1981; 38: 167-168.
33. Volavka J, Huis KS, Anderson B, Nemer Z, O'Donnell J, Lajtha A. Psychiatry Res 1983; 10: 243-252.
34. Kissling W, Möller HJ, Bürk F, Kraemer S. Abstract 14th CINP Congress, Florence, 1984: 146.
35. Azorin JM, Charbaut J, Granier F, et al. Abstract Symposium Neuropeptides and Brain Function, May 1986, Utrecht, The Netherlands: 52.
36. Schoemaker H, Davis ThP. Peptides 1984; 5: 1049-1054.
37. Davis TP, Schoemaker H, Culling A. Eur J Pharmacol 1984; 100: 249-251.
38. Höllt V, Bergmann M. Neuropharmacology 1982; 21: 147-154.
39. Ham J, Smyth DG. Neuropeptides 1985; 5: 497-500.
40. De Wied D. Progr Brain Res 1987; 72: 93-108.
41. Kovács GL, Telegdy G, De Wied D. Neuropharmacology 1982; 21: 451-454.
42. Bovenberg RAL, Burbach JPH, Wiegant VM, et al. Brain Res 1986; 376: 29-37.
43. De Jongh BM, Verhoeven WMA, Van Ree JM, De Wied D, Van Rood JJ. Immunogenet 1984; 9: 381-388.
44. Van Ree JM, Verhoeven WMA, Claas FHJ, De Wied D. Prog Brain Res 1986; 65: 221-235.

REGULATORY PEPTIDES IN THE CONTROL OF FOOD INTAKE

John E. Morley

Geriatric Research, Education and Clinical Center VA Medical Center, Sepulveda, CA and Department of Medicine, University of California, Los Angeles, CA

A number of peptides have been demonstrated to modulate feeding. Physiological studies have suggested that one or more humoral agents that are released from the gastrointestinal tract during the digestion of food are responsible for the termination of a meal. It appears that these humoral agents are gastrointestinal peptides. This system for the regulation of food intake has been termed the PERIPHERAL SATIETY SYSTEM. Regulatory peptides also interact with other neurotransmitters within the central nervous system to stimulate or inhibit feeding. This central regulatory system is known as the CENTRAL FEEDING SYSTEM.

CHOLECYSTOKININ AS A SATIETY FACTOR

A number of impure gastrointestinal preparations were originally shown to decrease feeding. Among these was enterogastrone which was subsequently shown to be rich in cholecystokinin (CCK). In 1972, Lars Sjodin (1) noted that an undesirable side-effect of CCK was to inhibit food intake in dogs. The following year Gibbs et al. (2) reported that CCK acted as a satiety agent in rats. Subsequently, CCK has been demonstrated to decrease feeding in a variety of species including humans (see 3 for a review). The active form of CCK necessary to produce its satiety action is the sulfated octapeptide (CCK-8S). Much controversy has existed concerning whether CCK is a true satiety agent or whether it produces its decrease in feeding as part of a general disruption of feeding. It appears that both options are true, depending on the dose of CCK employed. Low doses of CCK-8S appear to be truly satiating, while high doses (10 μg/kg and greater) appear to produce behavioral disruption (sickness). In humans, doses of CCK-8S that inhibit feeding are just below the threshold dose for nausea, but much higher than the doses necessary for gallbladder contraction or pancreatic enzyme secretion.

The mechanism by which CCK produces its satiety action has been extensively study in rodents. It appears that CCK produces its effect by a paracrine action within the peritoneum (4). Studies by Smith and his colleagues (5) have shown that CCK acts by stimulating the ascending fibers of the gastric vagus, sending its satiety message to the hindbrain where its message is relayed in the nucleus tractus solitarius (see 6 for a review). The mechanism by which the satiety message is relayed from the hindbrain is controversial. Crawley et al. (7) have suggested projections from the nucleus tractus solitarius to the paraventricular nucleus (PVN) carry the satiety message and that both midbrain knife cuts and lesions of the PVN will inhibit the CCK response. However, we and others (unpublished observations) have had difficulty in reproducing the inhibitory effect of knife cuts on the CCK effect. Within the central nervous system CCK may produce its effect by modulating the monoamine system. CCK has been shown to alter both catecholamine and serotonin turnover and vagotomy can alter the effects of CCK on these measurements (8).

The satiety effect of CCK is not the only effect that appears to be mediated by stimulation of the ascending vagal fibers. Feeding rodents has been shown to enhance memory and it has been suggested that this effect is mediated by release of gastrointestinal hormones, including CCK (9). The CCK enhancement of memory is blocked by vagotomy. CCK also causes the release of oxytocin from the posterior pituitary and this effect is vagally dependent (10). It is possible that the memory enhancement effect of CCK is secondary to the release of oxytocin.

In summary, CCK appears to play a role in the termination of a meal. Its effect, however, is of small magnitude (in humans the meal is reduced by approximately 10-16%). Behavioral tolerance develops rapidly to repeated injections, i.e., the animals eat less at a single meal but increase the number of meals eaten and thus the overall food intake remains unaltered (11). This has suggested that other gastrointestinal peptides may be involved in the termination of a meal.

THE PERIPHERAL SATIETY SYSTEM

It has been shown that when food is placed in a transplanted (denervated) stomach and its pylorus is closed, the animal will decrease its food intake. This suggests that a humoral factor released from the stomach is involved in satiety. A leading candidate for this effect is the frog skin, tetradecapeptide, bombesin and/or its mammalian form, gastrin releasing peptide. Bombesin is a potent inhibitor of feeding when injected peripherally (12). Bombesin appears to produce its effect by a mechanism independent of the vagus, as it still decreases feeding following vagotomy (13). Based on the differential starvation model, bombesin appears to have more illness-like effects (i.e., more similar to Lithium chloride) than does CCK-8 (14). This would fit with the common experience that excessive overeating is accompanied by overdistension of the stomach and a nauseous feeling! Bombesin has, however, been shown to decrease feeding in humans at doses that do not produce nausea (15).

A number of other gastrointestinal peptides have also been shown to decrease feeding following pharmacological administration peripherally. These peptides appear to produce their effects either by vagally dependent or vagally independent mechanisms. Glucagon appears to work through stimulating the coeliac fibers of the vagus nerve (16). Antibodies to glucagon administered peripherally enhance feeding, suggesting that the glucagon effect is physiological (16). Somatostatin also inhibits feeding through a vagally dependent mechanism (18).

It appears that gastrointestinal peptides produce additive effects on food intake (19). Thus it may be a combination of gastrointestinal peptides acting in concert that terminates a meal. However, all peptides are not additive; some show infra-additivity, e.g., bombesin and CCK; somatostatin while being additive with CCK inhibits bombesin's effects on feeding (Silver, Flood and Morley: unpublished observations). Further studies will be necessary to define the interaction of gastrointestinal peptides in the satiety cascade.

OPIOIDS AND FEEDING

In 1973, Holtzman (20) demonstrated that the opiate antagonist, naloxone, decreased feeding in rats. Subsequently, opiate antagonists have been demonstrated to decrease feeding in multiple mammalian species. Opioid antagonists appear to particularly affect the intake of highly palatable foods, especially those with a high-fat content (21). It has been suggested that the function of the opioid feeding system is to distinguish between toxic and non-toxic foods.

Numerous studies have addressed which endogenous opioid and which opioid receptor is involved in the regulation of feeding. It appears that dynorphin, the endogenous kappa ligand, is the most potent of the opioids (22). Dynorphin appears to produce its effect through a double-lock receptor consisting of a non-opioid "address" portion which allows access to the opioid receptor. Dynorphin appears to produce its feeding effect within the PVN, the same area where norepinephrine has been shown to stimulate feeding (23). It is possible that other opioids stimulate feeding in other areas of the central nervous system.

Opioid antagonists have been shown to decrease feeding over a single meal in humans (24). However, when opioid antagonists are administered for prolonged periods tolerance rapidly develops to their effects and they fail to reduce weight in obese humans (25). In addition, they have been shown to produce liver dysfunction.

Endogenous opioids produce a multitude of effects which would be useful to an animal while foraging for food. Opioids inhibit gastric acid secretion, enhance natural killer cell immunity, raise the blood glucose and are analgesic. Thus from an evolutionary perspective opioids could be seen as the co-ordinators of the behaviors necessary to allow for a successful hunt.

Of interest is that prolonged administration of opiate agonists does lead to weight loss. This appears to be secondary to peripheral effects of opiates on energy metabolism and not due to their effects on feeding. Norepinephrine, which similarly increases feeding when administered into the PVN, also leads to weight loss when given peripherally. This suggests to us that the ability of these neurotransmitters to stimulate the search for food is secondary to their ability to cause energy mobilization in the periphery. Teleologically, this would be an eminently sensible arrangement!

NEUROPEPTIDE Y - A POTENT OREXIGENIC AGENT

Neuropeptide Y (NPY) is a 36 amino acid member of the pancreatic polypeptide family. It is present in high concentrations in the hypothalamus, particularly within the arcuate nucleus and the PVN. Clark et al. (25) demonstrated that NPY was a potent stimulator of feeding. Subsequently, NPY has been shown to have a highly selective effect on carbohydrate intake (27). NPY appears to act by a mechanism that does not involve the alpha-2 norepinephrine feeding system. The major site of action of NPY appears to be within the hypothalamus either within the PVN (28) or in the anterior ventromedial hypothalamus (27). NPY also produces feeding when administered into the fourth ventricle, suggesting that it has a second site of action in a IV ventricle-related structure (Steinman, Gunion, Morley: unpublished observations). NPY also increases water intake and its effect on drinking can be disassociated from its effect on feeding by midbrain knife cuts.

NPY will reverse the weight loss seen in animals with lateral hypothalamic lesions (unpublished observations). The closely related, PYY (29) and NPY (30) will both continue to produce increased feeding and weight gain when administered chronically into the cerebroventricles. This increase in feeding is associated with overdistension of the gastrointestinal tract, suggesting a slowing of transit time through the gut. Like norepinephrine and the opioids, peripherally administered PYY causes weight loss rather than weight gain seen after central administration.

In preliminary studies we have shown that NPY levels are reduced in the cortex, but not in the hypothalamus, following 48-hour starvation and that these levels return towards normal with refeeding (Scarpace and Morley: unpublished observations). Thus, there is mounting evidence for a role for NPY in the physiological regulation of carbohydrate intake. It has been suggested that NPY may play a role in the pathogenesis of bulimia.

OTHER PEPTIDES AS STIMULATORS OF FOOD INTAKE

Besides the opioids and NPY, a number of other peptides have been reported to stimulate food intake. These include galanin (31), growth hormone releasing hormone (32) and motilin (33). None of these effects appears to be as potent as the opioids or NPY. Other neurotransmitters which increase food intake include norepinephrine, through an alpha-2 receptor mechanism, and gamma amino butyric acid and its analog, muscimol, when administered directly into the ventromedial hypothalamic area. Dopamine appears to have a biphasic effect on food intake. At low doses, it enhances food intake, while at higher doses it produces stereotypy and food intake is decreased (34).

CORTICOTROPIN RELEASING FACTOR AND ANOREXIA NERVOSA

Both psychological and physical stresses in animals modulate feeding behavior. Both duration and type of stressor appears to be important in determining the effect on food intake. Mild stressors, e.g., tail-pinch, tend to enhance feeding, while severe stressors e.g., immobilization, decrease food intake. Stress-induced overeating appears to involve the endogenous opioid system, while stress-induced anorexia may be mediated by corticotropin-releasing factor.

Corticotropin releasing factor (CRF) is the peptide which best fits Hans Selve's mythical stress co-ordinating factor in that it releases ACTH and cortisol as well as activating the autonomic nervous system. CRF is a potent inhibitor of feeding (35). The decrease in feeding it produces is accompanied by an increase in grooming - a normal satiety behavior. The alpha-helical form of CRF which acts as a CRF antagonist, has been shown to increase feeding, which has been reduced by stress (36). The satiety effect of CRF is partially dependent on the activation of the hypothalamic-adrenomodullary axis.

CRF produces its effect within the PVN nucleus (Krahn, Levine and Morley: unpublished observations). Lesions of the PVN result in an increase in feeding, suggesting that the major effect of the PVN on feeding is to inhibit it. Norepinephrine, which acts within the PVN, stimulates feeding by inhibiting the release of an inhibitor. Norepinephrine inhibits the release of CRF from the hypothalamus, suggesting that norepinephrine may

Table 1. Neurotransmitters that have been shown to alter feeding after pharmacological administration

A. PERIPHERAL SATIETY SYSTEM:

	Increase Feeding	Decrease Feeding
	Motilin Insulin	Cholecystokinin Insulin Somatostatin Bombesin Gastrin Releasing Peptide Calcitonin Glucagon Prostaglandins

B. CENTRAL FEEDING SYSTEM:

	Increase Feeding	Decrease Feeding
Monamines:	Norepinephrine (α-2) Dopamine (low concentrations) Serotonin	Dopamine (high concentrations) Serotonin
Peptides:	Opioids, e.g., dynorphin Neuropeptide Y Galanin Growth Hormone Releasing Hormone Motilin ?Somatostatin	Calcitonin Corticotropin Releasing Factor Calcitonin Gene Related Peptide Bombesin Cholecystokinin Neurotensin Thyrotropin Releasing Factor Cyclohistidyprolinediketo-piperazine Somatostatin Insulin Growth Factor
Others:	Gamma Amino Butyric Acid	Gamma Amino Butyric Acid Prostaglandins

mediate its effects on feeding through CRF. Similarly, serotinin, which inhibits feeding within the PVN, produces the release of CRF. Thus norepinephrine and serotonin may alter feeding by modulating the release of CRF from the PVN.

Patients with anorexia nervosa have long been recognized to have increased activity of the hypothalamic-pituitary-adrenal axis. Recently, two groups have shown that patients with anorexia nervosa have elevated levels of CRF in their CSF (37, 38). It seems reasonable to postulate that CRF plays a role in the pathophysiology of anorexia nervosa. Similarly, approximately two-thirds of patients with endogenous depression have evidence of activation of the hypothalamic-pituitary-adrenal axis and elevated levels of CRF in the CSF (39). Depressed patients often loss weight and this could be related to their elevated levels of CRF. CRF, in animals, produces a number of effects that could be considered similar to the DSM-III criteria for depression. Thus it is necessary to consider CRF as one of the "players" in the pathogenesis of depression.

CRF appears to have gained considerable stature as a physiological regulator of feeding. Evidence is also accumulating for its playing a role in the pathophysiology of the appetite disturbances seen in anorexia nervosa and depression.

THE CENTRAL FEEDING SYSTEM

Besides CRF, a variety of other peptides have been shown to decrease feeding following central administration. It is unclear which of these peptides are truly involved in feeding regulation and which are purely generalized disrupters of behavior. The neuropeptides which have been shown to alter feeding behavior are listed in Table 1.

Both calcitonin and calcitonin gene related peptide (CGRP) have been shown to decrease feeding after central administration (40, 41). Calcitonin is also an effective anorectic after peripheral administration. Calcitonin appears to produce its effects on feeding by altering calcium metabolism within the hypothalamus.

Present knowledge suggests that the regulation of feeding involves a delicate balance between a variety of neurotransmitters. At least two peptidergic feeding systems appear to exist. One is the dynorphin-dopaminergic system which is involved in the regulation of the intake of fatty or other highly palatable foods. The other is the NPY system which appears to specifically drive carbohydrate feeding. The serotonin-norepinephrine-CRF system appears to inhibit both of these feeding systems.

The realization that it is often anatomy that gives specificity to a peptide's behavioral effects, raises major questions concerning the development of anti-obesity drugs. It seems that any centrally acting peptide analog that decreases feeding is also likely to have a variety of other effects, such as altering memory and sexual function.

REFERENCES

1. Sjodin L. Effects of secretion, cholecystokinin and caerulein on gastric secretion in response to sham. Feeding in dogs. Acta Physiol Scand 1972; 85: 110-117.
2. Gibbs J, Young RC, Smith GP. Cholecystokinin decreases food intake in rats. J Comp Physiol Psychol 1973; 84: 488-495.
3. Morley JE. Neuropeptide regulation of appetite and weight. Endocrine Reviews 1987; in press.
4. Greenberg D, Smith CP, Gibbs J. Infusion of CCK-8 into hepatic-portal vein fails to reduce food intake in rats. Am J Physiol 1987; 252: R1015-1018.
5. Smith GP, Gibbs J. The satiety effect of cholecystokinin recent progress and current problems. In: Vanderhagen JJ, Crawley JN, eds. Neuronal Cholecystokinin. New York Acad Sci 1985; 448: 417-423.
6. Morley JE. Behavioral effects of peripherally administered cholecystokinin. ISI Atlas of Science (Pharmacology) 1987; 1: 49-51.
7. Crawley JN, Kiss JZ, Mezey E. Bilateral midbrain transections block the behavioral effects of cholecystokinin on feeding and exploration in rats. Brain Res 1984; 322: 316-321.
8. Orosco M, Rybarczyk MC, Rouch C, Cohen Y, Jacquot C. Cholecystokinin and bombesin in vagotomized or intact lean and obese rats: effects on neurotransmitters in brain. Neuropharmacol 1987; 26: 575-579.

9. Flood JF, Smith GE, Morley JE. Modulation of memory processing by cholecystokinin: dependence on the vagus nerve. Science 1987; 236: 832-834.
10. Verbalis JG, McCann MJ, McHale CM, Stricker EM. Oxytocin secretion in response to cholecystokinin and food differentiation of nausea from satiety. Science 1986; 232: 1417-1419.
11. West DB, Fey D, Woods SC. Cholecystokinin persistently suppresses meal size but not food intake in free-feeding rats. Am J Physiol 1984; 246: R776-787.
12. Gibbs J, Fauser DJ, Rowe EA, Rolls BJ, Rolls ET, Maddison SP. Bombesin suppresses feeding in rats. Nature 1979; 282: 208-210.
13. Morley JE, Levine AS, Kneip J, Grace M. The effect of vagotomy on the satiety effects of neuropeptides and naloxone. Life Sci 1982; 30: 1943-1947.
14. Billington CJ, Levine AS, Morley JE. Are peptides truly satiety agents? A method of testing for neurohumoral satiety effects. Am J Physiol 1983; 245: R920-926.
15. Muurhainen NE, Kisseleff MR, Thorton J, Pi-Sunyer FX. Bombesin: another peptide that inhibits feeding in man. Soc Neurosci Abstr 1983; 9: 183.
16. Geary N, Smith GP. Selective hepatic vagotomy blocks pancreatic glucagon's satiety effect. Physiol Behav 1983; 31: 391-394.
17. Langhans W, Zieger U, Scharrer E, Geary N. Stimulation of feeding in rats by intraperitoneal injection of antibodies to glucagon. Science 1982; 218: 894-895.
18. Levine AS, Morley JE. Peripherally administered somatostatin reduces feeding by a vagal mediated mechanism. Pharmacol Biochem Behav 1982; 16: 897-902.
19. Hinton V, Rosofsky M, Granger J, Geary N. Combined injection potentiates the satiety effects of pancreatic glucagon, cholecystokinin and bombesin. Peptides 1986; 17: 615-619.
20. Holtzman SG. Behavioral effects of separate and combined administration of naloxone and d-amphetamine. J Pharmacol Exp Ther 1974; 189: 51-60.
21. Marks-Kaufman R, Kanarek RB. Modifications of nutrient selection by naloxone in rats. Psychopharmacology (Berlin) 1981; 74: 321-324.
22. Morley JE, Levine AS. Involvement of dynorphin and the kappa opioid receptor in feeding. Peptides 1983; 4: 797-800.
23. Gosnell BA, Morley JE, Levine AS. Opioid-induced feeding: localization of sensitive brain sites. Brain Res 1986; 369: 177-184.
24. Atkinson RL. Naltrexone for weight loss in obesity. Abstr 4th Int Cong Obesity 1983; 30A: 81.
25. Mitchell JE, Morley JE, Levine AS, Hatsukami D, Gannon M, Pfohl D. High-dose naltrexone therapy and dietary counselling for obesity. Biol Psychiat 1987; 22: 35-42.
26. Clark JJ, Kalra S, Crowley R, Kalra SP. Neuropeptide Y and human pancreatic polypeptide stimulate feeding behavior in rats. Endocrinology 1984; 115: 427-429.
27. Morley JE, Levine AS, Gosnell BA, Kneip J, Grace M. Studies on the effect of neuropeptide Y on ingestive behaviors in the rat. Am J Physiol 1987; 252: R599-R609.
28. Stanley BG, Leibowitz SF. Neuropeptide Y: Stimulation of feeding and drinking by injection into the paraventricular nucleus. Life Sci 1984; 33: 2635-2642.
29. Morley JE, Willenbring ML, Krahn DD, et al. Opioid control of thyroid function. In: Frontiers in Neuroscience: The Opioid Modulation of Endocrine Function. New York: Raven Press, 1984: 267-275.
30. Stanley GG, Kyrkouli SE, Rampert S, Leibowitz SF. Neuropeptide Y chronically injected into the hypothalamus: a powerful neurochemical inducer of hyperphagia and obesity. Peptides 1986; 7: 1189-1192.

31. Kyrkouli SE, Stanley BG, Leibowitz SF. Galanin: stimulation of feeding induced by medical hypothalamic injection of this novel peptide. Eur J Pharmacol 1986; 122: 159-160.
32. Vaccarino FJ, Bloom FE, Rivier J. Stimulation of food intake in rats by the administration of hypothalamic growth hormone - releasing factor. Nature 1985; 314: 147-148.
33. Rosenfield DJ, Garthwaite TL. Central administration of motilin stimulates feeding in rats. Physiol Behav 1987; 39: 753-756.
34. Morley JE, Levine AS, Grace M, Kneip J. Dynorphin-(1-13), dopamine and feeding in rats. Pharmacol Biochem Behav 1982; 16: 701-705.
35. Morley JE, Levine AS. Corticotropin-releasing factor, grooming and ingestive behavior. Life Sci 1982; 31: 1459-1464.
36. Krahn DD, Gosnell BA, Grace M, Levine AS. CRF antagonist partially reverses CRF- and stress-induced effects on feeding. Brain Res Bull 1986; 17: 285-289.
37. Hotta M, Shibaski T, Masuda A, et al. The responses of plasma adrenocorticotropin and cortisol to corticotropin-reducing hormone and cerebrospinal fluid immunoreactive CRH in anorexia nervosa patients. J Clin Endocrinol Metab 1986; 62: 319-321.
38. Kaye WH, Gwirtsman HE, George DT, et al. Elevated cerebrospinal fluid levels of immunoreactive corticotropin-releasing factor in anorexia nervosa: relation to state of nutrition, adrenal function and intensity of depression. J Clin Endocrinol Metab 1987; 64: 203-208.
39. Nemeroff CB, Widerlov E, Bisette G. Elevated concentrations of corticotropin releasing factor-like immunoreactivity in depressed patients. Science 1984; 226: 1342-1343.
40. Levine AS, Morley JE. Reduction of feeding in rats by calcitonin. Brain Res 1982; 222: 187-191.
41. Krahn DD, Gosnell BA, Levine AS, Morley JE. Effects of calcitonin-gene-related peptide on food intake. Peptides 1984; 5: 861-864.

REGULATORY PEPTIDES IN HEALTH AND DISEASE

J.H.H. Thijssen and C.J.M. Lips

Department of Endocrinology and Department of Internal Medicine, Academisch Ziekenhuis, Catharijnesingel 101 3511 GV Utrecht, the Netherlands

INTRODUCTION

During the last ten years a large number of poly- and oligopeptides have been discovered, that have been recognised as intercellular messengers. These messengers are defined as substances, that are released by one cell, capable of modifying the functional activity of other cells in the immediate vicinity or at a distance. Many of these peptides have been characterised, released by different tissues throughout the body; they have been observed in the nervous system, the gastrointestinal tract and other organs, not only in man but also in vertebrate and invertebrate animals.

The detection of peptides, originally identified in nonneural tissues, within the central nervous system has raised questions regarding their origin and function (1). Because the majority of peptides are not capable of passing easily the blood-brain barrier, soon after their discovery local synthesis in the nervous system was supposed to take place and today much experimental evidence is available to support this supposition. Also their function has been clarified to some extent. These peptides are thought to play a role in the delicate interaction between cells, they appear to have neurotransmitter or neuromodulatory roles and appear to be involved in the regulation of a number of homeostatic systems (2).

Outside the central nervous system identical peptides are acting as "pure" hormones, after release into the blood they are influencing cells at some distance of the secreting cell e.g. cholecystokinin in periferal blood acts on gall bladder activity but locally in the brain its presence results in other effects. It is obvious that we are looking at compounds belonging to a control system which unifies the nervous and the endocrine regulation systems by virtue of the common active peptides. The variety of the peptides isolated points to the fact that there is an enormous diversity of transmitter signals, leading to the possibility of multiple transmitter-receptor interactions which permit unique specificity of biological responses.

TECHNOLOGY INVOLVED IN DISCOVERY

Many of these so-called regulatory peptides have been identified during the last ten years because only recently improved techniques for isolation, immunological detection, purification and microsequencing have enabled investigators to study successfully these peptides. In addition recombinant DNA technology has been applied, resulting in the discovery of many new peptides, the identity of the messenger RNAs, their amino acid sequence and their chromosomal organisation.
In this technology a number of individual processes or steps can be distinguished (3).

1. Use of restriction enzymes

Since the discovery of the non-overlapping triplet DNA-code, read from a fixed starting point, attempts have been made to establish nucleotide sequences in genes. The necessary specific breakage of the extremely large DNA-molecules into smaller fragments is made possible at defined points by the use of certain restriction enzymes. These enzymes have the ability of recognising rather short specific sequences of DNA as targets for cleavage. Each enzyme has a particular target in DNA, a specific sequence of usually 4 to 6 base pairs. Every place where this sequence occurs will be cleaved by the enzyme. Different restriction enzymes with different target sequences have been isolated from bacteria. Application of these enzymes will result in the formation of smaller DNA-fragments.

2. Hybridisation

The highly specific binding of RNA to DNA, by the pairing of base-pairs, makes it possible to use (synthetic) specific probes in the recognition of complementary DNA sequences. By the introduction of (radioactive) labels, the paired complementary sequence can be recognised, if present in one or more of the DNA-fragments, and can be isolated using chromatographic or electrophoretic techniques.

3. DNA cloning

The isolated DNA fragment can be inserted into bacterial plasmid and thus be used to produce almost indefinite amounts of this particular fragment out of even a single original DNA molecule.

4. Sequencing of DNA

The sequence of DNA can be obtained more easily and rapidly than protein sequences. The general principle is to generate a series of single stranded DNA molecules, each molecule being one base pair shorter than the last, and separating these molecules by electrophoresis. The generation of the series of single stranded molecules is done by subjecting the original fragment sequentionally to 4 different base-specific reactions and analysing the products in parallel bands during electrophoresis. This sequencing of DNA has shown to be a very accurate process.

Application of these techniques has resulted in the isolation and identification of a large number of peptides. It appeared that sometimes several of the isolated peptides showed structural homology, which means that a high percentage of the amino acids were in identical positions in peptides from different origins. This homology makes it possible to classify peptides in so-called families, some examples are given in Table 1.

Table 1. Polypeptide families showing homology between peptide hormones, receptors and oncogenes. Within each group substances have been given which show homology in their amino acid composition

Growth Hormone Releasing Factor Vasoactive Intestinal Peptide Secretin Glucagon Gastrin Inhibiting Peptide Corticotropin Releasing Factor	Arginin Vasopressin Neurophysin Oxytocin Neurophysin
	Prolactin Growth Hormone Human Placental Lactogen
Caerulein Phyllocaerulein Gastrin Cholecystokinin beta-Lipotropin Calcitonin I and II Calcitonin Gene Related Peptide I and II	Prothoracicotropic Hormone Insulin-like Growth Factor I and II human Insulin (a-chain)
	Platelet Derived Growth Factor Simian Sarcoma virus Derived Oncog.
Ranatensin Bombesin Gastrin Releasing Peptide	
Relaxin beta Nerve Growth Factor	human Insulin Receptor human Epidermal Growth Factor Rec. h.osteosarcoma derived Oncogene murine Abelson leukemia virus Onc. avian erythroblastosis virus Oncog. avian Rous sarcoma virus Oncogene murine Moloney sarcoma virus Oncog.

The homology of peptides differing so much in activity, together with the characterisation of a number of related polypeptides in multicellular invertebrates, in unicellular organisms and in plants, are arguments for a common origin of the regulatory peptides dating back far in the evolutionary development of living organisms (4, 5). Modern biochemistry offers a number of possible explanations for the diversity of the regulatory peptides and the observed homologies.

EXPLANATIONS FOR COMPLEXITY AND HOMOLOGY

At the biosynthetic level there are several mechanisms that may underly the diversity of peptides with homology at the same time. The basic events behind the mechanisms are supposed to have happened at various points of the physiological process of peptide replication:

1. At the DNA level

Soon after the discovery of the genetic code it was shown that the change of a single base in the DNA, a point mutation, can lead to a change in one single amino acid and that such a change can be the molecular basis of an inherited disease (6) or even of malignant transformation of a cell. Apart from being the origin of diseases, point mutations are essential to the diversity of the intercellular messengers in the evolution.

The mechanism of point mutation results in addition, deletion or substitution of a nucleotide. In nature mutations occur spontaneously in a low frequency under normal physiological conditions e.g. due to errors in DNA replication and/or low concentrations of mutagenic metabolites or agents in the cell or the environment.

In the genome of many species related proteins are often encoded by related but different genes which appear to have evolved from a single evolutionary ancestor following gene duplication. The duplication of the original gene is thought to could have resulted from a process called unequal crossing-over, happening during multiplication of DNA sequences in vivo. After such an event one of the DNA strands will be left with less DNA while the other will have acquired more.

Another process resulting in duplication of genes is the transposition of large pieces of DNA from one genetic locus to another one. This transposition may occur in different ways. Firstly genomic DNA may be incorporated into a retrovirus and transmitted to other individuals or species. Secondly it may be mobilized within the genome as part of a transposon (transposable DNA element), a DNA sequence which is able to replicate and to insert one copy at a new location in the genome. Thirdly a messenger RNA molecule may be reverse transcribed into DNA by viral enzymes and reinserted into the genome.

2. At the RNA level

The original RNA transcript in a cell nucleus is an exact copy of the transcribed DNA and this copy contains non-coding fragments, called introns. These introns will be eliminated during the so-called splicing process, which leads to m-RNA consisting of coding sequences, of exons only. This m-RNA finally will be translated into proteins. The removal of introns is not a nonspecific process, identical DNA sequences leading to identical RNA transcripts can be processed in an organ-specific way. Alternative splicing pathways can generate organ-specifically different m-RNAs and thus different proteins from one gene or, more accurately, from one transcript unit.

An example of this alternative splicing can be observed in the RNA coding for the peptide hormone calcitonin, normally produced in the thyroid gland. Although a large amount of this RNA coding for calcitonin can be found in the hypothalamus, very little of it is produced there. Instead another protein called "calcitonin gene related peptide (CGRP)" has been detected (7). Both calcitonin and CGRP are produced from the same primary transcript by different splicing routes. These routes generate two mature mRNAs which have a common start but different tails. It is completely ununderstood why cellular differentiation has lead to these alternative splicing pathways.

3. At the translational product, the peptide level

Many proteins are being synthesized at first as larger precursors containing some 15 to 25 additional amino acids at their NH_2-terminal ends.

These signal sequences are part of peptides that have to move through the cellular membrane before being secreted. This signal will be cleaved by specific proteolytic enzymes which will generate the functional peptide.

Several of the smaller biologically active peptides are derived from cleavage of larger precursor proteins. These precursors are processed in the cell after translation, they are modified from minimally biologically active polypeptides into biologically active peptides. An example: pro-opiomelanocortin will be processed and converted into several active peptides as ACTH, beta-lipoprotin and the endorphins (8). In addition to this intracellular processing, it has been observed in recent years that proteolysis after secretion can occur and that this pathway can be the biosynthetic origin of active peptides (9).

The additional peptide regions of the larger precursor molecules can have several functions in biosynthesis as for example the role of C-peptide in proinsulin, which contributes to the correct formation of the disulphide bridge in insulin, the final, mature product of the cell. In general this allows the conclusion that additional regions can be important for the formation of the tertiary structure of proteins and thus for the specific degradation into active forms.

PROBLEMS RELATED TO APPLICATIONS

As a consequence of the evolutionary events described before, a large diversity of related peptides can be found in the periferal circulation and in different organs of human subjects. This will cause serious problems in studies aiming at measuring one of these peptides specifically, as many of the antibodies used in these measurements do not possess sufficient specificity to be able to distinguish between (some of) the related substances, because of their homology.

In addition to the problems caused by the homology, it has been established for a number of the regulatory peptides that they do exist in different molecular sizes. For instance cholecystokinin (CCK) in blood of normal human subjects (10) was shown to occur in at least 4 different molecular forms, varying in size between 8, 22, 33 and 58 amino acids with identical sequences in the various molecular forms. Furthermore, evidence has been presented that the molecular nature of CCK, measured in periferal blood under different stimuli, may show other molecular compositions (11), the percentages of the individual molecular forms vary. In another recent publication (12) differences in the molecular composition of total CCK between normal and abnormal pituitaries were described, suggesting the production of altered forms of this hormone by pituitary adenomas.

With these observations the difficulties encountered in quantitative determinations of regulatory peptides in human subjects become very clear, the extreme heterogeneity of some of these peptides in combination with their very low concentrations in blood will make studies of their role in the pathophysiology of diseases very difficult. Otherwise their function in many of the important homeostatic processes becomes more and more evident and thus the requests for reliable estimations. Careful measurements, probably using combinations of chromatographic separations and highly specific antisera towards well-characterised parts of the molecule, will be necessary to gain more insight in their role in human health and disease.

CONCLUSIONS

Polypeptide hormones found their roots early in evolution, the original principles of intercellular communication appear to be highly preserved. Biochemical studies using recombinant DNA technology provide more insight into the evolutionary mechanisms behind the diversity of still related peptides. The diversity of transmitter substances has led to the possibility of multiple unique transmitter receptor interactions, allowing specificity of biological responses. Despite the diversity there appears to be an overlap between the endocrine and the nervous system because of the similarity of the messengers.

Biosynthesis by bacteria or fungi will provide us with new, now hardly available peptides such as growth factors, releasing hormones etc. The future will undoubtedly show the identification of new peptide hormones, tissue factors but also of oncogene products and their precursors with mutual homology. The use of determinations of these substances for diagnostic purposes will have to be established, but this process will be highly complicated because of the difficulties encountered in specific measurements of the regulatory peptides.

Improvements in diagnostic procedures will be achieved by the use of in-situ hybridisation techniques and/or by highly specific solution hybridisation assays. In this way the detection of constitutional gene defects as origin of inherited diseases and inherited predisposition to tumour growth will become possible.

REFERENCES

1. Krieger DT. Brain peptides: what, where, and why? Science 1983; 222: 975-85.
2. Lips CJM. The biosynthesis of polypeptide hormones in health and disease. Neth J Med 1986; 29: 53-61.
3. Genes III. Wiley and Sons, 1987.
4. LeRoith D, Liotta AS, Roth J, et al. Corticotropin and β-endorphin-like materials are native to unicellular organisms. Proc Natl Acad Sci USA 1982; 79: 2086-90.
5. Bleich HL, Moore NJ. Seminars in medicine of the Beth Israel Hospital, Boston. New Eng J Med 1982; 306: 523-7.
6. Ingram VM. A specific chemical difference between the globins of normal human and sickle-cell anaemia haemoglobin. Nature 1956; 178: 792-4.
7. Amara SG, Jonas V, Rosenfeld MG, Ong ES, Evans RM. Alternative RNA processing in calcitonin gene expression generates mRNAs encoding different polypeptide products. Nature 1982; 298: 240-4.
8. Nakanishi S, Inoue A, Kita T, et al. Nucleotide sequence of cloned cDNA for bovine corticotropin-β-lipotropin precursor. Nature 1979; 278: 423-7.
9. De Kloet R, De Wied D. The brain as target tissue for hormones of pituitary origin: behavioral and biochemical studies. In: Martini L, Ganong WF, eds. Front Neuroendocr 1980; 6: 157.
10. Cantor P, Rehfeld JF. The molecular nature of cholecystokinin in human plasma. Clin Chim Acta 1987; 168: 153-8.
11. Jansen JBMJ, Lamers CBHW. Molecular forms of cholecystokinin in human plasma during infusion of bombesin. Life Sci 1983; 33: 2197-205.
12. Rehfeld JF, Lindholm J, Andersen BN, et al. Pituitary tumors containing cholecystokinin. New Engl J Med 1987; 316: 1244-7.

b-ENDORPHIN AND REPRODUCTION: EVIDENCE FOR OVARIAN PARTICIPATION

A.R. Genazzani, F. Petraglia, F. Mercantini, A. Monzani
G. D'Andrea and F. Facchinetti

University of Modena, Dept. of Obstetrics and Gynecology
Modena, Italy

A. PHYSIOLOGICAL CHANGES OF PLASMA b-ENDORPHIN

Many studies have accumulated on the Proopiomelanocortin (POMC) family of peptides since they appear to be involved in various physiological processes. b-Endorphin (b-EP) in particular displays either neurotransmitter or hormonal actions, even though its targets remain to be established. The possible interactions of b-EP with reproductive processes could be operating at three different levels: the hypothalamus, the pituitary and the gonads.

In this chapter we will summarize firstly the plasma b-EP changes at the different stages of life and secondly attention will be focussed on the possible "local" opioid system existing in the ovary.

1. Infancy and adolescence

Early infancy (1-3 yrs) is characterized by b-EP plasma levels ranging from 2.2 to 3.9 fmol/ml with mean concentrations of 3.1 ± 0.7 in males and females, no sex-related differences are present at this age, nor in older children. With the progression of age b-EP plasma levels increase progressively and significantly, reaching adult levels just before the onset of puberty. This leads to the speculation that opioid peptides may be of importance in the activation of the adrenal glands, the adrenarche, as concomitant increases of dehydroepiandrosterone-sulphate (DHA-S) have been observed throughout prepuberty (1).

It is worth pointing out that despite such changes in b-EP and DHA in prepuberty, ACTH and cortisol levels remain constant from early infancy to adulthood. This discrepant pattern suggests a different control or a partially different origin of the peptides, although these peptides are believed to derive from the same pituitary precursor molecule. During infancy b-EP shows circadian changes.

A possible effect of POMC derived peptides, other than ACTH, on the adrenal gland is suggested by the finding that the N-terminal POMC fragment stimulates the granulosa cell mitotic index and potentiates the effects of ACTH on the cells (2).

Pubertal maturation in males and females is not accompanied by changes in the circulating levels of b-EP, leading to the conclusion that changes in gonadal steroid plasma levels occurring in this period do not interfere with the control of circulating b-EP levels. The similarity in behaviour with ACTH and cortisol could indicate a common Corticotropin Releasing Factor (CRF) mediated mechanism as the cause of such a chronobiological organization (3).

2. Menstrual cycle and pregnancy

The menstrual cyclicity is accompanied by important changes in the functioning of the hypothalamus-pituitary(HP)-target organs axis. In particular the HP-adrenal axis undergoes a marked activation during the late follicular and ovulatory periods. Detailed studies on the pattern of circulating ACTH patterns indicate the existence of a periovulatory increase with low levels in the early follicular phase.

Plasma b-EP levels show a peculiar pattern with high levels in the preovulatory phase, a fall in the ovulatory and high levels in the early luteal phase. This behaviour differentiates the pattern of b-EP from that of ACTH. The observation that the pattern of circulating b-EP from the late follicular to mid-luteal phase shows a similar pattern as that of circulating E_2 suggests that b-EP change during the menstrual cycle is an estrogen mediated mechanism.

Castrated females, animal and human, show a significant reduction of circulating b-EP levels. This phenomenon as well as the observation that estrogen replacement therapy results in increasing b-EP in castrated females support the hypothesis that estrogen may interfere in the control of peripheral b-EP levels (4).

Early pregnancy is characterized by a reduction of plasma b-EP; from the beginning of the second trimester, it progressively increases to reach the highest concentration at term (5). Labour is characterized by a significant and progressive increase until delivery. This last pattern seems to be related to the painful and stressing stimuli of the uterine contractions (6). After delivery the b-EP levels rapidly decline to the normal range and no peculiar pattern was observed during puerperium and under suckling stimulation.

3. Menopause

In addition to the general ageing processes, which are accompanied by a reduction of plasma and CSF b-EP levels both in male and females, menopause seems to represent a specific condition, negatively affecting the opioid system. Ovariectomy of women during reproductive life is accompanied by a significant decrease of circulating b-EP levels (7) and by the loss of LH increases induced by naloxone (8), which indicates a failure in the activity of central opiate receptors.

The changes in circulating plasma levels are accompanied by similar ones in both pituitary lobes as three weeks after gonadectomy the b-EP contents decreases in male and female rats (4). Interestingly, appropriate replacement treatment with testosterone propionate or estradiol benzoate is able to restore the opioid content in pituitary and plasma. These data therefore give convincing evidence that gonadal steroids influence opiatergic activity. In view of the time lapse between castration and the biochemical changes, one should conclude that sexual steroids play a modulatory rather than a direct role.

The decreased opioid content after castration is accompanied by important behavioural changes. In rats, the analgesic effects induced by morphine are attenuated by castration and the tail-flick latency is restored to normal after estrogen treatment (9). Behavioural changes can be documented also in postmenopausal women thus supporting the idea that the climacteric syndrome could partly originate also from an opioid disfunction. Indeed, endogenous opioids are thought to control affective equilibrium and are involved in the regulation of body temperature (10). It is worth noting that hot flushes are a typical neuroendocrine phenomenon (11) where opioids seem to play a fundamental role. During subjective hot flushes, in spite of a 20% increase of plasma LH, there is a sudden b-EP release leading to doubling of basal values (12).

In conclusion, all data collected in this period of life demonstrate that the reduction of opiate activity could be associated with the occurrence of some climacteric complaints, mainly postmenopausal pain syndromes, affective disorders and hot flushes. The positive role played by estrogens on endogenous opioids could stimulate the use of replacement treatments also for the so-called climacteric syndrome.

B. PROOPIOMELANOCORTIN-RELATED PEPTIDES CHANGES IN HUMAN FOLLICULAR FLUID

1. Introduction

In addition to pituitary and brain, opioid peptides have been found in several peripheral organs and compartments and in different fluids. Various review articles already report data on opioids in the adrenal (13), in the gastrointestinal tract (14), in other endocrine-related structures and in several peripheral organs (15).

Beta-Endorphin has been detected both by immunohistochemistry and radioimmunoassay in the cytoplasm of rat Leydig cells, and in epithelia of epididymis, seminal vesicle and vas deferens. The tissue contents were unaltered by hypophysectomy suggesting a non-pituitary dependent control (16). Further POMC-peptides such as a-EP, des-Tyr-g-EP, des-acetyl-a-MSH, were also found in rat testis homogenates. The prevalence of smaller, low molecular weight peptides indicates an extensive processing of POMC similar to that occurring in the brain, rather than that in the anterior pituitary (17). Acetyl-b-EP is absent from testicular homogenates (17), but extensive N-a-acetylation occurs in germ cells using a-EP and g-EP as substrate (18). Beta Endorphin has been detected also in human semen (19), where its levels are 4-12 times higher than in circulating plasma thus suggesting a local production (20).

In the female reproductive tract, ACTH, b-EP and some larger molecular weight forms containing the b-EP sequence have been first detected in follicular cells and corpus luteum of ewes (21). In the ovaries of mice as well as of rats the most intense staining was observed in the corpora lutea, while a lower immunoreactivity was found in granulosa cells. No staining was demonstrated in theca cells, in primary follicles or in ovaries of immature animals, thus linking the POMC-peptide presence to the ovarian endocrine activity (22).

Indeed, both HCG and PMSG increased b-EP staining in ovarian interstitium of cyclic mice (22). Total b-EP immunoreactivity in the rat ovary was lowest at estrous, progressively increased throughout the cycle and showed a peak at proestrous. These data suggest that gonadotropins positively regulate ovarian POMC-peptides. Recently, as a definitive proof, it was demonstrated that PMSG treatment of immature rats resulted in a large increase of POMC-like mRNA in the ovary (23).

In humans, elevated concentrations of b-EP, b-LPH and ACTH have been found in both peritoneal and follicular fluids (24). The peritoneal opioid concentrations are higher in the periovulatory period; they show very low values in postmenopausal women and in subjects taking oral contraceptives. Recently immunohistochemical data indicated the presence of b-EP in luteinized theca internal cells of normal ovarian tissue, confirming a reduced staining in postmenopausal tissue (25). In this paper, we describe the characterization of the POMC-related peptides content of follicular fluid by using HPLC coupled to RIA.

2. Materials and methods

Subjects. Twelve healthy menstruating women (21-35 years old) were recruited in the present study. They underwent laparoscopy for tubal sterilization (n = 6) or for our in vitro fertilization and embryo transfer (IVF-ET) program (n = 6). Subjects were selected on the basis of their regular menstrual cycle (28 ± 2 days).

Follicular fluids were collected during the follicular phase (6 cases) or on the preovulatory day in the patients submitted to IVF-ET. The occurrence of ovulation was monitored during the previous and the current cycle by echographic evaluation of the follicular development (twice a day in those participating to the IVF-ET) and the measurement of basal body temperature. No subject received drugs for at least 6 months; patients participating in the IVF-ET program received 100 mg/day of clomiphene citrate for 5 days (Serophene, Serono, Rome, Italy) from the 5th day of the cycle and human chorionic gonadotropin (5000 IU, im.; Profasi hp 5000, Serono) was given on the preovulatory day.

Laparoscopy was performed under general anesthesia and follicular fluid was aspirated from follicles through a special needle. Immediately after collection of the sample, aprotinine (1000 Kallikrein IU/ml, Bayer, Leverkusen, Fed. Rep. Germany) was added to the follicular fluid to reduce proteolytic activity; the sample was then centrifuged (3500 rpm x 15 min) and the supernatant stored at -20°C until assays.

Peptide Assays. The samples were extracted with octadodecacilylsilica columns (Sep-pak, Water Ass., Milford, MA), previously activated with urea and methanol. After washing the columns with 20% methanol in 0.5% acetic acid, POMC-related peptides were recovered with 4 ml of 90% methanol in acetic acid. The recovery of the different peptides ranged from 81 to 94%. The extracts were dried, redissolved and then injected into a reverse phase HPLC system. The column (RP-C18 u Bondapack, 3.9 x 300 mm, 10 u size) was eluted with a linear gradient from 18 to 40% of acetonitrile in 0.01 M hydrochloric acid in 22 min, at a flow rate of 1.5 ml/min. Fifty-two fractions, dried and redissolved in 1 ml of 0.125 M phosphate buffer containing 0.1% BSA, were evaluated for their contents of 1-39 ACTH (retention rime = 15.2 min), desacetyl-alpha-MSH (des-a-MSH) 8.8 min, a-MSH 11.4 min, b-LPH 15.8 min, b-EP 18.5 min, and a-EP 13.5 min, using specific RIAs. The characteristics of the RIAs for b-LPH, b-EP and 1-39 ACTH were previously reported elsewhere (26, 27).

For the a-MSH RIA, synthetic human a-MSH and rabbit anti-human a-MSH was kindly donated by Dr. V. Wiegant (Utrecht, NL). A ^{125}I labelling of the peptide was performed using the chloramine T method. The final dilution of the antiserum was 1 : 15,000. The cross reactions of anti a-MSH serum were: 100% with a-MSH, 22% with des-a-MSH, 0.01% with 1-39 ACTH, 0.3% with 1-24 ACTH. Des-a-MSH content of the eluted fractions could be measured in view of its cross reaction with anti a-MSH serum.

Synthetic human b-EP was provided by Organon (Oss, NL). Rabbit antihuman gamma-EP antiserum was kindly donated by Dr. V. Wiegant. Labelling with ^{125}I was also performed with the chloramine T method. The final dilution of anti gamma-EP serum was 1 : 25,000 and its cross reactions were: gamma-EP = 100%, b-EP = 4%, alpha-EP = 8%.

3. Results

The chromatograms shown in Figure 1 indicate that immunoreactivities, revealed by the different RIA systems, coelutes with reference peptides, thus suggesting the specificity of HPLC/RIA procedures. The quantitative evaluations of these peaks are reported in Figure 2.

In the immature follicles, gamma-EP (701 ± 190 fmol/ml, M ± SD) levels were higher than those of b-EP (173 ± 111, P <0.01) and b-LPH (70 ± 27, P <0.01).

This pattern is reversed in superovulated follicles where b-EP levels prevail (836 ± 594, P <0.05 vs immature) (Figure 2).

The b-EP/b-LPH ratio was different in both classes of follicles: immature had a ratio of 3.2 ± 2.9; superovulated of 13.5 ± 11.6. The gamma-EP/b-EP ratio was in favour of the former peptide in immature follicles (6.4 ± 5.5), while b-EP predominated in both superovulated ones.

As far as ACTH and its derived peptides are concerned, the levels measured in immature follicles (ACTH: 180 ± 158; des-a-MSH: 3916 ± 1297 and a-MSH: 742 ± 279 fmol/ml) are 10-50 times higher than in superovulated follicles (p <0.001) (Figure 2). The a-MSH/ACTH ratio was similar in both groups.

4. Conclusions

These data demonstate that almost all compounds related to the POMC-opioid system are present in follicular fluid and that their concentrations vary as a function of the endocrine activity of follicular cells. The specificity and sensitivity of radioimmunoassays in combination with the powerful chromatographic separation of HPLC permits the conclusion that the compounds measured are very similar if not identical to the reference

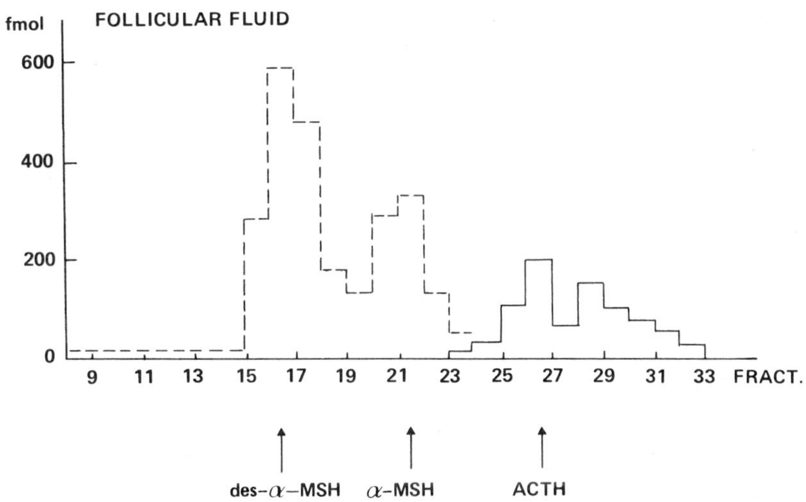

Fig. 1. Immunoreactivity measured with alpha-MSH (dotted line) and ACTH (solid line) antisera in human follicular fluid. Arrows refer to elution of standard peptides.

peptides. On these grounds, we report for the first time significant amounts of b-EP, a-MSH and its des-acetylated form in human follicular fluid.

In the follicular fluid, as observed in the brain, the concentrations of smaller molecules like a-MSH and b-EP largely prevails that of their intermediate precursors (ACTH and b-LPH, respectively) suggesting an extensive processing of POMC. Nothing is known about the possible role of locally produced ovarian opioids. However, it is worth noting that preliminary experiments done in vitro indicate that nanomolar concentrations of Met-Enk, but not of b-EP, enhance the FHS-dependent progresterone secretion from granulosa cells (28). Moreover, this effect was dose-related and partly prevented by naloxone, thus suggesting the mediation of specific opiate receptors.

As reported for POMC, the mRNA encoding for proenkephalin A has been detected in the ovary of three mammalian species i.e. rat, hamster and cattle (29). Moreover a small amount of immunoreactive Dynorfin has been found in rat ovarian homogenates (30) thus demonstrating that the three opioid systems are expressed also in the ovary.

Recent data from our laboratory demonstrated the presence of b-EP and Met-Enk in the urine fluid of fertile women and of normally cycling and superovulated cows. The two opioids are undetectable in the fluid of untreated postmenopausal women, whereas they are present following estrogen-progestagen treatment (31). As previously reported for b-EP (32), Met-Enk concentrations were higher in the secretory than in the proliferative phase of the menstrual cycle. Immunoreactive b-EP and Met-Enk are present also in the follicular, oviductal and uterine fluids of cows. In the uterine fluid Met-Enk concentrations are highest in the superovulated animals linking again ovarian endocrine activity to opioid content of female genital tract (personal communication).

The isolated increase of follicular b-EP in preovulatory follicles could be related to the isolated b-EP peak described in peripheral

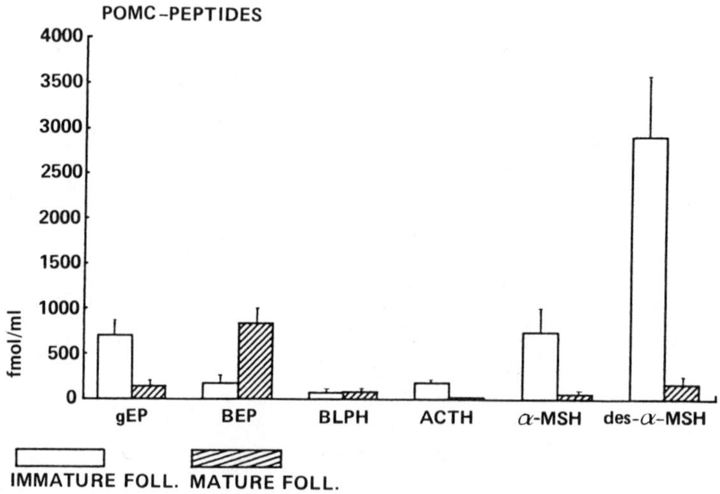

Fig. 2. Concentrations (M ± SEM) of different peptides in follicular fluids collected from immature and preovulatory follicles.

circulation around ovulation (33, 34). The absence of a concomitant b-LPH rise has been interpreted as a phenomenon independent of the anterior pituitary secretion (34). The present data and the absolute requirement of ovulation for the presence of a peak in plasma (34) led to the hypothesis that "ovarian endorphines" could contribute to peripheral b-EP levels in certain circumstances, such as at ovulation.

If the ovary could indeed be a source of b-EP, unexplained phenomena like the transient reduction of the peptide during the luteal phase of women suffering of premenstrual syndrome (35) could be explained (36, 37).

REFERENCES

1. Genazzani AR, Facchinetti F, Pintor C, et al. Proopiocortin-related peptide plasma levels throughout prepuberty and puberty. J Clin Endo Metab 1983; 57: 56-61.
2. Tilders FJH, Post M, Jackson S, Lowry PJ, Smelik PG. Beta-adrenergic stimulation of the release of ACTH- and LPH-related peptides from the pars intermedia of the rat pituitary gland. Acta Endocrinology 1981; 97: 343-351.
3. Sicuteri F. Is acute tolerance to 5-hydroxytryptamine opioid dependent? Its absence in migraine sufferers. Cephalalgia 1983; 3: 187-90.
4. Petraglia F, Penalva A, Locatelli V, et al. Effect of gonadectomy and gonadal steroid replacement on pituitary and plasma β-endorphin levels in the rat. Clin Endocrinol 1982; 111: 1224-9.
5. Genazzani AR, Facchinetti F, Parrine D. β-Lipotrophin and β-endorphin plasma levels during pregnancy. Clin Endocrinol 1981; 14: 409-18.
6. Facchinetti F, Centini G, Parrini D, et al. Opioid plasma levels during labour. Gynaecol Obstet Invest 1982; 13: 155-63.
7. Genazzani AR, Facchinetti F, Ricci-Danero MG, et al. Beta-lipotropin and beta-endorphin in physiological and surgical menopause. J Endocrinol Invest 1981; 4: 375-78.
8. Melis GB, Paoletti AM, Gambacciani M, Mais V, Fioretti P. Evidence that estrogens inhibit LH secretion through opioids in postmenopausal women using naloxone. Neuroendocrinology 1984; 39: 60-3.
9. Degli Uberti EC, Petraglia F, Trasforini G, et al. Effect of dermorphin on metyrapone-evoked release of acth, β-endorphin and β-lipotropin in man. J Endocrinol Invest 1985; 8: suppl 3.
10. Krieger DT, Martin JB. Brain peptides. New Engl J Med 1981; 304: 876-85.
11. Specific nonopiate receptors for β-endorphin. Science 1979; 205: 1033-5.
12. Genazzani AR, Petraglia V, Facchinetti F. Increase of proopiomelanocortin-related peptides during subjective menopausal flushes. Am J Obstet Gynaecol 1984; 149: 775-9.
13. Vermes I, Mulder GH, Smelik PG, Tilders FJH. Differential control of β-endorphin β-lipotropin secretion from anterior and intermediate lobes of the rat pituitary gland in vitro. Life Sci 1980; 27: 1761.
14. Abadia R, Boschat M, Opolon P. Endorphines et tube digestif. Ann Med Interne 1983; 134: 742-53.
15. Tson SD, Phillips D, Halmi N, et al. ACTH and β-endorphin-related peptides are present in multiple sites in the reproductive tract of the male rat. Endocrinology 1982; 110: 2204-6.
16. Margioris AN, Liotta AS, Vaudry H, Wayne Bardin C, Krieger DT. Characterization of immunoreactive proopiomelanocortin-related peptides in rat testes. Endocrinology 1982; 113: 663.
17. Cheng MC, Clements JA, Smit AI, Lolait SJ, Funder JW. N-Acetyl endorphin in rat spermatogonia and primary spermatocytes. J Clin Invest 1985; 75: 832.

18. Sharp B, Pekary AE. β-Endorphin 61-91 and other β-endorphin-immunoreactive peptides in human semen. J Clin Endocrinol Metab 1981; 52: 586.
19. Fabbri A, Gnessi L, Perricone R, et al. Human semen inhibits T rosette formation through an opiate mediated mechanism. J Clin Endocrinol Metab 1985; 60: 807-9.
20. Lolait SJ, Autelitano DJ, Lim ATW, Smith AI, Hock Toh B, Funder JW. Ovarian immunoreactive β-endorphin and estrous cycle in the rat. Endocrinology 1985; 117: 161-8.
21. Shaha C, Margioris A, Liotta AS, Krieger DT, Bardin CW. Demonstration of immunoreactive β-endorphin- and γ_3-melanocyte-stimulating hormone-related peptides in the ovaries of neonatal, cyclic, and pregnant mice. Endocrinology 1984; 115: 378-84.
22. Chen C-LC, Chang C-C, Krieger DT, Bardin CW. Expression and regulation of proopiomelanocortin-like gene in the ovary and placenta: comparison with the testis. Endocrinology 1986; 118: 2382.
23. Petraglia F, Segre A, Facchinetti F, Campanini D, Ruspa M, Genazzani A. β-Endorphin and met-enkephalin in peritoneal and ovarian follicular fluids of fertile and postmenopausal women. Fertil Steril 1985; 44: 615-21.
24. Alleem FA, Omar RA, ElTabbakh GH. Immunoreactive β-endorphin in human ovaries. Fertil Steril 1986; 45: 507-11.
25. Davis JM, Lowy MT, Yim GKW, Lamb DR, Malven PV. Relationship between plasma concentrations of immunoreactive beta-endorphin and food intake in rats. Peptides 1983; 4: 79-83.
26. Genazzani AR, Nappi G, Facchinetti F, et al. Central deficiency of β-endorphin in alcohol addicts. J Clin Endocrinol Metab 1982; 55: 583-6.
27. Facchinetti F, Ruspa M, Turci A, et al. Met-enkephalin enhances follicle-stimulating hormone-dependent progesterone production from cultured granulosa cells. J Clin Endocrinol Metab 1986; 63: 1222-4.
28. Kilpatrick DL, Howells RD, Noe M, Bailey LC, Udenfriend S. Expression of preproenkephalin-like mRNA and its peptide products in mammalian testis and ovary. Proc Natl Acad Sci USA 1985; 82: 7467-9.
29. Spampinato S, Goldstein A. Immunoreactive dynorphin in rat tissues and plasma. Neuropeptides 1983; 3: 193-212.
30. Petraglia F, Facchinetti F, M'Futa K, et al. Endogenous opioid peptides in uterine fluids. Fertil Steril 1986; 46: 247-51.
31. Wahlström T, Laatikainen T, Salminen K, Leppäluoto. Immunoreactive β-endorphin is demonstrable in the secretory but not in the proliferative endometrium. Life Sci 1985; 36: 987-90.
32. Veith JL, Anderson J, Slade SA, Thompson P, Laugel GR, Getzlaf S. Plasma β-endorphin, pain thresholds and anxiety levels across the human menstrual cycle. Physiol Behav 1984; 32: 31-4.
33. Petraglia F, Di Meo G, De Leo V, Nappi C, Facchinetti F, Genazzani AR. Plasma β-endorphin levels in anovulatory states: changes after treatment for the induction of ovulation. Fertil Steril 1986; 45: 185-90.
34. Chuong CJ, Coulam CB, Chih Kao P, Bergstrahl EJ, Go VLW. Neuropeptide levels in premenstrual syndrome. Fertil Steril 1985; 44: 760-65.
35. Fertil Steril 1987; in press.
36. Reid RL. Neuropeptides and PMS. Fertil Steril 1986; 46: 738-39.

CHAPTER 23
THE THYROID AND FUNCTIONING HORMONES

Cellular aspects of thyroid hormones
M.C. Sheppard and J.A. Franklyn

Methods of analysis of thyroid-related hormones
J.G. Ratcliffe

Thyrotropin secretion in health and diseases
G. Faglia, P. Beck-Peccoz, M. Nissim, G. Medri, and G. Piscitelli

CELLULAR ASPECTS OF THYROID HORMONES

M.C. Sheppard and Jayne A. Franklyn

Department of Endocrinology, Queen Elizabeth Hospital
Birmingham, B15 2TH, England

There is considerable evidence that the biological actions of thyroid hormones within the cell are mediated by binding to specific iodothyronine receptors found in close association with nuclear chromatin (1). There is an excellent correlation between the binding of T3 analogues to nuclear receptors and their thyromimetic potency. A second line of evidence is the correlation between the sensitivity of tissues to thyroid hormone action and the concentration of nuclear receptor sites in that tissue. Tissues such as the anterior pituitary and liver which are highly responsive to thyroid hormones have large numbers of nuclear receptors for T3, but tissues such as the testis and spleen, which are unresponsive to thyroid hormones, have little or no demonstrable nuclear binding. There is a clear relationship between nuclear occupation and nuclear response; when nuclear sites are fully saturated the measured biological response is maximal. The rapidity with which certain mRNA sequences change in response to T3 occupation of the nuclear sites further supports the concept of a causal relationship between the two. The receptor has been characterised as a non-histone protein of molecular weight 50,000 and a sedimentation coefficient of 3.5S. The receptors are of high affinity (Ka 5 x 10^{10} 1/mol) and are an integral component of a larger unit, the receptor-containing chromatin complex. Several hormone receptors may be localized to a nuclear matrix fraction in responsive tissues and hormone-responsive genes may be enriched in the nuclear matrix during hormonal stimulation. An example of this is the observation that 30-50% of nuclear T3 receptors in cultured GC cells are concentrated in the nuclear matrix which contains less than 1% of nuclear DNA, but which is enriched with hormone-responsive genes (2).

Thyroid hormones have a profound effect on the growth, development and metabolism of most tissues in higher organisms. Nuclear receptors are found in all mammalian tissues responsive to T3. These receptors are particularly abundant in the anterior pituitary and amongst the various actions of thyroid hormones that have been reported only the suppression of thyrotrophin (TSH) can be considered to be in any way "specific". Regulation of TSH gene transcription will therefore form the major part of this review. Other thyroid hormone-responsive genes include those coding for growth hormone (GH), prolactin (PRL), malic enzyme, α-glycerophosphate dehydrogenase, myosin heavy chain, sodium potassium ATPase, and S14.

TSH is a glycoprotein composed of two dissimilar non-covalently bound subunits, α and β. The β-subunit structure is unique to TSH and confers biological and immunological specificity. The subunits are encoded by different genes on different chromosomes. The mechanism whereby binding of T3 or T4 to the anterior pituitary nuclear receptor in turn regulates TSH β and α-subunit gene expression is not yet determined. Estimates of the rate of TSH gene transcription have demonstrated that T3 rapidly decreases transcription of the TSH β and α genes in thyrotrophic tumour explants, the change in transcription rate being directly proportional to the occupancy of nuclear receptors for T3, suggesting cause and effect (3). There is some evidence that the inhibitory effects of thyroid hormone on TSH gene expression are mediated via the synthesis of an intermediary inhibitory protein (4). Our own studies of T3 administration to the hypothyroid rat have revealed a biphasic time-dependent response of TSH mRNA (5). Using the technique of cytoplasmic dot-blot hybridization with complementary (c) DNA probes encoding the TSH β- and α-subunits we have demonstrated a significant increase in pituitary cytoplasmic levels of TSH β and α mRNA at 1 and 6 hours after administration of T3 to hypothyroid animals. The predicted inhibitory effect was evident only at 48 and 72 hours. This response may reflect direct but variable time-dependent influences of T3 on TSH gene transcription. Alternatively the early increase in TSH mRNA may be related to widespread stimulation of gene transcription after administration of T3 to the hypothyroid animal, T3-dependent synthesis of the putative inhibitory protein resulting in a delayed fall in TSH mRNA. It is of interest that a variable dose-dependent relationship for T3 and T4 and TSH production has been documented in studies of thyrotrophic tumour cells in vitro, low doses exerting a stimulatory effect and higher doses an inhibitory effect.

Our studies of the rat in vivo further defined the effects of thyroid status on cytoplasmic concentrations of TSH β and α mRNA, thus indicating regulation of the pre-translational stages of the TSH biosynthetic pathway by thyroid hormones. Additional studies performed in our laboratory have examined the role of other factors which may interact with T3 regulation of TSH gene transcription. We have shown for example that pretreatment of hypothyroid animals with oestrogen abolishes the early rise in TSH β and α mRNA seen after T3 administration (6).

It has been suggested that interaction of the thyroid hormone nuclear receptor complex with genomic DNA determines the regulatory actions of thyroid hormones. It is proposed that interaction of nuclear receptors with specific sequences of DNA close to structural genes regulates their transcriptional activity. Recent experiments have examined whether the 5′ flanking region of the rat GH gene contains a DNA element which could mediate thyroid hormone control (7). A chimeric gene was constructed comprising a 1.8 kilobase fragment of the 5′ flanking region of the GH gene ligated to DNA containing the structural gene for the enzyme xanthine-guanine phosphoribosyltransferase (XGPT). GH producing GC cells were transfected with this chimeric gene yielding stable transformants which demonstrated marked stimulation of XGPT mRNA and XGPT activity by T3. This study therefore provides evidence that the 5′ flanking region of the rat GH gene contains sequences which confer T3 regulation. Further studies have pointed to sequences conferring regulation of the human TSH α-subunit gene by T3 as well as cyclic AMP and dexamethasone (8). These studies used a gene construct comprising the 5′ sequences of the human α gene ligated to the bacterial gene encoding chloramphenicol acetyltransferase (CAT), and transfected rat GH3 cells. Cyclic AMP and dexamethasone increased CAT gene expression whereas T3 decreased CAT expression, indicating that the 5′ sequences mediate regulation of the α-gene promoter by these factors. Characterization of the nuclear-regulatory proteins, including the glucocorticoid and T3 receptors should help our understanding of DNA - protein interactions and of the control of gene activation and repression.

The human glucocorticoid receptor cDNA has been sequenced and shown to be functionally active. Sequence analysis of the receptor showed it to be related to the product of the v-erb-A oncogene of avian erythroblastosis virus. This led to the proposal that the steroid receptors and the erb-A oncogene products share a common origin, and that the erb-A products may also be proteins that bind to DNA enhancer elements. The similarity of steroid and thyroid hormone actions led to the examination of the possibility that the erb-A protein may itself be the thyroid hormone receptor. In a series of elegant experiments it was demonstrated that the translation product from the human c-erb-A cDNA possesses intrinsic thyroid hormone binding activity, characteristic of the native thyroid hormone receptor molecule (9). The evidence for the c-erb-A gene coding for a thyroid hormone receptor includes: (a) molecular weight of c-erb-A protein is the same as that of the previously characterised receptor, (b) the protein binds T3 and its analogues with affinities typical of the T3 receptor, (c) binding activity can be removed with anti-erb-A antibodies, and (d) both the thyroid hormone receptor and c-erb-A are expressed at low levels in many tissues. Further functional characterisation is awaited.

Our understanding of the molecular basis of thyroid hormone action has increased substantially with the recent developments in molecular biology and is an area of potential practical interest in both the diagnosis and management of thyroid disorders. It becomes clear from these observations that it is extremely unlikely that a single biochemical parameter will be identified that will reflect thyroid hormone function uniquely, and that a given level of serum T4 or T3 will always be associated with the same biochemical state of the tissues.

REFERENCES

1. Oppenheimer JH. Thyroid hormone action at the nuclear level. Annals of Internal Medicine 1985; 102: 374-384.
2. Kumara-Siri MH, Shapiro LE, Surks MI. Association of the T3 nuclear receptor with the nuclear matrix of cultured growth hormone producing rat pituitary tumour cells. Journal of Biological Chemistry 1986; 261: 2844-2852.
3. Shupnik MA, Ardisson LJ, Meskell MJ, Bornstein J, Ridgway EC. T3 regulation of thyrotropin subunit gene transcription is proportional to T3 nuclear receptor occupancy. Endocrinology 1986; 118: 367-371.
4. Melmed S, Park J, Hershman JM. Triiodothyronine induces a transferable factor which suppresses TSH secretion in cultured mouse thyrotropic tumour cells. Biochemical Biophysical Research Communications 1981; 98: 1022-1028.
5. Franklyn JA, Wood DF, Balfour NJ, et al. Effect of hypothyroidism and thyroid hormone replacement in vivo on pituitary cytoplasmic concentrations of thyrotropin (TSH) β and α-subunit mRNAs. Endocrinology 1987; 120: 2279-2290.
6. Franklyn JA, Wood DF, Balfour NJ, Ramsden DB, Docherty K, Sheppard MC. Modulation by oestrogen of thyroid hormone effects on thyrotrophin gene expression. Journal of Endocrinology 1987; in press.
7. Casanova J, Copp RP, Janocko L, Samuels HH. 5'-flanking DNA of the rat growth hormone gene mediates regulated gene expression by thyroid hormone. Journal of Biological Chemistry 1985; 260: 11744-11748.
8. Gurr JA, Kourides IA. Regulation of the transfected human glycoprotein hormone alpha subunit gene promoter. Endocrinology 1986; 119: T30 (abstract).

9. Weinberger C, Thompson CC, Ong ES, Lebo R, Gruol DJ, Evans RM. The c-erb-A gene encodes a thyroid hormone receptor. Nature 1986; 324: 641-646.

METHODS OF ANALYSIS OF THYROID-RELATED HORMONES

J.G. Ratcliffe

Wolfson Research Laboratories, Queen Elizabeth Medical Centre, Birmingham, B15 2TH, UK

The rapid development and commercial exploitation of new immunoassay techniques continues to have a major impact on serum assays for thyroid-related hormones. This revieuw will focus on methods for TSH and 'free' thyroid hormones with brief mention of methods for total Thyroxine (T4) and Triiodothyronine (T3).

THYROTROPHIN (TSH)

a. Radioimmunoassay (RIA)

Until recent years RIA was the only method for TSH suitable for routine clinical application. TSH RIA is based on established principles of limited reagent assays. Particular attention needs to be given to selection of high quality reagents, standardisation (including possible matrix effects) and optimised protocols.

Antisera must have high avidity, titre and specificity with respect to structurally related glycoprotein hormones. Thus, cross reaction with FSH and LH should be less than 1-2% and with HCG less than 0.1% to minimise specificity problems in post menopausal women and pregnancy. The requirement for high avidity is best met by polyclonal rather than monoclonal antisera. High quality labelled hormone requires iodination of highly purified natural human TSH with ^{125}I to a specific activity equivalent to one atom of carrier free iodine per molecule TSH (76 mCi/mg, 2.8 mBq/mg) which is a compromise between immunoactivity stability and count rate. Satisfactory methods include chloramine-T, solid phase lactoperoxidase, and N-bromosuccinimide. The ionidated peptide must be purified before use: suitable column materials are Sephadex G100 or Ultragel ACA 54.

Standardisation can be difficult. While the choice of TSH preparation is straightforward (2nd International Reference Preparation human TSH code 80/558), selection of matrix is not. Potentially suitable matrices are animal sera (e.g. horse serum), T3 - suppressed or thyrotoxic human sera, or TSH immunoadsorbed sera. Each individual donation of human serum must be tested for hepatitis antigen and human immunodeficiency virus antibody. Suitability of each batch should be assessed by comparison of zero antigen binding with that in thyrotoxic or T3-suppressed sera, parallellism of serial dilutions of test sera and standards in matrix and by quantitative recovery of added TSH from matrix. A useful external check is to compare values obtained on sera previously circulated in external quality assessment schemes which have reliable target values.

A wide variety of phase separation methods is available. 'In house' methods usually favour double antibody or polyethylene glycol (PEG)-assisted double antibody techniques, whereas commercial kits often employ magnetic solid phase or coated tube methods to avoid centrifugation. All TSH RIAs require extended incubation protocols and often delayed addition of tracer to maximise sensitivity. Overall assay time is usually at least 24 h with the more sensitive assays requiring several days. Acceptable precision usually extends only over a limited range (about 10 fold).

Despite refinements, the major limitations of TSH RIAs have not been overcome. These relate to unreliability in identifying as undetectable sera known to contain very low TSH levels (e.g. thyrotoxic) and imprecision near the limit of detection (e.g. low euthyroid range). Hence most RIAs cannot distinguish subnormal from normal levels. These analytical limitations coupled with prolonged assay times have provided a major impetus for the development of more sensitive, robust and rapid methods.

b. Immunometric assay (IMA)

Immunometric assays offer potential solutions to the problems of sensitivity, precision and speed (1). This approach differs fundamentally from RIA in that labelled specific antibody is used in excess to bind to TSH. In practice the two site protocol has advantages of specificity, convenience and ruggedness. In this, the labelled complex is insolubilised by reaction with a second antibody (also specific for TSH) linked to solid phase. This enables the solid phase linked complex to be separated and washed thoroughly before measuring the response, thus minimising background 'noise' and thereby enhancing sensitivity. Specificity is also improved since both labelled and solid phase antibody must bind to TSH to generate the signal. Fragments or subunits of TSH which do not process both binding epitopes are not detected. Because reagents are in relative excess, reaction times are short compared to RIA (hours vs days) and working ranges extended (1000 vs 10 fold).

Two site assays also offer considerable flexibility in design. TSH can be reacted with solid phase antibody followed by reaction of the product with labelled antibody. Alternatively TSH can be reacted first with labelled antibody followed by reaction with solid phase antibody or both antibodies can be incubated simultaneously with analyte.

Antibodies must be carefully selected for compatibility and specificity. One of the antibodies must be as specific for TSH as that required for RIA. A major factor in facilitating the practical implementation of immunometric assays is the availability of monoclonal antibodies. These can be readily purified in large amounts to provide a consistent specific reagent for labelling. Although specific antibodies can be prepared from polyclonal antisera by immunoselection on solid phase antigen, the procedure requires large volumes of high avidity antisera, is technically complex and requires rigorous quality control. In addition, non-specific bindig may be higher and more variable than with monoclonal antibodies with consequent adverse effects on assay sensitivity. However, the higher avidities of polyclonal antisera and increasingly refined methods of antibody purification may offer advantages for commercial reagent producers. For the solid phase reagent, both polycloncal and monoclonal antibodies are used successfully. Large volumes of either are required as the solid phase antibody must be present in excess to avoid reduction in binding at high analyte concentrations.

The use of monoclonal antibodies for both reagents may present a theoretical problem if their unique specificities distinguish between different molecular forms of TSH (isohormones). The assay may then recognise only a proportion of the molecular forms which it is desirable to detect. Depending upon the differences in isohormonal forms in endogenous and standard TSH, the assay may show consistent bias characteristic of the particular pair of monoclonal antibodies employed.

High sensitivity is required in TSH assays for clinical application since subnormal and normal levels must be distinguished. The ultimate sensitivity of any immunoassay is determined by four factors: antibody avidity, non specific binding (misclassification), experimental imprecision and specific activity of the label. The relative importance of these factors differs between IMAs and RIAs. Thus, high antibody avidity is not so critical in IMAs and it is possible to achieve good sensitivity in IMAs using antibodies which yield insensitive RIAs. Low and reproducible non specific binding is critical in IMA so that the signal is generated against an insignificant background. Solid phases with antibody capacity and consistent physical characteristics are required. Commercial kits show much ingenuity in their choice of suitable solid phases e.g. walls of tubes, microtitre wells, magnetisable particles or solid beads.

Given a suitable solid phase with low non-specific binding, assay sensitivity is then related to specific activity of tracer. While it is possible to label antibodies with ^{125}I to high specific activity, the immunoactivity of such labels deteriorates rapidly, thus preventing realisation of the high sensitivity of detection of ^{125}I (10^{-18} moles). This practical problem together with the requirement to avoid radioactivity, has encouraged the development of alternative non isotopic labels which are stable and can be measured at very low concentration.

An increasing choice of non isotopic labels is now available for TSH assays:

i. Chemiluminescence. Certain acridinium esters couple spontaneously to antibody under mild conditions and the conjugate has high specific activity and retains immunoactivity. Acridinium compounds can be stimulated to produce light via a chemiluminescent reaction. N-methyl acridone is produced in an electronically excited state which relaxes to its ground state with emission of photons which can be quantified by a photon counter. The chemiluminescent reaction, initiated by hydrogen peroxide, occurs after dissociation of label from antibody.

ii. Time resolved fluorescence. Antibodies can be coupled to lanthanides such as europium can be coupled to antibodies without loss of immunoactivity and stability. In the coupled state the label does not fluoresce, but an enhancement solution dissociates bound europium from the antibody linked chelate which then fluoresces under appropriate conditions. The europium label is chelated with B-diketone held in a triton X-100 micelle. In this hydrophobic environment, fluorescence is generated by exciting with light at 340 nm and measuring emission at 613 nm. The large Stokes shift and long half life of emission of the europium chelate (microseconds) in the excited state enables a pulsed light source to be used so that emitted light is measured after the non-specific background fluorescence has decayed. Each well can thus be measured many times in the one second measurement period.

iii. Enzyme amplification. An enzyme label is used to provide a trigger substance for a secondary enzyme system present in excess that generates a large quantity of coloured product. In the TSH assay, alkaline phosphatase has been employed as the enzyme label coupled to a redox amplifier involving cycling of reducing equivalents and generation of a formazan dye with a strong absorbance in the visible region of the spectrum. The amplifying system is added at a fixed time after the primary substrate in order to minimise non-specific background.

iv. Enhanced chemiluminescence. An enzyme (horse radish peroxidase) linked to antibody is used as label to oxidise luminol with consequent generation of a light signal. Light emission is massively enhanced and stabilised by addition of an enhancer (e.g. benzothiazole), while background is reduced.

The new generation of high sensitivity TSH assays provides a major advance with detection limits of 0.1 mU/L or less, extended working ranges, brief assay times and technical simplicity. Performance in the UK external quality assessment scheme (EQAS) has demonstrated quantitative recovery overall at low levels. This has allowed all specimens with TSH levels >1 mU/L to be used for the calculation of cumulative statistics. The majority of laboratories using these methods now achieve bias less than + 10% and variability of bias below 10%. However, there remain some method related differences in bias due to differences in calibration and possibly specificity arising from the use of particular monoclonal antibodies.

Disadvantages of present IMAs include the cost of commercial kits and dedicated instrumentation for non-isotopic assays. Depending upon local arrangements for reimbursement of laboratories, this may slow the rate at which non-isotopic assays are adopted. Occasionally, interfering factors have been identified. For example, antibodies in test sera which bind to mouse IgG can give falsely elevated results. Their effect can often be reduced by including animal sera in assay diluent.

Further experience is required in interpreting low TSH levels in patients with non-thyroidal illness, and in patients receiving centrally acting drugs. It remains to be established what sensitivity is required clinically and how reliable the new methods are in widespread routine practice. Nevertheless, there has been a significant move of laboratories to IMAs replacing in house and kit RIAs with IRMAs and, to a lesser extent, non isotopic IMAs.

'FREE' THYROID HORMONES (FT4 AND FT3)

Traditional methods for free thyroid hormones are indirect in that the free hormone level is inferred from two measurements (total hormone and a measure of serum protein binding (e.g. FTI, T4: TBG ratio) or the fraction free (e.g. equilibrium dialysis, or ultrafiltration)). Methods of quantitating the free fraction are technically complex and subject to error due to impurities in the tracer, and affected by ionic composition of buffers, pH and temperature. More recently a range of direct methods for free thyroid hormones has been introduced. Equilibrium dialysis of serum in specially constructed dialysis chambers followed by direct RIA of thyroid hormones in the dialysate (ED/RIA) avoids problems associated with tracer impurities but requires high avidity antisera and is technically complex. Newer techniques have concentrated on avoiding the dialysis step by a variety of methods such as sequential two stage back titration, microencapsulation of antibody or using a labelled hormone analogue (2). There has been much controversy about whether some of these approaches are analytically valid since conventional criteria such as recovery experiments are not applicable. The following tests of validity have been proposed: (1) dilution of serum. Although this perturbs the equilibrium originally present in undiluted serum, mathematical modelling of T4 binding equilibria with all the serum bindig proteins suggests that dilution up to 50 fold will not significantly alter FT4 concentrations. Apparent FT3 concentrations are reduced by up to 20% because T3 is less strongly protein bound. (2) The free hormone concentration should be accurately reflected in sera in which the normal relationship between bound and free moieties is disturbed e.g. when binding proteins are altered.

In the two stage sequential incubation method, a trace amount of solid phase thyroid hormone antibody is added to serum in the first stage after which serum is removed by washing. In the second stage, labelled hormone is added to back titrate the unoccupied antibody binding sites. The fraction of labelled hormone bound to solid phase is then inversely proportional to the free hormone concentration and can be quantitated by comparison with serum standards with known free hormone concentration determined independently. This method appears to be analytically valid, survives the dilution test and agrees well with ED/RIA methods in specimens with abnormal binding proteins (e.g. pregnancy, abnormal thyroxine binding globulin (TBG), albumin). However, the sequential incubation and washing steps are inconvenient and the procedure is subject to assay drift and errors arising from variations in timing of incubation and wash stages. Technical improvements may be expected to overcome these difficulties so that this method may become more suitable for clinical use.

In the microencapsulated antibody method, anti-T4 serum is enclosed in microspheres surrounded by a semi permeable membrane. Labelled T4 is bound to this antiserum and in the presence of serum the amount of labelled T4 displaced from the antibody to the outside of the microsphere, is proportional to the FT4 concentration in the sample. Although this method is little affected by serum dilution it does not appear to be completely unaffected by endogenous serum binding proteins and is relatively imprecise.

In the labelled analogue method a labelled hormone analogue and solid phase antibody are added concurrently to the serum sample. Provided the amount of antibody employed does not perturb the equilibrium significantly (less than 1% total T4 bound to antibody) and the analogue does not bind serum proteins but does bind to antibody, the proportion of label bound is inversely proportional to the free hormone concentration. Quantitation requires comparison with serum standards calibrated independently. Labelled analogue methods are widely available as commercial kits and are simple, quick and precise. However, there remains controversy about their analytical validity which centres on the extent to which the labelled analogues so far employed interact with serum binding proteins.

These methods are affected by serum dilution, and may not be reliable in sera with extremes of albumin concentration, and in pregnancy, non-thyroidal illness and heparin therapy. Results by current methods are artefactually raised in familial dysalbuminaemic hyperthyroxinaemia due to binding of the analogue to the abnormal high affinity binding site on albumin. However, in sera with normal binding proteins and conditions in which altered TBG is the major abnormality the method gives results which correlate well with ED/RIA methods. These effects suggest that labelled analogue methods do not measure free hormones per se but are another type of free thyroid hormones index with advantages over the conventional FTI (calculated from total hormone and an uptake test) in abnormal TBG states. These reservations about analytical validity and the misleading results in certain clinical situations suggest caution in recommending their suitability as sole first line tests in assessing thyroid function (3).

With the exception of the direct equilibrium methods (e.g. ED/RIA) the other direct methods require independent calibration. Estimates of free hormone levels in normal subjects by ED/RIA show reasonable agreement (mean FT4 15 pmol/L mean FT3 7 pmol/L). This contrasts with the considerable variation reported using indirect methods (mean FT4 25-50 pmol/l, mean FT3 3-7 pmol/L).

A major difficulty in obtaining useful information from EQAS is the lack of suitable target values which reflect the lack of a readily available reference method and the inapplicability of simple tests of validity (e.g. recovery). In euthyroid sera with or without added T4 the relationship of free to total T4 in EQAS specimens agrees well with that obtained using ED/RIA to measure FT4. However significant method-related differences are observed in serum pools with abnormal binding proteins and on modest dilution of serum reflecting the analytical invalidity of current labelled analogue methods.

TOTAL THYROID HORMONES (TOTAL T4, T3)

Since specific antisera for T4 and T3 became available in the early 1970's, RIA using ^{125}I labelled tracers has been the method of choice for clinical purposes. The assay of these hormones in unextracted serum requires their displacement or release from serum binding proteins by agents such as 8-anilino-1-naphthalene sulphonic acid, salicylate and/or alkaline buffers such as barbitone pH 8.6 or glycinate pH 10.5. Total hormone is then measured by conventional competitive immunoassay with separation of bound and free fractions and counting of the bound fraction. A wide variety of 'in-house' and commercial methods is available, differing principally in the source of reagents, choice of displacing agents, and phase separation.

Performance in the UK EQAS for total T4 is now generally satisfactory with few laboratories having a cumulative bias greater than ± 10% or variability of bias >15%. The All Laboratory Trimmed Mean appears to be a valid benchmark for assessing bias on the basis of its quantitative recovery and stability. 'In-house' double antibody, polyethylene glycol (PEG) or PEG assisted double antibody methods perform as well as the best commercial methods although some of the latter avoid centrifugation by use of magnetisable particles or other solid phases. A significant recent development is the availability of commercial methods using non-isotopic end points. Some are classical competitive assays requiring phase separation, but employing enzyme or fluorescent labels which generate colorimetric, fluorescent or chemiluminescent end points. Other more novel systems are homogeneous, in which the end point is measured without requiring phase separation (e.g., enzyme multiplied immunoassay, fluorescence transfer, fluorescence polarisation). Such methods are of particular interest to clinical laboratories since they are amenable to full automation.

Performance of total T3 assays is less satisfactory than for total T4 in UK EQAS, with 5-10% of laboratories having a cumulative bias greater than ± 20% or variability of bias >20%. There are significant method related differences which appear to be due mainly to the effect of different iodothyronine-free matrices and preparation of T3 standards. Because of uncertainties in the molar extinction coefficient of T3 and variable hydration of pure T3, standard material must be thoroughly desiccated before reconstitution and assignation of standard values by gravimetry. For 'in-house' methods phase separation by PEG alone is less satisfactory than by PEG-assisted double antibody or double antibody alone. Non-isotopic methods are as yet less available for total T3 than for total T4.

CONCLUDING COMMENTS

Recent trends in methods for thyroid-related hormones have been identified. Methods for TSH have progressed significantly following the development and widespread introduction of immunometric assays. This development was facilitated by the availability of monoclonal antibodies and has been increasingly exploited using non-isotopic endpoints. These methods offer high sensitivity, wide working ranges and brief assay times. Methods for 'free' thyroid hormones have also developed significantly towards convenient, simple, direct assays. However, the analytical validity of current labelled analogue methods remains uncertain. Methods for total thyroid hormones have developed towards homogeneous non-isotopic assays which are potentially suitable for full automation.

The new generation of TSH assays is particularly important in that it opens up for investigation areas of pathophysiology which were previously inaccessible. In the diagnostic arena, the new methods enable the cost effectiveness of alternative strategies for assessing thyroid status to be reappraised. It should be remembered however, that despite these advances, there has been no corresponding progress in the development of simple tests to assess tissue responses to circulating thyroid hormones. Results obtained with the new methods may thus be difficult to interpret in some circumstances (e.g. patients on thyroid replacement therapy, non-thyroidal illness).

ACKNOWLEDGEMENT

I am grateful to Mrs Wendy Baldwin for expert secretarial help.

REFERENCES

1. Woodhead JS, Weeks I. Circulating thyrotropin as an index of thyroid function. Ann Clin Biochem 1985; 22: 455-9.
2. Pearce CJ, Byfield PG. Free thyroid hormone assays and thyroid function. Ann Clin Biochem 1986; 23: 230-7.
3. Alexander NM. Free thyroxin in serum: labeled thyroxin-analog methods fall short of their mark. Clin Chem 1986; 32: 417.

THYROTROPIN SECRETION IN HEALTH AND DISEASE

G. Faglia, P. Beck-Peccoz, M. Nissim, G. Medri and G. Piscitelli

Dept. of Endocrinology, School of Medicine, University of Milan, Italy

Thyrotropin (TSH) is a glycoprotein hormone (m.w. 28,000 daltons) composed of two dissimilar, non covalently bound subunits (α and β). Within a species the α-subunit is virtually identical to that of the other glycoprotein hormones (LH, FSH, and hCG), while the β-subunit confers biological specificity. The two subunits are independently synthetized in the thyrotrophs, and then assembled to form complete TSH; only minute amounts of both subunits are secreted in the free form and are measurable in the peripheral blood. The individual subunits seem to be devoid of biological activity (1).
TSH secretion is regulated by the balanced action of stimulatory and inhibitory factors (2, 3). The inhibitory control is mainly exerted by thyroid hormones (TH), especially by triiodothyronine (T3) derived from the intrathyrotroph conversion of thyroxine (T4) into T3. Thyrotropin-releasing hormone (TRH) exerts its powerful stimulatory effect on TSH synthesis and release after binding to specific receptors on thyrotroph plasma membrane. Dopamine, somatostatin, opioid peptides, central nervous system, cholecystokinin, corticosteroids and arginine vasopressin on one hand, and norepinephrine and estrogens on the other hand, also contribute to the inhibitory or stimulatory control of TSH secretion, respectively. TSH secretion shows a circadian rhythm with a nadir in the afternoon and a zenith at night.
TSH is a heterogeneous molecule both in its physico-chemical characteristics and in its biological properties. Recently, evidence has been provided in humans that, under particular circumstances, circulating TSH may vary its biological activity (4).
Due to the dominant regulatory effect of TH, the thyroid status is generally reflected by opposite variations in TSH and TH concentrations in the peripheral plasma. However, hypothalamic-pituitary lesions, or alterations of TSH regulatory mechanisms, or changes in TSH bioactivity may also result in disorders of thyroid function.

The recent introduction of the ultrasensitive immunoradiometric (IRMA) assay for TSH has substantially improved the possibility of adequately investigating TSH secretion in health and disease. This paper will briefly review the clinical value of the estimation of serum concentrations of TSH and its subunits, as well as the assessment of TSH bioactivity.

IMMUNORADIOMETRIC ASSAY OF THYROTROPIN

TSH-IRMA is based on the use of two monoclonal antibodies, each reacting with specific epitopes of the TSH subunits. This confers high specificity to the assay. In particular, both α- and β-subunit do not crossreact in TSH-IRMA (Figure 1), enabling the correct estimation of serum TSH levels in conditions where the serum subunit levels are high (i.e patients with α-subunit-secreting pituitary tumors, choriocarcinomas, hepatocarcinomas, chronic renal failure, postmenopausal women, primary hypothyroidism, etc.). TSH-IRMA is not affected by the presence of circulating heterophilic antibodies (anti-rabbit gammaglobulins) which can cause overestimation of serum TSH levels by conventional RIA methods (Figure 2). In contrast, the presence of circulating anti-TSH antibodies may cause underestimation of serum TSH levels (Figure 3). TSH-IRMA is highly sensitive (0.07 μU/ml), accurate, and reproducible (intra- and interassay coefficients of variation (CV) <10% over the range from 0.3 to 200 μU/ml).

MEASUREMENT OF α- AND β-SUBUNITS OF TSH

Circulating free α- and β-subunits of TSH are usually measured by RIA, utilizing materials distributed by the National Pituitary Agency, NIDDK (Bethesda, MD, USA). Only recently, commercial kits have become available for α-subunit assay. The sensitivity is about 0.2 ng/ml and 0.1 ng/ml for α- and β-subunit, respectively. The intra- and interassay CV are <5% for both the subunits. The cross-reaction of TSH in the α-subunit assay is about 4%, whereas in the β-subunit assay is as low as 0.5%. α-subunit measurement and the calculation of the α-subunit/TSH molar ratio are particularly useful for discriminating neoplastic from non-neoplastic inappropriate secretion of thyrotropin (IST) (5). However, attention should be paid to compare the observed values with appropriate control groups matched for age and sex, as serum α-subunit levels show a wide variability throughout life, according to the gonadotropic activity.
The measurement of serum β-subunit levels may be of some importance in patients with hypothalamic idiopathic hypothyroidism and bioinactive TSH who have been reported to have high levels of circulating β-subunit (6).

TSH BIOASSAY

The methods for the evaluation of TSH bioactivity are technically difficult and time consuming. The cytochemical bioassay (7) is highly sensitive, but relatively non-specific, whereas methods based on the measurement of the adenylate cyclase activity in human thyroid membranes

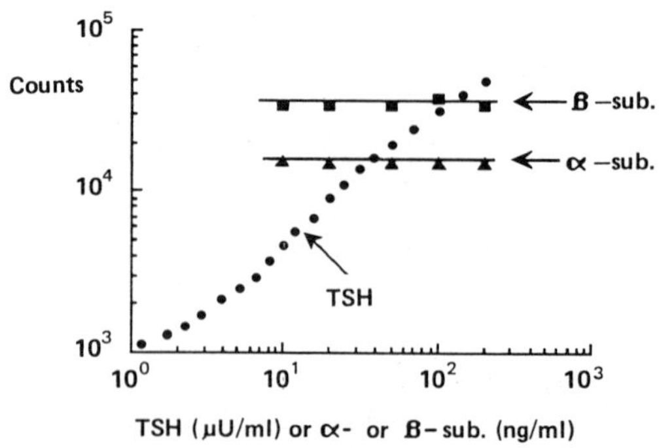

Fig. 1. Absence of cross-reactivity of α- and β-subunits of thyrotropin in TSH-IRMA.

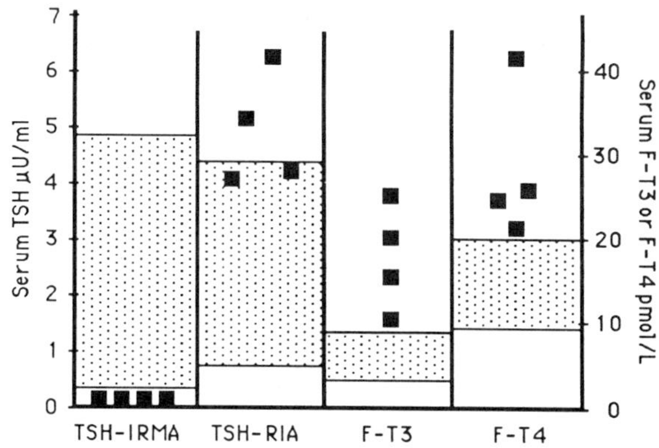

Fig. 2. The presence of heterophilic anti-rabbit antibodies in sera from 4 hyperthyroid patients, which caused TSH overestimation by RIA, did not affect the correct estimation of suppressed serum TSH levels by IRMA. It is worth noting that these patients would have been misdiagnosed by RIA as affected with "Inappropriate secretion of TSH".

are more specific, but far less sensitive, and require preliminary purification and concentration of serum TSH by immunoaffinity chromatography (4). The recently introduced bioassay methods, based on the measurement of cAMP formation and iodide trapping by a continuous line of functional rat thyroid cells (FRTL-5) (8), which appears to be very sensitive, specific and precise, are very promising (9).

CLINICAL VALUE OF TSH, α- AND β-SUBUNIT, AND TSH BIOACTIVITY MEASUREMENTS

As already mentioned, due to the dominant regulatory effect of TH, the thyroid status is generally reflected by opposite variations in TSH and TH concentrations in the peripheral plasma.
In this light, because of its high sensitivity that allows the discrimination of suppressed from non-suppressed serum TSH levels, TSH-IRMA has been proposed as "first line" test of thyroid function (10). Figure 4 shows serum IRMA-TSH levels recorded in normal individuals and in patients with various thyroid disorders. The measurement of circulating α- and β-subunit levels as well as the estimation of serum TSH bioactivity are of clinical value only in particular conditions and will be mentioned when appropriate.

Normal subjects

TSH-IRMA has substantially confirmed what already known about serum TSH in normals, including circadian fluctuations and changes after stimulatory or inhibitory maneouvres.

TRH test

It has been shown that serum TSH response to TRH is strictly correlated with basal serum TSH levels. Because of its high sensitivity TSH-IRMA makes the TRH test useless in most patients, as those who have suppressed basal serum TSH-IRMA invariably show no TSH response to TRH, and those who have slightly high basal serum TSH show exaggerated responses. The usefulness of TRH testing will be confined to a minority of patients including those with hypo- or hyperthyroidism secondary to hypothalamic-pituitary disorders where TRH test remains mandatory.

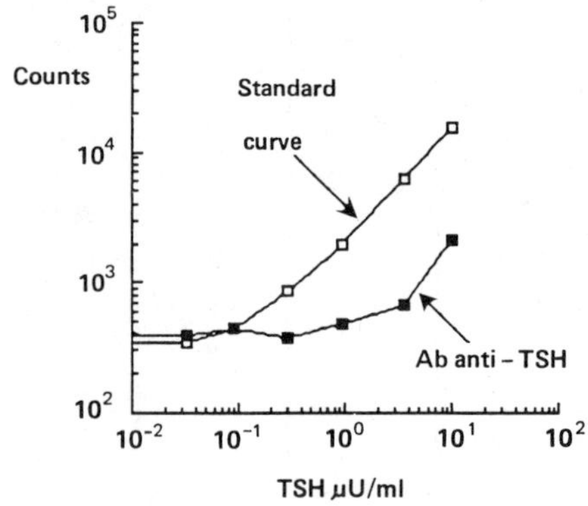

Fig. 3. Effect of the presence of circulating anti-TSH antibodies on serum TSH levels measured by IRMA.

Non toxic goiter

In euthyroid patients with non toxic goiter, serum TSH-IRMA levels, although generally normal, show a wider range of distribution, reflecting the tendency toward hypo- or hyperthyroidism.

Hyperthyroidism

Thyroid hyperfunction, either due to thyroid autonomy or stimulation by immunoglobulins, is characterized by high TH levels and suppressed TSH secretion. TRH administration does not elicit any TSH increase. TSH-IRMA clearly discriminates suppressed from non-suppressed serum TSH levels. Thus, the finding of unsuppressed serum TSH-IRMA levels in hyperthyroid patients, should suggest the diagnosis of inappropriate TSH-secretion. TSH-IRMA is particularly useful in the diagnosis of amiodarone-induced hyperthyroidism. In fact, the finding of suppressed serum TSH levels is crucial, as high serum T4 levels are usually found in euthyroid amiodarone-treated patients and serum T3 levels may not be clearly elevated in hyperthyroid patients for the drug-induced impairment of the enzymatic conversion of T4 into T3.

TSH-induced hyperthyroidism

"TSH-induced hyperthyroidism" or "Central hyperthyroidism" is part of the "Inappropriate Secretion of Thyrotropin" (IST) syndrome (11, 12). This disorder is characterized by elevated TH in concentrations and unsuppressed or inappropriately high serum TSH levels. IST may result from either selective resistance to TH action (non-neoplastic inappropriate secretion of TSH: nnIST) or TSH-secreting pituitary tumors (neoplastic IST: nIST). The recognition of this rare disorder and the differentiation of neoplastic from non-neoplastic form is of clinical importance, as their management differs from both each other and from that of classical hyperthyroidism.

Moreover, central hyperthyroidism must be distinguished from other conditions with apparently high TH levels and unsuppressed TSH, including those due to methodological problems with TSH or TH measurements, such as the presence of circulating heterophilic antibodies, anti-TH antibodies, qualitative and quantitative alterations of TH transport proteins, alterations of the peripheral conversion of T4 into T3, and generalized resistance to TH action. The current availability of ultrasensitive and highly specific TSH-IRMA which clearly distinguish between suppressed and unsuppressed TSH levels, the use of "direct and absolute" methods for the

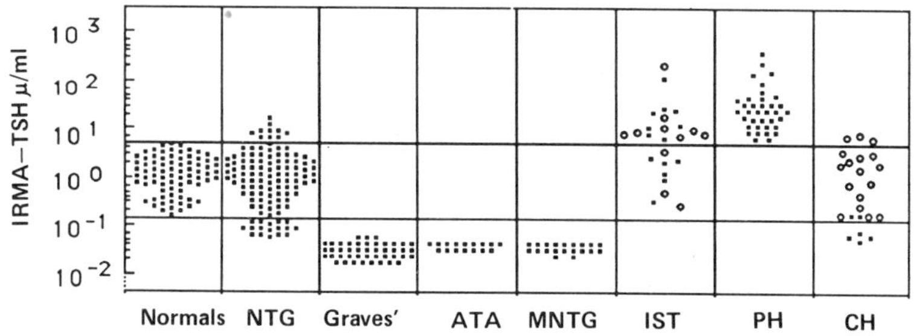

Fig. 4. Serum IRMA-TSH levels in normal individuals and in patients with altered thyroid function. NTG: non-toxic goiter; Graves': Graves' disease; ATA: hyperfunctioning autonomous thyroid adenoma; MNTG: multinodular toxic goiter; IST: inappropriate secretion of TSH (neoplastic: closed symbols, non-neoplastic: open symbols); PH: primary thyroid failure; CH: central hypothyroidism (of pituitary origin: closed symbols, of hypothalamic origin: open symbols).

measurement of circulating free TH concentration (13, 14, 15) and the evaluation of parameters of TH action at the peripheral tissue level, enable the definite recognition of central hyperthyroidism. The distinction between nnIST and nIST rests on several clinical, neuroradiological, and laboratory criteria, the most useful of which are the measurement of serum free α-subunit and the α-subunit/TSH molar ratio, both increased in nIST, and the TSH response to TRH or high doses of T3, abolished in nIST, but qualitatively normal in nnIST (Table 1). A careful evaluation of the hypothalamic-pituitary function is also important, as in about one third of patients with nIST, TSH hypersecretion is associated with other anterior pituitary hormone hypersecretion by mixed tumors. Increased bioactivity of circulating TSH has been documented in two patients with nIST who had basal serum TSH levels within the normal range (16, 17). The measurements of serum levels of TSH, α-subunit, and free TH are valid tools in assessing the effects of the different treatments proposed for IST. By these means, it has been shown that about 41% of patients with nIST are cured by pituitary surgery followed by radiant therapy (12), that the long-acting analog of somatostatin SMS 201 - 995 is effective in nIST, but not in nnIST because of escape (18), that bromocriptine is only seldom effective, and that triiodothyroacetic acid (TRIAC, a T3 derivative with potent TSH suppressive effect, but practically devoid of thyromimetic effects on peripheral tissues) is able to lower serum TSH in nnIST with restoration of euthyroidism (19).

Primary thyroid failure

Primary hypothyroidism is characterized by high serum TSH and low serum TH levels. However, the "thyrostat" is so finely regulated that even small diminutions in serum TH concentration, still within the normal range, may cause basal serum TSH to rise. The high specificity of TSH-IRMA renders more accurate the discrimination of normal subjects from patients with subclinical hypothyroidism. Serum α- and β-subunit levels are elevated in patients with primary thyroid failure. The α-subunit/TSH molar ratio is always normal, even in those patients who develop sella enlargements (so called "feedback pituitary TSH-secreting tumors"). It has been recently shown that in patients with thyroid failure there is an inverse correlation between circulating TH levels and the biological activity (assayed with the cAMP formation method on FRTL-5 cells) of serum TSH. This implies that TH can directly or indirectly modulate TSH bioactivity through conformational changes, probably related to modifications of the carbohydrate moiety of the TSH molecule (9).

Table 1. Clinical and biochemical profile, and serum TSH response to dynamic tests in patients with hyperthyroidism. Data are taken from the literature (12)

	Neoplastic (n = 76)	Non-neoplastic (n = 37)
Age (range)	17-69	4-80
Sex (F/M)	43/33	28/ 8
Goiter*	43/52	22/23
Sella alterations*	61/66	0/37
Exophthalmos*	5/76	0/37
Acropachy*	0/76	0/37
TSAb, TBI, or LATS*	1/26	1/11
TSH, μU/ml (range)	1.2-480	1.0-480
α-subunit, ng/ml (range)	0.5-105	0.5-7.7
α-subunit/TSH molar ratio (range)	0.5- 98	0.08-4.4
α-subunit/TSH molar ratio > 1*	35/38	5/20
TSH response to TRH*	17/66	31/31
T3*	6/38	25/25
ADA°*	1/12	4/ 4
ATD°*	14/20	9/11
DAA°*	7/31	6/ 6
CS°*	11/12	15/ 1

* Presence of the sign or response/reported cases. °ADA: antidopaminergic drugs; ATD: antithyroid drugs; DAA: dopamine agonists; CS: corticosteroids

L-T4-treated patients

TSH-IRMA is a valid tool for the evaluation of patients treated with L-T4. In hypothyroid patients on replacement therapy, the restoration of normal serum TSH levels is a good index of the adequacy of L-T4 dosage. In patients treated for thyroid cancer or simple goitre, the effectiveness of L-T4 doses in suppressing TSH secretion can be easily and precisely assessed by TSH-IRMA instead of TRH test.

Central hypothyroidism

Central hypothyroidism may result from either pituitary insufficiency or hypothalamic defects (20). The distinction between the two forms may be achieved by TRH test. Some patients with hypothalamic hypothyroidism have detectable and occasionally elevated basal serum TSH levels and a prolonged and/or exaggerated TSH response to TRH. In some of these cases, TSH lacks biologic activity because of impaired binding to thyroid receptor (4). The findings of elevated circulating free β-subunits and elevated β/α-subunit molar ratio, along with a larger apparent molecular weight of circulating TSH (6), may reflect abnormalities in glycosylation steps leading to alterations in the assembling of TSH subunits. Oral administration of TRH reverts to normal the TSH β/α-subunit molar ratio and the binding property of TSH to its specific thyroid receptor with recovery of biological activity. These findings imply that TRH regulates not only TSH secretion, but also its specific molecular and conformational features required for hormone action.

Non-thyroid illnesses

The knowledge of hypothalamic-pituitary control of thyroid function in non-thyroid illnesses is far to be complete. Serum TSH levels are usually in the normal range, but they may dramatically fall during severe illness (21). In a number of studies serum TSH levels are reported to be above the control mean (22). TSH response to TRH is generally impaired (23) suggesting an adaptation of the thyrotroph to the altered T4 to T3 metabolism and possible alterations of the inhibitory components of the TSH-regulating central mechanisms.

CONCLUSIONS

The investigation of thyrotropin regulation in health and disease has revealed that alterations of the components of hypothalamic-pituitary-thyroid axis, not only can cause quantitative, but also qualitative TSH changes. In this view, the evaluation of serum TSH levels with ultra-sensitive methods is mandatory in patients with disorders of thyroid function and, under particular conditions, should be associated with the measurement of TSH subunits and biological activity. TSH ultrasensitive methods should not be used alone as "first line" test of thyroid function, but always in association with free TH determinations by "direct and absolute" methods, as the sole TSH-IRMA measurement can cause patients with central hyper- or hypothyroidism to be misdiagnosed.

REFERENCES

1. Kourides IA, Vamvakopoulos NC, Maniatis GM. mRNA-directed biosynthesis of α-and β-subunits of thyrotropin. J Biol Chem 1979; 254: 11106-10.
2. Morley JE. Neuroendocrine control of thyrotropin secretion. Endocr Review 1981; 2: 336-396.
3. Hersham JM, Pekary AE. Regulation of thyrotropin secretion. In: Immura H. ed. The pituitary gland. Raven Press, 1985: 149.
4. Beck-Peccoz P, Amr S, Menezes-Ferreira MM, Faglia G, Weintraub BD. Decreased receptor binding of biologically inactive thyrotropin in central hypothyroidism. New Eng J Med 1985; 312: 1085-90.
5. Elewaut A, Mussche M, Vermeulen A. Familial partial target organ resistance to thyroid hormones. J Clin Endocr Metab 1976; 43: 575-81.
6. Faglia G, Beck-Peccoz P, Ballabio M, Nava C. Excess of β-subunit of thyrotropin (TSH) in patients with idiopathic central hypothyroidism due to the secretion of TSH with reduced biological activity. J Clin Endocr Metab 1983; 56: 908-14.
7. Bitensky L, Alaghband-Zadeh J, Chayen J. Studies on thyroid-stimulating hormone and the long-acting thyroid stimulating activity. Clin Endocrinol Oxf 1974; 3: 363.
8. Ambesi-Impiombato FS, Parks LAM, Coon HG. Culture of hormone-dependent functional epithelial cells from rat thyroids. Proc Natl Acad Sci USA 1980; 77: 3455-9.
9. Dahlberg PA, Petrick PA, Nissim M, Menezes-Ferreira MM, Weintraub BD. Intrinsic bioactivity of thyrotropin in human serum is inversely correlated with thyroid hormone concentrations. J Clin Invest 1987; 79: 1388-94.
10. Seth J, Kellett HA, Caldwell G, et al. A sensitive immunoradiometric assay for serum thyroid stimulating hormone: a replacement for the thyrotrophin releasing hormone test? Br Med J 1984; 289: 1334-6.
11. Brenner-Gati L, Gershengorn MC. Thyroid-stimulating hormone-induced hyperthyroidism. In: Immura H. ed. The Pituitary Gland. Raven Press, 1985: 467.

12. Faglia G, Beck-Peccoz P, Piscitelli G, Medri G. Inappropriate secretion of thyrotropin by the pituitary. Hormone Res 1987; 26: 79-99.
13. Immune assays Clinical Chemistry, Churchill Livingstone, 1983: 319.
14. Beck-Peccoz P, Romelli PB, Cattaneo MG, et al. Evaluation of free thyroxine methods in the presence of iodothyronine-binding autoantibodies. J Clin Endocrinol Metab 1984; 58: 736-9.
15. Spencer CA. The comparative clinical value of free T_4 estimation using different methodological approaches. Nuc Compact 1985; 16: 321-7.
16. Desoye G, Schweditsch MO, Pfeiffer KP, Zechner R, Kostner GM. Correlation of hormones with lipid and lipoprotein levels during normal pregnancy and postpartum. J Clin Endocrinol Metab 1987; 64: 704-12.
17. Evans PJ, Weeks I, Woodhead JS, Jones MK, Scanlon MF. Measurement of circulating thyrotrophin concentrations by an immunochemiluminometric assay. J Endocrinol 1985; 104: 42.
18. Massart C, Hody B, Condé D, Leclech G, Edan G, Nicol M. Effets de l'amiodarone in vitro sur les propriétés fonctionnelles du tissu thyroidien humain stimulé par la TSH. Annales d'Endocrinologie 1987; 48: 198.
19. Beck-Peccoz P, Piscitelli G, Cattaneo MG, Faglia G. Successful treatment of hyperthyroidism due to nonneoplastic pituitary TSH hypersecretion with 3,5,3'-triidothyroacetic acid (TRIAC). J Endocrinol Invest 1983; 6: 217-23.
20. Werner's The Thyroid. Lippincol Co, 1986: 1235.
21. Wehmann RE, Gregerman RI, Burns WH, Sarai R, Santos GW. Suppression of thyrotropin in the low-thyroxine state of severe nonthyroidal illness. New Eng J Med 1985; 312: 543.
22. Torella R, Giugliano D, Scognamiglio G, Passariello N, Tirelli A. Glucagon secretion in patients with hypoparathyroidism: effect of serum calcium on glucagon release. J Clin Endocrinol Metab 1982; 54: 229.
23. Maturlo SJ, Rosenbaum RL, Pan C, Surks MI. Variable thyrotropin response to thyrotropin-releasing hormone after small decreases in plasma free thyroid hormone concentrations in patients with nonthyroidal diseases. J Clin Invest 1980; 66: 451-6.

CHAPTER 24
VETERINARY CLINICAL CHEMISTRY

Biochemical disorders in nonhuman primates with hyperbilirubinemia
C.E. Cornelius

Biochemical alterations in liver diseases of domestic animals
E. Bogin

Clinical chemistry in liver-toxicity testing: differences between species
J.P. Braun, M. Bonnefoi, V. Burgat, and A.G. Rico

Lipid metabolism in large animals
Th. Wensing

BIOCHEMICAL DISORDERS IN NONHUMAN PRIMATES WITH HYPERBILIRUBINEMIAS

Charles E. Cornelius

School of Veterinary Medicine and California Primate Research Center, University of California, Davis, California, USA

Close similarities exist in the hyperbilirubinemic responses in inherited and acquired hepatic disorders between human and nonhuman primates (1); only little differences exist in the chemical structure of fetal and adult bile pigments and hepatic bilirubin conjugating enzymes (2). Analogies between different primate species suggest that nonhuman primates are in most cases the animal models of choice in which to study human hepatic disease mechanisms involving jaundice.

The purpose of this review is to (A) present recent advances and improvements in measuring the levels and chemical species of bile pigments in the serum; and (B) briefly summarize current information regarding various hepatic disorders in nonhuman primates which may result in jaundice.

SERUM BILIRUBIN

Most clinical laboratories still measure serum bilirubin levels by the van den Bergh test or modifications of it when more complex tests offer distinct advantages (3, 4). Increases in primarily "indirect-reacting" serum bilirubin levels suggest the presence of hemolysis or defective hepatic uptake, storage and/or conjugation of bilirubin, while primarily a "direct-reaction" with diazo-reagent points toward hepatobiliary disorders with cholestasis. It has been known for some time that the amounts of "direct and indirect-reacting" bilirubin are only rough estimates and may not accurately reveal the exact quantities of conjugated and unconjugated bilirubin, respectively. At lower normal serum bilirubin levels, this is particularly a problem since there is some "direct-reaction" of unconjugated bilirubin; this results in an overestimation of conjugated bilirubin levels. One usually finds that 5-10% of the total serum bilirubin reacts "directly" even if no conjugated pigment is present. Considerable conjugated pigment, however, may be present in the serum in hemolytic states due to the production of secondary hepatic disease or competition between unconjugated and conjugated pigment for hepatocellular uptake. Increases in unconjugated bilirubin in serum during cholestasis also occurs due to defective hepatic uptake or due to increased erythrocyte turnover in malignancies. When serum bilirubin levels are high, the diazo-tests can be clinically useful; however, other liver function tests may be needed to firmly confirm how much hemolytic and/or cholestatic jaundice may be present (1).

More tedious extraction procedures for the isolation of bilirubin metabolites and their subsequent diazotization and thin-layer chromatographic (TLC) analyses have been developed (5). Such procedures appear a bit too time consuming for routine analyses in the clinical laboratory. Recently, a method involving the separation of the mono- and dimethyl esters of bilirubin and its conjugates appears promising (6). Methyl esters of bilirubin and its conjugates can now be quickly extracted from serum and readily separated by either TLC or high performance liquid chromatography (HPLC) (6). This newer method uses a reverse-phase HPLC system and possesses a sensitivity which has been shown to even measure the low levels of the various bilirubins (<8.5 μmol/l) in normal human and canine sera (7, 8). Another HPLC method allows for the quantification of the various sugar conjugates of bilirubin in dogs and other species by measuring their azodipyrroles with ethyl anthranilate or p-iodoaniline (8).

With such new sensitive methods, it is possible to detect increases in the amounts of bilirubin conjugates in the serum of over 80 percent of normobilirubinemic human patients with biopsy-proven liver disease. It has recently been proposed that increases in the percent of conjugated bilirubin in patients with normal serum bilirubin levels is probably the single most sensitive marker for hepatobiliary disease. This new method was compared to a variety of other serum tests for sensitivity such as: alanine aminotransferase activity, γ-glutamyltransferase activity and fasting bile acid levels (9). Once such new chromatographic methods are fully automated, they may be diagnostically rewarding improvements over our current tests to measure serum pigment levels. Such HPLC methods are particularly useful in differentiating patients with Gilbert's syndrome from those with frank hepatopathy and in detecting early and moderate liver disorders (10).

It has long been confusing to the clinical pathologist that in certain cholestatic syndromes in animals and man, the level of "direct-reacting" serum bilirubin continues to remain high at a time when bilirubinuria has subsided and is absent. Since conjugated bilirubin, unlike unconjugated bilirubin, is lost to the urine, one normally observes increased concentrations of conjugated pigment in both serum and urine in hepatobiliary diseases. We finally have an explanation for this discrepancy. Only recently, an additional "direct-reacting" bilirubin fraction was observed in human and animal patients with cholestatic jaundice due to hepatobiliary disease or Dubin Johnson syndrome (8, 11). In cholestatic syndromes, total bilirubin concentrations are higher when measured by conventional diazo techniques than by HPLC techniques. When serum is denatured by alkaline methanolysis during its preparation for HPLC analysis, it was found to contain an alkali-stable pigment-protein complex. This fraction, referred to as "biliprotein" or "delta" bilirubin, reacts "directly" with diazo reagents and, unlike other conjugated bilirubins, is covalently bound to serum albumin (12). The abnormal accumulation of bilirubin glycosyl-conjugates in liver and serum leads to their covalent binding to protein. As jaundice subsides during recovery from cholestasis, bilirubin conjugates are normally lost to the urine but the covalently bound bilirubin conjugate remains in the serum and may persist long after the disappearance of bilirubinuria. The life span in the serum of this "biliprotein" is similar to that of serum albumin. It has been postulated that the ϵ-amino group of lysine forms an amide bond with the carboxyl groups of one propionic acid side-chain of bilirubin (13). Since no "biliprotein" is present in Gunn rats which lack UDP hepatic glucuronyl transferase for bilirubin conjugation, it is assumed that conjugated bilirubin must be present prior to its coupling to albumin (14).

Concentrations of serum "biliprotein" levels were first crudely estimated as the difference observed between the concentrations of bilirubin measured by diazo-methods and HPLC. A recent study has reported that such "biliprotein" can now be easily assayed without tedious pre-incubation or special equipment. This method includes first, the selective removal of other bilirubins by organic solvent extraction and secondly, the direct measurement of the remaining denatured "biliprotein" by a simple diazo method (15). This procedure may prove useful in the future in the clinical laboratory setting and allow for a quick quantification of this additional direct-reacting bilirubin fraction in cholestasis.

FETAL AND ADULT BILE PIGMENTS

Bile pigment composition is nearly identical in human and nonhuman primates (16). Bile pigment accumulation does not occur in the fetus since there is very limited fetal hepatic conjugation of pigment prior to birth. When unconjugated bilirubin, unlike conjugated bilirubin, is present in the fetal circulation of monkeys, it is rapidly transported across the placenta to the maternal circulation where it is ultimately conjugated and excreted into the maternal bile (16).

Bile of human and nonhuman primates contains over 80% bilirubin-IXα diglucuronide with the remainder composed of monoglucuronides and traces of other bilirubin isomers and sugars (17). Other bilirubin isomers and sugar conjugates increase in concentrations in serum and bile during intrahepatic or extrahepatic cholestasis (18). Only limited bilirubin-IXα conjugation occurs early in human and monkey fetal liver. Much of the bilirubin produced early in gestation is apparently the IX-β isomer which is replaced by the IX-α isomer at birth (19).

NEONATAL JAUNDICE

The only animal species which exhibit a marked neonatal unconjugated hyperbilirubinemia are primates. The rhesus monkey is currently recognized as the most appropriate animal model to use for studies regarding this potentially serious metabolic crisis that normally occurs in humans and monkeys at birth. Due to the smaller metabolic size of monkeys, the time course for elevated serum bilirubin levels in monkeys is three-times shorter than in humans. When neonatal jaundice persists with serum levels >340 μmol/l and associated with acidosis or hypoxia, bilirubin may penetrate the nervous tissues and produce mental retardation, deafness and even death (20). Neonatal jaundice occurs at birth due to the slow postnatal development of hepatic bilirubin conjugating enzymes and hepatocellular bilirubin uptake mechanisms; the presence of a large load of overproduced unconjugated bilirubin; and increased enterohepatic circulation of pigment (21, 22). In neonatal monkeys, the first phase of hyperbilirubinemia shows increases normally up to 85 μmol/l by 20 hours after birth; this is followed by a second phase of near 17 μmol/l by 48 hours which persists for up to 96 hours (21). Following these elevations, serum levels decline to normal levels of approximately 8.5 μmol/l. Adult hepatic conjugating activities (UDPGA-transferase) are attained by 48 hours after birth.

Therapeutic regimes to prevent marked neonatal hyperbilirubinemia have in the past included exchange transfusions of whole blood or albumin-primed solutions and/or phototherapy which produces photo-oxidative products of bilirubin that are easily excreted in bile without requiring conjugation (22). A new method is currently under study which inhibits the rate of heme oxidation to bilirubin through the use of competitive inhibitors of microsomal heme oxygenase, the rate limiting enzyme in bile pigment production (23). Two competitive inhibitors have been utilized: Sn and Zn protoporphyrin. Neonates quickly respond to as little as 10 micromoles of such inhibitors when injected subcutaneously: serum bilirubin levels rapidly decline. If these compounds are proven to be safe for human use, this serious neonatal disease may become only a medical threat of the past.

FASTING UNCONJUGATED HYPERBILIRUBINEMIAS IN GILBERTS-LIKE SYNDROMES

A marked fasting hyperbilirubinemia occurs following an overnight fast in humans with Gilbert's syndrome, normal horses, mutant Southdown sheep and mutant Gunn rats (24). Fasting unconjugated hyperbilirubinemia is also observed in all normal Bolivian and Peruvian squirrel monkeys as in Gilbert's syndrome (25). Fasting in Guyanese and Brazilian squirrel monkeys, however, does not result in marked elevations in serum bilirubin levels. Since all these squirrel monkeys are classified as of the same species (Saimiri sciureus), the Brazilian and Guyanese monkeys have been used as normal controls for comparative research studies. Bolivian squirrel monkeys do not exhibit decreased hepatic clearances of sulfobromophtalein (BSP) and indocyanine green and appear similar to humans with Gilbert's syndrome type I. They: exhibit decreased plasma clearance of unconjugated bilirubin during fasting; have 50% lower hepatic UDP-glucuronyl transferase activities; have a decreased bilirubin diglucuronide to monoglucuronide ratio in bile; have normal hepatic ligandin levels; and do not exhibit an increased erythrocyte turnover. Radiolabeled bilirubin kinetic studies suggest that overproduction of bilirubin also occurs in fasted Bolivian monkeys in addition to the defective bilirubin clearance from serum during fasting (26).

Recent unpublished data from our laboratory indicates that during fasting, no induction of hepatic UDP-glucuronyl transferase activity is present as in controls. The substrate UDP glucuronic acid, which is necessary for bilirubin conjugation, was not lower in concentration in fasted Bolivian monkey livers than in control monkey livers (27). Current studies are aimed at determining if the K_m and V_{max} of hepatic UDP glucuronyl transferase with such substrates could be altered in these monkeys and might account for the decreased conjugation. Whether the bilirubin plasma clearance defect could also be due in-part to alterations in the plasma and/or microsomal membranes should be investigated.

EXTRAHEPATIC BILIARY ATRESIA

A rhesus monkey was recently confirmed to have extrahepatic biliary atresia at the time of an exploratory laparotomy at 6 weeks of age (28). At birth, all clinical chemical parameters were normal including the icteric indices. By 6 days after birth, clinical jaundice and bilirubinemia were apparent; urinalysis revealed the absence of uribilinogen. At 9 days, total serum bilirubin levels were 770 μmol/l (70% conjugated); alkaline phosphatase levels were also extremely elevated. By 6 weeks of age, serum bilirubin levels stabilized at around 255 μmol/l (60% conjugated) and remained at this level until the animal was euthanatized at 10 months of age. The monkey exhibited all of the clinical signs that are present in most cases of biliary atresia in humans. These include: an atretic and fibrotic bile duct; biliary cirrhosis; rickets; and an elevated serum antibody titer for reovirus type III. The rhesus monkey will no doubt be useful as an animal model in future years in which to study the relationship of such viral infections as etiologic agents in this rare, but deadly, neonatal disease.

OTHER DISEASES EXHIBITING CONJUGATED HYPERBILIRUBINEMIA

A number of spontaneous hepatic disorders have been observed in nonhuman primates which result in conjugated hyperbilirubinemias. Hepatic necrosis and/or fibrosis can result from infections such as Leptospirosis or from ingested toxins such as monocrotaline (29). Biliary hyperplasia can also be observed in pyridoxine deficiency. Hepatic amyloidosis occurs secondarily to chronic debilitating disorders associated with enteric infections from Salmonella or Shigella organisms. Hemosiderosis of the liver and other organs occurs spontaneously in certain zoo animals such as Mynah birds, African rock hyrax and the lemur, a nonhuman primate. Hemosiderosis in lemurs is, however, not closely analogous to human idiopathic hemochromatosis due to its different patterns of iron deposition and absence of diabetes mellitus and heart disease (30). Hemoseridosis, with its resulting hepatic venous and periportal fibrosis as occurs in lemurs held in captivity for extended periods, is probably related to artificial diets fed at zoological parks.

Nonhuman primates have been recently used in the production of a variety of hepatobiliary diseases associated with jaundice such as cholelithiasis in squirrel monkeys (31); necrosis and chronic active hepatitis from hepatitis A (32); hepatitis B (33); non A-, non B- (34) and delta infections (35) in chimpanzees and marmosets; and cirrhosis in ethanol-fed baboons (36).

Nonhuman primates have recently been established as ideal animal models in which to study a variety of biochemical mechanisms present in human disorders. Their use in studying bile pigment metabolism and hepatic bilirubin transport is a good example. Advantages of nonhuman primates also includes the possible use of human immunologic reagents which usually cross-react with the counterpart nonhuman primate proteins important in bilirubin transport. Although other animal species may mimic certain human biochemical functions, nonhuman primates are generally more analogous due to their close phylogenetic relationship.

REFERENCES

1. Cornelius CE. The use of nonhuman primates in the study of bilirubin metabolism and bile secretion. American Journal of Primatology 1982; 2: 343-354.
2. Cornelius CE. Comparative bile pigment metabolism in vertebrates. In: Ostrow JD, ed. Bile pigments and Jaundice. (Liver: Normal Function and Disease Series, Volume 4). New York: Marcel Dekker, 1986; Chapter 23: 601-647.
3. Hootegem PV, Fevery J, Blanckaert N. Serum bilirubins in hepatobiliary disease: comparison with other liver function tests and changes in the postobstructive periods. Hepatology 1985; 5: 112-117.
4. Fevery J, Blanckaert N. What can we learn from analyses of serum bilirubin? Journal of Hepatology 1986; 2: 113-121.
5. Fevery J, Claes J, Heirwegh K, DeGroote J. Hyperbilirubinemia - significance of the ratio between direct-reacting total bilirubin. Clincal Chim Acta 1967; 17: 73-79.
6. Blanckaert N. Analysis of bilirubin and bilirubin mono- and diconjugates-determination of their relative amounts in biological samples. Biochem 1980; 185: 115-128.
7. Muraca M, Blanckaert N. Liquid-chromatographic assay and identification of mono- and diester conjugates of bilirubin in normal serum. Clin Chem 1983; 29: 1767-1771.
8. Rothuizen J. Hyperbilirubinemia in canine hepatobiliary disease. Proefschrift, Rijksuniversiteit, Utrecht, Netherlands 1985: 1-168.

9. Scharschmidt BF, Blanckaert N, Farina FA. Measurement of serum bilirubin and its mono- and diconjugates - applications to patients with hepatobiliary disease. Gut 1982; 23: 643-649.
10. Jansen PLM, Peters WHM, Janssens AR. Clinical value of serum bilirubin subfractionation by HPLC and conventional methods in patients with primary biliary cirrhosis. Meeting of the Dutch Gastroenterological Society and Dutch Society of Hepatology, Oct. 4-5, 1985, Noordwijkerhout, The Netherlands.
11. Weiss JS, Gautam A, Lauff JJ, et al. The clinical importance of a protein-bound fraction of serum bilirubin in patients with hyperbilirubinemia. New England Journal of Medicine 1983; 309: 147-150.
12. Lauff JJ, Kasper ME, Wu W, Ambrose RT. Isolation and preliminary characterization of a fraction of bilirubin in serum that is firmly bound to protein. Clinical Chemistry 1982; 28: 629-637.
13. Wu TW. Bilirubin analysis - the state of the art and future prospects. Clinical Chemistry 1984; 17: 221-229.
14. McDonnagh AF, Palma LA. Origin of mammalian biliprotein and rearrangement of bilirubin glucuronides in the rat. Journal of Clinical Investigation 1984; 74: 763-770.
15. Blanckaert N, Servaes R, Leroy P. Measurement of bilirubin-protein conjugates in serum and application to human and rat sera. Journal Laboratory Clinical Medicine 1986; 108: 77-87.
16. Cornelius CE. Bilirubin metabolism in fetus and neonate. In: Brans YW, Kuehl TJ, eds. Nonhuman Primates in Perinatal Research, Chapter 36; 1987; in press.
17. Fevery JM, Van de Vijver M, Michiels R, Heirwegh KPM. Comparison in different species of biliary bilirubin-IXα conjugates with the activities of hepatic and renal bilirubin-IXα uridine diphosphate glycosyltransferases. Biochem J 1977; 164: 737-746.
18. Fevery J, Van Damme B, Michiels R, DeGroote J, Heirwegh KPM. Bilirubin conjugates in bile of man and rat in the normal state and liver disease. J Clin Invest 1972; 51: 2482-2492.
19. Blumenthal SG, Taggart DB, Rasmussen RD, Ikeda RM, Ruebner BH. Conjugated and unconjugated bilirubins in humans and rhesus monkeys: structural identity of bilirubins from biles and meconiums of newborn humans and rhesus monkeys. Biochem J 1979; 179: 537-547.
20. Claireaux AE. Pathology of human kernicterus. In: Sass-Kortsak A, ed. Kernicterus. Toronto: University of Toronto Press, 1961: 140-149.
21. Gartner LM, Lee KS, Vaisman S, Lane D, Zarafu I. Development of bilirubin transport and metabolism in the newborn rhesus monkey. J Pediatr 1977; 90: 513-531.
22. Odel GB. Neonatal Hyperbilirubinemia. New York: Grune and Stratton, 1980: 35-134.
23. Cornelius CE, Rodgers PA. Prevention of neonatal hyperbilirubinemia in rhesus monkeys by tin-protoporphyrin. Pediatr Res 1984; 18: 728-730.
24. Cornelius CE. Animal Models (Chapter). In: Arias IM, Popper H, Schacter D, Shafritz DA, Jakoby B, eds. The Liver; Biology and Pathobiology, 2nd ed. New York: Raven Press, 1987; in press.
25. Cornelius CE, Rodgers PA, Bruss ML, Ahlfors CE. Characterization of a Gilbert-like syndrome in squirrel monkeys (Saimiri sciureus). J Med Primatol 1985; 14: 59-74.
26. Bruss ML, Myers BA, Portamn OW, Roy-Chowdhury J, Cornelius CE. Kinetics of plasma bilirubin in squirrel monkeys with a Gilbert-like syndrome. 1987; in press.
27. Myers B, Cornelius CE (Unpublished data). California Primate Research Center. Davis, CA: University of California; 1987.
28. Rosenberg DP, Morecki R, Lollini LO, Glaser J, Cornelius CE. Extrahepatic biliary atresia in a rhesus monkey (Macaca mulatta). Hepatology 1983; 3: 577-580.

29. Kruckenberg SM, Cornelius CE, Cook JE. Liver Function and Disease in Primates. In: T-W-Finennes RN, ed. Pathology of Simian Primates. Basel, Switzerland: S. Karge, 1972.
30. Gonzales J, Benirschke K, Saltman P, Roberts J, Robinson PT. Hemosiderosis in Lemurs. Zoo Biol 1984; 3: 255-265.
31. Portman OW, Alexander M, Tanaka N, Toshiaki O. Role of diet in normal biliary physiology and gallstone formation. In: Primates in Nutritional Research. New York: Academic Press, Inc., 1979.
32. Provost PJ, Conti PA, Giesa PA, et al. Studies in chimpanzees of live, attenuated hepatitis A vaccine candidates. Proc Soc Exp Biol Med 1983; 172: 357-363.
33. Gerin JL, Alexander H, Shih JW-K, et al. Chemically synthesized peptides of hepatitis B surface antigen duplicate the d/y specificities and induced subtype-specific antibodies in chimpanzees. Proc Natl Acad Sci (USA) 1983; 80: 2365-2369.
34. Bradley DW. The agents of non-A, non-B viral hepatitis. J Virol Meth 1985; 10: 307-319.
35. Govindarajan S, Fields HA, Humphrey CD, Margolis HS. Pathologic and ultrastructural changes of acute and chronic delta hepatitis in an experimentally infected chimpanzee. Am J Path 1986; 122: 315-322.
36. Lieber CS, DeCarli LM. An experimental model of alcohol feeding and liver injury in the baboon. J Med Primatol 1974; 3: 153-163.

BIOCHEMICAL ALTERATIONS IN LIVER DISEASES OF DOMESTIC ANIMALS

E. Bogin

Kimron Veterinary Institute, Tel Aviv University Medical School, P.O. Box 12, Bet Dagan, Israel

The liver is one of the most frequently damaged organs in the body of domestic animals. The list of diseases which affects the liver spans a wide range of vascular, metabolic, toxic, obstructive and neoplastic involvements. All extensive diseases of the liver which erode the large hepatic functional reserve tend to produce similar clinical signs and symptoms in the patient, with jaundice and liver failure being the most common pathophysiologic syndromes.

Hepatic pathology involves primarily the hepatic cells, bile ducts and vascular systems. Most hepatic disorders are observed pathologically as: 1) hepatic cell necrosis 2) intrahepatic or extrahepatic bile duct obstruction 3) hepatic atrophy and or fibrosis. Being such a dynamic organ the liver composition and enzymatic profile changes rapidly under the influence of nutritional, hormonal and toxic substances. Being in close equilibrium with the blood, these changes are reflected in the blood composition and enzymatic makeup especially under conditions that hepatocytes necrosis occurs.

This presentation describes some of the recent studies on the biochemical aspects of two commonly seen liver diseases in animals: fatty liver syndrome and bile duct obstruction.

Fatty liver syndrome, commonly seen in animals and man, has been widely studied. Steatosis, which is common and non-specific response of the liver to different forms of acquired injury or inherited metabolic derangement is characterized by the significant accumulation of triglycerides in the hepatocytes. Because of the interplay of the various pathological mechanisms in which fat is accumulated in the liver, it is hard to make a precise statement on the mechanisms which operate at any given case. It is however possible to group the pathological mechanisms into: a) those which involve an imbalance of nutritional or metabolic factors (starvation, low protein-high carbohydrate diet, hormonal imbalance) b) those due to toxic substances and c) result of anoxia (1-9).

The second study describes the biochemical aspects taking place during cholestasis, implicating obstruction of bile flow leading to hepatic and secondary renal failure.

Table 1. Mean body and liver weights, liver enzymes and serum composition of geese, at various times of forced feeding

Forced feeding (days)	Body weight (kg)	Liver weight (g)	Liver (U/g protein)						Serum (enzymes U/l; metabolites g/l)							
			ME	ICDH	LDH	MDH	ALP	AST	ME	ICDH	LDH	MDH	ALP	AST	ALB	TL
0	4.10	96	35	18	918	1073	2.70	152	2.5	18	289	276	50.7	17.8	15.4	6.24
19	5.83	143	134	69	1425	2934	2.07	215	24.8	19.5	137	366	51.8	19.5	18.5	11.92
26	6.05	312	242	86	1417	2412	1.29	273	72.5	33.3	185	572	44.6	41.3	18.6	-
30	7.01	575	196	83	2645	5101	1.05	496	161	70.7	604	1030	29.5	79.0	17.7	13.17
33	7.58	688	191	77	-	3374	1.10	358	-	37.3	547	958	20.3	49.5	13.0	-
41	8.32	900	214	79	5741	3271	1.08	391	143	43.0	643	1191	24.0	50.2	11.5	14.10

(n = 5 - 20)

In both studies, the biochemical changes occurring in the liver during the development of the disease and its relationship to the blood composition are described.

MATERIALS AND METHODS

Fatty liver in the following animals was induced: geese by force feeding for 30-35 days (1); chicken by estradiol injection (8.0 mg/kg bw) for 3-5 days; rats by sucrose rich diet (60%) for 4-6 weeks (2); dairy cow - from the slaughterhouse (metabolic and nutritional imbalance (3)). Obstructive jaundice was done by the ligation of the common bile duct (CBDL) in rats (4).

Blood and liver enzymes and metabolites were determined spectrophotometrically (1-10). Enzymes and antibiotics clearance rates ($T\frac{1}{2}$) were measured following their i.v. administration to control and fatty liver induced poults. Enzymes injected were obtained from tissues of poults of the same breed and age following their extraction and purification (6). Enzymes and drugs studied are listed in the results section.

Liver homogenization and fractionation was done immediately after the animals were sacrified. Three fractions were obtained: mitochondrial, microsomal and soluble (4). Enzymes measured are listed in the results section.

RESULTS

Mean body weight, liver weight, enzymes and metabolites in the liver and serum of geese at various times of force feeding are shown in Table 1. As seen, liver enzymes, except for ALP, significantly increased with time, paralleling levels of serum enzymes and liver size. Relatively good correlations (r) were seen between liver size and the enzymes ALP, LDH and MDH. The best r was obtained between liver size and MGH to albumin ratio.

Levels of enzymes, proteins and lipids in the sera of cows with normal and fatty liver are given in Table 2. Cows with fatty liver were divided into 2 groups, moderate and heavy fatty liver with 44 ± 2 and 139 ± 2 vs 30 ± 2 mg lipids/gr liver in the controls. The degree of enzyme changes in the serum corresponded to the degree of liver fattening. While liver total lipids increased during fattening, serum lipids decreased (Table 2).

Enzyme levels in the normal and fatty liver of cows are shown in Table 3. Except for SDH, all enzymes increased significantly. The change seen parelled the degree of liver fattening.

The relative distribution of the various lipid fractions in the normal and fatty liver are depicted in Figure 1. As seen, the relative distribution of phospholipids and unesterified cholesterol decreased, while levels of cholesterol ester triglycerides significantly increased.

Mean enzyme levels in control and fatty liver from rats are given in Table 4. In the rat, where fatty liver was induced by high carbohydrate diet supplemented with 1% orotic acid, liver size and liver to body ratio increased. Levels of enzymes in the sera are shown in Table 5 and liver enzymes are given in Table 6. Significant increases in the activities of ALT and MDH and decrease in the activity of GGT were seen in the sera of rats with fatty liver. In the liver, the changes seen were more dramatic with G6PDH, ME, GGT, aldolase, 6PGD, MDH, LDH and AST increasing by 235,

Table 2. Levels on enzymes, proteins and lipids in the sera of cows with normal and fatty liver

Group	Serum Constituents													
	ME	ICDH	G6PDH	LDH	MDH	GLDH	AST	ALT	ALP	ACP	TP	ALB	GLOB	TL
Normal	8.7	28.5	2.58	1200	600	2.8	54	8.0	26.9	6.7	70.6	38.8	31.7	6.86
	1.5	2.8	0.39	285	51	1.1	13	1.5	4.5	1.4	0.5	0.9	0.8	0.94
Moderate fatty liver	13.9	30.9	2.65	1362	979	3.1	96	12.6	22.0	8.4	69.3	35.1	34.1	3.99
	2.1	2.7	0.39	222	200	0.5	17	4.6	2.6	1.1	2.2	1.7	2.0	0.42
	(+60)	-	-	(+13)	(+63)	-	(+78)	-	-		-	-	-	-
Heavy fatty liver	12.8	49.8[a]	4.03[b]	1863	1281[b]	8.4[b]	129[b]	7.8	45.2[a]	12.1[b]	70.8	33.0	37.8	4.18[a]
	3.0	5.9	0.51	276	219	2.6	27	1.7	9.3	2.1	1.6	1.0	1.2	0.67
	(+47)	(+75)	(+56)	(+55)	(+113)	(+200)	(+139)	-	(+68)	(+81)	-	-	-	(-31)

Enzymes activity is expressed in U/l, proteins and lipids in gr/l
a = p <0.05 b = p <0.01 n = 7 - 17 (% change from control).

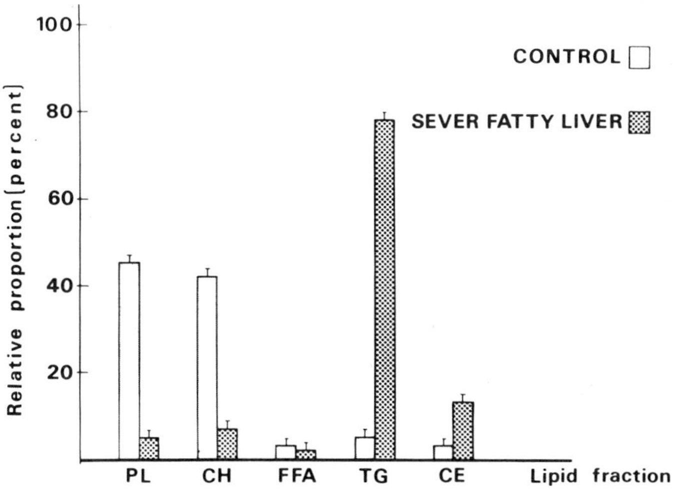

Fig. 1. Relative proportion of lipid constituents from normal and fatty liver in cow.

170, 113, 95, 63, 38, 37 and 36% respectively. Levels of SDH and ALT were lower.

The distribution of enzyme activities in the non-fatty parts of the fatty liver vs. the non fatty liver, control are shown in Table 4. Whenever a change in enzyme activity was seen in the fatty liver, is was larger in the non-fatty part than in the fatty part.

Obstructive jaundice, induced by the ligation of the common bile duct, greatly increased the blood levels of bile acids, urea, creatinine, bilirubin and the enzymes ALD, ICDH, LDH, ALP, AST and MDH. The peak of the disease, as evaluated by these changes, was at 3-6 days post ligation. There are also significant changes in the enzyme activities in the liver. Levels of ALD, ICDH, LDH, SDH, ME and MDH were lower, while that of ALP and AST were higher (Table 7). Liver microsomal Na-K ATPase was higher and MG-ATPase was lower, not shown in the table. Renal failure associated the hepatic insult, paralleled by lower enzyme activities and mitochondrial oxidative phosphorylation in the kidney. Studying the effect of bile acids and bilirubin, which were greatly elevated during liver obstruction, showed to greatly inhibit enzyme and oxidative phosphorylation.

Tables 8 and 9 showed that liver fattening in the chicken induced by estradiol significantly affected liver function as evaluated by clearance rate of enzymes and drugs from the blood. Both distribution and clearance rate were affected.

DISCUSSION

Steatosis is a common and non-specific response of the liver to different forms of acquired injury or metabolic derangement. Because of the interplay of the various pathological mechanisms by which fat is accumulated in the liver and because of the wide diversity of functions of the liver, it is hard to define the mechanisms which operate in a given case, leading to lipid accumulation.

Table 3. Levels of enzymes in normal and fatty liver from cows

Group	Enzymes (units/gr protein)										
	ME	ICDH	G6PDH	LDH	MDH	SDH	GLDH	AST	ALT	ACP	ALP
Normal	0.94	92	1.85	517	1016	79.1	12.4	239	2.71	11.3	3.60
Moderate fatty liver	1.67	112	2..53	814[a]	1171	69.6	18.9	231	2.21	12.2	4.56
	(+78)	(+22)	(+37)	(+57)	(+15)	(-12)	(+52)	-	-	-	(+27)
Heavy fatty liver	2.16[b]	133[a]	3.43[b]	1292[b]	1358[a]	27.5[b]	17.3	217	2.93	10.8	6.36[a]
	(+130)	(+45)	(+85)	(+150)	(+34)	(-65)	(+40)	-	-	-	(+77)

a = $p < 0.05$; b = $p < 0.01$ n = 8 - 17 (% change from control)

Table 4. Mean enzyme levels in control and parts of fatty liver from rats

Group	Enzyme (U/protein)										
	AST	ALT	ALD	ALP	ME	ICDH	G6PDH	6PDH	MDH	LDH	SDH
Control	246	129	35.5	0.60	9.8	135	12.7	35.0	2937	1459	91
Fatty liver Nonfatty part	320	137	48.7	0.55	22.6[b]	152	52.7[b]	59.7[b]	4326[a]	1946[a]	59.4
Fatty liver Fatty part	289	113	53.3[a]	0.51	16.2[b]	146	27.0[b]	44.0[b]	3682	1523	71.8

a. $p < 0.05$; b. $p < 0.01$; $n = 4 - 9$

Table 5. Mean enzyme and metabolite levels in the sera of control and fatty liver animals

Animal	Enzyme (units/l)													Metabolites (g/l)	
	ME	ICDH	LDH	MDH	G6PDH	ALP	AST	ALT	ACP	6PGDH	ALD	GLDH	GGT	ALB	TL
Goose	143 (2.5)	43 (18)	643 (289)	1191 (276)	0 (0)	24.0 (50.7)	50.2 (17.8)	-	-	-	-	-	-	11.5 (15.4)	14.1 (6.2)
Rat	12.4 (12.5)	4.8 (5.4)	905 (824)	560 (315)	2.8 (2.8)	84.7 (82.7)	60.4 (50.9)	13.7 (8.1)	-	14.2 (12.5)	59.7 (48.8)	-	8.1 (17.4)	- -	- -
Chicken	30.9 (3.0)	8.7 (11.3)	577 (670)	359 (590)	-	1170 (834)	108 (126)	-	-	-	-	-	-	-	34.5
Cow	12.8 (8.7)	49.8 (28.5)	1863 (1200)	1281 (600)	4.0 (2.8)	45.2 (26.9)	129 (54)	7.8 (8.0)	12.1 (6.7)	-	-	8.4 (2.8)	-	33.0 (38.8)	4.2 (6.9)

(control)

Table 6. Mean enzyme levels in the liver of control and fatty liver animals

Animal	Enzyme (units/g protein)													
	ME	ICDH	LDH	MDH	G6PDH	ALP	AST	ALT	ACP	6PGDH	ALD	SDH	GGT	GLDH
Goose	191 (35)	77 (60)	5741 (918)	3374 (1173)	127 (60)	1.08 (2.70)	391 (152)	-	-	-	-	-	-	-
Rat	27.8 (10.3)	165 (151)	2033 (1480)	3405 (2468)	57.6 (17.3)	0.72 (0.80)	404 (298)	125 (162)	42 (46)	43.1 (26.5)	69.1 (33.5)	72.5 (94.9)	2.89 (1.36)	-
Chicken	212 (174)	113 (149)	2439 (3304)	3479 (3200)	-	9.6 (8.6)	376 (407)	-	-	-	-	-	-	-
Cow	2.2 (0.9)	133 (92)	1292 (517)	1358 (1016)	3.4 (1.8)	6.4 (3.6)	217 (239)	2.9 (2.7)	10.8 (11.3)	-	-	27.5 (79.1)	-	17.3 (12.4)

control

Table 7. Enzyme levels in the blood and liver of control and CBDL rats at day 6 after ligation

ENZYME		ALD	ICDH	LDH	SDH	ALP	AST	ME	MDH	Na-K ATPase	Mg ATPase
Cont.	Blood (U/l)	37	1.8	138	-	43	62	-	141	-	-
Exper.		66[c]	9.1[b]	279[c]	-	148[c]	231[c]	-	297[b]	-	-
Cont.	Liver (U/gr)	192	240	1448	82	0.2	320	15	1603	64	12
Exper.		119[c]	130[c]	1295[a]	41[c]	2.2[c]	594[b]	12[b]	1139[c]	126[b]	7[b]

P was significantly different below 5% n = 10 - 20

Studying the biochemical aspects of fatty liver revealed several interesting observations which may shed some light on the sequence of events leading to liver fattening. Studying several models, inducing fatty liver by different methods, showed that enzymes induction occurs. Increased levels of NADPH generating enzymes were seen in models where fatty liver induction was accomplished by feeding surplus rich diet, for a long time (geese and rats). This was not seen in the chicken model, where fatty liver induction was acute and was obtained by hormonal imbalances. Of interest are the observations showing larger changes in the non-fatty part of the fatty liver, suggesting that during the liver fattening process, there is an induction of enzymes in the liver, reaching a peak prior to lipid accumulation, declining thereafter.

In the dairy cow, metabolism is geared toward milk production. Large metabolic changes take place around the calving period. One such event is the mobilization of lipids from the body reservoir to make up for the great need of energy required for milk production. Lipoproteins, the major carrier of lipids in the blood, have an important role in lipid metabolism. Of interest is the relative distribution of the various fractions HDL, LDL and VLDL. While in man the major fraction is LDL, in the dairy cow it is the HDL. Furthermore the absolute and relative levels of the triglycerides in the blood of the dairy cow is relatively low, ranging between 2-10% of the total lipids, in comparison to man, where it comprises 25-35%. The levels of this fraction are even lower in the blood of a cow having a fatty liver. The relationship of the low blood levels of the triglycerides and LDL and relatively high HDL levels, are being studied in connection to the development of fatty liver in dairy cattle.

The present series of studies suggest that lipid accumulation in the liver is a result of both increased de-novo synthesis of lipids in the liver and the failure to transport the lipids out of the liver. This is supported by the fact that triglycerides levels in the cow's blood during liver fattening is lower than prior to the onset of the syndrome. It is of interest that liver fattening is commonly seen in both the dairy cow and the laying hen during a period when the animals are under production stress of milk or eggs.

Table 8. Half life time of enzymes (minutes) in the blood of control and fatty liver, chicken

Enzyme	LDH		MDH		AST		ICDH		ME	
Group	α	β	α	β	α	β	α	β	α	β
Control	14.6±2.9	281±20	12.7±2.3	267±13	15.0±1.4	773±56	17.1±3.5	279±33	9.6±3.7	499±138
Fatty liver	8.5±2.4	248±16	30±1.9	394±47	9.6±0.4	820±67	39.5±4.0	294±44	12.9±5.7	611±78

(mean ± S.D.)

Table 9. Mean half life time ($T½$) and distribution (AUC) of antibiotics in the blood of control and fatty liver chicken

Group	Antibiotics									
	Erythromycin		Lincomycin		Penicillin		Streptomycin		Oxytetracycline	
	$T½$	AUC	$T½$	AUC	$T½$	AUC	$T½$	AUC	$T½$	AUC
Control	209	221	147	284	18	105	351	34933	169	492
Fatty liver	74	154	51	275	39	53	193	22458	615	1263

Another aspect of liver described is associated with toxic substances. In the rat model, where liver disease was induced by obstruction of the bile duct, leading to leakage of bile acids and bilirubin into the blood. Our studies (4) revealed that these substances are toxic to many enzymes thus affecting metabolic pathways and inhibit mitochondrial oxidative phosphorylation leading to lower production of high energy intermediates required for cellular function. This can also explain the renal failure seen during the induction of obstructive jaundice, known as the hepatorenal syndrome.

These studies revealed some of the biochemical aspects of liver function and the multifactorial involvement of liver diseases and the potential of some biochemical parameters to be used as genetic markers for sensitivity of fatty liver development.

ACKNOWLEDGEMENT

Parts of this research were supported by U.S.A.-Israel BARD fund No. I-76-79.

REFERENCES

1. Bogin E, Avidar Y, Merom M, et al. Biochemical changes associated with fatty liver in geese. Avian Pathol 1984; 13: 683.
2. Bogin E, Avidar Y, Merom M. Biochemical changes in liver and blood during liver fattening in rats. J Clin Chem Clin Biochem 1986; 24: 621.
3. Bogin E, Avidar Y, Merom M, Soback S, Brenner G. Biochemical changes associated with fatty liver syndrome in cows. J Comp Pathol 1987; in press.
4. Israeli B, Bogin E. Biochemical changes in liver, kidney and blood associated with common bile duct ligation. Clin Chim Acta 1986; 160: 211.
5. Rivetz B, Bogin E, Horenstein K. Half life time of LDH isoenzymes, MDH, GOT and CPK in chicken blood. Zbl Vet Med A 1977; 24: 345.
6. Avidar Y, Bogin E, Soback S. Fatty liver syndrome in farm animals, - biochemical and pharmacological aspects. Isr J Vet Med 1986; 42: 318.
7. Bogin E, Avidar J, Rivetz B, Israeli B. Fatty liver in fattened geese: enzyme profile of liver and serum. Zbl Vet Med 1978; 25: 727.
8. Bogin E, Avidar Y. Biochemical changes associated with liver fattening in poultry. Avian Pathol 1987; submitted.
9. Soback S, Ziv G, Bogin E, Cohen Z, Yearon Y. Pharmacokinetic changes of several antibiotics in chicken during induced fatty liver. Res Vet Sci 1987; in press.
10. Bergmeyer HU. Methods of Enzymatic Analysis, 2nd ed. Deerfield Beach, Florida: Verlag Chemie International, 1987.

CLINICAL CHEMISTRY IN LIVER-TOXICITY TESTING: DIFFERENCES BETWEEN SPECIES

J.P. Braun, M. Bonnefoi, V. Burgat and A.G. Rico

Laboratoire Associé INRA Toxicologie Biochimique et Métabolique, ENV, F-31076 Toulouse Cedex

Liver-toxicity testing is necessary for the development of all new drugs, as well for human as for veterinary uses. It must be performed in various animal species, mainly rodents but also primates, dogs and sometimes other domestical animals. The major criterion of liver toxicity remains microscopic pathologic examination, which unfortunately requires the sacrifice of a large number of small animals or repeated biopsies in larger ones. Thus alternative, mainly biochemical, methods were devised to investigate liver damages, liver cancerogenesis and liver metabolism alterations.

1. SERUM ENZYMES AND LIVER TOXICITY

Toxic compounds can produce different effects on the liver; the two most important ones are cytolysis and cholestasis, but liver toxicity can also result from some chronic damage or from enzyme induction. In experimental toxicology, patterns of enzymes were proposed to investigate these different troubles (1) but significant species differences were observed, which are the basis for a more cautious use of clinical chemistry in liver-toxicity testing.

When any cytolysis occurs in the liver, the activity of so-called liver specific enzymes is dramatically increased in the serum or the plasma. The measurement of enzyme activities can easily be repeated thus allowing to monitor the evolution of the toxic process which can greatly differ according to the chemical compound. Nevertheless, the choice of a sensitive and specific marker greatly differs according to species.

For routine liver toxic damage testing, both Alanine and Aspartate Aminotransferases are used. Alanine Amino Transferase (ALAT, EC 2.6.1.2) is too often considered a liver-specific enzyme for it has a high concentration in the liver of man, rat and dog (2-4) while its activity in the other tissues of these species is much lower. Moreover, its concentration in the liver of the dog is approximately twice as high as in the liver of the rat, making it a much more sensitive marker in the former. But in some other species such as horses, cows, sheep or goats, the concentration of ALAT in the liver is lower than in skeletal muscles (5).

Aspartate Amino Transferase (ASAT, EC 2.6.1.2) is not a liver-specific enzyme in most species, but in the rat its concentration in other organs is so low that it is a good marker for liver damage; moreover, its hepatic concentration being about five times higher than that of ALAT, it is more sensitive (Figure 1).

Other enzymes are more specific to test liver damage in larger animal species such as cattle, horses, sheep, ...: Glutamate Dehydrogenase (GLD, EC 1.4.1.2) and Sorbitol Dehydrogenase (SDH, EC 1.1.1.14) are the most widely used but they lack sensitivity (5, 6). Nevertheless, when a toxic compound is known or expected to induce a liver damage, unspecific but sensitive enzyme markers, such as the Aminotransferases, can also be used: for instance, good correlations were observed between their variations and those of SDH in carbon tetrachloride experimental intoxication of calves (7).

Such differences can also be observed for enzyme markers of cholestasis, for which Gamma-glutamyl transferase (GGT, EC 2.3.2.2) is widely used. This enzyme only has a low concentration in the liver of rats, mice, dogs and cats, thus being a poorly sensitive test in these species while its concentration is much higher in the liver of ruminants, horses or monkeys (8-10) in which it is a sensitive indicator of cholestasis.

The increases of serum GGT in hepatobiliary diseases of ruminants are so significant that its measurement has been proposed and is now routinely used as a test of detection of liver diseases such as sporidesmine intoxications (11).

In other species, such as dog, Alkaline Phosphatases (ALP, EC 3.1.3.1) are much more sensitive to indicate signs of cholestasis but they lack specificity; in this occurrence, serum GGT can contribute by its specificity to identify the origin of ALP increase (12).

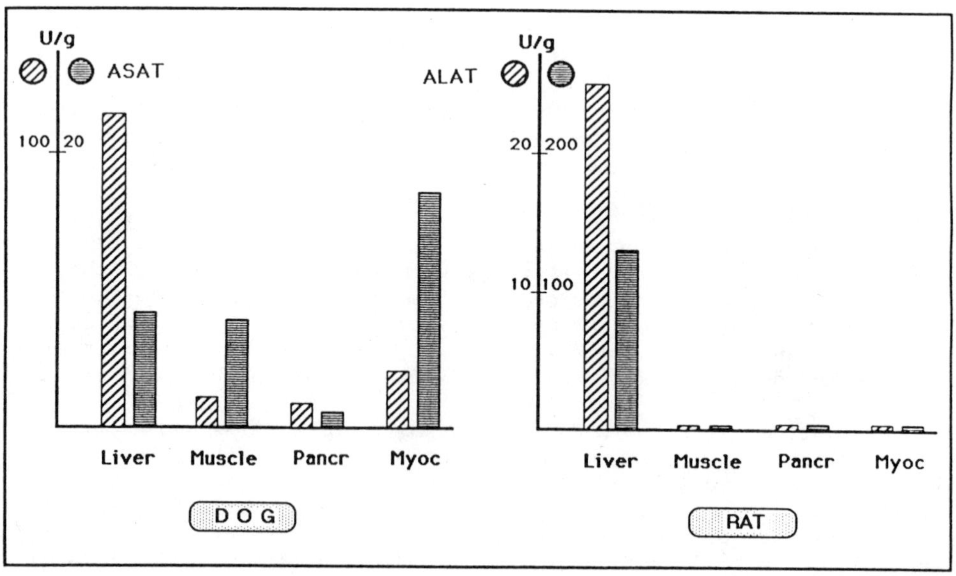

Fig. 1. Comparison of the distribution of alanine aminotransferase (ALAT) and aspartate aminotransferase (ASAT) in the liver, skeletal muscles, pancreas and myocardium of the dog and rat.

2. LIVER-CANCEROGENESIS TESTING

GGT is often used to support in vivo-testing of liver cancerogenesis in rodents. In the liver of the rat and the mouse, this enzyme has a relatively high activity in the foetus and the newborn but it rapidly decreases almost to zero in the adult; when a grafted or chemically-induced cancer develops, preneoplastic lesions and cancer nodules contain a very high GGT activity. This can be observed with a great variety of cancerogens such as 2-acetyl-amino-fluorene, ethionine, 3′-methyl diamino azobenzene (13, 14).

The increase of GGT activity within the liver depends on the compound, of the duration of the intoxication and on the various combinations of promoters which can be used. These changes of liver GGT with the tumor process can be monitored by the measurement of serum GGT which is also increased.

Inter-species differences can also be observed in liver chemical cancerogenesis. The preceding compounds, which induce GGT rich liver hepatomas in rats, are ineffective as liver-canceroigens in guinea-pigs which have a much higher GGT concentration in adults (15). Finally, neither species is a pertinent model of the situation observed in human spontaneous primary or metastatic liver cancers: some authors noted increases of GGT while others observed decreases of GGT in liver tumors. Meanwhile, serum GGT is reported to be increased in 50 to 95% of human liver cancers (18).

3. METABOLIC BASES FOR DIFFERENCES

The understanding of so numerous inter species differences requires the elucidation of the different metabolic pathways in the various species. These cannot easily be investigated and are not yet fully explained, with the exception of some models, as acetaminophene, a widely used analgesic and antipyretic.

Acetaminophen toxicity greatly differs between the rat, mouse and man (16): it is much more toxic for man and mice than for rats (Figure 4) and determines a more or less severe liver necrosis which can be evidenced by

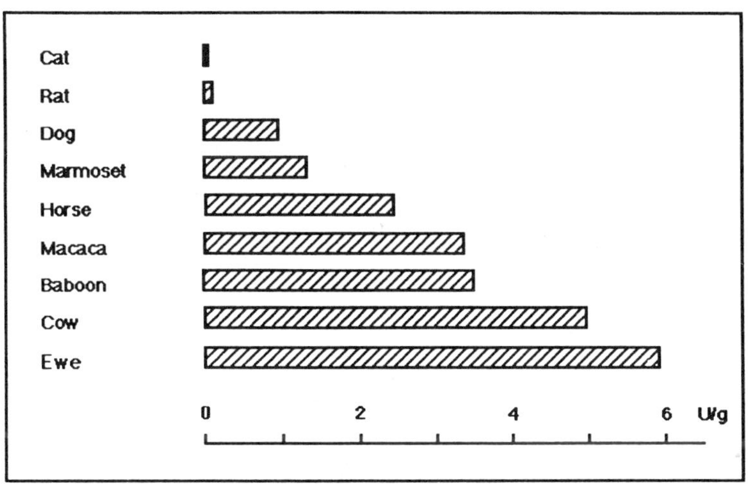

Fig. 2. Comparison of the concentration of gamma-glutamyl transferase in the liver of different animal species.

serum ALAT and ASAT increases (17). Acetaminophene itself is not the ultimate toxic molecule: it needs prior bioactivation by a cytrochrome P-450 oxydative system which leads to NAPQI; N-Acetyl Parabenzo Quinone Imine, which in turn can:

- either covalently bind to liver proteins, thus causing necrosis,
- or be detoxified by glutathione conjugation or reduction to acetaminophene, which can be directly detoxified as glucuronide or sulphate.

Thus, the amount of acetaminophene-mercapturate eliminated is directly related to the formation of NAPQUI, that is to potential toxic liver effects. In Figure 4, it can be observed that in the mouse, the proportion of mercapturate is about twice as high as in rats. So, for acetaminophen toxicity testing, the mouse, and even more the Syrian hamster, is a good model for man (16).

Similar differences are numerous and an animal species cannot be predicted to be a good model for another one in all opportunities. For instance, hexobarbital and phenylbutazone were both reported to have some toxic effect on the liver; the former is mainly an inducer of biotransformations, while the latter is mainly a cytotoxic agent. Their toxic effects differ between species and differences are often related to the biological half-life of these compounds (18). Hexobarbital half-life is very short in the mouse and rat but it is relatively similar in man and dog (respectively about 6 and 3 hours); on the contrary phenylbutazone half-life is about ten times longer in man (approximately 70 hours) than in the dog (approximately 7 hours).

Thus, the dog is both: a) a good model for man for hexobarbital, which is transformed by alipathic oxidation, b) a poor model for man for phenylbutazone, which is transformed by aromatic oxidation.

Such studies were performed about many different metabolic pathways in various species, as shown in Table 1.

In liver toxicity testing, the possibilities offered by clinical chemistry are so numerous and the results thus achieved are so accurate that it cannot be replaced, even by more specific pathologic examinations which are more tedious and require the sacrifice of large numbers of

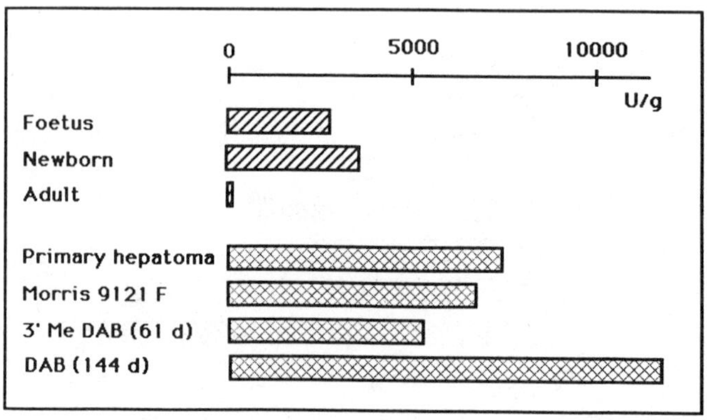

Fig. 3. Activity of gamma-glutamyl transferase in the liver of the normal rat and in grafted and chemically induced hepatomas (3'Me DAB = 3'-methyl diamino azobenzene; DAB = diamino azobenzene; Morris 9121F: grafted hepatoma; d = days).

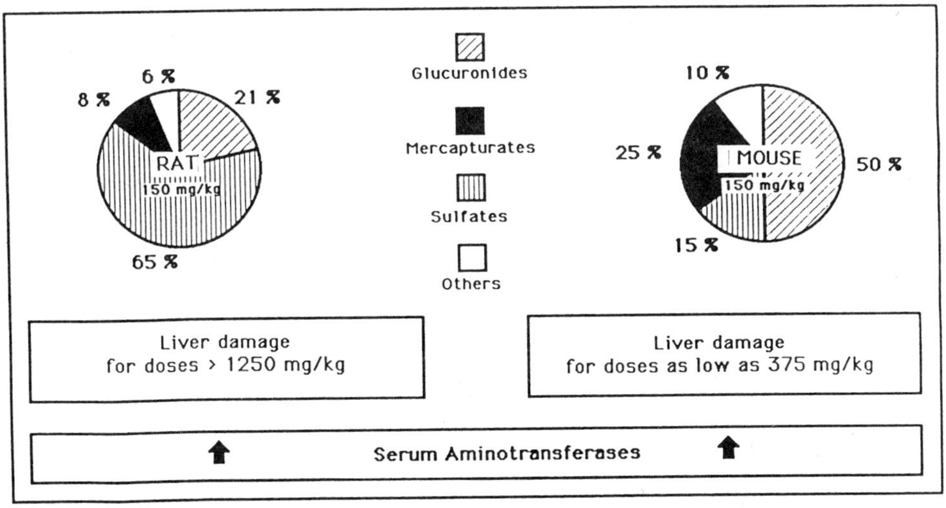

Fig. 4. Correlation between the various forms of elimination of acetaminophen and its toxicity in rats and mice.

animals. But interspecific differences rest mainly on the specific distribution of enzymes that determine the reaction of the animal to the toxic challenge. These differences have to be known to avoid some errors and also to understand more precisely the mechanisms of action of toxic compounds, thus to cure or at least to limit their effects.

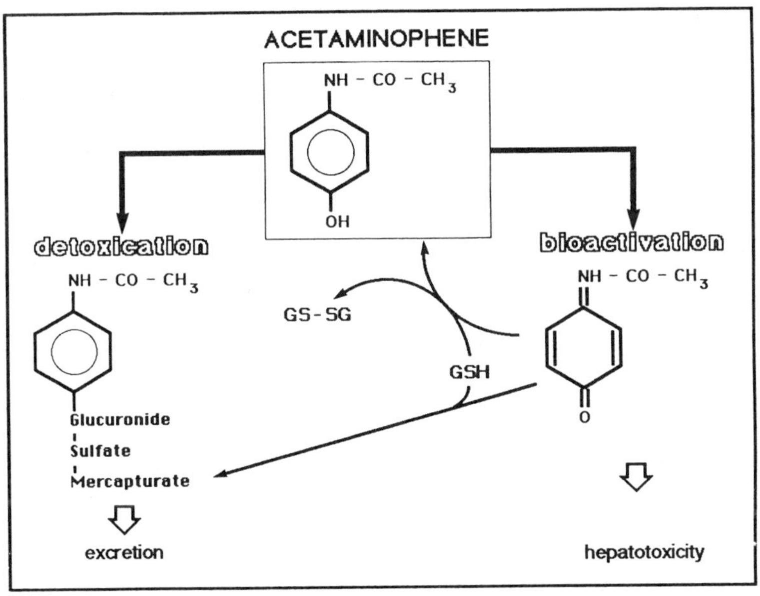

Fig. 5. Metabolic pathways of acetaminophene.

Table 1. Inter-specific differences for Phase 1 Reactions of drug metabolism (++: good model for man; +: intermediate model; ± poor model)

	Rat	Rabbit	Dog	Monkey
Aliphatic oxidation	±		++	
Aromatic oxidation	+		+	±
Dihydrodiol formation	±	++	++	
N and O dealkylation	±		++	
S dealkylation	++		±	
Deamination	±	++	±	
Dehalogenation	±		++	++
N hydroxylation	±		±	±
Ketone reduction	++			
Hydrolysis	++		++	

REFERENCES

1. Zimmerman HJ. Various forms of chemically induced liver injury and their detection by diagnostic procedures. Environ Hlth Persp 1976; 15: 3-12.
2. Knox WE. Enzyme patterns in fetal, adult, and neoplastic rat tissues. Basel: Karger, 1976.
3. Schmidt E, Schmidt FW. Enzyme activities in human liver. Enzym Biol Clin 1970; 11: 67-129.
4. Keller P. Enzymaktivitäten in Organen, Zellfraktionen und Körperflüssigkeiten des Hundes under spezielle Berücksichtigung klinisch-diagnostischer Aspekte. Schweiz Arch Tierheilk 1986; 128: 1-25.
5. Boyd JW. The comparative activity of some enzymes in sheep, cattle and rats. Normal serum and tissue levels and changes during experimental liver necrosis. Res Vet Sci 1962; 3: 256-268.
6. Keller P. Serumenzyme beim Rind: Organanalysen und Normalwerte. Schweiz Arch Tierheilk 1971; 113: 615-626.
7. Anwer MS, Engelking LR, Gronwall R, Klentz RD. Plasma bile acid elevation following CCl_4 induced liver damage in dogs, sheep, calves and ponies. Res Vet Sci 1976; 20: 127-130.
8. Braun JP, Soula G, Rico AG, Bénard P, Burgat-Sacaze V. Gamma-glutamyl transférase et cancer. Ann Biol Clin 1981; 39: 53-59.
9. Braun JP, Rico A, Bénard P. Evolution de la distribution de la gamma-glutamyl transférase dans les organes de la souris en fonction du développement. Paris: CR Acad Sci, 1978; 286: 1483-1485.
10. Braun JP, Rico A, Bénard P, Burgat V. Distribution of γ-glutamyl transferase in some domestic animals. In: Goldberg DM, Wilkinson JH. Enzymes in health and disease. Basel: Karger, 1978: 46-49.
11. Bézille P, Braun JP, LeBars J. Première identification de l'eczéma facial chez les ovins en France. Aspects épidémiologiques, cliniques et biologiques. Rec Méd Vét 1984; 12: 339-347.
12. Guelfi JF, Braun JP, Bénard P, Rico AG, Thouvenot JP. Value of so-called cholestasis markers in the dog: an experimental study. Res Vet Sci 1982; 33: 309-312.
13. Fiala S, Fiala AE, Dixon B. γ-glutamyl transpeptidase in transplantable, chemically induced rat hepatomas and "spontaneous" mouse hepatomas. J Natl Cancer Inst 1972; 48: 1393-1401.

12. Guelfi JF, Braun JP, Bénard P, Rico AG, Thouvenot JP. Value of so-called cholestasis markers in the dog: an experimental study. Res Vet Sci 1982; 33: 309-312.
13. Fiala S, Fiala AE, Dixon B. γ-glutamyl transpeptidase in transplantable, chemically induced rat hepatomas and "spontaneous" mouse hepatomas. J Natl Cancer Inst 1972; 48: 1393-1401.
14. Harada M, Okabe K, Shibata K, Masuda H, Miyata K, Enomoto M. Histochemical demonstration of increased activity of γ-glutamyl transpeptidase in rat liver during hepatocarcinogenesis. Acta Histochem Cytochem 1976; 9: 168-179.
15. Fiala S, Fiala ES. Activation by chemical carcinogens of γ-glutamyl transpeptidase in rat and mouse liver. J Natl Cancer Inst 1973; 51: 151-158.
16. Savides MC, Oehme FW. Acetaminophen and its toxicity. J Appl Toxicol 1983; 3: 96-111.
17. Dixon MF, Fulker MJ, Walker BE, Kelleher J, Losowsky MS. Serum transaminase levels after experimental paracetamol-induced hepatic necrosis. Gut 1975; 16: 800-807.
18. Adamson RH, Davies DS. Comparative aspects of absorption, distribution, metabolism, and excretion of drugs. In: Michelson MJ. Comparative Pharmacology. Oxford: Pergamon Press, 1973: 851-938.

LIPID METABOLISM IN LARGE ANIMALS

Th. Wensing*

State University, Veterinary Faculty, Utrecht
The Netherlands

Studying lipid metabolism in large animals has sense as the results either can be of help in elucidating the etiology of metabolic disorders that affect economic loss in large animal husbandry or can demonstrate how large animals can be used as models in fundamental research in the field of lipid metabolism.

In relation to items as the mobilisation of depot fat, either by starvation or by long term labor and of VLDL metabolism, the pony, especially the Shetland pony, has been found to be an attractive experimental animal. The results obtained in the study of milk fat depression in lactating cows suggest that ruminants can be used as a model in research that aims to study how the fatty acid composition in the blood can be influenced by changes in the diet. Finally evidence is obtained that sheep can be of help in the study of the processes that underlie fatty infiltration of the liver.

Though the results obtained in the research of lipid metabolism in ponies, lactating cows and sheep are important in relation to fundamental questions concerning lipid metabolism, first of all they have contributed to the solution of veterinary problems. In relation to hyperlipoproteinaemia of ponies results of experiments have lead to a better prevention and have widened the possibilities to treat patients suffering from hyperlipoproteinaemia. As far as milk fat depression in high producing dairy cows is concerned, it was found how milk fat depression can be prevented and what measures have to be taken to restore normal fat milk production. In relation to fatty infiltration of the liver in cows, a lot of information has been published about the consequences of this phenomenon as an increased susceptibility to metabolic disorders, a decreased fertility and a decreased milk production. Much less is known, however, about the processes that induce the increased uptake of fat by the liver.

---

* Dr. Th. Wensing, associate professor. Head of the department for clinical chemistry and toxicology. Department for Large Animal Medicine. Veterinary Faculty, State University Utrecht, Yalelaan 16, 3508 TD Utrecht, The Netherlands.

The first investigations into the backgrounds of hyperlipoproteinaemia in ponies started in the early sixties (1, 2, 3, 4). It has been found that the hyperlipoproteinaemia in ponies is a hypertrigyceridaemia caused by a tremendous increase of the mobilisation of depot fat. This disturbance of the fat metabolism goes together with increased serum activities of liver specific enzymes indicating live cell damage, Figure 1.

Evidence was obtained that the hyperlipoproteinaemia in ponies resembles the Frederickson types IV and V. At postmortem in the hyperlipoproteinaemic ponies a fatty degeneration of the liver was found. Till now the reason why the mobilised fatty acids are not metabolised as usual under circumstances of a negative energy balance but are used for the formation of triacyl glycerols in the liver is not known. Neither is the reason why the very low density lipoproteins circulating in the blood in very high concentrations are not metabolised. It has been demonstrated that the hyperlipoproteinaemia in ponies can be induced by starvation, that an inadequate metabolisation of free fatty acids may play a role in the development of this hyperlipoproteinaemia and that liver cell damage seems

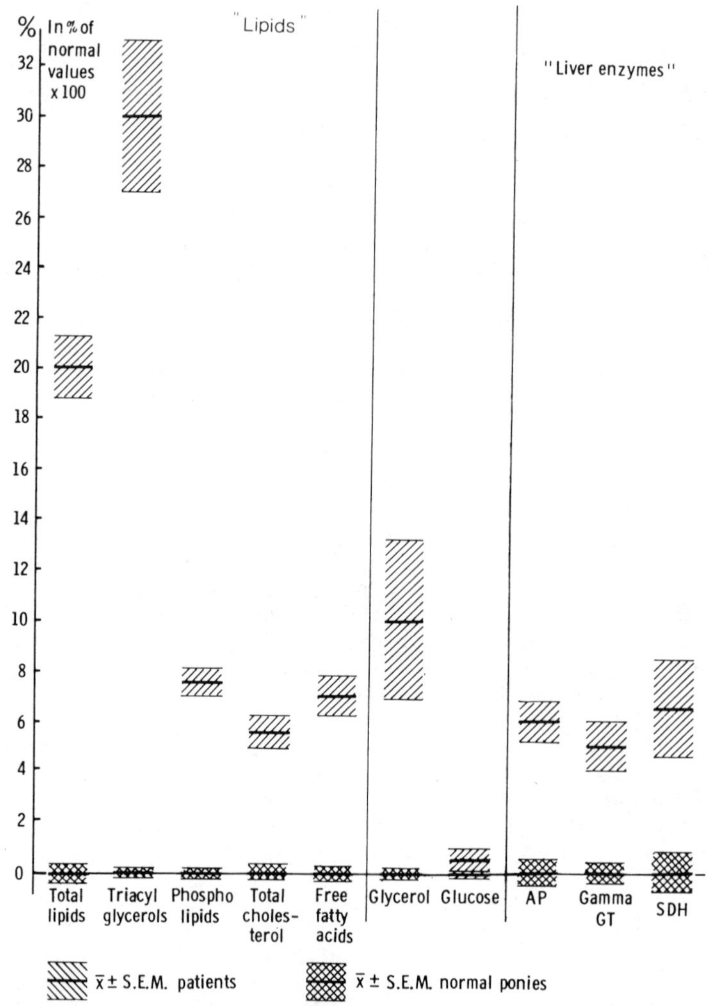

Fig. 1. The blood "profile" of hyperlipaemic ponies compared with that of clinically normal ponies.

to start in the same time as the FFA concentration in blood increases, Figure 2. The excretion of triacylglycerols from the liver into the blood increases later, suggesting that liver cell damage preceeds the onset of the increased TG excretion.

In order to check whether the very low density lipoproteins, VLDL's, circulating in the blood of hyperlipaemic ponies can be metabolised, lipaemic plasma originating from hyperlipaemic ponies has been infused into clinically normal ponies of the same breed and age. It was found that the infused VLDL were metabolised in the healthy pony, Figure 3. This observation suggests that the VLDL's circulating in the blood of hyperlipaemic ponies do not have a strongly deviating structure that makes a normal breakdown impossible. The results presented so far suggest that the Shetland pony is not capable to metabolise mobilised fatty acids adequately. This, however, is not true. It is well known that the Shetland pony can do long-term labour at least partly on the base of a breakdown of mobilised fatty acids.

Comparing the changes in the blood concentration of glucose and insulin in ponies during forced starvation with those occurring during long term forced labour shows that starvation effects a rather sudden decrease of the blood glucose concentration and a decrease of the insulin concentration which is not seen during long term labour, Figure 4. On the base of the results presented before one could suppose that a temporarily decrease of insulin as observed during starvation induced lipolysis to a degree that a complete breakdown of the mobilised fatty acids via the beta oxidation is not possible or, facing the actual need, is not completely necessary. Consequently the rather toxic fatty acids that are not broken down are used for triacyl glycerol synthesis and stored as such in the liver or are excreted in VLDL's. As VLDL's can inactivate insulin this again effects an increase of lipolysis. In that way a vicious circle is closed and lipolysis and therewith VLDL production and excretion are at least partly out of control. The observation that a treatment with insulin can effect clear improvement in the hyperlipaemic ponies, especially in the phase that the hyperlipoproteinaemia is developing, supports this hypothesis.

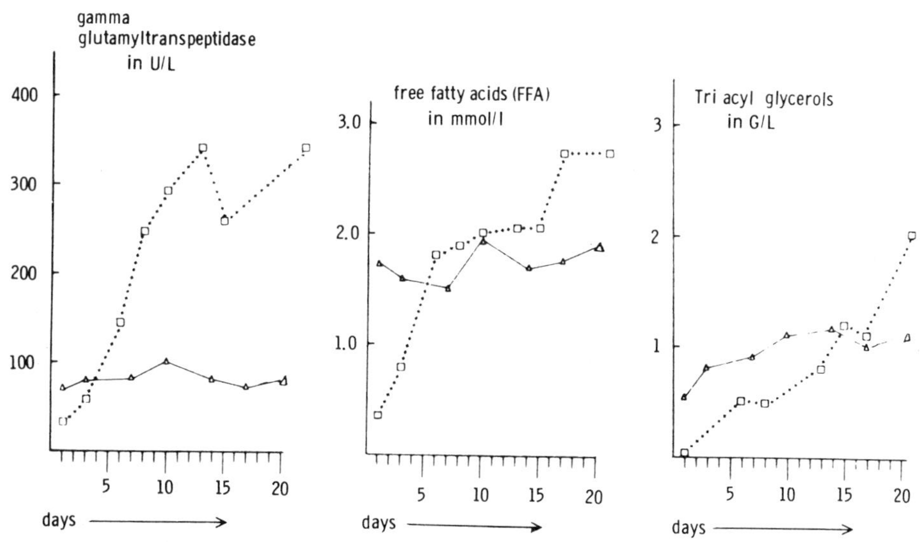

Fig. 2. The changes in the serum activity of gammaGT and serumtriacylglycerol and FFA concentration in two ponies during starvation.

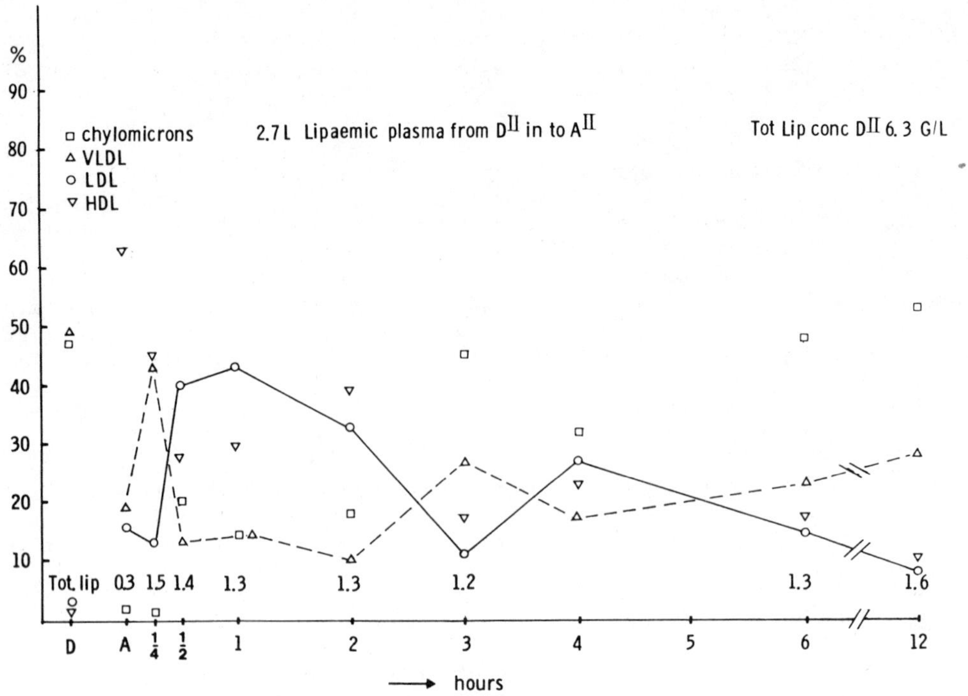

Fig. 3. The changes in the lipoprotein spectrum of a clinically normal pony after i.v. infusion of lipaemic plasma.

Since the clear increase in milk production that was made possible by genetic improvement, better nutrition, housing and management of dairy cows, the frequency of the occurrence of milk fat depression has increased clearly. Information has been collected indicating that milk fat depression can develop in cows fed a ration with a roughage to concentrate ratio smaller than 1 : 2 especially when roughage with a low amount of fibrous material is fed. In order to elucidate the etiology of milk fat depression the possible role of the physical form of the roughage, the composition of the concentrates and the feeding regime in the development of milk fat has been investigated intensively.

In connection with this presentation I should like to focus on experiments which aimed to study changes that occur during the development of milk fat depression in high yielding dairy cows:

- in the rumen fluid especially with regard to the pH and the pattern of the volatile fatty acids,
- in blood as far as parameters as insulin, glucose, glycerol, non esterified fatty acids (NEFA's) and the composition of the fatty acid fraction are concerned and
- in milk , especially with respect to milk production, fat content and the composition of the fatty acid fraction in the milk.

Increasing the amount of concentrates in de diet mostly resulted in a clear increase in the molar proportion of propionate in the rumen fluid. This effect was more clearly when flaked maize was incorporated in the concentrates.

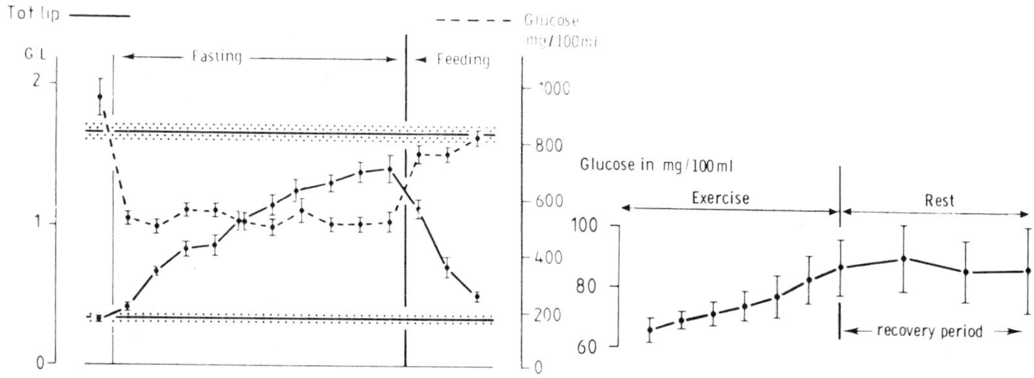

Fig. 4. The course of the blood glucose concentration in ponies during exercise and during starvation.

In literature it is suggested that milk fat depression results from depressed lipolysis caused by an increased insulin activity. In this theory, the glucogenic theory, the mentioned increase of the propionate concentration is thought to effect an increased blood glucose concentration with consequently an increased blood insulin activity (5).

The soundness of this theory was tested frequently. In some experiments increased blood glucose and insulin concentration have been found in milk fat depressed cows which indeed supports this theory. In other experiments, however, it was found that infusions of propionate into the rumen effected only minor changes in the blood insulin concentration (6, 7). In connection with milk fat depression experiments have been carried out at our institute by Van Beukelen. In an experiment with four fistulated high yielding dairy cows, milk fat depression was induced by feeding a high concentrate, low roughage ration. All cows developed a clear milk fat depression that went together with a clear increase of the propionate concentration in the rumen fluid. It was found that in the period that a milk fat depression existed serum insulin activity was increased at two hours after feeding (8). The cow that experienced the most severe milk fat depression showed the less severe weight loss. These results support the glucogenic theory.

In relation to changes in the fatty acid composition of the lipids in blood, milk and body fat stores during milk fat depression some information is present in literature. In our institute the effect of milk fat depression on this fatty acid composition has been investigated too. In experiments with high yielding dairy cows, milk fat depression was induced by feeding chopped hay or by feeding a low roughage, high concentrate ration. In the blood of the cows fed chopped hay a decrease in the percentage of linoleic acid was found which was most clearly in the cow that experienced the most severe milk fat depression, Figure 5.
In the milk the percentage of stearic acid and oleic acid were somewhat increased. During milk fat depression in cows fed a low roughage, high concentrate ration, in the blood a clear increase of the percentage of linoleic acid could be observed whereas the percentages of stearic acid and oleic acid decreased. In the milk of these cows the percentage of stearic acid was increased, that of oleic acid was decreased. The increase in the percentage of oleic acid could only be observed in cows that experienced a severe milk fat depression that was induced by a low roughage high concentrate ration. During milk fat depression, in the body fat stores the percentage of stearic acid was increased that of oleic acid was decreased.

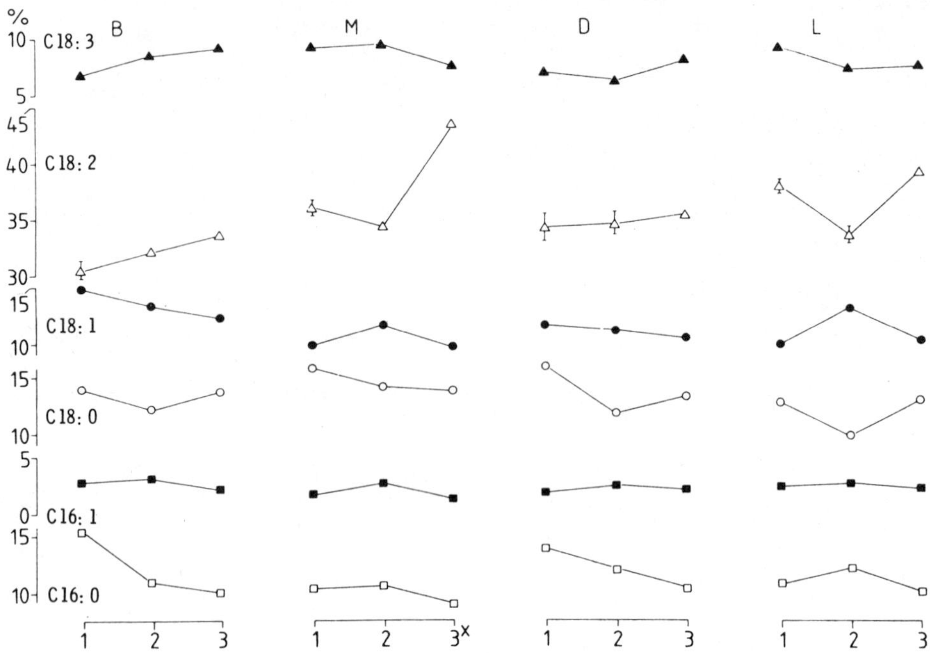

Fig. 5. Means of SEM, of some fatty acids in the blood lipid fraction of 4 cows during a control period, 1, during milk fat depression, 2 and during recovery from milk fat depression, 3.

In conclusion it can be stated that milk fat depression has its origin in the rumen and that during the developing of milk fat depression the fatty acid composition in blood, milk and body fat stores is changing. The observation that milk fat depression can go together with clearly increased serum activities of some "liver specific" enzymes, however, suggests that the liver is involved too.

Apart from milk fat depression, there is another disturbance of fat metabolism in dairy cows that can cause great economic loss, the fatty liver. Bovine fatty liver, due to accumulation of triacylglycerols is mostly seen in preparturient dairy cows. It is obvious that cows with a fatty liver are more frequently suffering from metabolic diseases as "fat cow syndrome", ketosis and milk fever, and from reduced fertility (9, 10, 11). The frequency that fatty liver occurs in dairy cows is not easy to document. Reid found that about 65% of the FH and about 35% of the Guernsey cows in England had a moderate or severe fatty liver at one week post partum. At 8 weeks post partum the incidence was found to be much less (11). As increased serum activities of "liver specific" enzymes are only weak indicators for fatty liver, investigation of liver biopsies is the only valid method to diagnose fatty liver definitely. In order to make taking a liver biopsy from a high producing dairy cow superfluous Reid tried to find from blood parameters that on its own, or in combination with others, correlates with liver fat content.

It was found that changes in the serum activity of AST and in serum NEFA and blood glucose concentrations indeed correlate with the fat percentage of the liver (11). For these investigations moderate yielding cows have been used. It is questionable, however, whether the mentioned correlation is valid for high yielding cows too. As a matter of fact, both the diagnosis of fatty liver and the etiology of the fatty infiltration of the liver are still under investigation. As far as the diagnosis is

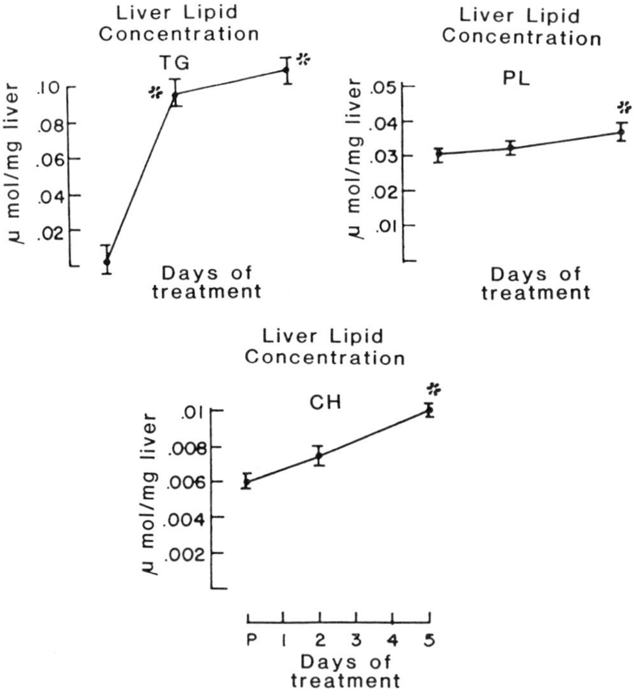

Fig. 6. The triacylglycerol (TG), cholesterol (CH), and phospholipid (PL) content in the liver of a sheep sampled before induced fat mobilisation, P, and on the second (2) and fifth day (5) of induced fat mobilisation.

concerned some investigations have been done last year at about 40 dairy farms in The Netherlands. Special attention was given to cows ante partum in order to find out whether fatty liver occurs at that time too. The results were very disappointing, no correlation was found between liver fat content and changes in the blood.

In respect to the etiology of fatty liver in dairy cows, some experiments have been carried out with sheep by Herdt at our institute (12). In these experiments attention was paid to both the possible factors that can result in accumulation of triacylglycerols in the liver: an increased TG synthesis in the liver and a decreased TG secretion from the liver into the blood. It is well established that VLDL synthesis and VLDL secretion rates are determined at the point of glycerolipid synthesis i.e. the synthesis of triacylglycerols and neutral phospholipids (13, 14). This rate, in turn, seems to be regulated by the relative activities of one or more enzymes: Glycerophosphate acyltransferase, GPAT, diacylglycerol acyltransferase, DGAT, and phosphatidate phosphohydrolase, PAP (5).

Under circumstances of phloridzin induced fat mobilisation it was found that a quick and clear accumulation of TG in the liver occurred, whereas the increase of cholesterol and phospholipids was moderate and seemed to be somewhat delayed, Figure 6. Preceeding the accumulation of TG in the liver, serum NEFA concentration increased dramatically with treatment. Evidence was obtained that the ratio of surface lipid to core lipid increased with time during the increased mobilisation of depot fat. In addition it was found that PAP and DGAT activities in liver biopsies increased clearly during accumulation of TG in the liver. These observations support the hypothesis that fatty liver at least partly results from both an increased uptake of fatty acids by the liver and a delayed production of VLDL's by the liver.

On the base of the experiences obtained till now it can be concluded that sheep can function very well as a model in the research of the etiology of fatty infiltration of the liver.

REFERENCES

1. Morris MD, Zilversmit DB, Hintz HF. J Lip Res 1972; 13: 383.
2. Schotman AJH, Wagenaar G. Zentralbl Vet Med A 1969; 16: 1.
3. Wegen P. The Netherlands: The State University Utrecht. Thesis 1979.
4. Wensing Th, Van Gent CM, Schotman AJH, Kroneman J. Clin Chim Acta 1974; 58: 1.
5. McClymont GL, Vallance S. Proc Nutr Soc 1962; 21: xli.
6. De Jong A. J Endocrin 1982; 92: 357.
7. Stern JS, Baile CE, Mayer J. Amer J Physiol 1970; 219: 84.
8. Van Beukelen P, Wensing Th, Breukink HJ. Z Tierphysiol Tierernärhrg u Futtermittelkde 1982; 48: 182.
9. Collins RA, Reid IM. Res Vet Sci 1980; 28: 273.
10. Gerloff BJ, Herdt TH, Emery. 8th International Symposium on Rumen physiology. Canada: Banff, 1984.
11. Reid IM. Vet Rec 1980; 107: 281.
12. Herdt et al. 1987; in press.
13. Beynen AC, Haagsman HP, Van Golde LMG, Geelen MJH. Biochim Biophys Acta 1981; 665: 1.
14. Haagsman HP, Van Golden LMG. Arch Bichem Biophys 1981; 208: 395.

CHAPTER 25
CLINICAL BIOCHEMISTRY IN DEVELOPING COUNTRIES

CLINICAL BIOCHEMISTRY IN DEVELOPING COUNTRIES IN ASIA

It-Koon Tan

Clinical Biochemistry Laboratories, Department of Pathology
Outram Road, Singapore 0316, Republic of Singapore

Asia covers a vast geographical area of some 20.75 million km^2 and contains more than half the world's total population of close to 5000 million. The region has about 25 countries/areas in various stages of socioeconomic development. There is great variation in land area, size of population as well as financial and manpower resources. More than 90% of the population live in the developing countries in the region. In 14 countries including China, India, Indonesia and Bangladesh which are 4 of the most populous in the region, 70% to 80% of the population live in rural areas where communication and transportation may present serious difficulty in health-care delivery and education. Of those who move to the cities, considerable number live in urban slums where the environment is not conducive to health. Even among the developing countries and between the urban and rural areas of the same country, there are considerable differences in the needs for health care. It is therefore not surprising that the diversity in requirements and national resources to meet these requirements give rise to wide variation in the priority and standard of health care and in the scope and sophistication of the clinical laboratories. Provision for laboratory services and facilities will have to be made according to the nature and magnitude of prevailing health problems which affect the population. Therefore it is pertinent to examine the major health problems confronting the developing countries/areas in Asia before taking a look at the current status of clinical biochemistry in these countries/areas.

REGIONAL HEALTH PROBLEMS

A high incidence of communicable diseases and malnutrition is found in countries at the lower levels of development. The principal communicable diseases in the region are diarrhoeal diseases, malaria, tuberculosis, acute respiratory infections, and leprosy. At the 39th World Health Assembly held in May 1986 in Geneva, national representatives from India, Laos, Nepal, Pakistan, and Sri Lanka highlighted the problems of these diseases (1).

Basic public health facilities such as supply of safe accessible drinking water and sanitation services which have been taken for granted in the developed countries are still inadequate in a number of developing countries, particularly in the rural areas. Although discernible progress has been made in the development of drinking-water supply facilities and their management in all countries in South East Asia, much remains to be done in the field of sanitation which is progressing very slowly. It has been estimated that by 1986, about 1200 million people in WHO's member states were still without safe drinking water and 1700 million without proper sanitation (2). An estimated 88% of the resources have been directed towards the towns. As over half of the world's population live in Asia and as the region comprises mainly developing countries/areas, it seems a reasonable estimate that 600-850 million people or more in Asia are without water and sanitation. This situation has contributed to the high endemicity of water-borne and diarrhoeal diseases, especially among children. Poor quality of potable water and inadequate waste disposal also favoured intestinal parasitism and skin diseases. The pattern of these diseases varies considerably both in magnitude and severity among the countries/areas. In Laos, the Philippines, Vietnam and Malaysia, acute diarrhoeal diseases were reported to be among the leading causes of morbidity and hospital admission (3). In Thailand, the problem is especially serious in children under 5 years (4).

In conjunction with diarrhoea, poverty and inadequate knowledge of sound nutritional habits have resulted in malnutrition being another major cause of illness in countries at the lower levels of development in the region. Malnutrition has been a main reason for the very high maternal, infant and young child morbidity and mortality, as well as retardation of growth and development, and lowered immunity to disease (3). Nutritional anaemia, iron deficiency and vitamin A deficiency have also been noted to be significant public health problems (5). Xerophthalmia is endemic in Bangladesh, Indonesia, India, Nepal, the Philippines and Sri Lanka and is also strongly suspected to be a problem in Burma, Kampuchea and Laos (6).

Despite the fact that there is an apparently simple and effective solution, iodine deficiency diseases affect a number of Asian countries. Communities with inadequate iodine suffered not only in educability but also in social development and ultimately in economic productivity. Extensive surveys had indicated that more people are affected and levels of severity are higher in South East Asia than anywhere in the world. In Bangladesh, Bhutan, Burma, India, Indonesia, Nepal, Sri Lanka, and Thailand, about 277 million people live in significantly iodine deficient areas, and about 102 million had goitre, about 6 million were cretins and another 36 million suffered from some degree of mental or motor impairment (7). In the northern parts of China and Pakistan and the mountainous regions of Vietnam, large number of people live in iodine deficient areas (8). It has been estimated that in China alone, approximately 300 million people are affected and only one third of that population had been adequately covered by control programmes.

In recent years, hepatitis B virus (HBV) infection and dengue fever have emerged as serious public health problems (5). The percentage of population carrying the HBV varies between 2 to 7% in South West Asia, and between 8 to 20% in South Asia and some parts of China (9). Considerable concern has also been expressed about the increasing prevalence of sexually transmitted diseases (3). Although the number of cases of acquired immune deficiency syndrome (AIDS) in Asia is still very low compared to other parts of the world, the deadly disease is now an increasing cause of worry because of the very high cost of detection, surveillance and medical care involved as well as the difficulty of mass education and control (10).

Even though communicable diseases are still the leading causes of morbidity and mortality in a number of countries, non-communicable and chronic degenerative diseases, environmental pollution, mental disorders, accidents, and diseases related to life-style (e.g. smoking and drug abuse) have reached such a proportion that it has become necessary for health authorities to give increasing support for the development and expansion of hospital and laboratory services required for these problems, especially in the urban areas.

CURRENT STATUS OF CLINICAL BIOCHEMISTRY

(A) General Situation

Table 1 provides comparative data on population, provision of hospital beds and availability of doctors in 20 countries (11-15) and Table 2 shows the degree of automation of biochemical analyses in 15 countries (16).

The most highly urbanised countries/areas are Singapore and Hong Kong where more than 92% of the population live in the city. These are followed by Taiwan, Brunei and South Korea where over 57% of the population live in the urban areas. These 5 countries/areas have the highest per capita income and also have relatively higher ratios of doctors to population and hospital beds to population. The main health problems are similar to those of the developed and industrialised countries, the principal causes of death and illness being: cardiovascular disease, hypertension, cancer, diabetes mellitus, respiratory disease, and accidents. Although the development of clinical biochemistry started some 10 years later than that in the developed countries, current status of practice, in terms of provision of routine and emergency analyses for patient care, is quite comparable to the trends in the more advanced countries. A wide variety of investigations ranging from the standard repertoire of tests to more specialised analyses such as those for endocrine investigations, study of cancer and inborn errors, as well as drug monitoring are provided for the routine diagnosis and management of patients. Biochemical screening of apparently healthy adults and infants is also conducted for the early detection and control of disease. The laboratories are highly automated and in a number of cases, also highly computerised. A variety of modern sophisticated multi-channel analysers are used for analytical work.

In the other countries where communicable diseases and various nutritional deficiencies are more important health problems, provisions for the hospital and laboratory have been more limited. The ratio of doctors to 1000 population is less than 0.5 in 14 countries and significantly less than 0.2 in half of these countries (Table 1). In 9 countries, only one or less than one bed is provided for 1000 persons. The number of doctors and hospital beds available for health care in the rural areas must be considerably less than the figures given for national average, as only 25% of the population in 12 countries live in the urban areas where the majority of the better and larger hospitals are located and where a large proportion of the trained personnel prefer to work. More than 75% of the population have to depend on the health centers/stations for their health care needs. Laboratory testing is not a usual part of the basic services in these centers. Where laboratory service is available, it is confined to simple haematological, microbiological and urine tests. Due to problem of distance and lack of transportation, large number of people do not make use of such centers. Instead, they rely on traditional medicine and self-care. In a recent television documentary film featuring health care in remote areas of Indonesia, attention was drawn to the acute shortage of medical personnel. One doctor would have to look after 40,000 population scattered among villages over a large area. For the doctor to get from one village to another requires several hours' journey by foot. There is no other form of transport because of the rough terrain. This means that while a small proportion of the urban population in many countries in Asia is served by a

Table 1. Comparative data for 20 developing countries in Asia (11)

Country/ Area	Population (million)	Percentages in Urban Areas	Hospital Beds/1000	Doctors per 1000	Life Expectancy
Afghanistan	15.4	16	0.32	0.06	37
Bangladesh	104.1	13	0.28	0.16	48
Bhutan	1.4	5	0.67	0.06	46
Brunei	0.2	64	3.15[a]	0.5	71
Burma	37.7	24	0.7	0.3	58
China	1050	20[b]	2.14	0.7[b]	64
Hong Kong	5.7	92	4.5	0.9	75
India	785	23	0.8	0.28	53
Indonesia	168.4	22	0.67	0.11	55
Korea (South)	43.3	57	2.4	1.3	66
Laos	3.7	16	1.0	0.05	44
Malaysia	15.8	32	1.61[c]	0.3[c]	67
Nepal	17.4	6	0.22	0.04	46
Pakistan	101.9	28	0.58	0.44	50
Philippines	58.1	37	1.45[c]	0.88[c]	62
Singapore	2.6[d]	100	3.8[d]	1.0[d]	71
Sri Lanka	16.6	22	2.9	0.2	68
Taiwan	19.6	67	3.7	0.9	73
Thailand	52.8	17	1.33	0.16	63
Vietnam	62	19	3.4	0.3	59

Note: For a, b, c, and d, please see respective references (12, 13, 14, 15).

Table 2. Comparative data on national per capita income and laboratory automation in 15 Asian-Pacific countries (16)

Country or area	Per capita income in US dollars	Automated instruments (number)	Automated instruments per million population	Automated instruments per 1000 hospital beds
Australia	8913	351	22.22	3.77
Bangladesh	135	3	0.03	0.11
Brunei	21600	5	22.73	7.25
Burma	172	4	0.11	0.15
Hong Kong	5937	73	13.27	2.95
India	193	61	0.08	0.07
Indonesia	566	40	0.24	0.37
Japan	9714	5097	42.19	2.91
Malaysia	1996	27	1.72	1.02
Pakistan	360	27	0.27	0.50
Philippines	603	39	0.69	1.14
Singapore	6922	46	17.70	4.66
Sri Lanka	340	3	0.18	0.07
Taiwan	3068	136	7.08	2.21
Thailand	646	28	0.53	0.42

few well-equipped and well-staffed hospitals and laboratories comparable to those of the developed countries, the majority do not enjoy the benefits of such modern health care facilities and expertise.

In general, health authorities have tended to be more pre-occupied with the establishment, provision and improvement of reliable services appropriate for the prevention, diagnosis and monitoring of the more prevalent communicable diseases caused by viruses, bacteria and parasites. WHO support has also been largely concentrating on the surveillance, control and eradication of such diseases. Therefore, much of the WHO programmes for the laboratory have been aimed at improving microbiological diagnostic services. Under such circumstances, it is understandable that the range of biochemical tests offered by public health and clinical laboratories must be quite limited. It is quite typical to find that clinical laboratories in most small district hospitals are unable to do even the first few tests listed in Repertoire (A) given below, while those in larger district/provincial hospitals offer (A), or (A) with part of Repertoire (B), or (A) and (B), depending on their size and availability of resources.

Repertoire (A): total protein, urea, glucose, bilirubin, ALP, AST, calcium, cholesterol, urinalysis especially for reducing sugar, protein, ketones, and microscopy.
Repertoire (B): sodium, potassium, chloride, total CO_2, albumin, creatinine, ALT, amylase, urate, phosphate.
In laboratories of major general hospitals in large cities, tests for other analytes may be added to the repertoire. The exact nature and number of tests are determined by the interest and status of the hospitals and the availability of financial support. The tests may include: CK, GGT, triglycerides, electrophoresis of proteins and enzymes, thyroid function tests, and tests for steroids and urinary pigments. Therapeutic drug monitoring (TDM) is only carried out in a few national central laboratories. Facilities for assessment of acid-base status is still not widely available. Investigations requiring special instruments, costly commercial reagents and highly skilled staff are also not widely performed.

A recent study on the status of automation in clinical biochemistry laboratories in Asia showed that most of the developing countries still rely on manual methods (16). When comparative assessment on utilisation of automated instruments was made on the common basis of "no. of instruments per million population" and "no. of instruments per 1000 hospital beds", laboratories in Brunei, Hong Kong, Singapore and Taiwan are found to be more highly mechanised. When figures for the 1985 per capita income of 13 countries were plotted against those for "no. of instruments per million population" and "no. of instruments per 1000 hospital beds", correlation coefficients (r) of 0.9930 and 0.9477 were obtained respectively, indicating a strong correlation between income status and number of automated equipment purchased. Owing to the high cost of imported automated instruments and low cost of manpower and problem of unemployment, governments would rather provide employment for their workforce whenever possible. Maintenance, repair and other problems make it difficult or impractical for any clinical laboratory outside major cities to use automated instruments. Many laboratories which cannot afford automation have introduced semi-automatic devices such as dispensers and dilutors for work-simplification.

In several countries such as India, Indonesia, Philippines, and to a lesser extent also Malaysia, Singapore, Taiwan, private hospital laboratories and independent clinical laboratories play an important role in providing fairly comprehensive range of biochemical analyses, especially for those able to pay for the service. In India, Indonesia and Philippines, such laboratories are often better equipped than those in public hospitals because of the absence of financial and bureaucratic constraints. In India, 80% of automated instruments are in the private laboratories. In Indonesia and the Philippines, senior professional staff of the university and public hospital laboratories are employed in the private sector as consultants to advise and supervise the laboratories on a part-time basis.

(B) Laboratory Staffing

Generally, a pathologist with or without training in clinical biochemistry is the overall administrator of the larger laboratory departments providing a service for the different branches of laboratory medicine. In countries or areas influenced by the British system, either a clinical pathologist with relevant training and experience or a clinical biochemist may be employed to supervise biochemical analyses. Various grades of biochemists and technicians with different levels of responsibility are engaged to work under the head of the biochemistry section. This is the current practice in Brunei, Hong Kong, Singapore and Malaysia. The majority of pathologists obtain their qualifications either from Britain or Australia. BSc(Hons) graduates in biochemistry who gained recognised academic/professional qualifications in clinical biochemistry such as MSc, PhD, MCB(UK), MRCPath(UK), MAACB(Australia) have the opportunity of being promoted to senior positions and to the headship of a biochemistry laboratory. However, since most universities do not have a distinct clinical biochemistry department, those wishing to pursue higher qualifications in the field can only do so abroad. Technical staff comprise of general degree science graduates, polytechnic graduates who attended a full-time course on medical technology and those who had completed their secondary school education and had sat for qualifying examinations after receiving 2 to 5 years of in-service training. The situation in India is similar. Laboratories attached to teaching and non-teaching institutions are adequately staffed with postgraduates in medicine and biochemistry, graduates in science and laboratory technologists. Length of training programmes for technologists ranges from 6 months to 3 years. Not all small hospitals, however, have adequately qualified staff.

In countries/areas where practice of laboratory medicine is influenced by the American system, a pathologist or a senior medical technologist may be given the responsibility of looking after the biochemistry section. This is the case in South Korea, the Philippines, Thailand and most laboratories in Taiwan. Generally, university graduates in biochemistry are not employed in clinical laboratories for the following reasons: (1) the job does not require high degree of specialisation or qualification, (2) the staffing structure does not cater for the employment of this category of staff, and (3) the salary and nature of work are not sufficiently attractive. The few biochemists known to be working in hospital laboratories, and even holding senior responsible positions appear to be the exceptions rather than the rule. Medical technologists in the Philippines, Thailand, and Taiwan are university/college graduates of a full time course on medical laboratory technology. Due to limited career advancement opportunities and the fact that medical technology is frequently not their first choice of study, many technologists in Thailand and Philippines are not contented to remain in their profession. When opportunity arises, they would work towards an academic teaching or research position. However, once in that position, they are lost to the laboratory as they would not contribute so significantly to the routine analytical work.

In Indonesia where laboratory practice was subjected to Dutch and American influence, pharmacists are permitted to carry out analytical work as well as supervise a clinical laboratory. In China, biochemists, biologists, pharmacologists and clinicians with some training in laboratory techniques work side by side in clinical laboratories and supervise technical staff the majority of whom received little or no formal training.

Without exception, small district hospitals employ only one or two laboratory technicians or assistants to carry out a small range of tests in the field of haematology, microbiology, and sometimes also biochemistry depending on local needs.

(C) Professional Organisations

Development and advancement of clinical biochemistry in Asia was hindered by two unfavourable situations: (1) the lack of communication among practitioners in the field even within the same country, and (2) the lack of opportunities for continuing education and participation at national, regional and international conferences. Cost of attending meetings which are mostly held in western countries is too prohibitive and discourages participation from developing countries in Asia. Unfortunately, locally there was neither a formal channel for dissemination of information and for communication among laboratory personnel, nor a representative association to organise meetings and relevant activities to satisfy their professional needs. The initiation of a continuing series of Asian-Pacific Congress of Clinical Biochemistry in 1979 was therefore a most timely and significant event. It offered the opportunity for meeting other laboratory workers and updating professional knowledge and practice on a regular basis.

Formation of the Asian and Pacific Federation of Clinical Biochemistry soon after the first regional congress provided a permanent framework for continuous collaboration among the countries and was a milestone in the development of clinical biochemistry in Asia. It provided the stimulus for the establishment of national societies or associations. Apart from India which had an association for clinical biochemists since the early 1970s, 5 other countries/areas have formed associations to represent the profession: Singapore (1978), Indonesia (1979), Taiwan/China (1982), Hong Kong (1983), and Korea (1986). The Singapore and Indonesian associations are members of the International Federation of Clinical Chemistry (IFCC). By the time this paper is presented, the other associations may have also been accepted as full members of IFCC. Since their formation, the various associations have been actively promoting the advancement of clinical biochemistry in their respective countries/areas, through the organisation of regular scientific meetings, educational and practical courses, and annual conferences or seminars. Manufacturers and suppliers of instruments and reagents have supported them by sponsoring and conducting educational workshops, teaching or demonstrating new methods and instruments. Some associations have been responsible for initiating national interlaboratory quality assurance programmes. Most associations produce a newsletter and one has also produced an annual publication for local scientific papers and reports.

(D) Quality Assurance

There is considerable variation in the practice of quality assurance. Internal quality control is not practised in some smaller laboratories. In most countries, large hospital laboratories in major cities generally have some form of internal quality control. However, the effectiveness of such programmes in some laboratories is questionable because the high cost of control materials and lack of expertise or raw materials to prepare the laboratories' own controls make it difficult for analyses of control materials to be done on a regular basis.

National central laboratories and laboratories of selected university and public hospitals participate in a variety of external quality assurance schemes. These include programmes offered by manufacturers of control materials and professional institutions/associations e.g. the Australian and American Colleges of Pathologists and the Australian Association of Clinical Biochemists. A few have been invited to participate in the WHO external quality assurance scheme. National interlaboratory programmes have been introduced in China, India, Indonesia, South Korea, Singapore, Taiwan and Thailand. Those in China and Indonesia are compulsory and are used by the health authorities for formal assessment of the laboratories. Due to cost constraint and logistic problems, the frequency of some programmes is unfortunately much lower than desired (17-28).

A significant dilemma encountered by many laboratories is the lack of ability to identify and solve problems when the quality control programmes indicate poor analytical performance. Most educational programmes on quality control have not taught staff how to trouble-shoot analytical problems and find the solutions. Another worrying situation is the relative lack of conscious effort in the practice of pre-analytical quality assurance. Although most laboratories have some knowledge about the usefulness and importance of quality control in analytical work even if they do not have a control programme, few realise the importance of quality assurance of non-analytical procedures i.e. collection, handling, preservation and storage of specimens and handling of patient/test data. There is therefore a need to educate both clinical and laboratory staff so that the necessary pre-analytical precautions are taken to ensure reliable test results. Better communication between clinical and laboratory staff would help to reduce erroneous results.

In some countries like Thailand, many laboratories use locally produced reagents for analytical work. Quality control for the local manufacture of reagents is not required by state law. Use of such reagents is an important reason for variability in test results.

At the specific requests of the IFCC President and the 13th IFCC Congress Committee, I will give a more detailed description on Philippines and China.

DEVELOPMENTS IN THE PHILIPPINES (29-32)

Philippines has a population of 58.1 million and a land area of about 300,000 km^2 consisting of 7107 islands and islets of which 16 are most important. Over 60% of the population live in rural areas where electricity supply is still non-existent. The major health problems of the country are: (1) communicable disease control, (2) malnutrition, (3) high population growth, (4) inequitable distribution of health resources, and (5) unsatisfactory sanitation of the environment. There is a total of 1839 hospitals, 1/3 of which belongs to the public sector and 2/3 in the private sector. Private hospitals are responsible for providing half the total number of hospital beds in the country. They are widely distributed, but with concentration in urban areas, providing medical care to a significant segment of the population on a fee-for-service basis. In the country's capital, Manila, where about 7 million people live, there are about 50 major hospitals with bed-capacity of 400 or more. Hospitals with laboratory service are only present in about 10 major cities. According to a managing director of an established local firm marketing laboratory diagnostic products, 70% of business in diagnostics are in Manila. This provides a good indication of the distribution of laboratory service in the Philippines.

The mode of practice of laboratory medicine is affected by 2 important factors: (1) a law which defines who is able to practise, and (2) attitude of hospital authority who determines how pathology should be practised. According to the law, the head of a clinical laboratory service must be a pathologist irrespective of the areas of training received. Most major hospitals treat the laboratory as an investment. Laboratory service is used to support other areas of service which do not generate sufficient revenue for the hospital. Whatever is done for the laboratory must be good for the publicity and income of the hospital but not necessarily beneficial for the laboratory discipline or profession. This has sometimes led to wasteful and absurd situation where hospitals compete to purchase new instruments which may not be appropriate for them. It is not uncommon to find that when one hospital buys an autoanalyser for 17 analytes, another hospital will want to purchase a machine with twice the capacity and requiring half the amount of specimen for analyses.

Pathology as a separate medical discipline started in the early 1960s. The Philippine Society of Pathologists has a current membership of 300-350, but only a third of them are active in society activities. About 120-150 of the members are trained pathologists, while the others are clinicians who went into laboratory work without specialised formal training, under the old practice. The "trained pathologists" receive some kind of certification upon satisfactory completion of 4-5 years of in-house training. The majority of the trained pathologists specialised in anatomic pathology. There are 3 reasons for the preference for anatomic pathology. The first 2 reasons are: (1) most public hospitals cannot afford to support purchase of laboratory equipment and medical graduates find that they would not be able to practise clinical pathology at a level that would give them adequate experience in the practice and application of tests for diagnosis, and (2) clinical pathology is not as financially rewarding as anatomic pathology. Payment for histopathology service is on a per case basis whereas a fixed fee is paid for clinical pathology service. The third reason is historical. Anatomic pathology was established as a distinct discipline much earlier than clinical pathology, which until recently was practised within the department of medicine. In recent years, more medical graduates are willing to specialise in clinical pathology. About 100 pathologists practise in Manila. Most pathologists look after at least 3 different hospitals' laboratory departments. They can at the same time open their own private laboratories. However, there are very few private laboratories which are not associated with a hospital. It would appear that pathologists can earn a reasonable income even though a senior member of the profession lamented that the Ministry of Health hardly encouraged doctors to specialise in pathology during a recent regional meeting on medical laboratory technology. About 80% of medical conferences and meetings held in the Philippines are either associated with relevant organisations in the USA or have significant American participation. Patients use the American practice of medicine as a standard for comparing the status of local specialists and tend to ask their doctors if they had been to the USA. Consequently, many clinicians and pathologists go to the USA to "improve their status".

Supervision and performance of biochemical analyses are largely carried out by graduates of full-time university or college courses in medical laboratory technology, pharmacy or biochemistry. In practice, few biochemistry graduates work in clinical laboratories even though they would be better paid than a medical laboratory technology graduate. The reason is the much more attractive salary and opportunities offered by the university/college, industry and other government departments. There are not more than 2 biochemists working in hospital laboratories, one of them holding the position of chief medical technologist for the biochemistry section.

A 4-year course for BSc in medical laboratory technology was first started in 1954. Today, 33 colleges and universities are offering the course, 14 of these are in Manila. On completion of the course, graduates have to sit for a government certification examination required for the licence to practise. Results of the examination would only be disclosed about one year later. During the period of waiting, graduates work in private laboratories to gain practical experience. Graduates in biochemistry are not required to take the examination before joining the laboratory service. In university courses, the ratio of male to female students is 1 : 2. On graduation, some 60% of all graduates join the degree course in medicine from first year. Most male graduates either continue their further study to become doctors or take up other better-paying jobs such as salesman for medical/diagnostic products. This accounts for the very small number of male staff in the laboratories. It is estimated that there are about 20,000 technologists. However, many have moved abroad because of better pay and career opportunities. They are unable to advance themselves in the field as few colleges or universities offer higher degree courses, continuing education programmes, or even short courses on specific techniques. Currently, the 23-year old Philippine Association of Medical Technologists has about 10,000 members of which only 15% are active. There are also laboratory technicians who do not have formal training and work at lower responsibility and salary. The Association is trying to discourage the further employment of such staff.

Recently, a nation-wide survey was conducted to assess the current status of analytical methods and instrumentation in hospital and clinical laboratories in the Philippines (32). Of the 350 questionnaires sent out, 76 responded. Of the respondents 75% came from the hospitals and 25% from clinical laboratories. The survey found that 75% relied on manual instruments, 23% used a combination of manual and semi-automated/automated instruments, 1% used semi-automated instruments and 1% fully automated instruments. 80% of respondents felt a need for addition or replacement of equipment with 33% opting for manual operations, 45% semi-automation, and 22% for automated systems. On the choice of analytical methods, 52% of respondents adhered to conventional methods, 41% employed both conventional and updated methods, and only 7% employed largely updated methods (i.e. specific, enzymatic reaction, kinetic uv measurement). When respondents were asked what would be their preference when given a choice, only 4% chose conventional procedures. 63% preferred a mix of conventional and updated methods and 33% chosed updated methods. It was noted that automated instruments and updated methods are used mainly in Manila. The criteria for selection and purchase of instruments are: (1) reliability of maintenance service and availability of spares, (2) cost, (3) open system, not confining to reagents by a single manufacturer, (4) knowledge of laboratory personnel or hospital administrator, listed in decreasing order of priority.

DEVELOPMENTS IN THE PEOPLE'S REPUBLIC OF CHINA (18, 33, 34)

China has a population of 1050 million and a vast land area of 9.6 million km^2. The administrative areas consist of 22 provinces, 5 autonomous regions and 3 municipalities (Beijing, Tianjin, Shanghai). There are 216 cities, 2085 counties and more than 50,000 rural communes. There are 200,866 health organisations and 67,000 hospitals of which 11,497 are provided with facilities equivalent to or above the level of county hospitals. Free medical service is given to all government employees, workers in major collective organisations, state-owned farms and college students. Individual farmers can join a cooperative medical care system by regular payment of a small sum of money.

Urban hospitals are divided into 3 classes: First Class Hospitals - hospitals under the control of provinces, municipalities, autonomous regions, and teaching hospitals of the medical colleges; Second Class Hospitals - hospitals of the prefectures; Third Class Hospitals - county hospitals. They are organised in a referral chain, so that hospitals at the higher levels are responsible for rendering medical directions and assistance as well as continuing education of the medical personnel for the hospitals at a lower level. Thus, clinical laboratories in hospitals at a lower level will benefit from the technical aid given by hospitals at a higher level. Departments for laboratory diagnosis have been established in all urban hospitals. Generally, they are responsible for laboratory work in biochemistry, bacteriology, immunology, serology and haematology. Some laboratories also run a blood bank. A range of routine examinations such as glucose, urea, creatinine, urate, sodium, potassium, chloride and CO_2 determinations, ALT, amylase and even protein electrophoresis are performed in hospital laboratories including those in Third Class hospitals. Other enzymes and blood gas analyses are added to the repertoire of major hospitals. More sophisticated tests are performed only in the First Class hospitals and those clinical laboratories which performed very well in the external quality assurance scheme sponsored by the WHO. In laboratories of lower level hospitals, most of the procedures are carried out manually using spectrophotometers, flame photometers, semi-automated instruments and electrophoresis equipment made by local manufacturers. However, in most of the large hospitals in major cities such as Beijing, Shanghai and Guangzhou, imported or indigenous semi-automated and automated analysers, densitometers, flame photometers, and imported spectrophotometers are used. Sophisticated discrete analysers like those made by Baker, Gilford, and Roche have been installed in some of the larger hospitals. Due to shortage of foreign exchange, prior application has to be made and approval sought for the purchase of imported instruments. Lack of foreign exchange makes it especially difficult for laboratories to rely on imported reagents for high volume routine analyses on an annual recurrent basis. Few if any laboratory would purchase a closed analytical system because they cannot be certain of the availability of the necessary foreign exchange required for the subsequent purchase of imported reagents. Therefore instruments proven to be workable with locally made reagents are preferred.

Prior to 1977, the services and facilities for laboratory investigations in China were very insignificant. Equipment were of the most basic type. The role of the laboratory in health care was not given proper recognition by the authorities. Little attention was paid to the activities and facilities of clinical laboratories. Consequently, for a long time laboratories suffer from shortage of qualified staff and modern equipment. Fortunately, since the end of the Cultural Revolution and the gradual introduction of new policies to promote or encourage economic, academic and technological developments, changes in laboratory medicine has been remarkably rapid. This is largely due to the use of the right people for the improvement of laboratory diagnostic service.

In 1986, there were 145,217 technical/professional staff in the clinical laboratories. Only 10% of them were clinical pathologists and technologists. A large number of the technicians had not received formal pre-job training. Realising the shortage of trained personnel and qualitative weakness of the existing staff, the Ministry of Health took the following measures to improve the manpower situation: (a) established departments of clinical laboratory in 18 medical schools to offer a 4-5 year course, (b) encouraged graduates in medicine and natural science to work in the clinical laboratories, and (c) established the National Clinical Laboratory Center (NCLC) to improve the level of practice in laboratory diagnosis on a nation-wide scale. Dr Y W Ye was appointed to the position of a central coordinator, the Director of NCLC. She has been most earnest and energetic in her efforts to improve the quality of analyses and to standardise methods and instruments where possible. She was able to pursuade the health authorities to place emphasis on excellence in laboratory work in order to achieve better health care. Fortunately, the current authorities are very enlightened and supported the laboratories with funds for equipment and quality control activities. The Health Ministry has now included quality of laboratory performance as an important criteria in the routine periodic evaluation of hospital services. If a hospital cannot show that its laboratory practises quality control and obtains satisfactory results, it would be penalised by allocation of less funds. Therefore, laboratories have great pressure and incentive to strive for good performance.

The NCLC was founded in December 1981, 2 years after it was proposed at the National Conference of Clinical Laboratory Diagnosis. It is now accommodated in the Beijing Hospital and has the following responsibilities: (1) to propagate quality control and ensure reliability of test results, (2) to recommend methods of routine examination, (3) to educate key laboratory personnel, (4) to introduce new technology and methods and to conduct research, (5) to solve laboratory problems, (6) to collaborate with industry in the production of apparatus and reagents, and (7) to encourage academic exchange, both at home and abroad. Experienced key personnel were selected to work in the center and college graduates were employed to form the main body of the staff. Currently, the center has 4 consultants of professor status, one of associate professor status, 5 staff in the rank of lecturer or visiting physician, 20 medical doctors and technologists and 3 attendants. Branch centers were also established in provinces, autonomous regions and municipalities forming a network in the country. Under the direct control of the Ministry of Public Health, the NCLC provides professional guidance to the branch centers. The latter in turn has professional and supervisory responsibility over hospital laboratories. Clinic laboratories in rural towns are under the supervision of hospital laboratories.

Continuing Education

Under a nationwide re-education programme, 25 training courses on a variety of topics have been held for key staff from 29 provinces and the autonomous regions. Each class consisted of an average of 80 persons. So far, about 2000 persons attended the courses. When they returned to their own branch centers and major laboratories, they had the duty to conduct similar refresher or educational courses for those under their supervision. Thus, throughout the country, about 100,000 laboratory staff had benefited from the training programmes.

Soon after its establishment, the NCLC was recognised by the National Committee for Academic Degree to award the Master degree in laboratory medicine and accepted postgraduate students for a course in clinical laboratory diagnosis. One doctor has already been awarded the degree and 7 others are undergoing training. Provincial centers also plan to award the degree with the approval of the NCLC. More recently, NCLC was also elected as the institute for the award of Doctorate degree in laboratory medicine.

Quality Assurance

In 1980, two national training courses were held. A total of 130 participants from 28 provinces attended. They were obliged to run a similar course in their own regions and to organise a local quality assessment scheme, including internal quality control activity. Four national interlaboratory surveys were carried out in 74 major laboratories to evaluate machine calibration and analytical performance. Results of the surveys had been published (17). In 1981, a more advanced course was held for senior staff in charge of regional quality control. Thereafter, the following actions were taken to eliminate some of the possible causes of errors: (1) provision of standard solutions for comparison against laboratory-prepared standards, (2) calibration of spectrophotometers (China-made model 721) using didymium filter and p-nitrophenol solution, and (3) supply of accurate pipettes. By 1986, the national quality assurance scheme was extended to 280 laboratories and processing of statistics was computerised. All participants are required to submit their internal quality control data for the month when returning their results. By the end of 1986, all hospitals above the county hospitals were expected to have a strict system of quality control for biochemical analyses. In 1987, about 500 laboratories will take part in the national scheme. The branch centers have been instructed to conduct their own regional quality control activities in the same manner. Since the institution of formal quality assessment schemes, there has been remarkable improvement in the reliability of analytical performance. Taking a variance index score (VIS) of 150 as acceptable performance, the percentage of acceptable results for bilirubin assay increased from 24.3 in 1980 to 93.2 in 1985. For glucose assay over the same period, the percentage increased from 59.2 to 92.6. Improvement in overall performance is shown by an increase from 39.1% to 71.1% of acceptable results. Quality control materials based on porcine sera are now produced in China by 6 biological products centers which are also responsible for the preparation of vaccines.

New Technology and Methods

New instruments and methods are first evaluated or developed in the NCLC before being introduced to the hospital laboratories. The more sophisticated or specialised machines and techniques are confined to few major centers. Thus determination of Mg, Zn and Cu are now performed by atomic absorption spectrometry in NCLC. Therapeutic drug monitoring (TDM) by HPLC and immunoassay has also been introduced in NCLC and selected laboratories. Nationwide training courses on TDM were conducted. Recently, NCLC produced reagents for myoglobin determination and provided kits to many hospital laboratories for the diagnosis of myocardial infarction.

Collaboration with Industry

The NCLC assists local reagent manufacturers in the production and supply of reagents so that they can be used by clinical laboratories at the grassroot levels. An example is the reagents for glucose assay by glucose oxidase method.

Professional Association

The Society of Laboratory Medicine is an association for laboratory specialists who have had 3 years practical experience after graduation from a medical school. Under the central society are the branch societies in different localities. National conference is held once or twice a year by the central society while branch societies organise their own scientific activities in every 1-2 months. Some of these are jointly organised with the NCLC. Hierarchy-wise, the Society of Laboratory Medicine is under the Chinese Medical Association which is in turn under a larger parent organisation, the China Association for Science and Technology. In March 1987, I was informed by Dr Y W Ye of the imminent formation of a Chinese Association for Clinical Chemistry.

Recent development of clinical biochemistry in China has been encouraging. The country now has one of the most organised comprehensive supervisory, checking and training systems among the developing countries.

However, those responsible for laboratory medicine feel that development in the field is lagging behind that of the clinical disciplines. They acknowledge the need for further improvements in analytical methods and performance, establishment of a reference laboratory, greater efforts in the area of continuing education and collaborative work in the domestic production of instruments and reagent kits.

SOME PROBLEMS IN THE DEVELOPING COUNTRIES

(1) Funding

Limited funding for laboratory facilities and activities is a major problem in the less developed countries. The small budget has been a main obstacle in the development of clinical laboratory services.

(2) Supply and Maintenance of Instrument

Laboratories face several major problems in their selection and purchase of instruments. The first is the relatively high cost of imported instruments and the inability of government clinical laboratories to obtain the necessary foreign exchange. The second problem is the inadequate or non-existent maintenance and repair service which is a prominent complaint and often the reason for instruments to become idle only a short time after their installation. Many local agents are only interested in selling instruments but not in providing after-sales service that requires employment of skilled technical personnel who are not readily available. Laboratory staff from several countries have complained that electrolytes measurements are not made because their flame photometers malfunctioned soon after the instruments were installed. Attempts to have them repaired were not successful. In the developed countries, there are many opportunities for assessment of instruments. Reports concerning the performance of a wide range of instruments are also readily available from journals, national associations or colleagues in other laboratories. This is not the case in many developing countries. Consequently, the lack of knowledge has resulted in purchasing an inappropriate instrument which is underutilised or one whose performance is short of expectation. Purchase of inappropriate or unworkable instruments could also be the result of: (1) decision by administrators who have little or no knowledge of laboratory operations and equipment, and (2) donations by organisations and developed countries who have no knowledge of local working environment and constraints.

(3) Lack of Training of Laboratory Staff

Laboratories other than those in premier hospitals are still short of trained staff. Knowledge in some important aspects of laboratory work, such as checking correctness of wavelength and absorbance reading calibration of spectrophotometer, accuracy of balance, and performance of pipettes and dilutors, is inadequate. Some staff do not carry out periodic checks on the temperature of water-baths used for enzyme assay and report activities for 25°C when neither their instrument nor working environment permits measurement at that temperature. There is a tendency for staff to have over-confidence in instruments, especially the semi-automated and automated machines. They are unaware that chemical aspects of the analyses can go wrong even though all the operational components of the machines appear to work normally. The consequence of this over-confidence is more serious if there is no adequate internal quality control programme to monitor routine laboratory performance.

(4) Unfavourable Laboratory Environment

Unexpected power cuts and sudden fluctuations in power supply contribute significantly to the instability and short life of equipment. Hot, humid and sometimes dusty environment in some countries is an important cause for malfunction of the more sophisticated instruments. Due to economic reasons, many laboratories are not provided with airconditioning.

(5) High Taxation

In some countries, import tax on instruments can be 100% higher than their sales price. Tax on reagents can even reach 200% or more if plastic disposable items are included in a reagent kit. For some, there is the added burden of an "unofficial tax" imposed by various officials. Prices of instruments and reagents can be further inflated by profit mark-up of local agents. These additional costs often push prices beyond the laboratory's budget.

(6) Poor communication between Clinical and Laboratory Staff

Due to shortage of laboratory personnel with higher academic/ professional status who are able to communicate with clinical colleagues on the same level, there is a general lack of contact and collaboration between clinical and laboratory staff.

REFERENCES

1. Thirty-Ninth World Health Assembly, Verbatim Records of Plenary Meetings and Reports of Committees, Geneva, May 1986, WHA39/1986/REC/2: 29, 34, 45-46, 198, 212.
2. Watters G. The Decade at Half-Time. In: World Health, the magazine of the WHO, Dec. 1986: 27-29.
3. The Work of WHO in the Western Pacific Region (1983-1985), Biennial Report of the Regional Director, WPR/RC36/2: 71-73, 127, 171.
4. Ramaboot S. Thailand: Attainable Targets. In: World Health, the magazine of the WHO, Apr. 1986: 23.
5. Evaluation of the Strategy for Health For All by the Year 2000, Seventh Report on the World Health Situation, Western Pacific Region, 1986; Vol. 7: 26-27.
6. Zeroing in on Xerophthalmia with Vitamin A. In: World Health, the magazine of the WHO, June 1985: 30-31 (news page).
7. Clugston GA, Bagchi K. Tackling Iodine Deficiency in South-East Asia. World Health Forum 1986; 7: 33-38.
8. Thirty-Ninth World Health Assembly, Summary Records of Committees, Geneva, May 1986, WHA39/1986/REC/3: 357-360.
9. Deinhardt F, Gust ID. Viral Hepatitis. Bulletin of the WHO, 1982; 60 (5): 661-691.
10. Thurston D. Deadly Serious. In: Asia Magazine (Singapore Edition, special issue on Aids in Asia) May 3, 1987: 8-15.
11. Asia Year Book, Far Eastern Economic Review 1987: 6-7.
12. Asia Year Book, Far Eastern Economic Review 1986: 6.
13. Shortage of Doctors in China (Report in China Daily, Beijing), Lianhe Zaobao (Sunday edition, Singapore). 1 March 1987: 25.
14. SEAMIC Health Statistics, 1985, compiled by Committee on Health Statistics, published by Southeast Asia Medical Information Center, Tokyo, 1985: 95, 100.
15. Yearbook of Statistics, Singapore, 1985/86, Chief Statistician, Department of Statistics, Singapore.

16. Tan IK, Jacob E, Png A, Yo SL, Gunasegaram R. Current Status of Clinical Laboratory Automation in the Asian-Pacific Region. Japanese J Clin Lab Automation 1986; 11 (6): (355) 9 - (371) 25.
17. Ye YW. Recent Experience in Interlaboratory Quality Assessment in Clinical Biochemistry in China. In: Proceedings of Plenary Lectures and Symposia, 2nd Asian-Pacific Congress of Clinical Biochemistry (2nd APCCB - Singapore), July 1983: 215-221.
18. Ye YW. Present Status of Hospital Laboratories in China. Japanese J Clin Lab Automation 1986; 11 (6): (373) 27 - (378) 32.
19. Sarkar AK. National Quality Control Project in Clinical Biochemistry in India. In: Proceedings of the 2nd APCCB, Singapore, July 1983: 231-236.
20. Kanagasabapathy AS. The Present Pattern of Clinical Biochemistry Service in India. In: Proceedings of the 2nd APCCB, Singapore, July 1983: 62-65.
21. Widyaharsana J. Laboratory Quality Assurance: The Indonesian Experience. In: Proceedings of the 2nd APCCB, Singapore, July 1983: 222-223.
22. Widyaharsana J. National Laboratory Quality Assurance in Indonesia. In: Proceedings of the 5th International Symposium on Quality Control - Tokyo. 5th I.S.Q.C. - Tokyo 1984: 542-545.
23. Lee SY. Current Status and Problems of Quality Control in Korea and the Activities of the Korean Society of Quality Control in Clinical Pathology. In: Proceedings of the 5th I.S.Q.C. - Tokyo 1984: 476-480.
24. Tan IK, Jacob E. The Status and Problems of Quality Control in Clinical Biochemical Analyses in Singapore. In: Proceedings of the 5th I.S.Q.C. - Tokyo 1984: 508-519.
25. Chen JS. Quality Control in Taiwan: A Review of Ten Years. In: Proceedings of the 5th I.S.Q.C. - Tokyo 1984: 481-483.
26. Whitehead TP. The Thailand Clinical Chemistry Project. In: Proceedings of the 2nd APCCB, Singapore, July 1983: 22-34.
27. Petchclai BP. Quality Control in Thailand. In: Proceedings of the 5th I.S.Q.C. - Tokyo 1984: 499-507.
28. Asian-Pacific Meeting on Quality Control, Hong Kong, December 1986: Presentations by representatives from Australia, Hong Kong, Indonesia, Japan, Korea, Malaysia, Singapore, Taiwan and Thailand.
29. Country Health Information Profile on Philippines. WHO/WPRO/HIN 1975-86.
30. Maramba T, Billote J, Cantos AC, Acedera CP, Ambrosio S, Suapenco R. Personal communications, Manila, November 1986.
31. Cruz J. Medical Laboratory Education: A Dilemma and a Challenge. Presented at the 2nd ASEAN Conference in Medical Laboratory Technology, Manila, Philippines, November 1986.
32. Cantos AC. Test Kits, Methods and Instrumentations. Presented at the 2nd ASEAN Conference in Medical Laboratory Technology, Manila, Philippines, November 1986.
33. Country Health Information Profile on People's Republic of China. WHO/WPRO/HIN 1984-86.
34. Ye YW. Personal communication, Beijing, China, October 1986.

CLINICAL BIOCHEMISTRY SERVICES IN TROPICAL AFRICA

Babatunde Osotimehin (U.K.)

F.M.C. Path, FWACP, Department of Chemical Pathology
University College Hospital, College of Medicine, University of Ibadan, Ibadan, Nigeria

The differentiation of clinical chemistry as a separate entity in the health care delivery system in most African countries is a recent phenomenon. Hitherto this area of clinical management was subsumed within the broad area of pathology which was heavily biased towards microscopic diagnosis i.e. parasitology, bacteriology etc. However with the development of medical faculties in the last quarter of a century, clinical chemistry has developed to the status of separate departments or administrative units within the context of large hospital establishments.

The pattern of development is however most uneven. Whereas countries like Nigeria have many viable clinical chemistry departments, there are African states (especially those without medical schools) that do not possess separate clinical chemistry departments. This is probably the dictates of the economy rather than desirability.

PRESENT STATUS

Clinical chemistry services are presently offered in Africa by two basic channels: (a) Hospital based services, (b) Private laboratories.

HOSPITAL BASED SERVICES

Depending on the size of the hospital clinical chemistry could be a separate unit or a wing of pathology. In Nigeria, clinical chemistry services are available in virtually all secondary and tertiary medical institutions. This is similar to what obtains in Ghana, Ivory Coast, Kenya, Tanzania and Zaire Republic. The services they render however differ from hospital to hospital and also across the continent. Predictably the tertiary centres which serve as teaching centres, tend to carry most conventional chemistry tests i.e.

- Monovalent Electrolytes - Na^+ K^+
- Divalent Electrolytes - Ca^{++} Mg^{++} (occasionally)
- Bicarbonate (as been excess HCO_3-)
- Urea, Uric Acid, Creatinine
- Glucose estimations (Blood cerebrospinal fluid)
- Serum Proteins and Protein electrophoresis

- Liver enzymes - Aspartate and Glutamate transaminases
- Bilirubin
- Alkaline phosphate, 5-Nucleotidase etc.

The smaller units tend to be restricted in their service potential and they usually offer such analytes like:

- Serum Na^+, K^+, Ca^{++}
- Serum Urea
- Serum Proteins
- Glucose estimation (Blood and CSF)

thus patients usually require referral to a secondary or tertiary centre to have a full work up in most instances.

The primary health clinics and district hospital usually do not have any capabilities for chemistry. They usually may do with urinalysis using dipsticks and blood dextrostix.

SPECIAL INVESTIGATIONS

Clinical care especially at the tertiary level often times requires more than the base investigations most of which have been listed above. This amorphous group I have referred to as special because of the organisation and logistics in our laboratories. Most of the 'special' investigations that we handled are endocrinological. These include:

(i) screening for catecholamine excess by the urinary V.M.A. estimations;
(ii) screening for carcinoid tumours by urinary 5 HIAA estimation;
(iii) Radioimmunoassay of various hormones
  (a) thyroxine
  (b) cortisol
  (c) reproductive hormones (peptides and steroid hormones).

The demand of these special investigations in our circumstance at Ibadan has necessitated the establishment of a separate (geographically) laboratory to handle these requests. Fortunately we get some of the reagents that we use for our clinical samples from the programme organised by the World Health Organisation Human Reproduction Programme. This is rather similar to what obtain in Kenya where there is also a WHO Clinical Centre for Reproduction.

Most of the other African centres that were surveyed had limited endocrine services within the hospital situation, Ghana offers a thyroid and cortisol service at the Medical School in Korle Bu.

PRIVATE LABORATORIES

The last decade has seen a proliferation of private laboratories offering clinical chemistry service in our countries mostly managed by technologists. These tend to be laboratories offering a wide range of services including clinical chemistry, bacteriology, haematology and blood transfusion and occasionally histopathology.

They provide services for usually the common analytes:

- Electrolytes (Monovalent and Divalent)
- Liver Function Tests
- Renal Function Tests etc.

Their services are targeted at the various private medical institutions and it would appear that they satisfy most of the needs of these establishments.

QUALITY CONTROL SERVICE

All laboratories (both hospital based and private) possess an internal quality control system which provides some sort of confidence for the laboratory performance. The big hospital based laboratories participate in external quality control schemes what are usually based in Europe or America. This situation demands constant foreign exchange payment and is fraught with the logistical problems of shipment and handling of quality control material.

Recently, the Association of Clinical Chemistry in Nigeria has proposed a quality control programme for the country which would be coordinated by a laboratory in Lagos. This we hope to extend to the West African subregion under the auspices of the West African Health Community. The East African subregion would also be encouraged to organise a similar scheme so that eventually we should have a continental programme.

TRAINING

There are different training schemes for the various levels of categories of workers in the clinical laboratories across the African continent. Our recent historical past seem to have a sustaining influence on the direction of our training programmes, books etc.

The pathologist

The pathologists in our hospitals are usually medical graduates who have opted to specialise in pathology after an initial medical training. There is a programme available for training in pathology under the auspices of the Nigerian National Postgraduate Medical College. A parallel programme also exists with the West African Post-graduate Medical College. The trainee in either of these programmes is expected to spend at least four years in pathology, the initial two years he is expected to rotate through all the branches of pathology (viz: Medical Microbiology, Histopathology, Haematology and Blood Transfusion, Virology) so that he can be adequately exposed.

The final two years are spent in his chosen speciality i.e. in this circumstance, clinical chemistry. There are prescribed examinations at various levels to assess the level of professional competence and in addition the candidate is also expected to produce a short well-researched dissertation at the end of his training programme.

I should like to stress that this training programme is limited to the English speaking West African states, the French West African states still obtain postgraduate training and experience in France or Canada. This is similar to the situation in the East African subregion where pathologists still have to go to the United Kingdom for training purposes. However, Kenya has recently established a Master of Medicine (M.Med.) programme in Human Pathology. The aim of this programme is to produce the general pathologists that will function at the district all provincial hospital level. This is a three year training programme during which period, they will be exposed to all branches of pathology. It is anticipated that in the very near future specific programmes for specialty training will be incorporated into the programme.

The Clinical Chemist

In Nigeria like some other African countries, academic institutions have programmes leading to the award of Masters degree in Chemical Pathology and some of the candidates can proceed to the Ph.D. level in their area of interest. This usually attracts candidates who have their first degrees in a related science, i.e. Chemistry, Biochemistry etc.

Unfortunately, a good number of the programmes are purely academic in content without adequate exposure on the 'shop floor' thus such graduates do not possess the necessary practical professional experience to perform as full fledged clinical chemists. However efforts are now been made in Nigeria to correct this anomaly and the necessary regulatory bodies are going to be established to certify this category of trainees in the near future.

The Medical Laboratory Technologist (Scientist)

Most African countries have established training systems for this cadre of professionals. The training programme is usually similar to that which can be obtained in the European linkage country i.e. the Nigerian programme of training is similar to what generally is obtained in Britain and the system in Ivory Coast to the French system.

The programmes are rather similar, however, in content and philosophy. The entry requirement is also similar i.e. General Certificate of Education passes (or equivalent) in English, Biology, Chemistry, Physics and Mathematics. Virtually all of the schemes strive to expose the trainee to all aspects of pathology and then they are made to spend a considerably longer period in the specialty of clinical chemistry where they are fuller exposed to available technology including specialised techniques. There is continuous assessment of performance and a final set of examinations to certify the candidate. In Nigeria and Ghana, there exists the National Institute of Medical Laboratory Technology or its equivalent which registers the new qualified trainee and regulate the professional conduct of technologists. Usually training takes place in a hospital setting except in Kenya where there is some training being conducted by the Kenya Polytechnic, however the practicals still takes place in a clinical setting.

Laboratory Assistants

The training of the laboratory assistant as personnel that looks after basic procedures in the laboratory i.e. the spinning and storage of samples, the preparation of some basic buffers etc. is an in-house thing in most hospitals. Usually the entre requirement is basic grade education.

EQUIPMENT (PROCUREMENT AND SERVICES)

Most clinical chemistry laboratories in tropical Africa still depend to a large extent on manual operations. There are a few tertiary centres in Nigeria, Kenya, Zaire that have limited automation. Thus clinical chemistry in our parts is still labour intensive. The major problems that we face relate to the fact that the major equipment manufacturing companies have limited outlets on the continent.

Usually, this is restricted to a small number one or two and situated at locations which in most circumstance are not accessible to the user. There is a predilection for the East African city of Nairobi or a North African location e.g. Cairo. Occasionally the supplier functions from a

European city base e.g. Geneva or London. The experience of the African consumer is a harrowing one when it comes to equipment.

The process of procurement and installation is probably the least traumatic that is not to say that problems are not encountered even at this early phase of the business contact. There are stories of faulty installations which eventually make the pieces of equipment totally useless or considerably shortens their life spans.

Subsequent to installation most of the suppliers disappear and even when the state government or the institution painstakingly enters into a prophylactic service agreement with the supplier these are usually not honoured and when they are the 'down-time' on the machines totally disrupts the service and sometimes questions the wisdom of the initial investment. It is also difficult to get the companies to accept to train some of our ground staff in at least preventive maintenance so that we only restrict our contact with them to major break-downs. It is also not unusual for some companies to ship equipment to us without the usual maintenance papers which are necessary to trace the fault. Thus most of the time we are reduced to a catch - 22 situation! This grim position is compoundedly by the fact that most African countries are relatively poor and thus the real investment when viewed against the capital base and the hard foreign exchange in which practically all those purchases have to be made is really large.

There is obviously a need for national policies on procurement i.e. procurement should be limited to one as two sources probably the same generation of equipment and an inducement for supplier performance. Probably there could be established regional cooperation in this kind of venture since it is possible that some of these countries might not be in a position to purchase more than a few pieces of equipment at a particular instance.

REAGENTS

The African continent with probably the exception of the Republic of South Africa is again at a great disadvantage with regards to reagents procurement and supply. There is no chemical company of any sort in these parts and thus all reagents either in part or whole have to be imported from Europe or the United States. These supplies are subjected to the vagaries of the economy i.e. adequate supply of foreign exchange at the appropriate time and the bureaucracy linked to orders etc. This discontinuous stream sometimes affects our ability to deliver services without interruption. Then of course there is the problem of preservation of the chemicals in transit and rapid clearance from the ports to ensure no deterioration. The same philosophy of regionalisation similar to the situation with equipment procurement might be the solution with reagent supply. The thesis is that if the market is large enough then suppliers would have sufficient motivation to maintain meaningful presence in the region. This will obviously ensure continuous service and guarantee the quality of the goods.

BUDGETARY ALLOCATIONS

It is difficult to isolate the specific amount expended on clinical chemistry at the national level in most African countries, it is subsumed in most circumstances in the budget of clinical services. Even though in relative terms the budget for clinical services vis-a-vis other health needs (preventive medicine, welfare services etc.) is rather large, but in absolute terms the amount of money that actually gets spent on pathology is small. This limits the range of investigations that are available in most hospitals, the tertiary centres being possibly the most well serviced. On a rough estimate, most African states spend about $ US 0.2 per individual on clinical chemistry annually. This gross underfunding makes it difficult to provide any meaningful service at our present scale of operations. Thus by inference when viewed on a global scale, the budgetary allocation to laboratory services is small and usually insufficient for optimal services at every level of the health service.

THE FUTURE

Economics and Population

The African continent offers a wide diversity when viewed in many perspectives. It is populated by different peoples with distinct cultural beliefs and attitudes. However in terms of economics, black African states have one thing in common - poverty. There are presently 31 independent black African states and 21 of these belong to the group which the United Nations refer to as least developed countries as defined by the different socio-economic indicators (Table 1). Logically, the financing of the health services of most African states is dismal, most of them expending less than 2% of their Gross National Product (G.N.P.) on health serving delivery (Table 2). Tragically, most of this meagre resources are spent on the running of central tertiary centres which offers health care to only a very small percentage of the population. The example of Malawi would elucidate this point (Table 3). Thus the primary and secondary health centres in most African states are chronically underfunded and consequently underutilised. Thus in rural Africa where 80% of the population live, it is estimated that only 20% of the health budget eventually percolate to that level ensuring the perpetuation of the vicious circle of ignorance, poverty and disease.

Table 1. Least developed countries - designated by the UN

Africa
1. Benin
2. Botswana
3. Burundi
4. Cape Verde
5. Central Africa Republic
6. Chad
7. Comoros
8. Ethiopia
9. Gambia
10. Guinea
11. Guinea Bissau
12. Lesotho
13. Malawi
14. Mali
15. Niger
16. Ruanda
17. Somalia
18. Sudan
19. Uganda
20. United Republic of Tanzania
21. Upper Volta

Table 2. Health and related socioeconomic indicators

	Least developed countries	Overdeveloping countries	Developed countries
Number of countries	31	89	37
Total population (millions)	283	3001	1131
Infant mortality rate (per 1000 live born)	160	94	19
Life expectancy (years)	45	60	72
% Infants with birth weight of 2500 g or more	70%	83%	93%
% Population with access to safe water supply	31%	41%	100%
Adult literacy rate	28%	55%	98%
GNP per capital	US$ 170	US$ 520	US$ 230
Per capital public expenditure on health	US$ 1.7	US$ 6.5	US$ 244
Public expenditure on health as % of GNP	1.0%	1.2%	3.9%
Population per doctor	17000	27000	520
Population per nurse	6500	1500	220
Population per health worker (any type, including traditional birth attendant)	2400	500	130

Note: Figures in the Table are weighted averages based on estimates for 1980 or for the latest year for which data are available. Source - World Bank

Table 3. Malawi unit costs of health service operations[a]

Health facility	K
Central hospitals	
Lilongwe	1980
Blantyre	2314
Government rural health services	
Primary health centres	25
Dispensaries with maternity	19
Dispensaries only	12
Maternities only	5
PHAM Institutions	
Hospitals	92
Primary health centres	12
Dispensaries with maternity	14
Dispensaries	3
Maternities	3
Local Government Health Services	
Dispensaries/Maternities	6
Dispensaries only	4
Maternities only	4

[a] Based on survey data for 1980-1981
K = Kwacha which is Malawi's currency unit

Unfortunately, the problem is compounded by high population growth rates in a significant number of African states (Table 4). It is estimated that by the year 2000, the population of the continent would be 870 m, an increase of 50% compared to 1987 figures. This situation can only further aggravate the health situation except there is a drastic change in policy, and direction of health care delivery.

Fortunately, the world health community is now committed to the principle of the Alma Ata Declaration of Health for All by the year 2000. This can only be achieved in the context of the African continent by laying emphasis on health care delivery at the primary level i.e. expending 80% of the health budget in the geographical location where 80% of the population live.

The epidemiology of disease in our continent dictates that most of the efforts should be directed at promotive and preventive health schemes as most of the health problems encountered are caused ignorance and poverty. The input of the other sectors of the economy and society need also be highlighted in this concerted effort to improve the living standard of the people. The availability of portable water supply (which hitherto only 30 - 40% of the population have had access to) and provision of proper housing for most of the people would significantly reduce the prevalence of some of the endemic diseases like - guinea worm, schistosomiasis etc. The agricultural outputs of these states also require a substantial boost to enable the population have access to adequate nutrition and probably some of these nations might earn enough hard foreign exchange from some of the excess food produced - this is different from the export earnings from the traditional cash crops like cocoa, coffee, tea etc.

Table 4. Demographic indicators

	Crude birth rate per 1000 population		Average percentage annual growth of population		
	1960-64	1980-84	1960-64	1980-84	2000-04
China	34	19	2.0	1.3	0.9
Ethiopia	51	50	2.3	2.7	2.5
India	44	33	2.3	1.9	1.3
Nepal	46	42	1.9	2.3	2.0
Pakistan	47	42	2.8	2.8	1.9
Zaire	48	45	2.4	2.9	2.7
Egypt	43	35	2.5	2.4	1.9
Guatamala	48	38	3.0	2.9	2.4
Indonesia	47	31	2.3	1.6	1.1
Ivory Coast	50	46	2.6	3.0	2.8
Nigeria	52	50	2.8	3.4	3.1

Where does Chemical Pathology or in a larger perspective Pathology services fit in this new scheme of things i.e. primary health care?

The directional change towards primary health care demands a reorientation in terms of technology and complexity of tests. I would like to propose that a task force be established to determine the appropriate attitude that should be adopted towards this development. This ideally should be a cooperative effort between the International Federation of Clinical Chemistry and the World Health Organisation. It will be the function of this task force to determine what are the basic requirements for pathology at the primary health care level. This will form the basic frame work on which to build recognising the fact that there will be variations in disease patterns depending on epidemiology that might dictate the modifications required in different geographical zones.

CLINICAL BIOCHEMISTRY SERVICES IN INDIA

A.S. Kanagasabapathy and S. Swaminathan

Department of Clinical Biochemistry, Christian Medical College and Hospital, Vellore, India

INTRODUCTION

India, with a population of 800 million is the second most populous country in the world. It is supporting 15% of world's population with only 2.5% of total land area. In 39 years after independence from the British, the population has been more than doubled, adding a second India, 75% of the people live in rural areas.

This largest democratic country in South Asia is a Union of 25 States and 9 Union territories. Under the constitution, the States are largely independent in matters relating to the delivery of health care to the people. The Central Government's responsibility consists mainly of policy making, planning, guiding, assisting, evaluating and coordinating the work of the States. After independence, the Government established Primary Health Centres (PHC) as peripheral institutions for providing comprehensive health care in an attempt to move away from the hospital-based curative services which were found inadequate to meet the health needs of the vast majority of rural population. Keeping in view the national commitment to attain the goal of "Health for All by 2000", the Government, over the years has built up a vast infrastructure of rural health services based on PHC and subcentres (SC). Today, there are over 11,000 PHC and 83,000 SC in the rural areas. Each SC serves a population of 5,000, while each PHC has been allotted for a community development block comprising about 100 villages with approximately 100,000 population.

There are 106 medical colleges with an annual admission capacity of 12,000 and about 8,000 hospitals with over 500,000 beds, 30% belonging to private, non-governmental institutions. Nearly 80% of these hospitals are situated in urban areas.

HEALTH CARE DELIVERY TO THE PEOPLE

The State Ministry of Health is headed by the Minister of Health and Family Welfare who is administratively supported by the Health Secretary. The Secretary carries out the work through three channels viz - Medical Service, Medical Education and Public Health. Each avenue is headed by a Director. The Director of Medical Services controls directly the State

Referral Hospitals, and through the District Medical officers controls the district headquarters hospitals, taluk headquarters hospitals, and dispensaries. It should be mentioned that there are private hospitals providing parallel service at all levels. The Director of Medical Education is responsible for medical education in the State. The Director of Public Health handles through Deputy Directors, several national projects funded by the Central Government for prevention of various diseases. Regional Assistant Directors assist the Director of Public Health through District Health officers at each District level to cater to the health needs of the people. The District Health officer in turn takes care of PHC and SC at the Village level with the help of several medical officers.

During the last five years, the Country has witnessed significant growth in laboratory facilities mainly in urban areas due to (1) the Government's liberalised policy for importing sophisticated equipment and reagents/kits from abroad and (2) the development of an increased number of private diagnostic centres. Details on various aspects of clinical biochemistry services were received through a questionnaire sent to over 400 laboratories all over the country - both urban and rural. The salient features of this questionnaire, for which approximately 75% response was received are:

1. Classification of the laboratory.
2. Staff and training of laboratory personnel.
3. Investigatory facilities and the use of commercial kits.
4. Equipment.
5. Equipment Maintenance.
6. Workload and Data processing.
7. Internal quality control and external quality assurance.

CLASSIFICATION OF THE LABORATORY

The laboratory is either attached to a teaching hospital, non-teaching hospital, a small hospital (Government or private in all these categories), a private clinic or private hospital/diagnostic centre. Small hospital laboratories include those in the Government hospitals at the District and Taluk levels, the PHC and a large number of Mission hospitals. While most of the Mission hospitals are situated in rural blocks, many private laboratories which are supported by private medical practitioners are in urban areas. The recent trend in the country is "corporatisation of medicare" by establishing hospitals and diagnostic centres through corporate funding by private agencies and individuals. It is envisaged that more units of this kind will be established in the near future in different regions of the country.

STAFF AND TRAINING OF LABORATORY PERSONNEL

Laboratories attached to large teaching and non-teaching institutions are adequately staffed with postgraduates in medicine (M.D.) and biochemistry (M.Sc. and Ph.D.), graduates in science (B.Sc.) and laboratory technologists. Not all small hospitals, however, have adequately qualified laboratory personnel.

All Universities in India offer a two year M.Sc. Biochemistry course to science graduates while very few medical colleges offer a 3 year M.Sc. Medical Biochemistry course. A medical graduate undergoes a course for 3 years for his M.D. degree in Biochemistry. Ph.D. programmes are offered to the above at University Biochemistry departments as well as medical college biochemistry departments. What is lacking in the country is a Mastership course as well as a postdoctoral training in clinical biochemistry. An M.D. or a Ph.D. is the Head of the Clinical Biochemistry section. This professional Clinical Biochemist is expected to have (i) analytical and data handling capability, (ii) consultant and educational capability and (iii) research ability, the latter being essential to the improvement of laboratory services through applied research and to the advancement of the discipline through fundamental research.

The questionnaire has revealed that the training of laboratory technologists is quite varied and it is provided by different organisations, both Government and private. (i) A three year B.Sc. Medical Laboratory Technology course conducted by the Government of India - 13 Institutions offer this course and 105 students are admitted every year, (ii) one year Laboratory Technology course offered by the Christian Medical Association of India and the Government of India - 78 Institutions provide this course with an annual intake of 1600 students, and (iii) short term training programmes (less than one year) conducted by a number of private laboratory schools. Although several courses are available for laboratory technologists, there is no uniformity in curriculum content and standard. It is a matter of concern that there is regular exodus of well trained technologists to foreign countries and it is hard to find suitable persons and retain them in the country.

INVESTIGATORY FACILITIES AND THE USE OF COMMERCIAL KITS

No biochemical test is as yet done by the PHC laboratories and the taluk headquarters hospital laboratories analyse only glucose; however, district headquarters hospital laboratories conduct a maximum of 9 important biochemical investigations (glucose, urea, creatinine, cholesterol, protein, bilirubin, AST, ALT and AP). While small private hospitals' and private clinics' laboratories analyse many more tests, laboratories of teaching and non-teaching hospitals and diagnostic centres carry out all routine tests along with a wide variety of other special investigations such as hormones, drugs, isoenzymes, etc.

Table 1 shows the list of biochemical analyses as recommended by the Government of India/WHO to be done by all laboratories throughout the country in a phased manner.

It is appropriate to mention that the Health Laboratory Technology Unit of the WHO has recently published a document (LAB/86-2) on "Identification of Essential clinical chemical and Haematological tests in Intermediate Labs". Out of the total of 34 essential tests recommended, 13 are biochemical investigations and these are included in the 20 tests recommended by Government of India/WHO. Certainly, during the last few years awareness of the diagnostic usefulness of these tests has increased among the rural laboratories and many of these tests are done routinely today. The Government is making earnest attempts to improve the laboratory facilities at the PHC level so that at least very basic tests could be done in these laboratories.

To do these tests, some of the large hospital laboratories use reagents prepared in their laboratories, employing local as well as imported chemicals. On the other hand, many small hospital laboratories employ commercial kits either imported or indigenous. This number of laboratories using commercial kits has significantly increased over the last five years or so. Due to the Government's liberalised import policy, foreign kits are easily available through the local distributors. This has helped many laboratories, particularly small hospital laboratories, to do more special tests, such as enzymes by kinetic assays, cholesterol and

Table 1. Biochemical tests

1. Haemoglobin	11. Potassium
2. Total protein	12. Chloride
3. Albumin	13. Amylase
4. Urea	14. Aspartate Transaminase
5. Glucose	15. Alanine Transaminase
6. Bilirubin	16. Cholesterol
7. Alkaline phosphatase	17. Urate
8. Calcium	18. Bicarbonate
9. Phosphorus	19. Creatinine
10. Sodium	20. Triglycerides

glucose by enzymatic methods. Few local companies now also offer indigenous kits for these. As more and more laboratories are switching over to commercial kits, evaluation of the quality of these preparations has become essential.

Over a decade ago radioimmunoassay was introduced by a few premier institutions in India, but during the last five years a large number of teaching/non-teaching institutions as well as private clinical laboratories have included RIA as one of the important analytical systems. While in the past mostly in-house reagent preparations were employed in carrying out the RIA, today there is easy availability of a wide range of RIA kits. Bhabha Atomic Research Corporation supplies a variety of radionucleides and RIA kits. In addition, a number of Indian agencies supply foreign-made kits, such as Serono (Italy), Diagnostic Products Corporation (U.S.A.), C.I.S. International (France), Amersham International (U.K.) and Leeco Diagnostics (U.S.A.).

Dry strips are increasingly used today in India for qualitative and semiquantitative screening tests due to the easy availability of imported strips from Boehringer Mannheim and AMES as well as local preparations from Span Diagnostics, Miles India and Bharat.

EQUIPMENT

In the past, there was a sharp contrast between the urban and rural hospital laboratories in the use of the range of equipment. The situation has certainly improved now, permitting the use of modern equipment even by the small hospital laboratories in the rural areas. The reasons are: (1) the manufacture of many indigenous equipment (some of these with foreign technical collaboration) like colorimeter/spectrophotometer, flame photometer, pH meter, centrifuge, electrophoretic unit, semiautomatic pipette and dispenser and manual gammacounter, and (2) relaxed import policy of the Government resulting in the easy availability of a wide range of sophisticated microprocessor controlled equipment.

During the last few years, foreign made discrete autoanalysers have been the main attraction for the Indian laboratories. Many laboratories have employed a variety of batch analysers. In the last two years, some of the large institutions have introduced discrete selective analysers - Hitachi-705 of Boehringer Mannheim and RA-1000 of Technicon. Many laboratories including small private laboratories employ microprocessor controlled photometers which have helped these laboratories to replace the conventional end point enzyme assays by kinetic assays.

EQUIPMENT MAINTENANCE

Equipment Maintenance is not quite satisfactory in India. While only very few hospitals have an Equipment Maintenance Section within the Institution, most of the laboratories hire local private personnel whenever necessary. Particularly for rural laboratories access to maintenance facilities is very difficult. Since foreign made equipment are now available in plenty in the country, the present trend is towards entering into service contracts with the local service agents of the foreign companies. Some of these agents provide satisfactory service, but some do not provide good servicing and they send their service engineers according to their convenience. Another problem is that some foreign companies change their local service agents very frequently causing a problem to the laboratory in taking time to re-establish rapport with the new local agent. Servicing of equipment by Government authorised institutions has yet to pick up well. The effort of the Central Scientific Instruments Organisation of Government of India is limited to nine cities and it does not cover the entire country.

The laboratory manager and the equipment manufacturer certainly have an important role, since unless remedial measures are taken towards equipment maintenance, the enormous money spent on equipment will contribute little to the improvement of health care. As users of equipment, we hold the responsibility to encourage local manufacturers towards designing and supplying appropriate laboratory equipment, taking into account power problem and environmental consideration, so that the equipment is reliable, easy to operate and robust. Thorough working knowledge on the equipment and preventive maintenance aspects is essential. Wide fluctuation in power supply and sudden power breakdown necessitate the use of voltage stabilisers and generators. Some leading institutions in India have started thinking about introducing uninterrupted power supply (UPS) in the place of generator for sophisticated equipment and we hope UPS will become popular very soon.

The equipment manufacturer has the foremost responsibility to ensure that his equipment once purchased is effectively installed and remains useable for a reasonable period of time. He should provide service manual, list of spares, advice on sources of breakdown and remedial measures, effective after sales services and training of the user on equipment operation and maintenance.

What we urgently need in the country are: (i) institution of a national advisory board consisting of national experts who could formulate guidelines and conduct equipment evaluations under prevalent conditions of climate and environment and (ii) the establishment of equipment maintenance centres in every region, with adequately trained biomedical and bioelectronic personnel, for the benefit of both urban and rural laboratories. Through the Government we should seek the cooperation of international agencies like WHO and IFCC towards achieving these goals.

WORKLOAD AND DATA PROCESSING

Table 2 shows the workload of laboratories in terms of total number of tests done annually against the bed strength of the hospitals.

Although laboratory data are at present manually processed, many large institutions will be introducing in the near future computerisation of laboratory data processing, due to the easy availability of a variety of indigenous computers. Already, computers have found their place in a number of hospitals for data processing of patients' diseases, epidemiology, etc.

Table 2. Workload

Number of beds	Tests done annually
< 200	up to 20,000
200 - 500	10,000 - 100,000
500 - 1000	50,000 - 200,000
> 1000	100,000 - 500,000

INTERNAL QUALITY CONTROL AND EXTERNAL QUALITY ASSURANCE

A pilot study conducted by our Vellore Laboratory in August 1977 for about 35 laboratories to check their performance on 8 routine biochemical parameters, which revealed alarming variations in the reported results, prompted us into organising an external quality assurance (EQA) programme in the same year under the aegis of the Association of Clinical Biochemists of India (ACBI).

Over the last ten years, the awareness on the quality of laboratory performance has certainly increased. The present questionnaire has therefore revealed that the concept of quality control (QC) is now an accepted clinical biochemistry practice among Indian laboratories. At present 360 laboratories participate in our ACBI programme and 34 laboratories participate in the Government of India/WHO programme. The number of laboratories employing internal QC for monitoring daily laboratory work has increased, however the situation must further improve and internal QC must be employed by all laboratories. Only some laboratories use their own pooled serum for this purpose. Although foreign made commercial accuracy control sera are used by a number of laboratories as part of internal QC, it is not economical to use precision control sera from these sources. Encouragement of local companies for the supply of QC materials will definitely improve the situation.

Under the ACBI EQA programme conducted from our Vellore Centre, we distribute during the first week of every month liquid QC sera for analysis of 10 biochemical constituents (glucose, urea, total protein, albumin, calcium, cholesterol, sodium, potassium, creatinine and lithium). Upon receipt of results from the participants we check their results and send our comments during the third week of every month. For evaluation, we follow the WHO method of comparison of performance on the basis of Variance Index Score (VIS) obtained by calculation using the Chosen Coefficient of Variation, reference laboratory's mean value and participant laboratory's value.

We have conducted so far 19 six-month programmes. In the first six-month programme conducted from October 1977 to March 1978 only 59 laboratories participated. The number increased significantly over the last ten years to 360 in the 19th programme ending June this year. As part of the programme, we help several laboratories improve their performance by providing reference materials, details of recommended procedures and standardisation techniques. It is heartening to note that the participating laboratories are spread over all the States of our vast country. The participants include laboratories of large teaching and non-teaching hospitals in urban areas and small hospital laboratories in rural areas.

Table 3. Ethylene glycol stabilised serum stability at different temperatures

Constituent		* Mean value at		
		-20°C	4°C	R.T. (30°C-35°C)
Glucose	(mg%)	152	153	152
Urea	(mg%)	66	65	66
Total protein	(g%)	5.5	5.5	5.6
Albumin	(g%)	2.6	2.6	2.6
Calcium	(mg%)	7.6	7.7	7.6
Creatinine	(mg%)	1.3	1.3	1.3
Cholesterol	(mg%)	139	138	139
AST	(U/l)	66	65	53
ALT	(U/l)	31	30	22
AP	(U/l)	83	83	82

* Mean of 16 values obtained during a period of 25 days

The chemicals used as preservatives in the pooled serum cause inhibition of enzymes in the QC sera thus preventing their inclusion in the EQA programme. This problem has been solved by us by the use of ethylene glycol stabilised serum based on a recent WHO recommendation. We have checked the stability of various constituents in this preparation at -20°C, 4°C and at room temperature (between 25°C and 35°C). All constituents are quite stable even at room temperature, particularly the three enzymes AST, ALT and AP (Table 3).

As part of the EQA programme, we conduct regularly QC workshops in different centres in the country. During the last two years we conducted such workshops in nine centres either for 1 day or 2 days in which the number of participants ranged from 20 to 50. The participants included personnel from the laboratories in the respective regions. The subjects covered in the workshop are: significance of QC in the laboratory, lab. errors, precision and accuracy, statistical calculations, methods of standardization, internal QC with demonstration, methods, and prevention maintenance of equipment. Further, one important aspect in the workshop is a special inter-laboratory programme conducted for the participants of the workshop and a discussion on their performance in relation to the methods and equipment employed.

CONCLUSION

It may be said that there has been improvement in biochemical services in our country - both in terms of the availability of tests and the quality of laboratory performance. Governments's financial assistance is essential for further progress in this direction and also for subsidising the cost of these services in such a way that the common man has easy access to them.

To improve laboratory services further in the country, the following are immediately required:

1. Adequate funding by Government for staff, space and equipment to laboratories.
2. Licensing for laboratory services.
3. Institution of a National Scheme for training in Laboratory Technology, with standardised training of personnel.
4. Encouragement of local production of quality equipment and reference materials.
5. Establishment of equipment maintenance centres in every region, with adequately trained biomedical and bioelectronic personnel.
6. Uniformity in the methods employed.
7. Evaluation of indigenous equipment and reagent kits and
8. Encouragement of more laboratories to adopt quality control measures.

CLINICAL BIOCHEMISTRY SERVICES IN INDONESIA

F.X. Budhianto Suhadi

Airlangga University School of Medicine, Dr. Soetomo General Hospital, Surabaya, Indonesia

Indonesia, the world's cross road position country consists of many islands. The population is 165 millions, 70% of which still live in villages. In order to the able to manage the development including health services, Indonesia has a Five-Year Development Plan; and as a strategy in health sector, the Government has developed a National Health System, in which the health services are mainly for the low income population including transmigrants and those who live in remote areas by introducing health delivery system as close as possible to the population.

Although the indicators for health status seem improving compared to the last Five-Year Development Plan, but since these are still far from what is expected, efforts during this Five-Year Development Plan are still emphasized on the health delivery system. Concrete actions to improve the health status is to improve the capability of the community to live healthily, to educate and train the health personnels and the community, to enable them to solve their own simple health problems.

Laboratory services in remote and rural areas are still very simple. There are two kinds of laboratories in Indonesia, the Government and the private ones. The Government labs are laboratories in the Government hospitals or Army hospitals and among them are laboratories attached to teaching hospitals. In the Fourth Five-Year Development Plan the main activities are to increase the capacity of examining bloodsmears from 20 to 25 each day or 6,000 specimens a year for a Peripheral Health Center Laboratories. This could be achieved by routine work and referral from other centers especially in the effort of controlling communicable diseases.

The capabilities of the Government laboratories are different according to the classification of the hospitals. There are type C_1, C_2, B_1, B_2, B_3 and type A. This classification is based on the medical services i.e., kind of specialist, subspecialist and supporting services departments and the total amount of beds. Standard tests that could be provided by the different type of laboratories has been determined.

Besides the clinical chemistry laboratories we also have Public Health Laboratories in the twenty seven Provinces of which there are six of type A in the Regions.

The laboratories in type A hospitals which are also University Hospitals are more sophisticated and serve as a National Top Referal Hospital Laboratories. Private labs also play an important role in enhancing health services.

To provide good laboratory results, we conduct a National Quality Assurance Program which is an interlaboratory program.
The participants consist of the Government laboratories which are free of charge and the private laboratories, since their participation is important for getting license to run a laboratory or as a prerequisite for renewal of laboratory licenses. There are two kinds of Quality Assurance Program, the Clinical Chemistry and Haematology. The amount of participating labs increases significantly each year. We are convinced that the clinical chemistry services will be improved in the near future.

INTRODUCTION

Indonesia is the world's largest archipelago extending between two oceans, the Pacific and Indonesia Ocean, and between two continents, Asia and Australia. Because of this in-between and strategic position, Indonesia is said to be the world's cross-road position.

Another name for Indonesia is Nusantara. The total area is 1,905,443 square kms (735,354 sq. miles), 84% of which is water and 16% are islands. There are more than 13,677 islands of which only 930 are inhabited. These islands stretch 5,152 kms (3,200 miles) from East to West, and 1,770 kms (1,100 miles) from North to South. The population of Indonesia is 165 millions and is the fifth most populous country in the world, exceeded only by China, India, the Soviet Union and the United States of America. Seventy percent of which still live in villages. The density is very variable, 63% of the total population are spread out in Java, which is the principal island in terms of population and land use and the capital of Indonesia, Jakarta, is located on its northwestern coast. The island of Bali, Indonesia's fabled tourist destination, is only 5,623 sq. kms (2,170 sq. miles) and has a population of less than 2% of the population.

In order to be able to manage the development, Indonesia has a Five-Year Development Plan. We started on the year 1969 with our first Five-Year Development Plan and now we are in the fourth period. As stated in the program, the health sector will be paid more attention to be more able in providing health services and to develop human resources.

THE STRATEGIES ARE AS FOLLOWS

1. In the health sector, the Government has developed a National Health System. The health services are mainly for the low income population including transmigrants and those who live in remote areas.
2. In promoting health the strategy is to introduce health delivery system as close as possible to the population.
3. Health Centres, hospitals and paramedical staffs including laboratory technicians will be increased.

CONDITION AND PROBLEMS

During the Third Five-Year Development Plan the health status improved considerably. As indicators for health status we use:

- Crude Death Rate.
- Infant Mortality Rate.

- Age Specific Death Rate (under five).
- Life Expectancy.
- Nutritional status.
- Prevalence.

These health indexes seem better compared to the last period. But since these are still far from what is expected, efforts during this Fourth Five-Year Development Plan are still emphasized on the health delivery system.

CONCRETE ACTIONS TO IMPROVE THE HEALTH STATUS

1. Improve the capability of the community to live healthily, to educate and train the health personnels and the community, to enable them to solve their own simple health problems.
2. Improve the environmental health.
3. Improve the nutritional status.
4. Decrease the morbidity and mortality rate due to communicable diseases.
5. Decrease the fertility rate significantly.

To come to the aims stated above good strategies and plannings should be organized.

LABORATORY SERVICES

There are 2 kinds of laboratories in Indonesia as in many countries: the Government and the private ones.

The Government laboratories are laboratories in the Government hospitals or Army hospitals and among them are laboratories attached to teaching hospitals.

In the Fourth Five-Year Development Plan the main activities are to increase the capacity of examining bloodsmears from 20 to 25 each day or 6,000 specimens a year for peripheral Health Center laboratories. This could be achieved by routine work and referal from other centers especially in the effort of controlling communicable diseases. In East Java for instance there are 814 Health Centers and only 724 out of those have simple laboratories and according to recent stratifications survey only 50% of them belong to the good group. The intermediate labs could be found in Kabupaten and Province level and the more developed in the Regional and National level. If a certain parameter or analysis cannot be performed in the laboratory, then the patient or the specimen may be referred to a larger or better equipped laboratory as far as possible.

The capabilities of the Government laboratories are different according to the classification of the hospitals, i.e. type C_1, C_2, B_1, B_2, B_3 and A. This classification is based on the medical services (kind of specialist, subspecialist, supporting service) and the amount of beds.

The standard of tests which could be done in the different type of laboratories are listed in Table 2. The development of dry chemistry system as a mean for getting immediate laboratory results and as a consequence possible for making early diagnosis and treatment is an advantage for smaller laboratories where small series of analyses have to be performed during the day and for larger laboratories in emergency cases and during night. These technologies are also widespread used in Indonesia. Beside the clinical chemistry laboratories we also have Public Health Laboratories in

Table 1. Health facilities (1983/1984 - 1988/1989)

	1983-1984	1988-1989
1. Peripheral Health Center	5,353	5,853
2. Peripheral Health Sub Center	13,636	19,636
3. Mobile Peripheral Health Center	2,479	4,000
4. Peripheral Health Center with inpatients	128	296
5. Government and Private Hospitals	1,246	1,329
6. Hospital and Health Center beds	103,505	119,385

the 27 Provinces of which there are six of type A in the Regions, which main activities are in the preventive health care and less clinically.

The laboratories in type A hospitals which are also University Hospitals are more sophisticated and serve as a National Top Referral Hospital Laboratory. Administrative works and statistical manipulations are performed with the help of computers.

Since the overall budget of the Government decreases, the participation of the community in health service are accepted as partner for the Government. Private labs in Indonesia still play an important role in enhancing health services. Some of them are well developed and equipped sophisticatedly.

QUALITY ASSURANCE PROGRAM

Reproducibility or precision and accuracy for each of the substances tested should be first set up as the top priority ranker before starting with laboratory work. Reproducibility is the ability to obtain a consistent result on sample from time to time and is best assessed within the laboratory either by measuring the same sample over a period of days, or by measuring samples in duplicates.

Accuracy, the ability to obtain result that approaches the true result, can only really be assessed by comparison of the method used in the laboratory with a reference method. Practically, a fair indication of accuracy can be obtained from national or international quality assurance schemes which list target values for each of the test. This target value may be the mean of all participants in the scheme (possibly minus some extreme outliers) or target value may be supplied by the analysis of samples by a few laboratories chosen because of some expertise in those particular tests.

Another tool which can be used as reference for accuracy is good assayed quality assurance material. It is important that the target values are relevant for the particular method of analysis in use at that laboratory.

Starting from the basic grounds for a good laboratory described above, and with the legal backing Legislation in the field of Laboratory Care (1978) we perform a ring trial beginning with a small amount of participating laboratories and few parameters. We called this period as the "Preconditioning of the Community" both professional as well as

Table 2. Standard clinical biochemistry services at different hospital levels

Hospital classification	C_1	C_2	B_1	B_2	B_3	A
Number of beds	100-200	200-400	400-600	600-800	800-1000	> 1000
Liver function tests	Alb,Glob,Bill ASAT,ALAT,ALP	Gamma GT Cholin-esterase	Serological markers			more sophisticated
Enzymes	Amylase, AC. Phosph, LDH	C P K	G6PD, CKMB			
Renal function tests	Ureum, Creat UCT, CCT, Uric acid					
Electrolyte	Na, K, Cl.	Ca, P	Mg, Fe			
Blood gas	-	+	+	+	+	
Electro-phorese protein	-	-	+	+	+	
Hormone			T4, T3U, TSH	+	TDM	

consumer. The idea is to motivate the participants and to understand the philosophy of Quality Assurance Program. Since 1980 the Directorate of Laboratory Services of the Indonesian Ministry of Health conduct a National Quality Assurance Program which is an interlaboratory program. The participants consist of the Government laboratories which are free of charge and the private laboratories, since their participation is important for getting license to run a laboratory or as a prerequisite for renewal of laboratory licenses. Later on besides for legal aspect there should be penalties for those who still belongs to the less satisfactory performers after guiding and counseling. So the program was started on a fully voluntary base but later after a certain grace period comprising several stages will gradually become compulsory by proper legislation.

Each participant will receive twice annually 2 bottles of blind lyophilized serum, instruction sheet, questionnaires about method, reagent and instrument used performing the tests and answer sheets. The participants are asked to reconstitute the serum with 5 ml aquabidest and to perform the test with routine method.

We realize that the so-called routine methods of analysis is particularly difficult to define since there are so many of them for a given analyte and laboratories differ so many in their workloads, instrumentation and special skills. The reference laboratories analyse the control serum with the same routine method used by most participants.

Table 3. Participation in Quality Assurance Program

Year	Parameter tested	Government labs	Private labs
1980	7	55	83
1981	7	81	123
1982	11	123	144
1983	11	165	160
1984	11	100	197
1985	11	133	217
1986	11	197	257

The participating laboratories are requested to treat control specimens in the same manner as specimens in the routine and not to treat them as special and non routine. After been filled the answer sheets are sent to the center. We hope that the participating laboratories do not send the materials to a better laboratory and then to send the results as their own.

At present the sheets are still processed and analysed in Germany and after some time every participant will get result sheet as shown in the appendix. As the reference labs we still use laboratories from outside the country, but in the near future we would organize and select out of the best participants group of our own.

The amount of parameters tested increases with time. Until now there are 11 parameters which should be done by the participants:

- Bilirubine
- Cholesterol
- Creatinine
- Glucose
- Total Protein
- Urea
- Uric Acid
- Triglycerides
- G O T
- GPT
- Calcium

Each participant will get a score indicating his performance during a certain scheme.

The formula used to calculate the score is as follows:

$$\text{score} = \frac{\Sigma\ 10 - (\text{deviation in SD})^2}{\Sigma\ \text{parameter}}$$

deviation in SD = deviation of each parameter between participant and mean of the reference labs' results expressed in SD of the reference labs' results.

There are two kinds of Quality Assurance Program, the Clinical Chemisty and Haematology. The program for haematology is just at the beginning phase. The amount of participating labs increases significantly each year. Suggestions, guidance and workshops on improving the result of the parameters tested are conducted by the committee by giving advice by letter and if there is available budget by making a visit to meet the local conditions.

Many laboratories also join a collaborative scheme with WHO, Australian College of Pathologist, Scottish Home and Health Department, etc.

CONCLUSION

Employing daily Internal Quality Assurance Program and participating actively in the External Quality Assurance Program, I believe undoubtedly that there will be a certain amount of progress among clinical chemistry laboratories in Indonesia and certainly will improve the patients care.

BIBLIOGRAPHY

- De Verdier CH, Aronsson T, Nyberg A. Quality Control in Clinical Chemistry - Efforts to find an efficient strategy. The Nordic Clinical Chemistry Project (Nordkem). Helsinki, Finland, 1984.
- Introducing Indonesia. Garuda magazine 1987; Vol. 7: 2.
- Garcia-Webb P. The small clinical laboratory. The clinician's requirements. Clinical Biochemistry, Principles and Practice. Third Asian and Pacific Congress of Clinical Biochemistry. Denpasar, Bali, 15 - 19 September 1985: 271-275.
- Symposium: Current status of laboratory service in urban and rural areas in developing countries. Second Asian and Pacific Congress of Clinical Biochemistry, Singapore, July 1983.
- Widyaharsana J. Laboratory Quality Assurance: The Indonesian experience. Clinical Biochemistry, Principles and Practice, Second Asian and Pacific Congress of Clinical Biochemistry. Singapore, July 1983.
- Workshop on National Health Service. Department of Health, Republic Indonesia. Jakarta, 14 - 18 March 1984.

APPENDIX Program nasional pemantapan kwalitas laboratorium klinik (PNPKLK)
Tabel hasil Ring trial Indonesien 2/86 Result sheet Ring trial Indonesien 2/86
Nomor laboratorium: 1101 Laboratory number : 1101

Parameter/parameter		NILAI REFEREN Reference values			HASIL PESERTA results of participants			NILAI ANDA your results	
Sample	Meth. No.	Target Value	CV%	Daerah Toleransi Tolerance limits	no nil	rata mean	CV%	Hasil Result	Def. STD Dev. in SD
Bilirubin				Jendrassik-grof					
A	11	2,620	5.00	2,200 to 3,020	337	2,525	15.48	2,470	-1,145
B	11	4,430	5.50	3,690 to 5,170	337	4,206	15.51	4,350	-0,328
Cholesterol				Chod-PAP					
A	21	113,000	7.00	89,300 to 137,000	220	112,745	10.64	99,000	-1,769
B	21	134,000	5.00	113,000 to 155,000	220	134,339	10.92	132,000	-0,298
Creatinine				Jaffe TCA deprot					
A	31	3,750	5.00	3,180 to 4,320	177	3,392	15.27	3,500	-1,333
B	31	2,350	5.00	1,990 to 2,710	177	2,355	19.27	2,000	-2,978
Glukosa				God-PAP					
A	41	133,000	4.10	116,000 to 150,000	251	125,500	13.31	129,000	-0,733
B	41	212,000	5.10	179,000 to 245,000	251	199,390	14.18	212,000	+0,000
Protein total				Biuret W/O blank					
A	51	5,000	3.90	4,410 to 5,590	325	5,243	9.57	5,100	+0,512
B	51	5,900	3.05	5,360 to 6,440	325	6,061	8.21	6,200	+1,667
Ureum				Berthelot-meth.					
A	61	90,000	4.60	77,500 to 102,500	271	86,760	19.85	96,700	+1,618
B	61	50,400	5.00	42,800 to 58,000	271	50,620	17.50	53,300	+1,150
Asam Urat				Uricase-colorim.					
A	72	5,230	5.10	4,420 to 6,040	196	5,287	14.95	5,500	+1,012
B	72	7,100	5.50	5,920 to 8,280	196	7,092	16.02	7,500	+1,024
Triglyserida				Enz. color test					
A	82	205,000	5.40	171,000 to 239,000	208	192,774	15.26	208,500	+0,316
B	82	118,000	6.50	94,000 to 142,000	208	116,542	18.94	117,500	-0,065
GOT				Opt. DGKC 25 C					
A	91	29,500	5.50	24,600 to 34,400	102	32,800	21.15	38,000	+5,238
B	91	45,000	6.50	36,200 to 53,800	101	48,548	22.05	53,200	+2,803
GPT				Opt. DGKC 25 C					
A	101	53,000	6.20	43,100 to 62,900	102	54,360	20.95	59,800	+2,069
B	101	38,600	6.50	31,000 to 46,200	101	39,614	23.38	43,800	+2,072
Calcium				Cresophth. compl					
A	112	9,500	5.30	7,990 to 11,000	134	9,851	9.40	4,970	-8,997
B	112	11,200	5.00	9,520 to 12,900	134	11,252	12.41	6,360	-8,642

PROBLEMS IN CLINICAL CHEMISTRY IN DEVELOPING COUNTRIES: SUMMARY AND POSSIBLE SOLUTIONS

P. Garcia-Webb

University of Western Australia, Dept. of Clinical Biochemistry, Perth, W.A., Australia

The concept of including a symposium on clinical chemistry in developing countries at an IFCC international congress is an excellent one. It is of course not new as the IFCC has been involved in this area for quite some time. In 1970, the then President of the IFCC, Professor Martin Rubin, suggested to WHO that there should be a joint meeting to discuss this very topic. That meeting was held in mid September 1971 at Geneva. Professor Rubin has kindly given me a report of that meeting; I shall refer to it a few times during this talk. I should mention that I have also been in correspondence with Professor Per Lous from Denmark and Dr Matt McQueen from Canada while preparing material for this presentation. Some of the ideas that I shall put forward therefore are not my own, although of course I agree with them.

The delegates at that meeting in Geneva 16 years ago considered the provision of clinical chemistry that might be appropriate to three levels of health care: the rural health centre, the intermediate health centre and the central laboratory. This of course continues to be the current approach. In 1971, and as I shall shortly show is still the case now, the provision of clinical chemistry services had to be taken in the context of the provision of total health care to the community. At that time it was calculated that the total amount of money available for health care in a typical developing country was unlikely to exceed 30 US cents per capita per year (calculated as no more than 5% of the national budget). Furthermore the primary health care problem at that time was communicable disease with parasites and infectious disease representing the greatest source of morbidity and mortality.

It is interesting to compare the current situation with the situation as described in the report of the 1971 meeting. Table 1 shows the government budget of a number of countries. The data are expressed in terms of 1985 US dollars and are taken from the Encyclopedia Britannica (1). The countries would normally be considered as being developing countries, at least from the point of view of laboratory medicine. Obviously there are large differences in the budget from country to country.

It is not easy to find out what percentage of the expenditure budget of a number of countries is spent on health care. The 1971 meeting assumed

Table 1. Money available for health care in countries from around the world. Data are from 1985. Budget for health care has been calculated as 5% of total budget

Region of World	Total Budget Per Capita (US$)	Health Care Budget Per Capita (US$)
Africa	38.0	1.9
Asia	46.5	2.3
Africa	76.3	3.8
South America	79.1	4.0
Africa	132	6.6
Asia	157	7.9
South America	296	14.8
Africa	309	15.0
Asia	743	37.2

the percentage would be no greater than 5%, I have used that figure again here. On that basis the second column in Table 1 shows the amount of money available for total health care in each country. You should note that these data refer to government spending and not to health care support provided by private or commercial concerns. The data therefore underestimate the amount actually spent on health care by an amount that will depend on the way in which private medical practice is funded in any particular country. In some countries such as Australia and, more particularly, the United States of America, ignoring the private arena would give a distorted view of health expenditure. However in the context of this presentation dealing with assistance to developing countries, the amount of money available for expenditure by government is the more relevant figure.

The table gives the amount of money available for total health care. In my state of Western Australia somewhat less than 1% of the money spent by the State Government on total health care is spent on provision of clinical chemistry. Let us assume then that a developing country might spend a maximum of 1% of its budget on clinical chemistry. It would follow then that in 1985, most developing countries would have less than 10 cents US per head to spend on clinical chemistry. Even the relatively well off developing countries would have had less than 40 cents to spend. This compares with a government expenditure on health for 1985 in Australia of US$ 154 per capita for total health care and US$ 1.54 for clinical chemistry. In fact of course much less than 10 to 40 cents may actually be available. Looked at from this point of view, the responsible attitude could well be to advise only minimal spending on clinical chemistry rather than the indefinite development of the discipline. An any event, any money should be spent wisely.

The question of how much should be spent on clinical chemistry can be approached in another way. The pattern of disease in developing countries has not changed greatly since 1971. Parasites and infectious disease still represent the greatest source of morbidity and mortality although in some countries malnutrition and starvation certainly play an important part. The Lancet in its section on notes and news commented last year that the lack of a clean water supply was still the biggest single cause of human disease (2). Some progress was being made particularly in bringing adequate sanitation to urban dwellers. However a recent WHO review stated that at the end of 1983, only 14% of rural dwellers had an adequate sanitation service. Presumably at least one of the factors responsible for this state of affairs is lack of finance. It is not totally inappropriate to compare the gain to the community likely to flow from a development of clinical chemistry with the gain likely to result from an improvement in rural sanitation. Perhaps the latter should have priority.

The 1971 joint IFCC/WHO meeting recognised this need for restraint in development of clinical chemistry. One of its recommendations was: "present investment in highly mechanised, expensive and complicated laboratory equipment including electronic data processing units was strongly discouraged." Unfortunately such advice was not accepted then, nor would it be accepted now. It is useful to consider why this might be so.

The developing countries see an ever increasing use of expensive technology in clinical chemistry laboratories in developed countries. If it is apparently all right for the one, why should it not be acceptable for the other? The journals and modern textbooks (where these are available in developing countries) continue to present clinical chemistry results as being no more unusual or difficult to obtain than the pulse rate. Bright young graduates from developing countries have travelled and continue to travel to centres of renown to work and study for higher degrees. Is it any wonder that on return to their own countries they should attempt to apply there the shiny new technology in which they had only recently been immersed?

With the possible exception of the last year or so, there has been no attempt on the part of developed countries to relate the expenditure on clinical chemistry to the real health care needs of their citizens. Indeed the United States Department of Commerce and Congressional Budget Office has estimated the total health care expenditure in the United States to continue to increase to a figure of US$ 511 billion for 1987. Commenting on that, an article in the New England Journal of Medicine concludes that the American society is not likely to favour the explicit rationing of proved health care services to the public (3). There is however little attempt to accurately define the word "proved". Presumably therefore the expenditure in developed countries will continue to increase. The tendency will be for developing countries to try and follow suit.

Now all this is not to say that there should be no clinical chemistry in developing countries. Nor is it to say that the developed countries do not have an important part to play in assisting with the organisation and development of clinical chemistry in developing countries. However the emphasis must be on the effective use of money in this area. Let us turn now to consider this question of cost effectiveness in more detail.

I shall start with examples of ineffective spending. Anyone who has visited laboratories in developing countries will be familiar with the sight of large and expensive items of equipment that are idle. Sometimes it is because there is no money for reagents or spare parts: on other occasions nobody bothered to find out if there were funds available to connect the necessary services to the instrument. I could borrow a more grandiose example from the field of chemistry. Recently the United States built a technical college in a city in a developing country. Unfortunately it seems that no one asked the government of that country if it had the funds available to run teaching courses in the college or indeed to maintain the building. As a result the college is underutilised and the building is crumbling. This type of approach is clearly unsatisfactory. It consumes both money and the time of the individuals involved and it builds up resentment on behalf of those offering the project and feelings of hopelessness and frustration in those on the receiving end.

Let us consider now, in a more positive frame of mind, how the developed countries have and can continue to play an important part in assisting the developing countries. There are of course many examples of such effective interactions. It would be impossible for me to list them all, nor is there space to do so. A few examples will have to suffice. The first Latin American Congress of Clinical Chemistry was held in Mar del Plata in 1968. Following the efforts of Professor Rubin at the congress, the law in Argentina governing the control and management of laboratories was changed to allow management by better trained personnel. Still in that part of the world the selection of Mexico City and later Rio de Janeiro as venues for IFCC congresses has had a major impact on the development of clinical laboratory practice in that region. This stimulatory effect of international congresses can be seen elsewhere also. Thus the Asian Pacific Federation of Clinical Chemistry has held a number of congresses. These congresses have played a crucial part in the development of national societies of clinical chemistry and in the progression of these societies to the status of IFCC membership.

The World Health Organisation has made and continues to make many excellent contributions of direct relevance to the practice of clinical chemistry in developing countries. Other international and national organisations, groups and individuals continue to provide excellent assistance and leadership in the areas of laboratory medicine.

To give two examples from my part of the world. Under the auspices of the Asian Pacific Federation of Clinical Biochemistry a regional quality control group has been formed. It is hoped that by correspondence and meetings this group will provide a forum for the distribution of cheap quality assurance material for use in national quality control programmes. The experience available in both developed and developing countries in the region will be pooled to tackle the problems that might exist. As the second example, the Pacific Paramedical Training Centre was set up in 1980 at the Wellington Hospital in New Zealand. This centre has developed and runs laboratory programmes that are appropriate for the health care needs of countries in the Oceania South Pacific region.

Thus there are examples of cooperation between developed and developing countries that are both satisfactory and unsatisfactory. Education clearly remains the corner stone to reduce that percentage of these interactions that are unsatisfactory. Broadly speaking there are a number of educational areas that should be addressed. These include criteria for test selection, methods of analysis including quality control and instrumentation and project design for interactions between developed and developing countries. I would like to highlight some of the important point in these three areas.

Elsewhere I have referred to criteria that might be used to decide whether the measurement of a particular analyte is likely to contribute to patient care (4). In essence I suggested that the value of a particular test was directly proportional to its clinical usefulness and inversely proportional to the difficulty of measurement.

There were a number of questions that should be decided before introducing a particular test into the laboratory's repertoire. First, would the availability of the test procedure an improvement in health care? Second, how many patients might be expected to derive benefit from the test per week or year? Third, would the test result affect the rapidity or completeness of recovery and finally, what consequences, both medical and financial, would be incurred if the test results were erroneous?

From the analytical point of view, there is one general rule for the selection of test methods: the test must be reliable in the particular laboratory environment. There should be a hierarchy of methods in laboratories depending on the complexity of the laboratory. This hierarchy moves from simple manual methods, to work simplification, to the use of mechanical equipment. However the same hierarchy applies within a single laboratory. As I have already implied I am not in favour of the indiscriminant use of large multi channel analysers in laboratories in developing countries or indeed in developed countries. However if such an instrument is used it should not be the only method for measuring the more important tests on the system since clearly a failure in supply of reagents or a crucial spare part could incapacitate the instrument for some considerable time.

I have given examples of some projects designed to enable developed countries to assist developing countries that have been unsuccessful. It is apparent that any exercise must be carefully planned if it is to avoid the pitfalls of the past. This planning should include a number of steps. First, it should be clear that the aim of the exercise is one that will indeed benefit the health care needs of the community. Second, the site or sites at which the exercise is to be mounted must be appropriate so that the people at whom the exercise is aimed will indeed be able to benefit by it. Third, it must be apparent that there are sufficient funds to run the exercise itself. And finally, and in many ways most importantly, it must be clear in advance that there will be sufficient resources to allow the benefit of the exercise to continue after the exercise itself has finished.

There is of course nothing particularly new in these ideas. The concept of teaching teachers so that they can then teach others embodies these principles. Furthermore in some countries this concept is applied very well. However in other countries mere lip service is paid to it. To be effective there needs to be considerable discussion between the donor and the recipient of the exercise. This discussion should take place at a professional level. It should be frank, so that grandiose ideas that bear little relevance to or have little contact with the reality of the situation can be tossed out by either party. For these discussions to be effective both parties must recognise and apply high professional standards. There is no place for the promoting of ideas that are really designed to promote the stature on an individual at what amounts to the expense of the community. It is only once a suitable plan has been drawn up that the authorities should be approached to seek the necessary financial support. The budgeting for this financial support must include provision for the effective continuation of the exercise after the assistance from the developed countries has ceased.

Now in my view the exercise should not be entertained by the donor country unless there is a clearly defined plan to allow for the effects of the exercise to be continued into the future. I recognise that this somewhat extreme view may not be accepted by some who claim that any education is better than none. However I would stress again that the amount of resource to assist in dealing with problems of clinical chemistry in developing countries is extremely limited. There are only so many individuals willing or able to spend time giving assistance. There is only so much money available to bring that expertise to the developing country. In the face of limited resource, it seems to me that the only effective approach is to use that resource wisely. Not to do so is to fritter the resource away; the only benefit will be to the egos of a few. The majority will not benefit.

So in conclusion, there have been many instances in the past of excellent contributions by developed countries to the health care and clinical chemistry needs of developing countries. However there have also been many disasters. It is essential to recognise that developing countries do not have the resources to adopt the apparent technological advances of developed countries without a very critical look at the actual contribution to the real health care needs of the community that would so result. Finally new projects should be effectively planned. That planning must include the identification of ways to extend the benefit of the project beyond the time frame of the project itself. Projects that do not include provision for this extension should probably not take place.

REFERENCES

1. Britannica Book of the year, Chicago 1986; 68: 618-9.
2. A government minister's view on alternative medicines and therapies. The Lancet 1986; 1: 1453.
3. Ginzberg E. Sounding board a hard look at cost containment. New Engl J Med 1987; 316: 1151-4.
4. World Health Forum, WHO, Geneva 1985; 6: 182-4.

AUTHOR INDEX

SUBJECT INDEX